The Future of Rice Demand: Quality Beyond Productivity

Antonio Costa de Oliveira
Camila Pegoraro • Vívian Ebeling Viana
Editors

The Future of Rice Demand: Quality Beyond Productivity

Springer

Editors
Antonio Costa de Oliveira
Department of Plant Sciences
Universidade Federal de Pelotas
Pelotas, Brazil

Camila Pegoraro
Department of Plant Sciences
Universidade Federal de Pelotas
Pelotas, Brazil

Vívian Ebeling Viana
Department of Plant Sciences
Universidade Federal de Pelotas
Pelotas, Brazil

ISBN 978-3-030-37509-6 ISBN 978-3-030-37510-2 (eBook)
https://doi.org/10.1007/978-3-030-37510-2

This Springer imprint is published by the registered company Springer Nature Switzerland AG
The registered company address is: Gewerbestrasse 11, 6330 Cham, Switzerland

Preface

Rice (*Oryza sativa* L.) is one of the most consumed cereals in the world and is associated with food security. Considering the increasing demand for this cereal, breeding programs focus mainly on the development of higher yielding genotypes. However, consumers have expressed interest regarding the quality of rice. This scenario creates a new challenge for breeders by directing programs for high yield and quality grain.

Rice grain quality is determined by four attributes, cooking and eating quality, milling quality, appearance quality, and nutritional quality. Cooking and eating quality depends on the properties of starch (Chap. 1). Milling quality and appearance quality are associated with rice processing and the characteristics of processed grain (Chaps. 2 and 3). Finally, nutritional quality is related to the achievements of rice consumption in human health. In specific cases, rice may have high concentrations of contaminants requiring strategies to mitigate this problem (Chaps. 4 and 5). On the other hand, rice is a source of health-beneficial compounds, making it a strategic food for the human diet (Chaps. 6, 7, 8, 9, 10, 11, and 12). The potential application of other *Oryza* species on the quality of cultivated rice is also explored (Chap. 13).

The accumulated knowledge can directly assist different breeding strategies, i.e., conventional breeding, transgenic mutation induction, and genome editing (Chaps. 14, 15, 16, and 17). The advent of omics tools such as genomics, transcriptomics, proteomics, and metabolomics (Chaps. 18, 19, 20, and 21) helps the different breeding strategies.

Thus, this book is designed for people interested in the topic to have access to relevant information directly associated with rice grain quality. In this book, we sought to make a collection of topics related to the quality of rice grain. This was made possible by the availability and reliability of experts from different countries with experience in different aspects related to rice grain quality.

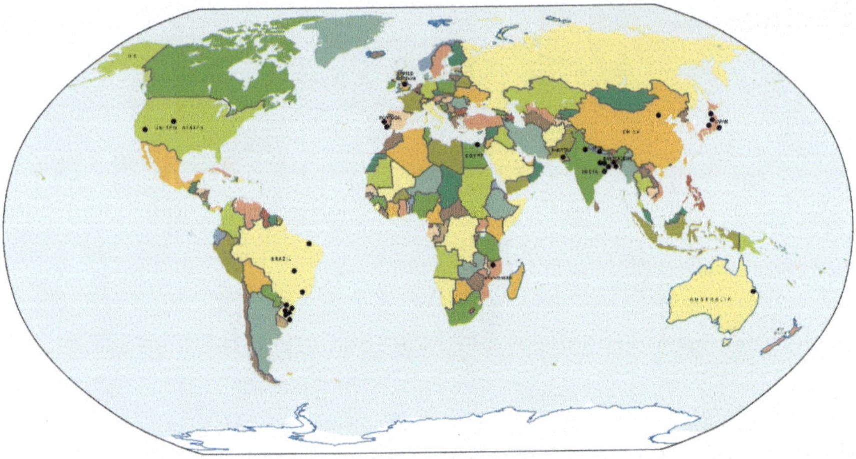

Worldwide map showing authors and coauthors who contribute with this book.

We hope the expert's brainstorm on rice quality can contribute to the state of the art and bring new insights into improving the quality of grain in rice and boosting the science of this cereal that means life to much of the world's population.

Pelotas, Brazil — Antonio Costa de Oliveira
Pelotas, Brazil — Camila Pegoraro
Pelotas, Brazil — Vívian Ebeling Viana

Contents

Contributors

Mohamed Abdelrahman National Key Facility for Crop Gene Resources and Genetic Improvement (NFCRI), Institute of Crop Science, Chinese Academy of Agriculture Sciences (CAAS), Beijing, China

Rice Research and Training Center, Field Crops Research Institute, Agricultural Research Center, Kafrelsheikh, Egypt

Aluana Gonçalves de Abreu Embrapa Arroz e Feijão, Santo Antônio de Goiás, Goiás, Brazil

Tanweer Fatah Abro Department Plant Breeding and Genetics, Faculty of Crop Production, Sindh Agriculture University Tandojam, Tandojam, Sindh, Pakistan

Gabriel Almeida Aguiar Instituto Federal Rio Grande do Sul, Campus Sertão, Sertão, RS, Brazil

Fahim Ahmed Department of Genetics and Plant Breeding, Bangladesh Agricultural University, Mymensingh, Bangladesh

May Sann Aung Department of Biological Production, Faculty of Bioresource Sciences, Akita Prefectural University, Akita, Japan

Research Institute for Bioresources and Biotechnology, Ishikawa Prefectural University, Ishikawa, Japan

Amrita Banerjee Central Rainfed Upland Rice Research Station, ICAR-National Rice Research Institute, Hazaribag, Jharkhand, India

N. Basak Crop Physiology and Biochemistry Division, ICAR-National Rice Research Institute, Cuttack, Odisha, India

Marouane Baslam Department of Applied Biological Chemistry, Faculty of Agriculture, Niigata University, Niigata, Japan

Priscila Zaczuk Bassinello Embrapa Arroz e Feijão, Santo Antônio de Goías, Goías, Brazil

Bruno Lemos Batista Universidade Federal do ABC, Santo André, SP, Brazil

Christine J. Bergman Food & Beverage and Event Management Department, University of Nevada Las Vegas, Las Vegas, NV, USA

Corey D. Broeckling Proteomics and Metabolomics Facility, Colorado State University, Fort Collins, CO, USA

Carlos Busanello Universidade Federal de Pelotas, Pelotas, RS, Brazil

Adriano Pereira de Castro Embrapa Arroz e Feijão, Santo Antônio de Goías, Goías, Brazil

Fabio Clasen Chaves Proteomics and Metabolomics Facility, Colorado State University, Fort Collins, CO, USA

Jeanaflor Crystal T. Concepcion School of Agriculture and Food Science, The University of Queensland, Brisbane, QLD, Australia

Department of Crop Sciences, University of Illinois at Urbana-Champaign, Urbana, IL, USA

Antonio Costa de Oliveira Universidade Federal de Pelotas, Campus Capão do Leão, Pelotas, RS, Brazil

Swapan K. Datta University of Calcutta, Laboratory of Translational Research on Transgenic Crops, Kolkata, India

Karabi Datta University of Calcutta, Laboratory of Translational Research on Transgenic Crops, Kolkata, India

Indra Deo Department of Genetics and Plant Breeding, College of Agriculture, G. B. Pant University of Agriculture and Technology, Pantnagar, India

Vívian Ebeling Viana Universidade Federal de Pelotas, Campus Capão do Leão, Pelotas, RS, Brazil

Paulo Henrique Karling Facchinello Embrapa Clima Temperado, Capão do Leão, RS, Brazil

Cristiano Dietrich Ferreira Laboratory of Post-Harvest, Industrialization and Grains Quality, Graduate Program in Food Science and Technology, Faculty of Agronomy Eliseu Maciel, Federal University of Pelotas, Brazil/Plant Science Department, Rothamsted Research, Harpenden, UK

José Manoel Colombari Filho Embrapa Arroz e Feijão, Santo Antônio de Goiás, Goiás, Brazil

Melissa A. Fitzgerald School of Agriculture and Food Science, The University of Queensland, Brisbane, QLD, Australia

Bruna Moreira Freire Universidade Federal do ABC, Santo André, SP, Brazil

Sharmin Hasan Queensland Alliance for Agriculture and Food Innovation, The University of Queensland, Brisbane, QLD, Australia

Department of Botany, Jagannath University, Dhaka, Bangladesh

Robert J. Henry Queensland Alliance for Agriculture and Food Innovation, The University of Queensland, Brisbane, QLD, Australia

Md. Shahjahan Kabir Bangladesh Rice Research Institute, Gazipur, Bangladesh

Takanori Kobayashi Research Institute for Bioresources and Biotechnology, Ishikawa Prefectural University, Ishikawa, Japan

J. Kumar Central Rainfed Upland Rice Research Station, ICAR-National Rice Research Institute, Hazaribag, Jharkhand, India

Ujjawal Kumar Singh Kushwaha Department of Genetics and Plant Breeding, College of Agriculture, G. B. Pant University of Agriculture and Technology, India/ Agriculture Botany Division, Nepal Agricultural Research Council, Khumaltar, Lalitpur, Nepal

Gustavo Heinrich Lang Laboratory of Post-Harvest, Industrialization and Grains Quality, Graduate Program in Food Science and Technology, Faculty of Agronomy Eliseu Maciel, Federal University of Pelotas, Brazil/Plant Science Department, Rothamsted Research, Harpenden, UK

Camila Neves Lange Universidade Federal do ABC, Santo André, SP, Brazil

Mohammad Abdul Latif Plant Pathology Division, Bangladesh Rice Research Institute, Gazipur, Bangladesh

Fernando C. Lidon Unidade de Geobiociências, Geoengenharias e Geotecnologias (GeoBioTec), Faculdade de Ciências e Tecnologia (FCT), Universidade NOVA de Lisboa (UNL), Monte de Caparica, Portugal

Viviane Kopp da Luz Plant Breeding and Genomics Center, Universidade Federal de Pelotas, Pelotas, Brazil

Ariano Martins de Magalhães Júnior Embrapa Clima Temperado, Capão do Leão, RS, Brazil

Latóia Eduarda Maltzahn Plant Breeding and Genomics Center, Universidade Federal de Pelotas, Pelotas, Brazil

N. P. Mandal Central Rainfed Upland Rice Research Station, ICAR-National Rice Research Institute, Hazaribag, Jharkhand, India

Hanna Paula Masuda Universidade Federal do ABC, Santo André, SP, Brazil

Hiroshi Masuda Department of Biological Production, Faculty of Bioresource Sciences, Akita Prefectural University, Akita, Japan

Research Institute for Bioresources and Biotechnology, Ishikawa Prefectural University, Ishikawa, Japan

Toshiaki Mitsui Department of Applied Biological Chemistry, Faculty of Agriculture, Niigata University, Niigata, Japan

Graduate School of Science and Technology, Niigata University, Niigata, Japan

Alex Ndayiragije International Rice Research Institute (IRRI) and Agricultural Research Institute of Mozambique (IIAM), Lichinga, Niassa, Mozambique

Naoko K. Nishizawa Research Institute for Bioresources and Biotechnology, Ishikawa Prefectural University, Ishikawa, Japan

Department of Global Agricultural Sciences, Graduate School of Agricultural and Life Sciences, The University of Tokyo, Tokyo, Japan

Maurício de Oliveira Laboratory of Post-Harvest, Industrialization and Grains Quality, Graduate Program in Food Science and Technology, Faculty of Agronomy Eliseu Maciel, Federal University of Pelotas, Pelotas, Brazil

Plant Science Department, Rothamsted Research, Harpenden, UK

Greiziany Soares Paulino Oliveira Universidade Federal do ABC, Santo André, SP, Brazil

Victoria Freitas de Oliveira Plant Breeding and Genomics Center, Universidade Federal de Pelotas, Pelotas, Brazil

Tereza Cristina de Oliveira Borba Embrapa Arroz e Feijão, Santo Antônio de Goías, Goías, Brazil

Isabel P. Pais Unidade de Investigação em Biotecnologia e Recursos Genéticos (UIBRG), Instituto Nacional de Investigação Agrária e Veterinária, I.P. (INIAV), Oeiras, Portugal

Fernanda Pollo Paniz Universidade Federal do ABC, Santo André, SP, Brazil

Tatiana Pedron Universidade Federal do ABC, Santo André, SP, Brazil

Camila Pegoraro Universidade Federal de Pelotas, Campus Capão do Leão, Pelotas, RS, Brazil

Rodrigo Mendes Pereira Universidade Federal do ABC, Santo André, SP, Brazil

José Almeida Pereira Embrapa Meio Norte, Teresina, Piauí, Brazil

Maria F. G. Pessoa Unidade de Geobiociências, Geoengenharias e Geotecnologias (GeoBioTec), Faculdade de Ciências e Tecnologia (FCT), Universidade NOVA de Lisboa (UNL), Monte de Caparica, Portugal

Christopher Proud School of Agriculture and Food Science, The University of Queensland, Brisbane, QLD, Australia

Vera Quecini Embrapa Uva e Vinho, Bento Gonçalves, RS, Brazil

José C. Ramalho Plant-Environment Interactions & Biodiversity Lab (PlantStress&Biodiversity), Linking Landscape, Environment, Agriculture and Food (LEAF), Instituto Superior de Agronomia (ISA), Universidade de Lisboa (UL), Oeiras and Lisbon, Portugal

Unidade de Geobiociências, Geoengenharias e Geotecnologias (GeoBioTec), Faculdade de Ciências e Tecnologia (FCT), Universidade NOVA de Lisboa (UNL), Monte de Caparica, Portugal

Fernando H. Reboredo Unidade de Geobiociências, Geoengenharias e Geotecnologias (GeoBioTec), Faculdade de Ciências e Tecnologia (FCT), Universidade NOVA de Lisboa (UNL), Monte de Caparica, Portugal

Ana I. Ribeiro-Barros Plant-Environment Interactions & Biodiversity Lab (PlantStress&Biodiversity), Linking Landscape, Environment, Agriculture and Food (LEAF), Instituto Superior de Agronomia (ISA), Universidade de Lisboa (UL), Lisbon, Portugal

Unidade de Geobiociências, Geoengenharias e Geotecnologias (GeoBioTec), Faculdade de Ciências e Tecnologia (FCT), Universidade NOVA de Lisboa (UNL), Monte de Caparica, Portugal

Faustino A. Roda Centro Zonal Noroeste, Instituto de Investigação Agraria de Moçambique (IIAM), Lichinga, Niassa, Mozambique

Centro de Biotechnologia (CB), Universidade Eduardo Mondlane (UEM), Maputo, Mozambique

Cesar Valmor Rombaldi Laboratory of Post-Harvest, Industrialization and Grains Quality, Graduate Program in Food Science and Technology, Faculty of Agronomy Eliseu Maciel, Federal University of Pelotas, Brazil/Plant Science Department, Rothamsted Research, Harpenden, UK

Somnath Roy Central Rainfed Upland Rice Research Station, ICAR-National Rice Research Institute, Hazaribag, Jharkhand, India

Railson Schreinert dos Santos Instituto Federal Farroupilha, Alegrete, RS, Brazil

Narendra Kumar Singh Department of Genetics and Plant Breeding, College of Agriculture, G. B. Pant University of Agriculture and Technology, Pantnagar, India

Eduardo Anibele Streck Instituto Federal Farroupilha, Campus Santo Augusto, Santo Augusto, RS, Brazil

Surendra Nath Tripathi Department of Entomology, College of Agriculture, G. B. Pant University of Agriculture and Technology, Pantnagar, India

Eduardo Venske Plant Breeding and Genomics Center, Universidade Federal de Pelotas, Pelotas, Brazil

Kaijun Zhao National Key Facility for Crop Gene Resources and Genetic Improvement (NFCRI), Institute of Crop Science, Chinese Academy of Agriculture Sciences (CAAS), Beijing, China

Part I
Aspects of Cooking Quality

Chapter 1
Rice Quality: Biochemical Composition, Eating Quality, and Cooking Quality

Fahim Ahmed, Tanweer Fatah Abro, Md. Shahjahan Kabir, and Mohammad Abdul Latif

1.1 Introduction

Rice is one of the most important food crops in the world as more than half of the world's population depends on rice as their major daily source of calories and protein (Parmar et al. 2011). Rice is regularly grown in almost all continents other than Antarctica and in over 100 countries, sea level to 3000 m in altitude, between 40°S and 53°N latitudes, and from plain land to under 1–2 m shallow land (Bhattacharya 2009).

Because of its adaptability in almost all types of climatic situation, so many varieties of rice are present in the world with different kind of eating and cooking qualities. A favorite type of rice for one cultural group is not suitable to others. An example of the variations of human taste for their native rice can be observed in different parts of Asia. The rice (*japonica*) variety available in the northern and eastern parts of Asia is sticky and soft when cooked, whereas the rice (*indica*) in South Asia is fluffy and dry, and the rice in Southeast Asia falls in between (Juliano and Pascual 1980). Similarly, Japanese prefer sticky rice which is suitable to make their favorite cuisine sushi, whereas consumers of South Asia and Middle East prefer aromatic rice which is dry, long, and slender when cooked (Suwannaporn and

F. Ahmed
Department of Genetics and Plant Breeding, Bangladesh Agricultural University, Mymensingh, Bangladesh

T. F. Abro
Department Plant Breeding and Genetics, Faculty of Crop Production, Sindh Agriculture University Tandojam, Tandojam, Sindh, Pakistan

Md. S. Kabir
Bangladesh Rice Research Institute, Gazipur, Bangladesh

M. A. Latif (✉)
Plant Pathology Division, Bangladesh Rice Research Institute, Gazipur, Bangladesh
e-mail: alatif1965@yahoo.com

A. Costa de Oliveira et al. (eds.), *The Future of Rice Demand: Quality Beyond Productivity*, https://doi.org/10.1007/978-3-030-37510-2_1

Linnemann 2007). The largest proportion of milled rice is composed of starch; thus, it plays a vital role in the processing quality of rice flour and milled rice.

Variation of its cooking quality is a small example of its total diversity within different varieties. Rice researchers around the world were fascinated about the diversity of eating and cooking quality of rice and want to clearly understand these attributes to use it for the development of new improved rice variety. It can be said that the study of eating and cooking quality of rice is one of the oldest topics of cereal chemistry.

Classification of rice into glutinous and non-glutinous types mainly depends on the composition of its endosperm. Almost whole consumption of rice in different countries is glutinous; thus, this type of rice is more important than non-glutinous type. Glutinous types of rice are sticky when cooked because of its dextrin content in the endosperm instead of starch. While cooked non- glutinous type of rice is not sticky. People prefer the glutinous type because of its softness after cooking. Non-glutinous rice contains higher fat content and less protein than glutinous type of rice.

The eating and cooking quality (ECQ) is the most important characteristic of rice which determines its price in the market and meets up the customer's satisfaction. Bao (2012) mentioned that the ECQs of rice are mainly controlled by some physicochemical properties of the starch in the endosperm such as gel consistency, gelatinization temperature (GT), pasting properties, and most importantly apparent amylose content. Thus, grain quality can be considered as a combination of physical, chemical, and eating and cooking qualities. Mechanisms controlling the traits of grain qualities are complex, controlled by many genes, and influenced by the environment. Several quantitative trait loci (QTL) controlling the grain properties were already identified using molecular markers (Fan et al. 2005; Lestari et al. 2009; Shao et al. 2010; Tabkhkar et al. 2012). Altogether, in this chapter, we discussed the biochemical properties of rice, its composition, and nutritive value. The eating and cooking quality of rice is also discussed here with details of their controlling factors and measurement systems.

1.2 Biochemical Properties of Rice

1.2.1 Nutritional Value of Rice

The rice grain is the main daily diet contributor. Cooked brown rice (195 g) consists of the following nutrients: calories (216.4 kcal), fat (1.76 g), monoun saturated fat (0.64 g), polyunsaturated fat (0.63 g), saturated fat (0.35 g), carbohydrates (44.8 g), protein (5.03 g), fiber (0.66 g), pantothenic acid (0.19 mg), vitamin B6 (0.28 mg), folic acid (7.80 mg), calcium (19.50 mg), iron (0.82 mg), and zinc (1.23 mg) (Atukorale 2002). However, despite their good nutritional value, rice loses some nutrients during milling.

1.2.2 Chemical Composition of Rice

Rice is a palatable high energy food as a result of its nutrient composition. Different fraction of rice was analyzed, and it was found that the rice bran consists of high protein and energy and the hull consists of low energy and nutrients. It has been also observed that all parts of brown rice are edible. As a result of intense friction milling to remove outer husk, pericarp, embryo, and aleurone layer, a significant amount of nutrient components, such as crude, protein, fat, ash, neutral detergent fiber, riboflavin, thiamine, α-tocopherol, and niacin, is lost (Table 1.1).

Starch or energy content in brown rice is lower than that of fined milled white rice. Heinemann et al. (2005) suggested that the nutrient composition of milled rice and brown rice, as a result of excessive analysis, is different from each other. Healthy dietary fiber remains in the hull (bran layer), and milled fine rice is low in fiber. Because of its low oil content, the bulk density and density of milled rice is high. Bulk density and density is low in hull and intermediate in bran.

The B vitamins are concentrated in the bran layers, as α-tocopherol (vitamin E) (Table 1.1). Although rice is a good source of the B vitamins, thiamine, riboflavin, and niacin, it contains little or no vitamin C, D, or A (Kennedy et al. 2002). The locational gradient in the whole rice grain is steeper for thiamine than for riboflavin and niacin, resulting in a lower percent retention of thiamine (vitamin B1) in milled rice (Table 1.1). The important B vitamins in rice are thiamine, riboflavin, and niacin, and they are largely found in the germ and bran. Rice milling to produce white rice removes about 80% of the thiamine alongside other nutrients in the bran, such as niacin, iron, and riboflavin. Following polishing, there are losses of 29% of the protein, 79% of the fat, and 67% of the iron (Atungulu and Pan 2014). Washing rice before cooking is a common practice, especially in Asia. The purpose is to remove bran, dust, and dirt from the processed rice; however, the washing procedure leads to the removal of most of the protein, ash, water-soluble vitamins, minerals, and up to two-thirds of crude fat. It is reported that when rice is washed before cooking, some 2–7% of protein, 20–41% of potassium, 22–59% of thiamine, 11–26% of riboflavin, and 20–60% of niacin could be lost (Juliano 1993).

Lipids are also known to influence viscoelastic properties by forming inclusion complexes with the helical structure of amylose. Juliano and Pascual (1980) found that defatting rice starch reduced both gelatinization temperature and gel viscosity of starch.

The protein content of milled rice is low in comparison with other cereals, although the whole rice grain content N × 5.95 % rice ranged from 7.0% to 10.8%, of which 70–80% is in the glutelin. The in vitro protein digestibility was relatively good (87.6–91.8%), as shown by Yousif (2000). Protein, as the other major constituent of rice, has not been thought to strongly influence cooking and eating qualities. When differences in gross protein content were examined in relation to texture of cooked rice, only a weak relationship was found, the higher-protein rice being somewhat less tender than low-protein rice. Because commonly eaten rice generally contains about 7% proteins and does not fluctuate widely form this level, protein

Table 1.1 Composition of rough rice milling fractions and vitamin and mineral content at 14% moisture

Rough rice contents and its milling fractions

Rice fraction	Crude protein (g N x 5.95)	Crude fat (g)	Crude fiber (g)	Crude ash (g)	Available carbohydrates (g)	Neutral detergent fiber (g)	Energy content (kJ)	(kcal)
Rough rice	5.8–7.7	1.5–2.3	7.2–10.4	2.9–5.2	64–73	16.4–19.2	1 580	378
Brown rice	7.1–8.3	1.6–2.8	0.6–1.0	1.0–1.5	73–87	2.9–3.9	1 520–1 610	363–385
Milled rice	6.3–7.1	0.3–0.5	0.2–0.5	0.3–0.8	77–89	0.7–2.3	1 460–1 560	349–373
Rice bran	11.3–14.9	15.0–19.7	7.0–11.4	6.6–9.9	34–62	24–29	670–1 990	399–476
Rice hull	2.0–2.8	0.3–0.8	34.5–45.9	13.2–21.0	22–34	66–74	1 110–1 390	265–332

Vitamin and mineral content of rough rice

Rice fraction	Thiamine (mg)	Riboflavin (mg)	Niacin (mg)	α-tocopherol (mg)	Calcium (mg)	Phosphorus (g)	Phytin P (g)	Iron (mg)	Zinc (mg)
Rough rice	0.26–0.33	0.06–0.11	2.9–5.6	0.90–2.00	10–80	0.17–0.39	0.18–0.21	1.4–6.0	1.7–3.1
Brown rice	0.29–0.61	0.04–0.14	3.5–5.3	0.90–2.50	10–50	0.17–0.43	0.13–0.27	0.2–5.2	0.6–2.8
Milled rice	0.02–0.11	0.02–0.06	1.3–2.4	75–0.30	10–30	0.08–0.15	0.02–0.07	0.2–2.8	0.6–2.3
Rice bran	1.20–2.40	0.18–0.43	26.7–49.9	2.60–13.3	30–120	1.1–2.5	0.9–2.2	8.6–43.0	4.3–25.8
Rice hull	0.09–0 21	0.05–0.07	1.6–4.2	0	60–130	0.03–0.07	0	3.9–9.5	0.9–4.0

References: Juliano (1985a, b), Pedersen and Eggum (1983), Eggum et al. (1982)

content is not considered an important indicator of quality (Hamaker and Griffin 1990).

Rough rice has higher fiber and ash content but lower protein and available carbohydrates (nitrogen-free extract, by difference) than brown rice.

The mineral composition of rice depends remarkably on soil nutrient availability during crop growth and on the diversity of sampling, preparation, and analytical methods used by different researchers. Higher levels of minerals are usually present in brown rice than in finely milled rice. A considerable portion of the rice caryopsis ash is accounted for by phosphorus. Potassium, magnesium, and silicon are present also in large amounts in brown and milled rice. By contrast, silica is the major element in hull ash (Champagne et al. 2004).

1.3 Eating and Cooking Quality of Rice

1.3.1 Grain Size, Shape, and Appearance

Choice of grain size, shape, and appearance differs from one group of people to another. Long slender grain is a prize commodity for one group of customers, other prefers medium types of grain, and short and bold type of grain is preferred by another group. In the Indian subcontinent, customers prefer long and slender grain, while customers of Southeast Asia prefer medium to medium long type rice. Short grain rice varieties are mainly cultivated in temperate regions. In international market, customers prefer long slender grain aromatic rice. Appearance of the grain depends on the size and shape of the kernel, chalkiness, and translucency. The shape of rice kernel is one of the main factors in determining its market value. Rice is classified into four categories based on its length and into three categories based on shape (ratios of grain length to grain width) (Cruz and Khush 2000) (Table 1.2).

Amount of chalkiness sometimes determines the appearance of a grain (Cruz and Khush 2000). Blurredness or opacity of the endosperm determines the chalkiness of rice and also classifies rice into waxy and non-waxy types (Rani et al. 2006). The difference between waxy and non-waxy rice is the presence or absence of amylose. Waxy rice contains trace amount of amylose and appears to be opaque, and

Table 1.2 Classification of rice according to grain size and shape

Type	Grain size (mm)	Grain shape (the ratios of grain length to grain width)
Very long	7.50	
Long	6.61–7.50	More than 3.0
Medium	5.51–6.60	Between 2.1 and 2.9
Short	Less than or equal to 5.50	Smaller than 2.0

References: Cruz and Khush (2000)

non-waxy rice contains amylose from a range of 2.1% to 32% and is translucent, dull, or hazy. Cruz and Khush (2000) classified rice on the basis of the percentage of chalkiness that covers the endosperm; they also mentioned that customer acceptability decreases with the increase of chalkiness in the grain.

1.3.2 Amylose Content (AC)

Yu et al. (2012) defined apparent amylose content, amylose to amylopectin ratio, or simply amylose content (AC), as the most important factor in determining the eating and cooking quality of rice. It is also known to be the main influencing factor in determining its texture upon cooking and processing. On the basis of AC, rice is classified into waxy, very low, low, intermediate, and high based on its amylose content (Kumar et al. 1987; Han et al. 2004) (Table 1.3).

The ECQs of rice are greatly influenced by its AC. Rice with variation of AC in their grain shows different texture upon cooking. Dry and fluffy texture is shown by rice with 25–30% AC (high), and often, it became hard after cooling; soft and sticky type of texture is shown by rice with 20–25% AC (intermediate), and sticky texture is shown by rice with less than 20% AC (low). Waxy rice contains almost zero amylose and is often called sticky rice. *Indica* rice belongs to all categories of amylose group; tropical *japonica* rice belongs to low to intermediate amylose group, and *japonica* belongs to low amylose group.

Amylose leaches out from the expanded starch content of the grain as a result of cooking. Those leached amylose form chain structure and bond together to produce gel when cooled down. The firmness of the cooked rice depends on its AC; if the AC is higher, then the cooked rice will be firmer. When the cooked rice cooled down in room temperature or beyond, the amylose chains formed during cooking crystallized, a phenomenon called retrogradation, which depends on protein content. A consumer always prefers rice with some characteristics like stickiness and firmness, which depends on AC.

Wang et al. (1995) suggested that the amylose of rice is produced by a single locus on chromosome 6 named *Wx* and the amount of protein produced by this locus determines its AC. With the increase of amylose on rice, the cooked rice loses its softness, viscosity, palatability, and luster. Palatability of cooked rice is not totally

Table 1.3 Types of rice based on their amylose content and cooking characteristics

Type	Percentage (%) of amylose content	Characteristics upon cooking
Waxy	0–2	Sticky rice
Very low	3–9	Sticky rice
Low	10–19	Quite soft and sticky
Intermediate	20–25	Softer and stickier
High	>25	Dry and fluffy

Reference: Han et al. (2004), Kumar et al. (1987)

determined by its amylose content. Different types of rice with the same amylose content taste different upon cooking. The Rapid Visco Analyzer (RVA) can find more palatable rice using its endosperm starch profile characteristics, which is not possible depending only on AC data. Researchers suggested that the information obtained from RVA, such as breakdown viscosity and setback viscosity, may be used to compare the palatability differences of different kinds of rice having same amylose content (Wu et al. 2001).

Kharabian-Masouleh et al. (2012) suggested that the properties of starch are controlled by two genes, granule-bound starch synthase I (*GBSSI*) and starch synthase IIa (*SSIIa*). Some physiochemical characteristics of starch such as gelatinization, pasting, cooked texture, retrogradation, amylose content, and the chain length distribution of short amylopectin chains are supposed to be influenced by *GBSSI* and *SSIIa* (Umemoto et al. 2002a, b; Chen et al. 2008; Lu et al. 2010; Yang et al. 2014).

1.3.3 Gelatinization Temperature (GT)

Cruz and Khush (2000) define GT as the range of temperature in which 90% of starch granules swell and lose their crystallinity and refraction quality. According to Tian et al. (2005), “GT is a physical-chemical property which directly reflects the cooking quality of rice grain in terms of energy and time needed for cooking.” Rice cooking starts at GT when starch granules of rice absorb water and lose their crystalline nature permanently. The GT indicates the time of cooking of a particular kind of rice. Depending upon types of rice, GT ranges from 55 to 79°C. The GT is classified into three categories: low, intermediate, and high at 55–69°C, 70–74°C, and more than 74°C, respectively (Graham 2002). The GT influences the cooking quality of rice greatly (Traore 2005). Traore (2005) also mentioned that GT plays an important role in water uptake, expansion of its volume, and elongation of the kernel. Cooking time for rice can be reduced by up to 4 min by lowering the GT of the grain (Fitzgerald et al. 2009). In case of overcooking, the high GT makes the rice too soft. If standard cooking procedure is applied, the grain can be less elongated and undercooked. Although rice grain quality preferences differ around the world, varieties with intermediate GT are generally preferred (Khush et al. 1979). High GT rice needs more energy than low GT rice upon cooking.

GT is predominantly determined by amylopectin structure (Umemoto et al. 2002a, b).Three classes of enzymes including starch synthase, starch branching, and starch debranching enzymes have been implicated in the synthesis of amylopectin (Smith et al. 1997; Myers et al. 2000; Nakamura et al. 2002). The gene *starch synthase IIa* (*SSIIa*) is the main contributor of rice GT which is located on chromosome 6 (Umemoto et al. 2002a, b, 2004; Waters et al. 2006). Nakamura et al. (2005) revealed that the function of *SSIIa* is to elongate the short A and B1 chains with degree of polymerization (DP) <10 to form long B1 chains of amylopectin. Genetic engineering by introduction of *indica* active *SSIIa* gene into *japonica* rice increases GT and gives longer amylopectin side chain length (Nakamura et al. 2005; Gao et al. 2011). Consequently, *japonica* starch granules have lower GT than *indica* starch granules (Little et al. 1958).

GT is highly correlated with alkali spreading value (ASV), which reflects the disintegration of milled rice in dilute KOH (Bergman et al. 2004; Little et al. 1958). Alkali spreading method is often used for the measurement of gelatinization. At present, differential colorimetry is used for the actual determination of gelatinization temperature and energy required for the cooking of a particular rice type.

1.3.4 Viscosity

Rice with similarly high AC may differ in cooking and eating quality due to potential effect of amylopectin structure and other factors. Gel consistency and viscosity are effective to differentiate rice with high AC (Bao 2014).

Viscosity indicates the cooking properties of rice, and it is measured by using Rapid Viscosity Analyzer (RVA) which mimics the cooking process of rice and monitors the change of the slurry made of water and rice flour during the test. In this test, the slurry is constantly stirred and heated to monitor the resistant of the slurry against the paddle. Starch granules take in water and become swollen, the slurry becomes thick, and viscosity begins to increase with the increase of temperature and constant stirring. The slurry is constantly heated and stirred until it reached 95°C and kept there for some time. At 95°C, the starch granules lose their structure and viscosity decreases. The hot slurry then cooled down to below gelatinization temperature (50°C), the amylose chain lines up and gels together, and again viscosity increases. Viscosity curve of different types of rice differs from each other and is a useful tool for rice variety development. This test is applicable in later stage of variety development as it requires certain amount of seed and the process is destructive.

Genetic studies show that the exon 10 SNP (single nucleotide polymorphisms) of *Wx* is responsible for the genetic basis for the gel consistency, the proportion of amylose bound to amylopectin, the proportion of amylose able to leach, gel hardness (Tran et al. 2011), and viscosity (Traore et al. 2011). Tran et al. (2011) indicated that the rice with SNP allele C at exon 10 produces soft, viscous gels and has a soft texture when cooked, but with high retrogradation, and the rice with SNP allele T gives a short, firm gel and has a firm texture when freshly cooked with little change in texture over storage. In a cross between two varieties having similar high AC, but with different viscosity properties, Traore et al. (2011) indicated that the exon 10 SNP marker is associated with most RVA pasting measurements and the proportion of soluble to insoluble apparent amylose.

1.3.5 Gel Consistency (GC)

Amylose content is the primary measure of the texture of cooked rice, but it fails to describe precisely the texture of certain types of rice (Juliano 2003). Gel consistency (GC) is, therefore, used to complement AC. Rice with the same amylose content can be classified as hard gel consistency (26–40 mm), medium gel consistency (41–60 mm), or soft gel consistency (61–100 mm) (Cagampange et al. 1973). The length of gel flow is inversely proportional to GC; therefore, the long gel length corresponds to soft GC and the short gel length means hard GC. Amylopectin, rather than AC, has been reported to be the major determinant of GC (Juliano and Perdon 1975). GC is reported to be affected by milling (lipid content), protein content, aging of milled rice (fat oxidation), and rice flour particle size (efficiency of dispersion) (Perez 1979). GC measurements are reported to have poor repeatability, and some laboratories have replaced them with starch paste viscosity parameters (Bergman et al. 2004). Cagampange et al. (1973) described gel consistency as a classification tool to differentiate rice varieties in which the gel length of previously cooked rice flour (with 0.2 M KOH) was measured. This method is used to differentiate between lower and intermediate classes of AC. Rohilla et al. (2000) mentioned that the rate of hardness and hardening differences of previously cooked rice correlate with GC. If the GC of a rice sample is hard, then it is supposed to be less sticky upon cooking, and if the GC of a rice sample is soft, then it is supposed to be tenderer upon cooking (Juliano 1985b).

Many QTLs for GC have been mapped using molecular markers (He et al. 2006; Wang et al. 2007; Septiningsih et al. 2003; Tan et al. 1999; Lanceras et al. 2000). Some researchers reported that a single locus in the *Wx* region and some modifier genes control GC (Tan et al. 1999, Lanceras et al. 2000). He et al. (2006) also reported that the *Wx* locus is the major determinant of GC, but it explained only 38.9% of the phenotypic variation. Using association analysis, the *Wx* was confirmed as the major gene controlling GC (Tian et al. 2009). A major QTL for GC at the *Wx* locus has been cloned (Su et al. 2011). In addition to the *Wx* gene, GC is reported to be affected by three other minor genes in the starch biosynthesis pathway—*AGPiso*, *SBE3*, *and ISA* (Tian et al. 2009). Two QTLs with minor effects on GC had earlier been reported (He et al. 1999; Bao 2012), and the *alk* locus has also been found to make minor contributions to GC (Wang et al. 2007).

1.3.6 Kernel Elongation

In addition to the amylose content, gelatinization temperature, gel consistency, and viscosity, other parameters, such as water absorption, volume expansion, and cooked rice elongation, have been set up to evaluate the cooking characteristics of rice (Bao et al. 2009).

The market demand of a particular rice remarkably depends upon its grain size and shape, and cooking quality largely depends on the characteristics of starch. Upon cooking, some varieties expand in size more than other varieties. The linear kernel elongation of rice while cooking is one of the main qualities of good aromatic rice. The genetic evaluation of the kernel length, shape, and its quality upon cooking plays a vital role in eating and cooking quality improvement program of rice grain (Bergman 2019). Elongation of kernel in length without the elongation in width is one of the most desirable characteristics of the consumer in high-quality aromatic rice (Rani et al. 2006; Khan et al. 2009). Rosniyana et al. (2006) suggested that elongation of milled rice in previously soaked condition seems to be more with low GT and intermediate amylose content. Increased storage life can remarkably improve cooking quality of rice (Faruq et al. 2014; Faruq et al. 2003a; b; Handoko et al. 2013). Cooking of aged rice results in more elongation of kernel and higher head rice. It is a common practice in South and Southeast Asian countries. However, storing of freshly harvested rice for more than 3–4 months affects its quality.

Ahn et al. (1993) identified a QTL on chromosome 8 for cooked rice elongation. Rani et al. (2011) found that a functional marker targeting an SNP in the *GS3* is associated with kernel elongation. Tian et al. (2005) detected three, two, and two QTLs for water absorption, volume expansion, and cooked rice elongation, respectively, in a DH population. While no QTL on chromosomes 3 and 8 was detected, one common QTL for all the traits is at the *Wx* locus on chromosome 6, suggesting that the *Wx* gene plays a major role in determining these cooking characteristics in addition to other cooking and eating quality traits (Tian et al. 2005).

1.3.7 Flavor and Aroma

Aroma or fragrance of the cooked rice is the most important feature of fragrance rice. Fragrance of cooked rice is the result of a combination of more than 100 volatile compounds, and among those, one chemical, 2-acetyl-1-pyrroline (2AP), is considered as the main contributor of cooked rice fragrance because of its low odor threshold value (Grosch and Schieberle 1997). The 2AP is considered as responsible for a variety of food flavors by different scientists, such as popcorn (Schieberle 1995), baguettes (Zehentbauer and Grosch 1998), cheese (Zehentbauer and Reineccius 2002), corn tortillas (Buttery and Louisa 1995), ham (Carrapiso et al. 2003), and green tea (Kumazawa and Hideki 2002). Among the volatile chemicals present in the aromatic rice, 2AP had lower threshold (Yoshihashi 2002).

The aroma of 2AP was evaluated and found to be popcorn-like, and the evaluation order of the amount of popcorn-like odor in 10 different rice varieties ranked them in general order of the concentration of 2AP (Tsugita 1985). The 2AP was found in different plant parts in different concentrations, usually high in aromatic types of rice, both aromatic and non-aromatic rice other than roots (Lorieux et al. 1996; Grosch and Schieberle 1997; Sakthivel et al. 2009). High demand for aroma among consumers makes this trait important among plant breeders.

The level of 2AP present in fragrance rice is much higher in aromatic rice than non-aromatic types (Fitzgerald et al. 2008a). Grosch and Schieberle (1997) described the aroma of 2AP as the freshly baked popcorn-like fragrance. It has been suggested that the presence of 2AP in aromatic rice depends on several factors, such as the location of cultivation, the temperature at the time of grain maturity and processing, the storage life of the grain, etc. (Itani et al. 2004; Fitzgerald et al. 2008b; Wongpornchai et al. 2004). Pleasant aroma, which we experience from the freshly cooked aromatic and non-aromatic rice, is a combination of different types of volatile compounds (Weber et al. 2000). The nice aroma we found in different freshly cooked rice is composed of similar types of volatiles but in different proportions. It has been suggested that a single recessive gene is responsible for the fragrance in rice (Lorieux et al. 1996). The gene responsible for the presence of 2AP in rice is called *fgr* or fragrance gene, located on chromosome 8, at a genetic distance of 4.5 *cM*, and is the principal compound of fragrance of rice (Cordeiro et al. 2002; Lorieux et al. 1996).

A histochemical test was developed by Nadaf et al. (2006) to confirm the presence of 2AP in aromatic rice. Nagaraju et al. (2002) developed a marker-assisted selection method for the confirmation of basmati type rice at the Center for DNA Fingerprinting and Diagnostics (CDFD), Hyderabad. The size and shape of the grain and the odor produced upon cooking have remarkable market interest of the consumers and exporters. The aroma of cooked rice has been studied by different scientists, and several aroma fractions were identified responsible for various fragrance (Widjaja et al. 1996; Jezussek et al. 2002; Champagne 2008; Yang et al. 2008; Zeng et al. 2009). Different components of rice aroma composed of saturated and unsaturated aldehydes, different alcohols, and cyclic compounds like hexanal, 1-octen-3-ol, and 2-pentylfuran are markers of both quality and ageing, while 2AP is the aroma quality marker for fragrance rice (Widjaja et al. 1996; Grimm et al. 2001; Mahatheeranont et al. 2001; Laguerre et al. 2007; Champagne 2008).

The aroma produced upon cooking of fragrance rice is the result of volatilization of different vaporous compound. The fragrance of freshly cooked rice can be described in different ways such as popcorn, alfalfa-grassy-green bean, floral, aromatic (fairy floss, caramel), vanilla, animal, grassy, hay-like (musty), corn, dairy, sewer, metallic, etc. Different variations in the aroma of rice can be a result of environmental conditions. The fragrance of rice can be measured by direct sniffing, or for more accuracy, gas chromatography can be used.

The aroma in fragrance rice is a complex character and determined by several chemical compounds. Two aromatic rice varieties, Jasmine 85 (Singh et al. 2000) and Della (Jodon and Sonnier 1973), were developed by researchers but criticized for their dull color and less noticeable aroma (Pinson 1994) and were found to contain much higher concentration of 2AP than Thai Jasmine variety (Weber et al. 2000). It has been suggested that the aroma of Della variety was controlled by a single recessive gene located on chromosome 8 as determined by Simple Sequence Repeat (SSR) method (Cordeiro et al. 2002).

Different types of aromatic rice (Dragon Eyeball 100, Amber, Jasmine 85, Della-X, PI45917, and A-301) were analyzed by Pinson (1994), and it was concluded that, among these varieties, four (Jasmine 85, Della-X, PI45917, and A-301) contained a single allelic gene and two (Dragon Eyeball 100, Amber) contained the allelic gene plus a novel gene.

Different scientists used different techniques for the determination of rice aroma. Berner and Hoff (1986) developed a method of determination of rice aroma by chewing half of a seed. Detecting rice aroma by chewing few seeds was also used (Dhulappanavar 1976). Reinke et al. (1991) developed a method involving testing all the seeds individually for the determination of rice aroma quality, and this method is still used worldwide in different breeding programs. An assay was developed involving adding of KOH or 12-KI to the plant sample for the release of aroma (Sood and Siddiq 1978). Gas chromatography was also used for the detection of 2AP in the plant sample, but this method requires large amount of seeds and is time consuming (Lorieux et al. 1996; Widjaja et al. 1996). Recently, marker-assisted selection was used for the detection of rice aroma using specific markers genetically linked to fragrance of rice (Kibria et al. 2008).

Leaf tissue and grain can be used to detect rice aroma. Nagaraju et al. (1975) developed a method of detecting rice aroma by heating leaves in water. Sood and Siddiq (1978) used 1.7% KOH on rice leaf and seed for the detection of aroma. Pinson (1994) mentioned that these methods can cause permanent damage to nasal passage, if carried out for a long period of time. Smelling or chewing individual seed for the detection of aroma was also being used by the researchers (Ghose and Butany 1952). Above-mentioned methods are subjective to individuals, and their ability to detect decreases with time. As a result of constant chewing and sniffing, it may cause damage to nasal and moth parts of researchers. These methods also require a panel of experts to detect aroma. Thus, sensory methods have limitations in case of detecting large samples (Reinke et al. 1991). Sensory methods for the detection of aroma in rice are currently being used by several researchers (Bounphanousay et al. 2008; Ahmad et al. 2009; Yi et al. 2009), and they are also following the standard protocol at the International Rice Research Institute (IRRI) grain quality laboratory.

The gas chromatography-mass spectrometry (GC-MS) method only detects the presence of compounds rather the actual amount often done by gas chromatography. Volatile profile of several rice cultivars was determined before and after storage using solid-phase microextraction (SPME) fibers in conjunction with gas chromatography mass spectrophotometer (Bryant and McClung 2011). Researchers found 93 volatile compounds among those plant samples, of which 64 were not previously identified in rice. New volatile compounds were identified in aromatic rice other than 2AP which is not present in non-aromatic rice. This research also suggested that storage life and aroma quality had minimal effect in aroma quality. Identification of 2AP using gas chromatography requires large tissue samples, and this process is time consuming (Lorieux et al. 1996; Widjaja et al. 1996).

Most of the scientists suggested that aroma of rice was mainly controlled by a single recessive gene (Lorieux et al. 1996; Jin et al. 2003; Bradbury et al. 2005; Datta 2011). It has been suggested that the production of 2AP on rice is mainly controlled by a single recessive gene, located on chromosome 8. Researchers also fine-mapped the chromosomal location of the aroma gene and developed both SSR (Jones et al. 2002) and SNP markers (Jin et al. 2003) markers for breeding program.

The discovery of genome sequence of rice provided the opportunity for the researchers to identify genes responsible for the aroma quality by comparing genome sequence between aromatic and non-aromatic rice (Bourgis et al. 2008). This will give future researchers opportunity to target the exact location on chromosome 8 for the resequencing of aromatic genes. Bradbury et al. (2005) designed an

Table 1.4 List of genes responsible for eating and cooking quality of rice

Genes	Controlling character	Position	References
Wx locus *SBEI and SBEIIb*	Amylose content Amylose content	Chromosome 6 Chromosome 2	Wang et al. (1995) Sun et al. (2017)
SSIIa	Gelatinization temperature	Chromosome 6	Umemoto et al. (2002a, b, 2004), Waters et al. (2006)
fgr	Rice aroma	Chromosome 8	Lorieux et al. (1996)
psy (Daffodil)+ *crt1 psy* (Maize)+ *crt1*	Vitamin A		Ye et al. (2000), Paine et al. (2005)
SoyferH1 (iron) *Ferritin* (iron) *HvNAS1* (iron) *OsNAS3* (iron) *OsYSL2* (iron) *OsIRT1* (iron+zinc)	Iron and zinc	Chromosome 6 Chromosome 7 Chromosome 5 Chromosome 7 Chromosome 2 Chromosome 3	Goto et al. (1999) Paul et al. (2012) Masuda et al. (2009) Lee et al. (2009) Ishimaru et al. (2010) Lee and An (2009)
(TKTKK) coding sequences	Protein	Chromosome 10	Jiang et al. (2016)

excellent marker for the genotypic selection of aromatic rice from non-aromatic types by using allele-specific amplification of a gene encoding betaine aldehyde dehydrogenase 2 (*BADH2*) on chromosome 8. Both aromatic and non-aromatic varieties contain a copy of gene encoding *BADH2*. However, non-aromatic rice variety carries a fully functional copy of the gene, while aromatic rice variety carries a copy that consists of an eight-base pair deletion and three SNPs in exon 7, which results in a frame shift that generates a premature stop codon that presumably stops the function of the gene.

Selection of aromatic properties of rice using maker-assisted selection has remarkable advantages over traditional methods. It is specific and less time consuming and requires small amount of leaf tissue for detection, which is suitable for the analysis of large size of plant population over relatively short period of time. Marker-assisted selection method allows researchers to screen the population for specific genotypes in early stage of plant growth other than conventional methods, which require seeds for the analysis of aroma (Zhou et al. 2003; Toojinda et al. 2005; Zhang 2007; Thomson et al. 2007). A list of different genes and their influence on the quality of rice is given in Table 1.4.

1.4 Nutraceutical Rice

1.4.1 Gamma Aminobutyric Acid (GABA) Rice

Gamma aminobutyric acid (GABA) is an amino acid found in mammalian brain and is also present in plants, supposed to be a typical inhibitory neurotransmitter in the central nervous system. It is also active and found in the peripheral nerve or tissues.

Researchers already developed different types of food intended to reduce hypertension using GABA as the main active ingredient (Inoue et al. 2003; Sugiyama et al. 2008). Kishine et al. (2008) reported some special properties of rice preventing diseases related to lifestyle. A high amount of vitamins and GABA was also reported in sprouted brown rice by Wu et al. (2013).

The production of GABA can be rapidly enhanced by adding water to brown rice. One Japanese company (Satake Co., Ltd.) takes advantage of that fact and produces a new variety of white rice enriched with GABA, and they called it GABA rice (Mizuno et al. 2012a, b). In GABA rice, they maintain the usual nutritional quality of white rice plus they add GABA to it. This rice can be consumed like normal rice which ensures high intake of GABA can lead to long-term health benefit. It was suggested that regular consumption of GABA rice can reduce hypertension and also can prevent early onset of high blood pressure and cardiovascular related diseases, if it was consumed for a long period of time (Nishimura et al. 2015).

1.4.2 Zinc Rice

Deficiency of zinc in diet can cause lack of development, stunted growth, and death from diarrheal disease in young children. Zinc deficiency among children can be reduced by regular consumption of zinc rice. Zinc rice was developed in 2013 by the researchers of Bangladesh Rice Research Institute (BRRI) using zinc-rich germplasm provided by IRRI and was released in Bangladesh. This was the world's first zinc-rich rice variety that contains 20–22 ppm (parts per million) of zinc, which was higher than their usual amount in normal rice (14–16 ppm).

1.4.3 Black Rice

The color of the outer part of seed kernel, which is called pericarp, of this rice is black due to the presence of anthocyanin (color pigment). Black rice is glutinous, rich in nutrients, and cultivated mainly in Asia. This rice is also called king's rice, forbidden rice, imperial rice, prized rice, heaven rice, and purple rice. This rice is believed to be the universal cure for a long range of diseases because of its high nutritive value and curative effects. This rice is also called long-life rice because of its supposed influence on life expectancy. There are several varieties of this rice that are available and traditionally cultivated in different parts of Asia, such as India, Thailand, and China (Kong et al. 2008). There are more than 200 varieties of black rice available in the world, and only China produces 62% of global production. China already developed more than 54 improved black rice varieties with higher production and multiple disease and insect resistance. Interest in black rice of different countries perhaps indicates by their collection of this rice germplasm, for example, China has 359, Sri Lanka has 50, Indonesia has 42, India has 30,

Bangladesh has 24, and Philippines has 25 (Chaudhary 2003). Compared to other kinds of rice, this rice is rich in dietary fiber, proteins, minerals (Fe, Zn, Ca, and P), and vitamins. The demand of black rice is increasing in Europe and the USA because of its attractive color and high nutritive value. It is also reported that black rice is double nutritious than blueberries in terms of antioxidant properties (Ichikawa et al. 2001).

1.5 Conclusion

The demand for high-quality rice is increasing day by day with the increase of total rice production. Criteria of high-quality rice for which consumers are willing to pay premium price depend upon its customers and their locality. They did not meet up with the criteria set for good quality rice in the laboratory. Different groups of people used to look for different types of attributes for their desired high-quality rice in the post-harvest system. Variety is the main contributor for grain quality which indicates its size, shape, color, and aroma. Post-harvest management of grain is very important to maintain or even improve it. Moisture content of finished rice is one of the main considerations in rough rice. Traders always look for good variety name, varietal integrity, good physical properties, low moisture content, and high head milled rice yield. Cooking and eating quality of rice is determined by physicochemical properties, particularly apparent amylose content. People around the world prefer more head rice yield and semitransparent grain than small and dull-looking rice. A rich customer is willing to pay premium price for their desired group of qualities which indicates their ability to pay. Although preference over high-quality rice did not vary between different income levels, lower-income-level consumer prefers grain that is more filling. Different countries around the world prefer different levels of amylose in their rice. Some Asian countries like Bangladesh, Vietnam, and Thailand prefer high amylose content. Pakistan, Bhutan, and Myanmar prefer intermediate amylose content, whereas low-amylose-content rice is preferred by Japan, the Republic of Korea, and Taiwan. Rice with very low amylose content is preferred by small proportions of people of Indonesia, Laos, Brunei, the Republic of Korea, Myanmar, Sarawak (Malaysia), Thailand, and the Philippines. Waxy types of rice are preferred in some portions of Indonesia, China, the Republic of Korea, Japan, West Malaysia, Laos, Thailand, Myanmar, and the Philippines. In north and northeast Thailand and Laos, waxy rice is used as the staple food. The GT of rice is also a very important factor for the determination of cooking and eating quality. It indicates the amount of energy required for cooking particular rice. Customers prefer low GT rice because it requires less energy and time to prepare. In case of gel consistency, customers prefer medium over hard gel consistency in non-waxy and waxy types of rice. In the international markets, demand is high for medium and high-quality long grain rice, short and medium grain rice, parboiled rice, Thai jasmine and aromatic rice, glutinous or waxy types of rice, etc., depending upon locality of the consumers (Efferson 1985). In Europe and the Middle East, demand is high for high-quality long grain rice, medium-quality long grain rice is mostly consumed in the underdeveloped countries of Asia, short grain rice is consumed in various areas, parboiled rice of high quality is demanded in the Near East and African countries, and lower-quality parboiled rice is demanded in Asia and Africa. The Near East countries mostly consume aromatic rice. Laos and some small parts of the world prefer waxy rice.

References

Ahmad SW, Ookawa T, Yoshihashi T, Khalid MA, Yosofzai W, Oikawa Y, Hirata Y (2009) Characterization of aroma and agronomic traits in Afghan native rice cultivars. Plant Prod Sci 12:63–69

Ahn SN, Bollich CN, McClung AM, Tanksley SD (1993) RFLP analysis of genomic regions associated with cooked-kernel elongation in rice. Theor Appl Genet 87:27–32

Atukorale DP (2002) *Nutritional value of rice*. Reproduced from the Daily News

Atungulu GG, Pan Z (2014) Rice industrial processing worldwide and impact on macro-and micronutrient content, stability, and retention. Ann N Y Acad. Sci 1324(1):15–28

Bao JS (2012) Toward understanding the genetic and molecular bases of the eating and cooking qualities of rice. Cereal Foods World 57:148–156

Bao J (2014) Genes and QTLs for rice grain quality improvement. In: InRice-Germplasm, genetics and improvement 2014 Apr 23. IntechOpen

Bao JS, Fjellstrom R, Bergman C (2009) Rice end-use quality. In: Datta SK (ed) Rice improvement in the genomics era. The CRC Press, Taylor & Francis Group, Florida 33487-2742, pp 269–305

Bergman CJ (2019) Rice end-use quality analysis. In: Rice 2019 Jan 1. AACC Inter Press, St. Paul, MN 55121 USA, pp 273–337

Bergman C, Bhattcharya K, Ohtsubo K (2004) Rice end-use quality analysis. In: Champagne E (ed) Rice chemistry and technology. AACC, St Paul, pp 415–472

Bhattacharya KR (2009) Physiochemical basis of eating quality of Rice. Cereal Foods World (CFW) 54(1):18

Bounphanousay C, Jaisil P, McNally KL, Sanitchon J, Sackville HNR (2008) Variation of microsatellite markers in a collection of Lao's black glutinous rice (*Oryza sativa* L.). Asian J Plant Sci 7(2):140–148

Bourgis F, Guyot R, Gherbi H, Tailliez E, Amabile I, Salse J, Lorieux M, Delseny M, Ghesquiere A (2008) Characterization of the major fragance gene from an aromatic japonica rice and analysis of its diversity in Asian cultivated rice. Theor Appl Genet 117:353–368

Bradbury LM, Fitzgerald TL, Henry RJ, Jin Q, Waters DL (2005) The gene for fragrance in rice. Plant Biotechnol J 3:363–370

Bryant RJ, McClung AM (2011) Volatile profiles of aromatic and non-aromatic rice cultivars using SPME/GC MS. Food Chem 124(2):501–513

Buttery RG, Louisa CL (1995) Volatile flavor components of corn tortillas and related products. J Agri Food Chem 43(7):1878–1882

Cagampange GB, Perez CM, Juliano BO (1973) A gel consistency test for eating quality in rice. J Sci Food Agri 24(12):1589–1594

Carrapiso AI, Bonilla F, García C (2003) Effect of crossbreeding and rearing system on sensory characteristics of Iberian ham. Meat Sci 65:623–629

Champagne ET (2008) Rice aroma and flavor: a literature review. Cereal Chem 85(4):445–454

Champagne ET, Wood DF, Juliano BO, Bechtel DB (2004) The rice grain and its gross composition. Rice chem Technol 3:77–107

Chaudhary RC (2003) Speciality rices of the world: Effect of WTO and IPR on its production trend and marketing. J Food Agric Environ 1(2):34–41
Chen M, Bergman CJ, Pinson SR, Fjellstrom RG (2008) Waxy gene haplotypes: associations with pasting properties in an international rice germplasm collection. J Cereal Sci 48(3):781–788
Cordeiro GM, Christopher MJ, Henry RJ, Reinke RF (2002) Identification of microsatellite markers for fragrance in rice by analysis of the rice genome sequence. Mol Breed 9(4):245–250
Cruz ND, Khush GS (2000) Rice grain quality evaluation procedures. Aromatic rices 3:15–28
Datta SK (2011) Rice improvement through application of biotechnological tools. Rice Knowl Manag Portal:1–35
Dhulappanavar CV (1976) Inheritance of scent in rice. Euphytica 25(1):659–662
Eggum BO, Juliano BO, Maningat CC (1982) Protein and energy utilization of rice milling fractions by rats. Plant Foods Hum Nutr 31:371–376
Fan CC, Yu XQ, Xing YZ, Xu CG, Luo LJ, Zhang Q (2005) The main effects, epistatic effects and environmental interactions of QTLs on the cooking and eating quality of rice in a double haploid line population. Theor Appl Genet 110:1445–1452
Faruq G, Mohamad O, Hadzim M, Meisner CA, Perai S (2003a) Optimization of aging time and temperature for four Malaysian rice cultivars. Pak J Nutr 2:125–131
Faruq GO, Hadjim MO, Meisner CA (2003b) Kernel ageing: an analysis in four Malaysian rice varieties. Int J Agri Biol 3:230–232
Faruq G, Prodhan ZH, Nezhadahmadi A (2014) Effects of ageing on selected cooking quality parameters of Rice. Int J Food Prop 18(4):922–933
Fitzgerald TL, Daniel LEW, Robert JH (2008a) The effect of salt on betaine aldehyde dehydrogenase transcript levels and 2-acetyl-1-pyrroline concentration in fragrant and non-fragrant rice (*Oryza sativa*). Plant Sci 175:539–546
Fitzgerald MA, Sackville HNR, Calingacion MN, Verhoeven HA, Butardo VM (2008b) Is there a second fragrance gene in rice? Plant Biotechnol J 6:416–423
Fitzgerald MA, McCouch SR, Hall RD (2009) Not just a grain of rice: the quest for quality. Trends Plant Sci 14(3):133–139
Gao Z, Zeng D, Cheng F, Tian Z, Guo L, Su Y, Yan M, Jiang H, Dong G, Huang Y, Han B, Li J, Qian Q (2011) *ALK*, the key gene for gelatinization temperature, is a modifier gene for gel consistency in rice. J Integr Plant Biol 53:756–765
Ghose RLM, Butany WT (1952) Studies on the inheritance of some characters in rice (*Oryza sativa* L.). Ind J. Genet Plant Breed 12:26–30
Goto F, Yoshihara T, Shigemoto N, Toki S, Takaiwa F (1999) Iron fortification of rice seed by the soybean ferritin gene. Nat Biotechnol 17(3):282–286
Graham R (2002) A proposal for IRRI to establish a grain quality and nutrition research center. Los Bafios, IRRI, Laguna, p 15
Grimm CC, Bergman C, Delgado JT, Bryant R (2001) Screening for 2-acetyl-1-pyrroline in the headspace of rice using SPME/GC-MS. J Agri Food Chem 49:245–249
Grosch W, Schieberle P (1997) Flavor of cereal products-a review. Cereal Chem 74(2):91–97
Hamaker BR, Griffin VK (1990) Changing the viscoelastic properties of cooked rice through protein disruption. Cereal Chem 67(3):261–264
Han Y, Xu M, Liu X, Yan C, Korban SS, Chen X, Gu M (2004) Genes coding for starch branching enzymes are major contributors to starch viscosity characteristics in waxy rice (*Oryza sativa* L.). Plant Sci 166:357–364
Handoko DD, Methven L, Elmore JS, Mottram DS (2013) Comparison of the Maillard-derived aroma volatiles of cooked milled and brown rice. In: Flavour science: proceedings from XIII Weurman Flavour Research Symposium. Academic Press, p 83
He P, Li SG, Qian Q, Ma YQ, Li JZ, Wang WM (1999) Genetic analysis of rice grain quality. Theor App Genet 98(3–4):502–508
He Y, Han Y, Jiang L, Xu C, Lu J, Xu M (2006) Functional analysis of starch-synthesis genes in determining rice eating and cooking qualities. Mol Breed 18(4):277–290

Heinemann RJ, Fagundes PD, Pinto EA, Penteado MD, Lanfer-Marquez UM (2005) Comparative study of nutrient composition of commercial brown, parboiled and milled rice from Brazil. J Food Comp Anal 18(4):287–296
Ichikawa H, Ichiyanagi T, Xu B, Yoshii Y, Nakajima M, Konishi T (2001) Antioxidant activity of anthocyanin extract from purple black rice. J Med Food 4(4):211–218
Inoue K, Shirai T, Ochiai H (2003) Blood-pressure-lowering effect of a novel fermented milk containing gamma-aminobutyric acid (GABA) in mild hypertensives. Eur J Clin Nutr 57:490–495
Ishimaru Y, Masuda H, Bashir K (2010) Rice metal-nicotianamine transporter, *OsYSL2*, is required for the long-distance transport of iron and manganese. Plant J 62:379–390
Itani T, Tamaki M, Hayata Y, Fushimi T, Hashizume K (2004) Variation of 2-acetyl-1-pyrroline concentration in aromatic rice grains collected in the same region in Japan and factors affecting its concentration. Plant Prod sci 7:178–183
Jezussek M, Juliano B, Schieberle P (2002) Comparison of key aroma compounds in cooked brown rice varieties based on aroma extract dilution analysis. J Agri Food Chem 50:1101–1105
Jiang SY, Ma A, Xie L, Ramachandran S (2016) Improving protein content and quality by over-expressing artificially synthetic fusion proteins with high lysine and threonine constituent in rice plants. Sci Rep. https://doi.org/10.1038/srep34427
Jin Q, Waters D, Cordeiro GM, Henry RJ, Reinke RF (2003) A single nucleotide polymorphism (SNP) marker linked to the fragrance gene in rice (*Oryza sativa* L.). Plant Sci 165:359–364
Jodon NE, Sonnier E (1973) Registration of Della Rice1 (Reg. No. 37). Crop Sci 13:773–773
Jones E, Dupal M, Dumsday J, Hughes L, Forster J (2002) An SSR-based genetic linkage map for perennial ryegrass (*Lolium perenne* L.). Theor Appl Genet 105:577–584
Juliano BO (1985a) Rice: chemistry and technology, 2nd ed. Am Assoc Cereal Chem, St Paul, p 774
Juliano BO (1985b) Criteria and tests for rice grain qualities. Rice chemistry and technology, 2nd edn. Am Assoc Cereal Chem, St Paul, pp p443–p513
Juliano BO (1993) Rice in human nutrition. FAO, Rome, p 162
Juliano BO (2003) Rice chemistry and quality. Philippine Rice Research Institute, Los Banos, Laguna
Juliano BO, Pascual CG (1980) Quality characteristics of milled rice grown in different countries. IRRI Res Paper Ser (Philippines) 48:1–25
Juliano BO, Perdon AA (1975) Gel and molecular properties of non-waxy rice starch. Starch Starke 27(4):115–120
Kennedy G, Burlingame B, Nguyen N (2002) Nutrient impact assessment of rice in major rice consuming countries. Korea 165:12–15
Khan AS, Muhammad I, Muhammad A (2009) Estimation of genetic variability and correlation for grain yield components in rice (*Oryza sativa* L.). J Agri Env Sci 6:585–590
Kharabian-Masouleh A, Waters DL, Reinke RF, Ward R, Henry RJ (2012) SNP in starch biosynthesis genes associated with nutritional and functional properties of rice. Sci Rep 2:1–9
Khush GS, Paule CM, DeLaCruz MN (1979) Rice quality evaluation and improvement at IRRI. In: Proceedings of the workshop on chemical aspects of rice grain quality. International Rice Research Institute, Los Banos, Laguna, pp 21–31
Kibria K, Islam MM, Begum SN (2008) Screening of aromatic rice lines by phenotypic and molecular markers. Ban J Bot 37:141–147
Kishine M, Suzuki K, Nakamura S, Ohtsubo K (2008) Grain qualities and their genetic derivation of 7 new rice for Africa (NERICA) varieties. J Agric Food Chem 56:4605–4610
Kong L, Wang Y, Cao Y (2008) Determination of Myo-inositol and D-chiro-inositol in black rice bran by capillary electrophoresis with electrochemical detection. J Food Compos 21(6):501–504
Kumar I, Khush GS, Juliano BO (1987) Genetic analysis of waxy locus in rice (*Oryza sativa* L.). Theor Appl Genet 73(4):481–488
Kumazawa K, Hideki M (2002) Identification of potent odorants in different green tea varieties using flavor dilution technique. J Agri Food Chem 50:5660–5663

Laguerre M, Mestres C, Davrieux F, Ringuet J, Boulanger R (2007) Rapid discrimination of scented rice by solid phase microextraction, mass spectrometry and multivariate analysis used as amass sensor. J Agri Food Chem 55:1077–1083

Lanceras JC, Huang Z-L, Naivikul O, Vanavichit A, Ruanjaichon V, Tragoonrung S (2000) Mapping of genes for cooking and eating qualities in Thai jasmine rice (kdml105). DNA Res 7(2):93–101

Lee S, An G (2009) Over-expression of *OsIRT1* leads to increased iron and zinc accumulations in rice. Plant Cell Environ 32:408–416

Lee S, Jeon US, Lee SJ (2009) Iron fortification of rice seeds through activation of the *nicotianamine synthase* gene. Proc Natl Acad Sci 106:22014–22019

Lestari P, Ham TH, Lee HH, Jiang W, Chu SH, Kwon SW, Ma K, Lee JH, Cho YC, Koh HJ (2009) PCR marker-based evaluation of the eating quality of Japonica Rice (*Oryza sativa* L.). J Agric Food Chem 57:2754–2762

Little RR, Hilder GB, Dawson EH (1958) Differential effect of dilute alkali on 25 varieties of milled white rice. Cereal Chem 35:111–126

Lorieux M, Petrov M, Huang N, Guiderdoni E, Alain G (1996) Aroma in rice: genetic analysis of a quantitative trait. Theor App Genet 93:1145–1151

Lu Y, Xiao P, Shao YF, Zhang G, Thanyasiriwat T, Bao JS (2010) Development of new markers to genotype the functional SNPs of *SSIIa*, a gene responsible for gelatinization temperature of rice starch. J Cereal Sci 52(3):438–443

Mahatheeranont S, Keawsard S, Dumri K (2001) Quantification of the rice aroma compound, 2acetyl-1-pyrroline, in uncooked Khao Dawk Mali 105 brown rice. J Agri Food Chem 49:773–779

Masuda H, Usuda K, Kobayashi T (2009) Overexpression of the barley nicotianamine synthase gene *HvNAS1* increases iron and zinc concentrations in rice grains. Rice 2:155–166. https://doi.org/10.1007/s12284-009-9031-1

Mizuno H, Fukumori T, Liu K, Sasaki Y, Ochiai S (2012a) The influence of heating and humidifying by humidified air on GABA generation and quality in paddy. J Jpn Soc Agric Mach 74:226–233

Mizuno H, Fukumori T, Liu K, Sasaki Y, Ochiai S (2012b) Development of GABA enriching equipment for brown rice by heating and humidifying of moist air. J Jpn Soc Agric Mach 74:234–243

Myers AM, Morell MK, James MG, Ball SG (2000) Recent progress toward understanding biosynthesis of the amylopectin crystal. Plant Physiol 122(4):989–998

Nadaf AB, Krishnan S, Wakte KV (2006) Histochemical and biochemical analysis of major aroma compound (2-acetyl-1-pyrroline) in Basmati and other scented rice (*Oryza sativa* L). Curr Sci 91(11):1533–1536

Nagaraju M, Chaudhary D, Balakrishna-Rao MJ (1975) Simple technique to identify scent in rice and inheritance pattern of scent. Curr Sci 44:599

Nagaraju J, Kathirvel M, Kumar RR, Siddiq EA, Hasnain SE (2002) Genetic analysis of traditional and evolved Basmati and non-Basmati rice varieties by using fluorescence based ISSR-PCR and SSR markers. Proc Natl Acad Sci USA 99(9):5836–5841

Nakamura Y, Sakurai A, Inaba Y, Kimura K, Iwasawa N, Nagamine T (2002) The fine structure of amylopectin in endosperm from Asian cultivated rice can be largely classified into two classes. Starch-Starke 54(3–4):117–131

Nakamura Y, Francisco PB, Hosaka Y, Sato A, Sawada T, Kubo A, Fujita N (2005) Essential amino acid of *starch synthase IIa* differentiate amylopectin structure and starch quality between *japonica* and *indica* rice varieties. Plant Mol Biol 58:213–227

Nishimura M, Yoshida S, Haramoto M (2015) Effects of white rice containing enriched gamma-aminobutyric acid on blood pressure. J Tradit Compl Med 6(1):66–71

Paine JA, Shipton CA, Chaggar S (2005) Improving the nutritional value of golden rice through increased pro-vitamin A content. Nat Biotechnol 23:482–487

Parmar RR, Jain KR, Modi CK (2011) Unified approach in food quality evaluation using machine vision. In: Int Con Ad Comp Comm Jul 22. Springer, Berlin/Heidelberg, pp 239–248

Paul S, Ali N, Gayen D (2012) Molecular breeding of *Osfer2* gene to increase iron nutrition in rice grain. GM Crops Food 3(4):310–316

Pedersen B, Eggum BO (1983) The influence of milling on the nutritive value of flour from cereal grains. 4. Rice. Plant Foods Hum Nutr 33(4):267–278

Perez CM (1979) Gel consistency and viscosity of rice. In: Proceeding of the workshop on chemical aspects of rice grain quality. International Rice Research Institute, Los Banos, Laguna, pp 303–311

Pinson SR (1994) Inheritance of aroma in six rice cultivars. Crop Sci 34(5):1151–1157

Rani NS, Manish KP, Prasad GSV, Sudharshan I (2006) Historical significance, grain quality features and precision breeding for improvement of export quality basmati varieties in India. Ind J Crop Sci 1:29–41

Rani NS, Sivaranjani AKP, Madhav MS, Sundaram RM, Prasad GSV, Srikanth S, Pandey M, Suneetha K, Sudharshan I (2011) Identification of molecular markers for cooking quality traits of rice. Indian J Genet Plan Breed 71:129–138

Reinke RF, Welsh LA, Reece JE, Lewin LG, Blakeney AB (1991) Procedures for quality selection of aromatic rice varieties. Int Rice Res Newslett 16:10–11

Rohilla R, Singh VP, Singh US, Singh RK, Khush GS (2000) Crop husbandry and environmental factors affecting aroma and other quality traits. Aromatic rices, Oxford and IBH Publishing Co. Pvt. Ltd, New Delhi, pp 201–216

Rosniyana A, Hashifah MA, Shariffah NSA (2006) Quality evaluation of retailed packed Malaysian milled rice sold in the market. J Tropical Agri Food Sci 34:45

Sakthivel K, Sundaram RM, Rani NS, Balachandran SM, Neeraja CN (2009) Genetic and molecular basis of fragrance in rice. Biotechnol Adv 27:468–473

Schieberle P (1995) Quantitation of important roast-smelling odorants in popcorn by stable isotope dilution assays and model studies on flavor formation during popping. J Agri Food Chem 43(9):2442–2448

Septiningsih EM, Trijatmiko KR, Moeljopawiro S, McCouch SR (2003) Identification of quantitative trait loci for grain quality in an advanced backcross population derived from the *Oryza sativa* variety ir64 and the wild relative *O. rufipogon*. Theor App Genet 107(8):1433–1441

Shao G, Tang S, Luo J, Jiao G, Wei X, Tang A, Wu j Z j, Hu P (2010) Mapping of *qGL7-2*, a grain length QTL on chromosome 7 of rice. J Genet Genom 37:523–531

Singh RK, Singh US, Khush GS, Rohilla R (2000) Genetics and biotechnology of quality traits in aromatic rices. Aromatic Rices (Edt. RH Singh, US Singh, GS Khush), Oxford & IBH Publishing Co. Pvt. Ltd., New Delhi, pp 47–70

Smith AM, Denyer K, Martin C (1997) The synthesis of the starch granule. Ann Rev Plant Physiol Plant Mol Biol 48(1):67–87

Sood BC, Siddiq EA (1978) A rapid technique for scent determination in rice. Ind J Genet Plant Breed 38(2):268–275

Su Y, Rao Y, Hu S, Yang Y, Gao Z, Zhang G (2011) Map-based cloning proves *qgc-6*, a major qtl for gel consistency of *japonica/indica* cross, responds by waxy in rice (*Oryza sativa* L.). Theor App Genet 123(5):859–867

Sugiyama S, Kishi M, Fushimi T, Oshima Y, Kajimoto O, Kaga T (2008) Hypotensive effect and safety of brown rice vinegar with high concentration of GABA on mild hypertensive subjects. Jpn Pharmacol Ther 36:429–444

Sun Y, Jiao G, Liu Z (2017) Generation of high-amylose rice through CRISPR/Cas9-mediated targeted mutagenesis of starch branching enzymes. Front Plant Sci. https://doi.org/10.3389/fpls.2017.00298

Suwannaporn P, Linnemann A (2007) Rice-eating quality among consumers in different rice grain preference countries. J Sens Stud. 23:1–13

Tabkhkar N, Rabie B, Sabouri A (2012) Genetic diversity of rice cultivars by microsatellite markers tightly linked to cooking and eating quality. Aus J Crop Sci 6:980–985

Tan YF, Li JX, Yu SB, Xing YZ, Xu CG, Zhang Q (1999) The three important traits for cooking and eating quality of rice grains are controlled by a single locus in an elite rice hybrid, shanyou 63. Theor App Genet 99(3–4):642–648

Thomson MJ, Septiningsih EM, Suwardjo F, Santoso TJ, Silitonga TS, McCouch SR (2007) Genetic diversity analysis of traditional and improved Indonesian rice (*Oryza sativa* L.) germplasm using microsatellite markers. Theor Appl Genet 114:559–568

Tian R, Jiang GH, Shen LH, Wang LQ, He YQ (2005) Mapping quantitative trait loci underlying the cooking and eating quality of rice using a DH population. Mol Breed 15(2):117–124

Tian Z, Qian Q, Liu Q, Yan M, Liu X, Yan C (2009) Allelic diversities in rice starch biosynthesis lead to a diverse array of rice eating and cooking qualities. Proc Natl Acad Sci 106(51):21760–21765

Toojinda T, Tragoonrung S, Vanavichit A, Siangliw JL, Pa-In N, Jantaboon J, Siangliw M, Fukai S (2005) Molecular breeding for rainfed lowland rice in the Mekong region. Plant Prod Sci 8:330–333

Tran NA, Daygon VD, Resurreccion AP, Cuevas RP, Corpuz HM, Fitzgerald MA (2011) A single nucleotide polymorphism in the Waxy gene explains a significant compomponent of gel consistency. Theor Appl Genet 123:519–525

Traore K (2005) Characterization of novel rice germplasm from West Africa and genetic marker associations with rice cooking quality. PhD diss., Texas A & M University, USA

Traore K, McClung AM, Chen MH, Fjellstrom R (2011) Inheritance of flour paste viscosity is associated with a rice waxy gene exon 10 SNP marker. J Cereal Sci 53:37–44

Tsugita T (1985) Aroma of cooked rice. Food Rev In 1(3):497–520

Umemoto T, Yano M, Satoh H, Shomura A, Nakamura Y (2002a) Mapping of a gene responsible for the difference in amylopectin structure between *japonica*-type and *indica*-type rice varieties. Theor App Genet 104(1):1–8

Umemoto T, Yano M, Satoh H, Shomura A, Nakamura Y (2002b) Mapping of validation of single nucleotide polymorphisms in relation to starch gelatinization temperature and other physicochemical properties in rice (*Oryza sativa* L.). Theor App Genet 113:1171–1183

Umemoto T, Aoki N, Lin HX, Nakamura Y, Inouchi N, Sato Y, Yano M, Hirabayashi H, Maruyama S (2004) Natural variation in rice starch synthase *IIa* affects enzyme and starch properties. Funct Plant Biolo 31:671–684

Wang ZY, Zheng F, Shen GZ, Gao JP, Snustad DP, Li MG, Zhang JL, Hong MM (1995) The amylose content in rice endosperm is related to the post-transcriptional regulation of the waxy gene. Plant J 7(4):613–622

Wang L, Liu W, Xu Y, He Y, Luo L, Xing Y (2007) Genetic basis of 17 traits and viscosity parameters characterizing the eating and cooking quality of rice grain. Theor App Genet 115(4):463–476

Waters DLE, Henry RJ, Reinke RF, Fitzgerald MA (2006) Gelatinization temperature of rice explained by polymorphisms in starch synthase. Plant Biotechnol J 4:115–122

Weber D, Rohilla R, Singh US (2000) Chemistry and biochemistry of aroma in scented rice. Aromatic Rices (Edt. RH Singh, US Singh, GS Khush), Oxford & IBH Publishing Co. Pvt. Ltd., New Delhi, pp 29–46

Widjaja RW, Craske JD, Wootton M (1996) Comparative studies on volatile components of non-fragrant and fragrant rices. J Sci Food Agri 70(2):151–161

Wongpornchai S, Dumri K, Jongkaewwattana S, Siri B (2004) Effects of drying methods and storage time on the aroma and milling quality of rice (*Oryza sativa* L.) cv. Khao Dawk Mali 105. Food Chem 87:407–414

Wu DX, Shu QY, Xia YW (2001) Assisted selection for early *indica* rice with good eating quality by RVA profile. Acta Agron Sin 27:165–172

Wu F, Yang N, Touré A, Jin Z, Xu X (2013) Germinated brown rice and its role in human health. Crit Rev Food Sci Nutr 53:451–463

Yang DK, Lee KS, Jeong OY, Kim KJ, Kays SJ (2008) Characterization of volatile aroma compounds in cooked black rice. J Agri Food Chem 56(1):235–240

Yang F, Chen Y, Tong C, Huang Y, Xu F, Li K, Corke H, Sun M, Bao J (2014) Association mapping of starch physicochemical properties with starch synthesisrelated gene markers in non-waxy rice (*Oryza sativa* L.). Molecular Breed 34(4):1747–1763

Ye X, Al-Babili S, Klöti A (2000) Engineering the provitamin a (β-carotene) biosynthetic pathway into (carotenoid-free) rice endosperm. Science 287:303–305. https://doi.org/10.1126/science.287.5451.303

Yi M, Nwe KT, Vanavichit A, Chai-arree W, Toojinda T (2009) Marker assisted backcross breeding to improve cooking quality traits in Myanmar rice cultivar Manawthukha. Field Crops Res 113:178–186

Yoshihashi T (2002) Quantitative analysis on 2-acetyl-1-pyrroline of an aromatic rice by stable isotope dilution method and model studies on its formation during cooking. J Food Sci 67:619–622

Yousif NE (2000) Effect of Fermentation and dry cooking following fermentation on protein fractions and invitro protein digestibility of sorghum, corn and rice. Ph.D. thesis, University of Khartoum-Sudan

Yu S, Ma Y, Menager L, Sun DW (2012) Physicochemical properties of starch and flour from different rice cultivars. Food Bioprocess Technol 5:626–637

Zehentbauer G, Grosch W (1998) Crust aroma of baguettes I. Key odorants of baguettes prepared in two different ways. J Cereal Sci 28(1):81–92

Zehentbauer G, Reineccius GA (2002) Determination of key aroma components of Cheddar cheese using dynamic headspace dilution assay. Flavour Fragrance J 17:300–305

Zeng Z, Zhang H, Zhang T, Tamogami S, Chen JY (2009) Analysis of flavor volatiles of glutinous rice during cooking by combined gas chromatography–mass spectrometry with modified headspace solid phase microextraction method. J Food Compos Anal 22(4):347–353

Zhang Q (2007) Strategies for developing green super rice. Proc Natl Acad Sci 104:16402–16409

Zhou HF, Xie ZW, Ge S (2003) Microsatellite analysis of genetic diversity and population genetic structure of a wild rice (*Oryza rufipogon* Griff.) in China. Theor Appl Genet 107:332–339

Chapter 2
Brown, White and Parboiled Rice

Maurício de Oliveira, Cristiano Dietrich Ferreira, Gustavo Heinrich Lang, and Cesar Valmor Rombaldi

2.1 Structure and Grain Composition

The genus *Oryza* comprehends many species which can present differences regarding grain quality properties. Here, we will focus on *Oryza sativa*, from which the industrial processing of the natural (non-parboiled) rice grain can be presented primarily in two forms: brown rice and milled rice.

To obtain brown, white, and parboiled rice for efficient processing, it is essential to know the physiological structure of a rice kernel. Paddy rice contains on average 6.7–8.3% protein, 2.1–2.7% lipid, 3.4–6.0% ash, and 70.5–84.2% carbohydrate. Bran obtained by milling of the brown rice consists of 13.0–14.5% lipids, 6.1–8.5% ash, and 48.3–55.4% carbohydrates.

The kernel outer structure (Fig. 2.1) is composed of the husk and a set of modified leaves called palea and larger lemma, with an inedible silky fibrous layer that protects the caryopsis from attack by insects, fungi, humidity, and atmospheric oxygen. Some genotypes have an awn at the tip of the lemma. This outer layer represents on average 23% of the kernel (Abaide et al. 2019); however, percentages ranging from 17 to 28% can be found between different cultivars and/or origins. The color and shade of the husk may vary depending on the cultivar and the region and/

M. de Oliveira (✉)
Laboratory of Post-Harvest, Industrialization and Grains Quality, Graduate Program in Food Science and Technology, Faculty of Agronomy Eliseu Maciel, Federal University of Pelotas, Brazil/Plant Science Department, Rothamsted Research, Harpenden, UK

Plant Science Department, Rothamsted Research, Harpenden, United Kingdom, Harpenden Common, Harpenden AL5 2JQ, UK
e-mail: mauricio@labgraos.com.br

C. D. Ferreira · G. H. Lang · C. V. Rombaldi
Laboratory of Post-Harvest, Industrialization and Grains Quality, Graduate Program in Food Science and Technology, Faculty of Agronomy Eliseu Maciel, Federal University of Pelotas, Brazil/Plant Science Department, Rothamsted Research, Harpenden, UK

A. Costa de Oliveira et al. (eds.), *The Future of Rice Demand: Quality Beyond Productivity*, https://doi.org/10.1007/978-3-030-37510-2_2

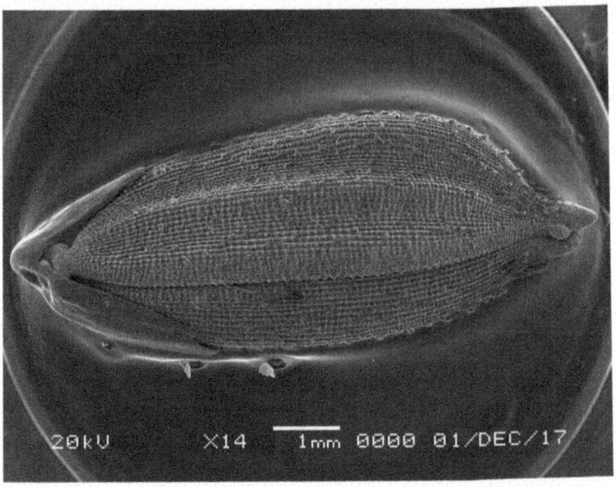

Fig. 2.1 Detail of a paddy rice kernel seen under electron microscope cryo-fractured and imaged using the JEOL JSM 6360 LV SEM. (Image courtesy of Ms. Hannah Walpole, Rothamsted Bioimaging, Rothamsted Research, Harpenden, UK)

or cultivation condition. The husk is mainly composed of cellulose (37–40%), hemicellulose (18–24%), lignin (12–22%), and others (17–33%) (Castro et al. 2016; Abaide et al. 2019).

The main edible portion of rice is the endosperm. Botanically and compositionally, different layers are present as outer cover to the endosperm. The first layer that is in unpigmented cultivars about 10 mm thick is called pericarp, composed of epicarp, cross cells, and cuticle. Beside the pericarp is the forehead, also called the grain coat, followed by the nucellar and aleurone layers (Fig. 2.2).

The aleurone layer (AL – Fig. 2.2) is strongly attached to the endosperm (EN – Fig. 2.2). In addition to these layers, the embryo is attached to one end of the endosperm, which is also similarly covered. This layer that is mainly composed of the bran covers the endosperm, whose purpose is to protect the inner layers from external aggressions (heat, water, oxygen, and others). These layers may therefore contain specialized and biologically active metabolites, including tocopherols, gamma-oryzanol, tocotrienols, and phenolic compounds (Goffman and Bergman 2004).

Studies show that there is a very significant variation in aleurone layer thickness between different rice genotypes, which can be grouped into numbers of 1–5 aleurone layers, and that Japonica rice normally has more aleurone cells than Indica rice. Differences larger than three times in the aleurone area, ranging from 2.3% to 7.5% among 321 rice varieties, were reported by Khin et al. (2013) who used a lipid staining method of a grain cross section to represent the aleurone area, demonstrating where much of the lipid fraction is stored. Among the different colors of the pericarp, studies show that white rice has a lower percentage of bran (3.57%) compared to red rice (10.13%) and purple bran rice (13.19%) (Ming-Hsuan and Anna 2019).

The appearance (color and size, among others) and quality (yield of whole, broken, and nutritional value, among others) of a rice grain as its color and size may be influenced by genetic and environmental factors, but its internal and external struc-

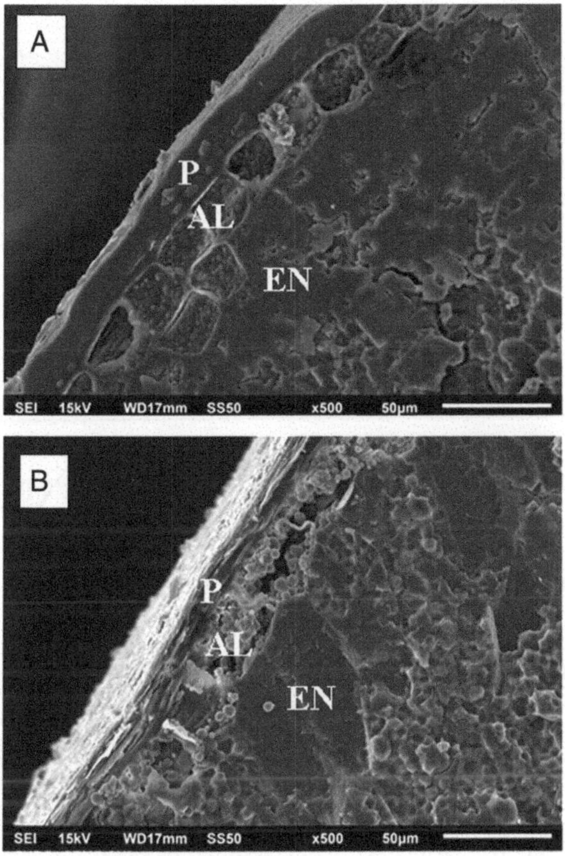

Fig. 2.2 Scanning electron microscopy of the cross section of black (**a**) and red (**b**) rice grains. Letters P, AL, and EN indicate the pericarp layers, aleurone layer, and endosperm region, respectively. (Images: Gabriela H. Alves and Maurício de Oliveira – Laboratory of Post-Harvest, Industrialization and Grains Quality, Graduate Program in Food Science and Technology, Faculty of Agronomy Eliseu Maciel, Federal University of Pelotas, Brazil)

ture remains the same. Proteins, lipids, and minerals are mostly accumulated in the aleurone layer, while starch and small amounts of stored protein are stored mainly in endosperm (Xiong et al. 2019).

In addition to the light pericarp rice grains (called brown rice or white rice), which are the best known and consumed, the *Oryza sativa* L. grains may have different pericarp colors, such as grains with black pericarp and grains with red pericarp. In addition to the genus *Oryza* – the most common – there is the so-called wild rice, which should not be confused with *Oryza* types because it is another botanically named *Zizania* sp.

The predominant layer of a rice grain is the endosperm composed mostly of starch. Besides starch, proteins also take part in the endosperm, and among the proteins found in rice caryopses, about 20–30% belong to glutelins and prolamines, respectively. Protein generally has a negative effect on rice food quality due to the involvement of starch granules by the protein network and the competitive absorption of water (Likitwattanasade and Hongsprabhas 2010; Zhu et al. 2019; Amagliani et al. 2016).

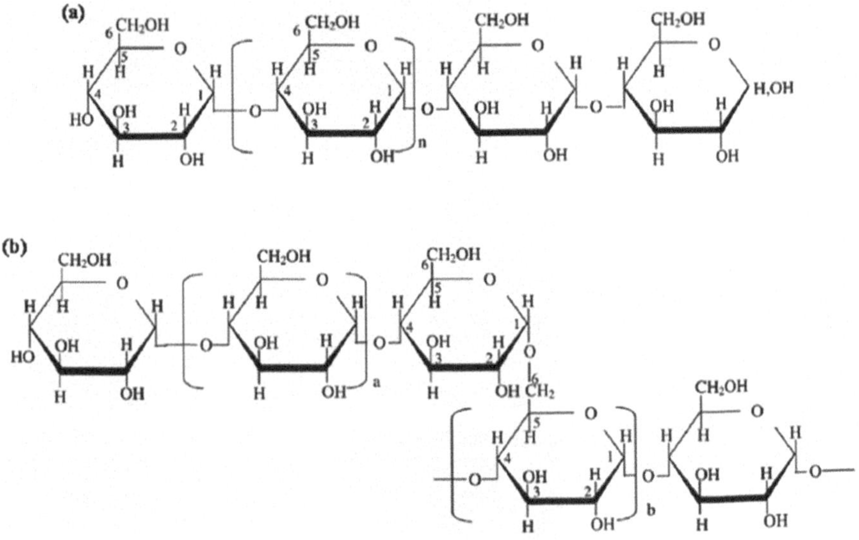

Fig. 2.3 Amylose (a) and amylopectin (b) structure. (Source: Amagliani et al. 2016)

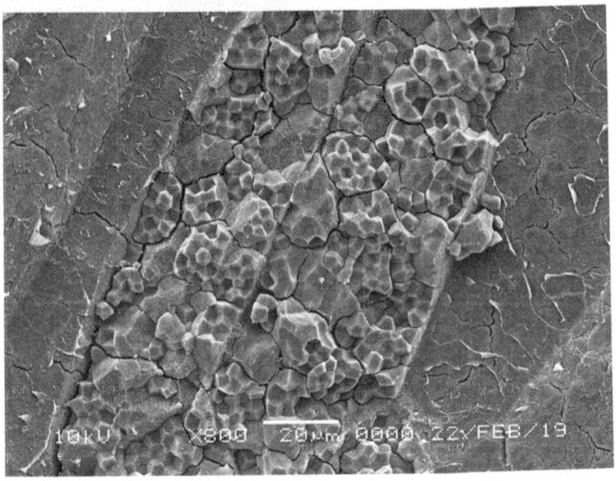

Fig. 2.4 Starch granule scanning electron micrograph of a mature imbibed rice grain cryo-fractured and imaged using the JEOL JSM 6360 LVSEM. (Image courtesy of Ms. Hannah Walpole, Rothamsted Bioimaging, Rothamsted Research, Harpenden, UK)

Starch is characterized by being the main reserve carbohydrate present in vegetables and is formed by two macromolecules, amylose and amylopectin. Amylose is a linear molecule composed of glucose units linked by α-1,4 glycosidic bonds (Fig. 2.3a), with a small number of branches, and amylopectin is composed of glucose units joined by α-1,4 and α-1,6 glycosidic bonds (Fig. 2.3b), forming a branched structure (Zavareze and Dias 2016).

In grains, starch is in the form of granules, which have varied shape and size according to the botanical origin. Rice starch granules have polyhedral shape (Fig. 2.4) and a diameter up to 150 μm, being the smallest among cereals (Amagliani et al. 2016). In granules, structural organization occurs in the form of growth rings, alternated by amorphous and crystalline regions, characterizing the formation of a semi-crystalline polymer (Denardin and Silva 2009).

The composition and structure of starch is fundamental for understanding the variations presented in the physicochemical properties of grains (Zhu 2018). The amylose content of a cultivar or grain is an important factor that influences thermal properties and directly affects the expansion volume and water absorption capacity during cooking as well as the hardness of cooked rice. According to the apparent amylose content, rice is classified as waxy (0.8–2.0%), low amylose (7–20%), medium amylose (20–25%), and high amylose (>25%) (Juliano 1985). In general, lower apparent amylose content results in higher viscosity peaks, lower retrogradation, higher gelatinization temperatures, and stickier texture.

As for its functional properties, rice starch is referred to as rapid digestion, although intrinsic factors such as physicochemical properties and starch-protein-starch lipid interactions alter the hydrolysis rate (Toutounji et al. 2019). Normally to be consumed, the grains undergo cooking processes or hydrothermal treatments that cause the starch granules to gelatinize, resulting in loss of crystallinity and amylose leaching. After gelatinization, the reassociation of amylose and amylopectin molecules promotes starch retrogradation, which implies a reduction in the digestive rate. Hydrothermal processing further favors the formation of amylose-lipid complexes, which hinders the access of hydrolytic enzymes, hindering and reducing starch digestibility (Ye et al. 2018).

According to Gomes and Magalhães Junior (2004), rice provides 20% of the energy and 15% of protein needed by man and stands out for its easy digestion. The most consumed rice can be classified as white (milled) or light brown (brown), depending on whether or not the aleurone layer is removed from the grain. In general, light brown rice can be considered to contain about 6–7% bran, 90% endosperm, and 2–3% embryo by weight (Chen et al. 2015). Genotypes with colored pericarp in the form of brown rice will be lighter when polished, as this process results in the removal of external pigmented structures that cover the grain. Among the molecules that give color to the pericarp of rice grains are anthocyanins in the case of black rice and proanthocyanidins in the case of red rice (Paiva et al. 2016).

2.2 Processing

Due to rice's own characteristics, grains cannot be consumed as nature produces them, and this means that the cereal's production chain has its own characteristics very different from grains of other species, that is, to be consumed, it needs to undergo processing. Over time, activities in the rice agribusiness, as throughout the food agroindustry, have undergone substantial modifications to cater to an increasingly demanding and informed consumer. Major advances include design and application of good manufacturing practices, hazard analysis and critical control points, criteria for raw material selection and processing, traceability, and diversification of production in pursuit of market niches and value increases aggregate. These facts

Fig. 2.5 Grains of the brown rice (**a**) and milled rice (**b**). (Image: Gabriela H. Alves and Maurício de Oliveira – Laboratory of Post-Harvest, Industrialization and Grains Quality, Graduate Program in Food Science and Technology, Faculty of Agronomy Eliseu Maciel, Federal University of Pelotas, Brazil)

are important because they are a food safety management system, which is based on controls of various stages of food production, analyzing potential health hazards of consumers, and determining preventive measures to control these hazards. For excellent processing and obtaining high-quality products, the grains should preferably be received and maintained separately by cultivar, trait, and/or quality until industrial processing.

With respect to industrial processing without additional process, natural (non-parboiled) rice grain can be presented primarily in two forms: brown rice (Fig. 2.5a), which is subjected to husking only, and milled rice by removing the bran (integument, pericarp, germ, and aleurone layer), obtaining the white rice (Fig. 2.5b) (Juliano 1985).

For a better understanding of the steps and terms that we will present below on three of the ways rice is processed, it is important to understand the terms defined below in accordance with Codex Standards 1988–1995, which will be used as a reference:

- *Paddy rice* is rice which has retained its husk after threshing.
- *Husked rice* (*brown rice*) is paddy rice from which the husk only has been removed.
- *Milled rice* (*white rice*) is husked rice from which all or part of the bran and germ has been removed by milling.
- *Parboiled rice* may be husked rice or milled rice processed from paddy or husked rice that has been soaked in water and subjected to a heat treatment so that the starch is fully gelatinized, followed by a drying process.

In addition, rice grains can be classified according to kernel length/width ratio, kernel length, or a combination of the kernel length and the length/width ratio:

Long grain rice: husked rice or parboiled husked rice with a length/width ratio of 3.1 or more, or milled rice or parboiled milled rice with a length/width ratio of 3.0 or more, or a kernel length of 6.6 mm or more, or a kernel length of more than

6.0 mm and with a length/width ratio of more than 2 but less than 3 or a kernel length of more than 6.0 mm and with a length/width ratio of 3 or more.

Medium grain rice: husked rice or parboiled husked rice with a length/width ratio of 2.1–3.0, or milled rice or parboiled milled rice with a length/width ratio of 2.0–2.9, or a kernel length of 6.2 mm or more but less than 6.6 mm, or a kernel length of more than 5.2 mm but not more than 6.0 mm and a length/width ratio of less than 3.

Short grain rice: husked rice or parboiled rice with a length/width ratio of 2.0 or less, or milled rice or parboiled milled rice with a length/width ratio of 1.9 or less, or a kernel length of less than 6.2 mm, or a kernel length of 5.2 mm or less and a length/width ratio of less than 2. The same standard classifies polished grains according to the following degrees of milling:

Undermilled rice: obtained by milling husked rice but not to the degree necessary to meet the requirements of well-milled rice

Well-milled rice: obtained by milling husked rice in such a way that some of the germ and all the external layers and most of the internal layers of the bran have been removed

Extra-well-milled rice: obtained by milling husked rice in such a way that almost all of the germ, all of the external layers and the largest part of the internal layers of the bran, and some of the endosperm have been removed

Several factors affect the nutritional value of rice, such as genotype, growing conditions and environment, crop management, storage and post-harvest, and especially the milling process. Content of starch, protein, lipids, fibers, and ashes according to processing is shown in Table 2.1.

Table 2.1 Average proximate composition (% in dry matter) of brown long, milled white, and milled parboiled rice grains

Component	Brown rice	White rice	Parboiled milled rice
Total starch	74.12	87.58	85.08
Protein (Nx5,95)	10.46	8.94	9.44
Lipids	2.52	0.36	0.69
Ash	1.15	0.30	0.67
Total fiber	11.76	2.87	4.15
Insoluble fiber	8.93	1.05	1.63
Soluble fiber	2.82	1.82	2.52

Source: Adapted from Denardin et al. (2004)

2.2.1 *Brown Rice*

The brown rice or husked rice (Fig. 2.5a) is an industrially simpler form of processing (Fig. 2.6) but the most noble in terms of nutritional value, since the process for obtaining it consists basically of dehusking, selection, and packaging. The grain of this simplified processing compared to other processes is called brown rice (or dehusked), which has a somewhat fibrous and greasy coating that prevents its easy cooking in water, with a layer that is collectively called bran.

There are three types of equipment used for husking: rubber roller, centrifugal sheller, and disc sheller. The detachment form is independent of the processing method used to obtain brown rice, white rice, or parboiled rice. In the first type of husking mentioned, the rubber roller, the rice grains are directed between two rubber rollers, which rotate in the opposite direction and at different speeds so that the shear force is increased, causing the husk to be removed. This is one of the most commonly used mismatching processes and one of the best performing industrial methods. Centrifugal sheller removes the husk by using centrifugal force. By means of a rotating rotor, the rice grains are subjected to centrifugal force and thrown with great force to the outer covering, which causes the palea and lemma to separate. The third type is the disc sheller, consisting of a set of two abrasive discs, one of which is static or moves in the opposite direction, and peeling occurs between the discs. Due to its characteristics, it can be adjusted for different materials and/or cultivars and/or formats.

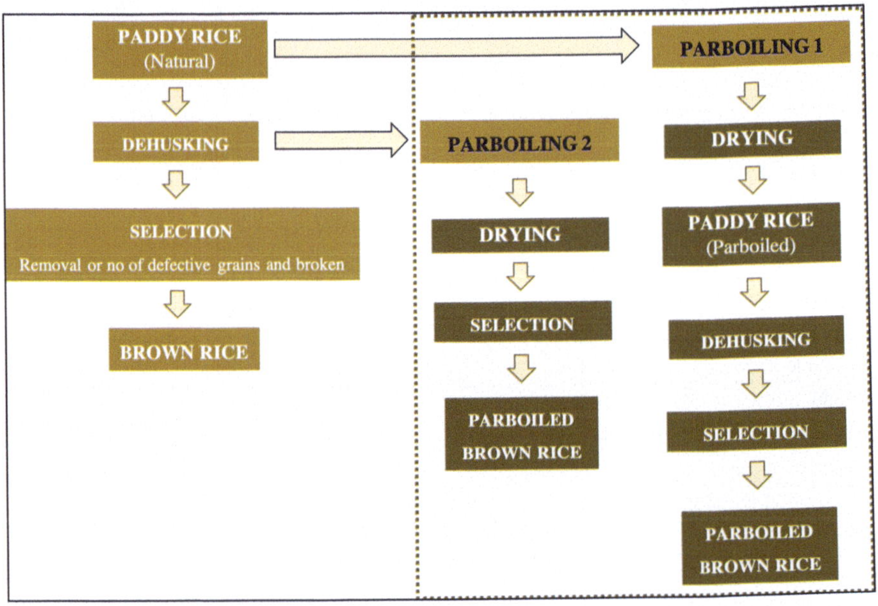

Fig. 2.6 Flow chart of processing brown rice from the rice husk

Fig. 2.7 Black (**a**) and red (**b**) brown rice grains. (Image: Alves, G.H. and Oliveira, M. – Laboratory of Post-Harvest, Industrialization and Grains Quality, Graduate Program in Food Science and Technology, Faculty of Agronomy Eliseu Maciel, Federal University of Pelotas, Brazil)

The husk is usually removed by suction, while the brown rice must be screened to separate the grains that remain husked after passing through the disc. The separation of the dehusked grains, which return to the sheller as many times as necessary to lose the husk, is usually done in separators whose principle of operation is the difference in density, as the grain density increases with dehusking. Worldwide, the production of husk from processing is estimated at 149 million tons, approximately 70% of which is destined for burning as fuel (Abaide et al. 2019).

Brown rice contains nutritional components such as dietary fiber, amino acids, minerals, proteins, vitamins B and E, and other nonessential nutrients concentrated in the germinal and external layers of starch endosperm, which are normally removed as bran during grain milling (Das et al. 2008). Even so it has less appeal and consumption preference due to its dark appearance and the longer time required for the cooking of the grains, as they resist more heat penetration and protect the cooking grain, which leads to greater energy loss during cooking. In addition, the fibrous layer of bran gives hard texture and increased firmness in cooked rice.

Brown rice grains, especially black and red pigmented integument genotypes (Fig. 2.7), are rich sources of phenolic compounds, while a smaller amount of phenolic compounds is found in non-pigmented brown rice and milled rice (Walter et al. 2008; Paiva et al. 2016). Black rice is an important source of anthocyanins and phenolic acids, while red rice is a source of phenolic acids and proanthocyanidins. According to Zaupa et al. (2015) and Zhang et al. (2015), the main phenolic acids found in both pigmented and non-pigmented rice are protocatechuic, sinapic, vanillic, p-coumaric, and ferulic acids.

In order to retain their appearance and maintain their beneficial health effects, pigmented rice grains are normally marketed and consumed in their entirety, in which the grains are husked and selected before packing. The storage of these grains mainly after processing, and especially in conditions of high temperatures (30–40 °C), is detrimental to the physicochemical, structural, texturometric, viscoamylographic, and sensory properties of brown rice grains (Chen et al. 2015), in addition to causing a decrease in phenolic content, flavonoids, proanthocyanidins, and antioxidant activity (Ziegler et al. 2018).

Black pericarp rice grains are considerable sources of phenolic and anthocyanin compounds, recognized for their antioxidant activity and reduced aging effects (Masisi et al. 2016). The phenolic acid content ranges from 1.47 to 2.08 mg g^{-1}, with ferulic, caffeic, and vanillic acids being the main compounds described in grains with this coloration (Goufo and Trindade 2014; Shao, et al. 2014; Zhang, et al. 2014). Among the anthocyanin types the two major forms found in these grains are cyanidin-3-O-glucoside and peonidin-3-O-glucoside, whose content ranges from 0.008 to 7.84 mg g^{-1} (Hou, et al. 2013; Lang et al. 2019).

Due to the higher lipid content, rice whole grains are more susceptible to chemical changes such as lipid oxidation during storage. And when the rice product is exposed to an unfavorable external environment (such as high temperature conditions typical of tropical countries), enzymatic activity can increase and accelerate lipid degradation, thereby reducing the sensory quality of rice (Zhou et al. 2016). Some technologies such as electron beam irradiation (EBI) have been shown to be effective for increasing the shelf life of stored rice as they reduce these effects (Luo et al. 2019).

2.2.2 White Rice

The white rice (Fig. 2.5b) is the most consumed and preferred worldwide, but its grains are poorer in nutritional terms. To obtain white rice, the grains need to go through the milling and selection steps. The polishing (milling) process is a rather drastic process in which the bran layer is removed, employing abrasion and/or friction. In abrasive type, the bran is removed by contact against a rough surface, while in friction type, removal takes place by rubbing one rice kernel against another while the grains are subjected to slight pressure (Fig. 2.8).

In some cases, industries use the milling process definition to refer also to cleaning and/or husking operations, but generally this definition refers only to the removal of bran or outer layers of brown rice to produce white rice. In this process, 7–12% of the outer layer of the grain and the germ is usually removed; this percentage can be called "milling degree." When this operation is performed in excess, there may be a reduction in whole grain yield (due to both the volume removed and grain breakage caused by prolonged friction time and excess bran generation); in contrast, when the grain is insufficiently milled, it presents darker hue (due to the higher content of the bran and aleurone layer) and requires longer cooking time. Besides the usual abrasion and friction milling processes, enzymatic polishing methods have been proposed by Das et al. (2008), using xylanases and cellulases for 2 h at 50°C. The enzymes act on the non-starch polysaccharides of the bran layers releasing their monomeric sugar constituents.

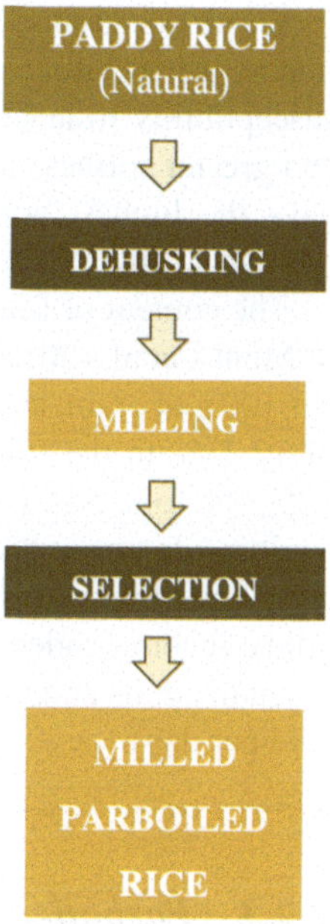

Fig. 2.8 Flow chart of processing white rice from the rice husk

Due to the removal of the pericarp from the grains, the layer on which the pigments are concentrated, they have a lighter color and tone (Fig. 2.5b), and this is one of the reasons that it is called white rice. This shade is associated with its greater acceptability by consumers.

Starch and protein are the two main components of milled rice. Its physicochemical properties are of great importance for the quality of cooked rice (Concepcion et al. 2015). For starch, the amylose content between 15 and 20% tends to have more desirable viscoelasticity than rice containing a higher (very hard) or lower (very soft) amylose content (Fukuda et al. 2014). Studies have reported a varying texture of rice, although they have similar starch content (Shinde et al. 2014). The ratio of long chains (average degree of polymerization, 92–98) was positively correlated with rice elasticity, but short chains may have a negative correlation (Ong and Blanshard 1995; Li et al. 2016). The process of milling rice grains causes a reduction in protein content, since in the bran portion, proteins are also found. Monks et al. (2013) verified 11.76% reduction in protein content when brown rice was milled with a milling degree of 8%.

The last step prior to packaging is selection using flat and circular sieve sets for size and shape separation, while color and/or defect separation is performed by selector machines using photoelectric cells and air jets (Paraginski et al. 2014). In addition to quantitatively reducing the selected grain mass, defects reduce yield when used for other purposes such as starch extraction, which is reduced as the defect index increases.

White rice grains, due to the removal of the bran layer, which hinders the absorption of water during the cooking process, have a shorter cooking time and greater susceptibility to leaching of compounds to the water used in the cooking process and greater softness and release after cooking. Although polished white rice grains have the highest consumption preference, as the bran and aleurone layers are removed, some special compounds are lost.

The content of folic acid decreases with the degree of milling (MD). According to Monks et al. (2013), the brown rice (non-milled) contained 188 mg of folic acid per 100 g of grain, and the removal of 8% of the outer layer caused a significant drop of 72.23% in the folic acid content. When the degree of polishing progressed to 10%, the folic acid content of the grains was 44.44% lower than those subjected to a milling degree of 8%. The difference between brown and polished rice (8% MD) is 135.8 mg of folic acid per 100 g of rice. This difference is greater than the additional intake recommended by the World Health Organization for lactating women compared to adults and/or pregnant women compared to lactating women (100 mg folic acid per day).

2.2.3 *Parboiled Rice*

Parboiled rice is husked or milled rice that goes through the parboiling process usually still in shell (Fig. 2.9). Parboiling is a hydrothermal process consisting of immersion, gelatinization, and drying. Around 20% of the rice produced in the world is processed as parboiled rice (Leethanapanich et al. 2016; Cheng et al. 2019). Parboiling improves rice storage stability with minimal changes in nutritional quality and reduces breakage rates during processing, especially polishing (Buggenhout et al. 2013; Min et al. 2014; Jayaraman et al. 2019). On the other hand, this process provokes a change in color conferring a yellowing that negatively affects the acceptability by consumers, especially Latin Americans (Bhattacharya and Subba 1996; Lamberts et al. 2006; Leethanapanich et al. 2016; Rockembach et al. 2019).

Parboiling may provide improvements in the technological properties and nutritional value of rice as it provides increased yield of whole grains, firm textured grains after cooking, and higher nutritional value due to the dissolution and transport of water-soluble substances into the grains during soaking step. In addition, the process implies reduced starch digestibility due to the occurrence of gelatinization and retrogradation phenomena (Dutta and Mahanta 2012).

During gelatinization, especially when the process is employed under pressure, enzymes present in the grain, such as lipases and peroxidases, are virtually inactivated, providing increased grain durability and stability of rice bran. Later, in the drying stage, the starch gelatinization and its retrogradation are completed, which makes the grains more resistant to friction operations (Elias et al. 2015), such as husking and milling.

Parboiling alters the physicochemical and organoleptic properties of rice grain, reduces rice stickiness, increases hardness, and darkens color. According to the lit-

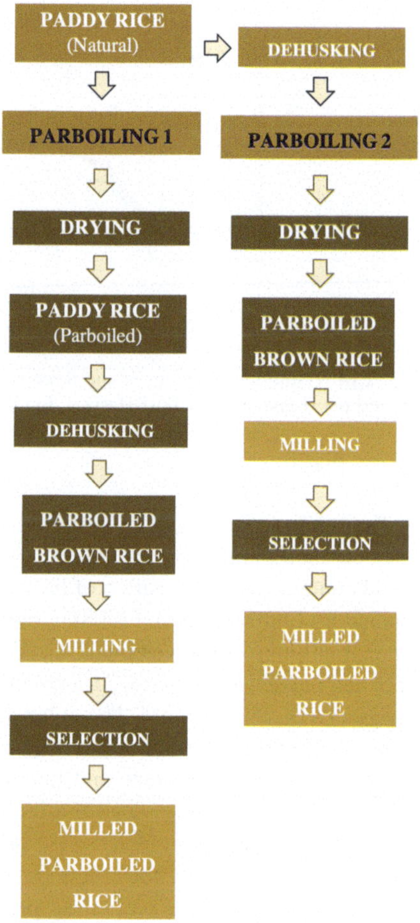

Fig. 2.9 Flow chart of processing parboiled rice from the rice husk

erature, color changes during parboiling may be due to the migration of shell and/or bran pigments and enzymatic browning and non-enzymatic browning of the Maillard type (Lamberts et al. 2008). In order to reduce grain darkening and the formation of Maillard compounds, which may have detrimental effects on consumer health, anti-darkening compounds such as glycine and reduced glutathione may be employed (Villanova et al. 2017).

The process conditions used in parboiling rice grains may affect the protein content. Thus, longer immersion and gelatinization times result in increased protein content of parboiled rice grains, although intensifying color changes. In addition to the processing conditions, rice genotype is considered a variable that affects the quality of parboiled grains, since different genotypes have different chemical composition and morphology, which, in turn, interferes with hydration rate, head yield, and intensity of Maillard reactions (Leethanapanich et al. 2016), causing different behaviors and susceptibility to browning.

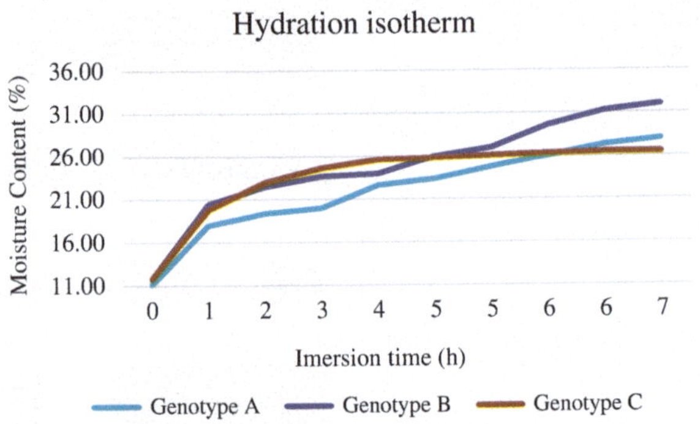

Fig. 2.10 Hydration isotherm of the three rice genotypes with distinct behaviors of the water absorption. (Laboratory of Post-Harvest, Industrialization and Grains Quality, Graduate Program in Food Science and Technology, Faculty of Agronomy Eliseu Maciel, Federal University of Pelotas, Brazil)

Research on the effect of parboiling on the phytochemical content of rice has been devoted mainly to the study of carotenoid stability and migration. The carotenoid content and composition of brown and ground brown and parboiled rice were studied by Lamberts et al. (2008). Color analyses of rice flour samples with different extraction rates showed that the yellow and red pigments are concentrated in the bran and outside endosperm. As a result, all pigments have been removed for grinding grades of 15% or more. The color-determining components present in the fractions of different rice cultivars were identified as β-carotene, lutein, and zeaxanthin carotenoids, with β-carotene and lutein being the predominant compounds. Thus, the decrease in brightness and the increase in red and yellow color intensities of parboiled rice are attributed to the Maillard reactions and/or physicochemical changes of the different rice components that occur during parboiling.

The first stage of the parboiling process, soaking (immersion), has as its main purpose the adequate hydration of rice to enable the subsequent complete starch gelatinization, usually occurring in tanks with heated water. Water absorption by the grain should be uniform and rapid until it reaches about 30% of the moisture, avoiding peel opening, grain breakage, and starch granules as much as possible. Among the three important stages of parboiling, immersion is the most critical step as it alters the chemical composition and nutrient distribution in the grains (Kale et al. 2015). In the immersion process, nutrient and amylose leaching into the immersion water also occur. The temperature and time to be used in this step depend on the hydrothermal behavior of each rice genotype and vary depending on factors such as amylose content, shape, and grain size, among others (Amato 2005), as shown in the example in Fig. 2.10.

In soaking the grains, heated water dissolves water-soluble substances such as vitamins and minerals, dragging them from the periphery to the center of the grains or allowing them to leach. Lipid molecules have their viscosity reduced with increasing temperature, favoring their transport into the grain. After the caryopsis has absorbed sufficient water, it should be heated to the starch gelatinization temperature, which fluctuates with the characteristics of the rice grain and the parboiling process, but generally requires a temperature higher than 58°C when the water conventional processes take from 4 to 7 hours. Most industries perform this operation through heated steam and in closed containers. Gelatinization by the use of an autoclave is the most usual method, using pressures ranging from 0.5 to 1.2 kgf cm^{-2} for 10–30 minutes. Before the rice is gelatinized, excess soaking water is eliminated.

Starch granules have limited capacity for absorption and expansion in cold water due to hydrogen bonds between the amylose and amylopectin fractions. During immersion in hot water, however, additional energy in the form of heat weakens the granular structure by disruption of hydrogen bonds increasing the surface for water absorption by starch granules. This increase allows for additional and irreversible hydration of the initially swollen granules by the higher dissociative activity of molecular water. This phenomenon, called starch gelatinization, begins in immersion and is complemented in the gelatinization step.

The rice grain must remain in the gelatinization step long enough for complete starch gelatinization to occur. Excessive times and/or temperatures at this stage increase the incidence of damaged grains and intensify the color of rice grains (yellowing). On the other hand, insufficient times and/or temperatures in this operation result in ungelatinized grains with poor mechanical strength. Mixing rice varieties can cause problems in parboiling operations, considering that the most suitable conditions for some varieties or some hybrids may not be for other genotypes.

In special situations, such as parboiling during the harvest period, when the rice is submitted to the immersion (soaking) operation directly from the harvest, without drying, the grains are processed without storage. This operation decreases the time required for grain hydration, since the recommended grain harvesting moisture degree for most climates is 20%. The degree of gelatinization of the grains, which directly interferes with susceptibility to breakage, is increased under higher pressure conditions. These conditions can reduce the presence of white center and the incidence of cracks in the grains, increasing their resistance (Buggenhout et al. 2014).

After the gelatinization step, the grains need to be dried to adequate moisture (12–13%) for safe processing and storage; moisture grade facilitates and confers friability to husk and grain storage stability.

Parboiling enhances the nutritional value and recovers mechanical damage from rice, eliminates chalked grains increasing their yield, but intensifies some biological defects such as stained and burnt grains. The steps of the soaking, gelatinization and drying promote the starch gelatinization and retrogradation, seals cracks and fissures, increasing grain resistance to friction operations, providing more whole grains, more nutritious rice than polished white and more stable than the whole grain. Sareepuang et al. (2008) observed an 18% increase in mineral content in polished parboiled rice compared to polished conventional rice due to phosphorus and potassium migration.

Nowadays, innovative parboiling methods have been developed, such as fast microwave parboiling which employs microwave waves for heating in the soaking and gelatinization steps. A method developed by Rockembach et al. (2019), which is ten times faster than the conventional parboiling process, provides lighter grains and shorter cooking time.

2.3 Rice Grain Cooking Properties

The consumption quality of rice grains has long been attributed to starch, or its interactions with other components such as lipids and proteins. Factors considered to determine the quality of cooking or consumption include amylose content (Juliano 1985), the solubility of amylose (Bhattacharya and Sowbhagya 1971), the length of α and β amylopectin chains (Hanashiro et al. 1996), gelatinization and melting temperatures of the amorphous and crystalline regions of amylopectin, and the amount of structure.

The grains are composed of a heterogeneous mixture with different levels of structure, so it changes with heating at the time of preparation. In the cooking process, water gradually penetrates from the surface to the center of the grain, laying the groundwork for the rapid hydration of each component (Horigane et al. 2014). Increased temperature promotes starch gelatinization and granular structure collapse (Bosmans et al. 2012); this is why uniform hydration is essential for high-quality cooked grain. One way to achieve adequate and uniform hydration is to use hydration prior to cooking.

The prolonged cooking required in the preparation of brown rice leads to the breaking and opening of grains which according to Bhattacharya (2011) is not acceptable. This is why in many cases and/or processes, a fairly mild polishing process is employed, around 2–3% for scratching the bran layer to facilitate hydration by the cooking water. This form of polishing minimizes nutritional losses and improves the preparation and consumption properties of rice.

Usually, rice grains are consumed as cooked grains, especially in the form of polished grains. Grain cooking is an important operation, usually done to provide the desired grain smoothness for direct feeding. The rice grains are cooked in water (limited or excess amount) (Ramesh 2001). During cooking, the starch content of the milled rice grain absorbs moisture and waves due to its gelatinization. In addition, some solid content of ground rice is dissolved in cooking porridge. Moisture absorption of rice grains during cooking is an important phenomenon to predict optimal cooking conditions (Kasai et al. 2005). Many studies have been directed to water absorption and gelatinization of rice starch during parboiling before polishing.

Parboiling provides low retrogradation values, which indicate that amylose is completely solubilized at the time of heating in the cooking and is not able to retrograde during the decreasing temperature cycle when the grains begin to cool. These occur due to a decrease in parboiled rice paste properties, regardless of the immersion and gelatinization temperatures used (Himmelsbach et al. 2008; Villanova et al. 2017).

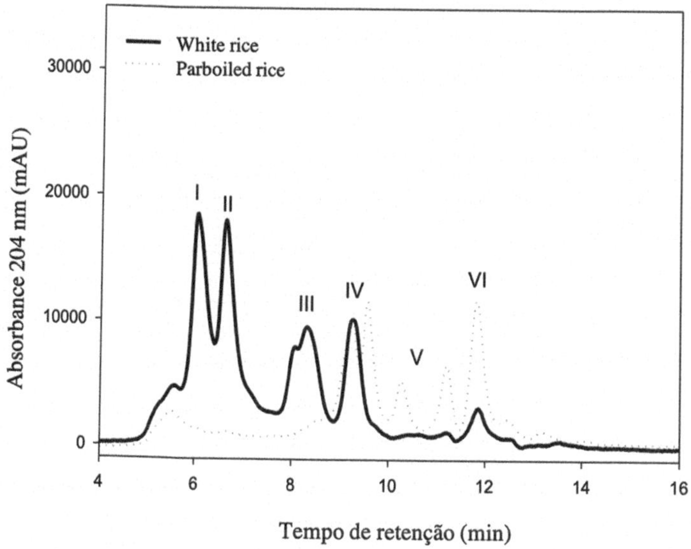

Fig. 2.11 Molecular weight distribution of proteins in white rice (non-parboiled) and parboiled rice. (Adapted from Villanova et al. 2017)

Fractions I and II, which have a molecular weight of approximately 97 and 66 kDa, respectively, indicate the region comprised of dimers, trimers, and more polymerized forms of proteins. Fraction III, with molecular weight between 12 and 34 KDa, indicates the chromatogram region comprised of glutelin subunit (α-β) pairs. Fraction IV indicates the region comprised of rice glutelin free acid (α) and basic (β) subunits, and fractions V and VI indicate the presence of low-molecular-weight proteins, mainly albumin, globulin, and/or prolamine.

Fractions I, II, and III, which presented a retention time of 6, 7, and 8 minutes, respectively, had reduced solubility in Sodium Dodecil Sulfate (SDS) due to the parboiling process (Fig. 2.11). On the other hand, this process increases the solubility of fractions IV, V, and VI. The reduction in solubility of fractions I, II, and III is due to the intensification of interactions between amino acid radicals, such as hydrophobic, electrostatic interactions and hydrogen bridges, which makes denaturing solvent accessibility to the most protected protein sites difficult.

References

Abaide ER, Tres MV, Zabot GL, Mazutti MA (2019) Reasons for processing of rice coproducts: reality and expectations. B Bio 120:240–256. https://doi.org/10.1016/j.biombioe.2018.11.032

Amagliani L, O'regan J, al K, O'Mahony JA (2016) Chemistry. Structure. Functionality and applications of rice starch. J Cer Sci 70:291–300. https://doi.org/10.1016/j.jcs.2016.06.014

Amato GW (2005) Elias MC Parboilização do arroz. Editora Ricardo Lenz Ziede, Porto Alegre, p 160

Bhat FM, Riar CS (2016) Effect of amylose, particle size & morphology on the functionality of starches of traditional rice cultivars. I J Bio Mac 92:637–644. https://doi.org/10.1016/j.ijbiomac.2016.07.078

Bhattacharya KR, Sowbhagya CM (1971) Water uptake by rice during cooking. C Sci 16:420–424. http://ir.cftri.com/7243

Bhattacharya KR, Subba RPV (1996) Processing conditions and milling yield in parboiling of rice. J Agr Foo Che 14:473–475. https://doi.org/10.1021/jf60147a008

Bhattacharya KR (2011) Rice quality: a guide to rice properties and analysis. Woodhead Publishing, Cambridge

Bosmans GM, Lagrain B, Deleu LJ, Fierens E, Hills BP, Delcour JA (2012) Assignments of proton populations in dough and bread using NMR relaxometry of starch, gluten, and flour model systems. J Agr Foo Che 60:5461–5470. https://doi.org/10.1021/jf3008508

Buggenhout J, Brijs K, Delcour JA (2013) Impact of processing conditions on the extractability and molecular weight distribution of proteins in parboiled brown rice. J Cer Sci 58:8–14. https://doi.org/10.1016/j.jcs.2013.05.002

Buggenhout J, Brijs K, Oevelen JV, Delcour JA (2014) Milling breakage susceptibility and mechanical properties of parboiled brown rice kernels. LWT – F Sci Tec 59:369–375. https://doi.org/10.1016/j.lwt.2014.05.001

Castro RCA, Fonseca BG, Santos HTL, Ferreira IS, Mussato SI, Roberto IC (2016) Alkaline deacetylation as a strategy to improve sugars recovery and ethanol production from rice straw hemicelluloses and cellulose. I Cro Pro 106:65–73. https://doi.org/10.1016/j.indcrop.2016.08.053

Chen Y, Jiang W, Jiang Z, Chen X, Cao J, Dong W, Dai B (2015) Changes in physicochemical, structural, and sensory properties of irradiated brown Japonica rice during storage. J Agr and F Che 63:4361–4369. https://doi.org/10.1021/jf5047514

Cheng KC, Chen SH, Yeh AI (2019) Physicochemical properties and in vitro digestibility of rice after parboiling with heat moisture treatment. J Cer Sci 85:98–104. https://doi.org/10.1016/j.jcs.2018.11.016

CODEXSTAN 198:1995, Codex Standard for rice

Concepcion JCT, Ouk M, Zhao D, Fitzgerald MA (2015) The need for new tools and investment to improve the accuracy of selecting for grain quality in rice. F Cro Res 182:60–67. https://doi.org/10.1016/j.fcr.2015.05.003

Das M, Gupta S, Kapoor V, Banerjee R, Bal B (2008) Enzymatic polishing of rice e A new processing technology. LWT - F Sci and Tec 41:2079–2084. https://doi.org/10.1016/j.lwt.2008.02.007

Denardin CC, Silva LP (2009) Estrutura dos grânulos de amido e sua relação com propriedades físico-químicas. C Rur 39:945–954. https://doi.org/10.1590/S0103-84782009005000003

Denardin CC, Silva LP, Storck CR, Nornberg JL (2004) Ineral composition of brown, parboiling and white rice cultivars. Ali e Nut 15:125–130

Dutta H, Mahanta CL (2012) Effect of hydrothermal treatment varying in time and pressure on the properties of parboiled rices with different amylose content. F Res Int 49:655–663. https://doi.org/10.1016/j.foodres.2012.09.014

Elias MC, Oliveira M, Nathan L (2015) Qualidade de arroz na pós-colheita e na agroindústria: análise, conservação e tipificação. Santa Cruz, Pelotas, p 221

Fukuda T, Takeda T, Yoshida S (2014) Comparison of volatiles in cooked rice with various amylose contents. Food Sci Technol Res 20:1251–1259

Goffman FD, Bergman CJ (2004) Rice kernel phenolic content and its relationship with antiradical efficiency. J Sci Food Agric 84:1235–1240. https://doi.org/10.1002/jsfa.1780

Gomes AS, Magalhães Junior AM (2004) Arroz Irrigado no Sul do Brasil. Embrapa Informação, Brasília. 899p

Goufo P, Trindade H (2014) Rice antioxidants: phenolic acids, flavonoids, anthocyanins, proanthocyanidins, tocopherols, tocotrienols, γ-oryzanol, and phytic acid. Foo Sci Nut 2:75–104. https://doi.org/10.1002/fsn3.86

Hanashiro I, Abe J, Hizukuri S (1996) A periodic distribution of the chain length of amylopectin as revealed by high performance anion exchange chromatography. C Pol 283:151–159. https://doi.org/10.1016/0008-6215(95)00408-4

Heinemann RJB, Fagundes PL, Pinto EA, Penteado MVC, Lanfer-marquez UM (2005) Comparative study of nutrient composition of commercial brown, parboiled and milled rice from Brazil. J Foo Com Ana 18:287–296. https://doi.org/10.1016/j.jfca.2004.07.005
Himmelsbach DS, Manful JT, Coker RD (2008) Changes in rice with variable temperature parboiling: thermal and spectroscopic assessment. Cereal Chem 85:384–390. https://doi.org/10.1094/CCHEM-85-3-0384
Horigane AK, Suzuki K, Yoshida M (2014) Moisture distribution in rice grains used for sake brewing analyzed by magnetic resonance imaging. J Cereal Sci 60:193–201. https://doi.org/10.1016/j.jcs.2014.02.011
Hou F, Zhang R, Zhang M, Su D, Wei Z, Deng Y, Zhang Y, Chi J, Tang X (2013) Hepatoprotective and antioxidant activity of anthocyanins in black rice bran on carbon tetrachloride-induced liver injury in mice. J Funct Foods 5:1705–1713. https://doi.org/10.1016/j.jff.2013.07.015
Jayaraman R, Uluvar H, Khanum F, Singh V (2019) Influence of parboiling of red paddy varieties by simple hot soaking on physical, nutrient, phytochemical, antioxidant properties of their dehusked rice and their mineral, starch, and antioxidant's bioaccessibility studies. J F Bio e12839:1–16. https://doi.org/10.1111/jfbc.12839
Juliano BO (1985) Criteria and tests for rice grain qualities. In: Rice chemistry and technology, 2nd edn. A A Cer Che, St. Paul Minnesota, pp 443–524
Kale SJ, Jha SK, Jha GK, Sinha JP, Lal SB (2015) Soaking induced changes in chemical composition, glycemic index and starch characteristics of basmati rice. R Sci 22:227–236. https://doi.org/10.1016/j.rsci.2015.09.002
Kasai M, Lewis A, Marica F, Ayabe S, Hatae K, Fyfe CA (2005) NMR imaging investigation of rice cooking. Food Res Int 38:403–410. https://doi.org/10.1016/j.foodres.2004.10.012
Khin OM, Sato M, Li-Tao T, Matsue Y, Yoshimura A, Mochizuki T (2013) Close association between aleurone traits and lipid contents of rice grains observed in widely different genetic resources of Oryza sativa. Plant Prod Sci 16:41–49. https://doi.org/10.1626/pps.16.41
Lamberts L, Bie E, Derycke V, Veraverbeke WS, De Man W, Delcour JA (2006) Effect of processing conditions on color change of brown and milled parboiled rice. Cereal Chem 83:80–85. https://doi.org/10.1094/CC-83-0080
Lamberts L, Brijs IR, Gebruers K, Delcour JA (2008) Impact of parboiling conditions on Maillard precursors and indicators in long-grain rice cultivars. Food Chem 110:916–922. https://doi.org/10.1016/j.foodchem.2008.02.080
Lang GH, Lindemann IS, Ferreira CD, Hoffmann JF, Vanier NL, Oliveira M (2019) Effects of drying temperature and long-term storage conditions on black rice phenolic compounds. 287:197–204. https://doi.org/10.1016/j.foodchem.2019.02.028
Leethanapanich K, Mauromoustakos A, Wang YJ (2016) Impacts of parboiling conditions on quality characteristics of parboiled commingled rice. J Cer Sci 69:283–289. https://doi.org/10.1016/j.jcs.2016.04.003
Li H, Prakash S, Nicholson TN, Fitzgerald MA, Gilbert RG (2016) The importance of amylose and amylopectin fine structure for textural properties of cooked rice grains. Food Chem 196:702–711. https://doi.org/10.1016/j.foodchem.2015.09.112
Likitwattanasade T, Hongsprabhas P (2010) Effect of storage proteins on pasting properties and microstructure of Thai rice. F Res Int 43:1402–1409. https://doi.org/10.1016/j.foodres.2010.04.011
Luo X, Li Y, Yang D, Xing J, Li K, Yang M, Wang R, Wang L, Zhang Y, Chen Z (2019) Effects of electron beam irradiation on storability of brown and milled Rice. J Sto Pro Res 81:22–30. https://doi.org/10.1016/j.jspr.2018.12.003
Martin M, Fitzgerald MA (2002) Proteins in rice grains influence cooking properties. J Cer Sci 36:285–294. https://doi.org/10.1006/jcrs.2001.0465
Masisi K, Beta T, Moghadasian MH (2016) Antioxidant properties of diverse cereal grains: a review on in vitro and in vivo studies. F Che 196:90–97. https://doi.org/10.1016/j.foodchem.2015.09.021

Min B, McClung A, Chen MH (2014) Effects of hydrothermal processes on antioxidants in brown, purple and red bran whole grain rice (Oryza sativa L.). F Che 159:106–115. https://doi.org/10.1016/j.foodchem.2014.02.164

Ming-Hsuan C, Anna MA (2019) Genotypic diversity of bran weight of whole grain rice and its relationship with grain physical traits. C Che 96:252–262. https://doi.org/10.1002/cche.10117

Monks JLF, Vanier NL, Casaril J, Berto RM, de Oliveira M, Gomes CB, de Carvalho MP, Dias ARG, Elias MC (2013) Effects of milling on proximate composition, folic acid, fatty acids and technological properties of rice. J F Com Ana 30:73–79. https://doi.org/10.1016/j.jfca.2013.01.009

Ong MH, Blanshard JMV (1995) Texture determinants in cooked, parboiled rice. I: rice starch amylose and the fine stucture of amylopectin. J Cereal Sci 21:251–260. https://doi.org/10.1006/jcrs.1995.0028

Paiva FF, Vanier NL, Berrios JJ, Pinto VZ, Wood D, Williams T, Pan J, Elias MC (2016) Polishing and parboiling effect on the nutritional and technological properties of pigmented rice. F Che 191:105–112. https://doi.org/10.1016/j.foodchem.2015.02.047

Paraginski RT, Evangelho JA, Colussi R, Marques e Silva R, Zavareze ER, de Oliveira M, Elias MC, ARG D (2014) Sta 66:729–737. https://doi.org/10.1002/star.201300291

Rockembach CT, Mello EHSL, Mesko MF, Gutkoski LC, Oliveira M (2019) Morphological and physicochemical properties of rice grains submitted to rapid parboiling by microwave irradiation. LWT F Sci Tec 103:44–52. https://doi.org/10.1016/j.lwt.2018.12.036

Sareepuang K, Siriamornpun S, Wiset l MN (2008) Effect of soaking temperature on physical, chemical and cooking properties of parboiled fragrant rice. W Jou Agr Sci 4:409–415

Shao Y, Xu F, Sun X, Bao J, Beta T (2014) Identification and quantification of phenolic acids and anthocyanins as antioxidants in bran, embryo and endosperm of white, red and black rice kernels (Oryza sativa L.). J Cereal Sci 59:211–218. https://doi.org/10.1016/j.jcs.2014.01.004

Shinde YH, Vijayadwhaja A, Pandit AB, Joshi JB (2014) Kinetics of cooking of rice: a review. J Food Eng 123:113–129. https://doi.org/10.1016/j.jfoodeng.2013.09.021

Tao K, Yu W, Prakash S, Gilberta RG (2019) High-amylose rice: starch molecular structural features controlling cooked rice texture and preference. C Pol. 219:251–260. https://doi.org/10.1016/j.carbpol.2019.05.031

Toutounji MR, Farahnaky A, Santhakumar AB, Oli P, Butardo VM, Blanchard CL (2019) Intrinsic and extrinsic factors affecting rice starch digestibility. T F Sci Tec 88:10–22. https://doi.org/10.1016/j.tifs.2019.02.012

Villanova FA, Vanier NL, De Avila MN, Pesek J, Matyska-Pesek M, Elias MC, de Oliveira M (2017) Improvement of the quality of parboiled rice by using anti-browning agents during parboiling process. F Che 235:51–57. https://doi.org/10.1016/j.foodchem.2017.05.053

Walter M, Marchezan E, Avila LA (2008) Arroz: composição e características nutricionais. C Rur, Santa Maria 38:1184–1192

Xiong Y, Ren Y, Li W, Wu F, Yang W, Huang X, Yao J (2019) NF-YC12 is a key multi-functional regulator of accumulation of seed storage substances in rice. J Exp Bot 70:3765–3780. https://doi.org/10.1093/jxb/erz168

Yadav BK, Jindal VK (2007) Water uptake and solid loss during cooking of milled rice (Oryza sativa L.) in relation to its physicochemical properties. J Foo Eng 80:46–54. https://doi.org/10.1016/j.jfoodeng.2006.05.011

Ye J, Hu X, Luo S, Mcclements DJ, Liang L, Liu C (2018) Effect of endogenous proteins and lipids on starch digestibility in rice flour. F Res Int 106:404–409. https://doi.org/10.1016/j.foodres.2018.01.008

Zaupa M, Calani L, Rio DD, Brighenti F, Pellegrini N (2015) Characterization of total antioxidant capacity and (poly)phenolic compounds of differently pigmented rice varieties and their changes during domestic cooking. Food Chem 187:338–347. https://doi.org/10.1016/j.foodchem.2015.04.055

Zavareze ER, Dias ARG (2011) Impact of heat-moisture treatment and annealing in starches: a review. Carbohydr Polym 83:317–328. https://doi.org/10.1016/j.carbpol.2010.08.064

Zavareze ER, Dias ARG (2016) Impact of heat-moisture treatment and annealing in starches: a review. C Pol 83:317–328. https://doi.org/10.1016/j.carbpol.2010.08.064

Zhang M, Tang S, Huang X, Zhang F, Pang Y, Huang Q, Yi Q (2014) Selenium uptake, dynamic changes in selenium content and its influence on photosynthesis and chlorophyll fluorescence in rice (Oryza sativa L.). Environ Exp Bot 107:39–45. https://doi.org/10.1016/j.envexpbot.2014.05.005

Zhang S, Dong Q, Zhang L, Xiong Y (2015) High quality syngas production from microwave pyrolysis of rice husk with char-supported metallic catalysts. Bioresour Technol 191:17–23. https://doi.org/10.1016/j.biortech.2015.04.114

Zhou ZK, Yang X, Su Z, Bu DD (2016) Effect of ageing-induced changes in rice physicochemical properties on digestion behaviour following storage. J S Prod Res 67:13–18. https://doi.org/10.1016/j.jspr.2016.01.004

Zhu F (2018) Relationships between amylopectin internal molecular structure and physicochemical properties of starch. T F Sci Tec 78:234–242. https://doi.org/10.1016/j.tifs.2018.05.024

Zhu D, Qian Z, Wei H, Guo B, Xu K, Dai Q, Zhang H, Huo Z (2019) The effects of field pre-harvest sprouting on the morphological structure and physicochemical properties of rice (Oryza sativa L.) starch. F Che 278:10–16. https://doi.org/10.1016/j.foodchem.2018.11.017

Ziegler V, Ferreira CD, Hoffmann JF, Chaves FC, Vanier LV, de Oliveira M, Elias MC (2018) Cooking quality properties and free and bound phenolics content of brown, black, and red rice grains stored at different temperatures for six months. F Che 242:427–434. https://doi.org/10.1016/j.foodchem.2017.09.077

Chapter 3
Industrial Quality

Ariano Martins de Magalhães Júnior, Eduardo Anibele Streck, Gabriel Almeida Aguiar, and Paulo Henrique Karling Facchinello

3.1 Introduction

Rice (*Oryza sativa* L.) plays an important strategic role at economic and social levels, as well as being fundamental in human nutrition, being responsible for supplying 20% of the calories consumed in the food of people worldwide (Sosbai 2018). It has an excellent nutritional balance, presenting 20% of the energy and 15% of the per capita protein needed by man, besides being a crop adapted to different environmental conditions, making it the species with the greatest potential to fight hunger in the world (Gomes and Magalhães Jr 2004).

Yield is the main attribute sought in rice breeding programs. However, the major increases in yield potential were obtained by replacing traditional cultivars with modern cultivars (high tillering, semi-dwarf with sturdy and short stalks, and erect leaves) (Magalhães Jr et al. 2003). However, the quality of grains in rice directly reflects its market value and consumer acceptance, where according to Walter (2008), a small amount of rice is consumed as an ingredient in milled products, mostly consumed as in its natural form (*in natura*) grains. Grain quality attributes in rice have become a priority in breeding programs, at the same level as yield, being decisive in adopting an improved line as a new commercial cultivar (Rangel et al. 2000).

Due to the increasing demand and concern of the consumer to purchase better quality products, there has been investment from research, industries, as well as

A. M. de Magalhães Júnior (✉) · P. H. K. Facchinello
Embrapa Clima Temperado, Capão do Leão, RS, Brazil
e-mail: ariano.martins@embrapa.br

E. A. Streck
Instituto Federal Farroupilha, Campus Santo Augusto, Santo Augusto, RS, Brazil

G. A. Aguiar
Instituto Federal Rio Grande do Sul, Campus Sertão, Sertão, RS, Brazil

A. Costa de Oliveira et al. (eds.), *The Future of Rice Demand: Quality Beyond Productivity*, https://doi.org/10.1007/978-3-030-37510-2_3

producers in items related to rice quality. The quality begins to be defined in the planning of the planting, when the variety to be cultivated is chosen; it continues in the management of the crop and is very critical in the phase that goes from the harvest to the storage. All of this characterizes the pre-industrialization stages, i.e., the raw material treatment phase. Since arrival in the industry, methods and processes have been refined to improve rice quality. Many industries are modernizing their facilities by investing in both technology and equipment to meet these new market demands.

In general, grain quality in rice can be divided into four aspects: nutritional value, industrial quality, product suitability to marketing standards, and culinary and sensory quality (Anacleto et al. 2015). Grain quality in rice influences market value of the product, consumer acceptance of the product, and the adoption of new cultivars. Attributes such as physical appearance and physicochemical and culinary properties characterize the product and drive decisions about cultivars in the consumer market (Castro et al. 1999).

The quality of grains in rice varies according to the eating habits of each person, thus varying worldwide according to consumer preference. There are countries in which high quality means short grain with low amylose, while in other countries, it is long grains with high amylose. The presence of chalkiness in the grain is also variable from country to country, desired in some and considered a defect for others.

In general, in Brazil, rice consumers prefer long, thin, vitreous grain with low cracked and/or damaged grain contents (Brasil 2009). Likewise, adequate milling performance, with good yields of whole grains, is also sought by producers and mills, since the breakage rate during grain milling affects the value of the product in the market and is a determining factor in the yield acceptance of new cultivars (Embrapa et al. 2013). From the Brazilian consumer's point of view, besides the appearance of the raw product, the preference is for a quality cooking rice that provides good pan yield, cooks fast, has dry and loose grains after cooking, and remains soft even after cooling. On the other hand, there is growing interest in the current market for other types of rice, such as pigmented or aromatic grains or for specific cuisines such as Italian or Japanese. They are the so-called special types of rice, with specific characteristics in terms of quality, covering mainly the nutritional and functional aspects (presence of beneficial health compounds) of the grain (Zhou et al. 2002).

Irrigated rice cultivars cultivated in Brazil belong mainly to the Indica group, consisting essentially of high amylose content (straight starch chain) and low amylopectin content (branched starch chain), becoming loose after cooking. Amylose content is associated with textural properties such as softness, cohesion, color, gloss, and water absorption and expansion volume (Mingotte et al. 2012). Most of these cultivars also have low or intermediate starch gelatinization temperature, where they refer to the cooking temperature where water is absorbed and 90% of starch granules are gelatinized and swollen irreversibly, with simultaneous breaking of crystallinity (Guimarães 1989). This feature directly implies less time and water, hence less energy for cooking (Mingotte et al. 2012).Thus, in Brazil, the pattern corresponds mainly to a so-called small needle grain, i.e., long and thin (length ≥6 mm, thickness ≤1.9 mm, and length/width ratio between 2.75 and 3.00 mm). In addition to the appearance of the product in natural form, the preference is for a quality cooking rice providing good pan yield with quick cooking, showing dry and loose grains after cooking, and remaining soft even after cooling (Castro et al. 1999; Usberti Filho et al. 1986; Guimarães 1989).

3.2 Classification of Rice

3.2.1 *Groups*

1. Paddy rice: grains derived from the species *Oryza sativa* L., in natural form, without industrial milling
2. Milled rice: the mature product that has undergone some beneficiation process and is at least devoid of its husk

3.2.2 *Subgroups*

1. Natural paddy rice: the product which does not undergo any industrial preparation or technological process prior to milling
2. Brown rice or husked rice: the product from which only the husk has been removed
3. Polished rice: the product from which the husk, the germ, the pericarp, and most of the inner layer (aleurone) are removed upon being milled
4. Parboiled rice: the product that has been subjected to the parboiling process, hydrothermal process in which the paddy rice is immersed in water at a temperature above 58 °C followed by partial or total starch gelatinization and drying
5. Parboiled brown rice: the product that has been subjected to the parboiling process and subsequently removed husk
6. Polished parboiled rice: the product that was submitted to the parboiling process and subsequent milling, where the husk, the germ, the pericarp, and most of the inner layer (aleurone) are removed

3.2.3 *Classes*

Rice should fall into five classes based on the size of whole grains after husking and polishing. Thus, it is necessary that at least 80% of the weight of the sample grains be within one of the first four classes (long thin, long, medium, and short) (Fig. 3.1); otherwise, it will go to a fifth class called mixed.

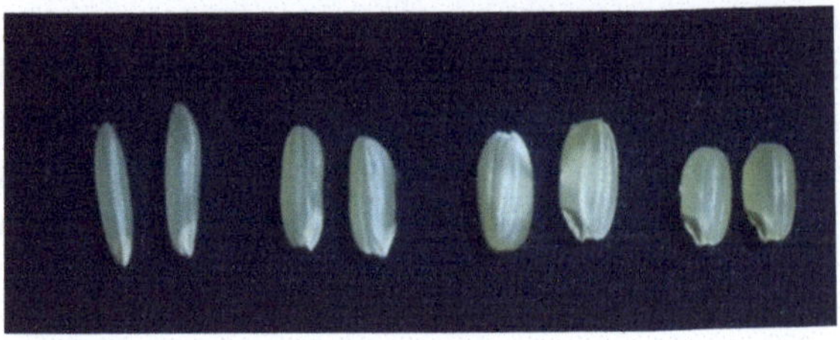

Fig. 3.1 Polished rice grains of the first four classes: long thin, long, medium, and short, respectively

Table 3.1 Classification of rice grains according to their length, thickness, and length/width ratio specifications

Grain classification	Specifications
Long thin	Length ≥ 6 mm
	Thickness ≤ 1.90 mm
	Length/width ratio > 2.75 mm
Long	Length ≥ 6 mm
Medium	Length from 5 mm to less than 6 mm
Short	Length < 5 mm
Mixed	Mixed presence of two or more classes

Source: Brasil (2009); MAPA (2012), adapted by the authors

I. Long thin: the product containing at least 80% of the weight of whole grains measuring 6.00 mm or more in length, the thickness less than or equal to 1.90 mm and the length/width ratio greater than or equal to 2.75 after polishing the grains (Table 3.1)
II. Long: the product containing at least 80% of the weight of whole grains measuring 6.00 mm or more in length after grain polishing (Table 3.1)
III. Medium: the product containing at least 80% of the weight of whole grains measuring from 5.00 mm to less than 6.00 mm in length after polishing the grains (Table 3.1)
IV. Short: the product containing at least 80% of the weight of whole grains measuring less than 5.00 mm in length after polishing the grains (Table 3.1)
V. Mixed: the product that does not fall into any of the previous classes (Table 3.1)

3.2.4 Types

In addition to class, all rice intended for marketing as grain for consumption must be classified into types. The types are numerically expressed and defined according to the percentage of defect occurrence and the percentage of cracked grains and special cracked type called "chirera" which are smaller grain parts, and they have maximum tolerance limits according to the Brazilian legislation with regard to the rice group and subgroup: natural paddy rice (Table 3.2) (Fig. 3.2), parboiled paddy rice (Table 3.3), whole benefited rice (Table 3.4), whole parboiled beneficiated rice (Table 3.5), polished beneficial rice (Table 3.6), and polished parboiled beneficial rice (Table 3.7).

Table 3.2 Maximum defect tolerance limits, expressed in %/weight, of paddy rice

Type	Spoiled	Chopped or stained	Chalky and green	Striped	Yellow
1	0.15	1.75	2.00	1.00	0.50
2	0.30	3.00	4.00	1.50	1.00
3	0.50	4.50	6.00	2.00	2.00
4	1.00	6.00	8.00	3.00	3.00
5	1.50	8.00	10.00	4.00	5.00

Source: Brasil (2009), MAPA (2012), adapted by the authors

Fig. 3.2 Natural husked rice of the first four classes, long thin, long, medium, and short, respectively

Table 3.3 Maximum defect tolerance limits, expressed in %/weight, of parboiled paddy rice

Type	Spoiled and dark	Nongelatinized	Damaged	Striped	Chopped or stained
1	0.20	20.00	0.50	1.00	1.75
2	0.40	25.00	1.00	1.50	3.00
3	0.60	35.00	1.50	2.00	4.50
4	0.80	45.00	2.00	3.00	6.00
5	1.00	55.00	3.00	4.00	8.00

Source: Brasil (2009), MAPA (2012), adapted by the authors
Note: The maximum allowable tolerance limit for non-parboiled grain is 0.30% for all types. Above this limit the product will be considered as Out of Type

Table 3.4 Maximum tolerance limits of defects, expressed in %/weight, of the milled whole rice

Type	Strange matters and impurities	Mold and spoiled	Chopped or stained	Chalky and green	Red and black	Yellow	Total cracked and "chirera"
1	0.10	0.15	1.75	2.00	1.00	0.50	4.00
2	0.20	0.30	3.00	4.00	1.50	1.00	7.50
3	0.30	0.50	4.50	6.00	2.00	2.00	12.50
4	0.40	1.00	6.00	8.00	3.00	3.00	15.00
5	0.50	1.50	8.00	10.00	4.00	5.00	20.00

Source: Brasil (2009), MAPA (2012), adapted by the authors
Note: The maximum tolerance allowed for unhusked is 10 grains in 1000 g for all types. Above this limit the product will be considered as Out of Type

Table 3.5 Maximum defect tolerance limits, expressed in %/weight, of parboiled whole beneficiated rice

Type	Strange matters and impurities	Mold, spoiled, and darkened	Nongelatinized	Damaged	Red and black	Chopped or stained	Total cracked and "chirera"
1	0.05	0.20	20.00	0.50	1.00	1.75	2.50
2	0.10	0.40	25.00	1.00	1.50	3.00	4.00
3	0.15	0.60	35.00	1.50	2.00	4.50	6.00
4	0.20	0.80	45.00	2.00	3.00	6.00	8.00
5	0.25	1.00	55.00	3.00	4.00	8.00	10.00

Source: Brasil (2009), MAPA (2012), adapted by the authors

Remarks: (1) The maximum allowable tolerance limit for non-parboiled grain is 0.30% for all types. Above this limit the product will be considered as Out of Type. (2) The maximum tolerance allowed for unhusked is 5 grains in 1000 g for all types. Above this limit the product will be considered as Out of Type

Table 3.6 Maximum defect tolerance limits, expressed in %/weight, of polished benefited rice

Type	Strange matters and impurities	Mold and spoiled	Chopped or stained	Chalky and green	Striped	Yellow	Total cracked and "chirera"	"Chirera" (maximum)
1	0.10	0.15	1.75	2.00	1.00	0.50	7.50	0.50
2	0.20	0.30	3.00	4.00	1.50	1.00	15.00	1.00
3	0.30	0.50	4.50	6.00	2.00	2.00	25.00	2.00
4	0.40	1.00	6.00	8.00	3.00	3.00	35.00	3.00
5	0.50	1.50	8.00	10.00	4.00	5.00	45.00	4.00

Source: Brasil (2009), MAPA (2012), adapted by the authors

Note: The maximum tolerance allowed for sailors is 10 grains in 1000 g for all types. Above this limit the product will be considered as Out of Type

Serious defects are those resulting from contamination of the product by foreign matter, mold, and burnt grains. The general defects refer to the presence of stained and chopped grains (mainly caused by the bed bugs), yellows, streaked (red rice), and chalk (grains usually harvested very immature) (Fig. 3.3).

3.3 Genetics of Physical Attributes of Rice Grain Quality

Rice grain composition and quality are subject to genetic differences, environmental, management, milling, and storage variations (Zhou et al. 2002). Therefore, the understanding of the genetic basis of quality traits is fundamental to obtain higher potential genetic gains and, consequently, genetically superior genotypes.

In recent years, the use of biotechnology has allowed the identification of several important QTLs (Quantitative Trait Loci) associated with the industrial quality of

Table 3.7 Maximum defect tolerance limits, expressed in %/weight, of polished parboiled benefited rice

Type	Strange matters and impurities	Mold and spoiled	Nongelatinized	Damaged	Striped	Chopped or stained	Total cracked and "chirera"	"Chirera" (maximum)
1	0.05	0.20	20.00	0.50	1.00	1.75	4.50	0.40
2	0.10	0.40	25.00	1.00	1.50	3.00	7.00	0.50
3	0.15	0.60	35.00	1.50	2.00	4.50	9.00	0.75
4	0.20	0.80	45.00	2.00	3.00	6.00	11.00	1.00
5	0.25	1.00	55.00	3.00	4.00	8.00	15.00	1.25

Source: Brasil (2009), MAPA (2012), adapted by the authors

Remarks: 1. The maximum allowable tolerance limit for non-parboiled grain is 0.30% for all types. Above this limit the product will be considered as Out of Type. 2. The maximum tolerance allowed for unhusked is 5 grains in 1000 g for all types. Above this limit the product will be considered as Out of Type

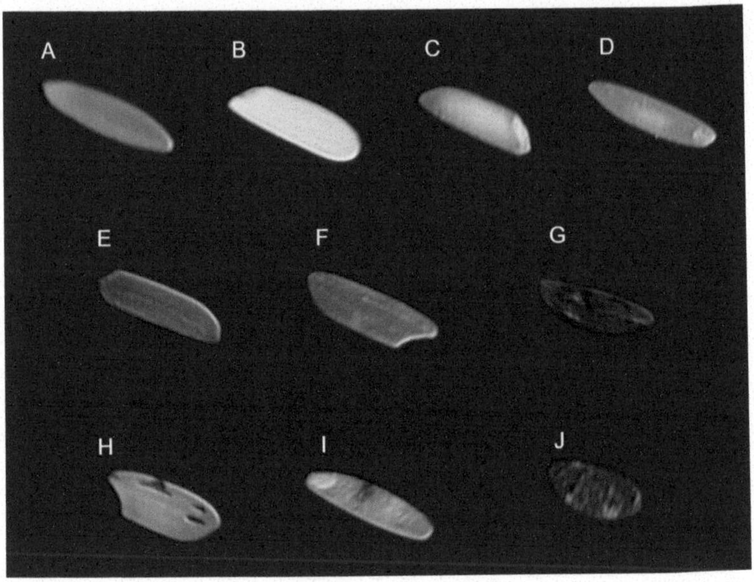

Fig. 3.3 Vitreous rice grain compared to grains with severe and general defects. Vitreous grain (**a**); chalky grain (**b**); grain with chalky center (**c**); grain with chalky (**d**); grain strong yellow (**e**); yellow grain (**f**); burnt/mold grain (**g**); streaked grain (**h**); chopped grain (**i**); invasive particles (**j**). (Images obtained through the S21 Statistical Rice Analyzer)

rice. However, the physical attributes of quality are very complex quantitative traits, so many genetic mechanisms and interactions with the environment still remain obscure (Shi et al. 2002; Zhou et al. 2009).

3.3.1 Whole Grain Yields

Whole grain yield after milling is a complex trait whose genetics are not fully understood. Several genes that control grain shape have been identified and cloned, such as *GW2* (Song et al. 2007), *GIF1* (Wang et al. 2008), *qSW5* (Shomura et al. 2008), *GS3* (Mao et al. 2010), *GS5* (Li et al. 2011), *qGL3* (Zhang et al. 2012), *GW8* (Wang et al. 2012), *GS6* (Sun et al. 2013), *GS2* (Hu et al. 2015), *GL7/GW7* (Wang et al. 2015), *OsMAPK6* (Liu et al. 2015), and *GLW7* (Si et al. 2016). In addition to these genes, many QTLs that affect grain size have been identified through linkage mapping and association studies (Zhao et al. 2011; Singh et al. 2012; Yang et al. 2014; Qiu et al. 2015; Edzesi et al. 2016; Feng et al. 2016) and some fine-mapping, such as *GW1-1* and *qGRL1.1* (Singh et al. 2012), *GW3* and GW6 (Guo et al. 2009), *qGL- 7* (Bai et al. 2010), and *qGRL7.1* (Singh et al. 2012).

These studies only allowed the verification of a strong correlation between grain shape and whole grain yield in long and thin rice. Recently, however, genetic mapping has allowed the identification and cloning of *qBRR-10* which was localized to a 39.5-kb region on chromosome 10, and two candidate genes were selected based on the difference in expression between parents, contributing to improve rice productivity and quality (Ren et al. 2016).These reports do not yet precisely target appropriate breeding techniques or methods; however, they improve the understanding of genetic control and provide molecular markers that are useful in increasing the chances of obtaining transgressive segregants and, consequently, genetic gains per breeding cycle.

3.3.2 Chalky Grains

In order to meet the most demanding rice consumption standards, the Brazilian rice market prefers long-grain, well-translucent grains. When this translucency is interrupted by opaque areas in the endosperm, we call it white center, white belly, and chalky grains. This grain opacity occurs due to the arrangement between the starch and protein granules developed under adverse weather and cultivation conditions, as well as due to genetic factors (Fitzgerald et al. 2008). This presence of chalky area in the grain negatively affects appearance, milling yield, cooking, and nutritional qualities. Therefore, these parameters are linked to the defects associated with rice grains, dictating the market value and playing a fundamental role in the adoption and, consequently, in the development of new cultivars. Regarding the genetics of these attributes, many QTLs have been identified; however, these attributes are highly influenced by the environment. Zhao et al. (2016) detected 78 and 43 QTLs in nine environments; however, few were stably expressed in different environments. A *QTL qPGWC-7* was accurately mapped to the 44-kb region on chromosome 7 (Zhou et al. 2009), and a QTL cluster for chromosome 4 on the *id4007289* and *RM252* flank was detected by single environment analysis and joint mapping in nine environments (Zhao et al. 2016). Another relevant feature is that many of the genes characterized for these attributes have pleiotropic effects, as evidenced by Woo et al. (2008) who found that the male sterility gene *ms-h* is recessive and has a pleiotropic effect on the grain chalkiness.

Only the *Chalk5* gene has been cloned, and its high expression affects protein synthesis and increases small vesicle-like structures, forming spaces between endosperm storage substances, increasing grain chalkiness, and affecting many other quality traits (Li et al. 2014). According to Li et al. (2014), two SNPs (Single Nucleotide Polymorphism) in the promoter region of the gene suppress *Chalk5* expression level, indicating that these two SNPs are the main reason for the variation in the length of chalk in long and thin rice (subspecies *indica*).Thus, some of the knowledge of chalkiness genetics can be explained by *Chalk5*; however, there are still many genes and their effects related to starch synthesis and grain formation that remain unexplored.

3.3.3 *Genetic Gains for Physical Attributes of Rice Grain Quality*

The development of genetically superior cultivars for grain yield and quality attributes is a major challenge for breeders to meet the requirements of developing technologies compatible with the reality of rice cultivation. At the level of yields, irrigated rice breeding programs are in advanced stages, with very significant genetic gains (Tabien et al. 2008; Breseghello et al. 2011; DoVale et al. 2012; Streck et al. 2018a). However, the recent intensification of selection pressure for grain quality traits in breeding programs, coupled with the complexity of the genetic involving these traits, shows the lack of consolidated selection strategies and methods for breeding.

Streck et al. (2018b) in a study of genetic progress for grain quality of rice cultivars released in southern Brazil estimated significant genetic gains for percentage of whole grains after milling (0.20% per year), percentage of chalkiness and white belly (−1.38% per year), total chalky area (−0.77% per year), total whiteness (−0.08% per year), and non-vitreous grain whiteness (−0.82% per year). Feng et al. (2017) studied the main quality characteristics of 635 Chinese rice varieties from 2000 to 2014, establishing that grain quality (milling yield and chalk) in Japanese rice and hybrid rice has improved, except for traditional Indian varieties with limited gains. These genetic advances are achieved from genetically superior cultivar releases over the years for these industrial quality attributes. Examples of "premium" industrial quality standard cultivars are the Brazilian irrigated rice cultivars IRGA 417 and BRS Pampa (Magalhães Jr et al. 2012; Streck et al. 2018b). Other cultivars are also noteworthy for their superior qualities, such as the cultivars BRS Pampeira (Magalhães Jr et al. 2017; Streck et al. 2018b) and BRS Primavera (Champagne et al. 2010). In addition, the cultivar BR-IRGA 409, despite being an old cultivar (launched in 1979), is still under cultivation in Brazil for its high grain quality (Streck et al. 2018b). Thus, in addition to yield, rice breeding programs have sought to select superior genotypes for the release of cultivars with excellent industrial grain quality.

3.4 Perspectives and Opportunities in the Genetics of Physical Attributes for Rice Grain Quality

Considering the complexity of these rice quality traits, there has been relative progress in research on its genetic controls. The progress of molecular genetics, such as selection assisted by molecular markers, has allowed significant advances in understanding various quality traits of rice grains. However, we still find a huge gap for further genetic studies. The CRISPR/Cas9 gene editing system, for example, has great untapped potential, enhancing research to understand these complex polygenic attributes, and is a promising approach to overcome barriers to the genetic improvement of rice quality. In contrast to this advance in molecular biology, few studies have been conducted in the field of quantitative genetics applied to conventional breeding. This fact highlights the need for exploration in this area, seeking to combine the advances of molecular biology with its application in the field improvement work field. This increases the likelihood of understanding these traits of physical grain quality and, consequently, the potential genetic gains per selection cycle.

3.5 Crop Management and Grain Quality

3.5.1 Cultivation System Management

Grain quality in rice cultivation is an extremely important condition for its classification at commercialization of the product since whole and flawless grains have higher market value and consumer acceptability. This condition is directly related to genotypic characteristics of cultivars and environmental factors such as management practices in cultivation and postharvest of grains. Environmental conditions during all phases of the crop phenological cycle directly reflect the quality and yield of rice. Thus, rice management practices aim to meet the necessary requirements for cultivars to express the maximum of their genetic potential.

One of the primary management practices to achieve high quality and grain yield in rice crop is sowing. Sowing time is directly related to the cultivation region and the cultivar cycle, which corresponds to the number of days from emergence until the physiological maturation of the plant. The decisive condition to establish the sowing time in rice is to make compatible the reproductive phase of the crop with the period of highest solar radiation. Thus, the most important periods between panicle differentiation and flowering, which influence the number of grains per panicle, and between flowering and maturation, which influence grain weight, occur at the moment of the highest solar radiation and temperature suitable for rice cultivation. The coincidence of these periods with the highest solar radiation and air temperature conditions establishes one of the main factors related to the yield of grain (Lack et al. 2012) as well as for grain quality in the crop.

Another important factor is the management of fertilization and mineral nutrition in rice cultivation, which are essential to provide proper growth and development of the crop, so that the cultivar expresses its genotypic potential for grain quality and yield. Each rice essential nutrient performs definite functions within the plant, and none can be replaced by another, all of which need to be present in the proper form and quantity to produce satisfactory results (Veloso et al. 2009). Likewise, water is indispensable for the absorption, transport, and translocation of nutrients between plant tissues and organs (Taiz and Zeigeir 2013) as well as from soil to plant. Based on the amount of nutrients required for rice cultivation, they are divided into macro- and micronutrients. The increased requirement for the macronutrient contents for plants is associated with their role in the composition of most plant cell carbohydrates, proteins, and lipids, while micronutrients participate primarily in the plant's enzymatic activation process (Fageria 2009). Nitrogen, potassium, and phosphorus are considered the main nutritional elements for the plant, so their use as fertilizer is fundamental; therefore, they are more widespread and used more than other elements.

Nitrogen is the nutrient most absorbed by the rice crop (Bianchet et al. 2015), being one of the limiting factors for crop yield (Fageria et al. 2011b). In addition to yield, the application of nitrogen fertilizers also has an important impact on improving rice quality, especially during the grain filling period (Tang et al. 2019). Chlorophyll component, nitrogen, has several physiological effects on the rice plant, such as increase in tiller, panicle, and grain number and grain size, directly reflecting the increase in grain yield and protein content (Lin et al. 2009). To this nutrient (nitrogen) is also responsible for obtained the higher whole (uncracked) grain yield in milling (Freitas et al. 2001), better weight of one thousand grains, and reduced spikelet sterility (Fageria 2007). The nitrogen recommendations for rice consider the soil organic matter content and the expected response of the crop to fertilization. In addition to the base fertilization, it is recommended to apply the coverage in the phenological stages of thinning and differentiation of the floral primordium.

Potassium plays many vital and essential roles related to physiological and biochemical processes in cultivated plants, such as enhancing root growth and improving water and nutrient absorption. In addition, it is required to activate at least 60 different enzymes involved in plant growth, aid in sugar and starch translocation, influence starch formation in grains, maintain turgor and reduce water loss and cell wilting, increase resistance of plant tissues, and reduce the fungal populations in the soil (Fageria 2009). Nutritional management of potassium in the rice crop is performed according to the interpretation of its contents, as a function of soil cation exchange capacity (CEC) from critical contents for the culture.

Potassium absorption by rice occurs mainly in the vegetative phase, being the best results obtained by the application before sowing, regardless of the cropping system. However, split application is indicated for crops in sandy soils, where nutrient losses are higher, and it is recommended to apply half of the dose before sowing and the remainder in panicle differentiation. Potassium applications significantly influence dry plant biomass, number of tiller per plant, number of grains per panicle, one thousand grain mass, and rice grain yield (Zaratin et al. 2004; Fageria et al. 2011a; Fageria 2015; Nascente and Kromocardi 2017).

Phosphorus is essential for plant growth and reproduction and is an important nutrient along with nitrogen and potassium. The flooded rice cultivation system provides significant increases in the availability of this nutrient to rice plants, demonstrating relatively small effects to phosphate fertilization, even in soils with low levels of available phosphorus. In other cropping systems, phosphorus is one of the main limiting factors for rice production. Its deficiency causes smaller number of tiller and panicles per plant and, consequently, low rice grain yields.

Phosphate fertilization contributes significantly to the increase of dry plant biomass, number of tillage per plant, one thousand grain mass, and grain yield (Nascente and Kromocardi 2017). Granule size, solubility, application season, and mode and frequency of application to soil have a significant influence on the relative efficiency of phosphate fertilizer. Due to the importance of phosphorus in the early phase of rice growth, the low mobility of the element in the soil, and the large translocation within the plant, the full application of the nutrient in sowing is indicated. Thus the nutritional analysis for the management of the fertilization of the crop, is based on the phosphorus content extracted from the soil.

In addition to the aforementioned factors, phytosanitary control in the rice crop is essential to obtain grain quality and yield, factors that are indispensable mainly for the consumer and the producer, respectively. Integrated management practices for weed control in rice, such as crop rotation, machine cleaning, certified seed adoption, and diversification in the use of herbicides in different modes of action, among others, should be effective in order to avoid high infestations that result in damage to grain quality and yield. The main common weeds of the rice crop in southern Brazil are *Echinochloa* spp., *Aeschynomene rudis*, *Cyperus* spp., *Sagittaria* spp., *Ipomoea grandifolia*, *Brachiaria* spp., *Digitaria* spp., *Paspalum* spp., and weed rice (*Oryza sativa* L.).

Rice is affected by more than 70 diseases through infection caused by bacteria, fungi, and viruses (Tamilarasan et al. 2018). Among the main fungal diseases of the rice crop, which impair grain quality and yield, *Pyricularia grisea* (Cooke) Saccardo, *Bipolaris oryzae* (Breda de Haan) Shoemaker, *Cercospora janseana* (Racib) O. Const., *Microdochium oryzae* (Hashioka & Yokogi) Samuels & Hallett, *Tilletia barclayana* (Bref.) Sacc. & Syd., *Ustilagino idea virens* (Cooke) Takahashi, and glum burning *Phoma sorghina* (Sacc.) Boerema, Dorenbosch & Van Kesteren stand out. These pathogens infect plants by reducing grain quality due to the appearance of spots and damages in the process of grain development and maturation. This fact occurs through the significant reduction of the moisture of the infected grains, predetermining in the field the higher incidence of cracks and the degradation in the constituent layers of the grains, which consequently will reduce the quality in the processing and the commercial classification of the cereal.

Throughout rice cultivation, grain quality and yield can be directly affected by the significant presence of various insect pests of the crop. Among these insect pests, noteworthy are the *Oryzophagus oryzae,* mainly in the larvae phase when cutting the roots reducing nutrient absorption, *Tibraca limbativentris* that occurs after emergence and when attacking the sheath leaves cause the death of the plant and after flowering when attacking the stalk causes sterility of the panicle spikelets. Another insect is the *Oebalus* spp., whose infestation occurs during the reproductive phase, and depending on the stage, it causes the formation of totally empty spikelets, and if the attack occurs after the formation of the grains, it causes partial loss of content, endosperm patches, smaller mass, irregular size, and chalky appearance. Also noteworthy is the *Pseudaletia* spp., whose attack starts from the emission phase of the panicles and can partially or totally destroy them.

Given this context, the rice farmer needs to be aware of the management practices technically recommended for rice cultivation, thus providing the best conditions of cultivation environment, consequently allowing the cultivar used to express its full genetic potential for grain quality and yield.

3.5.2 Postharvest Management

Just as the management of the rice cultivation system aims to provide the ideal conditions for high yielding and high quality grain plants, postharvest management involves a series of steps aimed at maintaining the quality and quantity of rice grain produced.

Postharvest operations involve a number of important steps such as grain transportation, reception, drying, storage, processing, and dispatch. Environmental conditions, especially during drying, storage, and processing, directly influence the postharvest quality of the grains and should be properly managed according to the characteristics, properties, and parameters related to the crop.

In the storage, rice grains are kept with approximately 13% moisture, which significantly reduces their metabolic activity, keeping them stable for long periods which consequently maintain their quality. However, at the time of harvest, the moisture content of rice grains is between 18% and 23%, and drying is essential.

The technical management of the drying of rice grains is extremely important because this cereal is characterized by having high mechanical resistance and high sensitivity to temperature variations. Thus, intermittent drying or dry-aeration systems are the most recommended, as they avoid cracks and breakage, providing greater amount of whole grains in the milling process. Maintaining the quality of rice grains after drying depends on the storage conditions, which should prioritize grains with uniform and relatively low humidity, low impurities, low microorganism index, absence of mycotoxins, and cooling of the grain mass.

The process of rice milling consists of the following steps: cleaning, husked, separation by the chamber of straw and unhusked grains, polishing, homogenization, and classification of grains. One of the most important quality parameters in rice processing is grain yield, which corresponds to the amount of whole grains obtained after processing, as well as vitreous grains.

3.6 Final Considerations on Rice Grain Quality

As discussed throughout this chapter, the quality of rice grains is influenced by several factors, due to the complexity of these characteristics, be they genetic and environmental factors, as well as their interaction. Table 3.8 summarizes the list of the main factors that specifically interfere with some of the important attributes that refer to industrial quality.

It is noteworthy to stress again that the quality of rice grains is currently considered a major character, along with grain yield, and that breeding, crop management, and postharvest studies have achieved very significant gains in the enrichment of

Table 3.8 Major factors that interfere on rice grain quality

Factors	Class	Total yield	Whole grains	Cooking	Spoiled	Mold	Chalky	Chipped	Stained	Streaked	Impurities
Soil preparation		x	x								x
Cultivar	x	x	x	x			x		x	x	
Seed quality		x	x	x			x			x	
Sowing (time, density, spacing)		x	x				x	x	x		
Diseases		x	x				x		x		
Pests		x	x				x	x	x		
Weeds					x					x	x
Fertilizer use	x	x	x				x				
Time of harvesting		x	x				x				
Harvesting											x
Transport		x	x								
Cleaning		x	x		x					x	x
Drying			x		x	x				x	x
Storage		x	x	x	x	x			x		
Storage pests		x	x	x							
Time of storage			x	x							

Source: Castro et al. (1999), adapted by the authors

this much-needed product for the world's population. However, much remains to be studied and technological advances (molecular biotechnology, industrial, and agrochemical, among others) will be fundamental tools in obtaining new cultivars, as well as in minimizing impacts and refining the milling of rice.

Appendix I

Technical terms (according to the Brazilian regulations, i.e., Instrução Normativa 2/2012/MAPA) (MAPA, 2019)

For a better understanding of terms associated with rice quality, below are the technical terms used in Brazil and their characteristics.

- Glutinous rice: the product belonging to the variety of the species *Oryza sativa* L. glutinous, whose white and opaque-looking grains tend to stick together because they are almost entirely made of amylopectin.
- Poorly polished rice: the product which after polishing has longitudinal streaks visible to the naked eye.
- Black rice: the product belonging to the variety of the species *Oryza sativa* L., whose grains have black-colored pericarp.
- Red rice: the product belonging to the variety of the species *Oryza sativa* L., whose grains have red-colored pericarp.
- Physiologically mature: the grain that reaches the full developmental stage of the cultivar and is in a condition to be harvested.
- Rice fragment: the product consisting of at least 90% of cracked and chirera grains.
- Granule: grain-shaped product reconstituted from rice flour or other cereals, subject to specific legislation.
- Yellow grain: the husked and polished grain, whole or cracked, yellow in whole or in part, varying from light yellow to dark yellow and in contrast to the working sample; the yellow grain found in the polished and parboiled rice mixture comes from the polished milled subgroup.
- Spoiled grain: any husked and polished grain, whole or cracked, showing all or part of its dark color from the fermentation process.
- Empty grain: the physiologically undeveloped husked grain leaking through a 1.75 × 20 to 22 mm oblong hole sieve.
- Damaged grain: whole or cracked husked and polished grain which, through parboiling, has cracks in the longitudinal direction, except the grain with small cracks, provided that its shape is not altered.
- Dark grain: the husked and polished whole or cracked grain that is blackened on its entire surface due to excessive heat action in the parboiling process.
- Gelatinized grain: whole or cracked grain presenting at least with its translucent gelatinized outer layer when observed under polarized light; gelatinized is also considered to be the grain which has, under polarized light, any vitreous (translucent) part irrespective of the size of the area.

- Chalky grain: grain which has been husked and polished, whole or cracked, and which has a completely opaque color similar to chalk; in rice of special varieties intended for national and international cuisine, the chalky grain shall not be considered as a defect.
- Whole grain: husked and polished grain with a length equal to or greater than three quarters (3/4) of the minimum length of the predominant class; in the specific case of short class rice, cracked grains shall be determined by reference to their maximum length, i.e., 4.99 mm and equal to or greater than 3.74 mm.
- Mold grain: husked and polished grain, whole or cracked, with fungal contamination (mold or mildew) visible to the naked eye.
- Ungelatinized grain: whole or cracked grain which does not have starch gelatinization due to poor parboiling and is totally opaque under polarized light.
- Unparboiled grain: husked and polished whole or cracked grain which has not undergone parboiling, corresponding to the polished processed rice found in parboiled rice.
- Chipped or stained grain: whole or cracked husked and polished grain showing a dark or whitish spot, punctures, or damage caused by pests or other agents, provided they are visible to the naked eye, and dark spots from the fermentation process in less than one quarter of the grain area.
- Black grain: husked grain, whole or cracked, which has a black pericarp.
- Cracked grain: the piece of husked and polished rice grain which is less than 3/4 (three quarter) parts of the minimum length of the predominant class and which is retained in the 1.60-mm-diameter circular hole sieve.
- Streaked grain: husked and polished grain, whole or cracked, showing any red or black spot or stripes, in contrast with the predominant variety; in rice of special varieties of red or black pericarp, the split grain shall not be considered as a defect.
- Green grain: husked and polished grain, whole or cracked, which is not physiologically developed and has a greenish coloration.
- Red grain: husked whole or cracked grain with a red pericarp.
- Impurities: debris from the product itself, e.g., rice husk, empty husks, and pieces of stem.
- Unhusked grain: the grain which preserves the husk after milling, corresponding to unhusked rice found in the milled rice.
- Strange matters: bodies or debris of every nature strange to the product, i.e., grains or seeds of other plant species, dirt, and dead insects.
- Macroscopic materials: those foreign to the product that can be detected by direct observation (naked eye), without the aid of optical instruments, and which are related to human health risk according to specific legislation.
- Microscopic materials: materials foreign to the product that can be detected with the aid of optical instruments that are related to human health risk according to specific legislation.
- Polished and parboiled rice mixture: the product consisting of the polished and parboiled processed rice mixture.

- Chirera: the rice fragment that passes through the circular holes of the sieve with 1.60 mm diameter.
- Milling yield: the percentage of beneficiated or processed and polished rice resulting from the processing of paddy rice.
- Grain yield: the percentage by weight of whole grains and cracked grains resulting from rice processing.
- Health hazardous substances: any substances or agents of biological, chemical, or physical origin that are harmful to health, such as mycotoxins, pesticide residues, or other contaminants, as described in specific legislation, whose value is within the prescribed maximum limits.

References

Anacleto R, Cuevas RP, Jimenez R, Llorente C, Nissila E, Henry R, Sreenivasulu N (2015) Prospects of breeding high quality rice using post genomic tools. Theor Appl Genet Rev 128:1449–1466. https://doi.org/10.1007/s00122-015-2537-6

Bai X, Luo L, Yan W, Kovi MR, Zhan W, Xing Y (2010) Genetic dissection of rice grain shape using a recombinant inbred line population derived from two contrasting parents and fine mapping a pleiotropic quantitative trait locus qGL7. BMC Genet 11:16. https://doi.org/10.1186/1471-2156-11-16

Bianchet P, Sangoi L, Souza CA, Klauberg-Filho O, Panison F (2015) Desenvolvimento vegetativo do arroz irrigado afetado pela inoculação com Azospirillum e aplicação de nitrogênio mineral. Rev Fac Agron La Plata 114:201–207

Brasil (2009) Ministério da Agricultura, Pecuária e Abastecimento. Instrução Normativa n° 6, de 16 e fevereiro de 2009. Aprova o regulamento técnico do arroz, definindo o seupadrão oficial de classificação, com os requisitos de identidade e qualidade, a amostragem, o modo de apresentação e a marcação ou rotulagem. Diário Oficial [da] República Federativa do Brasil, Brasília, DF, 17 fev. Seção 1:3

Breseghello F, Morais OP, Pinheiro PV, Silva ACS, Castro EM, Guimarães EP, Castro AP, Pereira JA, Lopes AM, Utumi MM, Oliveira JP (2011) Results of 25 years of upland rice breeding in Brazil. Crop Sci 51:914–923. https://doi.org/10.2135/cropsci2010.06.0325

Castro EDM, Vieira NRA, Rabelo RR, Silva SA (1999) Qualidade de grão sem arroz. Rev Embrapa Arroz e Feijão 34:1–30. http://www.infoteca.cnptia.embrapa.br/infoteca/handle/doc/208032

Champagne ET, Bett-Garber KL, Fitzgerald MA, Grimm CC, Lea J, Ohtsubo K, Jongdee S, Xie L, Bassinello PZ, Resurrección A, Ahmad R, Habibi F, Reinke R (2010) Important sensory properties differentiating premium rice varieties. Rice 3:270–281. https://doi.org/10.1007/s12284-010-9057-4

Dovale JC, Soares PC, Cornélio VMO, Reis MS, Borges V, Bisi RB, Soares AA, Fritsche-Neto R (2012) Contribuição genética na produtividade do arroz irrigado em Minas Gerais no período de 1998 a 2010. Bragantia 71(4):460–466. https://doi.org/10.1590/S0006-87052012000400002

Edzesi WM, Dang X, Liang L, Liu E, Zaid IU, Hong D (2016) Genetic diversity and elite allele mining for grain traits in rice (*Oryza sativa* L.) by association mapping. Front Plant Sci 7:787. https://doi.org/10.3389/fpls.2016.00787

Embrapa, Empresa B, De Pesquisa A (2013) Arroz: o produtor pergunta, a Embrapa responde, 2nd edn. DF, Ver. Ampl. Brasília, p 245

Fageria NK (2007) Yield physiology of rice. J Plant Nutr 30:843–879. https://doi.org/10.1080/15226510701374831

Fageria NK (2009) The use of nutrients in crop plants. CRC Press, Boca Raton, p 448

Fageria NK (2015) Potassium requirements of lowland rice. Commun Soil Sci Plant Anal 46:1459–1472. https://doi.org/10.1080/00103624.2015.1043444
Fageria NK, Baligar VC, Jones CA (2011a) Growth and mineral nutrition of field crops, 3rd edn. CRC Press, Boca Raton, p 586
Fageria NK, Moreira A, Coelho AM (2011b) Yield and yield components of upland rice as influenced by nitrogen sources. J Plant Nutr 34(3):361–370. https://doi.org/10.1080/01904167.2011.536878
Feng Y, Lu Q, Zhai R, Zhang M, Xu Q, Yang Y, Wang S, Yuan X, Yu H, Wang Y, Wei X (2016) Genome wide association mapping for grain shape traits in indica rice. Planta 244:819–830. https://doi.org/10.1007/s00425-016-2548-9
Feng F, Li Y, Qin X, Liao Y, Siddique K (2017) Changes in rice grain quality of Indica and Japonica type varieties released in China from 2000 to 2014. Front Plant Sci 8:1863. https://doi.org/10.3389/fpls.2017.01863
Fitzgerald MA, Mccouch SR, Hall RD (2008) Not just a grain of rice: the quest for quality. Trends Plant Science 14:133–139. https://doi.org/10.1016/j.tplants.2008.12.004
Freitas JG, Azzini LE, Cantarella H, Bastos CR, Castro LHSM, Gallo PB, Felício JC (2001) Resposta de cultivares de arroz irrigado ao nitrogênio. Scientia Agricola 58(3):573–579. https://doi.org/10.1590/S0103-90162001000300021
Gomes AS, Magalhães AM Jr (2004) Arroz Irrigado no Sul do Brasil. Embrapa Clima Temperado, Pelotas, p 899
Guimarães EP (1989) Qualidade de grão em arroz. In: Reunião da Comissão Técnica de Arroz da Região II, Campinas (1989). EMBRAPA-CNPAF, Goiânia, p 14
Guo L, Ma L, Jiang H, Zeng D, Hu J, Wu L, Gao Z, Zhang G, Qian Q (2009) Genetic analysis and fine mapping of two genes for grain shape and weight in rice. J Integr Plant Biol 51:45–51. https://doi.org/10.1111/j.1744-7909.2008.00793.x
Hu J, Wang Y, Fang Y, Zeng L, Xu J, Yu H, Shi Z, Pan J, Zhang D, Kang S, Zhu L, Dong G, Guo L, Zeng D, Zhang G, Xie L, Xiong G, Li J, Qian Q (2015) A rare allele of GS2 enhances grain size and grain yield in rice. Mol Plant 8:1455–1465. https://doi.org/10.1016/j.molp.2015.07.002
Lack S, Marani NM, Mombeni M (2012) The effects of planting date on grain yield and yield components of rice cultivars. Adv Env Biol 6(1):406–413. https://doi.org/10.1016/j.fcr.2013.05.028
Li Y, Fan C, Xing Y, Jiang Y, Luo L, Sun L, Shao D, Xu C, Li X, Xiao J, He Y, Zhang Q (2011) Natural variation in GS5 plays an important role in regulating grain size and yield in rice. Nat Genet 43:1266–1269. https://doi.org/10.1038/ng.977
Li Y, Fan C, Xing Y, Yun P, Luo L, Yan B, Peng B, Xie W, Wang G, Li X, Xiao J, Xu C, He Y (2014) Chalk5 encodes a vacuolar H+-translocating pyrophosphatase influencing grain chalkiness in rice. Nat Gene 46:398–404. https://doi.org/10.1038/ng.2923
Lin XQ, Zhu DF, Chen HZ, Cheng SH, Uphoff N (2009) Effect of plant density and nitrogen fertilizer rates on grain yield and nitrogen uptake of super hybrid rice (*Oryza sativa* L.). J Agric. Biotechnol Sustain Dev 1(2):44–53. https://doi.org/10.1016/S1672-6308(08)60070-0
Liu S, Hua L, Dong S, Chen H, Zhu X, Jiang JE, Zhang F, Li Y, Fang X, Chen F (2015) OsMAPK6, a mitogen-activated protein kinase, influences rice grain size and biomass production. Plant J. 84:672–681. https://doi.org/10.1111/tpj.13025
Magalhães AM Jr, Fagundes PR, Franco DF (2003) Melhoramento genético, biotecnologia e cultivares de arroz irrigado. In: Magalhães AM Jr, Gomes AS (eds) Arroz irrigado: melhoramento genético, manejo do solo e da água e prognóstico climático, vol 113. Embrapa Clima Temperado, Pelotas, pp 13–33
Magalhães AM Jr, Morais OP, Fagundes PRR, Moura Neto FP, Franco DF, Neves PCF, Nunes CDM, Rangel PHN, Petrini JA, Severo ACM (2012) BRS Pampa: cultivar de arroz irrigado de alta produtividade e excelência na qualidade de grãos. Pelotas: Embrapa Clima Temperado 282:8
Magalhães AM Jr, Morais OP, Fagundes PRR, Colombari Filho JM, Franco DF, Cordeiro ACC, Pereira JA, Rangel PHN, Moura Neto FP, Streck EA, Aguiar GA, Facchinello PHK (2017) BRS

Pampeira: new irrigated rice cultivar with high yieldpotential. Crop Breed Appl Biotechnol 17:78–83. https://doi.org/10.1590/1984-70332017v17n1c13

Mao H, Sun S, Yao J, Wang C, Yu S, Xu C, Li X, Zhang Q (2010) Linking differential domain functions of the GS3 protein to natural variation of grain size in rice. Proc Natl Acad Sci U S A 107(45):19579–19584. https://doi.org/10.1073/pnas.1014419107

Mingotte FLC, Hanashiro RK, Fornasieri Filho D (2012) Característicasfísico-químicas do grão de cultivares de arroz emfunção da adubaçãonigrogenada. Semina: CiênciasAgrárias 33(1):2605–2618. https://doi.org/10.5433/1679-0359.2012v33Supl1p2605

Ministério Da Agricultura Pecuária e Abastecimento (2012) Regulamento Técnico de Classificação do Arroz. Instrução Normativa n° 2, de 6 de fevereiro de 2012. Brasília, Disponível em: http://sistemasweb.agricultura.gov.br/sislegis/action/detalhaAto.do?method=visualizarAtoPortalMapa&chave=918108049. Acesso em: 15 de março 2019.

Nascente AS, Kromocardi R (2017) Genotype selection and addition of fertilizer increases grain yield in upland rice in Suriname. Acta Amazonica 47(3):185–194. https://doi.org/10.1590/1809-4392201603374

Qiu X, Pang Y, Yuan Z, Xing D, Xu J, Dingkuhn M, Li Z, Ye G (2015) Genome-wide association study of grain appearance and milling quality in a worldwide collection of Indica rice germplasm. PLoS ONE 10(12):e0145577. https://doi.org/10.1371/journal.pone.0145577

Rangel PHN, Pereira JA, Morais OP, Guimarães EP, Yokokura T (2000) Ganhos na produtividade de grãos pelo melhoramento genético do arroz irrigado no meio-norte do Brasil. PesquisaAgropecuáriaBrasileira 35(8):1595–1604. https://doi.org/10.1590/S0100-204X2000000800012

Ren D, Rao Y, Huang L, Leng Y, Hu J, Lu M, Zhang G, Zhu L, Gao Z, Dong G, Guo L, Qian Q, Zeng D (2016) Fine mapping identifies a new QTL for brown rice rate in rice (Oryza sativa L.). Rice 9:1–10. https://doi.org/10.1186/s12284-016-0076-7

Shi CH, Wu JG, Lou XB, Zhu J, Wu P (2002) Genetic analysis of transparency and chalkiness area at different filling stages of rice (*Oryza sativa* L.). Field Crops Res 76:1–9. https://doi.org/10.1016/S0378-4290(02)00011-4

Shomura A, Izawa T, Ebana K, Ebitani T, Kanegae H, Konishi S, Yano M (2008) Deletion in a gene associated with grain size increased yields during rice domestication. Nat Genet 40:1023. https://doi.org/10.1038/ng.169

Si L, Chen J, Huang X, Gong H, Luo J, Hou Q, Zhou T, Lu T, Zhu J, Shangguan Y, Chen E, Gong C, Zhao Q, Jing Y, Zhao Y, Li Y, Cui L, Fan D, Lu Y, Weng Q, Wang Y, Zhan Q, Liu K, Wei X, An K, An G, Han B (2016) OsSPL13 controls grain size in cultivated rice. Nat Genet 48:447–456. https://doi.org/10.1038/ng.3518

Singh R, Singh A, Sharma T, Singh A, Singh N (2012) Fine mapping of grain length QTLs on chromosomes 1 and 7 in Basmati rice (Oryza sativa L.). J Plant Biochem Biotechnol 21:157–166. https://doi.org/10.1007/s13562-011-0080-3

Sociedade Sul-Brasileira De Arroz Irrigado – SOSBAI (2018) Arroz irrigado: recomendações técnicas da pesquisa para o Sul do Brasil. SOSBAI, Farroupilha, p 205

Song XJ, Huang W, Shi M, Zhu MZ, Lin HX (2007) A QTL for rice grain width and weight encodes a previously unknown RING-type E3 ubiquitin ligase. Nat Genet 39:623. https://doi.org/10.1038/ng2014

Streck EA, Magalhães AM Jr, Aguiar GA, Facchinello PKH, Fagundes PRR, Franco DF, Nardino M, Oliveira AC (2018a) Genetic progress in 45 years of irrigated rice breeding in Southern Brazil. Crop Sci 58:1094–1105. https://doi.org/10.2135/cropsci2017.06.0383

Streck EA, Magalhães AM Jr, Aguiar GA, Facchinello PHK, Perin L, Fagundes PRR, Oliveira AC (2018b) Genetic progress of grain quality of flooded-irrigated rice cultivars in the state of Rio Grande do Sul, Brazil. Pesq Agrop Brasileira 53(4):453–463. https://doi.org/10.1590/s0100-204x2018000400007

Sun L, Li X, Fu Y, Zhu Z, Tan L, Liu F, Sun X, Sun X, Sun C (2013) GS6, a member of the GRAS gene family, negatively regulates grain size in rice. J Integr Plant Biol 55:938–949. https://doi.org/10.1111/jipb.12062

Tabien RE, Samonte SOPB, Mcclung AM (2008) Forty-eight years of rice improvement in Texas since the release of cultivar Bluebonnet in 1944. Crop Sci 48:2097–2106. https://doi.org/10.2135/cropsci2007.12.0680

Taiz L, Zeigeir E (2013) Fisiologia vegetal, 5th edn. Artmed, Porto Alegre, p 954

Tamilarasan G, Pillai MA, Kannan R, Kumari SMP (2018) Research Article Genetic divergence analysis for bacterial leaf blight (BLB) disease resistance in rice (Oryza sativa L.). Electron J Plant Breed 9(3):1194–1205. https://doi.org/10.5958/0975-928X.2018.00147.3

Tang S, Zhang H, Liu W, Dou Z, Zhou Q, Chen W, Wang S, Ding Y (2019) Nitrogen fertilizer at heading stage effectively compensates for the deterioration of rice quality by affecting the starch-related properties under elevated temperatures. Food Chem 277:455–462. https://doi.org/10.1016/j.foodchem.2018.10.137

Usberti Filho JA, Azzini LE, Soave J (1986) Características agronômicas e de qualidade de produtos de cultivares de sequeiro e irrigado. Pesquisa Agropecuária Brasil 21(12):1283–1296

Veloso CAC, Botelho SM, Lopes ADM, Carvalho EJM (2009) Nutrição mineral e adubação da cultura do arroz de sequeiro. Belém, PA: Embrapa Amazônia Oriental, 29p. (Documentos/ Embrapa Amazônia Oriental, ISSN 1983-0513; 360).

Wang H, Zhang W, Liu L, Shen Y, Wang J, Jiang L, Zhai H, Wan J (2008) Dynamic QTL analysis on rice fat content and fat index using recombinant inbred lines. Cereal Chem 85:769–775. https://doi.org/10.1094/CCHEM-85-6-0769

Wang S, Wu K, Yuan Q, Liu X, Liu Z, Lin X, Zeng R, Zhu H, Dong G, Qian Q, Zhang G, Fu X (2012) Control of grain size, shape and quality by OsSPL16 in rice. Nat Genet 44:950–954. https://doi.org/10.1038/ng.2327

Wang S, Li S, Liu Q, Wu K, Zhang J, Wang S, Wang Y, Chen X, Zhang Y, Gao C, Wang F, Huang H, Fu X (2015) The OsSPL16-GW7 regulatory module determines grain shape and simultaneously improves rice yield and grain quality. Nat Genet 47:949–954. https://doi.org/10.1038/ng.3352

Woo MO, Ham TH, Ji HS, Choi MS, Jiang W, Chu SH, Piao R, Chin JH, Kim JA, Park BS, Seo HS, Jwa NS, Mccouch S, Koh HJ (2008) Inactivation of the UGPase1 gene causes genic male sterility and endosperm chalkiness in rice (Oryza sativa L.). Plant J 54:190–204. https://doi.org/10.1111/j.1365-313X.2008.03405.x

Yang W, Guo Z, Huang C, Duan L, Chen G, Jiang N, Wei F, Feng H, Xie W, Lian X, Wang G, Luo Q, Zhang Q, Liu Q, Xiong L (2014) Combining high throughput phenotyping and genome-wide association studies to reveal natural genetic variation in rice. Nat Commun 5:5087. https://doi.org/10.1038/ncomms6087

Zaratin C, Souza SA, Pantano AC, Sá ME, Arf O, Buzetti S (2004) Efeitos de quatro doses de potássio em seis cultivares de arroz de sequeiro irrigados por aspersão. Científica 32:115–120. https://doi.org/10.15361/1984-5529.2004v32n2p121-126

Zhang X, Wang J, Huang J, Lan H, Wang C, Yin C, Wu Y, Tang H, Qian Q, Li J, Zhang H (2012) Rare allele of OsPPKL1 associated with grain length causes extra-large grain and a significant yield increase in rice. Proc Natl Acad Sci USA 109:21534–21539. https://doi.org/10.1073/pnas.1219776110

Zhao K, Tung CW, Eizenga GC, Wright MH, Ali ML, Price AH, Norton GJ, Islam MR, Reynolds A, Mezev J, McClung AM, Bustamante CD, McCouch SR (2011) Genome-wide association mapping reveals a rich genetic architecture of complex traits in Oryza sativa. Nat Commun 2:467. https://doi.org/10.1038/ncomms1467

Zhao X, Daygon V, Mcnally K, Hamilton R, Xie F, Reinke R, Fitzgerald MA (2016) Identification of stable QTLs causing chalk in rice grains in nine environments. Theor Appl Genet 129:141–153. https://doi.org/10.1007/s00122-015-2616-8

Zhou Z, Robards K, Helliwell S, Blanchard C (2002) Composition and functional properties of rice. Int J Food Sci Technol 37:849–868. https://doi.org/10.1046/j.1365-2621.2002.00625.x

Zhou LJ, Jiang L, Zhai HQ, Wan JM (2009) Current status and strategies for improvement of rice grain chalkiness. Hereditas 31(6):563–572

Part II
Contaminants

Chapter 4
Arsenic in Rice Grain

Camila Neves Lange, Tatiana Pedron, Bruna Moreira Freire, Rodrigo Mendes Pereira, and Bruno Lemos Batista

4.1 Arsenic: A Brief History

Arsenic (As) is a metalloid labeled in Group 5 of the Periodic Table of the Elements and, according to the Agency for Toxic Substances and Disease Registry (ATSDR), the most dangerous element for human health, considering its occurrence, toxicity, and exposure potential (ATSDR 2017). This element is located in the same group of nitrogen, phosphorus, antimony, and bismuth. Its name derives from the Greek *arsenikon*, which means bold, potent, or valiant (Cullen 2008). Arsenic is the 20th most abundant element in the earth's crust and occurs naturally in both inorganic and organic forms. The first appears in the environment combined with oxygen (O_2), sulfur (S), and chlorine (Cl), while the organic forms are present in animals and plants combined with hydrogen and carbon (ATSDR 2007; Mandal and Suzuki 2002). The most common oxidation states of As are the trivalent (arsenite, As^{3+}) and the pentavalent (arsenate, As^{5+}). Exposure to As may occur through the soil, air, water, and food (Hughes et al. 2011). Currently, As is the first element in the Substance Priority List from the Agency for Toxic Substances and Disease Registry (ATSDR 2017).

Minerals containing As were used by ancient civilizations. A common practice was the smelting of minerals As-containing to produce tools and weapons. When heating these minerals, a white smoke was produced and used for medicinal purposes. In the Western Mediterranean (3000 B.C.) and Europe (during Iron Age), As was used by smiths to increase the resistance of copper alloys. Only in 1250, As was first isolated by Albertus Magnus from As trioxide (Cullen 2008). Arsenic has been known as a poison for centuries because of its history and use in the past. It was widely applied as both suicidal and homicidal agent in the Renaissance and Middle Ages because of its discreetness and potency. Unintentional poisonings by As are

C. N. Lange · T. Pedron · B. M. Freire · R. M. Pereira · B. L. Batista (✉)
Universidade Federal do ABC, Santo André, SP, Brazil
e-mail: bruno.lemos@ufabc.edu.br

A. Costa de Oliveira et al. (eds.), *The Future of Rice Demand: Quality Beyond Productivity*, https://doi.org/10.1007/978-3-030-37510-2_4

related to its use as pigment mainly in wallpapers but also in candles and toys in the 1800s (Hughes et al. 2011; Scheindlin 2005; Vahidnia et al. 2007). Only in the middle of 1700s As was able to be analytically detected. In 1832, James Marsh developed a method for As detection, which was first used in 1840 (Cullen 2008; Hughes et al. 2011).

Therapeutic uses of As have been reported for more than 2400 years. For example, in 1786, a solution of potassium arsenite (K_3AsO_3) named Fowler's solution begins to be used for many diseases, such as syphilis. Another compound more recently used in medicine is As trioxide (As_2O_3), which is approved for the treatment of acute promyelocytic leukemia (Antman 2001; Frith 2013; Hughes et al. 2011; Scheindlin 2005).

Arsenic compounds were applied over the years as a pesticide. Paris Green, the same compound used as a pigment, was applied as an insecticide between 1867 and 1900. Lead arsenate was also used as a pesticide at the end of the 1800s. However, their uses were discontinued due to the severe adverse effects observed. Since 1940, chromated copper arsenate (CCA) started to be used. From 1950 until now, organic arsenical pesticides, less toxic than inorganic ones, emerged in agriculture (Cohen et al. 2006; Cullen 2008; Hughes et al. 2011).

4.2 Toxicology of Arsenic

Classified as carcinogenic substance 1 for humans, inorganic As (i-As) compounds (As^{3+} and As^{5+}) are about one hundred times more toxic than partially methylated organic forms (IARC 2012). In humans, trivalent and pentavalent forms of As are absorbed (about 90%) in the gastrointestinal tract (Klaaseen 2008). Detoxification occurs through the methylation of As by the liver where i-As^{3+} is oxidatively methylated to monomethyl As^{+5}, reduced to monomethyl As^{3+}, methylated again to dimethyl As^{5+}, and finally reduced to dimethyl As^{3+} (Cullen 2014; Watson and Yager 2007). There are pieces of evidence that only one enzyme itself may be capable of catalyzing both methylation steps. According to Li et al. (2005), arsenic(III) methyltransferase (AS3MT) is the enzyme responsible for arsenic species reduction as long as there is a source of reducing equivalents such as glutathione or thioredoxin (Li et al. 2005; Watson and Yager 2007).

The genotoxicity of i-As is directly associated with the chemical species and their methylation. When As^{3+} interacts with the mitochondria, they may alter the redox reactions of the cells, promoting the uncontrolled formation of a superoxide anion radical, which is harmful to the human organism (Cohen et al. 2006; Jomova et al. 2011; Valko et al. 2005). Besides, the presence of i-As in the organism may promote genotoxic effects such as single-strand breaks of DNA, chromosomal aberrations, and increased oxidative stress of DNA bases (Jomova et al. 2011). These are linked to the carcinogenic ability of i-As to induce various types of cancers such as lung, bladder, and kidney, in addition to other diseases that are not directly linked to carcinogenicity, such as vascular, skin, and diabetes mellitus diseases (Khairul et al. 2017).

In general, the intake of As for humans is about 20–300 μg day^{-1} from food and water, with inhalation being the smallest source of exposure, about 1 μg day^{-1} (Chatterjee et al. 1995; IARC 2012). Contaminated soils and water are the main contributors to As in the food chain. Among the foods, rice has more considerable notoriety. Consumed by more than half of the world's population, rice can absorb about ten times more As than other cereals such as wheat and barley (Meharg and Zhao 2012). This is due to several factors, such as cultivation system (usually flooded), genotype, and soil characteristics. Besides the damages caused in the health of the consumers, As also causes damages in plants. After uptake, As can cause several toxic effects in physiological, morphological, and growth processes of plants. Arsenic exposure of various plant species such as *Boehmeria nivea* L., *Oryza sativa* L., *Pisum sativum* L., and *Trigonella foenum-graecum* L. can cause increased production of reactive oxidative species, decreasing seed twinning, chlorophyll, and essential amino acids (Abbas et al. 2018).

4.3 Arsenic in the Environment and Places of the World

Environmental matrices such as water, soils, bottom sediments, and dust could contain As in various forms and levels (Wang et al. 2008). In nature, As can be found in four oxidation states, −3, 0, +3, and +5, in both organic and inorganic forms (Table 4.1), bound or complexed with other metals, such as silver, lead, nickel, and cobalt (Hughes et al. 2011; IARC 2012; WHO 2001).

A large number of natural and anthropogenic sources of As in the environment have been reported (Nriagu et al. 2007; Reimann et al. 2009), and high levels of this element in agricultural soil or irrigation water may induce an increment of its concentration in crop plants (Hettick et al. 2015; Punshon et al. 2017).

Anthropogenic activities play an important role in dispersing As contamination to the hydrosphere, pedosphere, and atmosphere (Masuda 2018). The primary anthropogenic sources of As are related to industrial activities (mining) as well as coal and petroleum extraction and refining (Han et al. 2003). Only the smelting of Cu has an estimated annual As emission of about 50,000 tonnes, which represents 80% of the emission of As by smelting of ores, for example (Bissen and Frimmel 2003; Sadler et al. 1994). As a consequence of this emission, the concentration of As in soil and water is considerably high where these activities take place. According to some studies, concentrations of up to 9 g kg^{-1} and 70 mg L^{-1} in soil and water, respectively, have already been determined at sites near mines (Sadler et al. 1994; Williams 2001). The burning of fossil fuels is another important anthropogenic source of arsenic in the environment, being the coal combustion on power plants the most important. Again, As emitted may create some environmental problems such as the contamination of natural waters and soils (Yudovich and Ketris 2005). On a smaller scale, the production, use, and disposal of some products containing

Table 4.1 Examples of organic and inorganic compounds of As

Compound	Formula
Arsine	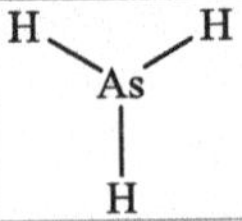
Arsenious acid (As^{3+})	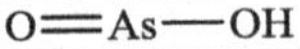
Arsenic acid (As^{5+})	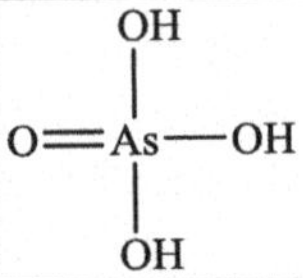
Monomethylarsonic acid (MMA) (V)	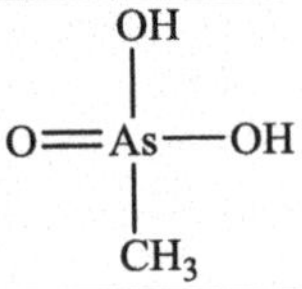
Dimethylarsinic acid (DMA) (V)	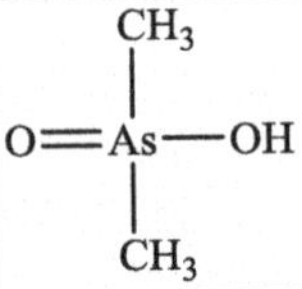
Arsenobetaine (AsB)	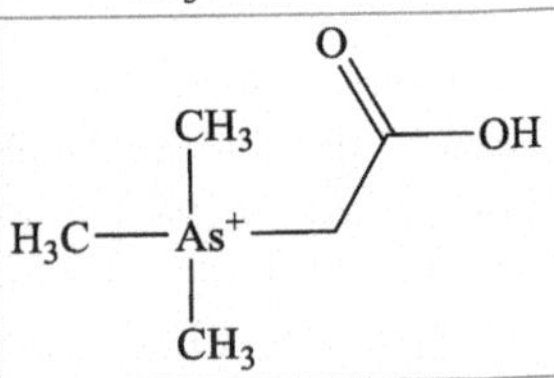
Arsenocholine (AsC)	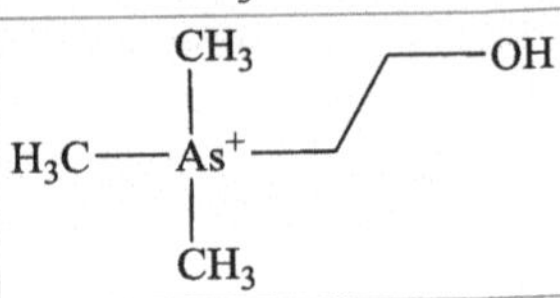

Extracted from Demesmay et al. (1994)

synthetic As compounds such as pesticides, herbicides, crop desiccants, and As-based additives in livestock feed, seaweed fertilizers, organic manures, and wood preservatives also contribute to the presence of arsenic in the environment (Bissen and Frimmel 2003; Punshon et al. 2017; Smedley and Kinniburgh 2013).

Besides As inputs due to human activities, several places of the world present natural high levels of this element, affecting mainly water supply and food chain. Environmental contamination of drinking water is the main primary concern of As exposure; due to that, the majority of worldwide regulatory agencies has set drinking water guideline values for As (Meharg and Raab 2010; Punshon et al. 2017). According to UNICEF (2008), geogenic groundwater contamination has been discovered in over 70 countries, and half of this was within the last 20 years. In Asia, Bangladesh, India, China, Pakistan, Myanmar, Afghanistan, and Cambodia are severely impacted by As-contaminated drinking water (Mukherjee et al. 2006). For example, groundwater As contamination is a significant concern in seven of India's 29 states because of their consumption, and it is also observed in other regions in the United States (Arizona and California), Argentina, Japan, Korea, and Australia (Chakraborti et al. 2016).

The element transference to groundwater occurs through complex biogeochemical interactions within the solid phase and water (Fig. 4.1). Arsenopyrite and As-substituted pyrite, among other sulfide minerals, are sensitive to oxidation in the near-surface environment gradually mobilizing quantities of As in the sediments and groundwater (Herath et al. 2016). Arsenic groundwater concentration is also regulated by adsorption/desorption reactions on metal ions like iron (Fe), manganese-mineral precipitation or dissolution, and physicochemical properties like pH, redox potential (Eh), organic matter content, the presence of interfering ions, depth of the aquifer, and reaction kinetics, among others (Sarkar and Paul 2016). Oxidation of minerals such as As_2O_3 releases As in the environment. Inorganic forms are interconverted by algae and microorganisms, both in sediments and flooded soils. In sediments, precipitation of As^{3+} may occur, forming As minerals again. In the

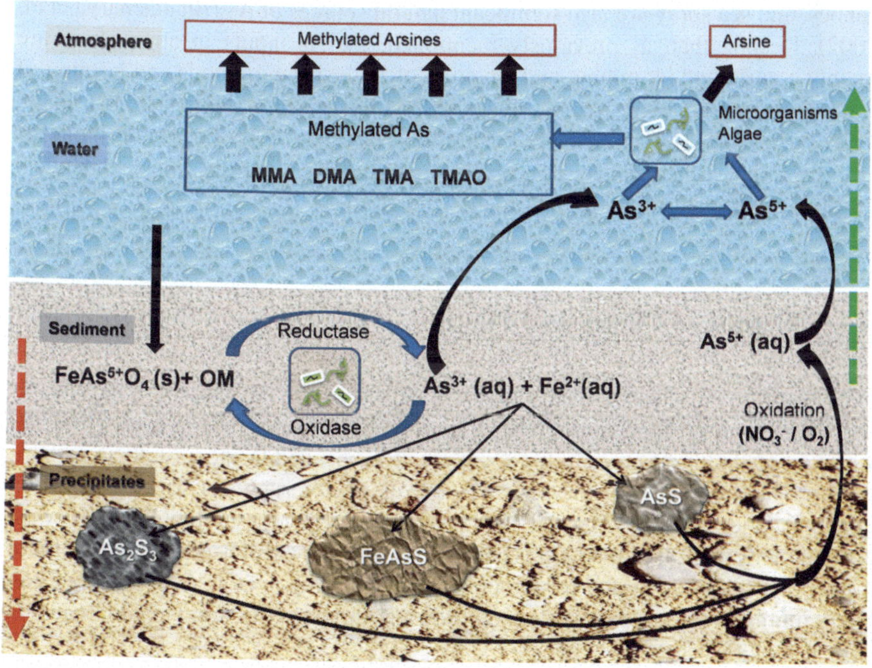

Fig. 4.1 Scheme of As forms and reactions in the environment (reduction, oxidation, and biotransformation). Dotted red line: increasing reductive environment; dotted green line: increasing oxidative environment. MMA monomethylarsonic acid, DMA dimethylarsinic acid, TMAO trimethylarsine oxide, TMAs trimethylarsine, OM organic matter. (Adapted from Guimarães et al. 2019)

flooded soil, inorganic species can either be methylated or form arsine, which, together with methylated arsines, is released in the atmosphere (Fig. 4.1).

Pedosphere is a significant sink of As; there are more than 568 naturally occurring minerals that contain As, of which approximately 60% are As^{5+}, 20% are sulfides and sulfosalts, and the remaining 20% includes arsenides, As^{3+}, oxides, silicates, and As^{o} (elemental arsenic) (IMA 2014; Mandal and Suzuki 2002; Onishi 1969). Arsenic is a redox-sensitive element; thus, seasonal variations influence its concentration and speciation in water and soil systems (Bradham et al. 2018). However, unlike other oxyanion-forming elements, As has relative mobility over a broad range of redox conditions and pH. Some key parameters also affect As mobility in soil systems such as soil acidity, organic matter content, clay mineralogy, redox-sensitive ions (Fe, Mn, and S), PO_4^{3-} (phosphates) and silicon content, and soil microorganisms (Kumarathilaka et al. 2018; Sarkar and Paul 2016). Arsenic anomalies in soils are caused by the chemical composition of local geology, biogeological factors, and human activities. Soil As contamination was reported in some parts of England, China, Thailand, Ghana, Zimbabwe, Mexico, Canada, the United States, and Brazil, among others (Asante et al. 2007; Bissen and Frimmel 2003; Borba et al. 2003; Li and Thornton 1993).

Windblown dust from crustal weathering, forest fires, vegetation emissions, volcanoes, and sea spray are also significant natural sources of As (Bhattacharya et al. 2002). Despite that, as previously mentioned, the As inputs in atmosphere are mainly contributed by mining and smelting of As-bearing minerals and burning of fossil fuels, mainly coal. Arsenic is generally adsorbed on polynucleated aromatic hydrocarbons (PAH) having a size range from 1 to 20 μm. Arsenic species in the air are generally washed out by rain or get precipitated by various physicochemical processes of the atmosphere (Sarkar and Paul 2016).

4.4 Arsenic in Plants and Food

Food security is a crucial issue for global development in a sustainable way, considering not only quantity but also the quality of food. However, the rapid growth of the world's population has meant that some countries needed to increase their agricultural productivity, often using low-cost irrigation as wastewater and leaving food quality behind. Thus, food security and human health have been threatened by some contaminants present in food crops, such as As, Pb, and Cd. The absorption of these metals or metalloids by the plant can occur through roots or leaves, and they can bioaccumulate and undergo biomagnification through the food chain, causing human health hazards.

In plants, As has no biological function and is potentially toxic. Moreover, arsenic in the soil can cause loss of plant fertility and decrease in harvest (Garg and Singla 2011; Zhao et al. 2010). It can be present as As^{3+} or As^{5+} depending on the soil redox status (Garg and Singla 2011; Smith et al. 2010; Verbruggen et al. 2009; Zhao et al. 2009). In aerobic soils, such as rainfed soils, there is a higher prevalence of As^{5+} due to their strong affinity with iron oxides and hydroxides. Considering that As^{5+} has a phosphate-like behavior, the transport of this species occurs by the same $PO_4^{3}-$ pathway in higher plants (Ma et al. 2008; Tripathi et al. 2007; Zhao et al. 2009). Studies are demonstrating that, in rice plants, As^{5+} may enter into the cell via phosphate transporters such as OsPHT1;1 and OsPHT1;8, the latter being considered the most important (Awasthi et al. 2017; Kamiya et al. 2013; Wu et al. 2011). Moreover, according to Wu et al. (2011), the regulators of phosphate transport such as OsPHF1 (phosphate transporter traffic facilitator 1) and PHR2 (phosphate starvation response 2) also contribute to As^{5+} uptake and transport in rice plants. After absorption by plant, As^{5+} can be reduced to As^{3+} in roots before species are transported to shoots.

On the other hand, in anaerobic soils, such as flooded soils, the presence of As^{3+} is favored over As^{5+} because of the development of reducing conditions. In general, As^{3+} levels vary from 0.01 to 3 μM, higher than non-contaminated aerobic soils. It is worth mentioning that inorganic species of As differ in their toxicity as well as their transport in plants since they have considerably different properties. In this sense, As^{3+} and some organic As species are transported mainly via nodulin 26-like intrinsic protein (NIP) class of aquaporin channels, among other aquaglyceroporins of various classes (Zhao et al. 2010; Mosa et al. 2012). A transporter that plays a very important role in As^{3+} uptake in plants is OsNIP2;1 (Lsi1), an NIP class transporter that is well known in the transport of silicic acid (Si) (Ma et al. 2008).

Methylated forms (MMA and DMA) may also be available in soils, having as sources pesticides used in the past or biomethylation performed by microorganisms and algae. The uptake of these species by plants has also been found to occur through Lsi1 (Awasthi et al. 2017). However, organic and inorganic As species have different mobilities. The mobility of inorganic As species is lesser than organic As species (Awasthi et al. 2017; Carey et al. 2010). Zhao et al. (2012) demonstrated that only 10% of total As^{3+} intake by rice plants reached the shoots and about 3% reached the grain. This fact may be attributed to the phytochelatins (PCs), which are oligomers of glutathione produced by the enzyme phytochelatin synthase. The PCs act as chelators and are very important for plant detoxification by metals or other elements, such as inorganic As (Raab et al. 2007).

The behavior of As into the soil-plant system is influenced by some factors such as climatic conditions, As availability in soils, habitat, and plant species (Bergqvist and Greger 2012; Smith et al. 2010; Vithanage et al. 2012; Zhao et al. 2010). Bergqvist and Greger (2012) evaluated plants from different habitats, finding predominantly the inorganic species As^{5+} and As^{3+}, but also MMA in almost all plant species. In submerged plants, As^{5+} was the predominant species, while in terrestrial plants, As^{3+} was predominant in shoots (Bergqvist and Greger 2012). According to Smith et al. (2010), there are small amounts of organic As in plants, such as DMA and MMA, which can be sourced from the soil or produced internally.

Studies have shown that mycorrhizas (symbioses between fungi and plant's roots) may influence the uptake of As^{5+} by the plant, retaining As in the fungi's region or bio-volatilized, reducing its absorption by the root. Some plants present restricted uptake and translocation of As, being called excluders (Smith et al. 2010;

Zhao et al. 2010). About the habitat, there are pieces of evidence that submerged plants accumulate more As than emergent and terrestrial plants (Bergqvist and Greger 2012).

Excessive As exposure generates responses and effects in the plant such as oxidative stress, disturbed respiration, and growth inhibition. The reduction in root growth can impact the absorption of water and nutrients due to the smaller contact surface, causing nutrient deficiencies. This may be somewhat compensated by mycorrhizal associations, which increase nutrients uptake. Certain species of terrestrial plants can accumulate substantial As levels. For example, ferns from the Pteridaceae family can absorb higher amounts of As than other plants and translocate this element from root to shoot more efficiently, and As hyperaccumulates in their tissues without toxicity symptoms. These plants can be used as an inexpensive tool to remediate soil contaminated with As, but some challenges should be considered, for example, post-harvest management of the contaminated biomass (EFSA 2009; Garg and Singla 2011; Smith et al. 2010; Verbruggen et al. 2009; Vithanage et al. 2012; Zhao et al. 2009). An alternative is its use for the generation of bioenergy. Gymnosperms also have potential to be used for As phytoremediation because they, as well as fern species, present a high relation between As in the shoot and the root indicating an efficient transport (Bergqvist and Greger 2012).

In general, crops, shrubs, and trees present relatively low As concentrations. However, Bergqvist and Greger (2012) have found high As concentrations in the edible parts of oats and alfalfa (290 and 370 μg kg^{-1}, respectively), indicating that high As levels could be ingested through those oats (Bergqvist and Greger 2012). Another crop that can uptake As from soil and accumulate it is rice. The flooded paddy cultivation causes mobilization of As^{3+}, which is efficiently absorbed and transported through silicon transporters, as previously mentioned (EFSA 2009; Verbruggen et al. 2009; Zhao et al. 2010).

Rice and rice-based products are known sources of total and i-As. They are widely consumed around the world, representing a potential health risk. The main source of As intake in humans in some regions of the world is the ingestion of rice that has been planted in soils contaminated with this element (Agency for Toxic Substances and Disease Registry (ATSDR) 2007, EFSA 2009, Gu et al. 2019, Islam et al. 2017, Silva et al. 2014, Zhao et al. 2010). The concentration of As in soil usually ranges from 1 to 40 mg kg^{-1}, but it may be much higher in contaminated areas (Agency for Toxic Substances and Disease Registry (ATSDR) 2007).

In food, total As concentrations typically range from 20 to 140 μg kg^{-1}, being influenced by the source and type of food. For example, husked rice and its derivatives have more As than white rice and white rice-based products. Inorganic As species are predominantly found in these types of food. Islam et al. (2017) have investigated As concentrations in rice and rice-based food (n=59) purchased from Australia, finding that 53% of analyzed products presented As concentrations higher than EU recommendation for young children (100 μg kg^{-1}) and 22% showed values higher than EU recommendation for adults (200 μg kg^{-1}). Rice-based foods (ready-to-eat rice, rice-based snacks, puffed rice, rice cakes, baby rice, and rice crackers) usually present higher levels of As. Children are particularly at risk due to the high consumption of rice products and lower body weight (Islam et al. 2017).

According to the Agency for Toxic Substances and Disease Registry (ATSDR), foods containing higher As levels are seafood, rice, mushrooms, and poultry (Agency for Toxic Substances and Disease Registry (ATSDR) 2007). The European Food Safety Authority (EFSA) gathered and assessed data regarding As in different food commodities, finding the highest levels for fish and seafood, food products or supplements based on algae (mostly hijiki), and cereal grains and derivatives, mainly rice, rice-based food, bran, and germ (EFSA 2009). Except for fish and seafood, which have higher concentrations of organic As, the percentage of i-As ranges from 50% to 100% of total As in the foods mentioned above. It is important to mention that one person ingests about 50 μg of As per day through dietary sources, being about 3.5 μg of i-As (Agency for Toxic Substances and Disease Registry (ATSDR) 2007). The EFSA has estimated i-As exposure from food and water ranging from 0.13 to 0.56 μg kg^{-1} body weight per day for average consumers, possibly reaching 1 μg kg^{-1} body weight per day in high-rice consumers. In addition to these specific consumer groups, children under 3 years are the most exposed to i-As (0.50–2.66 μg kg^{-1} body weight per day), due to the higher consumption of food in relation to body weight and the high consumption of rice-based products (EFSA 2009). Therefore, understanding the absorption, distribution, metabolisms, and effects of As in rice, it is possible to develop strategies for mitigation in rice.

4.5 Arsenic in Rice: Absorption, Translocation, Distribution, and Accumulation

In recent years, several studies have been carried out in order to understand the mechanisms of As uptake and translocation by rice (Abedin et al. 2002a; Abedin et al. 2002b; Awasthi et al. 2017; Carey et al. 2010; Kamiya et al. 2013; Liu et al. 2006; Lombi et al. 2009; Ma et al. 2008; Meharg et al. 2008; Rahman et al. 2007; Shri et al. 2009; Tripathi et al. 2007; Wu et al. 2011; Zhao et al. 2009; Zhang et al. 2011). In short, the arsenic uptake mechanisms in rice occur as follows: (i) As^{5+} uptake in the rice root by the same transporters of PO_4^{3-} (such as OsPHT1;1 and OsPHT1;8) due to its chemical similarity and (ii) As^{3+} and methylated As species (MMA and DMA) uptake by roots through the silicic acid transport system using transporters Lsi1 and Lsi2, which belong to the aquaporin family and the family of putative anion transporters, respectively (Bakhat et al. 2017; Chen et al. 2013; Kumarathilaka et al. 2018; Li et al. 2009; Ma et al. 2008; Guo et al. 2007; Wu et al. 2015; Zhao et al. 2013).

The translocation and accumulation of As species in rice tissues are driven by some metabolic pathways, such as arsenate reduction, arsenite efflux, and (As^{3+})-thiol complexation and sequestration. This translocation efficiency for different As species in rice plants is understood as follows: DMA > MMA > inorganic As species (Kumarathilaka et al. 2018). According to Rahman et al. (2007), who studied two rice varieties, As content was about 28- and 75-folds higher in root than that of the shoot and raw rice grain, respectively. The authors also evaluated As content in fractions of parboiled and non-parboiled rice grain of both varieties, and the order of As concentration was rice hull > bran-polish (the thin outer layer of milled rice) > brown rice > raw rice > polish rice. In short, there are no significant differences between the rice varieties; however, As concentrations were higher in non-parboiled rice grain than that of parboiled rice. Lombi et al. (2009) found that (As^{3+})–thiol complexes dominated bran and endosperm and that the translocation from the maternal to filial tissues may be a bottleneck for As accumulation in the grain. Similar findings were reported by Carey et al. (2010). The authors found that DMA speciation is altered in the plant, potentially through complexation with thiols. The same authors investigated the roles of phloem and xylem transport on As distribution in rice and concluded that phloem transport contributed 90% of As^{3+} unloaded into the grain and 55% of DMA in excised panicles. Ma et al. (2017) studied the uptake and translocation of As species in rice plant samples including roots, straws, and grains collected from paddy fields near a mine impacted city in Southern China. These authors observed a total As concentration in the descending order as root > soil > straw > grain, and the As species predominantly detected in rice plants were inorganic forms: As^{3+} and As^{5+}, but they also detected MMA and DMA and other two organoarsenicals, AsB and AsC (arsenocholine). DMA and AsB were mainly formed in rice roots with the assistance of microorganisms. MMA and AsC detected in straws might be derived from methylation and oxidation of As^{3+}. Batista et al. (2014) also reported the greater translocation efficiency of MMA and DMA in rice plants than inorganic forms.

Arsenic levels and species loading into the rice grains are severely dependent on rice plant genetic material, environmental conditions, and plant maturity stage. Loading mechanisms are also extensively affected by S, Fe, $PO_4{}^{3}-$, and silicates availability (Awasthi et al. 2017; Kumarathilaka et al. 2018). Under flooded environment, the rice roots release O_2 into the rhizosphere resulting in the formation of an Fe plaque on the root surface, and the effects of this plaque on As accumulation in the rice grains have been investigated by several authors (Hossain et al. 2009; Liu et al. 2004; Li et al. 2015; Seyfferth et al. 2011; Somenahally et al. 2011; Xu et al. 2008; Yamaguchi et al. 2014). The Fe plaque can sorb inorganic species of As^{3+} and As^{5+}, reducing As uptake and accumulation in rice grains, especially in rice cultivars with high root porosities, which allows faster rates of radial oxygen loss (ROL) and subsequently greater potential for the Fe plaque formation (Pan et al. 2014; Wu et al. 2017; Yu et al. 2016). However, this process is very dependent on temporal redox variations on soil matrix, since when soil is under anaerobic conditions during rice growth, Fe plaque is not present in the entire root; thus, As^{3+} is not the dominant arsenic species in the rhizosphere, and during draining season, the conditions shift to aerobic leading to an increase of As^{5+} species (Yamaguchi et al. 2014).

The formation of Fe plaque is also affected by the S and P availability in the soil (Hossain et al. 2009; Hu et al. 2005, 2007). Phosphate levels in the rice-growing environment also influence the dynamics of As accumulation, mainly due to the interaction of this dissolved anion with Fe plaques and sorption competition (by ligand exchange) of adsorbed As from these plaques (Hossain et al. 2009). Hu et al. (2005) reported that Fe plaque formation was beneficiated under low P conditions, thus leading to more As sequestration in the Fe plaque. Batista et al. (2014) demonstrated that S considerably decreased As concentrations in rice, since it influences on the formation of As–phytochelatin (As-PC), which is produced when rice plants are exposed to As stress, but this process is specific to certain varieties, while in others, different mechanisms such as As-PC release and transport seem to be decisive.

In a review study, Farooq et al. (2016) resumed the mechanism responsible for As entry into cell and detoxification pathways as follows: (i) As^{5+} and As^{3+} are uptake by plant cells through PO_4^{3-} transporters (Pi) and aquaporins (NIP), respectively. Then, As^{5+} is reduced to As^{3+} by arsenate reductase (AR), which uses glutathione (GSH) as a reductant. Inside the cell, As^{3+} are complexed with thiol compounds such as phytochelatins (PCs) and glutathione (GSH) and sequestered into the central vacuole; (ii) As^{5+} and As^{3+} are translocated to shoot tissue where a similar reduction and sequestration occurs; (iii) in seed, As^{3+} is loaded by phloem through INT genes (inositol transporter) and transported to seeds. The mechanisms of transport and detoxification of methylated As species (MMA and DMA) in plants are still not clarified. Figure 4.2 shows a schematic representation of these mechanisms based on Awasthi et al. (2017), Batista et al. (2014), and Farooq et al. (2016) studies.

The effects of As in rice grains have also been the subject of several studies. In a recent review, Bakhat et al. (2017) summed up 17 studies that reported a drastic reduction in rice grain yield due to an increase in soil As concentrations. The reduction in yield percentage in these studies ranged from 15% to 99%, depending on the rice-growing conditions. In this way, the mitigation of these effects is a significant concern for both producers and consumers to ensure high yield and food safety.

In a pot experiment, Hossain et al. (2009) studied the effect of different ratios of As, P, and Fe application in soil matrix under the flooded condition on grain and straw yields. The authors reported As grain concentration ranged from 194 to 708 $\mu g\ kg^{-1}$ and concluded that Fe application reduced As toxic effects resulting in increasing grain yields and decreasing As accumulation. On the other hand, the PO_4^{3-} application resulted in lower grain yields and higher As concentrations in both grain and straw. The effect of Fe and As combined stresses on yield-related parameters and As distribution and speciation were evaluated by Vromman et al. (2013). The study revealed that the combined stresses affected plant nutrition and significantly reduced the plant yield by limiting the number of grains per plant and the grain filling.

In rice fields cultivated for 2 years, Linquist et al. (2015) and Carrijo et al. (2018) evaluated various treatments and frequency and duration of alternate wetting and drying (AWD) on As concentration and rice grain yields. These authors found ranges of As concentration in rice grains from 114 to 433 $\mu g\ kg^{-1}$ and 38 to 120 $\mu g\ kg^{-1}$, respectively. Linquist et al. (2015) emphasized that depending on AWD water management conditions, it was possible to reduce As levels in rice while maintaining the same grain yields as under conventional submerged conditions. Carrijo et al.

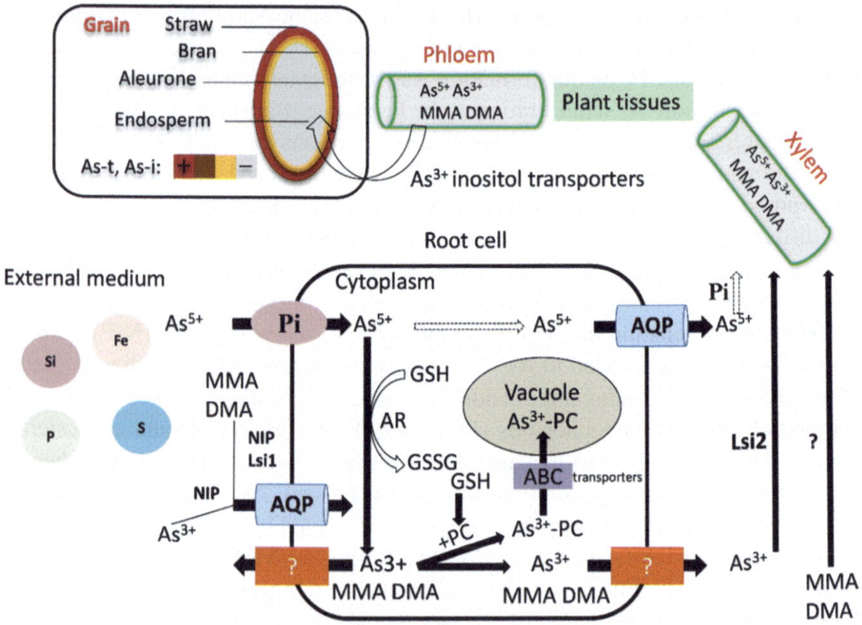

Fig. 4.2 Schematic representation of arsenic metabolism absorption, metabolism, and distribution on rice based on Awasthi et al. (2017), Batista et al. (2014), and Farooq et al. (2016) studies

(2018) observed that AWD safest condition was 2 days of drying, and this condition did not decrease grain As concentration; however, grain As concentration was decreased in the two driest AWD treatments. Norton et al. (2017) also concluded that AWD could increase grain production when compared to continuous submerged conditions. Besides, they presented pieces of evidence that AWD promotes leaf elongation and increases leaf abscisic acid and two cytokines, thus increasing the number of productive tillers. The mean concentration of As in rice grains found by these authors was around 226–304 μg $kg-^{1}$.

The losses of rice productivity and quality in Bangladesh, where the rice fields are well known to contain high concentrations of As, have been increasingly explored. No evidence from yield or panicle sterility data of As toxicity to rice was observed in a study performed in 150 paddy rice samples collected from Bangladesh, with As grain concentrations ranging from 10 to 420 μg $kg-^{1}$, but rice processing (parboiling and milling) was considered as an alternative to reduce As concentration in rice (Duxbury et al. 2003). In another study conducted in this country, farmer fields with soil-As levels ranging from about 10–70 mg $kg-^{1}$ and rice grain containing As concentrations from 300 to 600 μg $kg-^{1}$ indicated field yield losses over the years due to irrigation water contaminated with As. This study finding also reported the remarkable importance of removing rice straws, since their As content was 10–40 times higher than that of grain (Panaullah et al. 2009). Khan et al. (2010) found that As addition in the cultivation of *Boro* rice or *Aman* rice from Bangladesh resulted in yield losses from 21 to 74% and 8 to 80%, respectively. These authors reported that As concentrations in rice grains (220–810 μg kg^{-1}), straw (2640–12520 μg kg^{-1}) and husk (1200–2480 μg kg^{-1}) increased with an increase in As addition. They also described that the toxicity symptoms of rice plants growing in As highly contaminated soil (>60 mg kg^{-1}) are stunted growth, brown spots, and scorching on leaves.

According to Abedin et al. (2002a), who evaluated As in Bangladesh irrigation water, rice plants exposed to high arsenate levels produced low shoot biomass and reduced plant height; however, the rice straw biomass was significantly decreased with increasing arsenate dose.

The evaluation of the combined effect of cadmium and As concentration in rice grain on grain yield was carried out by Duan et al. (2017) in Chinese rice genotype selection study. Arsenic grain concentration varied from 70 to 500 μg kg^{-1}. It is worth noting that the authors found a significant and positive correlation between grain yield and grain Cd concentration, but a negative correlation between grain yield and grain As concentration. In the European context, Moreno-Jiménez et al. (2014) studied the effects of water management over 7 successive years of a field experiment carried out in Spain on rice production and on As and Cd concentrations. By comparison of sprinkler and flood irrigation systems, the authors mentioned that inorganic and organic As concentration in grain was reduced (from 550 to 90 μg kg^{-1}) and that Cd concentration increased under sprinkler system, but both irrigation techniques resulted in similar grain yields.

To sum up, the mentioned studies presented in this section among others helped to bring attention to scientific knowledge and awareness about some key parameters to be considered to control rice grain As contamination and also to improve rice productivity and quality even under As stressed growing environment to reduce the negative impact of this element in human food chain.

4.6 Strategies of Mitigation of Arsenic Uptake in Plants

There are different strategies to reduce the As concentration in foods. Among them are plant breeding, cultivation practices (irrigation management), phytoremediation, and postharvest processing (Mitra et al. 2017; Pedron et al. 2019; Souza et al. 2015; Zhao et al. 2010). Although there are several forms of mitigation, these are not always effective or applicable in all situations due to several factors such as viability and economic conditions.

Most rice cultivars in the world are produced under flood conditions, due to several benefits, such as greater availability of nutrients and reduction of pathogens and weeds (Li et al. 2009; Peng et al. 2006). However, in flooding conditions, anaerobic soils with low redox potential and high pH favor the absorption of As^{3+}, about 80% of the total As in the solid fraction of soil. On the other hand, upland (dryland) conditions favor a higher absorption of cadmium (Arao 2019; Zhao et al. 2010). Therefore, intermittent irrigation techniques that are used to minimize the bioavailability of As in the plant through water management were developed (Li et al. 2009; Mitra et al. 2017; Souza et al. 2015). Honma et al. (2016) observed a reduction of 62% and 56% of i-As and cadmium, respectively, in rice grains cultivated under intermittent irrigation. The study consisted of intermittent irrigation for 3 days of flooding followed by 5 days of drainage. However, prolonged drainage may favor aerobic conditions and consequently an increase in the concentration of cadmium available in soil (Honma et al. 2016).

Another form of mitigation involves plant breeding through the selection and development of resistant/tolerant cultivars to As. The selection of genotypes or varieties determines characteristics such as the blockade, translocation, and accumulation of As (Dasgupta et al. 2004; Norton et al. 2014; Syu et al. 2015; Zhang et al. 2008). Norton et al. (2009) measured the As content in grain of 76 cultivars irrigated with As-contaminated tube well water in two localities in Bangladesh. The authors observed highly significant cultivar differences in the grain As between the two field sites and a significant correlation ($r = 0.802$), indicating stable genetic differences in As accumulation. The content of i-As decreased linearly with increasing total As, indicating that i-As is regulated by the total concentration of As. However, genetic variation within this trend was identified, indicating that there is a genetic component regulating the species. Phytoremediation is also worth noting, in which several bioaccumulating plants (*Cistus ladanifer*, *Lavandula stoechas*, *Plantago subulata*, *Thymus mastichina*) can be used to remove soil contaminants (Lazaro et al. 2006). The species *Pteris vittata* is widely used for the removal of As in contaminated soils. Studies with *P. vittata* in pots containing soils contaminated with As used in rice plantations showed the removal of about 3.5–11.4% of As total in soils (Ye et al. 2011). Although plants such as *P. vittata* can remove 11–176 times more As of soil than *Oryza sativa* L, the removal efficiency takes into account factors such as rate of transfer depending on the availability of the metal in the soil and natural or anthropogenic contamination, aside from the region's climate (Ye et al. 2011; Zhao et al. 2010).

4.7 Final Considerations and Perspectives

Arsenic is the most dangerous element to human health considering its occurrence, toxicity, and exposure potential, especially the inorganic As forms (As^{3+} and As^{5+}). While organic As forms present moderate to low toxicity, inorganic forms are harmful to human health in very low concentration. There are many sources of As in the environment (natural and anthropogenic), among which stand out industrial activities (mining as well as coal and petroleum extraction and refining). As a result, high levels of As have been reported in agricultural soil and irrigation water, which may induce an increment of its concentration in cereal crops such as rice. Moreover, natural high levels of As are common in several places of the world, and it mainly affects water supply and food chain. Thus, the need for an adequate quality control as well as surveillance agencies monitoring to ensure food safety of the population regarding consumption of rice and other crop plants is evident.

On the other hand, it is worth noting that many studies have been performed aiming to reduce the impact related to the presence of toxic elements such as As on soils and contaminated waters on crop production. Plant breeding, cultivation practices (irrigation management), phytoremediation, the use of natural As chelators, hyperaccumulating plants, and postharvest processing are some of the most common strategies for As mitigation in rice. In addition, there are more recent strategies such as bioremediation by microbes resistant to As, which may influence the uptake of As^{5+} by the plant, reducing its absorption by the root. In this way, it is necessary to know more about As uptake, mobility, translocation, accumulation, and resistance in crop plants grown in contaminated environments. Thus, it will be possible to develop strategies that are probably very useful in the future to mitigate As in crop plants and ensure the quality and food safety avoiding health problems caused by the almost inevitable exposure to toxic elements.

References

Abbas G, Murtaza B, Bibi I, Shahid M, Niazi N, Khan M, Amjad M, Hussain M (2018) Arsenic uptake, toxicity, detoxification, and speciation in plants: physiological, biochemical, and molecular aspects. Int J Environ Res Public Health 15:59. https://doi.org/10.3390/ijerph15010059

Abedin MJ, Cresser MS, Meharg AA, Feldmann J, Cotter-Howells J (2002a) Arsenic accumulation and metabolism in rice (*Oryza sativa L.*). Environ Sci Technol 36(5):962–968. https://doi.org/10.1021/es0101678

Abedin MJ, Cotter-Howells J, Meharg AA (2002b) Arsenic uptake and accumulation in rice (*Oryza sativa* L.) irrigated with contaminated water. Plant Soil 240(2):311–319

Agency for Toxic Substances and Disease Registry (ATSDR) (2007) Toxicological Profile for Arsenic (Update). U.S. Department of Health and Human Services. Public Health Service, Atlanta

Agency for Toxic Substances and Disease Registry (ATSDR) (2017) 2017 Substance Priority List. Atlanta, GA: U.S. Division of Toxicology and Human Health Sciences. Available at: https://www.atsdr.cdc.gov/spl/#2017spl. Accessed 21 Mar 2019

Antman KH (2001) Introduction: the history of arsenic trioxide in cancer therapy. Oncologist 6:1–2. https://doi.org/10.1634/theoncologist.6-suppl_2-1

Arao T (2019) Mitigation strategies for cadmium and arsenic in rice. In: Cadmium toxicity. Springer, Singapore, pp 125–138. https://doi.org/10.1007/978-981-13-3630-0_10

Asante KA, Agusa T, Subramanian A, Ansa-Asare OD, Biney CA, Tanabe S (2007) Contamination status of arsenic and other trace elements in drinking water and residents from Tarkwa, a historic mining township in Ghana. Chemosphere 66:1513–1522. https://doi.org/10.1016/j.chemosphere.2006.08.022

Awasthi S, Chauhan R, Srivastava S, Tripathi RD (2017) The journey of arsenic from soil to grain in rice. Front Plant Sci 8:1007. https://doi.org/10.3389/fpls.2017.01007

Bakhat HF, Zia Z, Fahad S, Abbas S, Hammad HM, Shahzad AN, Abbas F, Alharby H, Shahid M (2017) Arsenic uptake, accumulation and toxicity in rice plants: possible remedies for its detoxification: a review. Environ Sci Pollut Res 24(10):9142–9158. https://doi.org/10.1007/s11356-017-8462-2

Batista BL, Nigar M, Mestrot A, Rocha BA, Barbosa Júnior F, Price AH, Raab A, Feldmann J (2014) Identification and quantification of phytochelatins in roots of rice to long-term exposure:

evidence of individual role on arsenic accumulation and translocation. J Exp Bot 65(6):1467–1479. https://doi.org/10.1093/jxb/eru018

Bergqvist C, Greger M (2012) Arsenic accumulation and speciation in plants from different habitats. Appl Geochem 27(3):615–622. https://doi.org/10.1016/j.apgeochem.2011.12.009

Bhattacharya P, Frisbie SH, Smith E, Naidu R, Jacks G, Sarkar B (2002) Arsenic in the environment: a global perspective. Handbook of heavy metals in the environment. Marcell Dekker Inc, New York, pp 147–215

Bissen M, Frimmel FH (2003) Arsenic—a review. Part I: occurrence, toxicity, speciation, mobility. Acta Hydrochim Hydrobiol 31(1):9–18. https://doi.org/10.1002/aheh.200390025

Borba RP, Figueiredo BR, Rawlins B, Matschullat J (2003) Geochemical distribution of arsenic in waters, sediments and weathered gold mineralized rocks from Iron Quadrangle. Brazil. Environ Geol 44(1):39–52

Bradham KD, Diamond GL, Burgess M, Juhasz A, Klotzbach JM, Maddaloni M, Nelson C, Scheckel K, Serda SM, Stifelman M, Thomas DJ (2018) In vivo and in vitro methods for evaluating soil arsenic bioavailability: relevant to human health risk assessment. J Toxicol Environ Hlth 21(2):83–114. https://doi.org/10.1080/10937404.2018.1440902

Carey AM, Scheckel KG, Lombi E, Newville M, Choi Y, Norton GJ, Charnock JM, Feldmann J, Prince AH, Meharg AA (2010) Grain unloading of arsenic species in rice. Plant Physiol 152(1):309–319. https://doi.org/10.1104/pp.109.146126

Carrijo DR, Akbar N, Reis AF, Li C, Gaudin AC, Parikh SJ, Green PG, Linquist BA (2018) Impacts of variable soil drying in alternate wetting and drying rice systems on yields, grain arsenic concentration and soil moisture dynamics. Field Crop Res 222:101–110. https://doi.org/10.1016/j.fcr.2018.02.026

Chakraborti D, Rahman MM, Chatterjee A, Das D, Das B, Nayak B, Pal A, Chowdhury UK, Ahmed S, Biswas BK, Sengupta MK, Lodh D, Samanta G, Chakraborty S, Roy MM, Dutta RN, Saha KC, Mukherjee SC, Pati S, Kar PB (2016) Fate of over 480 million inhabitants living in arsenic and fluoride endemic Indian districts: Magnitude, health, socio-economic effects and mitigation approaches. J Trace Elem Med Biol 38:33–45. https://doi.org/10.1016/j.jtemb.2016.05.001

Chatterjee A, Das D, Mandal BK, Chowdhury TR, Samanta G, Chakraborti D (1995) Arsenic in ground water in six districts of West Bengal, India: the biggest arsenic calamity in the world. Part I. Arsenic species in drinking water and urine of the affected people. Analyst 120(3):643

Chen XW, Wu FY, Li H, Chan WF, Wu C, Wu SC, Wong MH (2013) Phosphate transporters expression in rice (*Oryza sativa* L.) associated with arbuscular mycorrhizal fungi (AMF) colonization under different levels of arsenate stress. Environ Exp Bot 87:92–99. https://doi.org/10.1016/j.envexpbot.2012.08.002

Cohen SM, Arnold LL, Eldan M, Lewis AS, Beck BD (2006) Methylated arsenicals: the implications of metabolism and carcinogenicity studies in rodents to human risk assessment. Crit Rev Toxicol 36:99–133. https://doi.org/10.1080/10408440500534230

Cullen WR (2008) Medicinal Arsenic: Toxic Arsenic. In: Cullen WR (ed) Is arsenic an aphrodisiac? the sociochemistry of an element. Royal Society of Chemistry, Cambridge, pp 1–55

Cullen WR (2014) Chemical mechanism of arsenic biomethylation. Chem Res Toxicol 27(4):457–461. https://doi.org/10.1021/tx400441h

Dasgupta T, Hossain SA, Meharg AA, Price AH (2004) An arsenate tolerance gene on chromosome 6 of rice. New Phytol 163(1):45–49. https://doi.org/10.1111/j.1469-8137.2004.01109.x

Demesmay C, Olle M, Porthault M (1994) Arsenic speciation by coupling high-performance liquid chromatography with inductively coupled plasma mass spectrometry. Fresenius J Anal Chem 348(3):205–210

Duan G, Shao G, Tang Z, Chen H, Wang B, Tang Z, Yang Y, Liu Y, Zhao FJ (2017) Genotypic and environmental variations in grain cadmium and arsenic concentrations among a panel of high yielding rice cultivars. Rice 10(1):9. https://doi.org/10.1186/s12284-017-0149-2

Duxbury JM, Mayer AB, Lauren JG, Hassan N (2003) Food chain aspects of arsenic contamination in Bangladesh: effects on quality and productivity of rice. J Environ Sci Hlth 38(1):61–69. https://doi.org/10.1081/ESE-120016881

EFSA (2009) Panel on Contaminants in the Food Chain (CONTAM); Scientific Opinion on Arsenic in Food. EFSA Journal 2009 7(10):1351. [199 pp.]. doi:10.2903/j.efsa.2009.1351. Available online: www.efsa.europa.eu. Acceessed 26 Mar 2019

Farooq MA, Islam F, Ali B, Najeeb U, Mao B, Gill RA, Guijun Y, Siddiquer KHM, Zhou W (2016) Arsenic toxicity in plants: cellular and molecular mechanisms of its transport and metabolism. Environ Exp Bot 132:42–52. https://doi.org/10.1016/j.envexpbot.2016.08.004

Frith J (2013) Arsenic-the "Poison of Kings" and the "Saviour of Syphilis". J Mil Veterans Health 21:11–17

Garg N, Singla P (2011) Arsenic toxicity in crop plants: physiological effects and tolerance mechanisms. Environ Chem Lett 9:303–321. https://doi.org/10.1007/s10311-011-0313-7

Gu JF, Zhou H, Tang HL, Yang WT, Zeng M, Liu ZM, Peng PQ, Liao BH (2019) Cadmium and arsenic accumulation during the rice growth period under in situ remediation. Ecotox Environ Safe 171:451–459. https://doi.org/10.1016/j.ecoenv.2019.01.003

Guimarães LHS, Segura FR, Tonani L, von-Zeska-Kress MR, Rodrigues JL, Calixto LA, Silva FF, Batista BL (2019) Arsenic volatilization by Aspergillus sp. and Penicillium sp. isolated from rice rhizosphere as a promising eco-safe tool for arsenic mitigation. J Environ Manag 237:170–179

Guo W, Zhu YG, Liu WJ, Liang YC, Geng CN, Wang SG (2007) Is the effect of silicon on rice uptake of arsenate (AsV) related to internal silicon concentrations, iron plaque and phosphate nutrition? Environ Pollut 148(1):251–257. https://doi.org/10.1016/j.envpol.2006.10.021

Han FX, Su Y, Monts DL, Plodinec MJ, Banin A, Triplett GE (2003) Assessment of global industrial-age anthropogenic arsenic contamination. Naturwissenschaften 90(9):395–340. https://doi.org/10.1007/s00114-003-0451-2

Herath I, Vithanage M, Bundschuh J, Maity JP, Bhattacharya P (2016) Natural arsenic in global groundwaters: distribution and geochemical triggers for mobilization. Curr Pollut Rep 2:68–89. https://doi.org/10.1007/s40726-016-0028-2

Hettick BE, Canas-Carrell JE, French AD, Klein DM (2015) Arsenic: a review of the element's toxicity, plant interactions, and potential methods of remediation. J Agric Food Chem 63(32):7097–7107. https://doi.org/10.1021/acs.jafc.5b02487

Honma T, Ohba H, Kaneko-Kadokura A, Makino T, Nakamura K, Katou H (2016) Optimal soil Eh, pH, and water management for simultaneously minimizing arsenic and cadmium concentrations in rice grains. Environ Sci Technol 50(8):4178–4185. https://doi.org/10.1021/acs.est.5b05424

Hossain MB, Jahiruddin M, Loeppert RH, Panaullah GM, Islam MR, Duxbury JM (2009) The effects of iron plaque and phosphorus on yield and arsenic accumulation in rice. Plant Soil 317(1-2):167–176

Hu Y, Li JH, Zhu YG, Huang YZ, Hu HQ, Christie P (2005) Sequestration of As by iron plaque on the roots of three rice (*Oryza sativa* L.) cultivars in a low-P soil with or without P fertilizer. Environ Geochem Hlth 27(2):169–176. https://doi.org/10.1007/s10653-005-0132-5

Hu ZY, Zhu YG, Li M, Zhang LG, Cao ZH, Smith FA (2007) Sulfur (S)-induced enhancement of iron plaque formation in the rhizosphere reduces arsenic accumulation in rice (Oryza sativa L.) seedlings. Environ Pollut 147(2):387–393. https://doi.org/10.1016/j.envpol.2006.06.014

Hughes MF, Beck BD, Chen Y, Lewis AS, Thomas DJ (2011) Arsenic exposure and toxicology: a historical perspective. Toxicol Sci 123:305–332. https://doi.org/10.1093/toxsci/kfr184

IARC (2012) Working Group on the Evaluation of Carcinogenic Risks to Humans. Arsenic, metals, fibres, and dusts. IARC monographs on the evaluation of carcinogenic risks to humans, v. 87, n. PTeC, p. 11, 2012

IMA (2014) The new International Mineralogical Association list of minerals. A work in progress. International Mineralogical Society (2014) Available at: http://www.ima-mineralogy.org/Minlist.htm. Accessed 06 Mar 2019

Islam S, Rahman MM, Rahman MA, Naidu R (2017) Inorganic arsenic in rice and rice-based diets: health risk assessment. Food Control 82:196–202. https://doi.org/10.1016/j.foodcont.2017.06.030

Jomova K, Jenisova Z, Feszterova M, Baros S, Liska J, Hudecova D, Rhodes CJ, Valko M (2011) Arsenic: toxicity, oxidative stress and human disease. J Appl Toxicol 31(2):95–107. https://doi.org/10.1002/jat.1649

Kamiya T, Islam R, Duan G, Uraguchi S, Fujiwara T (2013) Phosphate deficiency signaling pathway is a target of arsenate and phosphate transporter OsPT1 is involved in as accumulation in shoots of rice. Soil Sci Plant Nutr 59:580–590. https://doi.org/10.1080/00380768.2013.804390

Khairul I, Wang QQ, Jiang YH, Wang C, Naranmandura H (2017) Metabolism, toxicity and anticancer activities of arsenic compounds. Oncotarget 8(14):23905. https://doi.org/10.18632/oncotarget.14733

Khan MA, Islam MR, Panaullah GM, Duxbury JM, Jahiruddin M, Loeppert RH (2010) Accumulation of arsenic in soil and rice under wetland condition in Bangladesh. Plant Soil 333(1-2):263–274. https://doi.org/10.1007/s11104-010-0340-3

Kumarathilaka P, Seneweera S, Meharg A, Bundschuh J (2018) Arsenic speciation dynamics in paddy rice soil-water environment: sources, physico-chemical, and biological factors-a review. Water Res 140:403–414. https://doi.org/10.1016/j.watres.2018.04.034

Lazaro JD, Kidd PS, Martinez CM (2006) A phytogeochemical study of the Trás-os-Montes region (NE Portugal): possible species for plant-based soil remediation technologies. Sci Total Environ 354(2-3):265–277. https://doi.org/10.1016/j.scitotenv.2005.01.001

Li X, Thornton I (1993) Arsenic, antimony and bismuth in soil and pasture herbage in some old metalliferous mining areas in England. Environ Geochem Hlth 15(2-3):135–144. https://doi.org/10.1021/es803643v

Li J, Waters SB, Drobna Z, Devesa V, Styblo M, Thomas DJ (2005) Arsenic (+3 oxidation state) methyltransferase and the inorganic arsenic methylation phenotype. Toxicol Appl Pharmacol 204:164–169. https://doi.org/10.1016/j.taap.2004.12.002

Li RY, Stroud JL, Ma JF, McGrath SP, Zhao FJ (2009) Mitigation of arsenic accumulation in rice with water management and silicon fertilization. Environ Sci Technol 43(10):3778–3783. https://doi.org/10.1021/es803643v

Li N, Wang J, Song WY (2015) Arsenic uptake and translocation in plants. Plant Cell Physiol 57(1):4–13. https://doi.org/10.1093/pcp/pcv143

Linquist BA, Anders MM, Adviento-Borbe MAA, Chaney RL, Nalley LL, Da Rosa EF, Van Kessel C (2015) Reducing greenhouse gas emissions, water use, and grain arsenic levels in rice systems. Glob Chang Biol 21(1):407–417. https://doi.org/10.1111/gcb.12701

Liu WJ, Zhu YG, Smith FA, Smith SE (2004) Do phosphorus nutrition and iron plaque alter arsenate (As) uptake by rice seedlings in hydroponic culture? New Phytol 162(2):481–488. https://doi.org/10.1111/j.1469-8137.2004.01035.x

Liu WJ, Zhu YG, Hu Y, Williams PN, Gault AG, Meharg AA, Charnock JM, Smith FA (2006) Arsenic sequestration in iron plaque, its accumulation and speciation in mature rice plants (*Oryza sativa* L.). Environ Sci Technol 40(18):5730–5736. https://doi.org/10.1021/es060800v

Lombi E, Scheckel KG, Pallon J, Carey AM, Zhu YG, Meharg AA (2009) Speciation and distribution of arsenic and localization of nutrients in rice grains. New Phytol 184(1):193–201. https://doi.org/10.1111/j.1469-8137.2009.02912.x

Ma JF, Yamaji N, Mitani N, Xu XY, Su YH, McGrath SP, Zhao FJ (2008) Transporters of arsenite in rice and their role in arsenic accumulation in rice grain. Proc Natl Acad Sci U S A 105(29):9931–9935. https://doi.org/10.1073/pnas.0802361105

Ma L, Wang L, Jia Y, Yang Z (2017) Accumulation, translocation and conversion of six arsenic species in rice plants grown near a mine impacted city. Chemosphere 183:44–52. https://doi.org/10.1016/j.017.05.089

Mandal BK, Suzuki KT (2002) Arsenic round the world: a review. Talanta 58:201–235. https://doi.org/10.1016/S0039-9140(02)00268-0

Masuda H (2018) Arsenic cycling in the Earth's crust and hydrosphere: interaction between naturally occurring arsenic and human activities. Prog Erth Plan Sci 5(1):68. https://doi.org/10.1186/s40645-018-0224-3

Meharg AA, Raab A (2010) Getting to the bottom of arsenic standards and guidelines. Environ Sci Technol 44(12):4395–4399. https://doi.org/10.1021/es9034304

Meharg AA, Zhao FJ (2012) Arsenic & rice. Springer Science & Business Media, Berlin. https://doi.org/10.1007/978-94-007-2947-6

Meharg AA, Lombi E, Williams PN, Scheckel KG, Feldmann J, Raab A, Zhu Y, Islam R (2008) Speciation and localization of arsenic in white and brown rice grains. Environ Sci Technol 42(4):1051–1057. https://doi.org/10.1021/es702212p

Mitra A, Chatterjee S, Moogouei R, Gupta D (2017) Arsenic accumulation in rice and probable mitigation approaches: a review. Agronomy 7(4):67. https://doi.org/10.3390/agronomy7040067

Moreno-Jiménez E, Meharg AA, Smolders E, Manzano R, Becerra D, Sánchez-Llerena J, Albarrán A, López-Piñero A (2014) Sprinkler irrigation of rice fields reduces grain arsenic but enhances cadmium. Sci Total Environ 485:468–473. https://doi.org/10.1016/j.scitotenv.2014.03.106

Mosa KA, Kumar K, Chhikara S, Mcdermott J, Liu Z, Musante C, White JC, Dhankher OP (2012) Members of rice plasma membrane intrinsic proteins subfamily are involved in arsenite permeability and tolerance in plants. Transgenic Res 21:1265–1277. https://doi.org/10.1007/s11248-012-9600-8

Mukherjee A, Sengupta MK, Hossain MA, Ahamed S, Das B, Nayak B, Lodh D, Rahman MM, Chakraborti D (2006) Arsenic contamination in groundwater: a global perspective with emphasis on the Asian scenario. J Health Popul Nutr 24:142–163

Norton GJ, Islam MR, Deacon CM, Zhao FJ, Stroud JL, McGrath SP, Meharg AA (2009) Identification of low inorganic and total grain arsenic rice cultivars from Bangladesh. Environ Sci Technol 43(15):6070–6075. https://doi.org/10.1021/es901121j

Norton GJ, Douglas A, Lahner B, Yakubova E, Guerinot ML, Pinson SR, Islam MR (2014) Genome wide association mapping of grain arsenic, copper, molybdenum and zinc in rice (Oryza sativa L.) grown at four international field sites. PLoS One 9(2):e89685. https://doi.org/10.1371/journal.pone.0089685

Norton GJ, Shafaei M, Travis AJ, Deacon CM, Danku J, Pond D, Cochrane N, Lockhart K, Salt D, Zhang H, Dodd IC, Hossain M, Islam MR, Price AH (2017) Impact of alternate wetting and drying on rice physiology, grain production, and grain quality. Field Crop Res 205:1–13. https://doi.org/10.1016/j.fcr.2017.01.016

Nriagu JO, Bhattacharya P, Mukherjee AB, Bundschuh J, Zevenhoven R, Loeppert RH (2007) Arsenic in soil and groundwater: an overview. Trace Metals Contam Environ 9:3–60

Onishi H (1969) Arsenic. In:Wedepohl KH (ed) Handbook of Geochemistry. Springer, New York, vol. II-2, Chapter 33

Pan W, Wu C, Xue S, Hartley W (2014) Arsenic dynamics in the rhizosphere and its sequestration on rice roots as affected by root oxidation. J Environ Sci 26(4):892–899. https://doi.org/10.1016/S1001-0742(13)60483-0

Panaullah GM, Alam T, Hossain MB, Loeppert RH, Lauren JG, Meisner CA, Ahmed ZU, Duxbury JM (2009) Arsenic toxicity to rice (Oryza sativa L.) in Bangladesh. Plant Soil 317(1-2):31. https://doi.org/10.1007/s11104-008-9786-y

Pedron T, Segura FS, Paniz FP, Souza FM, Santos MC, Magalhães AM Jr, Batista BL (2019) Mitigation of arsenic in rice grains by polishing and washing: Evidencing the benefit and the cost. J Cereal Sci 87:52–58. https://doi.org/10.1016/j.jcs.2019.03.003

Peng S, Bouman B, Visperas RM, Castañeda A, Nie L, Park HK (2006) Comparison between aerobic and flooded rice in the tropics: agronomic performance in an eight-season experiment. Field Crop Res 96(2-3):252–259. https://doi.org/10.1016/j.fcr.2005.07.007

Punshon T, Jackson BP, Meharg AA, Warczack T, Scheckel K, Guerinot ML (2017) Understanding arsenic dynamics in agronomic systems to predict and prevent uptake by crop plants. Sci Total Environ 581:209–220. https://doi.org/10.1016/j.scitotenv.2016.12.111

Raab A, Williams PN, Meharg A, Feldmann J (2007) Uptake and translocation of inorganic and methylated arsenic species by plants. Environ Chem 4:197–203. https://doi.org/10.1071/EN06079

Rahman MA, Hasegawa H, Rahman MM, Rahman MA, Miah MAM (2007) Accumulation of arsenic in tissues of rice plant (Oryza sativa L.) and its distribution in fractions of rice grain. Chemosphere 69(6):942–948. https://doi.org/10.1016/j.chemosphere.2007.05.044

Reimann C, Matschullat J, Birke M, Salminen R (2009) Arsenic distribution in the environment: The effects of scale. Appl Geochem 24(7):1147–1167. https://doi.org/10.1016/j.apgeochem.2009.03.013

Sadler R, Olszowy H, Shaw G, Biltoft R, Connell D (1994) Soil and water contamination by arsenic from a tannery waste. Water Air Soil Pollut 78:189–198. https://doi.org/10.1007/BF00475677

Sarkar A, Paul B (2016) The global menace of arsenic and its conventional remediation-A critical review. Chemosphere 158:37–49. https://doi.org/10.1016/j.chemosphere.2016.05.043

Scheindlin S (2005) The duplicitous nature of inorganic arsenic. Mol Interv 5:60–64. https://doi.org/10.1124/mi.5.2.1

Seyfferth AL, Webb SM, Andrews JC, Fendorf S (2011) Defining the distribution of arsenic species and plant nutrients in rice (Oryza sativa L.) from the root to the grain. Geochim Cosmochim Ac 75(21):6655–6671. https://doi.org/10.1016/j.gca.2011.06.029

Shri M, Kumar S, Chakrabarty D, Trivedi PK, Mallick S, Misra P, Shukla D, Mishra S, Srivastava S, Tripathi RD, Tuli R (2009) Effect of arsenic on growth, oxidative stress, and antioxidant system in rice seedlings. Ecotox Environ Safe 72(4):1102–1110. https://doi.org/10.1016/j.ecoenv.2008.09.022

Silva JMB, Barrio RJ, Moreira JC (2014) Arsênico-saúde: uma relação que exige vigilância. Vigil Sanit Debate 2:57–63. https://doi.org/10.3395/vd.v2i1.130

Smedley PL, Kinniburgh DG (2013) Arsenic in groundwater and the environment. In: Selinus O (ed) Essentials of medical geology. Springer, Dordrecht, pp 279–310

Smith SE, Christophersen HM, Pope S, Smith FA (2010) Arsenic uptake and toxicity in plants: integrating mycorrhizal influences. Plant Soil 327:1–21. https://doi.org/10.1007/s11104-009-0089-8

Somenahally AC, Hollister EB, Yan W, Gentry TJ, Loeppert RH (2011) Water management impacts on arsenic speciation and iron-reducing bacteria in contrasting rice-rhizosphere compartments. Environ Sci Technol 45(19):8328–8335. https://doi.org/10.1021/es2012403

Souza JM, Carneiro MF, Paulelli ACC, Grotto D, Júnior AMM, Júniora FB, Batista BL (2015) Aersênio e arroz: toxicidade, metabolismo e segurança alimentar. Quim Nova 38(1):118–127. https://doi.org/10.5935/0100-4042.20140279.

Syu CH, Huang CC, Jiang PY, Lee CH, Lee DY (2015) Arsenic accumulation and speciation in rice grains influenced by arsenic phytotoxicity and rice genotypes grown in arsenic-elevated paddy soils. J Hazard Mater 286:179–186. https://doi.org/10.1016/j.jhazmat.2014.12.052

Tripathi RD, Srivastava S, Mishra S, Singh N, Tuli R, Gupta DK, Maathuis FJM (2007) Arsenic hazards: strategies for tolerance and remediation by plants. Trends Biotechnol 25:158–165. https://doi.org/10.1016/j.tibtech.2007.02.003

UNICEF (2008) UNICEF Policy Brief: Mitigating arsenic in drinking water. Available from: https://www.unicef.org/wash/files/UNICEF_Mitigating_arsenic_in_drinking_water_policy_brief_2018.pdf. Accessed 3 Mar 2019.

Vahidnia A, van der Voet GB, de Wolff FA (2007) Arsenic neurotoxicity—a review. Hum Exp Toxicol 26:823–832. https://doi.org/10.1177/0960327107084539

Valko MMHCM, Morris H, Cronin MTD (2005) Metals, toxicity and oxidative stress. Curr Med Chem 12(10):1161–1208. https://doi.org/10.2174/0929867053764635

Verbruggen N, Hermans C, Schat H (2009) Mechanisms to cope with arsenic or cadmium excess in plants. Curr Opin Plant Biol 12:364–372. https://doi.org/10.1016/j.pbi.2009.05.001

Vithanage M, Dabrowska BB, Mukherjee AB, Sandhi A, Bhattacharya P (2012) Arsenic uptake by plants and possible phytoremediation applications: a brief overview. Environ Chem Lett 10:217–224. https://doi.org/10.1007/s10311-011-0349-8

Vromman D, Lutts S, Lefèvre I, Somer L, De Vreese O, Šlejkovec Z, Quinet M (2013) Effects of simultaneous arsenic and iron toxicities on rice (Oryza sativa L.) development, yield-related parameters and As and Fe accumulation in relation to As speciation in the grains. Plant Soil 371(1–2). https://doi.org/10.1007/s11104-013-1676-2199-217

Wang J, Wang T, Burken JG, Chusuei CC, Ban H, Ladwig K, Huang CP (2008) Adsorption of arsenic (V) onto fly ash: a speciation-base approach. Chemosphere 72(3):381–388. https://doi.org/10.1016/j.chemosphere.2008.02.045

Watson WH, Yager JD (2007) Arsenic: extension of its endocrine disruption potential to interference with estrogen receptor-mediated signaling. Toxicol Sci 98(1):1–4. https://doi.org/10.1093/toxsci/kfm111

WHO (2001) World Health Organization. In: Environmental health criteria 224: arsenic and arsenic compounds. World Health Organization, Geneva, pp 1–108

Williams M (2001) Arsenic in mine waters: an international study. Environ Geol 40:267–278. https://doi.org/10.1007/s002540000162

Wu Z, Ren H, McGrath SP, Wu P, Zhao FJ (2011) Investigating the contribution of the phosphate transport pathway to arsenic accumulation in rice. Plant Physiol 157:498–508. https://doi.org/10.1104/pp.111.178921

Wu C, Zou Q, Xue S, Mo J, Pan W, Lou L, Wong MH (2015) Effects of silicon (Si) on arsenic (As) accumulation and speciation in rice (Oryza sativa L.) genotypes with different radial oxygen loss (ROL). Chemosphere 138:447–453. https://doi.org/10.1016/j.chemosphere.2015.06.081

Wu C, Huang L, Xue SG, Huang YY, Hartley W, Cui MQ, Wong MH (2017) Arsenic sorption by red mud-modified biochar produced from rice straw. Environ Sci Pollut Res 24(22):18168–18178. https://doi.org/10.1007/s11356-017-9466-7

Xu XY, McGrath SP, Meharg AA, Zhao FJ (2008) Growing rice aerobically markedly decreases arsenic accumulation. Environ Sci Technol 42(15):5574–5579. https://doi.org/10.1021/es800324u

Yamaguchi N, Ohkura T, Takahashi Y, Maejima Y, Arao T (2014) Arsenic distribution and speciation near rice roots influenced by iron plaques and redox conditions of the soil matrix. Environ Sci Technol 48(3):1549–1556. https://doi.org/10.1021/es402739a

Ye WL, Khan MA, McGrath SP, Zhao FJ (2011) Phytoremediation of arsenic contaminated paddy soils with Pteris vittata markedly reduces arsenic uptake by rice. Environ Pollut 159(12):3739–3743. https://doi.org/10.1016/j.envpol.2011.07.024

Yu HY, Ding X, Li F, Wang X, Zhang S, Yi J, Liu C, Xu X, Wang Q (2016) The availabilities of arsenic and cadmium in rice paddy fields from a mining area: the role of soil extractable and plant silicon. Environ Pollut 215:258–265. https://doi.org/10.1016/j.envpol.2016.04.008

Yudovich YE, Ketris MP (2005) Arsenic in coal: a review. Int J Coal Geol 61:141–196. https://doi.org/10.1016/j.coal.2004.09.003

Zhang J, Zhu YG, Zeng DL, Cheng WD, Qian Q, Duan GL (2008) Mapping quantitative trait loci associated with arsenic accumulation in rice (Oryza sativa). New Phytol 177(2):350–356. https://doi.org/10.1111/j.1469-8137.2007.02267.x.

Zhang J, Zhao QZ, Duan GL, Huang YC (2011) Influence of sulphur on arsenic accumulation and metabolism in rice seedlings. Environ Exp Bot 72(1):34–40. https://doi.org/10.1016/j.envexpbot.2010.05.007

Zhao F-J, Ma J-F, Meharg AA, McGrath SP (2009) Arsenic uptake and metabolism in plants. New Phytol 181:777–794. https://doi.org/10.1111/j.1469-8137.2008.02716.x

Zhao F-J, McGrath SP, Meharg AA (2010) Arsenic as a food chain contaminant: mechanisms of plant uptake and metabolism and mitigation strategies. Annu Rev Plant Biol 61:535–559. https://doi.org/10.1146/annurev-arplant-042809-112152

Zhao F-J, Stroud JL, Khan MA, McGrath SP (2012) Arsenic translocation in rice investigated using radioactive 73As tracer. Plant Soil 350(1–2):413–420

Zhao F-J, Harris E, Yan J, Ma J, Wu L, Liu W, McGrath SP, Zhou J, Zhu Y-G (2013) Arsenic methylation in soils and its relationship with microbial arsm abundance and diversity, and as speciation in rice. Environ Sci Technol 47:7147–7154. https://doi.org/10.1021/es304977m

Chapter 5
Lead in Rice Grain

Rodrigo Mendes Pereira, Camila Neves Lange, Tatiana Pedron, Fernanda Pollo Paniz, Greiziany Soares Paulino Oliveira, Hanna Paula Masuda, and Bruno Lemos Batista

5.1 Lead: A Brief Historical

According to the Agency for Toxic Substances and Disease Registry (ATSDR), lead (Pb) is the second most dangerous element for human health due to its occurrence, toxicity, exposure and high environmental accumulation (ATSDR 2019; Wani et al. 2015). Lead in the environment comes from natural and anthropogenic sources. Indeed, the anthropogenic sources are the most troublesome, since they occur in a greater extent, being generally responsible for Pb contamination in foods (cereals, legumes, fruits, vegetables, and other), the primary source of human exposure (European Food Safety Authoriry (EFSA) (2010)).

One of the earliest Pb mining records was around 6500 B.C in Turkey. Its properties as malleability, ductility and low electrical conductivity were fundamental for its use as kitchen utensils, pots, water pipes, coins, food additives and cosmetics. About 10^5 tonnes of Pb per year were extracted from mines during the Roman Empire. Its extensive use was only overcome during the industrial revolution (Boulakia 1972; Needleman 1999; Hong et al. 1994). Lead toxicity has been recognized since 2000 B.C., with the register of colic and anemia after exposure (Gilfillan 1965; Smith 1978; Needleman 2004).

In 1921, the search for higher motor efficiency led to the addition of Pb tetramethyl as anti-detonating in automobile fuels. Its use was extended until 1970 when researchers proved the Pb toxic effects in the human body. In 1996, the United States completely banned the use of this additive in fuels (Needleman 1999, 2004).

Although widespread use of Pb has been discontinued in several countries, it is still widely used in many industrial activities (Wani et al. 2015). In fact, the primary anthropogenic sources of Pb in the environment (air, soil, dust and water) are related

R. M. Pereira · C. N. Lange · T. Pedron · F. P. Paniz ·
G. S. P. Oliveira · H. P. Masuda · B. L. Batista (✉)
Universidade Federal do ABC, Santo André, SP, Brazil
e-mail: bruno.lemos@ufabc.edu.br

A. Costa de Oliveira et al. (eds.), *The Future of Rice Demand: Quality Beyond Productivity*, https://doi.org/10.1007/978-3-030-37510-2_5

to these activities. Industrial activities related to mining, smelting and refining of mineral ores and concentrates; refining leaded petrol activities; production of lead-containing products (batteries, sheets, pipes, paints, makeups, grids, etc.) and the management of waste electrical and electronic equipment (WEEE) are the most predominant forms of introduction of this element into the environment (Evans et al. 2018; Norton et al. 2014).

5.2 Toxicology of Lead in Humans

In general, Pb is present in soils, water, food, medicines, cosmetics, toys and packaging, and exposure to Pb is associated with physiological and biochemical problems including neurological, cardiovascular, reproductive, developmental and musculoskeletal diseases (ATSDR 2019).

Body fluids, urine and bones are used as biological biomarkers; however, blood is the most commonly used (ATSDR 2019). Blood Pb detection is possible up to 3 months after exposure. This is due to the high affinity of Pb for porphobilinogen synthase, a hemoglobin synthesis enzyme (Howarth 2012). Several mechanisms maintain the balance between the passive influx and the active extrusion of Ca^{2+} in the red blood cells. Lead interacts with these transport mechanisms, mimicking the Ca^{2+}, where Pb^{2+} have a passive permeability 10^5 times higher than Ca^{2+} and the ability to form anionic complexes (Simons 1985, 1988). The deficiency of Fe in children may increase Pb-absorption due to the divalent cation pump, where Pb occupies the place of Fe during intestinal absorption (Howarth 2012).

Children are generally more exposed to Pb due to their direct contact with the ground (soil, dust) during the 1st years of life and the frequent habit of bringing the hand to the mouth. For this reason, children are 4 times more exposed than adults. The higher intestinal absorption and the central nervous system in development bring children to a high Pb-susceptibility (Needleman 2004; Howarth 2012; ATSDR 2019). In 2012, the US Centers for Disease Control and Prevention (CDC) set the reference value of 5 $\mu g\ dL^{-1}$ of Pb in children's blood (CDC 2012). For exposed workers, the index of biological exposure (BEI) for blood Pb is 20 $\mu g\ dL^{-1}$ (Acgih 2012).

5.3 Lead in the Environment and Places of the World

Lead naturally isotopic abundance is 52.4%, 24.1%, 22.1% and 1.4% for ^{208}Pb, ^{206}Pb, ^{207}Pb and ^{204}Pb, respectively. This isotope composition has been used as "fingerprints" for environment pollution (Wang et al. 2013a, b). The oxidation states of this element are 2^+ and 4^+ and the most common in nature is 2^+. The main natural sources of Pb are weathering of geological materials, windblown dust, emissions to the atmosphere from volcanoes, forest fires, biogenic materials, meteoritic dust (extraterrestrial dust) and sea spray. Some studies have reported the content of lead in the Earth Upper Crust (Goldschmidt 1954; Rudnick and Gao 2014; Shaw et al. 1986; Taylor 1964; Wedepohl 1995) and these values varied from 8 to 16 mg kg^{-1}. Globally, the mean predicted natural emissions of Pb to the atmosphere is 1.8×10^9 kg y^{-1} (Richardson et al. 2001).

Lead in nature is rarely found as a metal. In general, Pb ores are found in igneous rocks associated with Zn but may be associated with other ores such as Cu, As, Sb and Ag. There are more than 200 Pb minerals, with galena ores (PbS), anglesite ($PbSO_4$), cerussite ($PbCO_3$) and minium (Pb_3O_4) being the most abundant (Caito et al. 2017). Certain sedimentary rocks, in particular in black shales, which have very small particles, are rich in organic matter and may contain high Pb-levels. Shales, clays, granites, and sandstones are also geological materials with considerable levels of lead. The average concentration of Pb in these rocks range from 12 to 30 mg kg^{-1} (Li and Thornton 2001).

Soils contain a variety of Pb-compounds. Under natural conditions, these compounds are not readily soluble; thus, its mobility is very low (Rahman and Singh 2019). Lead mobility in soils is associated with soil characteristics such as acidity; redox potential; organic matter, oxides and clay content and others properties (McBride 1994; Sjöstedt et al. 2018). Under acid and oxidizing environment, the Pb mobility is increased and in reducing soil its solubility is very low, especially due to its precipitation by sulfide generation from sulfate reduction (Tabelin et al. 2018). Due to the differences in bedrock composition and widespread anthropogenic emissions, it is challenging to establish typical values of Pb in soils. Kabata-Pendias and Pendias (2010) reported worldwide averages of Pb in surface soils ranging from 10 to 67 mg kg^{-1}. They also considered soils with Pb concentration over 100 mg kg^{-1} as potentially contaminated.

Lead solubility in natural surface water and groundwater is generally low. The distribution of Pb-species in water is dependent on pH, salinity, competing ions and organic matter. Lead is present at $pH<7.5$ in freshwater as Pb^{2+}, but it easily complexes with dissolved carbonates at $pH>5.4$, forming $PbCO_3$ and $Pb_2(OH)_2CO_3$ and Pb-hydroxides, sulfates and phosphates (Reeder et al. 2006). Lead solubility increases at low pH. Therefore, if minerals Pb-containing are exposed to acidic waters ($pH<4$), a high Pb-amount could be detected in these waters (HEM 1976). The World Health Organization (World Health Organization (WHO) 2011) maximum limit for Pb in drinking water is 10 μg L^{-1}. Lead contamination of soils and waters are from several sources such as vehicle exhaust and mining activities.

5.3.1 *Mining*

The volume of metals consumed in recent years around the world has increased considerably. Thus, the mining activities have also increased in order to supply the world demand. The production of the six major base metals (aluminum, copper, lead, nickel, tin and zinc) in the latest decades has increased by up to eight times, while the world population has only doubled (Azcue 2012). Despite the economic benefits of this activity, it is also responsible for severe environmental pollution problems resulting from mineral excavation, ore transportation, smelting and refining, and disposal of tailings and wastewaters around mines (Navarro et al. 2008; Zhuang et al. 2009b).

Based on recent data from the U. S. Geological Survey, the estimated Pb production in China was 2100 tonnes, which accounted for approximately 50% of the total produced (US Geological Survey 2019). Other countries that stand out for the production of Pb are Australia (450 tonnes), Peru (300 tonnes), United States (260 tonnes), Mexico (240 tonnes), Russia (200 tonnes), India (170 tonnes), Bolivia (100 tonnes), Kazakhstan (100 tonnes), Sweden (70 tonnes) and Turkey (60 tonnes).

In the same way that China has the highest lead production (Fig. 5.1), there are also many reports in the literature demonstrating the negative impact of these activities related to environmental pollution by potentially toxic metals in this country (Cao et al. 2009b; Fang et al. 2015; Lee et al. 2008; Ma et al. 2015; Sun et al. 2011; Wen et al. 2015; Williams et al. 2009; Wu et al. 2011; Xue et al. 2017; Yang et al. 2004; Yang et al. 2010; Yanqun et al. 2005; Zhang et al. 2012; Zhang et al. 2008; Zhou et al. 2007).

The contamination of soils and water with potentially toxic metals, especially Pb, affects the biodiversity of impacted regions (Cao et al. 2009b; Ma et al. 2015). Moreover, it is affecting the production of foods such as cereals, legumes, fruits, vegetables and other, since high concentrations of Pb in soil cause several problems for the growth and development of plants (Ernst et al. 1992). On the other hand, Pb has high transfer rates (from soil to plant); several studies showed that foods produced in nearby regions of mining activities is being contaminated with potentially toxic metals (Ashraf et al. 2015; Cao et al. 2009b; Fang et al. 2015; Iqbal et al. 2017;

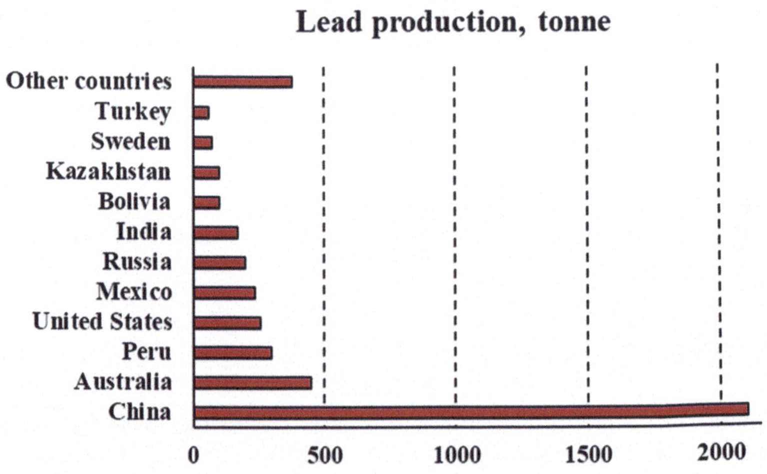

Fig. 5.1 Estimated worldwide lead production (tonne) in 2018 (US Geological Survey 2019)

Lu et al. 2015; Ma et al. 2015; Xue et al. 2017; Yu et al. 2016; Zhang et al. 2015; Zhuang et al. 2013; Zhuang et al. 2009a; Zhuang et al. 2009b).

Mining activities are indirectly contaminating a large part of the food produced around the world. In this way, these activities may be considered as one of the primary sources of Pb exposure to humans. Thus, the need and importance of actions that help in decreasing the impacts caused by mining activities and other environmental problems are remarkable. In this sense, some studies have focused on the development of plants (e.g. cereals, legumes, fruits, etc.) tolerant to toxic elements and capable of storing these species in non-edible parts, such as the roots (Iqbal et al. 2017; Liu et al. 2013; Yang et al. 2000).

5.3.2 Lead Derived from Vehicle Exhaust

Throughout history, the most famous and widespread environmental contamination by Pb was the addition of this element in gasoline (Landrigan and Garg 2002). The element was added in the early 1920s as Pb alkyls, trimethyl lead chloride (TML), triethyl lead chloride (TEL), and tetraethyl lead (TTEL), to boost octane (Nriagu 1990). At the very beginning ages of Pb alkyls production, several workers from Pb alkyls manufacturing plants died due to the occupational exposure (Oudijk 2010).

Patterson (1965) reported the first scientific data to show that high Pb levels in industrial nations are human-made and endemic. Murozumi et al. (1969) studied concentrations of Pb in Greenland snow and reported 200 times higher than the ancient times. They concluded that this increase was related to alkyl-leaded petrol. These two studies were decisive on reducing the use of Pb alkyls in gasoline. Since then, several other scientific reports were published worldwide, indicating the catastrophic consequences to the environment and for public health. The majority of these studies applied Pb isotopes determination to provide valuable information about the input of Pb from car exhaust by showing evidence of changes in Pb isotope ratios (Bollhöfer and Rosman 2002; Brännvall et al. 1999; Kylander et al. 2009; Shotyk et al. 2001).

The main Pb chemical compounds emitted from vehicle exhaust were $PbBr_2$, PbBrCl, Pb(OH)Br, $(PbO)_2PbBr_2$ and $(PbO)_2PbBrCl$. Bromide and chloride compounds were readily dissociated from these compounds and lead fallout in soils was transformed in $PbSO_4$ or associated with Fe-Mn oxides, organic matter or carbonates (McBride 1994).

Despite all strong evidence of Pb alkyls health effects, these pieces of evidence were rejected for more than 60 years, since Pb alkyls had robust economic importance and implications for the chemical and automotive corporations (Hernberg 2000). Fortunately, technical advances in automotive engineering and petroleum industry allowed the elimination of Pb alkyls from gasoline. By 1986, leaded gasoline was banished from the U. S. A. market; and by the 2000's it had been almost banned all over the world (Oudijk 2010). After evaluating epidemiologic data, Shy

(1990) concluded that leaded gasoline resulted in several public health and environmental issues and considered it as "The mistake of 20th century".

Even after the end of the use of leaded petrol, its consequences are still present in the environment and might represent health problems (Mielke et al. 2011). Wang et al. (2013a) and Wang et al. (2013b) evaluated Pb isotopes mobility in river water–cereal–atmosphere system (Yangtze River delta region). They determined Pb isotopic ratios of pedogenic parent rocks, fluvial suspended particulate matter, tillage soils, soil profiles, irrigated river water, fertilizer, Pb ore, cereal roots and grains (wheat and rice). The conclusion was that a considerable amount of Pb in cereal grains comes from the atmosphere. The estimated model shows an average of 33.5% of Pb in rice grain is airborne, mainly from vehicle exhausts.

5.3.3 Lead-Containing Products

Although the use of Pb in petrol is widely recognized, this element has also been used in the production of other various products. Its application in the production of batteries, Pb parts (sheets, pipes, grids, vehicle parts, tanks, and other), paints, cosmetics, among other products, is economically critical and has been carried out for many years around the world. This is due to some of its characteristics added to its relatively low cost when compared to other materials. Therefore, the production, the use and disposal of Pb-containing products are also important anthropogenic sources of this element, which may contaminate food.

Its application occurs mainly on the production of Pb-acid batteries, plumbing materials and alloys (World Health Organization (WHO) 2010). In 2004, about 70% of the total world Pb consumption was in Pb-acid batteries (Casas and Sordo 2006). This is a result of the high demand for batteries in various industry sectors, especially in the automotive industry, which has not yet found an alternative to Pb-acid batteries that is economically feasible for large-scale production.

The most common problems related to the exposure of Pb in the environment and human beings occur during the manufacture and recycling of Pb-acid batteries. Considering the large number of small and medium-scale factories without production permission, several pollution incidents have frequently been occurring, especially in countries with high Pb production such as China (Chen et al. 2012). Besides, thousands of people are exposed to the waste discharge from the Pb-acid battery factories.

An alarming study that was conducted in China showed that about 24% of children ($n = 94{,}778$) were Pb poisoned with levels exceeding 10 μg dL^{-1} in blood. The authors attribute these results to the high growth of the Pb-acid battery industry, which increased human exposure to this toxic element (van der Kuijp et al. 2013). This fact shows the significant impact that the Pb-acid battery industry has on the environment and humans, proving to be a dangerous source of Pb exposure as well as an important anthropogenic source.

Similarly, the production of Pb parts such as sheets, grids, vehicle parts, tanks has also an impact on the environment and humans. Lead sheets are the most widely produced and used metal parts in the world. They are used in the building and construction industry due to its characteristics, as above mentioned. Moreover, it is used as radiation shielding to protect against shortwave radiation (gamma and X-ray) in the nuclear power industry and medical installations (Casas and Sordo 2006).

Lead water pipes are still a problem. Over the years several studies have verified that the lead contained in these pipes is solubilized by water, becoming an essential source of human exposure to this element (Bjørklund et al. 2018, Casas and Sordo 2006, World Health Organization (WHO) 2010). In the late 1980s, several regulatory agencies, that perform quality control of public water supply around the world, banned the use of pipes that were not Pb free (Rabin 2008). Because of this, the use of Pb water pipes has decreased considerably and other materials such as copper, iron and PVC have been used for replacement. Despite this, there are still several water piping systems around the world that have not yet been replaced; also the use of Pb pipes in the industry is still widespread given its high mechanical and corrosive resistance.

Lead in the production of paints is also very common since it can be used as pigments and paint additives. These additives are used to increase the resistance to corrosion, accelerates drying, increases the durability as well as to give other characteristics to the paint (Casas and Sordo 2006). The use of Pb in paint has been restricted and eliminated in many countries, given the several problems caused by exposure to Pb, especially in children. However, several countries still use Pb in the production of paints (Clark et al. 2009, Gottesfeld 2015, Whitehead and Buchanan 2019, World Health Organization (WHO) 2010). Among these countries, some of the world's largest Pb producers, such as China, Peru, Mexico, Russia, India and Kazakhstan (Kessler 2014).

Cosmetics is another source of Pb exposure to humans and they have grown significantly in recent years. People from different social classes consume cosmetics since before the invention of writing (Draelos 2015; Mesko et al. 2017). Potentially toxic metals are widespread contaminants in cosmetics because of raw materials, especially those of mineral origin (Al-Saleh et al. 2009; Chauhan et al. 2010). Several studies are demonstrating high levels of Pb in various cosmetics even though there are guidelines for the safe manufacture of cosmetics, and the maximum limits for Pb in these products are well established in many countries (Al-Saleh et al. 2009; Chauhan et al. 2010; Draelos 2015).

5.3.4 *Waste Electrical and Electronic Equipment (WEEE)*

The vast amount of waste electrical and electronic equipment generated in the last years is very expressive. This waste may be a source of hazardous substances that pose a risk to the environment and health of the population (Babu et al. 2007). In this regard, the management of WEEE plays a significant role in minimizing these impacts. Despite this, in most countries, some of the waste is recycled and the rest is incinerated or sent to landfills, resulting in serious environmental problems in the future. Few countries can recycle or dispose of the WEEE generated. At the same time, a considerable amount of WEEE generated in several countries of the world is exported to developing countries, especially China and India, which are the popular destinations of WEEE. In these countries, the WEEE is recycled in informal units or dumped without any concern for the environment (Huo et al. 2007; Premalatha et al. 2014).

The high-tech recycling processes of WEEE are clean and safe, however, are very expensive and the methods used during the recycling activities in most countries are primitive (Huo et al. 2007). These recycling processes in addition to WEEE incineration and dumped are one of the major causes of the emission of Pb into the environment (water, soil and, specifically, air) (Ashraf et al. 2015; Babu et al. 2007; Huo et al. 2007; Papanikolaou et al. 2005; Premalatha et al. 2014; Wani et al. 2015). The WEEE is a severe problem and they have grown faster than the effectiveness of management methods. Besides, some predictions suggest that this scenario does not seem to reverse soon (Premalatha et al. 2014). As a consequence of the considerable emission of Pb due to precarious waste management, several are the sources of exposure to this element. Among these, it should be highlighted the intake of contaminated food given its production in an impacted environment. In other words, it is perceived that regardless of the anthropogenic source the final result is always the contamination of the environment (soil, water and air) and, consequently, of all the food that is produced in it.

5.4 Applied Lead Compounds in Agriculture

Lead arsenate (acid: $PbHAsO_4$ and basic: $Pb_4[PbOH][AsO_4]_3$) was the most extensively used insecticide for crops before the introduction of dichlorodiphenyltrichloroethane (DDT) in 1947 (Peryea and Creger 1994; Murphy and Aucott 1998). Especially in apples, Pb arsenate was recommended to control apple scab (*Venturia inaequalis*) and codling moth (*Cydia pomonella*) for routine spraying throughout the growing season. For other crops like asparagus, tomato, potato, sweet potato, eggplant, pepper, cucumber, squash, melons, carrots, celery, corn, spinach and beets, Pb arsenate (in formulations such sprays, dusts and baits) was recommended annually as a pesticide. For beans, the use of Pb arsenate was not indicated because of the toxicity of arsenic in grains (Codling et al. 2015; Merwin et al. 1994; Murphy and Aucott 1998). In general, the recommended annual application was very high, reaching, for example, 215 kg ha^{-1} of Pb for apple plantations (Peryea and Creger 1994).

In New Jersey is estimated that the amount of Pb arsenate (acid and basic) applied from 1900 to 1980 in cropland, turf and golf courses is about 2.22×10^7 kg (Murphy and Aucott 1998). In the first half of the 20th century, it is estimated an average yearly application of 59 kg ac^{-1} of Pb in apple and pear plantations in fields of the state of Washington (Wolz et al. 2003). New York state registered about 3300 kg ha^{-1} of the pesticide (Merwin et al. 1994).

A study conducted at the Hanford site (Washington State) evaluated 100 areas of historical plantations (sample depth from 10 to 50 cm). The results demonstrated that the Pb concentrations varied from 1100 mg kg^{-1} (10 cm) to 30 mg kg^{-1} (50 cm) (Yokel and Delistraty 2003). A similar study was carried out by Peryea and Creger (1994), where the result demonstrated that absolute soil enrichment with Pb occurred to depths between 15 and 50 cm. This difference of Pb mobility could be attributed to soil characteristics like coarse soil texture, low organic matter content and use of irrigation (Peryea and Creger 1994).

5.5 Lead in Plants: Absorption, Distribution and Accumulation

Lead uptake by plants occurs mainly by the roots and its availability is affected by several factors that include soil pH, redox conditions, cation exchange capacity, soil composition (including organic and inorganic ligand levels) and biological and microbiological factors. Lead precipitates with phosphate, sulfate and other chemicals in the rhizosphere that decrease its solubility and availability for plant uptake (Zeng et al. 2017). The presence of other metals in the soil such as Cd, Zn, Cr, Cu and Ni also decrease its availability to plants whereas the combination of Zn and Cu in the presence of Pb made it more available for uptake (Orroño et al. 2012). However, roots and soil/root's microorganisms produce exudates and other metabolites that change the rhizosphere environment. These metabolites, for example, modify the soil pH interfering with dissolution processes and soluble metal-organic complexes formation, such as fumaric, citric and uronic acid that chelate metal ions including Pb (Leyval and Berthelin 1991; Mench et al. 1987; Pourrut et al. 2011). Thus, the amount of Pb absorbed by the plants depends on the overall combination of these factors, Pb concentration in soils and plant species/variety (Liu et al. 2003; Meyers et al. 2008; Krzesłowska et al. 2009, 2010; Uzu et al. 2009).

Once soluble in the soil/soil solution, part of the Pb^{2+} is adsorbed and immobilized on the surface of the roots through the binding to carboxyl groups of uronic acids and other polysaccharides of rhizodermal cells (Seregin and Ivanov 2001, for review Kohli et al. 2019). A higher amount of Pb^{2+} is passively absorbed by the root cells and translocated to the aerial part of the plant, following the water-route. When roots absorb water or nutrients from soils, ions and molecules move towards that organ both by mass flow with the water from the soil and by diffusion (Fig. 5.2). Thus, the plant transpiration rate affects Pb uptake – the higher the transpiration rate, higher the Pb absorption (Tanton and Crowdy 1971; Liao et al. 2006).

Lead uptake by plant cells is a nonselective process and is dependent on H^+ ATPase activity (Wang et al. 2007). Uptake of cationic solutes, including Pb^{2+}, is thought to be driven by the negative membrane potential across plasma membrane

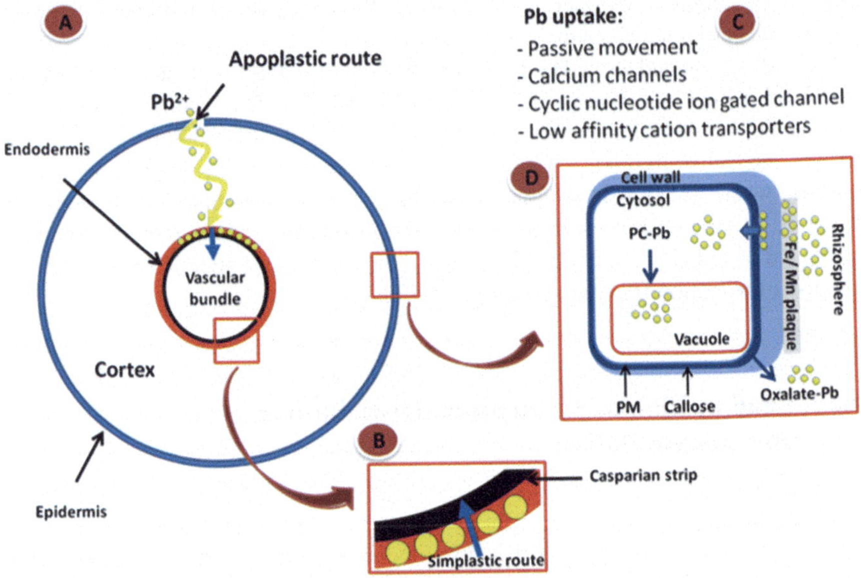

Fig. 5.2 Schematic representation of the different forms of absorption and accumulation of Pb in plant roots. (**a**, **b**) Pathways of Pb^{2+} from soil to the vascular bundle. (**a**) Pb travels from the epidermis and cortex up to the endodermis via apoplastic route. (**b**) In the endodermis, due to the physical barrier of the Casparian strip, the route of the Pb becomes simplastic and only part of Pb^{2+} reaches the vascular bundle. (**c**) List of possible Pb absorption by the cells of plant roots. (**d**) Forms of Pb entry and accumulation in the root cells. Formation of Fe/Mn plaques in the surface of the roots adsorbs part of Pb^{2+} soluble in the soil/soil solution. The cell walls become thicker and more callose is deposited in outer surface of the cells, sequestering Pb^{2+}. Oxalate is secreted and chelates Pb outside the plant root (Fahr et al. 2013)

partly generated by H^+ ATPase H^+ extrusion. Further evidence pointed that Ca^{2+} channels also mediate Pb^{2+} uptake. Reports of Pb inhibition of voltage-gated Ca^{2+} channel activity and Ca^{2+} inhibition of Pb^{2+} uptake were already shown (Huang and Cunningham 1996; Kim et al. 2006; Pourrut et al. 2008). Finally, cyclic nucleotide ion-gated channels or low-affinity cation transporters also mediate Pb^{2+} uptake (Arazi et al. 1999; Kim et al. 2002; Wojas et al. 2007).

In rice (*Oryza sativa* L.), rhizosphere acidification promoted by plasma membrane H^+ ATPase and H^+ pump under P deficiency was already shown but information of how and whether it affects Pb uptake is still missing (Zhang et al. 2011). On the other hand, the addition of Ca^{2+} and Mg^{2+} decreased Pb content in the root of rice seedlings via competition for the cation transporters that typically carries Ca^{2+} and Mg^{2+} (Kim et al. 2002). It is still not known whether there is a primary transporter or pathway for Pb uptake in rice for each physiological condition and cultivar, but it is known that Pb is not uniformly absorbed in the roots. The main site for Pb uptake is the root tip of rice seedlings, similar to what was already described in other plant species (Kim et al. 2002; Eun et al. 2000).

After the uptake, Pb follows an apoplastic route within the root and accumulates in the endodermis (Lane and Martin 1977). In the endodermis, the Casparian strips act as a partial physical barrier to the movement of the Pb between the root and the aerial part, forcing the Pb to enter the protoplast in a simplastic path (Fig. 5.2). This could partially explain the higher Pb accumulation in roots compared to shoots in several plant species, including rice (Ashraf et al. 2017a, 2018; Maodzeka et al. 2017; Leal-Alvarado et al. 2016; Hattab et al. 2016). Other possible explanations for Pb accumulation in the roots are (1) intercellular precipitation of Pb inorganic salts, (2) accumulation in the plasma membrane, (3) immobilization of Pb in the negatively charged cell walls and (4) vacuolar sequestration of Pb (Islam et al. 2007; Kopittke et al. 2007; Jiang and Liu 2010; Arias et al. 2010). Once in the roots, Pb eventually reaches the xylem and moves into the vascular bundle to the shoot in a transpiration rate-dependent manner whereas the movement from shoot to leaves occurs via vascular flow (Tanton and Crowdy 1971; Liao et al. 2006; Krzesłowska et al. 2010). Although roots are able to absorb large amounts of Pb, they also restrict its translocation to the aerial part.

In mature rice plants, highest and lowest amounts of Pb are found in the roots and grains, respectively, forming a gradient of Pb accumulation along the plant axis, where the amounts of Pb in the plant follows the decreasing sequence: roots > stem > leaf > grain (Liu et al. 2003). Within the cells, about 80% and 70% of the absorbed Pb are found in the cell wall fraction of root and shoot cells, respectively (Liu et al. 2015). Less than 10% of Pb accumulated in each organelle-rich, membrane-rich and soluble fraction of root cells, whereas about 10% accumulated in each fraction (soluble, organelle-rich and membrane-rich) of shoot tissues.

The translocation of Pb to the aerial part can be more or less efficient depending on the rice cultivar and subspecies (Liu et al. 2003; Liu et al. 2013). It was shown that the average of Pb content in *indica* milled rice was higher than those in *japonica* (Jiang et al. 2012a, b). Accordingly, in another experiment using hybrid *indica* as well as *indica* and *japonica* cultivars, the translocation factor from shoots to grains were in the following order: hybrid *indica* > *indica* > *japonica* (Liu et al. 2013). Finally, the correlation between transfer factor from shoots to ear/grains and grain Pb levels in all cultivars studied were positive and significant, suggesting that the Pb content in the shoot, but not in the root, is possibly the main source of the Pb transferred to the grains (Liu et al. 2015).

5.6 Effects and Tolerance of Lead in Rice

One of the most significant problems of Pb contamination in the soil is the rapid absorption by plants, entering the food chain. Besides human health issues, Pb contamination decreases grain yielding and growth of rice crops. The effects of Pb contamination in rice plants are evident at all growth stages from germination to maturity. Lead impairs germination and overall growth, affecting crucial physiological and biochemical processes such as respiration and photosynthesis. The higher the Pb concentration in the soil or medium, the stronger the detrimental effects on the plant growth.

Rice seeds germinated under 10 μM or 100 μM Pb showed a decrease in the germination percentage and germination index (Mishra and Choudhuri 1999). Several studies reported length and biomass decreasing of both roots and shoots rice seedlings upon treatment with Pb in different concentrations (Mukherji and Maitra 1977; Verma and Dubey 2003; Mishra and Choudhuri 1999; Srivastava et al. 2014). In older plants, it was reported a decrease of 12% in rice yielding of plants grown in red paddy soil containing 1000 mg kg^{-1} Pb (Gu et al. 1989). Another experiment found a significant decrease in rice growth and yielding at 1200 mg of Pb kg^{-1} of soil (Li et al. 2007). In pot experiments, the grain yield/pot of aromatic rice cultivars grown in 1200 mg kg^{-1} Pb were at least 50% of the control plants without Pb (Ashraf et al. 2017a). Conversely, two studies reported that the plant growth/biomass of some rice genotypes is either not impaired or even stimulated upon Pb treatment (Ashraf et al. 2015). Nonetheless, most of the rice genotypes are negatively impacted upon Pb exposure.

The decrease in the biomass and grain yield is probably due to the alteration of several critical physiological and biochemical processes. For example, Pb impairs plant nutrition affecting the plant-nutrient relationship and changes internal nutrient ratios among the plant tissues (Chatterjee et al. 2004; Sharma and Dubey 2005; Gopal and Rizvi 2008). Although data is still missing to allow a definite conclusion of the Pb toxicity and nutrient uptake, it was shown that Pb restricts the uptake of divalent cations such as Ca^{2+}, Mg^{2+}, Fe^{2+}, Zn^{2+} and Mn^{2+} and anions like NO_3^- in many plant species including rice (Chatterjee et al. 2004). The decreased absorption of nutrient in the presence of Pb may be due to the competition with ions with an atomic size similar to Pb or changes in physiological plant activities (Pourrut et al. 2011). It was reported that exposure of 4 and 8 mmol kg^{-1} Pb significantly decreased shoot nitrate content and nitrate reductase activity in *Brassica pekinensis*, suggesting that the effect of lead in nitrate reductase activity might be responsible for the decreased nitrate absorption (Xiong et al. 2006).

Another physiological effect well established for Pb toxicity is the inhibition of photosynthesis in many plants, including rice (Xiong et al. 2006; Liu et al. 2008; Piotrowska et al. 2009; Cenkci et al. 2010; Li et al. 2012a, b; Ashraf et al. 2017a, b; Chen et al. 2017). The decrease in photosynthetic efficiency is probably due to (1) affinity of Pb for protein N and S ligands causing damage to chloroplastic ultrastructure (Islam et al. 2007), (2) disrupted electron transport chain (Qufei and Fashui 2009), (3) degradation of chlorophyll by enhanced activity of chlorophyllase (Liu et al. 2008), (4) inhibition and substitution of metal nutrients such as Mg^{2+} and Fe^{2+} by Pb^{2+}, (5) decreased CO_2 levels due to stomatal closure (Romanowska et al. 2006), (6) lowered activity of enzymes such as ferredoxin-$NADP^+$ reductase and delta-aminolevulinic acid dehydratase (Gupta et al. 2009), (7) inhibition of catalytic enzymes of the Calvin cycle (Liu et al. 2008) and (8) decreased content of carotenoids and plastoquinones (Jing et al. 2007, Liu et al. 2008).

Lead toxicity also influences plant respiratory chain, resulting in decreased ATP content and energy-dependent processes (Romanowska et al. 2002). Divalent cations such as Pb^{2+}, Cd^{2+}, Ni^{2+}, Zn^{2+} and Co^{2+} may bind to mitochondrial membranes and cause decoupling of phosphorylation by disrupting electron transport chain reactions. According to Romanowska et al. (2005, 2006), Pb accelerates respiratory processes by increasing dark or mitochondrial respiration without affecting photorespiration. Although these studies of carbon fixation C3 were in other species, rice also has a C3 metabolism and is likely behaving in a similar way.

Finally, the generation of oxidative stress in response to the production of reactive oxygen species (ROS) alters the cell redox status and is considered one of the most significant phytotoxic effects of Pb. Several forms of ROS are produced including superoxide anion, hydrogen peroxide (H_2O_2), hydroxyl ions and singlet oxygen species. These ROS pose a severe challenge to maintain cellular metabolism when produced in large quantities (Liu et al. 2008; Jiang and Liu 2010). They rapidly oxidize several important biomolecules including lipids, proteins and nucleic acid. Lead divalent ions are known to induce lipid peroxidation, decrease the level of saturated fatty acids and increase the unsaturated fatty acid content of membranes changing its composition in several plant species (Piotrowska et al. 2009; Singh et al. 2010). As a result, these changes in lipid membranes cause cell abnormalities by disrupting cellular membranes (Gupta et al. 2009) and organelles like mitochondria, peroxisomes and chloroplast (Małecka et al. 2008; Hu et al. 2007). In rice, lipid peroxidation occurs upon Pb treatment in a dose-dependent manner, as observed with the increase in malondialdehyde (MDA) content, a lipid peroxidation marker (Verma and Dubey 2003; Ashraf et al. 2017a, b; Ashraf and Tang 2017).

Because of Pb toxic effects, plants developed various ways to respond to this metal stress. The primary defense, termed passive mechanism, involves physical avoidance of metal entry into the plant system (Fig. 5.2) (Pourrut et al. 2011). One of these mechanisms is the deposition of thicker cell walls in the root cells as a response to Pb exposure (Krzesłowska et al. 2010; Krzesłowska 2011). Pectin is one of the cell wall components. Lead divalent ions form complexes with pectin's carboxylic group and are immobilized in the cell wall. The pectin-Pb complex formation is a significant method by which plant declines Pb toxicity (Meyers et al. 2008; Jiang and Liu 2010). Another cell wall response upon Pb exposure is the deposition of callose, which sequesters Pb from the protoplast and deposit in the cell wall (Samardakiewicz et al. 1996; Krzesłowska et al. 2010; Krzesłowska 2011). Finally, the deposition of iron/manganese plaques on the surface of roots also protects the plant from the entry of Pb and other potentially toxic metals. These plaques increase the sequestration of Pb on the root surface and in the rhizosphere and restrain Pb absorption (Hansel et al. 2002). In some rice cultivars, Fe/Mn plaques were shown to decrease Pb uptake and accumulation. Lower concentrations of Pb were found in the roots of rice plants with plaque compared to higher Pb amount in the plants without the plaque (Liu et al. 2011; Cheng et al. 2014; Yang et al. 2016).

The second line of defense is the induction of various metal transporters, metal-chelating compounds and antioxidants. Metal transporters bind to metals and carry them in or out of the cytoplasm. Metal-efflux transporters are considered detoxification proteins and help in the efflux of the excess and/or toxic metals from the cytoplasm (Qian et al. 2016). *AtPDR12* and *AtATM3* from *Arabidopsis thaliana* are examples of ATP-binding cassette (ABC) carriers that conferred resistance or enhanced tolerance to Pb (Lee et al. 2005; Kim et al. 2006). Another example is the *Arabidopsis* knockout line of *AtAPX1*, which encodes an ascorbate peroxidase that resulted in an increased Pb tolerance due to the activation of ABC-type transporter genes expression (Jiang et al. 2017).

The synthesis of specific low-molecular-weight chelators, such as reduced glutathione (GSH) and phytochelatins (PCs), is another way to detoxify plant cells from potentially toxic elements. Metal ions such as Cd^{2+} and Pb^{2+} bind to thiol groups of phytochelatins (PC) and glutathione (GSH) and these chelators facilitate metal transport into vacuoles avoiding denaturation of important proteins (Verbruggen et al. 2009; Chen et al. 2015; Jacquart et al. 2017). Phytochelatins play an important role in the accumulation of Cd, As and Cr in rice (Batista et al. 2014; Yamasaki et al. 2018, Das et al. 2017, Uraguchi et al. 2017, Mukta et al. 2019). Most of the literature of PC in rice refers to these three elements and information of PC in lead detoxification is poor. Nonetheless, since Pb binds to thiol groups of PC and PC are present and functional for other toxic metal tolerance in various rice cultivars, it is plausible to think that it would confer tolerance to Pb as well. Oxalate was also reported to chelate Pb outside the plasma membrane of rice roots (Yang et al. 2000). The authors reported that rice genotypes Pb-tolerant produced new adventitious roots that secreted oxalate and conferred tolerance to Pb^{2+}, by excluding this ion from the plant cell.

Besides playing an important role in the defense against potentially toxic elements, glutathione is also a classical nonenzymatic components of plant defense system against ROS, together with cysteine (Noctor et al. 2012). *Arabidopsis* plants under Pb stress showed increased levels of cysteine, *GSH* genes expression and GSH (Liu et al. 2008, 2009). Glutathione protects plants from Pb toxicity by scavenging Pb-induced ROS generation. In rice, reduced glutathione levels are higher in plants under Pb stress, maintaining GSH pool (Verma and Dubey 2003; Ashraf et al. 2017a, b). Similarly, the activity of antioxidative enzymes (third line of defense) such as superoxide dismutase (SOD), ascorbate peroxidase (APX), catalase (CAT), glutathione S-transferase (GST) and glutathione reductase (GR) are enhanced in rice plants of various cultivars grown in the presence of Pb (Verma and Dubey 2003; Srivastava et al. 2014; Ashraf et al. 2017a, b; Thakur et al. 2017).

5.7 Mitigation of Lead in Rice Crops

Once in soils, Pb remains for years (Pehlivan et al. 2009). Plants play an important role in Pb-removal from contaminated soils affecting its global transport. However, when Pb is excessively accumulated in edible parts of crops, it enters the food chain and becomes a safety concern for human health. Thus, several mechanisms for soil management are used to decrease Pb toxicity and bioavailability to crops. Agricultural practice (e.g., water management), the addition of amendments (chemical and biochar) and bioremediation are some of these mechanisms (Pehlivan et al. 2009; Ashraf et al. 2015; Jiang et al. 2012a, b; Yuan et al. 2015).

Several studies have shown that water management is a promising technique that prevents the uptake of potentially toxic metal by rice plants (Arao et al. 2009, Hu et al. 2013, Wang et al. 2014). Interestingly, flooding in acidic soil markedly decreased the Pb content in the aboveground parts of rice when compared to aerobic condition (Wan et al. 2019). In alkaline soil, Pb content in the shoots and grains were indifferent to the water regime. Alternate wetting and drying (AWD) water regime decreased in 35–50% of Pb content in the grains, depending on the cultivars and compared to continuous ponding (Ashraf et al. 2018).

Another mechanism for soil remediation is the addition of chemical amendments that inhibit the uptake of toxic metals by plants such as Ca, P and Si-containing materials. These materials selectively decrease the mobility and/or availability of toxic metals in soils (Udeigwe et al. 2011). They have been used as immobilizing agents for the treatment of substantial contaminated areas due to its cost-effectiveness and speed of implementation (Kim et al. 2016; Lee et al. 2009).

Many studies have demonstrated that the Pb-immobilization in contaminated soil using phosphate amendments is a highly-efficient and environmentally-positive remediation method (Cao et al. 2009a; Silvano et al. 2012; Li et al. 2017). Other studies using Ca-containing materials as soil amendments, such as dolomite, slaked lime and limestone, showed a reduction of Pb uptake and translocation in paddy rice of various cultivars (Dong et al. 2016; Kim et al. 2006; Guo et al. 2018). The use of oyster shell waste, rich in calcium carbonate and calcium oxide, as a soil ameliorant, also increased Pb immobilization in the soil (Ok et al. 2010). The addition of Ca-based amendments decreased the soil redox potential and increased the pH. In an anaerobic environment that contains H_2S and S_2, Pb solubility in water was reduced due to the formation of PbS, which is less soluble than Pb^{2+} (Yang et al. 2016). The increase of soil pH caused deprotonating on the soil surface and enhanced the adsorption of Pb^{2+} in the negatively charged soil (Grybos et al. 2009). Also, the precipitation of Pb as $Pb(OH)_2$, $PbCO_3$ and $Pb(CO_3)_2(OH)_2$ contributed to the decreased Pb uptake by rice plants (Guo et al. 2018).

Studies using silicon-based materials to decrease the Pb absorption by plants are poor. Nonetheless, an experiment using conventional silicon and silicon nanoparticles (nano-Si) indicated that both materials decreased Pb-uptake and root-to-shoot translocation in two rice cultivars (Liu et al. 2015). Interestingly, nano-Si was more efficient than conventional silicon to ameliorate the toxic effects of Pb on rice growth by preventing Pb transfer from roots to shoot, especially in high Pb accumulating cultivars. Nanoparticles of hydroxyapatite (nHAP) also decreased the mobility and toxicity of Pb in rice under Pb stress by internalizing these nanoparticles in root cells and decreasing the translocation to shoots (Ye et al. 2018).

Other environmental technologies, such as biochar soil amendment and bioremediation, are also being used to mitigate Pb soil pollution. Biochar is a solid and porous carbonaceous product obtained from the burning of biomass under oxygen-free/oxygen-deficient conditions and is increasingly being used to immobilize potentially toxic metals in soils (Paz-Ferreiro et al. 2014). Adsorption of Pb^{2+} to biochar occurs mainly through non-electrostatic mechanism via the formation of surface complexes between Pb^{2+} and biochar functional groups (Yuan et al. 2015). Most of the Pb^{2+} removal is due to the inorganic fraction of the biochar (Xu et al. 2014). Studies have shown that rice plants grown in biochar-amended soils showed a reduction of Pb content in the shoot and grain (Zheng et al. 2012; Li et al. 2016). These and other studies also showed that the amount of biochar in the soil and the type of biomass in the biochar production affects Pb content in Fe plaques of the roots (Zheng et al. 2012; Bian et al. 2014; Li et al. 2016).

Bioremediation is also being used for soil decontamination. One of the types of bioremediation is phytoremediation that has as main agents plants that can extract this contaminant from soils and accumulate in their tissues. Phytoremediation has several mechanisms of action such as phytoaccumulation, phytoextraction, phytofiltration, phytostabilization, phytovolatilization, rhizofiltration and other that vary according to the biological behavior of each species (Kim et al. 2002; Ashraf et al. 2015; Costa et al. 2012). In rice-growing experiments, other bioremediation strategies to decrease Pb contaminants were tested, such as the use of fungi and bacteria (Lin et al. 2014; Zhou et al. 2016). In these studies, both microorganisms were able to decrease Pb availability and uptake by rice plants, resulting in lower Pb content in the grains.

Finally, combined strategies have been used to ameliorate and decrease Pb uptake in rice crops. Examples of some of these strategies are the combination of different chemical amendments (Zhou et al. 2014, Xiao et al. 2017, Yang et al. 2017) and combined inorganic and organic additives (Yang et al. 2016; Hamid et al. 2018). Still, other strategies and combinations should be tested to check for the efficiency or not in the mitigation of Pb in the plant, especially in grains.

5.8 Lead Concentration in Cereals and Foods

Cereal grains are widely consumed as staple food worldwide. Wheat, rice, and maize can be contaminated by residues of Pb present at the environment as a result of anthropogenic processes cited before. For plant development, Pb can decrease growth, photosynthesis, nutritional value, yield and grain quality. In humans, their exposure during childhood cause anemia, abdominal pain and neurological and adverse developmental effects because their systems are not fully developed and consequently, they are more vulnerable than adults to the effects of Pb. For adults, depends on the level in the blood, the effects can vary from decreased vitamin D absorption to damage of severe central nervous system and death (Hernández-Martínez and Navarro-Blasco 2012; Rizwan et al. 2018; Tong et al. 2000).

Food is the primary source of Pb exposure to humans; the World Health Organization (WHO) has established the maximum allowed level at cereals of 0.2 mg kg^{-1} (European Commission 2006). Rice is a staple food for more than 40% of the world's population. The elemental composition in this cereal is of great importance for its nutritional value as well as estimation of toxic levels. For food safety, it is necessary to keep contaminants within permissible levels (Parengam et al. 2010). In this sense, many studies have been conducted around the world to verify the nutritional quality of rice, as well as the risks associated with its consumption due to the presence of toxic elements.

Table 5.1 shows studies carried out from 1989 to 2019 in several countries around the world. Most studies are from Asian countries, probably because it is the continent that most produces and consumes this cereal. The concentration of Pb in these studies varies over a wide range (from <1 to 98,300 ng g^{-1}). Despite this vast variation, the concentration of Pb in several of these studies was lower than the maximum allowed level at cereals by WHO (Batista et al. 2012; Bennett et al. 2000;

Table 5.1 Concentrations of Pb determined in rice grains from several countries around the world

Origin	Rice type/sampling	Concentration range or mean concentration (ng g^{-1})	Reference
Argentina	White rice/Purchased at markets	<2.0–139	Londonio et al. (2019)
Bangladesh	N.T.[a]/ Harvested in field	2600–15,890	Alam et al. (2002)
	N.T./Collected from households near the studied area	1.0–89	Bergkvist et al. (2010)
	N.T./Purchased at markets	713	Ahmed et al. (2015)
Brazil	N.T./N.S.[b]	124–156	Krełowska-Kułas (1992)
	White, parboiled white and parboiled brown/ Purchased at markets	0.40–14.5	Batista et al. (2012)
	White, parboiled and brown rice/Purchased at markets	37–150	Poletti et al. (2014)
	N. T./ Harvested in field	<40	Corguinha et al. (2015)
	White, parboiled and brown rice/Purchased at markets	<40	de Lima et al. (2015)
	White, parboiled and brown rice, Purchased at markets	120–263	de Oliveira et al. (2016)
	White rice/Harvested in field	0.14–21	Maione et al. (2016)
	White, parboiled and brown rice/ Purchased at markets	<30–127	Mataveli et al. (2016)

(continued)

Table 5.1 (continued)

Origin	Rice type/sampling	Concentration range or mean concentration (ng g^{-1})	Reference
Canada	Husked, parched wild rice/Purchased at markets	<10–6700	Pip (1993)
China	Wild rice/Harvested in field	<10	Zhai et al. (1996)
	Glutinous and non-glutinous rice/Purchased at markets	4.0–419	Zhang et al. (1998)
	Unpolished rice/Harvested in field	1010–9290	Yang et al. (2004)
	Unpolished rice/Harvested in field	240–800	Liu et al. (2005)
	Japonica rice/Harvested in field	<10–1136	Fangmin et al. (2006)
	N.T./Purchased at markets that sell rice from farmland near the studied area	65–197	Zheng et al. (2007)
	Unpolished rice/Harvested in field and purchased at markets	167–2602	Fu et al. (2008)
	White rice/Harvested in field	<150–957	Hang et al. (2009)
	N.T./Harvested in field	1300–1900	Zhuang et al. (2009)
	Polished rice/Purchased at markets that sell rice from farmland near the studied area	7.6–120	Cao et al. (2010)
	N.T./Purchased at markets	5–400	Qian et al. (2010)
	Brown rice/Harvested in field	280–1630	Sun et al. (2010)
	N.T./Harvested in field	150	Zhao et al. (2012)
	Polished rice/Purchased at markets	5–220	Huang et al. (2013)
	N.T./Cultivated at laboratory using soils from uncontaminated fields	290–550	Liu et al. (2013)

(continued)

Table 5.1 (continued)

Origin	Rice type/sampling	Concentration range or mean concentration (ng g^{-1})	Reference
China	N.T./Harvested in field	380	Wang et al. (2013a, b)
	Brown rice/Cultivated at laboratory using soils from contaminated fields	300	Hu et al. (2014)
	Brown rice/Harvested in field	3.0–103	Zeng et al. (2015)
	N.T./Purchased at markets that sell rice from farmland near the studied area	34–76	Yu et al. (2016)
	Polished and brown rice/ Harvested in field	6.0–183	Mu et al. (2019)
India	Parboiled rice/Purchased at markets	156–280	Srikanth et al. (1995)
	N.T./Purchased at markets	19	Tripathi et al. (1997)
	N.T./Collected from households, in which rice is produced from farmland near the studied area	1107–4325	Kumar et al. (2016)
Iran	N.T./Harvested in field	520–974	Bakhtiarian et al. (2001)
	N.T./Collected in rice farms when farmers harvested their crops	1600–2600	Khaniki and Zazoli (2005)
	N.T./Collected in rice farms near the studied area	23–170	Falahi et al. (2010)
	N.T./Purchased at markets	387	Malakootian et al. (2011)
	N.T./Purchased at markets	41–128	Shakerian et al. (2012)
	N.T./Purchased at markets	137–785	Yousefi et al. (2016)
	N.T./Harvested in field	10,200–28,500	Rahimi et al. (2017)
	Review article that brings 21 references on the concentration of arsenic and lead in Iranian rice	70–35,000	Fakhri et al. (2018)

(continued)

Table 5.1 (continued)

Origin	Rice type/sampling	Concentration range or mean concentration (ng g^{-1})	Reference
Japan	N.T./Purchased at markets	14	Saito et al. (1988)
	Unpolished rice/ Collected in rice farms near the studied area	11	Zhang et al. (1997)
	N.T./Harvested in field	1.8–3.0	Shimbo et al. (2001)
Macedonia	N.T./Harvested in field	11,000–98,3000	Rogan et al. (2009)
Portugal	N.T./Harvested in field	12,000–35,000	Fernandes and Henriques (1990)
Russia	Unpolished rice/ Collected from households and farms near the studied area	1580	Gorbunov et al. (2003)
South Korea	Polished white rice/ Purchased at markets and collected from households and farms near the studied area	81–374	Jung et al. (2005)
Taiwan	N.T./Purchased at markets and collected from households and farms near the studied area	<100–200	Lin et al. (2004)
Turkey	N.T./Purchased at markets	54	Oymak et al. (2009)
Turkey	N.T./Purchased at markets	9.0–162	Gunduz and Akman (2013)
United States	N.T./Harvested in field	<2.0–70	Wolnik et al. (1985)
	Wild rice/Purchased at markets	5.0–115	Nriagu and Lin (1995)
	Wild rice/Harvested in field	151–171	Bennett et al. (2000)
Venezuela	N.T./Purchased at markets	<1.2–595	Buscema et al. (1997)
Asia and others	Glutinous, non-glutinous, polished and unpolished rice/Purchased at markets	<1.0–269	Watanabe et al. (1989)
	Glutinous, non-glutinous and polished rice/ Purchased at markets	1.6–58	Zhang et al. (1996)

(continued)

Table 5.1 (continued)

Origin	Rice type/sampling	Concentration range or mean concentration (ng g^{-1})	Reference
Australia, Egypt, India, Thailand and United States	N.T./Purchased at markets	23–1529	Al-Saleh and Shinwari (2001)
Australia, India, Italy, Thailand, United States and others	White, parboiled and brown rice/Purchased at markets	<2–16	Jorhem et al. (2008)
Bangladesh, China, France, Ghana, India, Italy, Japan, Nepal, Pakistan, Philippines, Spain, Sri Lanka, Thailand, United States and Vietnam	White and brown rice/Purchased at markets	1.0–4185	Norton et al. (2014)
Australia, Bangladesh, Canada, India, Italy, Pakistan, Thailand and Vietnam	N.T./Purchased at markets	2–1248	Rahman et al. (2014)
China, Philippines and South Korea	Brown rice/Harvested in field	410–1200	Chung et al. (2015)
India, Iran and Pakistan rice	N.T./Purchased at markets	57	Dehghani and Mosaferi (2016)

[a]not informed type; [b]not informed sampling

Bergkvist et al. 2010; Cao et al. 2010; Corguinha et al. 2015; Dehghani and Mosaferi 2016; Fakhri et al. 2018; Falahi et al. 2010; Gunduz and Akman 2013; Jorhem et al. 2008; Krełowska-Kułas 1992; de Lima et al. 2015; Lin et al. 2004; Londonio et al. 2019; Maione et al. 2016; Mataveli et al. 2016; Mu et al. 2019; Nriagu and Lin 1995; Oymak et al. 2009; Poletti et al. 2014; Saito et al. 1988; Shakerian et al. 2012; Shimbo et al. 2001; Tripathi et al. 1997; Watanabe et al. 1989; Wolnik et al. 1985; Yu et al. 2016; Zeng et al. 2015; Zhai et al. 1996; Zhang et al. 1997; Zhang et al. 1996; Zhao et al. 2012; Zheng et al. 2007). It should be noted that those studies conducted in Argentina, Brazil, Japan, Taiwan, Turkey and the United States had Pb concentrations always below 270 ng g^{-1}.

Although China is the most environmentally impacted country by anthropogenic activities in the world, the higher concentrations of Pb were observed in samples of rice collected in Europe, more specifically in Macedonia and Portugal (Fernandes and Henriques 1990; Rogan et al. 2009). In both cases, the authors investigated possible contaminations by mining areas near padding fields. Fernandes and Henriques (1990) investigated the Pb concentration in rice cultivated near the Sado River, in the Alcácer do Sal region, Portugal. The authors analyzed soil, root, shoot and grain samples collected in six rice fields near the Sado river. The Pb concentration in grains ranged from 12,000 to 35,000 ng g^{-1}. The concentrations of Pb were relatively high, showing the negative impact of mining activities (Fernandes and

Henriques 1990). Rogan et al. (2009) evaluated the concentration of Pb in rice cultivated near Bregalnica River and its tributaries, in Kočani Field (eastern Macedonia). According to the authors, there are several Pb-Zn ore deposits in this region, which has a long history of mining. The rice samples analyzed in this study showed high concentrations of Pb (from 12,000 to 98,300 ng g^{-1}) and undoubtedly indicated contamination. The authors related these results to mining activities and acid mine drainage-impacted riverine water, which is used for irrigation purposes (Rogan et al. 2009).

There are other studies presented in Table 5.1 that also aimed to evaluate the impact of anthropogenic sources near rice fields. Alam et al. (2002) also attribute the high concentrations of Pb determined in rice grown in Bangladesh, which ranged from 2600 to 15,890 ng g^{-1}, to possible irrigation of paddy fields with the use of water contaminated by mining wastes or waste from geothermal power stations. Yang et al. (2004) investigated the concentration distribution of Pb in paddy soil–rice system and its potential health risk to animal/human in paddy soils around a Pb/Zn mine in Lechang of Guangdong Province, South China. The authors observed that Pb concentration in unpolished rice (up to 9290 ng g^{-1}) was about 5 times higher than the environmental quality guideline for Pb in cereals in China, considering the mean concentration that was approximately 5000 ng g^{-1} (Yang et al. 2004). In the same way, Liu et al. (2005) also evaluated the concentration of Pb in rice cultivated in an environmentally impacted region. In 1985, the collapse of the tailing dam in Chenzhou Pb/Zn mine (Hunan, southern China) led to spread of mining waste spills on the farmland along the Dong River. After the accident, an urgent soil cleaning up was carried out in some places and in 2002, samples of rice and maize were collected to evaluated four sites near the old mine. Areas that cultivated rice and maize reported Pb concentration was 0.80–0.24 mg kg^{-1} for rice (grain) and 0.18–0.29 mg kg^{-1} for maize. Especially in rice, the difference between grain and husk could be 28 times higher for Pb (Liu et al. 2005).

In summary, most of the studies presented in Table 5.1 that show Pb concentrations in rice above 1000 ng g^{-1} are due to contamination of soils or water used for irrigation by anthropogenic activities related to mining, environmental accidents or the inappropriate waste disposal (Alam et al. 2002, Fangmin et al. 2006, Fernandes and Henriques 1990, Gorbunov et al. 2003, Hang et al. 2009, Khaniki and Zazoli 2005, Kumar et al. 2016, Rahimi et al. 2017, Rogan et al. 2009, Sun et al. 2010, Yang et al. 2004, Zhuang et al. 2009). These studies show the considerable impact of anthropogenic activities has on the production of contaminated food, especially rice. In this sense, it is evident the importance of adequate quality control of these foods to ensure the food safety of the population and minimize the risks to human health due to the ingestion of foods containing high Pb concentrations.

5.9 Lead Speciation in Food

Lead toxicity in food depends not only on its quantity but is also related to its oxidation states and/or chemical form. The chemical form of Pb not only dominates its bioavailability and toxicity but also controls its mobility and persistence (He et al.

2016; Lee and Jiang 2005). Generally, organometallic compounds are much more toxic than the ions of the corresponding inorganic compounds. Lead, mercury and tin obey this general rule, whereas arsenic and selenium represent an exception (Kot and Namieśnik 2000). It is important to notice that Pb speciation in food remains poorly understood and studies related to this subject are scarce (Crews 2005). Lower precision of the existing analytical techniques, few applications in the complex food matrix and lower extraction efficiency are some of the challenges to be overcome in the detection and quantification of Pb speciation in food (Xia et al. 2017).

He et al. (2016) investigated Pb species in Chinese seafood, including fish, shellfish and shrimp. About half of the samples were detected to have only Pb^{2+}, and another half of the samples were detected to have both Pb^{2+} and trimethyl Pb (TML). In comparison with shellfish/shrimp samples, fish samples have higher concentrations of TML and Pb^{2+}. Fish has a stronger ability to uptake and accumulate TML. However, total Pb levels were within allowable levels for human consumption.

Xia et al. (2017) reported speciation of Pb in two rice samples focusing on the separation, detection and quantification of the following species: Pb^{2+}, TML, TEL, and TTEL using reversed-phase chromatography–inductively coupled plasma mass spectrometry (RP-HPLC-ICP-MS). The method was conducted with a C18 column in gradient elution using methanol and acetic acid/ammonium acetate. Moreover, a combination of 1% ethylenediaminetetraacetic acid (EDTA) in 2% methanol with the assistance of ultrasound was chosen as the rice sample extraction solution. Only Pb^{2+} was detected in both samples analyzed. However, the authors reported also instability and natural decomposition of TML, TEL and TTEL when the environment changed.

5.10 Final Considerations and Perspectives

It is unquestionable that Pb is an element of significant public health concern due to its toxicity, persistence and occurrence. Moreover, it has no known biological function in the human organism and may cause adverse health effects at very low concentrations. The presence of this element in the environment is mainly related to anthropogenic sources, which are generally responsible for Pb contamination in foods. Among the activities that contribute to this scenario, mining, smelting and refining of mineral ores and concentrates are the most critical. Nevertheless, although widespread use of Pb has been discontinued in several countries, the production of lead-containing products and the management of WEEE also contribute significantly to the introduction of Pb in the environment. Even so, high Pb concentrations in rice are mainly caused by the use of soils or water contaminated by anthropogenic activities related to mining, environmental accidents or inappropriate waste disposal. In addition, moderate concentrations of Pb have been observed in rice around the world, but in most cases the concentrations were lower than the maximum allowed level at cereals by the World Health Organization. Nevertheless, it is important to mention that the consumption of foods with moderate concentrations of Pb over the years will probably cause problems to human health.

Unlike other elements that have a toxicity of organic and inorganic species well-defined, there is not much information for lead species. In this sense, the quality control of foods regarding total concentration of Pb is indispensable to ensure food safety, especially for foods widely consumed worldwide such as rice. Thus, the development of actions that help in decreasing the impacts caused by mining activities and other environmental problems are remarkable. Last years, several studies have been focused on understanding the uptake and translocation mechanisms of lead by plants. In addition, research projects aimed at the development of plants tolerant to toxic elements and capable of storing these species in non-edible parts have been performed in the last years. These studies may be an interesting starting point in the development of strategies for the production of safe foods in contaminated environments.

References

Acgih M (2012) TLVs and BEIs, threshold limit values for chemical substances and physical agents and biological exposure indices. Advisory Committee on Childhood Lead Poisoning Prevention (ACCLPP)

Ahmed MK, Shaheen N, Islam, MS, Habibullah-Al-Mamun M, Islam S, Banu CP (2015) Trace elements in two staple cereals (rice and wheat) and associated health risk implications in Bangladesh. Environ Monit Assess 187:326.

Alam MGM, Allinson G, Stagnitti F, Tanaka A, Westbrooke M (2002) Metal concentrations in Rice and Pulses of Samta Village, Bangladesh. B Environ Contam Tox 69:323–329. https://doi.org/10.1007/s00128-002-0065-y

Al-Saleh I, Shinwari N (2001) Report on the levels of cadmium, lead, and mercury in imported rice grain samples. Biol Trace Elem Res 83:91–96. https://doi.org/10.1385/bter:83:1:91

Al-Saleh I, Al-Enazi S, Shinwari N (2009) Assessment of lead in cosmetic products. Regul Toxicol Pharmacol 54:105–113. https://doi.org/10.1016/j.yrtph.2009.02.005

Arao T, Kawasaki A, Baba K, Mori S, Matsumoto S (2009) Effects of water management on cadmium and arsenic accumulation and dimethylarsinic acid concentrations in Japanese Rice. Environ Sci Technol 43:9361–9367. https://doi.org/10.1021/es9022738

Arazi T, Sunkar R, Kaplan B, Fromm H (1999) A tobacco plasma membrane calmodulin-binding transporter confers Ni2+ tolerance and Pb2+ hypersensitivity in transgenic plants. Plant J 20(2):171–182. https://doi.org/10.1046/j.1365-313x.1999.00588.x

Arias JA, Peralta-Videa JR, Ellzey JT, Ren M, Viveros MN, Gardea-Torresdey JL (2010) Effects of Glomus deserticola inoculation on Prosopis: enhancing chromium and lead uptake and translocation as confirmed by X-ray mapping, ICP-OES and TEM techniques. Environ Exp Bot 68(2):139–148. https://doi.org/10.1016/j.envexpbot.2009.08.009

Ashraf U, Tang X (2017) Yield and quality responses, plant metabolism and metal distribution pattern in aromatic rice under lead (Pb) toxicity. Chemosphere 176:141–155. https://doi.org/10.1016/j.chemosphere.2017.02.103

Ashraf U, Kanu AS, Mo Z, Hussain S, Anjum SA, Khan I, Abbas RN, Tang X (2015) Lead toxicity in rice: effects, mechanisms, and mitigation strategies – a mini review. Environ Sci Pollut R 22:18318–18332. https://doi.org/10.1007/s11356-015-5463-x

Ashraf U, Kanu AS, Deng Q, Mo Z, Pan S, Tian H, Tang X (2017a) Lead (Pb) toxicity; physio-biochemical mechanisms, grain yield, quality, and Pb distribution proportions in scented rice. Front Plant Sci 8(259):1–17. https://doi.org/10.3389/fpls.2017.00259

Ashraf U, Kanu AS, Deng Q, Mo Z, Pan S, Tian H, Tang X (2017b) Lead (Pb) toxicity; physio-biochemical mechanisms, grain yield, quality and Pb distribution proportions in scented rice. Front Plant Sci 8:259. https://doi.org/10.3389/fpls.2017.00259

Ashraf U, Hussain S, Akbarb N, Anjumb SA, Hassan W, Tang X (2018) Water management regimes alter Pb uptake and translocation in fragrant rice. Ecotox Environ Safe 149:128–134. https://doi.org/10.1016/j.ecoenv.2017.11.033

ATSDR (2019) Agency for Toxic Substances and Disease Registry (ATSDR). 2019. Toxicological profile for lead. (Draft for Public Comment). Atlanta: U.S. Department of Health and Human Services, Public Health Service.

Azcue JM (eds) (2012) Environmental impacts of mining activities: emphasis on mitigation and remedial measures. https://doi.org/10.1007/978-3-642-59891-3

Babu BR, Parande AK, Basha CA (2007) Electrical and electronic waste: a global environmental problem. Waste Manag Res 25:307–318. https://doi.org/10.1177/0734242x07076941

Bakhtiarian A, Gholipour M, Ghazi-Khansari M (2001) Lead and cadmium content of Korbal rice in Northern Iran. Iran J Public Health 30:129–132

Batista BL, Nacano LR, De Freitas R, De Oliveira-Souza VC, Barbosa F (2012) Determination of essential (Ca, Fe, I, K, Mo) and toxic elements (Hg, Pb) in Brazilian rice grains and estimation of reference daily intake. Food Nutr Sci 3(1):129–134. https://doi.org/10.4236/fns.2012.31019

Batista BL, Nigar M, Mestrot A, Rocha BA, Barbosa-Júnior F, Price AH, Raab A, Feldmann J (2014) Identification and quantification of phytochelatins in roots of rice to long-term exposure: evidence of individual role on arsenic accumulation and translocation. J Exp Bot 65:1467–1479. https://doi.org/10.1093/jxb/eru018

Bennett JP, Chiriboga E, Coleman J, Waller DM (2000) Heavy metals in wild rice from northern Wisconsin. Sci Total Environ 246:261–269. https://doi.org/10.1016/S0048-9697(99)00464-7

Bergkvist C, Kippler M, Hamadani JD, Grandér M, Tofail F, Berglund M, Vahter M (2010) Assessment of early-life lead exposure in rural Bangladesh. Environ Res 110:718–724. https://doi.org/10.1016/j.envres.2010.07.004

Bian R, Joseph S, Cui L, Pan G, Li L, Liu X, Zhang A, Rutlidge H, Wong S, Chia C, Marjo C, Gong B, Munroe P, Donne S (2014) A three-year experiment confirms continuous immobilization of cadmium and lead in contaminated paddy field with biochar amendment. J Hazard Mater 272:121–128. https://doi.org/10.1016/j.jhazmat.2014.03.017

Bjørklund G, Dadar M, Chirumbolo S, Aaseth J (2018) High content of lead is associated with the softness of drinking water and Raised Cardiovascular Morbidity: a review. Biol Trace Elem Res 186(2):384–394. https://doi.org/10.1007/s12011-018-1336-8

Bollhöfer A, Rosman KJR (2002) The temporal stability in lead isotopic signatures at selected sites in the Southern and Northern Hemispheres. Geochim Cosmochim Acta 66(8):1375–1386. https://doi.org/10.1016/s0016-7037(01)00862-6

Boulakia JDC (1972) Lead in the Roman world. Am J Archaeol 76(2):139–144. https://doi.org/10.2307/503857

Brännvall ML, Bindler R, Renberg I, Emteryd O, Bartnicki J, Billström K (1999) The medieval metal industry was the cradle of modern large-scale atmospheric lead pollution in northern Europe. Environ Sci Technol 33(24):4391–4395. https://doi.org/10.1021/es990279n

Buscema I, Prieto A, Araujo L, Gonzalez G (1997) Determination of lead and cadmium content in the Rice Consumed in Maracaibo, Venezuela. B Environ Contam Tox 59:94–98. https://doi.org/10.1007/s001289900449

Caito S, Lopes ACBA, Paoliello MMB, Aschner M (2017) Chapter 16: toxicology of lead and its damage to Mammalian Organs. In: Sigel A, Sigel H, Sigel RKO (eds) Lead – its effects on environment and health. De Gruyter, Berlin, pp 501–534. https://doi.org/10.1515/9783110434330-016

Cao X, Wahbi A, Ma L, Li B, Yang Y (2009a) Immobilization of Zn, Cu, and Pb in contaminated soils using phosphate rock and phosphoric acid. J Hazard Mater 164:555–564. https://doi.org/10.1016/j.jhazmat.2008.08.034

Cao H-c, Luan Z-q, Wang J-d, Zhang X-l (2009b) Potential ecological risk of cadmium, lead and arsenic in agricultural black soil in Jilin Province, China. Stoch Env Res Risk A 23:57–64. https://doi.org/10.1007/s00477-007-0195-1

Cao H, Chen J, Zhang J, Zhang H, Qiao L, Men Y (2010) Heavy metals in rice and garden vegetables and their potential health risks to inhabitants in the vicinity of an industrial zone in Jiangsu, China. J Environ Sci 22:1792–1799. https://doi.org/10.1016/S1001-0742(09)60321-1

Casas JS, Sordo J (2006) Chapter 1: An overview of the historical importance, occurrence, isolation, properties and applications of lead. In: Casas JS, Sordo J (eds) Lead. Elsevier Science, Amsterdam, pp 1–40. https://doi.org/10.1016/B978-044452945-9/50001-4

CDC (2012) CDC response to Advisory Committee on Childhood Lead Poisoning Prevention. Recommendations in "Low level lead exposure harms children: a renewed call for primary prevention." Atlanta: Centers for Disease Control and Prevention, Department of Health and Human Services. Available in: https://www.cdc.gov/nceh/lead/acclpp/final_document_030712.pdf. Accessed 1 Jul 2019

Cenkci S, Cigerci IH, Yildiz M, Özay C, Bozdag A, Terzi H (2010) Lead contamination reduces chlorophyll biosynthesis and genomic template stability in Brassica rapa L. Environ Exp Bot 67(3):467–473. https://doi.org/10.1016/j.envexpbot.2009.10.001

Chatterjee C, Dube BK, Sinha P, Srivastava P (2004) Detrimental effects of lead phytotoxicity on growth, yield, and metabolism of rice. Commun Soil Sci Plant Anal 35(1–2):255–265. https://doi.org/10.1081/css-120027648

Chauhan AS, Bhadauria R, Singh AK, Lodhi SS, Chaturvedi DK, Tomar VS (2010) Determination of lead and cadmium in cosmetic products. J Chem Pharm Res 2(6):92–97

Chen L, Xu Z, Liu M, Huang Y, Fan R, Su Y, Hu G, Peng X, Peng X (2012) Lead exposure assessment from study near a lead-acid battery factory in China. Sci Total Environ 429:191–198. https://doi.org/10.1016/j.scitotenv.2012.04.015

Chen J, Yang L, Gu J, Bai X, Ren Y, Fan T, Cao S (2015) MAN3 gene regulates cadmium tolerance through the glutathione-dependent pathway in Arabidopsis thaliana. New Phytol 205(2):570–582. https://doi.org/10.1111/nph.13101

Chen Q, Zhang X, Liu Y, Wei J, Shen W, Shen Z, Cui J (2017) Hemin-mediated alleviation of zinc, lead and chromium toxicity is associated with elevated photosynthesis, antioxidative capacity; suppressed metal uptake and oxidative stress in rice seedlings. Plant Growth Regul 81(2):253–264. https://doi.org/10.1007/s10725-016-0202-y

Cheng H, Wang MY, Wong MH, Ye ZH (2014) Does radial oxygen loss and iron plaque formation on roots alter Cd and Pb uptake and distribution in rice plant tissues? Plant and Soil 375:137–148. https://doi.org/10.1007/s11104-013-1945-0

Chung I-M, Kim J-K, Lee J-K, Kim S-H (2015) Discrimination of geographical origin of rice (Oryza sativa L.) by multielement analysis using inductively coupled plasma atomic emission spectroscopy and multivariate analysis. J Cereal Sci 65:252–259. https://doi.org/10.1016/j.jcs.2015.08.001

Clark CS, Rampal KG, Thuppil V, Roda SM, Succop P, Menrath W, Chen CK, Adebamowo EO, Agbede OA, Sridhar MKC, Adebamowo CA, Zakaria Y, El-Safty A, Shinde RM, Yu J (2009) Lead levels in new enamel household paints from Asia, Africa and South America. Environ Res 109:930–936. https://doi.org/10.1016/j.envres.2009.07.002

Codling EE, Chaney RL, Green CE (2015) Accumulation of lead and arsenic by carrots grown on lead-arsenate contaminated orchard soils. J Plant Nutr 38(4):509–525. https://doi.org/10.1080/01904167.2014.934477

Corguinha APB, de Souza GA, Gonçalves VC, Carvalho CA, de Lima WEA, Martins FAD, Yamanaka CH, Francisco EAB, Guilherme LRG (2015) Assessing arsenic, cadmium, and lead contents in major crops in Brazil for food safety purposes. J Food Compos Anal 37:143–150. https://doi.org/10.1016/j.jfca.2014.08.004

Costa ETS, Guilherme LRG, Melo EEC, Ribeiro BT, Inácio ESB, Severiano EC, Faquin V, Hale BA (2012) Assessing the tolerance of castor bean to Cd and Pb for phytoremediation purposes. Biol Trace Elem Res 145:93–100. https://doi.org/10.1007/s12011-011-9164-0

Crews HM (2005) Speciation of Lead in Food and Wine. Speciation of lead in food and wine. In: Cornelis R, Caruso J, Crews H, Heumann K (eds) Handbook of elemental speciation II – species in the environment, food, medicine and occupational health. https://doi.org/10.1002/0470856009.ch2j(ii)

de Lima CVS, Hoehne L, Meurer EJ (2015) Cádmio, cromo e chumbo em arroz comercializado no Rio Grande do Sul. Cienc Rural 45(12):2164–2167. https://doi.org/10.1590/0103-8478cr20140654

de Oliveira RM, Antunes ACN, Vieira MA, Medina AL, Ribeiro AS (2016) Evaluation of sample preparation methods for the determination of As, Cd, Pb, and Se in rice samples by GF AAS. Microchem J 124:402–409. https://doi.org/10.1016/j.microc.2015.09.018

Das N, Bhattacharya S, Bhattacharyya S, Maiti MK (2017) Identification of alternatively spliced transcripts of rice phytochelatin synthase 2 gene OsPCS2 involved in mitigation of cadmium and arsenic stresses. Plant Mol Biol 94(1-2):167–183. https://doi.org/10.1007/s11103-017-0600-1

Dehghani M, Mosaferi F (2016) Determination of heavy metals (cadmium, arsenic and lead) in Iranian, Pakistani and Indian rice Consumed in Hormozgan Province, Iran. J Mazandaran Univ Med Sci 25:363–367

Dong HX, Tang SY, Ye SQ, Zhao ML, Li HH, Wang G (2016) Effect of lime on the transfer of Cd and Pb in the soil-rice cultivation system and their accumulation in the rice grains. J Safety Environ 16:226–231

Draelos ZD (2015) Cosmetics: the medicine of beauty. J Cosmet Dermatol-US 14(2):91. https://doi.org/10.1111/jocd.12146

Ernst WHO, Verkleij JAC, Schat H (1992) Metal tolerance in plants. Acta Botanica Neerlandica 41(3):229–248

Eun SO, Youn HS, Lee Y (2000) Lead disturbs microtubule organization in the root meristem of Zea mays. Physiol Plant 110(3):357–365. https://doi.org/10.1111/j.1399-3054.2000.1100310.x

European Commission (2006) Commission Regulation (EC) No 1881/2006 of 19 December 2006 setting maximum levels for certain contaminants in foodstuffs. Off J Eur Union 364:365–324

European Food Safety Authoriry (EFSA) (2010) Scientific Opinion on Lead in Food. EFSA J 8(4):1570

Evans J, Pashley V, Madgwick R, Neil S, Chenery C (2018) Tracking natural and anthropogenic Pb exposure to its geological source. Sci Rep-UK 8(1):1969. https://doi.org/10.1038/s41598-018-20397-y

Fahr M, Laplaze L, Bendaou N, Hocher V, MEl M, Bogusz D, Smouni A (2013) Effect of lead on root growth. Front Plant Sci 4:175. https://doi.org/10.3389/fpls.2013.00175

Fakhri Y, Bjørklund G, Bandpei AM, Chirumbolo S, Keramati H, Hosseini Pouya R, Asadi A, Amanidaz N, Sarafraz M, Sheikhmohammad A, Alipour M, Baninameh Z, Mohseni SM, Sarkhosh M, Ghasemi SM (2018) Concentrations of arsenic and lead in rice (Oryza sativa L.) in Iran: a systematic review and carcinogenic risk assessment. Food Chem Toxicol 113:267–277. https://doi.org/10.1016/j.fct.2018.01.018

Falahi E, Hedaiati R, Ghiasvand AR (2010) Survey of iron, zinc, calcium, copper, lead, and cadmium in rice samples grown in Iran. Food Addit Contam B 3:80–83. https://doi.org/10.1080/19440041003671288

Fang T, Liu G, Zhou C, Lu L (2015) Lead in soil and agricultural products in the Huainan Coal Mining Area, Anhui, China: levels, distribution, and health implications. Environ Monit Assess 187(3):152. https://doi.org/10.1007/s10661-015-4368-y

Fangmin C, Ningchun Z, Haiming X, Yi L, Wenfang Z, Zhiwei Z, Mingxue C (2006) Cadmium and lead contamination in japonica rice grains and its variation among the different locations in southeast China. Sci Total Environ 359:156–166. https://doi.org/10.1016/j.scitotenv.2005.05.005

Fernandes J, Henriques FS (1990) Heavy metal contents of paddy fields of Alcácer do Sal. Portugal Sci Total Environ 90:89–97. https://doi.org/10.1016/0048-9697(90)90188-Z

Fu J, Zhou Q, Liu J, Liu W, Wang T, Zhang Q, Jiang G (2008) High levels of heavy metals in rice (Oryza sativa L.) from a typical E-waste recycling area in southeast China and its potential risk to human health. Chemosphere 71:1269–1275. https://doi.org/10.1016/j.chemosphere.2007.11.065

Gilfillan SC (1965) Lead poisoning and the fall of Rome. J Occup Environ Med 7(2):53–60

Goldschmidt VM (1954) Geochemistry. Oxford University Press, London, p 730

Gopal R, Rizvi AH (2008) Excess lead alters growth, metabolism and translocation of certain nutrients in radish. Chemosphere 70(9):1539–1544. https://doi.org/10.1016/j.chemosphere.2007.08.043

Gorbunov AV, Frontasyeva MV, Kistanov AA, Lyapunov SM, Okina OI, Ramadan AB (2003) Heavy and toxic metals in Staple foodstuffs and Agriproduct from Contaminated Soils. J Environ Sci Health B 38:181–192. https://doi.org/10.1081/pfc-120018448

Gottesfeld P (2015) Time to ban lead in industrial paints and coatings. Front Public Health 3:144. https://doi.org/10.3389/fpubh.2015.00144

Grybos M, Davranche M, Gruau G, Petitjean P, Pédrot M (2009) Increasing pH drives organic matter solubilization from wetland soils under reducing conditions. Geoderma 154(1–2):13–19. https://doi.org/10.1016/j.geoderma.2009.09.001

Gu SH, Zhu JZ, Gu ZL (1989) Study on the critical lead content of red paddy soil. Agro Environ Prot 8:17–22

Gunduz S, Akman S (2013) Determination of lead in rice grains by solid sampling HR-CS GFAAS. Food Chem 141:2634–2638. https://doi.org/10.1016/j.foodchem.2013.05.020

Guo J, Li Y, Hu C, Zhou S, Xu H, Zhang Q, Wang G (2018) Ca-containing amendments to reduce the absorption and translocation of Pb in rice plants. Sci Total Environ 637–638:971–979. https://doi.org/10.1016/j.scitotenv.2018.05.100

Gupta D, Nicoloso F, Schetinger M, Rossato L, Pereira L, Castro G, Srivastava S, Tripathi R (2009) Antioxidant defense mechanism in hydroponically grown Zea mays seedlings under moderate lead stress. J Hazard Mater 172(1):479–484. https://doi.org/10.1016/j.jhazmat.2009.06.141

Hamid Y, Tang L, Wang X, Hussain B, Yaseen M, Aziz MZ, Yang X (2018) Immobilization of cadmium and lead in contaminated paddy field using inorganic and organic additives. Sci Rep-UK 8(1):17839. https://doi.org/10.1038/s41598-018-35881-8

Hang X, Wang H, Zhou J, Ma C, Du C, Chen X (2009) Risk assessment of potentially toxic element pollution in soils and rice (Oryza sativa) in a typical area of the Yangtze River Delta. Environ Pollut 157:2542–2549. https://doi.org/10.1016/j.envpol.2009.03.002

Hansel CM, La Force MJ, Fendorf S, Sutton S (2002) Spatial and temporal association of As and Fe species on aquatic plant roots. Environ Sci Technol 36:1988–1994. https://doi.org/10.1021/es015647d

Hattab S, Hattab S, Flores-Casseres ML, Boussetta H, Doumas P, Hernandez LE, Banni M (2016) Characterisation of lead-induced stress molecular biomarkers in Medicago sativa plants. Environ Exp Bot 123:1–12. https://doi.org/10.1016/j.envexpbot.2015.10.005

He Y, Chen Z, Mo F, Huang L, Xu L, Wu Y, Xue Z, Fu F (2016) Species characteristics of lead in sea foods collected from coastal water of Fujian, Southeastern of China. Sci Rep-UK 6:33294. https://doi.org/10.1038/srep33294

Hem JD (1976) Geochemical controls on lead concentrations in stream water and sediments. Geochim Cosmochim Acta 40(6):599–609. https://doi.org/10.1016/0016-7037(76)90106-X

Hernández-Martínez R, Navarro-Blasco I (2012) Estimation of dietary intake and content of lead and cadmium in infant cereals marketed in Spain. Food Control 26(1):6–14. https://doi.org/10.1016/j.foodcont.2011.12.024

Hernberg S (2000) Lead poisoning in a historical perspective. Am J Ind Med 38(3):244–254. https://doi.org/10.1002/1097-0274(200009)38:3%3C244::AID-AJIM3%3E3.0.CO;2-F

Hong S, Candelone JP, Patterson CC, Boutron CF (1994) Greenland ice evidence of hemispheric lead pollution two millennia ago by Greek and Roman civilizations. Science 265(5180):1841–1843. https://doi.org/10.1126/science.265.5180.1841

Howarth D (2012) Lead exposure: implications for general practice. Aust Fam Physician 41(5):311. https://doi.org/10.1007/s11270-018-3865-3

Hu P, Li Z, Yuan C, Ouyang Y, Zhou L, Huang J, Huang Y, Luo Y, Christie P, Wu L (2013) Effect of water management on cadmium and arsenic accumulation by rice (Oryza sativa L.) with different metal accumulation capacities. J Soils Sediments 13(5):916–924

Hu Y, Norton GJ, Duan G, Huang Y, Liu Y (2014) Effect of selenium fertilization on the accumulation of cadmium and lead in rice plants. Plant Soil 384:131–140. https://doi.org/10.1007/s11104-014-2189-3

Hu J, Shi G, Xu Q, Wang X, Yuan Q, Du K (2007) Effects of Pb2+ on the active oxygen scavenging enzyme activities and ultrastructure in Potamogeton crispus leaves. Russ J Plant Physiol 54(3):414–419. https://doi.org/10.1134/s1021443707030181

Huang JW, Cunningham SD (1996) Lead phytoextraction: species variation in lead uptake and translocation. New Phytol 134:75–84. https://doi.org/10.1111/j.1469-8137.1996.tb01147.x

Huang Z, Pan X-D, Wu P-G, Han J-L, Chen Q (2013) Health risk assessment of heavy metals in rice to the population in Zhejiang. China PLoS One 8(9):e75007. https://doi.org/10.1371/journal.pone.0075007

Huo X, Peng L, Xu X, Zheng L, Qiu B, Qi Z, Zhang B, Han D, Piao Z (2007) Elevated blood lead levels of children in Guiyu, an Electronic Waste Recycling Town in China. Environ Health Perspect 115(7):1113–1117. https://doi.org/10.1289/ehp.9697

Iqbal MM, Murtaza G, Naz T, Niaz A, Mehdi SM (2017) Genotypic differences in rice cultivars for lead (pb) tolerance in salt affected soil. J Agric Res 55(4):627–637

Islam E, Yang X, Li T, Liu D, Jin X, Meng F (2007) Effect of Pb toxicity on root morphology, physiology and ultrastructure in the two ecotypes of Elsholtzia argyi. J Hazard Mater 147(3):806–816. https://doi.org/10.1016/j.jhazmat.2007.01.117

Jacquart A, Brayner R, El Hage JM, Ha-Duong NT (2017) Cd2+ and Pb2+ complexation by glutathione and the phytochelatins. Chem-Bio Interact 267:2–10. https://doi.org/10.1016/j.cbi.2016.09.002

Jiang W, Liu D (2010) Pb-induced cellular defense system in the root meristematic cells of Allium sativum L. BMC Plant Biol 10(1):40. https://doi.org/10.1186/1471-2229-10-40

Jiang S, Shi C, Wu J (2012a) Genotypic differences in arsenic, mercury, lead and cadmium in milled rice (Oryza sativa L.). Int J Food Sci Nutr 63(4):468–475. https://doi.org/10.3109/09637486.2011.636343

Jiang T Y, Jiang J, Xu R-K, Li Z (2012b) Adsorption of Pb(II) on variable charge soils amended with rice-straw derived biochar. Chemosphere 89(3):249–256. https://doi.org/10.1016/j.chemosphere.2012.04.028

Jiang L, Wang W, Chen Z, Gao Q, Xu Q, Cao H (2017) A role for APX1 gene in lead tolerance in Arabidopsis thaliana. Plant Sci 256:94–102

Jing C, Cheng Z, Li-ping L, Zhong-yang S, Xue-bo P (2007) Effects of exogenous salicylic acid on growth and H2O2-metabolizing enzymes in rice seedlings under lead stress. J Environ Sci 19(1):44–49

Jorhem L, Åstrand C, Sundström B, Baxter M, Stokes P Lewis J, Grawé KP (2008) Elements in rice from the Swedish market: 1. Cadmium, lead and arsenic (total and inorganic). Food Addit Contam A 25(3):284–292. https://doi.org/10.1080/02652030701474219

Jung MC, Yun S-T, Lee J-S, Lee J-U (2005) Baseline study on essential and trace elements in polished Rice from South Korea. Environ Geochem Hlth 27(5–6):455–464. https://doi.org/10.1007/s10653-005-4221-2

Kabata-Pendias A, Pendias H (2010) Trace elements in soils and plants. CRC Press Inc., Boca Raton, p 548. https://doi.org/10.1201/b10158

Kessler R (2014) Lead-based decorative paints: where are they still sold – and why? Environ Health Perspect 122(4):A96–A103. https://doi.org/10.1289/ehp.122-a96

Khaniki GRJ, Zazoli MA (2005) Cadmium and lead contents in rice (Oryza sativa) in the North of Iran. Int J Agric Biol 7:1026–1029. https://doi.org/10.1016/j.toxlet.2008.06.251

Kim Y-Y, Yang Y-Y, Lee Y (2002) Pb and Cd uptake in rice roots. Physiol Plant 116(3):368–372. https://doi.org/10.1034/j.1399-3054.2002.1160312.x

Kim S-C, Kim HS, Seo B-H et al (2016) Phytoavailability control based management for paddy soil contaminated with Cd and Pb: Implications for safer rice production. Geoderma 270:83–88

Kim DY, Bovet L, Kushnir S, Noh EW, Martinoia E, Lee Y (2006) AtATM3 is involved in heavy metal resistance in Arabidopsis. Plant Physiol 140(3):922–932. https://doi.org/10.1104/pp.105.074146

Kohli SK, Handa N, Bali S, Khanna K (2019) Current scenario of Pb toxicity in plants: unraveling plethora of physiological responses. In: de Voogt P (eds) Reviews of environmental contamination and toxicology. https://doi.org/10.1007/398_2019_25

Kopittke PM, Asher CJ, Kopittke RA, Menzies NW (2007) Toxic effects of Pb2+ on growth of cowpea (Vigna unguiculata). Environ Pollut 150(2):280–287. https://doi.org/10.1016/j.envpol.2007.01.011

Kot A, Namieśnik J (2000) The role of speciation in analytical chemistry. TrAC-Trend Anal Chem 19(2–3):69–79. https://doi.org/10.1016/S0165-9936(99)00195-8

Krełowska-Kułas M (1992) Studies on trace elements content in some cereal products. Food/Nahrung banner 36(3):269–272. https://doi.org/10.1002/food.19920360309

Krzesłowska M (2011) The cell wall in plant cell response to trace metals: polysaccharide remodeling and its role in defense strategy. Acta Physiol Plant 33(1):35–51. https://doi.org/10.1007/s11738-010-0581-z

Krzesłowska M, Lenartowska M, Mellerowicz EJ, Samardakiewicz S, Woźny A (2009) Pectinous cell wall thickenings formation – a response of moss protonemata cells to lead. Environ Exp Bot 65(1):119–131. https://doi.org/10.1016/j.envexpbot.2008.05.006

Krzesłowska M, Lenartowska M, Samardakiewicz S, Bilski H, Woźny A (2010) Lead deposited in the cell wall of Funaria hygrometrica protonemata is not stable – a remobilization can occur. Environ Pollut 158(1):325–338. https://doi.org/10.1016/j.envpol.2009.06.035

Kumar M, Rahman MM, Ramanathan AL, Naidu R (2016) Arsenic and other elements in drinking water and dietary components from the middle Gangetic plain of Bihar, India: health risk index. Sci Total Environ 539:125–134. https://doi.org/10.1016/j.scitotenv.2015.08.039

Kylander ME, Weiss DJ, Kober B (2009) Two high resolution terrestrial records of atmospheric Pb deposition from New Brunswick, Canada, and Loch Laxford, Scotland. Sci Total Environ 407(5):1644–1657. https://doi.org/10.1016/j.scitotenv.2008.10.036

Landrigan PJ, Garg A (2002) Chronic effects of toxic environmental exposures on children's health. J Toxicol Clin Toxicol 40(4):449–456. https://doi.org/10.1081/CLT-120006747

Lane SD, Martin ES (1977) A histochemical investigation of lead uptake in Raphanus sativus. New Phytol 79:281–286. https://doi.org/10.1111/j.1469-8137.1977.tb02206.x

Leal-Alvarado DA, Espadas-Gil F, Sáenz-Carbonell L, Talavera-May C, Santamaría JM (2016) Lead accumulation reduces photosynthesis in the lead hyper-accumulator Salvinia minima Baker by affecting the cell membrane and inducing stomatal closure. Aquat Toxicol 171:37–47. https://doi.org/10.1016/j.aquatox.2015.12.008

Lee TH, Jiang SJ (2005) Speciation of lead compounds in fish by capillary electrophoresis-inductively coupled plasma mass spectrometry. J Anal At Spectrom 20(11):1270–1274. https://doi.org/10.1039/b505779e

Lee M, Lee K, Lee J, Noh EW, Lee Y (2005) The Arabidopsis pxa1 Mutant is defective in an ATP-Binding Cassette Transporter-Like Protein Required for Peroxisomal Fatty Acid β-Oxidation. Plant Physiol 138(3):827–836. https://doi.org/10.1104/pp.127.3.1266

Lee CSL, Qi S-H, Zhang G, Luo C-L, Zhao LYL, Li X-D (2008) Seven thousand years of records on the mining and utilization of metals from Lake Sediments in Central China. Environ Sci Technol 42:4732–4738. https://doi.org/10.1021/es702990n

Lee SH, Lee JS, Choi YJ, Kim JG (2009) In situ stabilization of cadmium-, lead-and zinc-contaminated soil using various amendments. Chemosphere 77:1069–1075. https://doi.org/10.1016/j.chemosphere.2009.08.056

Leyval C, Berthelin J (1991) Weathering of a mica by roots and rhizospheric microorganisms of pine. Soil Sci Soc Am J 55(4):1009–1016. https://doi.org/10.2136/sssaj1991.03615995005500040020x

Li X, Thornton I (2001) Chemical partitioning of trace and major elements in soils contaminated by mining and smelting activities. Appl Geochem 16 (15):1693–1706

Li JX, Yang XE, He ZL, Jilani G, Sun CY, Chen SM (2007) Fractionation of lead in paddy soils and its bioavailability to rice plants. Geoderma 141(3–4):174–180. https://doi.org/10.1016/j.geoderma.2007.05.006

Li B, Wang X, Qi X, Huang L, Ye Z (2012a) Identification of rice cultivars with low brown rice mixed cadmium and lead contents and their interactions with the micronutrients iron, zinc, nickel and manganese. J Environ Sci 24(10):1790–1798. https://doi.org/10.1016/s1001-0742(11)60972-8

Li X, Bu N, Li Y, Ma LA, Xin S, Zhang L (2012b) Growth, photosynthesis and antioxidant responses of endophyte infected and non-infected rice under lead stress conditions. J Hazard Mater 213–214:55–61. https://doi.org/10.1016/j.jhazmat.2012.01.052
Li H, Liu Y, Chen Y, Wang S, Wang M, Xie T, Wang G (2016) Biochar amendment immobilizes lead in rice paddy soils and reduces its phytoavailability. Sci Rep 6:31616. https://doi.org/10.1038/srep31616
Li H, Guo X, Ye X (2017) Screening hydroxyapatite for cadmium and lead immobilization in aqueous solution and contaminated soil: the role of surface area. J Environ Sci 52:141–150. https://doi.org/10.1016/j.jes.2016.04.005
Liao Y, Chien SC, Wang M, Shen Y, Hung P, Das B (2006) Effect of transpiration on Pb uptake by lettuce and on water soluble low molecular weight organic acids in rhizosphere. Chemosphere 65(2):343–351. https://doi.org/10.1016/j.chemosphere.2006.02.010
Lin H-T, Wong S-S, Li G-C (2004) Heavy metal content of rice and shellfish in Taiwan. J Food Drug Anal 12:167–174
Lin A, Zhang X, Yang X (2014) Glomus mosseae enhances root growth and Cu and Pb acquisition of upland rice (Oryza sativa L.) in contaminated soils. Ecotoxicology 23(10):2053–2061. https://doi.org/10.1007/s10646-014-1368-6
Liu J, Li K, Xu J, Zhang Z, Ma T, Lu X, Yang J, Zhu Q (2003) Lead toxicity, uptake, and translocation in different rice cultivars. Plant Sci 165(4):793–802. https://doi.org/10.1016/s0168-9452(03)00273-5
Liu H, Probst A, Liao B (2005) Metal contamination of soils and crops affected by the Chenzhou lead/zinc mine spill (Hunan, China). Sci Total Environ 339(1–3):153–166. https://doi.org/10.1016/j.scitotenv.2004.07.030
Liu D, Li T, Jin X, Yang X, Islam E, Mahmood Q (2008) Lead induced changes in the growth and antioxidant metabolism of the lead accumulating and non-accumulating ecotypes of Sedum alfredii. J Integr Plant Biol 50(2):129–140. https://doi.org/10.1111/j.1744-7909.2007.00608.x
Liu J, Duan CQ, Zhang XH, Zhu YN, Hu C (2009) Subcellular distribution of chromium in accumulating plant Leersia hexandra Swartz. Plant and Soil 322(1–2):187–195. https://doi.org/10.1007/s11104-009-9907-2
Liu J, Leng X, Wang M, Zhu Z, Dai Q (2011) Iron plaque formation on roots of different rice cultivars and the relation with lead uptake. Ecotoxicol Environ Saf 74(5):1304–1309. https://doi.org/10.1016/j.ecoenv.2011.01.017
Liu J, Ma X, Wang M, Sun X (2013) Genotypic differences among rice cultivars in lead accumulation and translocation and the relation with grain Pb levels. Ecotoxicol Environ Saf 90:35–40. https://doi.org/10.1016/j.ecoenv.2012.12.007
Liu JG, Mei C, Cai H, Wang M (2015) Relationships between subcellular distribution and translocation and grain accumulation of Pb in different rice cultivars. Water Air Soil Pollut 226(4):93. https://doi.org/10.1007/s11270-015-2380-z
Londonio A, Morzán E, Smichowski P (2019) Determination of toxic and potentially toxic elements in rice and rice-based products by inductively coupled plasma-mass spectrometry. Food Chem 284:149–154. https://doi.org/10.1016/j.foodchem.2019.01.104
Lu Y, Song S, Wang R, Liu Z, Meng J, Sweetman AJ, Jenkins A, Ferrier RC, Li H, Luo W, Wang T (2015) Impacts of soil and water pollution on food safety and health risks in China. Environ Int 77:5–15. https://doi.org/10.1016/j.envint.2014.12.010
Ma L, Sun J, Yang Z, Wang L (2015) Heavy metal contamination of agricultural soils affected by mining activities around the Ganxi River in Chenzhou, Southern China. Environ Monit Assess 187(12):731. https://doi.org/10.1007/s10661-015-4966-8
Maione C, Batista BL, Campiglia AD, Barbosa F, Barbosa RM (2016) Classification of geographic origin of rice by data mining and inductively coupled plasma mass spectrometry. Comput Electron Agr 121:101–107. https://doi.org/10.1016/j.compag.2015.11.009
Malakootian M, Yaghmaeian K, Meserghani M, Mahvi AH, Danesh Pajouh M (2011) Determination of Pb, Cd, Cr and Ni concentration in Imported Indian Rice to Iran. Iran J Health Environ 4:77–84

Małecka A, Piechalak A, Morkunas I, Tomaszewska B (2008) Accumulation of lead in root cells of Pisum sativum. Acta Physiol Plant 30(5):629–637. https://doi.org/10.1007/s11738-008-0159-1
Maodzeka A, Hussain N, Wei L, Zvobgo G, Mapodzeke JM, Adil MF, Jabeen S, Wang F, Jiang L, Shamsi IH (2017) Elucidating the physiological and biochemical responses of different tobacco (Nicotiana tabacum) genotypes to lead toxicity. Environ Toxicol Chem 36(1):175–181. https://doi.org/10.1002/etc.3522
Mataveli LRV, Buzzo ML, De Arauz LJ, Carvalho MFH, Arakaki EEK, Matsuzaki R, Tiglea P (2016) Total arsenic, cadmium, and lead determination in Brazilian Rice samples using ICP-MS. J Anal Methods Chem 2016:1–9. https://doi.org/10.1155/2016/3968786
McBride MB (1994) Environmental chemistry of soils. Oxford University Press, New York
Mench M, Morel JL, Guchert A (1987) Metal binding properties of high molecular weight soluble exudates from maize (Zea mays L.) roots. Biol Fertil Soils 3:165–169. https://doi.org/10.1007/bf00255778
Merwin I, Pruyne PT, Ebel JG Jr, Manzell KL, Lisk DJ (1994) Persistence, phytotoxicity, and management of arsenic, lead and mercury residues in old orchard soils of New York State. Chemosphere 29(6):1361–1367. https://doi.org/10.1016/0045-6535(94)90267-4
Mesko MF, Novo DLR, Rondan FS, Pereira RM, Costa VC (2017) Sample preparation of lipstick for further Cd and Pb determination by ICP-MS: is the use of complexing acids really necessary? J Anal Atom Spectrom 32:1780–1788. https://doi.org/10.1039/c7ja00139h
Meyers DE, Auchterlonie GJ, Webb RI, Wood B (2008) Uptake and localisation of lead in the root system of Brassica juncea. Environ Pollut 153(2):323–332. https://doi.org/10.1016/j.envpol.2007.08.029
Mielke HW, Gonzales CR, Mielke PW Jr (2011) The continuing impact of lead dust on children's blood lead: comparison of public and private properties in New Orleans. Environ Res 111(8):1164–1172. https://doi.org/10.1016/j.envres.2011.06.010
Mishra A, Choudhuri MA (1999) Monitoring of phytotoxicity of lead and mercury from germination and early seedling growth indices in two rice cultivars. Water Air Soil Pollut 114(3–4):339–346. https://doi.org/10.1023/A:1005135629433
Mu T, Wu T, Zhou T, Li Z, Ouyang Y, Jiang J, Zhu D, Hou J, Wang Z, Luo Y, Christie P, Wu L (2019) Geographical variation in arsenic, cadmium, and lead of soils and rice in the major rice producing regions of China. Sci Total Environ 677:373–381. https://doi.org/10.1016/j.scitotenv.2019.04.337
Mukherji S, Maitra P (1977) Growth and metabolism of germinating rice (Oryza sativa L.) seeds as influenced by toxic concentrations of lead. Z Pflanzenphysiol 81(1):26–33. https://doi.org/10.1016/s0044-328x(77)80034-2
Mukta RH, Khatun MR, Nazmul Huda AKM (2019) Calcium induces phytochelatin accumulation to cope with chromium toxicity in rice (Oryza sativa L.). J Plant Interact 14(1):295–302. https://doi.org/10.1080/17429145.2019.1629034
Murozumi M, Chow TJ, Patterson CC (1969) Chemical concentrations of pollutant lead aerosols, terrestrial dusts and sea salts in Greenland and Antarctic snow strata. Geochim Cosmochim Acta 33:1247–1294. https://doi.org/10.1016/0016-7037(69)90045-3
Murphy EA, Aucott M (1998) An assessment of the amounts of arsenical pesticides used historically in a geographical area. Sci Total Environ 218(2–3):89–101. https://doi.org/10.1016/S0048-9697(98)00180-6
Navarro MC, Pérez-Sirvent C, Martínez-Sánchez MJ, Vidal J, Tovar PJ, Bech J (2008) Abandoned mine sites as a source of contamination by heavy metals: A case study in a semi-arid zone. J Geochem Explor 96:183–193. https://doi.org/10.1016/j.gexplo.2007.04.011
Needleman HL (1999) History of lead poisoning in the world. In: International conference on lead poisoning prevention and treatment, Bangalore
Needleman H (2004) Lead poisoning. Annu Rev Med 55:209–222. https://doi.org/10.1146/annurev.med.55.091902.103653
Noctor G, Mhamdi A, Chaouch S, Han Y, Neukermans J, Marquez-Garcia B, Queval G, Foyers CH (2012) Glutathione in plants: an integrated overview. Plant Cell Environ 35(2):454–484. https://doi.org/10.1111/j.1365-3040.2011.02400.x

Norton GJ, Williams PN, Adomako EE, Price AH, Zhu Y, Zhao F-J, McGrath S, Deacon CM, Villada A, Sommella A, Lu Y, Ming L, PMCS DS, Brammer H, Dasgupta T, Islam MR, Meharg AA (2014) Lead in rice: analysis of baseline lead levels in market and field collected rice grains. Sci Total Environ 485–486:428–434. https://doi.org/10.1016/j.scitotenv.2014.03.090
Nriagu JO (1990) The rise and fall of leaded gasoline. Sci Total Environ 92:13–28. https://doi.org/10.1016/0048-9697(90)90318-O
Nriagu JO, Lin T-S (1995) Trace metals in wild rice sold in the United States. Sci Total Environ 172:223–228. https://doi.org/10.1016/0048-9697(95)04809-X
Ok YS, Oh S-E, Ahmad M, Hyun S, Kim K-R, Moon DH, Lee SS, Lim KJ, Jeon W-T (2010) Yang JE (2010) Effects of natural and calcined oyster shells on Cd and Pb immobilization in contaminated soils. Environ Earth Sci 61(6):1301–1308. https://doi.org/10.1007/s12665-010-0674-4
Orroño DI, Schindler V, Lavado RS (2012) Heavy metal availability in Pelargonium hortorum rhizosphere: interactions, uptake and plant accumulation. J Plant Nutr 35:1374–1386. https://doi.org/10.1080/01904167.2012.684129
Oudijk G (2010) The rise and fall of organometallic additives in automotive gasoline. Environ Foren 11(1–2):17–49. https://doi.org/10.1080/15275920903346794
Oymak T, Tokalıoğlu Ş, Yılmaz V, Kartal Ş, Aydın D (2009) Determination of lead and cadmium in food samples by the coprecipitation method. Food Chem 113:1314–1317. https://doi.org/10.1016/j.foodchem.2008.08.064
Papanikolaou NC, Hatzidaki EG, Stamatis B, Tzanakakis GN, Tsatsakis AM (2005) Lead toxicity update. A brief review. Med Sci Monit 11(10):329–336
Parengam M, Judprasong K, Srianujata S, Jittinandana S, Laoharojanaphand S, Busamongko A (2010) Study of nutrients and toxic minerals in rice and legumes by instrumental neutron activation analysis and graphite furnace atomic absorption spectrophotometry. J Food Compos Anal 23(4):340–345. https://doi.org/10.1016/j.jfca.2009.12.012
Patterson CC (1965) Contaminated and natural lead environments of man. Arch Environ Health 11(3):344–360. https://doi.org/10.1080/00039896.1965.10664229
Paz-Ferreiro J, Fu S, Méndez A, Gasco G (2014) Interactive effects of biochar and the earhworm Pontoscolex corethrurus on plant productivity and soil enzymes activities. J Soil Sediment 14:483–494. https://doi.org/10.1007/s11368-013-0806-z
Pehlivan E, Ozkan AM, Dinc S, Parlayici S (2009) Adsorption of Cu2+ and Pb2+ ion on dolomite powder. J Hazard Mater 167(1–3):1044–1049. https://doi.org/10.1016/j.jhazmat.2009.01.096
Peryea FJ, Creger TL (1994) Vertical distribution of lead and arsenic in soils contaminated with lead arsenate pesticide residues. Water Air Soil Pollut 78(3–4):297–306. https://doi.org/10.1007/bf00483038
Piotrowska A, Bajguz A, Godlewska-Żyłkiewicz B, Czerpak R, Kamińska M (2009) Jasmonic acid as modulator of lead toxicity in aquatic plant Wolffia arrhiza (Lemnaceae). Environ Exp Bot 66(3):507–513. https://doi.org/10.1016/j.envexpbot.2009.03.019
Pip E (1993) Cadmium, copper and lead in wild rice from Central Canada. Arch Environ Contam Toxicol 24:179–181. https://doi.org/10.1007/bf01141345
Poletti J, Pozebon D, Barcellos de Fraga MV, Dressler VL, de Moraes DP (2014) Toxic and micronutrient elements in organic, brown and polished rice in Brazil. Food Addit Contam B 7(1):63–69. https://doi.org/10.1080/19393210.2013.845249
Pourrut B, Perchet G, Silvestre J, Cecchi M, Guiresse M, Pinelli E (2008) Potential role of NADPH-oxidase in early steps of lead-induced oxidative burst in Vicia faba roots. J Plant Physiol 165(6):571–579. https://doi.org/10.1016/j.jplph.2007.07.016
Pourrut B, Shahid M, Dumat C, Winterton P, Pinelli E (2011) Lead uptake, toxicity, and detoxification in plants. In: Whitacre D (ed) Reviews of environmental contamination and toxicology. Springer, New York, pp 113–136. https://doi.org/10.1007/978-1-4419-9860-6_4
Premalatha M, Tabassum A, Abbasi T, Abbasi SA (2014) The generation, impact, and management of E-Waste: state of the Art. Crit Rev Environ Sci Technol 44:1577–1678. https://doi.org/10.1080/10643389.2013.782171

Qian Y, Chen C, Zhang Q, Li Y, Chen Z, Li M (2010) Concentrations of cadmium, lead, mercury and arsenic in Chinese market milled rice and associated population health risk. Food Control 21(12):1757–1763. https://doi.org/10.1016/j.foodcont.2010.08.005

Qian JIN, Shan XQ, Wang ZJ, Tu Q (2016) Distribution and plant availability of heavy metals in different particle-size fractions of soil. Sci Total Environ 187(2):131–141. https://doi.org/10.1016/0048-9697(96)05134-0

Qufei L, Fashui H (2009) Effects of Pb2+ on the structure and function of photosystem II of Spirodela polyrrhiza. Biol Trace Elem Res 129(1–3):251–260. https://doi.org/10.1007/s12011-008-8283-8

Rabin R (2008) The lead industry and lead water pipes "a modest campaign". Am J Public Health 98(9):1584–1592. https://doi.org/10.2105/ajph.2007.113555

Rahimi G, Kolahchi Z, Charkhabi A (2017) Uptake and translocation of some heavy metals by rice crop (Oryza sativa) in paddy soils. Agric (Poľnohospodárstvo) 63(4):163–175. https://doi.org/10.1515/agri-2017-0016

Rahman Z, Singh VP (2019) The relative impact of toxic heavy metals (THMs) (arsenic (As), cadmium (Cd), chromium (Cr)(VI), mercury (Hg), and lead (Pb)) on the total environment: an overview. Environ Monit Assess 191(7):419. https://doi.org/10.1007/s10661-019-7528-7

Rahman MA, Rahman MM, Reichman SM, Lim RP, Naidu R (2014) Heavy metals in Australian grown and imported rice and vegetables on sale in Australia: health hazard. Ecotox Environ Safe 100:53–60. https://doi.org/10.1016/j.ecoenv.2013.11.024

Reeder RJ, Schoonen MA, Lanzirotti A (2006) Metal speciation and its role in bioaccessibility and bioavailability. Rev Mineral Geochem 64(1):59–113. https://doi.org/10.2138/rmg.2006.64.3

Richardson GM, Garrett R, Mitchell I, Mah-Poulson M, Hackbarth T (2001) Critical review on natural global and regional emissions of six trace metals to the atmosphere, report. Risklogic Sci Serv., Ottawa, ON, Canada.

Rizwan M, Ali S, ur Rehman MZ, Javed MR, Bashir A (2018) Lead toxicity in cereals and its management strategies: a critical review. Water Air Soil Pollut 229(6):211. https://doi.org/10.1007/s11270-018-3865-3

Rogan N, Serafimovski T, Dolenec M, Tasev G, Dolenec T (2009) Heavy metal contamination of paddy soils and rice (Oryza sativa L.) from Kočani Field (Macedonia). Environ Geochem Hlth 31:439–451. https://doi.org/10.1007/s10653-008-9197-2

Romanowska E, Igamberdiev AU, Parys E, Gardeström P (2002) Stimulation of respiration by Pb2+ in detached leaves and mitochondria of C3 and C4 plants. Physiol Plant 116(2):148–154. https://doi.org/10.1034/j.1399-3054.2002.1160203.x

Romanowska E, Pokorska B, Siedlecka M (2005) The effects of oligomycin on content of adenylates in mesophyll protoplasts, chloroplasts and mitochondria from Pb2+ treated pea and barley leaves. Acta Physiol Plant 27(1):29–36. https://doi.org/10.1007/s11738-005-0033-3

Romanowska E, Wróblewska B, Drozak A, Siedlecka M (2006) High light intensity protects photosynthetic apparatus of pea plants against exposure to lead. Plant Physiol Biochem 44(5–6):387–394. https://doi.org/10.1016/j.plaphy.2006.06.003

Rudnick RL, Gao S (2014) Composition of the continental crust. Treatise on Geochemistry (Second Edition) 4:1–51. https://doi.org/10.1016/B978-0-08-095975-7.00301-6

Saito I, Oshima H, Kawamura N, Yamada M (1988) Screening method for determination of high levels of cadmium, lead, and copper in foods by polarized Zeeman atomic absorption spectrometry using discrete nebulization technique. J Assoc Off Anal Chem 71:829–832

Samardakiewicz S, Strawiński P, Woźny A (1996) The influence of lead on callose formation in roots of Lemna minor L. Biol Plantarum 38(3):463–467. https://doi.org/10.1007/bf02896682

Seregin IV, Ivanov VB (2001) Physiological aspects of cadmium and lead toxic effects on higher plants. Russ J Plant Physiol 48(4):606–630. https://doi.org/10.1023/A:1016719901147

Shakerian A, Rahimi E, Ahmadi M (2012) Cadmium and lead content in several brands of rice grains (Oryza sativa) in central Iran. Toxicol Ind Health 28(10):955–960. https://doi.org/10.1177/0748233711430979

Sharma P, Dubey RS (2005) Lead toxicity in plants. Braz J Plant Physiol 17(1):35–52. https://doi.org/10.1590/S1677-04202005000100004

Shaw DM, Cramer JJ, Higgins MD, Truscott MG (1986) Composition of the Canadian Precambrian Shield and the continental crust of the earth. In: Dawson JB, Carswell DA, Hall J, Wedepohl KH (eds) The nature of the lower continental crust, vol 24. Geological Society Special Publication, pp 275–282 https://doi.org/10.1144/gsl.sp.1986.024.01.24

Shimbo S, Zhang Z-W, Watanabe T, Nakatsuka H, Matsuda-Inoguchi N, Higashikawa K, Ikeda M (2001) Cadmium and lead contents in rice and other cereal products in Japan in 1998–2000. Sci Total Environ 281:165–175. https://doi.org/10.1016/S0048-9697(01)00844-0

Shotyk W, Weiss D, Kramers JD, Frei R, Cheburkin AK, Gloor M, Reese S (2001) Geochemistry of the peat bog at Etang de la Gruère, Jura Mountains, Switzerland, and its record of atmospheric Pb and lithogenic trace metals (Sc, Ti, Y, Zr, and REE) since 12,370 14C yr BP. Geochim Cosmochim Acta 65(14):2337–2360. https://doi.org/10.1016/S0016-7037(01)00586-5

Shy CM (1990) Lead in petrol: the mistake of the XXth century. Wld Hlth Statist Quart 43:163

Silvano M, Alessia C, Vincenzo F (2012) Evaluation of the effectiveness of phosphate treatment for the remediation of mine waste soils contaminated with Cd, Cu, Pb, and Zn. Chemosphere 86(4):354–360. https://doi.org/10.1016/j.chemosphere.2011.09.050

Simons TJ (1985) Influence of lead ions on cation permeability in human red cell ghosts. J Membr Biol 84(1):61–71. https://doi.org/10.1007/BF01871648

Simons TJ (1988) Active transport of lead by the calcium pump in human red cell ghosts. J Physiol 405(1):105–113. https://doi.org/10.1113/jphysiol.1988.sp017323

Singh R, Tripathi RD, Dwivedi S, Kumar A, Trivedi PK, Chakrabarty D (2010) Lead bioaccumulation potential of an aquatic macrophyte Najas indica are related to antioxidant system. Bioresour Technol 101(9):3025–3032. https://doi.org/10.1016/j.biortech.2009.12.031

Sjöstedt C, Löv Å, Olivecrona Z, Boye K, Kleja DB (2018) Improved geochemical modeling of lead solubility in contaminated soils by considering colloidal fractions and solid phase EXAFS speciation. Appl Geochem 92:110–120. https://doi.org/10.1016/j.apgeochem.2018.01.014

Smith DD (1978) Lead poisoning and the fall of Rome. J Chem Educ 55(5):334. https://doi.org/10.1021/ed055p334.3

Srikanth R, Ramana D, Rao V (1995) Role of rice and cereal products in dietary cadmium and lead intake among different socio-economic groups in south India. Food Addit Contam 12(5):695–701. https://doi.org/10.1080/02652039509374359

Srivastava RK, Pandey P, Rajpoot R, Rani A, Dubey RS (2014) Cadmium and lead interactive effects on oxidative stress and antioxidative responses in rice seedlings. Protoplasma 251:1047–1065. https://doi.org/10.1007/s00709-014-0614-3

Sun H-f, Li Y-h, Ji Y-f, Yang L-s, Wang W-y, Li H-r (2010) Environmental contamination and health hazard of lead and cadmium around Chatian mercury mining deposit in western Hunan Province, China. T Nonferr Metal Soc 20:308–314. https://doi.org/10.1016/S1003-6326(09)60139-4

Sun G-X, Wang X-J, Hu Q-H (2011) Using stable lead isotopes to trace heavy metal contamination sources in sediments of Xiangjiang and Lishui Rivers in China. Environ Pollut 159(12):3406–3410. https://doi.org/10.1016/j.envpol.2011.08.037

Tabelin CB, Igarashi T, Villacorte-Tabelin M, Park I, Opiso EM, Ito M, Hiroyoshi N (2018) Arsenic, selenium, boron, lead, cadmium, copper, and zinc in naturally contaminated rocks: a review of their sources, modes of enrichment, mechanisms of release, and mitigation strategies. Sci Total Environ 645:1522–1553. https://doi.org/10.1016/j.scitotenv.2018.07.103

Tanton TW, Crowdy SH (1971) The distribution of lead chelate in the transpiration stream of higher plants. Pestic Sci 2:211–213. https://doi.org/10.1002/ps.2780020507

Taylor SR (1964) Trace element abundances and the chondritic earth model. Geochim Cosmochim Acta 28(12):1989–1998. https://doi.org/10.1016/0016-7037(64)90142-5

Thakur S, Singh L, Zularisam AW, Sakinah M, Din MFM (2017) Lead induced oxidative stress and alteration in the activities of antioxidative enzymes in rice shoots. Biol Plantarum 61(3):595–598. https://doi.org/10.1007/s10535-016-0680-9

Tong S, Schirnding YEV, Prapamontol T (2000) Environmental lead exposure: a public health problem of global dimensions. B World Health Organ 78(9):1068–1077

Tripathi RM, Raghunath R, Krishnamoorthy TM (1997) Dietary intake of heavy metals in Bombay city, India. Sci Total Environ 208:149–159. https://doi.org/10.1016/S0048-9697(97)00290-8

Udeigwe TK, Eze PN, Teboh JM, Stietiya MH (2011) Application, chemistry, and environmental implications of contaminant- immobilization amendments on agricultural soil and water quality. Environ Int 37:258–267. https://doi.org/10.1016/j.envint.2010.08.008

Uraguchi S, Tanaka N, Hofmann C, Abiko K, Ohkama-Ohtsu N, Weber M, Kamiya T, Sone Y, Nakamura R, Takanezawa Y, Kiyono M, Fujiwara T, Clemens S (2017) Phytochelatin Synthase has Contrasting effects on Cadmium and Arsenic Accumulation in Rice Grains. Plant Cell Physiol 58(10):1730–1742. https://doi.org/10.1093/pcp/pcx114

US Geological Survey (2019) Mineral commodity summaries 2019: U.S. Geological Survey, 200 p. https://doi.org/10.3133/70202434

Uzu G, Sobanska S, Aliouane Y, Pradere P, Dumat C (2009) Study of lead phytoavailability for atmospheric industrial micronic and sub-micronic particles in relation with lead speciation. Environ Pollut 157(4):1178–1185. https://doi.org/10.1016/j.envpol.2008.09.053

van der Kuijp TJ, Huang L, Cherry CR (2013) Health hazards of China's lead-acid battery industry: a review of its market drivers, production processes, and health impacts. Environ Health 12(1):61. https://doi.org/10.1186/1476-069x-12-61

Verbruggen N, Hermans C, Schat H (2009) Molecular mechanisms of metal hyperaccumulation in plants. New Phytol 181(4):759–776. https://doi.org/10.1111/j.1469-8137.2008.02748.x

Verma S, Dubey RS (2003) Lead toxicity induces lipid peroxidation and alters the activities of antioxidant enzymes in growing rice plants. Plant Sci 164:645–655. https://doi.org/10.1016/s0168-9452(03)00022-0

Wan Y, Huang Q, Camara AY, Wang Q, Li H (2019) Water management impacts on the solubility of Cd, Pb, As, and Cr and their uptake by rice in two contaminated paddy soils. Chemosphere 228:360–369. https://doi.org/10.1016/j.chemosphere.2019.04.133

Wang H, Shan X, Wen B, Owens G, Fang J, Zhang S (2007) Effect of indole-3-acetic acid on lead accumulation in maize (Zea mays L.) seedlings and the relevant antioxidant response. Environ Exp Bot 61(3):246–253. https://doi.org/10.1016/j.envexpbot.2007.06.004

Wang C, Wang J, Yang Z, Mao C, Ji J (2013a) Characteristics of lead geochemistry and the mobility of Pb isotopes in the system of pedogenic rock–pedosphere–irrigated riverwater–cereal–atmosphere from the Yangtze River delta region, China. Chemosphere 93(9):1927–1935. https://doi.org/10.1016/j.chemosphere.2013.06.073

Wang X, Zhou C, Liu G, Dong Z (2013b) Transfer of Metals from Soil to Crops in an Area near a Coal Gangue Pile in the Guqiao Coal Mine, China. Anal Lett 46:1962–1977. https://doi.org/10.1080/00032719.2013.777923

Wang X, Ye ZH, Li B, Huang LN, Meng M, Shi JB, Jiang GB (2014) Growing rice aerobically markedly decreases mercury accumulation by reducing both Hg bioavailability and the production of MeHg. Environ Sci Technol 48(3):1878–1885. https://doi.org/10.1021/es4038929

Wani AL, Ara A, Usmani JA (2015) Lead toxicity: a review. Interdiscip. Toxicol 8(2):55–64. https://doi.org/10.1515/intox-2015-0009

Watanabe T, Nakatsuka H, Ikeda M (1989) Cadmium and lead contents in rice available in various areas of Asia. Sci Total Environ 80:175–184. https://doi.org/10.1016/0048-9697(89)90073-9

Wedepohl KH (1995) The composition of the continental crust. Geochim Cosmochim Acta 59(7):1217–1232. https://doi.org/10.1016/0016-7037(95)00038-2

Wen H, Zhang Y, Cloquet C, Zhu C, Fan H, Luo C (2015) Tracing sources of pollution in soils from the Jinding Pb–Zn mining district in China using cadmium and lead isotopes. Appl Geochem 52:147–154. https://doi.org/10.1016/j.apgeochem.2014.11.025

Whitehead LS, Buchanan SD (2019) Childhood lead poisoning: a perpetual environmental justice issue? J Public Health Man 25:S115–S120. https://doi.org/10.1097/phh.0000000000000891

Williams PN, Lei M, Sun G, Huang Q, Lu Y, Deacon C, Meharg AA, Zhu Y-G (2009) Occurrence and partitioning of cadmium, arsenic and lead in mine impacted paddy rice: Hunan, China. Environ Sci Technol 43(3):637–642. https://doi.org/10.1021/es802412r

Wojas S, Ruszczyńska A, Bulska E, Wojciechowski M, Antosiewicz DM (2007) Ca2+-dependent plant response to Pb2+ is regulated by LCT1. Environ Pollut 147(3):584–592. https://doi.org/10.1016/j.envpol.2006.10.012

Wolnik KA, Fricke FL, Capar SG, Meyer MW, Satzger RD, Bonnin E, Gaston CM (1985) Elements in major raw agricultural crops in the United States. 3. Cadmium, lead, and eleven other elements in carrots, field corn, onions, rice, spinach, and tomatoes. J Agric Food Chem 33:807–811. https://doi.org/10.1021/jf00065a010

Wolz S, Fenske RA, Simcox NJ, Palcisko G, Kissel JC (2003) Residential arsenic and lead levels in an agricultural community with a history of lead arsenate use. Environ Res 93(3):293–300. https://doi.org/10.1016/s0013-9351(03)00064-1

World Health Organization (WHO) (2010) Preventing disease through healthy environments – exposure to lead: a major public health concern. Switzerland, Geneva

World Health Organization (WHO) (2011) Guidelines for drinking-water quality. Chronicle 38:104–108

Wu Y, Xu Y, Zhang J, Hu S, Liu K (2011) Heavy metals pollution and the identification of their sources in soil over Xiaoqinling gold-mining region, Shaanxi, China. Environ Earth Sci 64(6):1585–1592. https://doi.org/10.1007/s12665-010-0833-7

Xia J, Fang Y, Chen Y, Pan Y, Li P, Xue M, Hu Q (2017) Lead speciation analysis in rice by reversed phase chromatography with inductively coupled plasma mass spectrometry. J Food Compos Anal 60:74–80. https://doi.org/10.1016/j.jfca.2017.03.002

Xiao R, Huang Z, Li X, Chen W, Deng Y, Han C (2017) Lime and phosphate amendment can significantly reduce uptake of Cd and Pb by field-grown rice. Sustain-Basel 9(3):430(1–10). https://doi.org/10.3390/su9030430

Xiong Z, Zhao F, Li M (2006) Lead toxicity in Brassica pekinensis Rupr.: effect on nitrate assimilation and growth. Environ Toxicol 21(2):147–153. https://doi.org/10.1002/tox.20167

Xu X, Cao X, Zhao L, Zhou H, Luo Q (2014) Interaction of organic and inorganic fractions of biochar with Pb(II) ion: further elucidation of mechanisms for Pb(II) removal by biochar. RSC Adv 4(85):44930–44937. https://doi.org/10.1039/c4ra07303g

Xue S, Shi L, Wu C, Wu H, Qin Y, Pan W, Hartley W, Cui M (2017) Cadmium, lead, and arsenic contamination in paddy soils of a mining area and their exposure effects on human HEPG2 and keratinocyte cell-lines. Environ Res 156:23–30. https://doi.org/10.1016/j.envres.2017.03.014

Yamazaki S, Ueda Y, Mukai A, Ochiai K, Matoh T (2018) Rice phytochelatin synthases OsPCS1 and OsPCS2 make different contributions to cadmium and arsenic tolerance. Plant Direct 2(1):e00034

Yang Y-Y, Jung J-Y, Song W-Y, Suh H-S, Lee Y (2000) Identification of Rice varieties with High Tolerance or sensitivity to lead and characterization of the mechanism of Tolerance. Plant Physiol 124:1019–1026. https://doi.org/10.1104/pp.124.3.1019

Yang QW, Shu WS, Qiu JW, Wang HB, Lan CY (2004) Lead in paddy soils and rice plants and its potential health risk around Lechang Lead/Zinc Mine, Guangdong, China. Environ Int 30(7):883–889. https://doi.org/10.1016/j.envint.2004.02.002

Yang Y, Li S, Bi X, Wu P, Liu T, Li F, Liu C (2010) Lead, Zn, and Cd in slags, stream sediments, and soils in an abandoned Zn smelting region, southwest of China, and Pb and S isotopes as source tracers. J Soil Sediment 10(8):1527–1539. https://doi.org/10.1007/s11368-010-0253-z

Yang X, Liu J, McGrouther K, Huang H, Lu K, Guo X, He L, Lin X, Che L, Ye Z, Wang H (2016) Effect of biochar on the extractability of heavy metals (Cd, Cu, Pb, and Zn) and enzyme activity in soil. Environ Sci Pollut Res 23:974–984. https://doi.org/10.1007/s11356-015-4233-0

Yanqun Z, Yuan L, Jianjun C, Haiyan C, Li Q, Schvartz C (2005) Hyperaccumulation of Pb, Zn and Cd in herbaceous grown on lead–zinc mining area in Yunnan, China. Environ Int 31(5):755–762. https://doi.org/10.1016/j.envint.2005.02.004

Ye XX, Wang GZ, Zhang YX, Zhao HJ (2018) Hydroxyapatite nanoparticles in root cells: reducing the mobility and toxicity of Pb in rice. Environ. Sci.: Nano 5(2):398–407

Yokel J, Delistraty DA (2003) Arsenic, lead, and other trace elements in soils contaminated with pesticide residues at the Hanford site (USA). Environ Toxicol 18(2):104–114. https://doi.org/10.1002/tox.10106

Yousefi N, Meserghani M, Bahrami H, Mahvi AH (2016) Assessment of human health risk for heavy metals in imported rice and its daily intake in Iran. Res J Environ Toxicol 10:75–81. https://doi.org/10.3923/rjet.2016.75.81

Yu Y, Li Q, Wang H, Wang B, Lu Q, Yan Z, Ding A (2016) Exposure risk of young population to lead: a case study in Le'an River Basin in Jiangxi Province, China. Environ Pollut 209:140–146. https://doi.org/10.1016/j.envpol.2015.11.017

Yuan H, Lu T, Huang H, Zhao D, Kobayashi N, Chen Y (2015) Influence of pyrolysis temperature on physical and chemical properties of biochar made from sewage sludge. J Anal Appl Pyrolysis 112:284–289

Zeng F, Wei W, Li M, Huang R, Yang F, Duan Y (2015) Heavy metal contamination in Rice-Producing Soils of Hunan Province, China and Potential Health Risks. Int J Env Res Pub He 12:15584–15593. https://doi.org/10.3390/ijerph121215005

Zeng G, Wan J, Huang D, Hu L, Huang C, Cheng M, Xue W, Gong X, Wang R, Jiang D (2017) Precipitation, adsorption and rhizosphere effect: The mechanisms for Phosphate-induced Pb immobilization in soils—A review. J Hazard Mater 339:354–367

Zhai CK, Tang WL, Jang XL, Lorenz KJ (1996) Studies of the safety of chinese wild rice. Food Chem Toxicol 34:347–352. https://doi.org/10.1016/0278-6915(96)00117-2

Zhang ZW, Moon CS, Watanabe T, Shimbo S, Ikeda M (1996) Lead content of rice collected from various areas in the world. Sci Total Environ 191:169–175. https://doi.org/10.1016/0048-9697(96)05257-6

Zhang Z-W, Moon C-S, Watanabe T, Shimbo S, Ikeda M (1997) Contents of pollutant and nutrient elements in rice and wheat grown on the neighboring fields. Biol Trace Elem Res 57(1):39–50. https://doi.org/10.1007/bf02803868

Zhang Z-W, Watanabe T, Shimbo S, Higashikawa K, Ikeda M (1998) Lead and cadmium contents in cereals and pulses in north-eastern China. Sci Total Environ 220:137–145. https://doi.org/10.1016/S0048-9697(98)00252-6

Zhang XY, Lin FF, Wong MTF, Feng XL, Wang K (2008) Identification of soil heavy metal sources from anthropogenic activities and pollution assessment of Fuyang County, China. Environ Monit Assess 154(1–4):439–449. https://doi.org/10.1007/s10661-008-0410-7

Zhang R, Liu G, Wu N, Gu M, Zeng H, Zhu Y, Xu G (2011) Adaptation of plasma membrane H+ ATPase and H+ pump to P deficiency in rice roots. Plant and Soil 349:3–11. https://doi.org/10.1007/s11104-011-0774-2

Zhang X, Yang L, Li Y, Li H, Wang W, Ye B (2012) Impacts of lead/zinc mining and smelting on the environment and human health in China. Environ Monit Assess 184(4):2261–2273. https://doi.org/10.1007/s10661-011-2115-6

Zhang X, Zhong T, Liu L, Ouyang X (2015) Impact of soil heavy metal pollution on Food Safety in China. Plos One 10(8):1–14. https://doi.org/10.1371/journal.pone.0135182

Zhao H, Xia B, Fan C, Zhao P, Shen S (2012) Human health risk from soil heavy metal contamination under different land uses near Dabaoshan Mine, Southern China. Sci Total Environ 417–418:45–54. https://doi.org/10.1016/j.scitotenv.2011.12.047

Zheng N, Wang Q, Zhang X, Zheng D, Zhang Z, Zhang S (2007) Population health risk due to dietary intake of heavy metals in the industrial area of Huludao city, China. Sci Total Environ 387(1–3):96–104. https://doi.org/10.1016/j.scitotenv.2007.07.044

Zheng R-L, Cai C, Liang J-H, Huang Q, Chen Z, Huang Y-Z, Arp HPH, Sun G-X (2012) The effects of biochars from rice residue on the formation of iron plaque and the accumulation of Cd, Zn, Pb, As in rice (Oryza sativa L.) seedlings. Chemosphere 89(7):856–862. https://doi.org/10.1016/j.chemosphere.2012.05.008

Zhou J-M, Dang Z, Cai M-F, Liu C-Q (2007) Soil heavy metal pollution around the Dabaoshan Mine, Guangdong Province, China. Pedosphere 17(5):588–594. https://doi.org/10.1016/S1002-0160(07)60069-1

Zhou H, Zhou X, Zeng M, Liao BH, Liu L, Yang WT, Wu YM, Qiu QY, Wang YJ (2014) Effects of combined amendments on heavy metal accumulation in rice (Oryza sativa L.) planted on contaminated paddy soil. Ecotoxicol Environ Saf 101:226–232. https://doi.org/10.1016/j.ecoenv.2014.01.001

Zhou G, Xia X, Wang H, Li L, Wang G, Zheng S, Liao S (2016) Immobilization of Lead by Alishewanella sp. WH16-1 in Pot Experiments of Pb-Contaminated Paddy Soil. Water Air Soil Pollut 227(9):339. https://doi.org/10.1007/s11270-016-3040-7

Zhuang P, McBride MB, Xia H, Li N, Li Z (2009a) Health risk from heavy metals via consumption of food crops in the vicinity of Dabaoshan mine, South China. Sci Total Environ 407(5):1551–1561. https://doi.org/10.1016/j.scitotenv.2008.10.061

Zhuang P, Zou B, Li NY, Li ZA (2009b) Heavy metal contamination in soils and food crops around Dabaoshan mine in Guangdong, China: implication for human health. Environ Geochem Hlth 31(6):707–715. https://doi.org/10.1007/s10653-009-9248-3

Zhuang P, Li Z-A, McBride MB, Zou B, Wang G (2013) Health risk assessment for consumption of fish originating from ponds near Dabaoshan mine, South China. Environ Sci Pollut Res 20(8):5844–5854. https://doi.org/10.1007/s11356-013-1606-0

Part III
Rice and Health

Chapter 6
Golden Rice

Swapan K. Datta and Karabi Datta

6.1 Introduction

Rice (*Oryza sativa* L.), a food crop, is consumed by more than 3 billion people of the world as a staple food that lacks adequate nutrition. Globally, malnutrition is a serious problem causing millions of death, under-development in children, and poor health of adult men and women (Datta and Bouis 2000). So far, our intensive screening of exotic rice germplasm has shown the absence of β-carotene in polished seeds (Tan et al. 2004).

All green tissues of plants, including rice plants, possess carotenoids in photosynthetic tissues but not in the edible part, the endosperm. The genes for key enzymes in β-carotene (provitamin-A) biosynthesis pathway, *phytoene synthase* (*psy*), and *phytoene desaturase* (*crt*I) were isolated and characterized from daffodil (*Narcissus pseudonarcissus*) and the bacteria (*Erwinia uredovora*), respectively (Burkhardt et al. 1997; Misawa et al. 1990, 1994; Sandmann 1994; Scheldz et al. 1996; Hirschenberg 2001).

Genetic engineering of metabolic pathway for β-carotene biosynthesis in endosperm was demonstrated in *japonica* type model rice cultivar, and the transgenic rice produced 1.6 μg/g total carotenoids (Ye et al. 2000). The *indica* rice variety is consumed by about 90% of the world's rice-consuming population (Khush 2001). Hence, the emphasis was given to develop Golden *indica* rice and its further improvement (Datta 1999; Datta et al. 2003, 2006, 2007, 2014; Parkhi et al. 2005; Rai et al. 2007; Rehana et al. 2018; Ghosh et al. 2019). The improvement of carotenoids accumulation in rice endosperm has been reported (Paine et al. 2005; Bai et al. 2016).

S. K. Datta (✉) · K. Datta
University of Calcutta, Laboratory of Translational Research on Transgenic Crops, Kolkata, India
e-mail: skdbot@caluniv.ac.in

A. Costa de Oliveira et al. (eds.), *The Future of Rice Demand: Quality Beyond Productivity*, https://doi.org/10.1007/978-3-030-37510-2_6

The present review will cover the development, breeding, expression of carotenoids in introgressed lines; the stability of carotenoids in Golden *indica* rice lines; and performance of Golden Rice in limited field conditions, human feeding experiments, and future direction.

6.2 Golden Rice (Background)

Vitamin A plays an important role in many physiological functions in human health. The biosynthesis of carotenoids in plants takes place within the plastids, chloroplasts of photosynthetic tissues, and chromoplasts of fruits and flowers. Chlorophyll, tocopherols, plastoquinone, phylloquinone, gibberellins, and carotenoids share a common biosynthetic precursor, geranylgeranyldiphosphate (GGPP). The genes necessary for the four enzymes (phytoene synthase, phytoene desaturase, z-carotene desaturase, and lycopene cyclase) catalyze to complete the pathway toward β-carotene (provitamin-A) biosynthesis from GGPP (Misawa et al. 1990) (Fig.6.1a). Conventional interventions (distribution, fortification, diet diversification) have been helpful in defeating vitamin-A deficiency (VAD) but are not effective enough. Conventional plant breeding to alter, modify, or introduce this biosynthetic machinery into the target endosperm tissues in rice has been impossible as of now, as no endosperm-active carotenoid biosynthetic genes have been found thus far in the available rice gene pool (Tan et al. 2004). Thus, the transgenic approach offers the most realistic way to improve rice by incorporating genes without any species barrier into rice and showing β-carotene expression in rice seeds (Ye et al. 2000; Datta et al. 2003; Rehana et al. 2018).

Considering the problems of malnutrition and particularly VAD, a project proposal on developing engineered rice with β-carotene biosynthesis based on the pioneering publication of stable transgenic *indica* rice (Datta et al. 1990a) was considered by the Rockefeller Foundation, USA to support financial grant to the ETH-Zurich, Ins. of Plant Sciences headed by Prof. Ingo Potrykus. During 1987–1993, a group of scientists started working on trait improvement of crop plants using microinjection/microculture of young zygotic and microspore-derived embryos. Dr Ingo Potrykus stated the background story: "*After many years of enormous effort, we found much to our disappointment that the microinjection system worked fine with protoplasts, but was extremely bad with cells surrounded by a cell wall...Swapan Datta asked for permission to work part-time (over the weekends) on the embryogenic protoplast transformation approach. And the Dattas (*Swapan K Datta and Karabi Datta*) made it*" (Potrykus 2001, http://www.agbioworld.org/biotech-info/topics/goldenrice/tale.html).

There came the first transgenic *indica* rice (Datta et al. 1990b, 1992) all from embryogenic protoplasts and direct gene transfer. During a brainstorming session in New York (1992), organized by the Rockefeller Foundation (organized by Dr.

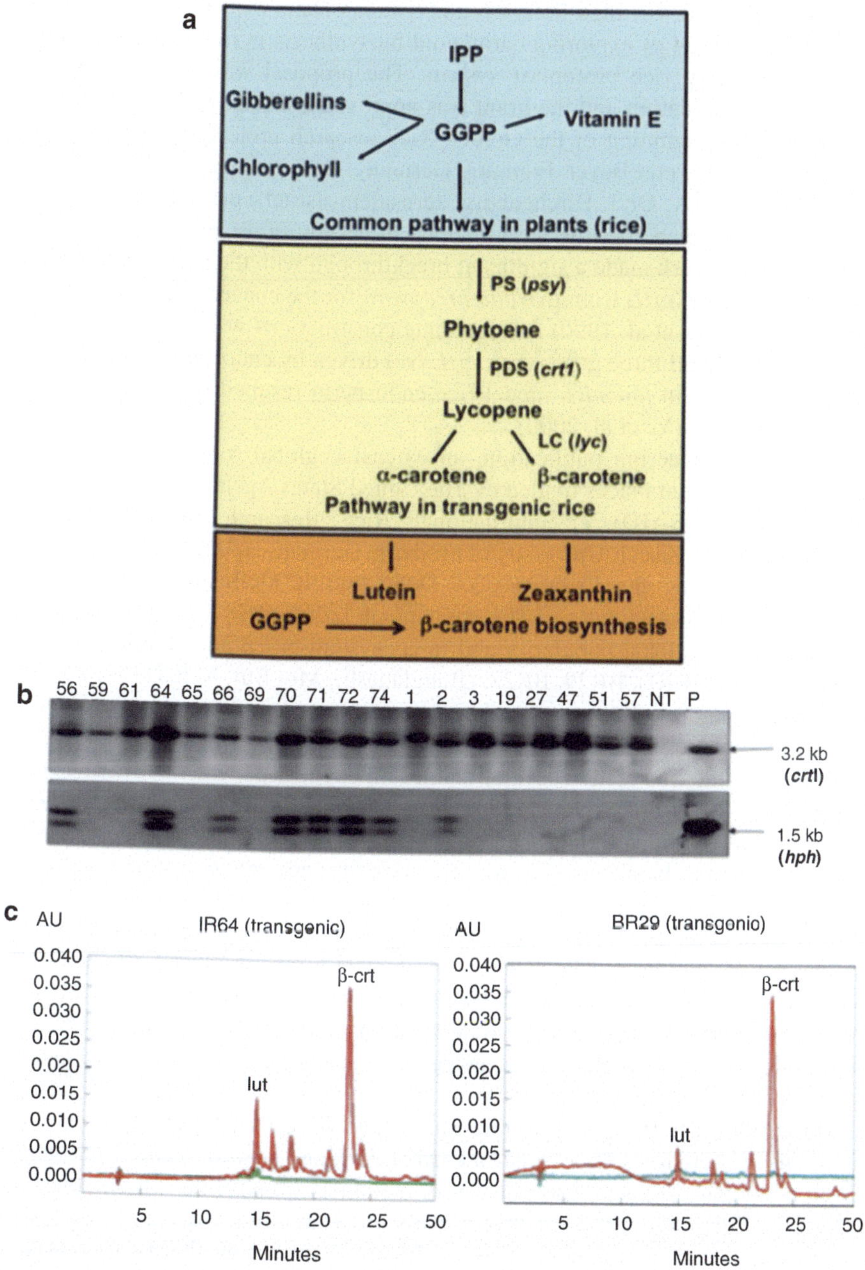

Fig. 6.1 Schematic figure showing carotenoid biosynthetic pathways (**a**); Southern blot showing the segregation pattern of gene of interest (*crtI*) and marker gene (*hph*) (Source: Parkhi et al. 2005) (**b**); HPLC profile showing carotenoid levels in transgenic Golden Rice line (Source: Datta et al. 2007) (**c**)

Gary Toenniessen), Dr. Ingo Potrykus and Swapan Datta requested for consideration of the project of exploring carotenoid biosynthesis in rice seed by genetic engineering using rice protoplast system. The proposal was accepted by the Rockefeller Foundation and the grant was approved considering its importance, which was the beginning of the Golden Rice research project. Several international groups (Dr Peter Bayer, Freiburg, Germany; Dr E.T. Wurtzel, NY, USA; Dr Li Li, Ithaca, USA, Dr J. Hirchenburg, Jerusalem, Israel), many concept notes, and review articles across the world helped in designing the research approach. Miswa's group work made a significant breakthrough with the publication showing only one gene (*crtI*) from *Erwinia uredovora* for the conversion of phytone to lycopene (Misawa et al. 1990) and the same construct *crtI* was available for this work and finally all three genes (*psy, crtI, lyc*) driven by endosperm specific promoters expressed in *japonica* model rice endosperm resulted in the yellow-colored Golden Rice (Ye et al. 2000).

After this pioneering publication, an extensive global research project was launched at different places funded by The United States Agency for International Development (USAID) at International Rice Research Institute (IRRI), Philippines, ETH-Zurich/University of Freiburg, Germany, and several Institutions in India, including the University of Delhi and ICAR-Institutes funded by Department of Biotechnology (Government of India). The rice cultivars were selected based on local popularity and in ecosystem-based adaptability, such as IR64, IR68144 (IRRI), BR29, BR28, (Bangladesh), Mot Bui, NHCD (Myanmar), and Swarna (India) and were used for genetic transformation for developing Golden *indica* rice.

Anther culture was used to develop stable homozygous genetic lines (Datta et al. 1990b, 2006, 2014; Baisakh et al. 2001; Datta 2005). Rice transformation methods (biolistic and *Agrobacterium)*, plant regeneration, positive plant (event) selection, and molecular and biochemical characterization of transgenic rice were conducted as described earlier (Datta and Bouis 2000; Datta et al. 2003; Vasconcelos et al. 2003; Parkhi et al. 2005; Rai et al. 2003). Two key genes, *psy* and *crt*I, driven by endosperm-specific promoters (*Glutelin* promoter) were introduced effectively to synthesize β-carotene in the target endosperm tissue. Embryogenic calli derived from scutellum of immature embryos were used as explants for gene transformation. Molecular analysis showed the evidence of stable integration of genes and its expression in edible endosperm tissue.

Three other reports are available on Golden Rice development. Hoa et al. (2003) used Vietnamese rice materials (MTL250) and a single gene construct to develop golden rice via Agrobacterium-mediated transformation. In another report, the Syngenta group enhanced levels of carotenoids with modified genes (from maize) in non-commercial rice cultivars (Paine et al. 2005). Recently, a combination of Arabidopsis thaliana genes AtDXS / AtOR with maize phytoene synthase gene (ZmPSY1) and bacterial desaturase gene (PaCRTI) in the presence of endogenous lycopene β-cyclase was reported to support the accumulation of significantly high levels of carotenoids in rice endosperm (Bai et al. 2016). The Syngenta-developed Golden Rice lines are known and referred to (here) as Golden Rice-2 (GR-2) and used as donor parents in different breeding programs.

The high levels of carotenoids in some Golden Rice lines degrade sharply during storage. Lipoxygenase (LOX) catalyzes lipid peroxidation of seed and causes seed quality deterioration. LOX activity reduces nutrition quality of seed by the process called "co-oxidation" because it produces hydroperoxy fatty acids which co-oxidize and decolorize the carotenoids (Gayen et al. 2015). To ameliorate the degradation of nutrition quality in Golden Rice seed, RNAi LOX driven by the Oleosin-18 promoter (aleurone specific promoter) was introduced in Golden Rice. This resulted in significant reduction in carotenoids degradation and supported further stabilization of Golden Rice (Gayen et al. 2014, 2015, 2016).

6.2.1 Development of Homozygous Lines

Golden Rice lines with the stable level of carotenoids are highly desirable. The development of homozygous lines is a primary prerequisite for the selection of dihaploid (DH) lines and for use in cross-breeding programs to transfer the traits or gene of interest (in the case of transgenic materials) into other backgrounds (Datta 2005). The selection of homozygous lines using transgene/s requires more than three generations, depending on the copy number and position of the inserted genes. The anther culture technique reduces the breeding cycle and rapidly fixes homozygosity via the production of homozygous dihaploid plants (Datta et al. 1990a).

Several factors, such as the quality of the donor plants, genotypic variations, media composition, and the handling of the cultures, play a crucial role in plant regeneration from anther cultures (Datta and Wenzel 1987; Datta et al. 1990b; Datta 2005). Dihaploid (homozygous) transgenic plants can be obtained in the following ways: (1) development of anther culture–derived dihaploid (pollen-derived) embryogenic cell lines that can be used for transformation, i.e., dihaploid cell lines are developed first; and (2) development of transgenic plants (T_1) with transgene of interest and development of microspore derived next generation plants (T_2) eventually which would be homozygous (dihaploid) transgenic plants. As a case study, dihaploid Golden Rice plants in addition to haploid and tetraploid plant were generated from anther cultures of these primary transgenic plants (Datta et al. 2014). Molecular characterization of the introduced transgenes (*psy, crtI, lyc*) and HPLC analysis confirmed the events of transgenic lines with levels of carotenoids including β-carotene in Golden Rice plants (Datta et al. 2014).

6.2.2 Introgression Breeding and Phenotyping of Golden Rice

Taipei 309 (T309) *japonica* Golden Rice when crossed with IR64 (*indica*) produced only a few seeds in BC_1F_1 with subsequent generation of backcrossing, the restitution of IR64 became prominent with better fertility in BC_2F_2 progenies. A few near isogenic lines (NILs) were identical to IR64 in most agronomic features (Baisakh et al. 2006). A similar approach has been utilized at IRRI, Philippines in converting GR-2 lines to BR29, IR64, Swarna. A similar type of work has been conducted at ICAR-Indian Agricultural Research Institute, New Delhi but discontinued further due to instability of the transgene function (Dr A.K. Singh, IARI, personal communication). However, it is reported that a few BR29 lines derived from the introgression of GR-2 line has been successfully grown in the field of Bangladesh. The earlier study with homozygous Golden Rice BR29, IR64, and control plants showed similar agronomic characteristics. The study was compared with Golden Rice IR64 with the control using 12 universal rice primers (URPs from SRILS Uniprimer™ Kit, Seolin Sci. Co. Ltd. Korea).

Another Golden Rice line, NHCD reported earlier was also included (Datta et al. 2003). Polymorphic bands were observed between IR64 and NHCD (for both transgenic and control). No polymorphism was found between the transgenic and control of both IR64 and NHCD. This study indicated that the intact and/or rearranged copies of the transgenes are clustered in a limited part of the genome, which could not be detected by these RAPD markers derived from the repetitive sequence of rice (Rai et al. 2003). As desirable from breeding perception, efforts have been made to select a transgenic event with a simple and single copy transgene for a Mendelian 3:1 segregation pattern and eventually to develop homozygous lines with acceptable agronomic performance. In addition, Anther culture–derived homozygous lines showed better gene stability and stable expression of the carotenoids in the progeny and selected for further breeding program (Datta et al. 2007).

6.2.3 Inheritance and Possible Transgeneration Carotenoid Levels in Subsequent Progenies

To study the inheritance and transgenerational of carotenoids, T_1 progenies from each selected line were grown to study the inheritance pattern of the integrated genes and their differential carotenoid expression level in the seeds. Molecular analysis of individual plants of each line showed single-locus Mendelian segregation (3:1) along with variable segregation pattern. This suggests the possibility of transgenes insertion in one or more than one locus (Fig 7.1b). A simple gene integration pattern at T_0 is the best selection choice to continue the subsequent progeny for attaining the homozygous material as well as the possibility of marker-free transgenic line development (Parkhi et al. 2005) (Fig. 6.1b).

After harvesting, the dried seeds of individual advanced progeny (T_2/T_3) of each line were polished to assess visual yellow color. The spectrophotometer and HPLC analysis showed a wide variation in the amount of carotenoid profiling in seeds (T_2) of individual progeny (T_1) of each line (Fig. 6.1c).

Interestingly, enhanced carotenoid levels were observed in many of the T_2 seeds when compared with their respective T_1 seeds. Such enhanced expression could be attributed due to the post-translational positive effect on carotenoids biosynthesis in rice seeds (Datta et al. 2006). Transgenic plant contained total carotenoids 9.34 μg/g of polished seeds and 3.92 μg/g of β-carotene, the highest value obtained from BR29 (Datta et al. 2006). Earlier papers reported the amount of carotenoids in transgenic rice were 1.6 μg/g in a *japonica* rice (Ye et al. 2000) and 1.05 μg/g in an *indica* Golden Rice (Datta et al. 2003).

About 4–5 μg/g carotenoids may provide more than 60% of the RDA (Recommended Daily Allowances) by the ICMR/WHO (B. Sivakumar, NIN, Hyderabad, personal communication; Sivakumar 1998). The carotenoid profile of the transgenic seeds is shown to contain lutein, β-cryptoxanthin, and α-carotene (Datta et al. 2006). Enhanced levels of carotenoids (37.0 μg/g) and (31.78 μg/g) were reported by Paine et al. 2005 and Bai et al. 2016, respectively. A few selected articles on Golden Rice development have been included in Table 6.1.

6.2.4 Golden Rice, Evaluation in Screenhouse and in Field: Phase I

BR29 is a well-adapted commercial rice cultivar in Bangladesh and is also gaining popularity in West Bengal, India. This cultivar has been evaluated and accepted successfully in several Asian countries through the International Network for Genetic Evaluation of Rice (INGER) and is also being adapted in some countries including India. Two BR29 Golden Rice lines (Datta et al. 2006) were transferred to Bangladesh for field evaluation. DH Golden Rice (BR29) plants grown along with control BR29 in the containment greenhouse were found similar without any phenotypic cost and showed no segregation in yellow color seed containing carotenoids and with haploid sterile plants as expected (Fig. 6.2a).

Golden Rice was found similar in different agronomic characters such as plant height, days to flowering, seed set, 100-seed weight, and total biomass as compared to similar control varieties grown side by side. Similarly, Golden Rice (IR64) and control IR64 grown in the containment greenhouse facility at IRRI, Philippines showed substantial equivalence (Fig. 6.2b). Another experiment of field evaluation at AgCenter Rice Research Station at Louisiana, USA with Golden Rice (US variety) developed by Syngenta showed encouraging results. However, it did not move further perhaps due to non-availability of regulatory approval and the study was not conclusive.

Table 6.1 A few selected publications indicating genes, source, and products related to Golden Rice development

Sl. No.	Genes involved	Rice variety	Result	References
1.	Daffodil (*Narcissus pseudonarcissus*) phytoene synthase, *psy* gene	Japonica Taipei 309 variety	Formation of carotenoid-specific intermediate phytoene	Burkhardt et al. (1997)
2.	Phytoene synthase (*psy*) and *β*–cyclase (*lcy*) from *Narcissus pseudonarcissus* and bacterial (*Erwinia uredovora*) phytoene desaturase (*CRTI*)	Japonica rice cultivarTP 309	Total carotenoids −1.6 μg/g	Ye et al. (2000)
3.	Phytoene synthase (*psy*) from daffodil, with seed-specific glutelin promoter (Gt-1 P), bacterial phytoene desaturase (*crtI*) fused to the transit peptide sequence of the pea-Rubisco small subunit were driven by the constitutive promoter CaMV35S P and lycopene β–cyclase (*lcy*)	IR 64, BR 29 Nang Hong Cho Dao and Mot Bui	Total carotenoids −1.05 μg/g in T_1 seeds in Nang Hong Cho Dao	Datta et al. (2003)
4.	Phytoene synthase (*psy*) from *Narcissus pseudonarcissus* and bacterial phytoene desaturase (*crt*I)	Japonica rice cultivar (Taipei 309) and Indica rice cultivars (IR64 and MTL 250)	Total carotenoids in T_2 grains of Taipei 309 and IR64 are 1.2 μg/g and 0.8 μg/g	Hoa et al. (2003)
5.	Maize phytoene synthase (*Zmpsy*) and bacterial (*Erwinia uredovora*) phytoene desaturase (*crt*I) gene	Asanohikari rice cultivar	Total carotenoids −37 μg/g	Paine et al. (2005)
6.	Introgression line with phytoene synthase (*psy*) and phytoene desaturase (*crt*I) and lycopene β–cyclase (*lcy*)	Indica rice cultivar IR64	Total carotenoids −1.06 μg/g	Baisakh et al. (2006)
7.	Phytoene synthase (*psy*) and phytoene desaturase (*crt*I)	Indica rice cultivars (IR 64 and BR 29)	Total carotenoids in BR 29 and IR 64 are 9.34 μg/g and 2.32 μg/g	Datta et al. (2006)
8.	Phytoene synthase (*psy*) and phytoene desaturase (*crt*I)	Indica rice cultivar	Total carotenoids −6.77 μg/g	Rai et al. (2007)
9.	Anther culture–derived line with phytoene synthase (*psy*) *Narcissus pseudonarcissus* and phytoene desaturase (*crt*I) *Erwinia uredovora*	Indica rice cultivar BR 29	Total carotenoids −3.188 μg/g	Datta et al. (2014)
10.	(a) co-expressed *AtDXS*, *ZmPSY*1, and *PaCRTI* (b) co-expressed *AtOR*, *ZmPSY*1, and *PaCRTI*	Wild-type rice (*Oryza sativa* L. cv. EYI105)	(a)Total carotenoids −31.78 μg/g (b) Total carotenoids −25.83 μg/g	Bai et al. (2016)

Fig. 6.2 Anther culture–derived homozygous double haploid (DH) Golden Rice plants (left), haploid plants (right), and non-segregating DH Golden Rice seeds (top): (Source: Datta et al. 2007) (**a**); Transgenic Golden Rice IR64 and Control IR64 plants showing phenotypically similar appearance grown in containment greenhouse facility (**b**)

6.2.5 Golden Rice, Freedom to Operate (FTO) and Potential Use: Phase II

Syngenta Golden Rice, GR2 (Paine et al. 2005): Six transgenic lines developed in *javanica* Kaybonnet background, and their respective null lines were hybridized with the genetic background of a mega rice variety “Swarna,” through marker-assisted backcross breeding. Homozygous Golden Rice lines showed surprising phenotypes of dwarf with pale green leaves and drastically reduced panicle size, grain number, and yield as compared to the recurrent parent, Swarna (Bollinedi et al. 2017). It evidently showed that the transgene disrupted *OsAux1* (native gene), resulting in disbalance of plant growth regulators, *viz.*, auxins, gibberellic acid, and abscisic acid, leading to the abnormalities in the growth and development of the lines homozygous for the transgene (Bollinedi et al. 2017).

Commercial rights of Golden Rice remained with Syngenta although there was an understanding and agreement with the owners of Intellectual Property Rights (IPR) of the technologies involved with the GR that freedom to operate would be available for the use of GR. Hopefully, with the support from the donors of IPR (Intellectual Property Rights), Syngenta, and the Funding agencies including the Melinda and Gates Foundation, FTO for the Golden Rice use in India, Bangladesh, and at the Southeast Asia could expedite the research, breeding, and its availability for the consumers.

6.2.6 Golden Rice, Feeding Trials in China and Evaluation: Phase III

Each group of two dozen or so children (aged 6–8) was given meals containing Golden Rice, spinach, or β-carotene capsules for lunch every week during the three-week trial (Qiu 2012; Tang 2013) which created controversy as adequate regulatory permission and ethics principles were not followed as reported in Nature (Qiu 2012).

6.2.7 Bioavailability and Future Study

Vitamin A is immensely required for the benefit of human health. The bioavailability of β-carotene present in Golden Rice is well accepted and conversion to retinol may vary due to different factors elaborated in the recent review (Dubock 2019) which may require further study.

β-carotene-enriched rice helps reduce VAD and protect against iron deficiency anemia (IDA). A few recent publications clearly showed that the high bioconversion efficiency of Golden Rice β-carotene to vitamin A shows that Golden Rice can be used as a source of vitamin A (Tang 2013, 2015). A dream rice biofortified with β-carotene, iron, zinc, and lysine along with higher yield and superior agronomic characters may support the food health and livelihood of the millions of people who need them most.

6.3 Conclusions

Golden Rice rich in provitamin-A may serve better life for human. It took long time since its perception at ETH-Zurich in 1992 led by Dr Ingo Potrykus when the authors (SKD & KD) established rice genetic transformation system at ETH-lab., Zurich and moved through IRRI, Philippines, Syngenta-USA, IARI-Delhi, University of Calcutta, India, BRRI-Bangladesh, and many laboratories including Children feeding trials in China. Phenotypically stable Golden Rice is safe (Owens 2018) and must be available to the people who need them most (Datta et al. 2007; Alberts et al. 2013) and it does not have additional cost except standard plant breeding system (Stein et al. 2006).

Recently in 2018, the Food Standards Australia New Zealand, the Health Canada, the United States Food and Drug Administration, and renowned international food safety regulatory agencies recommended 'Golden Rice' with a positive feedback on it. Regulatory package, selection of right event, and desirable cultivars could expedite further move for its utilization. However, freedom to operate (FTO) must be ensured and the compliance and management of the regulatory system need to be left to the country concerned for better utilization of Golden Rice.

Acknowledgments We thank the Department of Biotechnology (DBT) for financial and research support at the University of Calcutta and Dr Ingo Potrykus (Former, ETH-Professor) and Prof. Peter Beyer (Univ of Freiurg, Germany) for providing us with the initial support to begin with the Golden Rice project. We gratefully thank Dr. M Mahiul Haque from Bangladesh for selecting the BR29 lines and its availability for developing Golden BR29 line; IRRI (Dr Gurdev S. Khush) and a few research scholars/scientists (Dr Nilanjan Baisakh, Dr. Vilas Parkhi, Dr Mayank Rai, Dr Sayeda Rehana, Dr Md. Alamgir, Dr Krishnan, Dr Anindya Bandyopadhyay, Dr Khalekuzzaman, Dr Moumita Ganguly, Editha Abrigo, Norman Oliva, Lina Torrizo, Dr Gayetri Sahoo and Shuvobrata Majumder) significantly contributed in developing and expanding Golden Rice research. We are grateful to The Rockefeller Foundation, USA for funding support to research and particularly Dr Gary Toenniessen of Rockefeller Foundation who contributed significantly for the successful global project on the Rice Biotechnology program and also supporting the grant for Golden Rice project to ETH-Zurich and IRRI, Philippines along with USAID, USA.

References

Alberts B, Beachy R, Baulcombe D, Blobel G, Datta SK, Federoff N, Kennedy D, Khush G, Peacock J, Rees M, Sharp P (2013) Standing up for GMOs. Science 341:1320

Bai C, Capell T, Berman J, Medina V, Sandmann G, Christou P, Zhu C (2016) Bottelnecks in carotenoid biosynthesisand accumulation in rice endosperm are influenced by the precursor-product balance. Plant Biotechnol J 14:195–205

Baisakh N, Datta K, Rai M, Rehana S, Beyer P, Potrykus I, Datta SK (2001) Development of dihaploid transgenic "golden rice" homozygous for genes involved in the metabolic pathway of ß-carotene biosynthesis. Rice Gene Newslett 18:91–94

Baisakh N, Rehana S, Rai M, Oliva N, Tan J, Mackill D, Khush GS, Datta K, Datta SK (2006) Marker-free transgenic (MFT) near-isogenic introgression lines (NILs) of 'golden' indica rice (cv IR64) with accumulation of provitamina in the endosperm tissue. Plant Biotechnol J 4:467–475

Bollinedi H, Krishnan GS, Prabhu KV, Singh NK, Mishra S, Khurana JP (2017) Molecular and functional characterization of GR2-R1 event based backcross derived lines of Golden Rice in the genetic background of a mega rice variety Swarna. PLoS One 12(1):e0169600

Burkhardt PK, Beyer P, Wunn J, Kloti A, Armstrong GA, Schledz M, von LJ, Potrykus I (1997) Transgenic rice (*Oryza sativa*) endosperm expressing daffodil (*Narcissus pseudonarcissus*) phytoene synthase, a key intermediate of provitamin A biosynthesis. Plant J 11:1071–1078

Datta SK (1999) Transgenic Cereals: *Oryza sativa* (rice). In: Vasil I (ed) Molecular improvement of cereal crops. Springer, Dordrecht, pp 149–187

Datta SK (2005) Androgenic haploids: factors controlling development and its application in crop improvement. Curr Sci 89:1870–1878

Datta SK, Bouis HE (2000) Application of biotechnology to improving the nutritional quality rice. Food Nutr Bull 21:451–456

Datta SK, Wenzel G (1987) Isolated microspore derived plant formation via embryogenesis in *Triticum aestivum*. Plant Sci 48:49–54

Datta SK, Peterhans A, Datta K, Paskowski J, Potrykus I (1990a) Genetically engineered indica rice recovered from protoplasts. Bio/Technol 8:736–740

Datta SK, Datta K, Potrykus I (1990b) Embryogenesis and plant regeneration from microspores of both *Indica* and *Japonica* rice (*Oryza sativa*). Plant Sci 67:83–88

Datta SK, Datta K, Soltanifar N, Donn G, Potrykus I (1992) Herbicide resistant indica rice plants from IRRI breeding line IR72 after PEG mediated transformation of protoplasts. Plant Mol Biol 20:619–629

Datta K, Baisakh N, Oliva N, Torrizo L, Abrigo E, Tan J, Rai M, Rehana S, Al-Babili S, Beyer P, Potrykus I, Datta SK (2003) Bioengineered 'golden' indica rice cultivars with beta-carotene metabolism in the endosperm with hygromycin and man-nose selection systems. Plant Biotechnol J 1:81–90

Datta K, Rai M, Parkhi V, Oliva N, Tan J, Datta SK (2006) Improved 'golden' indica rice and post-transgeneration enhancement of metabolic target products of carotenoids (ß-carotene) in transgenic elite cultivars (IR64 and BR29). Curr Sci 91:935–939

Datta SK, Datta K, Parkhi V, Rai M, Baisakh N, Sahoo G, Rehana S, Bandyopadhyay A, Alamgir M, Ali MS, Abrigo E, Oliva N, Torrizo L (2007) Golden Rice: introgression, breeding, and field evaluation. Euphytica 154:271–278

Datta K, Sahoo G, Krishnan S, Ganguly M, Datta SK (2014) Genetic stability developed for β-carotene synthesis in BR29 rice line using dihaploid homozygosity. PLoS One 9:e100212

Dubock A (2019) Golden Rice: To Combat Vitamin A deficiency for public health. Intech Open 1-16

Gayen D, Ali N, Ganguly M, Paul S, Datta K, Datta SK (2014) RNAi mediated silencing of lipoxygenase gene to maintain rice grain quality and viability during storage. Plant Cell Tiss Org Cult 118:229–243

Gayen D, Ali N, Sarkar SN, Datta SK, Datta K (2015) Down regulation of lipoxygenase gene reduces degradation of carotenoids of golden rice during storage. Planta 252:353–363

Gayen D, Ghosh S, Paul S, Sarkar SN, Datta SK, Datta K (2016) Metabolic regulation of carotenoid-enriched golden rice line. Front Plant Sci 7:1622

Ghosh S, Datta K, Datta SK (2019) Rice vitamins. In: Bao J (ed) Rice, chemistry and technology, 4th edn. Elsevier, Rice, pp 195–220. https://doi.org/10.1016/978-0-12-811508-4.00007-1

Hirschenberg J (2001) Carotenoid biosynthesis in flowering plants. Curr Opin Plant Biol 4:210–218

Hoa TTC, Al-Babli S, Schaub P, Potrykus I, Bayer P (2003) Golden indica and *japonica* rice lines amenable to deregulation. Plant Physiol 133:161–169

Khush GS (2001) Green revolution: the way forward. Nat Rev Genet 2:815–821

Misawa N, Nakagawa M, Kobayashi K, Yamano S, Izawa Y, Nakamura K, Harashima K (1990) Elucidation of the *Erwinia uredovora* carotenoid biosynthetic pathway by functional analysis of gene products expressed in *Escherichia coli*. J Bacteriol 172:6704–6712

Misawa N, Yamano S, Linden H, de Felipe MR, Lucas M, Ikenaga H, Sandmann G (1994) Functional expression of the *Erwinia uredovora* carotenoid biosynthesis gene crt1 in transgenic

plants showing an increase of ß-carotene biosynthesis activity and resistance to the bleaching herbicide norflurazon. Plant J 4:833–840
Owens B (2018) Golden Rice is safe to eat, says FDA. Nat Biotechnol 36:559–560
Paine JA, Shipton CA, Chagger S, Howles RM, Kennedy MJ, Vernon G, Wright SY, Hincliffe E, Adams JL, Silverstone AL, Drake R (2005) Improving the nutritional value of Golden rice through increased pro-vitamin A content. Nat Biotechnol 23:482e487
Parkhi V, Rai M, Tan J, Oliva N, Rehana S, Bandyopadhyay A, Torrizo L, Ghole V, Datta K, Datta SK (2005) Molecular characterization of marker free transgenic indica rice lines that accumulate carotenoids in seed endosperm. Mol Genet Genomics 274:e325–e336
Potrykus I (2001) Golden Rice tale. In vitro Cell Dev Biol-Plant 37:93. https://doi.org/10.1007/s11627-001-9-9
Qiu J (2012) China sacks officials over controversy of Golden Rice trials. Nature 12. https://doi.org/10.1038/nature.2012.11998
Rai M, Datta K, Baisakh N, Abrigo E, Oliva N, Datta SK (2003) Agronomic performance of Golden indica rice (cv. IR64). Rice Genet Newsl 20:30–33
Rai M, Datta K, Parkhi V, Tan J, Oliva N, Chwla HS, Datta SK (2007) Variable T-DNA linkage configuration affects inheritance of carotenogenic transgenes and carotenoid accumulation in transgenic indica rice. Plant Cell Rep 26:1221–1231
Rehana S, Baisakh N, Datta K, Oliva N, Abrigo E, Mazil MA, Alam SA, Uddin MR, Datta SK (2018) Engineering Provitamin A synthesis pathway with ß-carotene metabolism in rice endosperm of a restorer line BR827R. Arc Agri Environ Sci 3(2):103–108
Sandmann G (1994) *Phytoene desaturase* genes, enzymes and phylogenetic aspects. J Plant Physiol 143:444–447
Scheldz M, Al-Babli S, von Lintig J, Haubruk H, Rabbani S, Klening H, Beyer P (1996) Phytoene synthase from *Narcissus pseudonarcissus* functional expression, galactolipid requirement, topological distribution in chromatoplasts and induction during flowering. Plant J 10:781–792
Sivakumar B (1998) Current controversies in carotene nutrition. Indian J Med Res 108:157–166
Stein A, Schdev H, Qaim M (2006) Potential impact and cost effectiveness of Golden Rice. Nat Biotechnol 24:200–201
Tan J, Baisakh N, Oliva N, Torrizo L, Abrigo E, Datta K, Datta SK (2004) The screening of rice germplasm including those transgenic rice lines which accumulate ß-carotene in their polished seeds for their carotenoid profile. Int J Food Sci Technol 40:563–569
Tang G (2013) Using plant foods rice in ß-carotene to combat vitamin A deficiency. Sight and Life 27(1):20–28
Tang G (2015) ß -carotene in Golden Rice is as good as ß -carotene in oil at providing vitamin A to children. Am J Clin Nutr 102:715
Vasconcelos M, Datta K, Oliva N, Khalekuzzaman M, Torrizo L, Krishnan S, Oliveira M, Goto F, Datta SK (2003) Enhanced iron and zinc accumulation in transgenic rice with the *ferritin* gene. Plant Sci 64(3):371–378
Ye X, Al-Babili S, Kloti A, Zhang J, Lucca P, Beyer P, Potrykus I (2000) Engineering the provitamin A (ß -carotene) biosynthetic pathway into (carotenoid-free) rice endosperm. Science 287:303–305

Chapter 7
Iron Biofortification: The Gateway to Overcoming Hidden Hunger

Hiroshi Masuda, May Sann Aung, Takanori Kobayashi, and Naoko K. Nishizawa

7.1 Introduction

7.1.1 The Problem of Iron Deficiency Anemia

Iron (Fe) is an essential micronutrient for most living organisms, including all plants and humans. Micronutrient deficiency is known as the "hidden hunger," and Fe deficiency in particular is one of the most prevalent micronutrient deficiencies in the world. Fe deficiency affects approximately two billion people (Stoltzfus and Dreyfuss 1998) and causes 0.8 million deaths per year worldwide (WHO 2002). Fe

Authors Hiroshi Masuda and May Sann Aung have been contributed equally to this work.

H. Masuda · M. S. Aung
Department of Biological Production, Faculty of Bioresource Sciences, Akita Prefectural University, Akita, Japan

Research Institute for Bioresources and Biotechnology, Ishikawa Prefectural University, Ishikawa, Japan

T. Kobayashi
Research Institute for Bioresources and Biotechnology, Ishikawa Prefectural University, Ishikawa, Japan

N. K. Nishizawa (✉)
Research Institute for Bioresources and Biotechnology, Ishikawa Prefectural University, Ishikawa, Japan

Department of Global Agricultural Sciences, Graduate School of Agricultural and Life Sciences, The University of Tokyo, Tokyo, Japan
e-mail: annaoko@mail.ecc.u-tokyo.ac.jp

A. Costa de Oliveira et al. (eds.), *The Future of Rice Demand: Quality Beyond Productivity*, https://doi.org/10.1007/978-3-030-37510-2_7

deficiency ranks as the sixth highest risk factor for death in developing countries with high mortality rates (WHO 2002), and it is the cause of over 60% of global anemia cases in both developed and developing countries (WHO 2008). Iron deficiency anemia (IDA), which is considered the most important contributing factor to the global burden of disease, causes disruptions in brain development in young children, maternal mortality in pregnancy and childbirth, and coma and death in severe anemia (WHO 2002). Consequently, it affects the social and economic development of nations by lowering the intellectual capacity of the population and the productivity of the adult workforce (UNICEF and Micronutrient Initiative 2004). In normal individuals, the amount of Fe lost per day is approximately 1.0 mg. These losses are balanced by ingesting an equivalent amount of Fe from the diet. However, because of the increased Fe needs of infants and young children for growth, in adolescents, in pregnant and breastfeeding women, and during blood loss due to menstruation in females, these populations frequently suffer from Fe deficiency anemia, and especially if they ingest inadequate amounts of dietary Fe (WHO 2002; Maeda 2005).

The three major chemical forms of nutritional Fe are heme, nonheme ferritin Fe, and nonheme low–molecular-weight Fe (as in Fe salts and chelates) (Theil 2011). Heme Fe is derived mainly from hemoglobin and myoglobin, found in animal foods such as red meats, fish, and poultry (Abbaspour et al. 2014). Most Fe in plant foods is derived from nonheme Fe (Hurrell 1997). The Fe absorption mechanisms of these three Fe forms in humans differ (Theil et al. 2012). Heme Fe is more easily absorbed than nonheme low-molecular-weight Fe. Ferritin Fe from food is readily bioavailable to humans *in vivo*, as indicated by cell culture experiments (Martin et al. 2008; Kalgaonkar and Lonnerdal 2008, 2009).

7.1.2 How to Solve Fe Deficiency Anemia?

Several options to mitigate against micronutrient deficiency in humans exist, including micronutrient supplementation, food fortification, dietary diversification, and biofortification. Nutritional supplementation and industrial food fortification were occasionally carried out by public health interventions of governments, non-governmental organizations (NGOs), and non-profit organizations (NPOs). These have successfully reduced micronutrient deficiencies worldwide. However, these interventions require continuous financial input to produce and distribute the supplements or the fortified foodstuffs. Also, purchasing power or access to markets are often not viable options for people living in remote rural areas (Mayer et al. 2008). Moreover, food fortification with Fe is difficult as it makes food less appetizing or changes the color, especially when using the most soluble and absorbable $FeSO_4$ compounds, and it is not as readily absorbed by the human body if the least soluble $Fe_4(P_2O_7)_3$ compounds are used instead (Hurrell 2002; Singh et al. 2017). Thus, projects of food fortification with Fe are not always sustainable in developing countries. Fe deficiency is frequently caused by dietary habits that have inadequate Fe intake. The bioavailability of nonheme Fe (except for Fe stored in ferritin) is lower than that of heme Fe (Miret et al. 2003). Thus, the diets of vegetarians are low in bioavailable Fe, and vegetarians have lower Fe stores (Hunt 2003). Greater dietary diversity (e.g., increasing consumption of legumes and nuts, especially if meat, eggs, or fish are either unavailable or too expensive) is one of the ways to solve Fe deficiency anemia. However, greater dietary diversity is not always successful because it depends on personal effort, finances, eating habits, food culture, or religion.

7.1.3 *What Is Biofortification?*

Biofortification is the process by which nutritional quality of the food crops is improved in the edible parts of the crops on a biological basis. Biofortified food can be produced via agronomic intervention such as improvement in cultivation methods and/or fertilizer application, or through genetic selection such as the development of new varieties in which the nutritional value is increased (White and Broadley 2005). In contrast to supplementation or food fortification, biofortification approaches do not require specific processing after harvesting or a special infrastructure (Grusak and DellaPenna 1999; Mayer et al. 2008). Once mineral-dense varieties have been produced, it requires no additional cost to improve the nutritional status of those consuming the crops. It is also beneficial for people with dietary habits that are difficult to change due to financial, cultural, regional, or religious restrictions. Thus, the biofortification approach is beneficial to both individuals and governments (Mayer et al. 2008) and is sustainable, especially in developing countries. Biofortification can complement existing supplementation and food fortification methods to alleviate micronutrient deficiencies.

The HarvestPlus Program (http://www.harvestplus.org/) set up by the Consultative Group for International Agricultural Research (CGIAR) focuses on various biofortification programs of major crops, including rice, wheat, maize, beans, sweet potato, cassava, and pearl millet, to mitigate micronutrient malnutrition such as Fe, Zinc (Zn), and provitamin A (Hotz and McClafferty 2007; Bouis et al. 2011). Fe biofortification is a current hot topic in plant nutrition research.

7.1.4 *Why Rice Is An Important Target for Fe Biofortification*

Rice is one of the most important target crops for biofortification, as it is one of the highest production crops in the world. The top four crops produced in the world are sugarcane, maize, wheat, and rice, with yields of 1840, 1135, 772, and 770 million tons, respectively, in 2017 (FAOSTAT 2017). Furthermore, 268 million tons of rice was produced in low-income food-deficit countries, compared to 98 million tons of maize and 148 million tons of wheat. Among the world regions, the number of individuals affected by IDA is the highest in Asia and Southeast Asia (WHO 2008), where rice is the main staple food. Sugarcane is mostly used on an industrial level to produce sugar or bioethanol. Maize is mainly used as silage for animal feed

(46%) and for bioethanol (37%), with only 2.6% used as food for human consumption (National Corn Growers Association 2018). Thus, sugarcane and maize are not so suitable targets for biofortification, whereas rice is an important crop in developing countries where Fe deficiency is an important issue to be resolved. Even though the nutritional value of some other crops is higher than that of rice, it is difficult to change people's dietary habits or their culture, especially if the only reason for doing so is to provide better nutrition. Therefore, biofortification approaches to increase the nutritional value of rice, which is currently the main staple diet, are the most efficient and sustainable approaches.

Another important reason is the way of consuming. Wheat and maize are normally processed into a powder after flour milling and used for cooking. This makes food fortification simpler, as it can take place during the milling process in wheat and maize by the simple addition and mixing of micronutrients with the wheat or corn flour. To this end, the Food Fortification Initiative created an agreement between governments and the private sector to promote the fortification of wheat flour in developing countries. As a result, the usage of Fe-fortified wheat flour increased from 18% in 2004 to 27% in 2007, helping 540 million people avoid Fe deficiency (Food Fortification Initiative 2013). On the other hand, rice is just polished to make white rice before cooking, and it is not often used in ground form. Furthermore, it is difficult and costly to add a micronutrient coating to the surface of polished (white) rice, and this rice is also usually washed before cooking. Thus, biofortification of rice, and especially biofortification with Fe, is the most suitable method to add micronutrients to this staple food.

Rice is generally eaten in the polished form. However, the Fe content is very low in polished rice (Grusak and Cakmak 2004; Masuda et al. 2008). People who mainly eat rice as their staple food are at risk of Fe deficiency, making Fe biofortification of rice especially important. The rice whole genome sequence is known, and many genetic databases exist for rice. The biofortification process is difficult to achieve in some crops. For example, the genetics of wheat is more complex than that of rice. Original wheat possesses two sets of chromosomes (diploid); however, many modern wheat species contain four (tetraploid) or six (hexaploid) sets of chromosomes (Hancock 2004), leading to difficulties in making functional mutations. Thus, although Fe biofortification of wheat has been attempted (Borrill et al. 2014), it remains more difficult to breed biofortified wheat than rice. Additionally, the production of transgenic rice is tequnically unproblematic than that of wheat.

7.2 Iron Nutrition in Rice

7.2.1 What Is Needed to Consider for Fe Biofortification in Rice?

The primary consideration before starting Fe biofortification in rice is to establish the Fe concentration in rice seeds. Brown rice includes the embryo, endosperm, and aleurone layers, and it has a high nutritional value. However, polishing removes the aleurone layers and the embryo (called bran) which are rich in Fe. The remaining endosperm, the commonly eaten part, contains low Fe levels (Grusak and Cakmak 2004). This is one reason why Fe deficiency is a serious issue in rice-consuming countries. Generally, Fe translocation efficiency to the endosperm is very low in rice. Indeed, out of the total Fe content per plant (above ground parts), the Fe content accounts for only 0.5% in the endosperm tissue, whereas it is 3.6% in brown seeds, 3.4% in the husk, 2.4% in the rachis, and 90.6% in straw of the Tsukinohikari rice variety (Masuda et al. 2008). As polished seeds are consumed the most, Fe biofortification in rice should be targeted to increase Fe in them rather than in brown seeds.

Fe contamination during the analysis for Fe concentration in polished rice seeds often occurs due to contamination by the remained bran or, the metal tubes or containers used for the analysis. It is vital to avoid any Fe contamination during analyses of Fe content, and it is also important to ensure the uniformity of the polishing levels of all rice seeds. The use of plastic tubes is recommended to avoid metal contamination. For example, the polishing method using a multi-beads shocker machine (Yasui Kikai, Osaka, Japan) and 2-mL multi-beads plastic tubes can be applied for a small number of seeds. Using this machine and a total of 30 seeds per tube for short and round japonica varieties (Masuda et al. 2009, 2012, 2013a, b) and about 15 seeds per tube for long and slender indica varieties (Aung et al. 2013, 2016), uniformly polished seeds can be obtained.

There is a positive correlation between the Zn and Fe content in grains, indicating that simultaneous selection for high Zn and Fe densities could prove to be highly efficient (Bürger et al. 2014; Aung et al. 2016). Zn biofortification of rice is also an important issue. Thus, a breeding program to select for both high Fe and Zn mineral contents is recommended.

7.2.2 *Target Fe Concentration for Biofortification of Rice*

The Fe intake per person is approximately 7.5–10.3 mg day^{-1} (FAO, 2002). The recommended dietary allowance for Fe in adult females is approximately 15–18 mg day^{-1} (Food and Nutrition Board 2001). The recommended daily intake of Fe varies between 8–18 mg day^{-1} according to age, body weight, and gender, and 30 mg day^{-1} is recommended for pregnant women (Vasconcelos et al. 2017). Thus, the per capita shortfall is estimated to be 5–10 mg day^{-1} (Aung et al. 2016) for females and 20 mg day^{-1} for pregnant women. The target amount of Fe biofortification in rice should be considered based on the target population (per capita rice consumption) and the target rice variety (baseline Fe concentration of the variety used). For example, Myanmar has a high prevalence of IDA; over 70% of children and pregnant women are reported as being Fe deficient (MOH 2003; WHO 2008). Myanmar has one of the highest rice consumption rates in Asia, with a per capita rice consumption of approximately 578 g day^{-1} (Maclean et al. 2002; FAOSTAT 2009). Based on this high consumption rate, we estimated that the target amount of Fe needed by the people of Myanmar is 8.6 μg g^{-1} (Aung et al. 2013). Thus, Myanmar represents one of the best target countries for Fe biofortification.

7.2.3 *Fe Uptake System in Plants*

The mechanisms of Fe uptake, translocation, and homeostasis regulation in higher plants, including rice, have been extensively investigated. In higher plants, Fe acquisition is mediated by two major mechanisms: Strategy I (Fe reduction-based mechanism) and Strategy II (Fe chelation-based mechanism) (Römheld and Marschner 1986). Under low Fe availability in the soil, non-graminaceous species employ the Strategy I Fe uptake system. First, the H^+-ATPase (AHA2 in *Arabidopsis thaliana*) acidifies the rhizosphere to enhance Fe solubilization (Santi and Schmidt 2009). The plant roots secrete coumarin-phenolic type compounds via ABCG (AtABCG37/AtPDR9 in *Arabidopsis*) transporters to chelate and mobilize Fe^{3+} (Fourcroy et al. 2014; Sisó-Terraza et al. 2016). The 2-oxoglutarate-dependent oxidase F6′H1 (Feruloyl-CoA 6′-hydroxylase 1) and two hydroxylase enzymes (S8H and CYP82C4) are involved in coumarin biosynthesis, including fraxetin and sideretin (Schmid et al. 2014; Rajniak et al. 2018). Second, roots induce Fe (III)-chelate reductase (FRO) and reduce Fe (III)-chelate to Fe^{2+} (Robinson et al. 1999), and this process is likely the rate-limiting step for the acquisition of Fe from the soil (Connolly et al. 2003). Third, the plant roots also directly take up Fe^{2+} through the major Fe^{2+} transporter, the Fe-regulated transporter (IRT) (Eide et al. 1996).

Rice (*Oryza sativa* L.) also employs a direct Fe^{2+} uptake by the Fe^{2+} transporter OsIRT1, which is a partial Strategy I Fe uptake mechanism (Bughio et al. 2002; Ishimaru et al. 2006). A possible reason for this is that rice is normally cultivated in submerged paddy fields where Fe^{3+} is easily reduced to Fe^{2+}, which is readily available to plants due to the low redox potential (Bughio et al. 2002; Ishimaru et al. 2006). However, due to a lack of a complete Strategy I system and also its low Fe(III)-chelate reductase activity (Ishimaru et al. 2006), rice has limited Fe uptake efficiency.

All graminaceous plants, including rice, employ a Fe chelation strategy (Strategy II) to acquire Fe from the soil. In Strategy II, plants synthesize and secrete mugineic acid family phytosiderophores (MAs) into the soil to acquire Fe (Takagi 1976; Marschner et al. 1986). MAs have a high capacity to chelate Fe^{3+} in the rhizosphere, especially under high pH (von Wirén et al. 2000). The plant takes up the resulting Fe(III)-MAs by YS1 and YSL transporters (Curie et al. 2001). In the MAs biosynthetic pathway (Mori and Nishizawa 1987; Shojima et al. 1990), three S-adenosyl-L-methionines are the precursors of nicotianamine (NA), which is synthesized by NA synthase (NAS) (Higuchi et al. 1999). Then, NA is further converted to 2′-deoxymugineic acid (DMA) by NA aminotransferase (NAAT) (Takahashi et al. 1999; Inoue et al. 2008) and DMA synthase (DMAS) (Bashir et al. 2006). Again, DMA is further converted to various types of MAs by dioxygenases in many graminaceous species, for example, IDS3 and IDS2 in barley (Nakanishi et al. 2000; Kobayashi et al. 2001).

Rice also synthesizes and secretes DMA from the roots to the rhizosphere to chelate Fe^{3+} through the transporter of the mugineic acid family phytosiderophores 1 (TOM1) (Nozoye et al. 2011). Then, the rice roots absorb the resulting Fe(III)–DMA complexes via Fe(III)-DMA transporters, such as OsYSL15 (Inoue et al. 2009; Lee et al. 2009a). However, rice usually does not secrete mugineic acid (Kobayashi et al. 2001); it secretes DMA at lower levels than other graminaceous species (Marschner et al. 1986). Thus, rice is less tolerant of Fe limitation than other graminaceous plants.

7.2.4 Genes Involved in Fe Transport in Plants

NA has a dual function: (1) it acts as a precursor of DMA, and (2) it acts as a chelator of metal cations, such as Fe^{2+} and Zn^{2+} within plants. NA is synthesized and utilized by all higher plants for the internal transport of metals, including Fe (Hell and Stephan 2003; Takahashi et al. 2003). Indeed, the NA-defective tomato mutant *chloronerva* (Rudolph et al. 1985) has a phenotype that is indicative of Fe deficiency (Pich and Scholz 1996; Stephan et al. 1996). Rice has three *NAS* genes: *OsNAS1*, *OsNAS2*, and *OsNAS3*. Although these genes are differentially regulated by Fe, all are expressed in cells involved in the long-distance transport of Fe (Inoue et al. 2003). Foliar application of NA-Fe to rice grown in the field increased the Fe and Zn concentrations in brown rice (Yuan et al. 2012). This indicates that NA contributes to the transport of Fe and Zn to the grains within the rice plant. The rice OsYSL2 transports Fe(II)-NA and manganese (Mn)-NA (Koike et al. 2004). This transporter is expressed specifically in leaf phloem cells, the vascular bundles of flowers, in developing seeds, suggesting its role in internal Fe transport (Koike et al. 2004). In addition, *OsYSL2* knockdown rice has a lower Fe concentration compared with non-transgenic rice: 18% and 39% lower in brown and polished seeds, respectively (Ishimaru et al. 2010). This suggests that OsYSL2 plays a vital role in Fe translocation to the seeds (Ishimaru et al. 2010).

Yoneyama et al. (2015) reviewed the role of chelators for Fe transportation within rice plants. The chemical forms of Fe in xylem sap were found to be bound largely to citrate (~65%) and DMA (~5%) (Ariga et al. 2014). Within the xylem, Fe is transported mainly as Fe(III)-citrate (Yoneyama et al. 2015). A Fe(III)-DMA transporter, *OsYSL15*, is expressed in the root epidermis and stele and participates in both Fe uptake and the internal translocation of Fe (Inoue et al. 2009; Lee et al. 2009a). Disruption of *OsYSL15* leads to lower accumulation of Fe in the seeds (Lee et al. 2009a). This suggests that OsYSL15 may act in Fe loading to the phloem to transport Fe from the root to the leaves and seeds. Another Fe(III)-DMA transporter, *OsYSL16*, is expressed in the root epidermis and vascular bundles of whole plants, including the nodes, suggesting that OsYSL16 transports Fe(III)-DMA in rice xylem sap via the xylem-to-phloem at the nodes (Kakei et al. 2012). *OsYSL18* is another Fe(III)-DMA transporter that is expressed in the reproductive organs and phloem of the lamina joints (Aoyama et al. 2009). Thus, OsYSL18 may act in unloading Fe(III)-DMA from the phloem at the panicles. Fe in the phloem sap binds to DMA, citric acid, and proteins (Nishiyama et al. 2012). In rice phloem sap, Fe(III)-DMA complexes were present as low-molecular-weight chemical forms of Fe, suggesting that DMA may participate in the long-distance transport of Fe in the phloem sap (Nishiyama et al. 2012). This indicates that likewise NA, DMA also participates in Fe distribution from the roots to the seeds in rice.

Yokosho et al. (2009) reported that OsFRDL1, a member of the multidrug and toxin efflux (MATE) family, acts as a citrate transporter that loads citrate into the xylem and is required for the efficient translocation of Fe in rice. Based on the results of *OsFRDL1* knockout lines, Yokosho et al. (2016) reported that *OsFRDL1* is expressed in the upper nodes and is required for the distribution of Fe in the panicles through solubilizing Fe deposited in the apoplastic part of the nodes, and is also required for the distribution of Fe to the grains in rice.

Permease in chloroplasts 1 (PIC1) protein in *Arabidopsis* transports Fe from the cytoplasm into the plastid (Duy et al. 2007). AtNRAMP3 and AtNRAMP4, members of the natural resistance-associated macrophage proteins (NRAMP), are assumed to participate in the release of Fe^{2+} from the vacuole in *Arabidopsis* (Thomine et al. 2003; Gross et al. 2003; Hall and Williams 2003; Lanquar et al. 2005; Grotz and Guerinot 2006; Puig et al. 2007). OsNRAMP5 contributes to Fe and Mn transport and is important for plant growth and development (Tang et al. 2017). Taken together, these genes may also function in Fe transportation in rice plants.

7.2.5 Fe Homeostasis Regulation in Rice

Fe is an essential element; however, excess Fe causes toxicity to plants. Thus, Fe homeostasis, including uptake, translocation, usage, and storage within plants must be strictly regulated throughout the plant lifespan. In rice, the ABI3/VP1 transcription factor, which is known as iron deficiency-responsive element-binding factor 1 (IDEF1), regulates the expression of most Fe deficiency-induced genes via the bHLH transcription factor OsIRO2 (Kobayashi et al. 2007; Ogo et al. 2007). IDEF1 is a central transcriptional regulator of Fe uptake- and utilization-related genes, especially at the early stages of Fe deficiency (Kobayashi et al. 2009). The hemerythrin motif-containing RING Zinc finger (HRZs) and BRUTUS (BTS) are E3 ubiquitin ligases possessing Fe-binding properties and function as potent negative regulators of Fe deficiency responses, which suggests that HRZs are important Fe sensors (Kobayashi and Nishizawa 2014, Kobayashi et al. 2019).

7.2.6 Fertilizer Effects for Iron Nutrition in Rice

Lyons and Cakmak (2012) reviewed the potential of Fe biofortification through the application of fertilizer. They claimed that biofortification by fertilizer application is possible for Zn, selenium, and iodine. However, it has been difficult to biofortify Fe using fertilizer (Rengel et al. 1999; Welch 2001). Inorganic Fe fertilizers applied to the soil are usually ineffective due to the rapid conversion of Fe^{2+} into Fe^{3+} forms, which are easily precipitated and are not available to the plants (Rengel et al. 1999; Frossard et al. 2000; Zhang et al. 2008). Fe provided in the chelate form is more available, but is expensive and is only marginally better than inorganic Fe for biofortification. Foliar Fe application has been slightly more effective than soil application (Rengel et al. 1999). The application of Fe(III)-DTPA to the rice during anthesis led to a 20% increase in Fe in the polished grains (He et al. 2013). Agronomic inputs such as fertilizer or foliar application require additional costs and effort from farmers during cultivation and have not effectively attained the target level of Fe biofortification.

7.2.7 Genetic Diversity of Fe Concentration in Rice Seeds

Mining of high-Fe rice varieties among extant rice varieties and then the identification of novel target genes from those varieties are important for the Fe biofortification of rice. Attaining higher Fe levels using transgenic approaches or conventional breeding is partly dependent on the base Fe concentration level of the host rice variety. Thus, to produce Fe biofortified rice, original high Fe rice varieties should be chosen as the host variety. This highlights the importance of surveying, screening, or mining for high Fe varieties among extant rice varieties. There are several previous research outcomes from the screening of high Fe rice among existing varieties. For examples, International Rice Research Institute (IRRI) screened a total of 1,138 rice genotypes collected from different countries with diverse genetic variation, and a natural variation of 6.3–24.4 μg g^{-1} of Fe concentrations in brown rice was reported (Gregorio et al. 2000). Banerjee et al. (2010) reported 4.8–22.7 mg kg^{-1} of seed Fe concentrations among the 46 tested rice genotypes. We also analyzed Fe and Zn concentrations in grains of 60 indigenous and popular Myanmar rice varieties, and Fe and Zn concentrations in polished rice ranged from 0.8–3.8 μg g^{-1} and 10–47 μg g^{-1}, respectively (Aung et al. 2016).

Grusak and Cakmak (2004) reported that aromatic rice genotypes contain consistently higher levels of Fe and Zn than the non-aromatic genotypes. They indicated that the aromatic rice germplasm was a promising genetic resource for improving micronutrient levels in rice. The natural variation of Fe content in grains suggests that the genetic potential for developing new varieties with high Fe exists. However, questions regarding polishing methods or the risk of Fe contamination remain because some varieties were reported to contain too high Fe concentrations.

This data should be carefully considered and investigated before any rice species are used as a breeding source.

7.3 Fe Biofortification Approaches

7.3.1 Biofortification by Conventional Breeding, QTLs, or Screening of Mutant Lines

Fe nutrition in rice has been improved by traditional breeding or marker-assisted breeding based on quantitative trait loci (QTLs). IRRI developed the cultivar IR68144 by conventional breeding. This cultivar contains approximately twice the Fe concentration in polished seeds compared with local varieties used in the Philippines (Gregorio et al. 2000). This IR68144 high Fe rice strain outperformed other, commercially available, low-iron rice in improving the Fe status of women (Haas et al. 2005). Anuradha et al. (2012) reported the mapping of QTLs and candidate genes for Fe biofortification, including *OsYSL1*, *OsMTP1*, *OsNAS3*, *OsNRAMP1*, *OsIRT1*, *OsZIPs*, and *OsAPRT* in Madhukar × Swarna indica rice varieties recombinant inbred lines. However, even though high Fe concentrations were present in some rice varieties, the majority of cultivated varieties generally contain low Fe levels, leading to insufficient daily Fe intake.

Sperotto et al. (2010) analyzed the gene expression profiles of 25 metal-related genes in eight rice varieties with different Fe and Zn concentrations in the seeds. They revealed that higher concentrations of Fe in the grains were positively correlated with the expression of genes such as *OsFER1*, *OsNRAMP4*, *OsNRAMP5*, *OsNRAMP6*, *OsYSL4*, *OsYSL6*, *OsYSL12*, *OsYSL14*, *OsZIP8*, *OsZIP10*, and *OsNAC5*, and also investigated putative target genes that contribute to increasing the grain Fe and Zn concentrations (Sperotto et al. 2010). Jeng et al. (2012) screened mutant lines with higher Fe or Zn concentrations in polished seeds among NaN_3-induced mutant lines (*Oryza sativa* cv. IR64). Ruengphayak et al. (2012) identified 76 mutant lines that contained higher Fe densities in the grains by screening 12,000 fast neutron-irradiated M4 mutant lines (*Oryza sativa* cv. Jao Hom Nin). Thus, there is the potential to detect novel candidate genes from these high Fe mutant rice lines to improve the Fe biofortification of rice.

Stangoulis et al. (2007) detected three QTLs for Fe on chromosomes 2, 8, and 12, two QTLs for Zn on chromosomes 1 and 12, with a QTL for Fe and Zn accounting for 13–18% of the variation. Neelamraju et al. (2012) identified seven QTLs for Fe and six QTLs for Zn that accounted for >30% of the phenotypic variance in rice accessions. Garcia-Oliveira et al. (2009) investigated two QTLs for Fe on chromosomes 2 and 9 and three QTLs for Zn on chromosomes 5, 8, and 12. There is potential to improve those characteristics with high heritability in a breeding program by marker-assisted selection.

Du et al. (2013) investigated three QTLs for Fe accumulation in grains on chromosomes 1, 6, and 8, which accounted for 15.7, 10.6, and 22.3% of the phenotypic variation, respectively, by mapping a population resulting from the cross Chunjiang 06 (japonica) × TN1 (indica). Furthermore, a QTL related to Fe concentration on chromosome 8 from a population derived from the cross cv. Lemont (japonica) × cv. TeQing (indica) was also identified by Zhang et al. (2014). To identify QTLs responsible for the mineral content in grains, Nawaz et al. (2015) performed genome-wide association mapping of QTLs for 221 accesses of *O. sativa*, five accesses of *O. glaberrima*, two accesses of *O. rufipogon* Griff., and one of *O. nivara*, and identified the genes *OsZIP1*, *OsHMA4*, *OsACA2*, *OsZIP2*, *OsCNGC*, *OsZIP3*, *OsZIP5*, *OsZIP9*, *OsHMA2* (an ABC transporter gene), *OsNAS3*, a CHY Zn finger gene, and *OsACA9* from 11 genetic regions. However, rice that is high in Fe, and that meets the target level for Fe content, has not yet been produced by conventional breeding or QTL breeding approaches. Genetic engineering may prove to be the best method to increase bioavailable Fe in food crops, including rice.

7.3.2 Fe Biofortification by Genetic Engineering Approaches

As discussed above, various genes involved in Fe uptake, translocation, and homeostasis in rice have been identified. Regeneration and transformation methods have also been established in rice (Hiei et al. 1994). Among the biofortification methods, we focus on genetic engineering approaches because they are the most efficient means to increase the Fe concentration in rice seeds. Previous studies have shown that various transgenic approaches can be applied for Fe biofortification of rice, as described below (Fig. 7.1).

7.3.2.1 Enhancing Fe Accumulation in Seeds

Ferritin is a ubiquitous Fe storage protein that sequesters as many as 4,000 Fe atoms in a complex (Theil 2003). Fe stored in ferritin is bioavailable and safe and is a good source of Fe that will prevent Fe deficiency (Zielińska-Dawidziak 2015). Expressing the ferritin gene under the control of endosperm-specific promoters leads to higher Fe accumulation in rice seeds. The rice glutelin promoter (*OsGluB1*) drives expression in the outer layers of the endosperm, and the rice globulin promoter (*OsGlb1*) drives expression in the inner layers of the endosperm (inner membranes of the endosperm cells) (Qu and Takaiwa 2004). Goto et al. (1999) introduced the soybean ferritin gene (*SoyferH1*) under the control of the rice endosperm-specific 1.3-kb *OsGluB1* promoter for expression in endosperms. The Fe concentration of polished seeds was increased 2 fold. This biofortification approach of endosperm-specific expression of ferritin has been used in previous research, as follows: a 2.2-fold increase in brown seeds of japonica cv. Taipei 309 (Lucca et al. 2002), a 3.7-fold

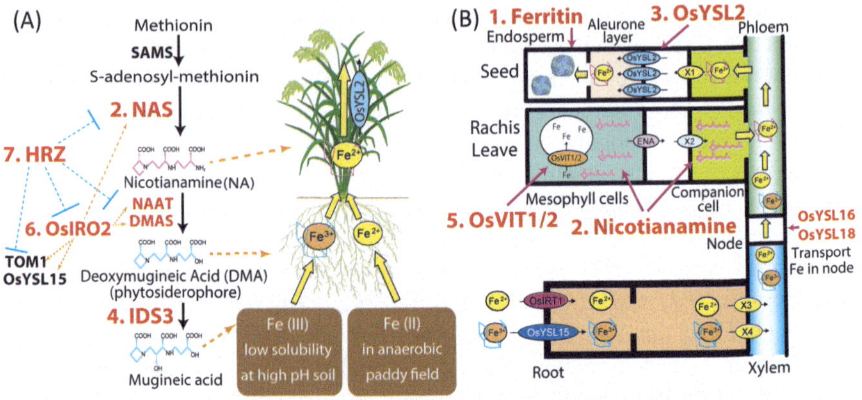

(C) Approaches with targeting genes to iron biofortification of rice

Approaches	Aims	Introduced gene or construct	Fold increase of Fe in grains	References
1. Expression of Ferritin	Enhancement of Fe accumulation in seeds	Endsperm specific promoter-*Ferritin*	2 to 3 times	Goto et al. (1999)
2. Production of NA	Enhancement of Fe translocation	*35S-HvNAS1, 35S-OsNAS3*	2 to 3 times	Masuda et al. (2009) Lee et al. (2009b) Johnson et al. (2011)
3. Expression of OsYSL2	Enhancement of Fe uptake and translocation	*OsSUT1* promoter-*OsYSL2*	2 times	Ishimaru et al. (2010)
4. Production of MA	Enhancement of Fe uptake and translocation	*IDS3* genome	40 %	Masuda et al. (2008) Suzuki et al. (2008)
5. Knockdown/out of OsVIT1/2	Suppression of Fe accumulation into vacuole	*OsVIT1/2 RNAi / T-DNA insertion*	2 times	Zhang et al. (2012) Bashir et al. (2013)
6. Expression of OsIRO2	Enhancement of Fe uptake and translocation	*35S-OsIRO2*	3 times (Calcareous soil)	Ogo et al. (2011)
7. Knockdown of HRZ	Enhancement of Fe uptake and translocation	*OsHRZ2-RNAi*	3 times	Kobayashi et al. (2013)

Fig. 7.1 Strategies for Fe biofortification of rice. **(A)** The biosynthetic pathway for mugineic acid family phytosiderophores (MAs) in graminaceous plants (modified from Masuda et al. 2013a). **(B)** The transporters and Fe translocation flow within the rice plant. (C) The approaches with targeting genes to Fe biofortification of rice. SAMS: S-adenosyl-methionine synthase, NAS: NA synthase, NAAT: NA aminotransferase, DMA: 2′-deoxymugineic acid, DMAS: DMA synthase, IDS3: MA synthase (dioxygenase that catalyzes the hydroxylation of DMA and epi-HDMA at the 2′ position), Ferritin: iron storage protein, OsYSL2: Fe(II)-NA and Mn(II)-NA transporter, OsIRO2: Fe deficiency–inducible bHLH transcription factor positively regulating DMA-based Fe uptake in rice, OsIRT1: Fe^{2+} transporter, OsYSL15: Fe(III)-DMA transporter, TOM1: efflux transporter of MAs, ENA: efflux transporter of NA, HRZ: RING ubiquitin ligase negatively regulating Fe uptake and translocation, VIT: vacuolar Fe transporter, X1, X2: unknown transporters related to Fe homeostasis. Numbers inside the figure show the target or introduced genes for Fe biofortification of rice

increase in polished seeds of indica cv. IR68144 (Vasconcelos et al. 2003), a 2.4-fold increase in indica cv. BR29 (Khalekuzzaman et al. 2006), and a 2.1-fold increase in polished seeds of indica cv. Pusa Sugandhi II (Paul et al. 2012).

Furthermore, Qu et al. (2005) introduced the *SoyferH1* gene under the control of two endosperm-specific promoters, *OsGlb1* and 1.3-kb *OsGluB1*, which increased the level of ferritin accumulation in rice seeds much higher than using single endosperm-specific promoters. However, the Fe concentration did not significantly increase compared to the expression driven by a single endosperm-specific promoter (Qu et al. 2005). Also, Fe deficiency symptoms appeared in the leaves of the transgenic plants when the *ferritin* gene alone was introduced (Qu et al. 2005; Masuda et al. 2013b). These results suggest that enhancement of the expression of *ferritin* alone may be insufficient to further increase Fe concentration in rice seeds.

It is also conceivable that rice plants may not actively translocate Fe to the endosperm to avoid the over-accumulation of Fe in their seeds. A strict regulatory system may exist in rice plants to prevent uptake and translocation of excess Fe into the endosperm. This may explain why there is less diversity for Fe concentration in current rice varieties and also the limits of traditional breeding to produce Fe-biofortified rice.

Using a positron-emitting tracer imaging system (PETIS), Tsukamoto et al. (2009) showed that more Fe is translocated from the basal parts of the plant body to the rice seeds than is remobilized from the flag leaf. Sperotto et al. (2013) also suggested that flag leaves are not necessarily the major source of Fe in rice seeds. Thus, in addition to increasing the amount of Fe stored in the seeds, it is necessary to enhance the Fe uptake ability from the soil and its translocation to the seeds rather than enhancing its translocation from the flag leaves. Thus, the following approaches are considered possible ways to increase Fe levels in the rice endosperm.

7.3.2.2 Enhancing Fe Uptake and Transport Within the Plant Body by NA, DMA, and MA

NAS-overexpressing transgenic rice showed 5–10-fold higher endogenous NA levels in the shoots and seeds (Masuda et al. 2009). *NAS* overexpression also increased the Fe concentration in polished rice seeds by 3–4-fold (Masuda et al. 2009; Lee et al. 2009b; Johnson et al. 2011). Thus, Fe transport within the plant body and to the edible part of the seeds may be improved by NA produced by overexpressed *NAS* (Masuda et al. 2009; Lee et al. 2009b; Johnson et al. 2011). Moreover, Lee et al. (2009b, 2012) reported that Fe-biofortified rice overexpressing *OsNAS3* or *OsNAS2* mitigated IDA in mice better than non-transgenic rice. Zheng et al. (2010) also showed that the bioavailability of Fe in the high NA-accumulating grain (*OsGluB1* promoter-*OsNAS1*) was higher than that in normal grains in an *in vitro* Caco-2 cell model that simulates the human digestive system. Nozoye (2018) also suggested that the potential of NA for the biofortification of crops, including rice, and the combined enhanced expression of the *NAS* and NA transporter genes could further elevate NA levels in the edible parts of the plant.

Interestingly, the DMA content was also increased by overexpressing *HvNAS1* in rice (Masuda et al. 2009; Lee et al. 2009b). Overexpression of *AtNAS1* in rice enhanced the expression of the genes involved in NA and DMA synthesis such as *OsSAMS2*, *OsNAS1*, *OsNAS3*, and *OsDMAS1* in the roots (Wang et al. 2013). Thus, higher expression of *NAS* and *DMAS* in *NAS* overexpression might enhance DMA productivity in rice. The DMA produced also contributes to the transport of Fe from roots to seeds. Díaz-Benito et al. (2018) reported the relationship between NA and DMA productivity and Fe content in the embryo and endosperm of rice seeds by analyzing transgenic rice highly expressing *OsNAS1*, *HvNAAT-B*, or both. The higher DMA content within the plant body may lead to additional DMA secretion from the roots, which subsequently enhances the uptake of Fe from the soil. Increase in DMA content by *OsNAS3* overexpression enhanced the Fe deficiency tolerance of rice plants (Lee et al. 2009b).

We proposed the next approach to Fe biofortification as enhancing Fe uptake and translocation by the introduction of the genome fragment of barley mugineic acid synthase gene, *IDS3* (Masuda et al. 2013a). Introduction of *IDS3* enables the rice plant to secrete MA (Kobayashi et al. 2001). The Fe concentrations in the seeds of transgenic rice expressing *IDS3* were analyzed. The Fe concentrations were 1.4- and 1.3-fold higher in the polished and brown seeds than in non-transgenic rice, respectively, after field cultivation in Fe-sufficient soil (Masuda et al. 2008). After field cultivation in soil with low Fe availability, Fe concentrations in this transgenic rice line were also 1.3-fold higher in the brown seeds (Suzuki et al. 2008). MA showed greater Fe(III) complex stability than DMA at a slightly acidic pH (von Wirén et al. 2000), and this is important because rice xylem sap is acidic. Thus, the MA synthesized by IDS3 may boost Fe translocation in rice.

7.3.2.3 Enhancing Fe Transport to Seeds by *YSLs*

Membrane transporters can be used to increase Fe concentrations in crops (Schroeder et al. 2013). However, it is necessary to consider the type of promoters that are used to enhance the expression of membrane transporters. Ishimaru et al. (2010) reported that simple overexpression of *OsYSL2* under the control of the cauliflower mosaic virus (*CaMV*) *35S* promoter did not increase Fe in the seeds. It may disrupt Fe translocation within the rice plant. Interestingly, the enhanced *OsYSL2* expression under the control of the promoter of *OsSUT1*, which encodes a rice sucrose transporter dominantly expressed in the panicle and immature seeds during seed maturation (Aoki et al. 2003), increased the Fe concentration up to 4-fold in polished rice seeds (Ishimaru et al. 2010).

Transgenic rice constitutively expressing *HvYS1* under the control of the maize *Ubi-1* promoter showed increased tolerance under alkaline soil cultivation and increased Fe concentration in leaves but not in seeds (Gómez-Galera et al. 2012). This indicates that constitutive expression of *HvYS1* mobilizes Fe into the vegetative parts of the plant but not into the seeds, similar to overexpression of *OsYSL2*.

Senoura et al. (2017) reported that *OsYSL9* is an important Fe(III)-DMA transporter expressed in immature seeds for Fe distribution within the seeds. Additionally, the knockdown of *OsYSL9* increased the Fe concentration in the endosperm (polished seeds). Thus, these *OsYSL* transporters can be used to genetically modify rice to efficiently increase the Fe content of the grain.

7.3.2.4 Enhanced Fe Translocation by Knockdown of the Vacuolar Fe Transporter Gene, *OsVITs*

AtVIT1 is reported as a vacuolar Fe transporter that transports Fe and Mn into the vacuole in *Arabidopsis* (Kim et al. 2006). The crystal structure and detailed molecular mechanism of the VIT1 transporter from rose gum have been recently reported (Kato et al. 2019). Disruption of the rice *OsVIT1* and *OsVIT2* increased Fe concentrations by 1.4-fold in brown seeds, but decreased Fe concentrations by 0.8-fold in flag leaves (Zhang et al. 2012). Bashir et al. (2013) also reported that an *OsVIT2* knockdown mutant showed 1.3-fold and 1.8-fold increases in Fe concentrations in brown and polished rice seeds, respectively. WT rice does not translocate Fe from the flag leaf to the seed (Tsukamoto et al. 2009), and this may be because *VIT* genes are highly expressed in rice flag leaves and Fe in the flag leaf is accumulated inside the vacuole by VITs. Thus, disruption of OsVIT1 or OsVIT2 prevents Fe accumulation in the vacuoles in flag leaves and allows extra Fe to translocate from flag leaves to the seeds. Therefore, *OsVIT1* and *OsVIT2* knockdowns are potential candidates for Fe biofortification approaches.

7.3.2.5 Negative Regulators of Fe Homeostasis in Plants

HRZ proteins in rice and its homologous BTS and BTSL in *Arabidopsis* are RING ubiquitin ligases that work as Fe uptake repressors and negative regulators of Fe deficiency responses in both graminaceous and non-graminaceous plants. (Kobayashi et al. 2013; Rodríguez-Celma et al. 2019). Moreover, rice HRZs also play a vital role in preventing Fe overload (Aung et al. 2018a). Kobayashi et al. (2013) reported that *OsHRZ* knockdown plants exhibited substantial tolerance to Fe deficiency and accumulated more Fe in their shoots and grains when cultivated in the field. This suggests that rice HRZs (OsHRZ1 and OsHRZ2) are also potent regulators of Fe accumulation. These genes may provide new approaches for Fe biofortification in rice.

7.3.2.6 Other Potential Genes for Fe Biofortification

There are other novel approaches that have the potential to increase Fe in rice grains or enhance Fe uptake/translocation within the plant. Lee and An (2009) produced transgenic rice that expressed *OsIRT1* under the control of the *Ubiquitin* promoter. This rice showed a 1.7-fold increase in Fe concentration in leaves and a 13% increase in Fe in brown seeds. Tan et al. (2015) also reported that overexpression of *MxIRT1* increased the Fe and Zn content in rice seeds. Additionally, Shin et al. (2013) reported that IRT1 degradation factor1 (IDF1), a RING E3 ubiquitin ligase, regulates the degradation of IRT1 in *Arabidopsis*. It is possible that the knockdown of the rice *IDF1* homolog will increase Fe uptake and Fe accumulation in rice grains similar to *OsHRZ*. Ogo et al. (2011) reported that overexpression of *OsIRO2* under the control of the *CaMV 35S* promoter resulted in Fe deficiency tolerance and a 3-fold higher Fe concentration in the seeds under calcareous soil cultivation, which is a low Fe availability condition. Yang et al. (2016) engineered fusion polypeptides composed of a carbohydrate-binding module family 11 (CBM11) and a Fe-binding peptide (IBP) for secretion into rice cell walls. The Fe concentration in rice polished seeds with *CBM-IBP* increased 2.5-fold. Grillet et al. (2018) and Hirayama et al. (2018) reported a ubiquitous family of peptides that control Fe transport in plants, which they designated as IRON MAN (IMA)/Fe-Uptake-Inducing Peptide (FEP). Expression of *Arabidopsis IMA/FEP* genes was induced under Fe-deficient conditions in the vascular tissues and mesophyll cells of leaves and in the vascular tissues in roots. Deletion of *Arabidopsis IMA3/FEP1* exhibited impaired Fe accumulation in shoots. Overexpression of *IMA/FEP* in *Arabidopsis* induced Fe uptake genes in roots, leading to the accumulation of Fe and Mn in all plant parts, including the seeds. Furthermore, some microRNAs are involved in maintaining Fe homeostasis in plants (Waters et al. 2012; Agarwal et al. 2015). Fe deficiency-inducible *IMA/FEPs* and their functions are conserved in rice and *Arabidopsis* (Grillet et al. 2018). These recent findings on Fe homeostasis-related genes and knowledge of Fe nutrition may provide novel approaches to further improve Fe biofortification of rice.

7.3.2.7 Combined Approaches

Combined approaches have been attempted for Fe biofortification of rice because the target level of Fe is difficult to achieve solely by introducing single genes. Wirth et al. (2009) introduced the *CaMV 35S* promoter-*AtNAS1* and *Globulin* promoter-*Pvferritin* genes into *Japonica* cv. Taipei 309 together with the *Globulin* promoter-*Afphytase* gene cassette to reduce the phytate content and improve Fe bioavailability in rice seeds. Fe concentrations were increased up to 6-fold in hydroponic cultures under various Fe nutrient conditions. Singh et al. (2017) produced wheat expressing both the rice *OsNAS2* gene and the bean ferritin gene (*PvFerritin*). Fe and Zn mineral concentrations were increased 2-fold in the transgenic line compared with a non-transgenic line of wheat flour.

We introduced 2.3-kb *OsGluB1* promoter-*SoyferH2 (soybean ferritin H2)* , *OsGlb1* promoter–*SoyferH2*, *OsActin1* promoter–*HvNAS1*, *OsSUT1* promoter-*OsYSL2*, and *OsGlb1* promoter–*OsYSL2* genes into *Japonica* cv. Tsukinohikari and produced Fer-NAS-YSL2 rice lines (Masuda et al. 2012). The Fe concentration in polished seeds was increased up to 6-fold under soil cultivation in a greenhouse. Moreover, the Fe concentration in polished seeds was 4.4-fold higher than in non-transgenic seeds under paddy field cultivation, without any loss in yield (Masuda et al. 2012). As described above, Myanmar is one of the important target regions for Fe biofortification of rice. Thus, we introduced this Fer-NAS-YSL2 gene into the Paw San Yin variety, which is a popular high-quality rice variety in Myanmar (Aung et al. 2013). The Fe concentration of the Paw San Yin Fer-NAS-YSL2 line increased

3.4-fold in polished seeds, which was close to the dietary target level (Aung et al. 2013).

We also introduced the *SoyferH2* gene under the control of the *OsGluB1* promoter and the *OsGlb1* promoter to enhance Fe accumulation in seeds, and introduced three fragments of the barley genome (5.9 kb of *HvNAS1*, 11 kb of *HvNAAT-A,-B*, and 6.0 kb of *IDS3*) to concomitantly enhance MA production in the rice plants (Masuda et al. 2013b). Transgenic lines showed tolerance to Fe deficiency in the calcareous soil where Fe availability was low due to the high pH of the soil. Concentrations of Fe in the polished seeds of this transgenic line were increased in both normal and calcareous soils, but not in the ferritin-introduced line, compared to the non-transgenic lines in calcareous soil. These results indicated that the concomitant introduction of the *ferritin* and MA biosynthetic genes effectively increased Fe levels in seeds without inducing Fe sensitivity under Fe-limited conditions.

Boonyaves et al. (2016, 2017) also produced Fe-biofortified rice by introducing the *NOD* promoter-*AtIRT1* together with the *CaMV 35S* promoter-*AtNAS1* and the *OsGlb1* promoter-ferritin genes. This demonstrated the synergetic effect of increasing Fe concentrations through a combination of multiple approaches. Trijatmiko et al. (2016) produced Fe-biofortified indica rice by introducing the *CaMV 35S* promoter-*OsNAS2* and the rice *glutelinB1* or rice *glutelinA2* promoter-*SoyferH1* and confirmed that the grain increased in Fe when grown in the field. Additionally, high Fe IR64 rice lines were backcrossed with agronomically important varieties such as Ciherang, BR29, and NSIC Rc222 for the purpose of achieving high Fe concentrations in those popular varieties (Trijatmiko et al. 2016). These results highlighted that the combined introduction of multiple Fe homeostasis genes was more effective than introducing a single gene to achieve Fe biofortification.

7.4 Research into the Fe Excess Response Mechanisms in Rice

Although Fe is an essential element for plants and animals, it causes toxic when accumulating excess amount within cells. Among graminaceous plants, rice is mainly cultivated in a paddy field, where ferrous ion is abundant. Interestingly, rice has a very wide range of Fe acceptance under Fe excess cultivation. Aung et al. (2018b) showed that there were no growth defects even when grown under 30× Fe excess compared to normal conditions. It is thought that a strict regulatory system, such as HRZ (Aung et al. 2018a), may exist in rice plants to prevent the uptake of excess Fe and translocate it into the seeds for the survival of plants under Fe excess condition. Thus, understanding the underlying mechanisms of the Fe excess response and controlling Fe excess regulation in rice by genome editing will enable Fe biofortification in rice. However, compared to the Fe deficiency response, the Fe excess response in rice is currently not well understood at the molecular level. We performed microarray analyses in various tissues of rice under various levels of Fe excess, which showed that many genes were induced under Fe excess conditions (Aung et al. 2018b). For example, in this study, we found that similar to the Fe excess–inducible genes *OsFerritin* and *OsVIT2*, *OsNAS3* was also highly induced in all tissues tested. Further investigation reveals that *OsNAS3* knockout showed sensitive in Fe excess, suggesting that *OsNAS3* is important for mitigating Fe excess stress in rice (Aung et al. 2019). It is speculated that rice has a strict regulatory mechanism for Fe uptake because it is typically cultivated in paddy fields that are rich in soluble ferrous Fe. Low pH acidic paddy conditions also cause damage to rice plants due to excess Fe. In this case, a tolerant rice genotype would have a molecular mechanism to avoid taking up excess Fe or to isolate the excess Fe within the plant body (Aung et al. 2018b). We also found that HRZ knockdown rice was sensitive to Fe excess (Aung et al. 2018a). Thus, studies of rice Fe excess regulators, such as HRZ, or Fe uptake/translocation limitation systems will provide new candidate genes for Fe biofortification.

7.5 Genome Editing Approaches

Biofortification through transgenic approaches is a successful methodology, but it remains impractical to apply in the field due to strict regulatory hurdles and low consumer acceptance worldwide. Vasconcelos et al. (2017) described the future possibility of biofortification by the application of TALEN and CRISPR/Cas9 (Schaart et al. 2015). Gene editing is a new and powerful tool for future biofortification programs by mutating a single or several DNA nucleotides, replacing existing alleles with new ones, or inserting new gene(s) at targeted sites in the genome (Wang et al. 2014). For example, Tang et al. (2017) reported that a knockout of *OsNramp5* using the CRISPR/Cas9 system produces low Cd-accumulating indica rice without compromising the yield. Provided that the gene-edited traits cannot be distinguished from the same traits derived from conventional breeding approaches or mutation approaches, new crop varieties generated by gene editing may not be subject to the strict controls that are commonly applied to genetically modified crops (Abdallah et al. 2015; Huang et al. 2016). Thus, gene-edited crops produced through recent technological advances will have greater consumer acceptance compared to current transgenic approaches.

7.6 The Gateway from the Laboratory to the People

The gateway of Fe biofortification of rice from the laboratory to the fields and the people are shown in Fig. 7.2. Transgenic approaches are efficient and have proven successful, but issues remain regarding adoption by farmers and consumer acceptance. To mitigate this, Masuda et al. (2013b) used a marker-free vector to obtain Fe-biofortified rice, in which the selection marker gene *HPT* can be removed by

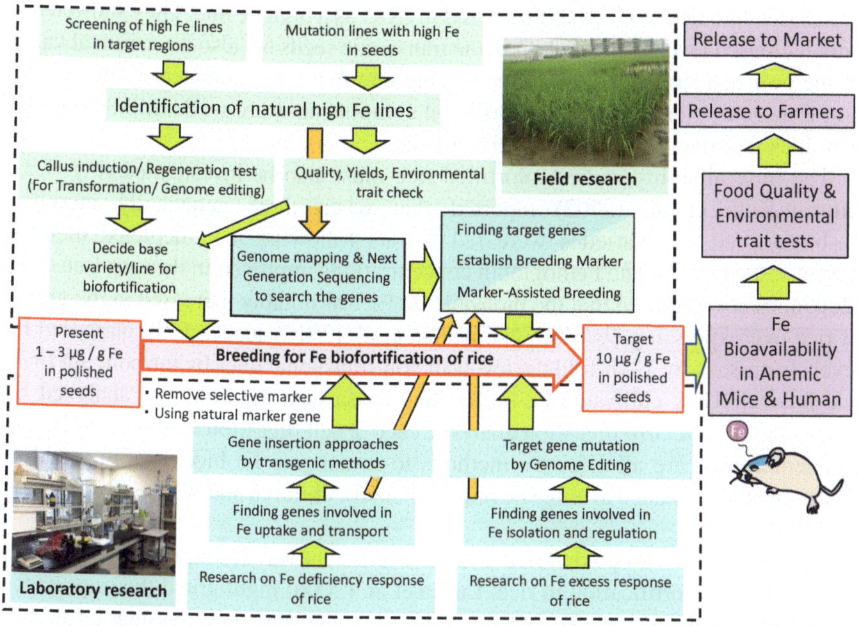

Fig. 7.2 The gateway of iron biofortification of rice: from the laboratory to the people. Flowchart of the necessary steps of rice Fe biofortification programs from laboratory research to practical applications, including breeding, field research, distribution to farmers, and release on the market

estradiol treatment after transformation (Nishizawa et al. 2006; Usuda et al. 2009). Oliva et al. (2014) generated a phytoferritin- overexpressing-popular indica variety without inducing a marker gene by the co-transformation approach (Miki and McHugh 2004). Additionally, the rice *OsALS* mutant gene is a naturally selectable marker gene by bispyribac-sodium to produce transgenic rice (Kawai et al. 2007). These marker-free systems or marker genes derived from the rice genome can be applied to create transgenic rice that is more readily accepted by the general public.

It is also very important to test biofortified lines in the fields in the target region (country, province, or prefecture) for Fe biofortification during several growing seasons to ensure that the increased Fe levels achieved are reproducible and robust (Fig. 7.2). Masuda et al. (2008, 2012) and Trijatmiko et al. (2016) confirmed improved Fe nutrition in rice grains grown both in greenhouses and in field cultivation trials.

Fe-biofortified rice produced by transgenic approaches can be a source of Fe biofortification of other agronomically important rice varieties. Paul et al. (2014) performed crossbreeding analyses based on *OsGluB1-SoyFerH1* rice lines together with the important indica rice cultivar Swarna.

It is also important to check other agronomical traits such as yield components, other nutritional components, and qualities of the Fe-biofortified variety before releasing it to farmers. Gayen et al. (2013) introduced the ferritin gene into the indica rice line IR68144, which was bred in IRRI as a high Fe line. They showed not only increased levels of Fe and Zn in the transgenic seeds but also a nutritional value of the polished rice grains in the transgenic line that was equivalent to that of the non-transgenic rice plants. The nutritional components were well within the range of values reported for other commercial lines.

The bioavailability of Fe-biofortified rice must also be investigated (Fig. 7.2). Murray-Kolb et al. (2002) reported that when three genetically modified Fe-biofortified rice varieties were fed to rats following a Fe-depleted diet for 2 weeks, the liver Fe and hemoglobin concentrations improved in the rats. Lee et al. (2009a, b) also reported that the bioavailable Fe was notably increased in the grains of rice overexpressing *OsNAS3*. Narayanan et al. (2019) successfully increased Fe levels in cassava by 7–18-fold the levels in non-transgenic lines by introducing *IRT1* and *ferritin*. They cultivated Fe-biofortified cassava in the field and analyzed Fe bioavailability. *In vitro* digestion analysis, caco-2 cell digestion analysis, or the use of anemic mice are all effective methods to investigate Fe bioavailability and to determine the estimated average Fe requirement for children and women (Narayanan et al. 2019).

Combining Fe with golden rice is another efficient way to improve the bioavailability of Fe biofortification in rice. Lucca et al. (2006) highlighted the importance of biofortification of both Fe and provitamin A, and thus the combination of rice containing high β-carotene (provitamin A), known as golden rice, is also effective. Dias et al. (2015) showed that the combination of rice and common beans with high carotenoid content increases Fe bioavailability and leads to increased hemoglobin regeneration efficiency in the human body. Singh et al. (2017) produced rice with high Fe, Zn, and β-carotene in the grains. They developed rice lines expressing *AtNAS1*, *PvFerritin*, *CRT1*, and *ZmPSY* in a single genetic locus to increase the Fe, Zn, and β-carotene content. Researchers involved with the VIRCA Plus Project (cassavaplus.org) are currently integrating high Fe and Zn traits into cassava varieties that are popular in Nigeria, with more field evaluations and assessments planned for 2019 (Danforth 2019). Their success may pave the way for Fe biofortification in rice.

Steur et al. (2015) reported that biofortification offers a complementary, cost-effective intervention for governments to improve public health, with market potential for producing biofortified crops by transgenic approaches. They also insisted that researchers continue to produce important new transgenic biofortification traits, while economists demonstrate their market potential.

7.7 Conclusion

We conclude that Fe biofortification in rice is the most sustainable and promising solution to mitigate Fe deficiency in humans. Fe biofortification using genetic engineering technologies alongside alternative solutions, such as marker-free transgenic approaches or those using markers of the rice genome itself, or with recent advanced approaches such as gene-editing, may offer complementary, cost-effective methods for Fe biofortification in the near future.

References

Abbaspour N, Hurrell R, Kelishadi R (2014) Review on iron and its importance for human health. J Res Med Sci 19:164–174

Abdallah N, Prakash C, McHughen A (2015) Genome editing for crop improvement: challenges and opportunities. GM Crops Food 6:183–205

Agarwal S, Mangrauthia SK, Sarla N (2015) Expression profiling of iron deficiency responsive microRNAs and gene targets in rice seedlings of Madhukar×Swarna recombinant inbred lines with contrasting levels of iron in seeds. Plant and Soil 396:137–150

Anuradha K, Agarwal S, Rao YV, Rao KV, Viraktamath BC, Sarla N (2012) Mapping QTLs and candidate genes for iron and zinc concentrations in unpolished rice of Madhukar×Swarna RILs. Gene 508:233–240

Aoki N, Hirose T, Scofield NG, Whitfeld RP, Furbank TR (2003) The sucrose transporter gene family in rice. Plant Cell Physiol 44:223–232

Aoyama T, Kobayashi T, Takahashi M, Nagasaka S, Usuda K, Kakei Y, Ishimaru Y, Nakanishi H, Mori S, Nishizawa NK (2009) OsYSL18 is a rice iron(III)-deoxymugineic acid transporter specifically expressed in reproductive organs and phloem of lamina joints. Plant Mol Biol 70:681–692

Ariga T, Hazama K, Yanagisawa S, Yoneyama T (2014) Chemical forms of iron in xylem sap from graminaceous and non-graminaceous plants. Soil Sci Plant Nutr 60:460–469

Aung MS, Masuda H, Kobayashi T, Nakanishi H, Yamakawa T, Nishizawa NK (2013) Iron biofortification of Myanmar rice. Front Plant Sci 4:158

Aung MS, Masuda H, Kobayashi T, Kakei Y, Tun YT, Nakanishi H, Yamakawa T, Nishizawa NK (2016) Identification of mineral-rich rice varieties in Myanmar. Myanmar Agri Res J 3:14–29

Aung MS, Kobayashi T, Masuda H, Nishizawa NK (2018a) Rice HRZ ubiquitin ligases are crucial for response to excess iron. Physiol Plant 163:282–296

Aung MS, Masuda H, Kobayashi T, Nishizawa NK (2018b) Physiological and transcriptomic analysis of responses to different levels of iron excess stress in various rice tissues. Soil Sci Plant Nutr 64:370–385

Aung MS, Masuda H, Nozoye T, Kobayashi T, Jeon J-S, An G, Nishizawa NK (2019) Nicotianamine synthesis by OsNAS3 is important for mitigating iron excess stress in rice. Front Plant Sci 10:660

Banerjee S, Sharma DJ, Verulkar SB, Chandel G (2010) Use of in silico and semi quantitative RT-PCR approaches to develop nutrient rich rice (*Oryza sativa* L.). J Biotechnol 9:203–212

Bashir K, Inoue H, Nagasaka S, Takahashi M, Nakanishi H, Mori S, Nishizawa NK (2006) Cloning and characterization of deoxymugineic acid synthase genes from graminaceous plants. J Biol Chem 281:32395–32402

Bashir K, Takahashi R, Akhtar S, Ishimaru Y, Hiromi N, Nishizawa NK (2013) The knockdown of OsVIT2 and MIT2 affects iron localization in rice seed. Rice 6:1–7

Boonyaves K, Gruissem W, Bhullar NK (2016) *NOD* promoter-controlled *AtIRT1* expression functions synergistically with *NAS* and *FERRITIN* genes to increase iron in rice grains. Plant Mol Biol 90:207–215

Boonyaves K, Wu T-Y, Gruissem W, Bhullar NK (2017) Enhanced grain iron levels in rice expressing an iron-regulated metal transporter, nicotianamine synthase and ferriting gene cassette. Front Plant Sci 8:130

Borrill P, Connorton JM, Balk J, Miller AJ, Sanders D, Uauy C (2014) Biofortification of wheat grain with iron and zinc: Integrating novel genomic resources and knowledge from model crops. Front Plant Sci 5:53

Bouis HE, Hotz C, McClafferty B, Meenakshi JV, Pfeiffer WH (2011) Biofortification: a new tool to reduce micronutrient malnutrition. Food Nutr Bull 32:202–215

Bughio N, Yamaguchi H, Nishizawa NK, Nakanishi H, Mori S (2002) Cloning an iron-regulated metal transporter from rice. J Exp Bot 53:1677–1682

Bürger A, Høgh-Jensen H, Gondah J, Hash CT, Haussmann BIG (2014) Micronutrient density and stability in West African pearl millet: Potential for biofortification. Crop Sci 54:1709–1720

Connolly EL, Campbell NH, Grotz N, Prichard CL, Guerinot ML (2003) Overexpression of the FRO2 ferric chelate reductase confers tolerance to growth on low iron and uncovers posttranscriptional control. Plant Physiol 133:1102–1110

Curie C, Panavience Z, Loulergue C, Dellaporta SL, Briat JF, Walker EL (2001) Maize yellow stripe1 encodes a membrane protein directly involved in Fe (III) uptake. Nature 409:346–349

Danforth D (2019) Cassava high in iron and zinc could improve diets and health in west Africa. Plant Science Center. https://www.eurekalert.org/pub_releases/2019-01/ddps-chi012819.php

Dias DM, Moreira MEC, Gomes MJC, Toledo RCL, Nutti MR, Sant'Ana HMP, Martino HSD (2015) Rice and bean targets for biofortification combined with high carotenoid content crops regulate transcriptional mechanisms increasing iron bioavailability. Nutrients 7:9683–9696. https://doi.org/10.3390/nu7115488

Díaz-Benito P, Banakar R, Rodríguez-Menéndez S, Capell T, Pereiro R, Christou P, Abadía J, Fernández B, Álvarez-Fernández A (2018) Iron and zinc in the embryo and endosperm of rice (*Oryza sativa L.*) seeds in contrasting 2′-deoxymugineic acid/nicotianamine scenarios. Front Plant Sci 9:1190

Du J, Zeng D, Wang B, Qian Q, Zheng S, Ling HQ (2013) Environmental effects on mineral accumulation in rice grains and identification of ecological specific QTLs. Environ Geochem Health 35:161–170

Duy D, Wanner G, Meda AR, von Wirén N, Soll J, Philippar K (2007) PIC1, an ancient permease in arabidopsis chloroplasts mediates iron transport. Plant Cell 19:986–1006

Eide D, Broderius M, Fett J, Guerinot ML (1996) A novel iron-regulated metal transporter from plants identified by functional expression in yeast. Proc Natl Acad Sci U S A 93:5624–5628

FAO. (2002). The State of Food Insecurity in the World 2002. Rome: Food and Agriculture Organization of the United Nations (FAO).

FAOSTAT (2009) Food and Agriculture Organization Statistical Database. Available from: http://www.fao.org/faostat/en/#data/QCFAOSTAT (2017) Food and Agriculture Organization Statistical Database. Available from: http://www.fao.org/faostat/en/#data/QC

Food and Nutrition Board (2001) Dietary reference intakes from vitamin A, vitamin K, arsenic, boron, chromium, copper, iodine, iron, manganese, molybdenum, nickel, silicon, vanadium, and zinc. Institute of Medicine. National. Academic Press, Washington, DC

Food Fortification Initiative (2013) Impact of fortifying wheat flour with iron, food fortification initiative. http://www.ffinetwork.org/

Fourcroy P, Sisó-Terraza P, Sudre D, Savirón M, Reyt G, Gaymard F, Abadía A, Abadia J, Álvarez-Fernández A, Briat JF (2014) Involvement of the ABCG37 transporter in secretion of scopoletin and derivatives by Arabidopsis roots in response to iron deficiency. New Phytol 201:155–167

Frossard E, Bucher M, Machler F, Mozafar A, Hurrell R (2000) Potential for increasing the content and bioavailability of Fe Zn and Ca in plants for human nutrition. J Sci Food Agric 80:861–879

Garcia-Oliveira AL, Tan L, Fu Y, Sun C (2009) Genetic identification of quantitative trait loci for contents of mineral nutrient in rice grain. J Integr Plant Biol 51:84–92

Gayen D, Sarkar SN, Datta SK, Datta K (2013) Comparative analysis of nutritional compositions of transgenic high iron rice with its non-transgenic counterpart. Food Chem 138:835–840

Gómez-Galera S, Sudhakar D, Pelacho AM, Capell T, Christou P (2012) Constitutive expression of a barley Fe phytosiderophore transporter increases alkaline soil tolerance and results in iron partitioning between vegetative and storage tissues under stress. Plant Physiol Biochem 53:46–53

Goto F, Yoshihara T, Shigemoto N, Toki S, Takaiwa F (1999) Iron fortification of rice seed by the soybean ferritin gene. Nat Biotechnol 17:282–286

Gregorio GB, Senadhira D, Htut H, Graham RD (2000) Breeding for trace mineral density in rice. Food Nutr Bull 21:382–386

Grillet L, Lan P, Li W, Mokkapati G, Schmidt W (2018) IRON MAN is a ubiquitous family of peptides that control iron transport in plants. Nat Plants 4:953–963

Gross J, Stein RJ, Fett-Neto AG, Fett JP (2003) Iron homeostasis related genes in rice. Genet Mol Biol 26:477–497

Grotz N, Guerinot ML (2006) Molecular aspects of Cu, Fe and Zn homeostasis in plants. Biochim Biophys Acta 1763:595–608

Grusak MA, Cakmak I (2004) Methods to improve the crop delivery of minerals to humans and livestock. In: Broadley MR, White PJ (eds) Plant nutritional genomics. Blackwell, Oxford, pp 265–286

Grusak MA, DellaPenna D (1999) Improving the nutrient composition of plants to enhance human nutrition and health. Annu Rev Plant Physiol Plant Mol Biol 50:133–161

Haas JD, Beard JL, Murray-Kolb LE, del Mundo AM, Felix A, Gregorio GB (2005) Iron-biofortified rice improves the iron stores of nonanemic Filipino women. J Nutr 135:2823–2830

Hall JL, Williams LE (2003) Transition metal transporters in plants. J Exp Bot 54:2601–2613

Hancock JF (2004) Plant Evolution and the Origin of Crop Species. CABI Publishing. ISBN 0-85199-685-X

He W, Shohag MJ, Wei Y, Feng Y, Yang X (2013) Iron concentration, bioavailability, and nutritional quality of polished rice affected by different forms of foliar iron fertilizer. Food Chem 141:4122–4126

Hell R, Stephan UW (2003) Iron uptake, trafficking and homeostasis in plants. Planta 216:541–551

Hiei Y, Ohta S, Komari T, Kumashiro T (1994) Efficient transformation of rice (*Oryza sativa L.*) mediated by Agrobacterium and sequence analysis of the bounderies of the T-DNA. Plant J 6:271–282

Higuchi K, Suzuki K, Nakanishi H, Yamaguchi H, Nishizawa NK, Mori S (1999) Cloning of nicotianamine synthase genes, novel genes involved in the biosynthesis of phytosiderophores. Plant Physiol 119:471–479

Hirayama T, Lei GJ, Yamaji N, Nakagawa N, Ma JF (2018) The Putative Peptide Gene FEP1 Regulates Iron Deficiency Response in Arabidopsis. Plant Cell Physiol 59:1739–1752

Hotz CC, McClafferty B (2007) From harvest to health: Challenges for developing biofortified staple foods and determining their impact on micronutrient status. Food Nutr Bull 28:271–279

Huang S, Weigel D, Beachy R, Li J (2016) A proposed regulatory framework for genome-edited crops. Nat Genet 48:109–111

Hunt JR (2003) Bioavailability of iron, zinc, and other trace minerals from vegetarian diets. Am J Clin Nutr 78:633–639

Hurrell RF (1997) Preventing iron deficiency through food fortification. Nutr Rev 55:210–222

Hurrell R (2002) How to ensure adequate iron absorption from iron-fortied food. Nutr Rev 60:7–15

Inoue H, Higuchi K, Takahashi M, Nakanishi H, Mori S, Nishizawa NK (2003) Three rice Nicotianamine synthase genes, *OsNAS1*, *OsNAS2*, and *OsNAS3* are expressed in cells involved in long-distance transport of iron and differentially regulated by iron. The Plant J 36:366–381

Inoue H, Takahashi M, Kobayashi T, Suzuki M, Nakanishi H, Mori S, Nishizawa NK (2008) Identification and localization of the rice nicotianamine aminotransferase gene *OsNAAT1* expression suggests the site of phytosiderophore synthesis in rice. Plant Mol Biol 66:193–203

Inoue H, Kobayashi T, Nozoye T, Takahashi M, Kakei Y, Suzuki K, Nakazono M, Nakanishi H, Mori S, Nishizawa NK (2009) Rice *OsYSL15* is an iron-regulated iron(III)-deoxymugineic acid transporter expressed in the roots and is essential for iron uptake in early growth of the seedlings. J Biol Chem 284:3470–3479

Ishimaru Y, Suzuki M, Tsukamoto T, Suzuki K, Nakazono M, Kobayashi T et al (2006) Rice plants take up iron as an Fe^{3+}-phytosiderophore and as Fe^{2+}. Plant J 45:335–346

Ishimaru Y, Masuda H, Bashir K, Inoue H, Tsukamoto T, Takahashi M, Nakanishi H, Aoki N, Hirose T, Ohsugi R, Nishizawa NK (2010) Rice metal-nicotianamine transporter, *OsYSL2*, is required for the long-distance transport of iron and manganese. Plant J 62:379–390

Jeng TL, Lin YW, Wang CS, Sung JM (2012) Comparisons and selection of rice mutants with high iron and zinc contents in their polished grains that were mutated from the indica type cultivar IR64. J Food Compos Anal 28:149–154

Johnson AAT, Kyriacou B, Callahan DL, Carruthers L, Stangoulis J, Lombi E, Tester M (2011) Constitutive overexpression of the *OsNAS* gene family reveals single gene strategies for effective iron- and zinc-biofortification of rice endosperm. PLoS One 6:e24476

Kakei Y, Ishimaru Y, Kobayashi T, Yamakawa T, Nakanishi H, Nishizawa NK (2012) *OsYSL16* plays a role in the allocation of iron. Plant Mol Biol 79:583–594

Kalgaonkar S, Lonnerdal B (2008) Effects of dietary factors on iron uptake from ferritin by Caco-2 cells. J Nutr Biochem 19:33–39

Kalgaonkar S, Lonnerdal B (2009) Receptor-mediated uptake of ferritinbound iron by human intestinal Caco-2 cells. J Nutr Biochem 20:304–311

Kato T, Kumazaki K, Wada M, Taniguchi R, Nakane T, Yamashita K, Hirata K, Ishitani R, Ito K, Nishizawa T, Nureki O (2019) Crystal structure of plant vacuolar iron transporter VIT1. Nat Plants 5:308–315

Kawai K, Kaku K, Izawa N, Fukuda A, Tanaka Y, Shimizu T (2007) Functional analysis of transgenic rice plants expressing a novel mutated ALS gene of rice. J Pestic Sci 32:385–392

Khalekuzzaman M, Datta K, Olival N, Alam MF, Joarder I, Datta SK (2006) Stable integration, expression and inheritance of the ferritin gene in transgenic elite indica rice cultivar BR29 with enhanced iron level in the endosperm. Indian Dent J 5:26–31

Kim SA, Punshon T, Lanzirotti A, Li L, Alonso JM, Ecker JR, Kaplan J, Guerinot ML (2006) Localization of iron in Arabidopsis seed requires the vacuolar membrane transporter VIT1. Science 314:1295–1298

Kobayashi T, Nishizawa NK (2014) Iron sensors and signals in response to iron deficiency. Plant Sci 224:36–43

Kobayashi T, Nakanishi H, Takahashi M, Kawasaki S, Nishizawa NK, Mori S (2001) In vivo evidence that *IDS3* from *Hordeum vulgare* encodes a dioxygenase that converts 2′-deoxymugineic acid to mugineic acid in transgenic rice. Planta 212:864–871

Kobayashi T, Ogo Y, Itai RN, Nakanishi H, Takahashi M, Mori S, Nishizawa NK (2007) The transcription factor IDEF1 regulates the response to and tolerance of iron deficiency in plants. Proc Natl Acad Sci 104:19150–19155

Kobayashi T, Itai RN, Ogo Y, Kakei Y, Nakanishi H, Takahashi M, Nishizawa NK (2009) The rice transcription factor IDEF1 is essential for the early response to iron deficiency, and induces vegetative expression of late embryo genesis abundant genes. Plant J 60:948–961

Kobayashi T, Nagasaka S, Senoura T, Itai RN, Nakanishi H, Nishizawa NK (2013) Iron-binding haemerythrin RING ubiquitin ligases regulate plant iron responses and accumulation. Nat Commun 4:2792

Kobayashi T, Nozoye T, Nishizawa NK (2019) Iron transport and its regulation in plants. Free Radical Biology and Medicine 133:11–20

Koike S, Inoue H, Mizuno D, Takahashi M, Nakanishi H, Mori S, Nishizawa NK (2004) *OsYSL2* is a rice metal-Nicotianamine transporter that is regulated by iron and expressed in the phloem. Plant J 39:415–424

Lanquar V, Lelièvre F, Bolte S, Hamès C, Alcon C et al (2005) Mobilization of vacuolar iron by AtNRAMP3 and AtNRAMP4 is essential for seed germination on low iron. EMBO J 24:4041–4051

Lee S, An G (2009) Over-expression of OsIRT1 leads to increased iron and zinc accumulations in rice. Plant Cell Environ 32:408–416

Lee S, Chiecko JC, Kim SA, Walker EL, Lee Y, Guerinot ML, An G (2009a) Disruption of *OsYSL15* leads to iron inefficiency in rice plants. Plant Physiol 150:786–800

Lee S, Jeon US, Lee SJ, Kim YK, Persson DP, Husted S, Schjørring JK, Kakei Y, Masuda H, Nishizawa NK, An G (2009b) Iron fortification of rice seeds through activation of the nicotianamine synthase gene. Proc Natl Acad Sci U S A 106:22014–22019

Lee S, Kim Y-S, Jeon US, Kim Y-K, Schjoerring JK, An G (2012) Activation of rice nicotianamine synthase 2 (*OsNAS2*) enhances iron availability for biofortiication. Mol Cells 33:269–275

Lucca P, Hurrell R, Potrykus I (2002) Fighting iron deficiency anemia with iron-rich rice. J Am Coll Nutr 21:184–190
Lucca P, Poletti S, Sautter C (2006) Genetic engineering approaches to enrich rice with iron and vitamin A. Physiol Plant 126:291–303
Lyons G, Cakmak I (2012) Agronomic biofortification of food crops with micronutrients. In: Bruulsema TW, Heffer P, Welch RM et al (eds) Fertilizing crops to improve human health: a scientific review. International Plant Nutrition Institute, Paris, France, pp 97–122
Maclean JL, Dawe DC, Hardy B, Hettel GP (eds) (2002) Rice almanac, third edition, source book for the most important economic activity on earth. CABI Publishing, Oxfordshire
Maeda M (2005) Trace elements related to anemia. Biomed Res Trace Elem 16:19–24
Marschner H, Römheld V, Kissel M (1986) Different strategies in higher plants in mobilization and uptake of iron. J Plant Nutr 9:695–713
Martin CDS, Garri C, Pizarro F, Walter T, Theil EC, Núñez MT (2008) Caco-2 intestinal epithelial cells absorb soybean ferritin by mu2 (AP2)-dependent endocytosis. J Nutr 138:659–666
Masuda H, Suzuki M, Morikawa KC, Kobayashi T, Nakanishi H, Takahashi M, Saigusa M, Mori S, Nishizawa NK (2008) Increase in iron and zinc concentrations in rice grains via the introduction of barley genes involved in phytosiderophore synthesis. Rice 1:100–108
Masuda H, Usuda K, Kobayashi T, Ishimaru Y, Kakei Y, Takahashi M, Higuchi K, Nakanishi H, Mori S, Nishizawa NK (2009) Overexpression of the barley nicotianamine synthase gene *HvNAS1* increase iron and zinc concentrations in rice grains. Rice 2:155–166
Masuda H, Ishimaru Y, Aung MS, Kobayashi T, Kakei Y, Takahashi M, Higuchi K, Nakanishi H, Nishizawa NK (2012) Iron biofortification in rice by the introduction of multiple genes involved in iron nutrition. Sci Rep 2:534
Masuda H, Aung MS, Nishizawa NK (2013a) Iron biofortification of rice using different transgenic approaches. Rice 6:40
Masuda H, Kobayashi T, Ishimaru Y, Takahashi M, Aung MS, Nakanishi H, Mori S, Nishizawa NK (2013b) Iron-biofortification in rice by the introduction of three barley genes participated in mugineic acid biosynthesis with soybean *ferritin* gene. Front Plant Sci 4:132
Mayer JE, Pfeiffer WH, Beyer P (2008) Biofortified crops to alleviate micronutrient malnutrition. Curr Opin Plant Biol 11:166–170
Miki B, McHugh S (2004) Selectable marker genes in transgenic plants: applications, alternatives and biosafety. J Biotechnol 107:193–232
Miret S, Simpson RJ, McKie AT (2003) Physiology and molecular biology of dietary iron absorption. Annu Rev Nutr 23:283–301
MOH (2003) National Haemoglobin Survey. National plan of action for food and nutrition. Nutrition Section, Department of Health, Ministry of Health (MOH), Yangon, Myanmar, pp 2006–2010
Mori S, Nishizawa N (1987) Methionine as a dominant precursor of phytosiderophore in graminaceae plant. Plant Cell Physiol 28:1081–1092
Murray-Kolb LE, Takaiwa F, Goto F, Yoshihara T, Theil EC, Beard JL (2002) Transgenic Rice is a source of Iron for Iron-Depleted Rats. J Nutr 132:957–960
Nakanishi H, Yamaguchi H, Sasakuma T, Nishizawa NK, Mori S (2000) Two dioxygenase genes, *Ids3* and *Ids2*, from *Hordeum vulgare* are involved in the biosynthesis of mugineic acid family phytosiderophores. Plant Mol Biol 44:199–207
Narayanan N, Beyene G, Chauhan RD, Gaitán-Solís E, Gehan J, Butts P, Siritunga D, Okwuonu I, Woll A, Jiménez-Aguilar DM et al (2019) Biofortification of field-grown cassava by engineering expression of an iron transporter and ferritin. Nat Biotechnol 37:144–151
National Corn Growers Association (2018) World of Corn 2018: Data is available from: http://www.worldofcorn.com/#/
Nawaz Z, Kakar KU, Li X, Li S, Zhang B, Shou H, Shu Q (2015) Genome-wide association mapping of quantitative trait loci (QTLs) for contents of eight elements in brown rice (*Oryza sativa* L.). J Agric Food Chem 63:8008–8016
Neelamraju S, Swamy BPM, Kaladhar K et al (2012) Increasing iron and zinc in rice grains using deep water rices and wild species – identifying genomic segments and candidate genes. Qual Ass Saf Crops Foods 4:138

Nishiyama R, Kato M, Nagata S, Yanagisawa S, Yoneyama T (2012) Identification of Zn-nicotianamine and Fe-2′-Deoxymugineic acid in the phloem sap from rice plants (*Oryza sativa* L.). Plant Cell Physiol 53:381–390

Nishizawa NK, Mori S, Nakanishi H (2006) Vector for plant transformation. Japan Patent, Patent Number 2006-191906, Tokyo

Nozoye T (2018) The Nicotianamine Synthase Gene is a useful Candidate for Improving the Nutritional Qualities and Fe-Deficiency Tolerance of various crops. Front Plant Sci 9:340

Nozoye T, Nagasaka S, Kobayashi T, Takahashi M, Sato Y, Sato Y, Uozumi N, Nakanishi H, Nishizawa NK (2011) Phytosiderophore Efflux Transporters are crucial for Iron Acquisition in Graminaceous Plants. J Biol Chem 286:5446–5454

Ogo Y, Itai RN, Nakanishi H, Kobayashi T, Takahashi M, Mori S, Nishizawa NK (2007) The rice bHLH protein OsIRO2 is an essential regulator of the genes involved in Fe uptake under Fe deficient conditions. Plant J 51:366–377

Ogo Y, Itai RN, Kobayashi T, Aung MS, Nakanishi H, Nishizawa NK (2011) *OsIRO2* is responsible for iron utilization in rice and improves growth and yield in calcareous soil. Plant Mol Biol 75:593–605

Oliva N, Chadha-Mohanty P, Poletti S, Abrigo E, Atienza G, Torrizo L et al (2014) Large-scale production and evaluation of marker-free indica rice IR64 expressing phytoferritin genes. Mol Breed 33:23–37

Paul S, Ali N, Gayen D, Datta SK, Datta K (2012) Molecular breeding of *Osfer2* gene to increase iron nutrition in rice grain. GM Crops Food 3:310–316

Paul S, Ali N, Datta SK, Datta K (2014) Development of an iron-enriched high-yieldings indica rice cultivar by introgression of a high-iron trait from transgenic iron-biofortified rice. Plant Foods Hum Nutr 69:203–208

Pich A, Scholz G (1996) Translocation of copper and other micronutrients in tomato plants (Lycopersicon esculentum Mill.): nicotianamine-stimulated copper transport in the xylem. J Exp Bot 294:41–47

Puig S, Andrés-Colás N, García-Molina A, Peñarrubia L (2007) Copper and iron homeostasis in *Arabidopsis*: responses to metal deficiencies, interactions and biotechnological applications. Plant Cell Environ 30:271–290

Qu LQ, Takaiwa F (2004) Evaluation of tissue specificity and expression strength of rice seed component gene promoters in transgenic rice. Plant Biotechnol J 2:113–125

Qu LQ, Yoshihara T, Ooyama A, Goto F, Takaiwa F (2005) Iron accumulation does not parallel the high expression level of ferritin in transgenic rice seeds. Planta 222:225–233

Rajniak J, Giehl RFH, Chang E, Murgia I, von Wirén N, Sattely ES (2018) Biosynthesis of redox-active metabolites in response to iron deficiency in plants. Nat Chem Biol 14, 442–450

Rengel Z, Batten GD, Crowley DE (1999) Agronomic approaches for improving the micronutrient density in edible portions of field crops. Field Crop Res 60:27–40

Robinson NJ, Procter CM, Connolly EL, Guerinot ML (1999) A ferric-chelate reductase for iron uptake from soils. Nature 397:694–697

Rodríguez-Celma J, Chou H, Kobayashi T, Long TA, Balk J (2019) Hemerythrin E3 ubiquitin ligases as negative regulators of iron homeostasis in plants. Front Plant Sci 10:98

Römheld V, Marschner H (1986) Evidence for a specific uptake system for iron phytosiderophore in roots of grasses. Plant Physiol 80:175–180

Rudolph A, Becker R, Scholz G, Procházka Z, Toman J, Macek T, Herout V (1985) The occurrence of the amino acid nicotianamine in plants and microorganisms: a Reinvestigation. Biochem Physiol Pflanz 180:557–563

Ruengphayak S, Ruanjaichon V, Phromphan S, Chaichumpoo E, Saensuk C, Toojinda T, Tragoonrung S, Kongkachuichai R, Vanavichit A (2012) Reverse and forward genetics approaches to identify mutants conferring high grain Fe density and tolerance to Fe toxicity. Paper presented at the 10th International Symposium on Rice Functional Genomics, Lotus Hotel Pang Suan Kaew, Chiang Mai

Santi S, Schmidt W (2009) Dissecting iron deficiency-induced proton extrusion in *Arabidopsis* roots. New Phytol 183:1072–1084

Schaart JG, van de Wiel CCM, Lotz LAP, Smulders MJM (2015) Opportunities for products of new plant breeding techniques. Trends Plant Sci 20:1–12

Schmid NB, Giehl RFH, Döll S, Mock HP, Strehmel N, Scheel D, Kong X, Hider RC, Von Wirén N (2014) Feruloyl-CoA 6'-hydroxylase1-dependent coumarins mediate iron acquisition from alkaline substrates in *Arabidopsis*. Plant Physiol 164:160–172

Schroeder JI, Delhaize E, Frommer WB, Guerinot ML, Harrison MJ, Herrera-Estrella L, Horie T, Kochian LV, Munns R, Nishizawa NK, Tsay YF, Sanders D (2013) Using membrane transporters to improve crops for sustainable food production. Nature 497:60–66

Senoura T, Sakashita E, Kobayashi T, Takahashi M, Aung MS, Masuda H, Nakanishi H, Nishizawa NK (2017) The iron-chelate transporter OsYSL9 plays a role in iron. Plant Mol Biol 79:1–13

Shin L-J, Lo J-C, Chen G-H, Callis J, Fu H, Yeh K-C (2013) IRT1 Degradation Factor1, a RING E3 Ubiquitin Ligase, regulates the Degradation of Iron-Regulated Transporter1 in Arabidopsis. The Planst Cell 25:3039–3051

Shojima S, Nishizawa NK, Fushiya S, Nozoe S, Irifune T, Mori S (1990) Biosynthesis of phytosiderophores - invitro biosynthesis of 2'-deoxymugineic acid from l-methionine and nicotianamine. Plant Physiol 93:1497–1503

Singh SP, Gruissem W, Bhullar NK (2017) Single genetic locus improvement of iron, zinc and β-carotene content in rice grains. Sci Rep 7:6883

Sisó-Terraza P, Luis-Villarroya A, Fourcroy P, Briat J-F, Abadía A, Gaymard F, Abadía J, Álvarez-Fernández A (2016) Accumulation and secretion of coumarinolignans and other coumarins in *Arabidopsis thaliana* roots in response to iron deficiency at high pH. Front Plant Sci 7:1711

Sperotto RA, Boff T, Duarte GL, Santos LS, Grusak MA, Fett JP (2010) Identification of putative target genes to manipulate Fe and Zn concentrations in rice grains. J Plant Physiol 167:1500–1506

Sperotto RA, Ricachenevsky FK, Waldow VA, Müller ALH, Dressler VL, Fett JP (2013) Rice grain Fe, Mn and Zn accumulation: How important are flag leaves and seed number? Plant Soil Environ 59:262–266

Stangoulis JCR, Huyn B, Welch RM, Choi E, Graham RD (2007) Quantitative trait loci for phytate in rice grain and their relationship with grain micronutrient content. Euphytica 54:289–294

Stephan UW, Schmidke I, Stephan VW, Scholz G (1996) The nicotianamine molecule is made-to-measure for complexation of metal micronutrients in plants. Biometals 9:84–90

Steur HD, Blancquaert D, Strobbe S, Lambert W, Gellynck X, Straeten DVD (2015) Status and market potential of transgenic biofortified crops. Nat Biotechnol 33:25–29

Stoltzfus RJ, Dreyfuss ML (1998) Guidelsines for the use of iron supplements to prevent and treat iron deficiency anemia. ILSI Press, Washington DC

Suzuki M, Morikawa KC, Nakanishi H, Takahashi M, Saigusa M, Mori S, Nishizawa NK (2008) Transgenic rice lines that include barley genes have increased tolerance to low iron availability in a calcareous paddy soil. Soil Sci Plant Nutr 54:77–85

Takagi S (1976) Naturally occurring iron-chelating compounds in oat-and rice-root washings. Soil Sci Plant Nutr 22:423–433

Takahashi M, Yamaguchi H, Nakanishi H, Shioiri T, Nishizawa NK, Mori S (1999) Cloning two genes for nicotianamine aminotransferase, a critical enzyme in iron acquisition (Strategy II) in graminaceous plants. Plant Physiol 121:947–956

Takahashi M, Terada Y, Nakai I, Nakanishi H, Yoshimura E, Mori S, Nishizawa NK (2003) Role of nicotianamine in the intracellular delivery of metals and plant reproductive development. Plant Cell 15:1263–1280

Tan S, Han R, Li P, Yang G, Li S, Zhang P, Wang W-B, Zhao W-Z, Yin L-P (2015) Over-expression of the MxIRT1 gene increases iron and zinc content in rice seeds. Transgenic Res 24:109–122

Tang L, Mao B, Li Y, Lv Q, Zhang LP, Chen C, He H, Wang W, Zeng X, Shao Y, Pan Y, Hu Y, Peng Y, Xiqin F, Li H, Xia S, Zhao B (2017) Knockout of *OsNramp5* using the CRISPR/Cas9 system produces low Cd-accumulating *indica* rice without compromising yield. Sci Rep 7:14438

Theil EC (2003) Ferritin: at the crossroads of iron and oxygen metabolism. J Nutr 133:1549–1553

Theil EC (2011) Iron Homeostasis and Nutritional Iron deficiency. J Nutr 141:724S–728S

Theil EC, Chen H, Miranda C, Janser H, Elsenhans B, Núñez MT, Pizarro F, Schümann K (2012) Absorption of iron from ferritin is independent of heme iron and ferrous salts in women and rat intestinal segments. J Nutr 142:478–483

Thomine S, Lelièvre F, Debarbieux E, Schroeder JI, Barbier-Brygoo H (2003) AtNRAMP3, a multispecific vacuolar metal transporter involved in plant responses to iron deficiency. Plant J 34:685–695

Trijatmiko KR, Dueñas JC, Tsakirpaloglou N, Torrizo L, Arines FM et al (2016) Biofortified indica rice attains iron and zinc nutrition dietary targets in the field. Sci Rep 6:19792

Tsukamoto T, Nakanishi H, Uchida H, Watanabe S, Matsuhashi S, Mori S, Nishizawa NK (2009) ^{52}Fe translocation in barley as monitored by a positron-emitting tracer imaging system (PETIS): Evidence for the direct translocation of Fe from roots to young leaves via phloem. Plant Cell Physiol 50:48–57

UNICEF and Micronutrient Initiative (2004) Vitamin and Mineral Deficiency: A Global Progress Report. Ottawa: The Micronutrient Initiative; New York: UNICEF.

Usuda K, Wada Y, Ishimaru Y, Kobayashi T, Takahashi M, Nakanishi H et al (2009) Genetically engineered rice containing larger amounts of nicotianamine to enhance the anti-hypertensive effect. Plant Biotechnol J 7:87–95

Vasconcelos M, Datta K, Oliva N, Khalekuzzaman M, Torrizo L, Krishnan S, Oliveira M, Goto F, Datta SK (2003) Enhanced iron and zinc accumulation in transgenic rice with the ferritin gene. Plant Sci 164:371–378

Vasconcelos MW, Gruissem W, Bhullar NK (2017) Iron biofortification in the 21st century: setting realistic targets, overcoming obstacles, and new strategies for healthy nutrition. Curr Opin Biotechnol 44:8–15

von Wirén N, Khodr H, Hider RC (2000) Hydroxylated phytosiderophore species from rye and barley possess an enhanced chelating efficiency and affinity for iron(III). Plant Physiol 124:1149–1157

Wang M, Gruissem W, Bhullar NK (2013) Nicotianamine synthase overexpression positively modulates iron homeostasis-related genes in high iron rice. Front Plant Sci 4:156

Wang Y, Cheng X, Shan Q, Zhang Y, Liu J, Gao C, Qiu J-L (2014) Simultaneous editing of three homoeoalleles in hexaploidy bread wheat confers heritable resistance to powdery mildew. Nat Biotechnol 32: 947–951

Waters BM, McIntruf SA, Stein RJ (2012) Rosette iron deficiency transcript and microRNA profiling reveals links between copper and iron homeostasis in *Arabidopsis thaliana*. J Exp Bot 63:5903–5918

Welch RM (2001) Micronutrients, agriculture and nutrition; linkages for improved health and well being. In K. Singh, S. Mori, and R.M. Welch (eds.). Perspectives on the Micronutrient Nutrition of Crops. Jodhpur, India: Scientific Publishers (India), 247–289

White PJ, Broadley MR (2005) Biofortifying crops with essential mineral elements. Trends Plant Sci 10:586–593

WHO (2002) The world health report. Reducing Risks, promoting healthy life. Word Health Organization (WHO), Geneva

WHO (2008) Worldwide prevalence of anemia 1993-2005: WHO Global Database of Anemia. WHO, Geneva, pp 12–13

Wirth J, Poletti S, Aeschlimann B, Yakandawala N, Drosse B, Osorio S, Tohge T, Fernie AR, Günther D, Gruissem W, Sautter C (2009) Rice endosperm iron biofortification by targeted and synergistic action of nicotianamine synthase and ferritin. Plant Biotechnol J 7:1–14

Yang H, Wei H, Ma G, Antunes MS, Vogt S et al (2016) Cell wall targeted *in planta* iron accumulation enhances biomass conversion and seed iron concentration in Arabidopsis and rice. Plant Biotechnol J 14:1998–2009

Yokosho K, Yamaji N, Ueno D, Mitani N, Ma JF (2009) *OsFRDL1* is a citrate transporter required for efficient translocation of iron in rice. Plant Physiol 149:297–305

Yokosho K, Yamaji N, Ma JF (2016) OsFRDL1 expressed in nodes is required for distribution of iron to grains in rice. J Exp Bot 67:5485–5494

Yoneyama T, Ishikawa S, Fujimaki S (2015) Route and regulation of zinc, cadmium, and iron transport in rice plants (*Oryza sativa* L.) during vegetative growth and grain filling: metal transporters, metal speciation, grain Cd reduction and Zn and Fe biofortification. Int J Mol Sci 16:19111–19129

Yuan L, Wu L, Yang C, Lv Q (2012) Effects of iron and zinc foliar applications on rice plants and their grain accumulation and grain quality. J Sci Food Agric 93:254–261

Zhang F, Fan M, Gao X, Zou C, Zuo Y (2008) Soil and crop management for improving iron and zinc nutrition of crops. In: Banuelos G, Lin Z-Q (eds) Development and uses of biofortified agricultural products. CRC Press, Boca Raton, USA, pp 71–93

Zhang Y, Xu YH, Yi HY, Gong JM (2012) Vacuolar membrane transporters *OsVIT1* and *OsVIT2* modulate iron translocation between flag leaves and seeds in rice. Plant J 72:400–410

Zhang M, Pinson SRM, Tarpley L, Huang XY, Lahner B, Yakubova E, Baxter I, Guerinot ML, Salt DE (2014) Mapping and validation of quantitative trait loci associated with concentrations of 16 elements in unmilled rice grain. Theor Appl Genet 127:137–165

Zheng L, Cheng Z, Ai C, Jiang X, Bei X et al (2010) Nicotianamine, a Novel enhancer of Rice Iron bioavailability to humans. PLoS One 5(4):e10190

Zielińska-Dawidziak M (2015) Plant Ferritin—a source of iron to prevent its deficiency. Nutrients 7:1184–1201

Chapter 8
Selenium Agronomic Biofortification in Rice: Improving Crop Quality Against Malnutrition

José C. Ramalho, Faustino A. Roda, Maria F. G. Pessoa, Fernando H. Reboredo, Isabel P. Pais, Alex Ndayiragije, Fernando C. Lidon, and Ana I. Ribeiro-Barros

8.1 Rice Importance in a Climate-Changing World

8.1.1 Global Changes and Warming Implications to Agriculture

Global warming is predicted to have severe impact on the Earth's ecosystems and to constitute a serious threat to agricultural sustainability (IPCC (Intergovernmental Panel on Climate Change) 2014; Tack et al. 2015). The increase of global air temperature by the end of the twenty-first century is expected to fall between 0.3 and 4.8 °C, depending on the actual and future trend of greenhouse gases (GHG) emission scenarios. Additionally, the number and severity of extreme weather events, including altered inter- and intra-annual rainfall patterns is also expected to increase (IPCC (Intergovernmental Panel on Climate Change) 2014).

Alex Ndayiragije, Fernando C. Lidon, Ana I. Ribeiro-Barros and José C. Ramalho contributed equally with all other contributors.

J. C. Ramalho (✉) · A. I. Ribeiro-Barros
Plant-Environment Interactions & Biodiversity Lab (PlantStress&Biodiversity), Linking Landscape, Environment, Agriculture and Food (LEAF), Instituto Superior de Agronomia (ISA), Universidade de Lisboa (UL), Oeiras and Lisbon, Portugal

Unidade de Geobiociências, Geoengenharias e Geotecnologias (GeoBioTec), Faculdade de Ciências e Tecnologia (FCT), Universidade NOVA de Lisboa (UNL), Monte de Caparica, Portugal
e-mail: cochichor@mail.telepac.pt; cochichor@isa.ulisboa.pt

F. A. Roda
Centro Zonal Noroeste, Instituto de Investigação Agraria de Moçambique (IIAM), Lichinga, Niassa, Mozambique

Centro de Biotechnologia (CB), Universidade Eduardo Mondlane (UEM), Maputo, Mozambique

A. Costa de Oliveira et al. (eds.), *The Future of Rice Demand: Quality Beyond Productivity*, https://doi.org/10.1007/978-3-030-37510-2_8

Plant cell metabolism is largely sensitive to warming, including the crucial photosynthesis and respiration pathways. Elevated temperature is known to reduce stomatal conductance and, consequently, the diffusion of CO_2 to the carboxylation sites in the chloroplast, while proportionally stimulates respiration more than photosynthesis (Wise et al. 2004; Wahid et al. 2007; Rodrigues et al. 2016), therefore with a direct impact on energy and carbohydrate skeletons availability that support growth. For instance, warming is expected to affect crop cereal development from sowing to grain maturity, including flower set, and grain-developmental stages (among them, filling), which are crucial to yield (Barnabás et al. 2008). Stronger impacts are expected with the cumulative occurrence of limited water availability, and additionally to the impact on yield also the quality of products could be significantly affected (Chaturvedi et al. 2017).

The world population is predicted to face an increase from 7600 to 9800 million people by 2050 (FAO (Food and Agriculture Organization of the United Nations) 2009, 2011). This implicates that global food production must be enhanced by ca. 60–70% (Powell et al. 2012; FAO (Food and Agriculture Organization of the United Nations) 2016), including a ca. 43% rise in staple crops (Powell et al. 2012). As regards specifically rice (*Oryza sativa* L.), the global production should present an additional 116 million tons already by 2035 as compared to 2010, representing about 26% increase. At the same time, in Africa, about 30 million tons more rice will be consumed, reflecting an expressive 130% greater value than in 2010 (Seck et al. 2012). Having this in mind, such strong increase in food and energy demand will require a profound transformation of food and agriculture systems worldwide in order to guarantee global food security (FAO (Food and Agriculture Organization of the United Nations) 2016). Taking into account that food production depends directly on natural resources closely dependent to climate conditions, an environmentally sustainable increase of agricultural productivity in a context of climate changes is an enormous and complex challenge (FAO (Food and Agriculture Organization of the United Nations) 2016; Semedo et al. 2018).

Cereals represent close to 58% of the area harvested annually, occupying 215 (wheat), 166 (rice), and 152 (maize) million ha (Dawe et al. 2010), and are responsible to supply about half of the world's food calories (Fischer et al. 2014). During the twentieth century, the rise up in mean annual temperatures of ca. 1 °C

M. F. G. Pessoa · F. H. Reboredo · F. C. Lidon
Unidade de Geobiociências, Geoengenharias e Geotecnologias (GeoBioTec), Faculdade de Ciências e Tecnologia (FCT), Universidade NOVA de Lisboa (UNL), Monte de Caparica, Portugal

I. P. Pais
Unidade de Investigação em Biotecnologia e Recursos Genéticos (UIBRG), Instituto Nacional de Investigação Agrária e Veterinária, I.P. (INIAV), Oeiras, Portugal

A. Ndayiragije
International Rice Research Institute (IRRI) and Agricultural Research Institute of Mozambique (IIAM), Lichinga, Niassa, Mozambique

was observed in areas of these cereals. This warming trend is expected to continue (and to be aggravated) in the coming years, with potential yield and quality losses under the new climatic scenarios (Chaturvedi et al. 2017; Zhao et al. 2017). Additionally, in several plant species, including crops as wheat and rice, it was reported a decline (dilution) effect of, among others, protein and mineral nutrient concentrations in the biomass under enhanced [CO_2] exposure, therefore foreseen potential quality implications to human and animal nutrition (McGrath and Lobell 2013; Loladze 2014; Pleijel and Hogy 2015). The dilution in mineral content could be related, namely, to higher growth rates, increased amounts of nonstructural carbohydrates (mainly starch), lower leaf transpiration rates, or changes in the nutrient allocation patterns (Thiec et al. 1995; McGrath and Lobell 2013; Houshmandfar et al. 2018). Additionally, changes in climate and in the associated soil organic carbon content might implicate altered soil micronutrient availability to plants, as for selenium (Se) that was estimated to suffer an overall reduction in soil concentrations, likely increasing the risk of Se deficiency under future climate change conditions (Jones et al. 2017). The referred lowered biomass content and soil availability of minerals would be even more important, since a significant fraction of the human population currently suffers from micronutrient malnutrition in one or more essential dietary nutrients (McGrath and Lobell 2013). This is usually referred as "hidden hunger," which is defined as a chronic lack in any of the dietary micronutrients and vitamins (Kennedy et al. 2003; Stein 2015; Rehman et al. 2018). Interestingly, in the coffee bean, the increase in air temperature might offset the elevated [CO_2] dilution effect on mineral content (Martins et al. 2014), while high [CO_2] mitigated the heat impact and preserved coffee bean quality characteristics (Ramalho et al. 2018), showing that much is still to be unveiled regarding the impacts of ongoing and future climate changes in crop product quality. In fact, by contrast, it should be also taken into account rice sensitivity to elevated CO_2 and heat stress (Chaturvedi et al. 2017). Moreover, climate changes were predicted to negatively affect global rice production (Teixeira et al. 2013), with estimates of yield reductions reaching 33%, 32%, and 23%, in India, Thailand, and Vietnam, respectively (Teng et al. 2016). Additionally, actual suitable cropped areas are expected to be reduced (Castro-Llanos et al. 2019), and grain quality parameters compromised, including opacity, amylose content, and cooking quality (Lin et al. 2010; Uprety and Reddy 2016). In fact, considering that grain development and total yield is maximal at the optimum temperature of 25 °C, a yield reduction of 10% is expected per each increased °C above this temperature, until negligible productions at 35–36 °C (Hatfield et al. 2011). Besides, productivity is highly impaired by spikelet sterility under temperatures above 33–35 °C, which will decrease grain formation and size, reduce the growth period, while increase maintenance respiration (Wassmann et al. 2010; Hatfield et al. 2011), all of which will impact yield. Even some less dramatic impact estimates (but still important) pointed to global yield reductions of 3.2% per each increased °C, although some regions (e.g., India) might present greater declines (up to 6%) in the future due to temperature rise (Zhao et al. 2017). Additionally, drought (Pandey et al. 2007), floods that may result in plants submergence (Redfern et al. 2012), and salinity, namely, affecting coastal and delta regions due to the rise in sea levels (Wassmann et al. 2010), will also contribute to a growing uncertainty regarding the future performance of this crop. As for other crops, some shifts of the traditional areas of production could be unavoidable, in order to guarantee good environmental conditions to rice crop. Therefore, production might have to move to colder regions in higher latitudes in countries like China or to higher altitudes in smaller countries, like Colombia (Wang et al. 2016; Castro-Llanos et al. 2019). In a few cases the benefit of warmer conditions, shifting rice crop northward, could even increase the suitable rice cropping area for irrigated rice production (Zhang et al. 2017).

Although these modeling approaches did not take into account the C-fertilization effect related to the enhancement of air CO_2 concentration, or the mitigation and genetic improvement measures that might be implemented, it is quite clear that climate changes and global warming will impose severe constraints to the amount and quality (including mineral nutrients) of this most important staple crop, rice. This, together with the predicted strong population increase, will demand the implementation of strategies that guarantees food security and safety at a global scale.

8.1.2 *Rice Importance as a Global Crop*

Rice is cultivated worldwide in more than 100 countries but its production and consumption are dominated by Asian countries (from Pakistan to the West to Japan to the East), which account for ca. 91% of world production, followed by the Americas with about 5.5%, Africa (3.3%), Europe (0.6%), and Oceania (0.1%) (FAOSTAT 2018). India, China, Indonesia, Thailand, Bangladesh, and Vietnam are the countries with the largest areas of rice cultivation (FAOSTAT 2017). Currently the world average rice yield is estimated at about 4.3 t ha^{-1}, with Australia presenting the highest productivity of about 10.2 t ha^{-1}, followed by Egypt with 9.6 t ha^{-1}, while Africa showed values of about 2.7 t ha^{-1}.

Rice is unquestionably one of the world's most important food crop (FAO (Food and Agriculture Organization of the United Nations) 2016), ranking third in production (741.5 million tons) in 2014, after sugar cane and maize (FAOSTAT 2017). Although being one of the three major staple crops (together with wheat and maize), rice is, by far, the most important in terms of worldwide human consumption, particularly in low- and lower-middle-income countries (Dawe et al. 2010). On average, this crop accounts for about 20% of calorie intake (considering 536 kcal/person/day) due to the high concentration of starch, 14% of protein, 2% of fat, and essential micronutrients and vitamins for human health (Kennedy and Burlingame 2003; Hirschi 2009; Lidon et al. 2015). Rice is the staple food for near half of the world's population (Hu et al. 2002; Mohanty 2013; Harvest Plus 2014), assuming an even higher percentage in Asia (Mohanty 2013), and for the poor, supplying about 27% of the calories in low- and lower-middle-income countries, above any other crop (Dawe et al. 2010). Therefore, taking into account such massive importance of rice, as well as the potential negative impacts of climate changes and global warming, it will be a major challenge to concurrently increase yield, and maintain quality, to feed an expected growing population, at affordable prices worldwide.

8.2 Selenium

8.2.1 Selenium Importance to Human Health

Selenium was identified as an essential trace element since Se-responsive diseases in animals were reported in the late 1950s and early 1960s (Schwarz and Foltz 1957). In fact, Se is crucial for humans, animals, and plants' health, being estimated that inadequate Se dietary intakes and Se deficiency affects ca. 15% of world's population (White and Broadley 2009).

The scarcity of micronutrients, such as Se, may lead to serious malnutrition problems, particularly in the most vulnerable populations. Among the several human health problems associated with Se deficiency are the increasing rates of premature mortality in adults, pregnant women, and children between 0 and 10 years old and adverse effects on thyroid hormone metabolism, asthma, male sterility, increased incidence of heart disease (Keshan disease), bone and joint condition (Kashin-Beck disease), cognitive decline, reduced immune capacity, certain forms of cancer, and AIDS (Baum et al. 2001; Mayer et al. 2008; Hirschi 2009; Fairweather-Tait et al. 2011; Garcia-Bañuelos et al. 2011; Rayman 2012; Fordyce 2013; Yang et al. 2016; Lyons 2018). These health disturbances would be related to the already 25 essential selenoproteins identified in humans, many of which are involved in catalytic functions, among them glutathione peroxidase, iodothyronine deiodinase, P and W selenoproteins, and thioredoxin reductase (Papp et al. 2007; Rayman 2012), playing important biological roles in the activation of several enzymes, pancreatic function, DNA repair and antioxidant defense (e.g., against Cd and As toxicity), cell differentiation, hormone metabolism, and testosterone biosynthesis (Combs Jr 2001; Rayman 2012; Zwolak and Zaporowska 2012). However, despite Se importance for human and animal health, the consumption in excess may also be toxic. Also, recent works pointed that the beneficial dose range for Se for human health is fairly narrow, that the adequate amount is much lower than previously accepted, and that high intake could provoke health disorders (Jablonska and Vinceti 2015; Manzanares and Hardy 2016). The Se daily intake recommended for adults was identified within the range of 70–400 $\mu g\ day^{-1}$, with the variations closely depending on each country guidelines and Se soil availability, as well as on factors such as age, sex, and health status of the population (Oldfield 1999; Institute of Medicine 2000; Thomson 2004; Padovani et al. 2006; Papp et al. 2007). Although research increasingly shows the essential nature of Se to human health, some uncertainty arose with observations that the upper safe limit of organic and inorganic Se intake could be much lower. In fact, adverse health effects (namely, in the endocrine function, diabetes incidence, prostate cancer) were reported with dietary exposure to ca. 300 μg day^{-1} (see Jablonska and Vinceti 2015), clearly showing the need for an accurate assessment of safe Se levels through diet, while promoting its access to entire populations (Garcia-Bañuelos et al. 2011).

Currently it is used an approach to identify the Se content necessary to optimize the function of glutathione peroxidase (GPx) selenoenzymes, which regulate cellular redox processes by catalyzing the reduction of organic hydroperoxides. Using the criteria of maximization of plasma GPx, recommended intakes of Se for adults were proposed, and although presenting important variations (e.g., due to soil Se content in each region), the doses varied between 30 and 85 μg day^{-1} for adults (Institute of Medicine 2000; Thomson 2004), with some reports pointing to a somewhat lower value in women as compared to men for the same region (Thomson 2004). More recently, an intake value close to 30 μg day^{-1} for organic Se was suggested to be safe (Jablonska and Vinceti 2015), in line with the Recommended Nutrient Intake set in 2004 by the World Health Organization (WHO) that pointed Se values between 26 (women) and 34 (men) μg day^{-1}, with ages between 19 and 65 years.

8.2.2 *Selenium Uptake by Humans via Food Crops*

Selenium is incorporated into the food chain mainly via crops, although a wide range of values can be found in food products. This variability regarding plant products is largely related, namely, with plant species, soil (and water) Se content and speciation, and the agronomic use, dose, and application form of Se fertilizers (Hartikainen 2005; Fordyce 2013; Li et al. 2017; Filippini et al. 2018; Natasha et al. 2018).

The presence of Se in agricultural products is particularly strongly related with the Se content and availability in the soil, as shown by the differences in mean Se concentration in bread wheat grain, which in the UK (soils with low Se content) was 28 ng g^{-1} FW (Adams et al. 2002), while in the USA and Canada (rich Se soils) the values reached as much as 457 and 760 ng g^{-1}, respectively (Broadley et al. 2010). This wide range of Se in agricultural products dependent on regions was also confirmed for a large variety of food products (cereals, vegetables, fruits), with Se contents from 6 ng g^{-1} DM in the deficient areas of Sweden and New Zealand to about 500 ng g^{-1} or even up to 3 μg g^{-1} in some parts of Canada (Carvalho et al. 2003; Navarro-Alarcon and Cabrera-Vique 2008; Mehdi et al. 2013; Gupta and Gupta 2017).

Plant species dependence is also important, since even at a given seleniferous site (1–100 μg Se g^{-1} soil) Se content in plants may vary over 100-fold between plant species (Malagoli et al. 2015). Additionally, it should also be taken into account that the intake of different food products promotes the availability of diverse Se forms, from inorganic (the most toxic) to organic Se species (e.g., selenoprotein-bound Se, selenocysteine-bound Se) (Filippini et al. 2018).

Using soil Se fertilization, Se contents can be enhanced throughout the entire food chain (soil-plants-animals-humans), while increasing plant yields in some cases (Hartikainen 2005). In edible parts of vegetables, Se values are usually below 100 ng g^{-1} DW, and fruits generally contain even lower Se amounts than vegetables. Cereals can also store Se in the seeds, mainly in the form of selenomethionine (SeMeth). This Se compound is the main uptake form from plants to animals and will be associated with protein tissues in the body, whereas inorganic Se is absorbed into other tissues (WHO-World Health Organization 1987, 1996; Rayman 2008). In fact, in humans it was suggested that Se concentrates in the internal organs which are rich in protein. In normal conditions, Se concentration is higher in the liver and lower in kidneys and in muscle tissues. However, the largest total amount of Se in the human body is allocated to muscles, since they represent the major body mass (Lyons et al. 2007).

8.3 Crop Mineral Biofortification to Fight Malnutrition

8.3.1 Biofortification Approach

As referred, malnutrition related to deficiencies of micronutrients, including essential trace elements as Se, constitutes a serious problem of public health, being estimated that from one third (FAO, IFAD, WFP (Food and Agriculture Organization of the United Nations, International Fund for Agricultural Development, World Food Program) 2015) until more than half of the world's population is affected, particularly those from developing countries (Fairweather-Tait et al. 2011; Stein 2015). In this context, the ultimate goal of biofortification is to provide nutritious and safe foods, sufficiently and sustainably, to improve the lives of people around the world that have limited access to diverse diets and micronutrient (Bouis and Saltzman 2017; Garg et al. 2018). Healthy human mineral nutrition is only possible through an adequate food supply of quality products. Therefore, mineral dietary deficiencies, namely of Se, can be addressed through diet diversification, food supplementation, and enhancement of micronutrient bioavailability by manipulating the levels of pro-nutrient or antinutrient components in foods and by crop biofortification approaches (e.g., Cakmak 2008; White and Broadley 2009; Stein 2015; Garcia-Casal et al. 2017; Lyons 2018; Rehman et al. 2018). The latter, however, has two crucial comparative advantages related to its long-term cost-effectiveness and the ability to reach more vulnerable rural populations (Bouis and Saltzman 2017). Biofortification can be implemented through soil and/or foliar fertilization (agronomic biofortification) and crop improvement (genetic biofortification) either by means of conventional plant breeding and selection or modern biotechnology related to genetic engineering/transgenic approaches, to add nutritional value by increasing the content of a micronutrient or improving its bioavailability (White and Broadley 2005; Brinch-Pedersen et al. 2007; Pfeiffer and McClafferty 2007;

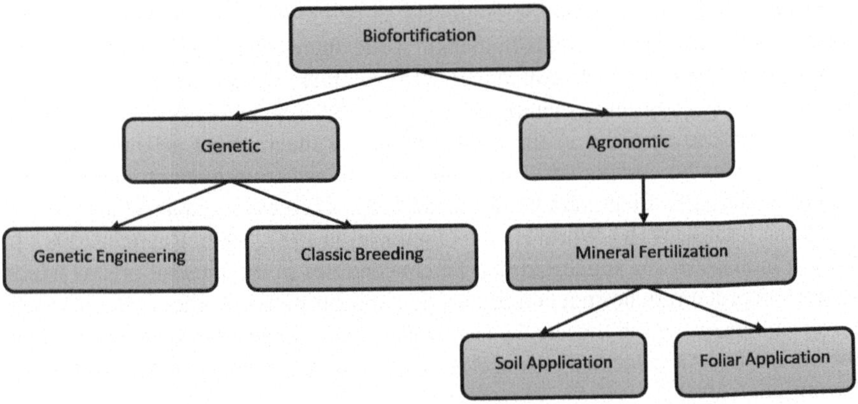

Fig. 8.1 Summary of biofortification approaches. (Modified from Nakandalage et al. 2016)

Cakmak 2008; Oliveira et al. 2015; Nakandalage et al. 2016; Bouis and Saltzman 2017; Garcia-Casal et al. 2017; Garg et al. 2018) (Fig. 8.1).

Most of the crops targeted by transgenic, conventional breeding, and agronomical approaches include staple crops like rice, wheat, maize, sorghum, lupine, common bean, potato, sweet potato, and tomato (Garg et al. 2018). Genetic biofortification can explore the natural existing genetic diversity within species, by selecting genotypes genetically predisposed to accumulate more Se, in order to increase Se intake in human diets (Lyons et al. 2004), but in the longer term, it will be possible to breed crops for increased Se concentration (Broadley et al. 2006). In fact, conventional breeding would provide an alternative to agronomic biofortification, in order to minimize the use of Se fertilizers (Lyons et al. 2004; Broadley et al. 2006). When genetic diversity is unavailable or the target nutrient does not naturally exist at the required levels, transgenic plant breeding could be an option to produce biofortified crops with the desired nutrient and agronomic traits (Bouis and Saltzman 2017; Garg et al. 2018). Among other advantages, transgenic approaches can be utilized for targeting multiple crops once a useful gene was identified, and allows the concomitant incorporation of genes involved in the enhancement of micronutrient concentration and bioavailability, as well as in the reduction in the concentration of antinutrients (Garg et al. 2018). Under this approach, transgenic rice lines were obtained, namely, as regards the increase of provitamin A (Golden rice) and the increase of Fe and Zn (Garg et al. 2018), but no information was found regarding Se.

However, the implementation of genetic biofortification programs demands a technological level that is absent in many countries. Furthermore, a strategy based solely in genetic biofortification would be largely dependent on Se pools in agricultural soils of each region, which is often low (Sun et al. 2010; Fordyce 2013; Jones et al. 2017), and its biological availability in soil, which is limited, namely, by the presence of $CaCO_3$, organic matter, and silt-clay grains (Zhao et al. 2005). As an example, a study using a large number of polished rice samples showed an average difference of 32-fold of Se between the highest values from the USA (seleniferous soils) and the lowest values from Egypt samples (Williams et al. 2009). Therefore, in order to address micronutrient malnutrition, food insecurity, and reach a large part of the population, particularly rural ones in developing countries where inadequate intake problem is most prevalent, emerges agronomic biofortification of staple crops as an efficient, fast, reliable, sustainable, more accessible, and cost-effective way to improve the intake of nutrient-rich foods (Cakmak 2008; Mayer et al. 2008; Meenakshi et al. 2010; Bouis et al. 2011; Garcia-Bañuelos et al. 2011; FAO (Food and Agriculture Organization of the United Nations) 2014; Bouis and Saltzman 2017; Garcia-Casal et al. 2017). This strategy has also the great advantage of allowing a rapid dissemination to be used in rural areas where agriculture is the basis of subsistence (Hartikainen 2005; Mudema and Manjate 2014), partly overcoming deficient distribution infrastructure and health systems (Forsman 2014). This approach has been extensively promoted to increase several mineral contents on the edible parts of widely used food and feed crops (particularly in vegetables and cereals). These included rice, wheat, maize, pearl millet, carrots, beans, turnip, brassicas, potato, and considered a variety of mineral nutrients, among them, Se, I, Zn, and Fe (Cakmak 2008; White and Broadley 2009; Bouis et al. 2011; Malagoli et al. 2015; Garcia-Casal et al. 2017; Li et al. 2018; Lyons 2018; Natasha et al. 2018; Oliveira et al. 2018). Nonetheless, it is highly desirable to complement the readily agronomic biofortification with longer-term implementation of genetic biofortification, namely, by searching for genetic variability in global germplasm banks and at the farmers, or to obtain new breed varieties with the ability to accumulate higher contents of the required micronutrients and substances (Cakmak 2008).

8.3.2 *Selenium Metabolism, Accumulation, and Beneficial Effects in Plants*

Selenium has gained importance due to its essential role as a trace element in animals and humans (Rayman 2008, 2012; Schomburg 2016). Still, unlike for humans and animals, Se is not considered an essential element for plants (Pilon-Smits and Quinn 2010), where its role remains partly unveiled (Djanaguiraman et al. 2010). Due to its chemical properties, similar to sulfur (S), Se is absorbed and metabolized through the same assimilation pathway and is present in S amino acids, such as selenomethionine (SeMeth) and selenocysteine (SeCys), as well as in Se-methylselenocysteine (MeSeCys) (Sors et al. 2005; Malagoli et al. 2015; Gupta and Gupta 2017; White 2018).

Plants greatly vary in their capability to accumulate Se in root and shoot tissues. Some can accumulate high amounts of Se in their tissues (Terry et al. 2000; Ellis and Salt 2003), namely, hyperaccumulator plants, which could reach values as high as between 100 and 1000 $\mu g\ g^{-1}$ DW without displaying toxicity symptoms. Regarding, non-hyperaccumulator plants, as food crops, vegetables, and grasses, much lower Se contents are present, but with large differences in the Se accumulation, with *Allium* spp. and *Brassica* spp. showing quite high Se concentrations (Galeas et al. 2007; Bañuelos et al. 2015; Natasha et al. 2018; White 2018).

A wealth of reports pointed for beneficial Se impacts in plants. In fact, at moderate concentrations, Se can, among others, (A) stimulate plant growth and development (Bitterli et al. 2010; Feng et al. 2013; González-Morales et al. 2017; Natasha et al. 2018), linked to an (B) increase crop yield and quality (Turakainen et al. 2004; Jiang et al. 2015b; Thavarajah et al. 2015; Oliveira et al. 2018). Moreover, it can (C) improve cell integrity (Mozafariyan et al. 2017) and delay leaf senescence (Xue et al. 2001; Djanaguiraman et al. 2004, 2010), (D) contribute to preserve the structure and fluidity of chloroplast membranes affected by Cd (Filek et al. 2010; Zembala et al. 2010), (E) enhance photosynthesis (Feng et al. 2015; Xiaoqin et al. 2009b; Jiang et al. 2015a), and (F) increase respiratory activity, all of which might contribute to greater seed production (Lyons et al. 2009). Finally, Se confer (G) protection to biotic (insect herbivory, pathogen fungi) (Freeman et al. 2007) and abiotic (Kuznetsov et al. 2003; Taspinar et al. 2009; Feng et al. 2013; Natasha et al. 2018) stresses, likely related, among others, with the (H) reinforcement of the defense against oxidative stress caused by such environmental constraints (Xue et al. 2001; Pennanen et al. 2002; Xiaoqin et al. 2009a; Zembala et al. 2010; Rosa et al. 2012).

Nevertheless, it should be noted that Se toxicity by high concentrations has also been reported in plants, namely, growth reduction and disturbance of the metabolic processes (Bitterli et al. 2010; Fu et al. 2011; Longchamp et al. 2015; González-Morales et al. 2017; Natasha et al. 2018). In fact, the several Se compounds found in plants have different toxicity levels and nutritional values, with higher biofortification efficiency usually found in organic forms (Malagoli et al. 2015). Selenate (Se(VI)) is the major inorganic Se compound found in both animal and plant tissues, whereas SeMeth and MeSeCys are the major Se-containing compounds. SeMeth is the major Se compound in cereal grains, grassland legumes, and soybeans, whereas MeSeCys is the major Se compound in Se-enriched plants such as garlic, onions, broccoli florets and sprouts, and wild leeks (Whanger 2002).

8.3.3 Selenium Translocation to Plant Shoot and Accumulation

Plants Se uptake from soil is performed in the form of selenate, selenite (Se(IV)), and organic Se (SeCys and SeMeth) (Ellis and Salt 2003; White et al. 2004). These Se forms differ regarding their mobility and absorption within the plant, and only a low portion (<5%) of soil-applied Se is estimated to be used by plants (Haug et al. 2007). The presence of co-occurring competing ions in soil, especially sulfate and phosphate anions, alters Se uptake. The addition to the soil of different forms of S (sulfur, sulfate, sulfite, sulfide, and methionine) reduces Se uptake by plants (Mackowiak and Amacher 2008), particularly in acid soils (González-Morales et al. 2017), whereas Se biofortification reduces the S content in biomass, namely, in rice grain, reflecting some S vs. Se antagonism (Mangueze et al. 2018). This might be related with the observations that Se presence affects the expression of key genes encoding S transporters and enzymes regulating S metabolism and that high Se concentration in plant tissues might reduce S metabolites' presence (González-Morales et al. 2017; White 2018).

Plants were reported to usually show greater affinity for selenate than to selenite, either by soil (Tian et al. 2016; Gupta and Gupta 2017) or foliar application (Boldrin et al. 2013), and Se cannot be directly absorbed as metallic selenide or elemental Se due to their water insolubility (Fernandez-Martınez and Charlet 2009). However, organic Se compounds, such as seleno-amino acids, have relatively higher phytoavailability (Kikkert and Berkelaar 2013) and can be used, namely, through the soil amendment with the Se-hyperaccumulator plant *Stanleya pinnata* (Pursh) Britton for Se enrichment of broccoli and carrots (Bañuelos et al. 2015). Selenate, which is the most prevalent form of bioavailable Se in agricultural soils (Gupta and Gupta 2017), enters root plasma membrane cells through high-affinity sulfate transporters and channels (Sors et al. 2005; Bitterli et al. 2010; Cabannes et al. 2011; Gupta and Gupta 2017). This Se compound is readily uploaded to xylem sap, and hereafter transported from root to shoot (Zhu et al. 2009; Winkel et al. 2015), and then to reproductive organs by phloem (Brown and Shrift 1982). Selenate forms (Na_2SeO_4 and K_2SeO_4) provide plant available Se for immediate plant uptake, while selenite and less soluble selenates (e.g., $BaSeO_4$) provide long-term effects (Broadley et al. 2006). Selenite uptake in roots is performed by phosphate transporters (Li et al. 2008), and then redistributed into the plant via the phloem (White et al. 2004). This turns it less translocated in plants (Terry et al. 2000; Cartes et al. 2005; Ramos et al. 2010) and is less accumulated in shoot, as compared to selenate, with most Se being transported to shoot in the latter form (Cartes et al. 2005). Besides, in the plant, selenite is quickly transformed in organic forms, limiting its translocation to shoot, whereas selenate conversion into its organic form is much slow (Li et al. 2008).

Foliar fertilization with Se has been successfully applied, namely, in rice, wheat, maize, garlic, onions, broccoli florets, turnip, and sprouts (Chen et al. 2002; Whanger 2002; Ramos et al. 2010; Malagoli et al. 2015; Oliveira et al. 2015; Li et al. 2018; Lidon et al. 2019 ; Mangueze et al. 2018), with reduced costs when compared to soil application (Wang et al. 2013). Also, Se foliar application is sometimes considered more efficient than soil application due to the mobility of Se in plants (Ducsay and Ložek 2006; Mao et al. 2014; Winkel et al. 2015; Ros et al. 2016) and not only obviates the soil factors that can reduce the effectiveness of soil agronomic biofortification of Se, but also reduces possible environmental Se accumulation as less Se is applied per hectare (Lyons et al. 2004).

After the transport to plant shoot, selenate is converted in Se organic forms, beginning with the action of sulfate assimilating enzymes in the chloroplast (Terry et al. 2000; Natasha et al. 2018). Selenate is firstly transformed to adenosine phosphoselenate by ATP sulfurylase, and then reduced to selenite by adenosine phosphosulfate reductase (Sors et al. 2005; Pilon-Smits and Quinn 2010; Gupta and

Gupta 2017). Afterwards, selenite is transformed to selenide by enzyme sulfite reductase or non-enzymatically by reduced glutathione (GSH) (Terry et al. 2000; Wallenberg et al. 2010) and then linked to O-acetyl serine by cysteine synthase to produce the amino acid SeCys (Gupta and Gupta 2017). This Se compound can be incorporated in proteins (instead of cysteine) causing toxicity. Thus, its transformation to SeMeth (less toxic), elemental Se (mostly not harmful), or methylated to MeSeCys (safe, since it will not be incorporated in proteins) can decrease such toxicity. Additionally, a more efficient strategy is the conversion to volatile molecules dimethylselenide (DMSe) and dimethyldiselenide (DMDSe), usually in non-accumulator and hyperaccumulator species, respectively, as promoted by the action of several enzymes, depending on plant species and environmental conditions (Pilon-Smits and Quinn 2010; Gupta and Gupta 2017; Natasha et al. 2018; White 2018).

8.3.4 Selenium Agronomic Biofortification in Rice

Cereals are the dominant Se source for those people with diets characterized by low protein content, common among malnourished populations (Williams et al. 2009). Rice has usually low content of Se in many world regions (Fig. 8.2) limiting its nutritional availability, what is particularly important as rice feeds about half of the world's population (Hu et al. 2002; Harvest Plus 2014). In this context, rice biofortification practices can constitute a strategic vehicle to reduce Se (and other micronutrients, as Zn) malnutrition, particularly in populations living in Se-deficient regions (Chen et al. 2002; Hu et al. 2002).

Agronomic biofortification in Se is clearly feasible (Combs Jr 2001; Hawkesford and Zhao 2007; Rayman 2008; Broadley et al. 2010; Mangueze et al. 2018), and the

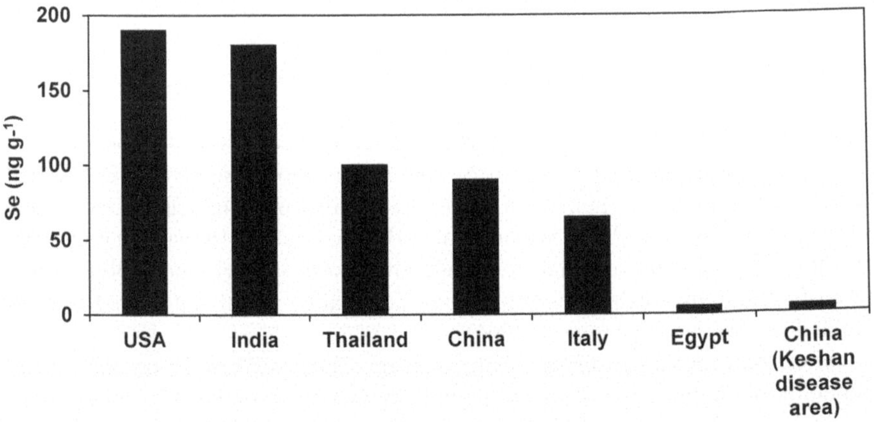

Fig. 8.2 Mean average Se levels in rice grains (ng g^{-1}) in key rice-producing countries. (Modified from Zhu et al. 2009)

increase of this essential nutrient concentration in the edible portions of agricultural crops can be performed using soil (with root uptake) and/or foliar (entering by the cuticle or stomata) applications of inorganic (sodium selenite and selenate) and organic (SeMeth or SeCys) Se compounds (Carvalho et al. 2003; Li et al. 2008; Fu et al. 2011; Kikkert and Berkelaar 2013; Longchamp et al. 2015; Lidon et al. 2018). This has allowed to effectively increase Se contents in diets, improving the Se status and people's health, with some successful nationwide programs, as in Finland and New Zealand, using selenate (White and Broadley 2005, 2009; Alfthan et al. 2015). Additionally, for several crops, Se biofortification was observed to improve other nutrient components, including mineral elements, polysaccharides, proteins, amino acids, and vitamin C (Turakainen et al. 2004; Oliveira et al. 2018).

With the wider use of biofortification, an effort was made to optimize applied doses as regards (cereal crop) fertilizers. The initial Se level in fertilizers of 16 μg g^{-1} was reduced in 1991 to 6 μg g^{-1}, but since this alteration led to adverse effect on the crop quality, Se concentration was raised to 10 μg g^{-1} in 1998 (Hartikainen 2005).

The reported rice agronomic biofortification results showed some variability that arise from several sources, among them the used genotypes, the way of Se application (soil and/or foliar), the chemical Se form, the developmental stage of the plant when the application was done, the partitioning of the dose along the plant life cycle, and trails performed under greenhouse or field conditions. The promising results obtained with rice are in line with the significant increases of Se observed in other cereals, namely, in winter wheat in China (Mao et al. 2014) and the UK (Broadley et al. 2010) and maize in Malawi (Chilimba et al. 2012) and China (Wang et al. 2013). In rice, Se accumulation is mostly found to be as organic SeMeth, followed by Se-MeSeCys and SeCys (Carey et al. 2012), and that Se is preferentially accumulated in bran (Gupta and Gupta 2017; Mangueze et al. 2018). Several works provides evidence that it was possible to significantly increase Se levels in grain, without negative (or even positive) effects on grain yield (see, e.g., Chen et al. 2002; Boldrin et al. 2012, 2013; Lidon et al. 2018; Mangueze et al. 2018), as shown in the successful case studies reported below.

Using a large number of samples of cropped rice in China, the average Se content was 25 ng g^{-1}. This value increased to 471 and 640 ng g^{-1} in field trails through foliar application of 20 g Se ha^{-1}, as sodium selenite and sodium selenate, respectively (Chen et al. 2002). The latter Se molecule showed a greater efficiency, reflected in a higher 36% accumulation of Se in the grain than that obtained by selenite. For a consumption of 180 g of rice, this will promote a daily intake of 85 or 115 μg Se. However, since rice is a staple food in China, a usual 400 g daily consumption would raise Se intake values to 188 and 256 μg day^{-1} (Chen et al. 2002).

Another study, in Brazil where most agricultural soils present low Se concentration, applied optimized Se doses of 0.75 μg g^{-1} for soil, and 50 μmol L^{-1} for foliar spraying, of both sodium selenate (Na_2SeO_4) and selenite (Na_2SeO_3), to potted plants under greenhouse conditions at the flowering stage (Boldrin et al. 2013). Noteworthy, this allowed to increase grain yield in 22% for selenate and 17% for selenite, accompanied by the Se enrichment of rice grain to ca. 1.16 μg g^{-1} and 0.27 μg g^{-1} (foliar application) and to 6.4 μg g^{-1} and 0.42 μg g^{-1} (soil application), for selenate and selenite, respectively. The higher values obtained for soil application were explained by the longer plant-Se contact time. Considering again a consumption of 180 g of rice per day, these grains could contribute to a dietary Se intake of 210 and 48 μg day^{-1} via foliar application, and 1150 and 75 μg day^{-1} via soil application, by using selenate and selenite, respectively (Boldrin et al. 2013).

In these two previous works, the higher accumulation obtained with selenate (as compared to selenite) could be justified by the possible lower conversion of selenate into Se-organic forms, namely, in the roots, what would favor its mobility in the xylem (Li et al. 2008).

A third selected study was performed in Portugal, a country that ranks first (by far) in rice consumption per capita in Europe. On field trials it was studied the effect of a large number of doses (from 4 to 300 g Se ha^{-1}) of selenate and selenite, applied through foliar spraying in four cropped genotypes (Lidon et al. 2019). One interesting difference from previous studies was the partitioning of Se dose application, which was done in the stages of booting, heading, and during grain filling, allowing a prolonged contact of Se with the plant, and a more gradual leaf uptake. In contrast with the works referred above, a higher Se accumulation in the grain was obtained with sodium selenite, relatively to sodium selenate, with the maximal dose of 300 g Se ha^{-1} leading to large rice grain Se concentrations up to 30–40 μg Se g^{-1} (selenite) and 10–13 μg Se g^{-1} (selenate), depending on genotype. Interestingly, the 30 g Se ha^{-1} dose led to values of ca. 1–1.2 μg Se g^{-1} (selenite) and 0.3–0.5 μg Se g^{-1} (selenate) in the whole grain (Lidon et al. 2018). Therefore, 30 g Se ha^{-1} foliar application was enough to promote biofortification for an adequate Se daily intake, together with a low or negligible impact in other grain characteristics, including lipid, sugars, and protein contents, irrespective of selenite or selenate form (Lidon et al. 2018).

Finally, it should be highlighted two field works performed at the experimental fields of the International Rice Research Institute (IRRI), Agricultural Research Institute of Mozambique (IIAM), Mozambique, in a country who faces serious micronutrient deficiencies (among them, Se), which affects about 66% of children, women, and the elders, and where rice is the second consumed cereal (Mangueze et al. 2018). The two local cropped genotypes (Mak and IR) used in the trials showed grain Se values below the detection limit, highlighting the need for Se biofortification. These genotypes were then submitted to foliar application of sodium selenite (0, 150, 300 g Se ha^{-1}) at the stage of beginning of grain filling (Mangueze et al. 2018). A second experiment used as well sodium selenite (0, 250, 500 g Se ha^{-1}), but parcelling Se applications, at the stages of beginning of heading, flowering, and grain formation (Roda et al. unpublished data). In this second trial, all Se treatments increased yield in both genotypes, related to a higher number of tillers per plant, as well as a higher number and bigger grains per panicle. Substantial differences were observed between the genotypes capability to incorporate Se into the grain, as reflected in the flour produced from whole (Mak – 1.97 μg g^{-1}; IR – 6.09 μg g^{-1}) or polished (Mak – 0.65 μg g^{-1}; IR – 4.46 μg g^{-1}) grains, considering the 300 g Se ha^{-1} dose. Among other reasons, this difference could be related to the higher leaf area index (LAI) of IR genotype, since a higher surface for mineral absorption would be likely an important advantage when promoting biofortification through foliar application. These results also confirmed that an important part of the Se was located at the husk (Zhang et al. 2014) and in the outer part of the grain (Mangueze et al. 2018).

One interesting common output of these case studies is that the obtained rice Se-enriched grains showed values higher than needed for a safe daily intake, assuming an average rice consumption of ca. 180 g day^{-1} per person (Mohanty 2013), and the Se intake values established by the WHO in 2004 as adequate for women (26 μg day^{-1}) and man (34 μg day^{-1}) (Jablonska and Vinceti 2015). Therefore, lower Se-fertilizer doses can be efficiently used in these rice plants, or, alternatively, the flour obtained from biofortified grain could be mix with other from non-biofortified rice flour to obtain a balanced chosen value. This would also reduce some undesirable impacts on grain compounds (Lidon et al. 2018) and decrease the antagonism to other minerals, namely, S (due to the chemical and physical similarities of these elements regarding plant uptake), Ca, K, and Zn (Boldrin et al. 2012; Lidon et al. 2015; Mangueze et al. 2018). Furthermore, positive (with low Se dose) and negative (high Se dose) yield impacts were reported with both selenite and selenate, pointing to a dose vs. genotype effect and, therefore, highlighting the importance of a fine tuned dose and genotype selection (Boldrin et al. 2012; Zhang et al. 2014; Mangueze et al. 2018).

In alternative to inorganic Se compounds, some authors looked to the use of organic Se compounds SeMeth, MeSeCys, and SeCys. Although we did not notice rice field trials using these organic Se compounds, this option was used in other plants species, therefore deserving some attention. In fact, rice plants grown hydroponically in growth chambers were exposed for 1 week to 20 μM Se as selenate, selenite, or SeMeth. Results showed that the shoots of selenate supplied plants accumulated the greatest amount of Se, followed by those supplied with SeMeth (Zayed et al. 1998). Still in rice, to elucidate the translocation of Se species to the filling grain, excised stems were feed with Se forms (selenate, selenite, SeMeth, MeSeCy), being observed that SeMeth and MeSeCys were transported to the grain more efficiently than both inorganic Se forms. SeMeth and SeMeSeCys were translocated exclusively via the phloem, while inorganic Se was transported via both the phloem and xylem. Notably, SeMeth and MeSeCys were observed to promote Se dispersion in the external grain layers, into the endosperm and, for SeMeSeCys, into the embryo. In contrast, selenite was retained at the point of grain entry (Carey et al. 2012). Nevertheless, selenate was observed to promote in a quite uniform Se accumulation in some species, as was the case in the caryopsis, embryo, and endosperm in wheat grain (Rodrigo et al. 2014).

In another work, hydroponically grown durum wheat (*Triticum turgidum* L.) and spring canola (*Brassica napus* L.) seedlings were exposed to 0.5 or 5.0 μM Se as selenate, selenite, SeMeth, or SeCys for ≤300 min, resulting in a greater uptake rate of organic Se forms than inorganic ones. For the 5.0 μM Se dose, selenite presented a 1.5- (canola) or fivefold (wheat) greater uptake rate than selenate. However, SeMeth was uploaded up to 40- (canola) or 100-fold (wheat) faster and SeCys 2- (wheat) or 20-fold (canola) faster than selenate. In fact, the plants treated with SeMeth presented the greatest Se concentrations in the shoot, although selenate was more mobile once taken up (Kikkert and Berkelaar 2013).

Finally, organic Se compounds were applied by amending the soil with the Se-hyperaccumulator plant *Stanleya pinnata* (Pursh) Britton, which resulted in Se enrichment of broccoli and carrots cropped in those soils (Bañuelos et al. 2015).

Altogether, these results point that organic Se forms could be more effective than inorganic forms for some plant species, including also broccoli and garlic (Malagoli et al. 2015), likely related to a quicker loading into the phloem and transport to the grain. These organic forms can also promote a more uniform Se distribution within the rice grain (in the outer layers and in the endosperm), when compared with inorganic species (Carey et al. 2012). Therefore, the use of organic Se forms should be looked, mostly for paddy rice.

8.4 Conclusion

Selenium is an essential element for men and animals, having also important roles in plant metabolism. Still, it is estimated that inadequate Se dietary intakes and Se deficiency affect about 15% of the world's population, with higher incidence in poor populations. Rice is responsible for feeding near half of the world's population, being the most important crop in terms of worldwide human consumption. In this context, it should be taken into account the expected growing pressure for food and feed, largely related to the significant increase of world population along this century, together with stronger impacts of the ongoing climate changes in the agricultural crops, among them, rice. Diet is the most important source of Se, and, although usually poor in Se content, rice can be a vehicle to fight Se (and other minerals) deficitary intake due to its widespread human consumption. To accomplish this goal, rice agronomic biofortification with Se stands out as a feasible, efficient, fast, reliable, sustainable, more accessible, and cost-effective strategy to respond to such widespread human malnutrition, particularly in rural areas and in developing countries, having the additional advantage of overcoming deficient distribution infrastructure and health systems in these regions. Therefore, the use of biofortification Se programs would greatly contribute to guarantee adequate food security and safety in a difficult and unfavorable context of growing population pressure and ongoing (and increasing) climate changes.

Acknowledgments This work was supported through the project GO-Fortification of Rice with selenium (Operation – 1.0.1 – Grupos Operacionais, PDR2020-101-030672 and 030672, IFAP and PDR2020), and by Portuguese national funds of Fundação para a Ciência e a Tecnologia to the research units UID/AGR/04129/2019 (LEAF) and UID/GEO/04035/2019 (GeoBioTec), and the contribution to the International Rice Research Institute, as well as by funding from the Portuguese Cooperation via Camões, Instituto da Cooperação e da Língua, I.P.

Author Contributions Taking into account their expertise, all authors contributed to the writing of the original draft, review, and editing.

Conflict of Interest Statement The authors declare that there are not any potential conflicts of interest.

References

Adams ML, Lombi E, Zhao F-J, McGrath SP (2002) Evidence of low selenium concentrations in UK bread-making wheat grain. J Sci Food Agric 82:1160–1165. https://doi.org/10.1002/jsfa.1167

Alfthan G, Eurola M, Ekholm P, Venalainen E-R, Root T, Korkalainen K, Hartikainen H, Salminen P, Hietaniemi V, Aspila P, Aro A (2015) Effects of nationwide addition of selenium to fertilizers on foods, and animal and human health in Finland: from deficiency to optimal selenium status of the population. J Trace Elem Med Biol 31:142–147. https://doi.org/10.1016/j.jtemb.2014.04.009

Bañuelos GS, Arroyo I, Pickering IJ, Yang SI, Freeman JL (2015) Selenium biofortification of broccoli and carrots grown in soil amended with Se-enriched hyperaccumulator *Stanleya pinnata*. Food Chem 166:603–608. https://doi.org/10.1016/j.foodchem.2014.06.071

Barnabás B, Jäger K, Fehér A (2008) The effect of drought and heat stress on reproductive processes in cereals. Plant Cell Environ 31:11–38. https://doi.org/10.1111/j.1365-3040.2007.01727.x

Baum MK, Campa A, Miguez-Burbano MJ, Burbano X, Shor-Posner G (2001) Role of selenium in HIV/AIDS. In: Hatfield DL (ed) Selenium: its molecular biology and role in human health, Chapter 20. Springer, Boston, pp 247–255. https://doi.org/10.1007/978-1-4615-1609-5_20

Bitterli C, Bañuelos G, Schulin R (2010) Use of transfer factors to characterize uptake of selenium by plants. J Geochem Explor 107:206–216. https://doi.org/10.1016/j.gexplo.2010.09.009

Boldrin PF, Faquin V, Ramos SJ, Guilherme LRG, Bastos CEA, Carvalho GS, Costa ETS (2012) Selenate and selenite on yield and agronomic biofortification with selenium in rice. Pesq Agropec Bras 47:831–837. https://doi.org/10.1590/S0100-204X2012000600014

Boldrin PF, Faquin V, Ramos SJ, Boldrin KVF, A Vila FW, Guilherme LRG (2013) Soil and foliar application of selenium in rice biofortification. J Food Compos Anal 31:238–244. https://doi.org/10.1016/j.jfca.2013.06.002

Bouis HE, Saltzman A (2017) Improving nutrition through biofortification: A review of evidence from HarvestPlus, 2003 through 2016. Global Food Secur 12:49–58. https://doi.org/10.1016/j.gfs.2017.01.009

Bouis HE, Hotz C, McClafferty B, Meenakshi JV, Pfeiffer WH (2011) Biofortification: A new tool to reduce micronutrient malnutrition. Food Nutr Bull 32(1 Suppl):S31–S40. https://doi.org/10.1177/15648265110321S105

Brinch-Pedersen H, Borg S, Tauris B, Holm PB (2007) Molecular genetic approaches to increasing mineral availability and vitamin content of cereals. J Cereal Sci 46:308–326. https://doi.org/10.1016/j.jcs.2007.02.004

Broadley MR, White PJ, Bryson RJ, Meacham MC, Bowen HC, Johnson SE, Hawkesford MJ, McGrath SP, Zhao FJ, Breward N, Harriman M, Tucker M (2006) Biofortification of UK food crops with selenium. Proc Nutr Soc 65:169–181. https://doi.org/10.1079/PNS2006490

Broadley MR, Alcock J, Alford J, Cartwright P, Foot I, Fairweather-Tait SJ, Hart DJ, Hurst R, Knott P, McGrath SP, Meacham MC, Norman K, Mowat H, Scott P, Stroud JL, Tovey M, Tucker M, White PJ, Young SD, Zhao F-J (2010) Selenium biofortification of high-yielding winter wheat (*Triticum aestivum* L.) by liquid or granular Se fertilisation. Plant Soil 332:5–18. https://doi.org/10.1007/s11104-009-0234-4

Brown T, Shrift A (1982) Selenium: toxicity and tolerance in higher plants. Biol Rev 57:59–84. https://doi.org/10.1111/j.1469-185X.1982.tb00364.x

Cabannes E, Buchner P, Broadley MR, Hawkesford MJ (2011) A comparison of sulfate and selenium accumulation in relation to the expression of sulfate transporter genes in *Astragalus* species. Plant Physiol 157:2227–2239. https://doi.org/10.1104/pp.111.183897

Cakmak I (2008) Enrichment of cereal grains with zinc: agronomic or genetic biofortification? Plant Soil 302:1–17. https://doi.org/10.1007/s11104-007-9466-3

Carey A-M, Scheckel KG, Lombi E, Newville M, Choi Y, Nort GJ, Price A, Meharg A (2012) Grain accumulation of selenium species in rice (*Oryza sativa* L.). Environ Sci Technol 46:5557–5564. https://doi.org/10.1021/es203871j

Cartes P, Gianfreda L, Mora M (2005) Uptake of selenium and its antioxidant activity in ryegrass when applied as selenate and selenite forms. Plant Soil 276:359–367. https://doi.org/10.1007/s11104-005-5691-9

Carvalho KM, Gallardo-Williams MT, Benson RF, Martin DF (2003) Effects of selenium supplementation on four agricultural crops. J Agric Food Chem 57:704–709. https://doi.org/10.1021/jf0258555

Castro-Llanos F, Hyman G, Rubiano J, Ramirez-Villegas J, Achicanoy H (2019) Climate change favors rice production at higher elevations in Colombia. Mitig Adapt Strat Global Change:1–30. https://doi.org/10.1007/s11027-019-09852-x

Chaturvedi AK, Bahuguna RN, Pal M, Shah D, Maurya S, Jagadish KSV (2017) Elevated CO_2 and heat stress interactions affect grain yield, quality and mineral nutrient composition in rice under field conditions. Field Crop Res 206:149–157. https://doi.org/10.1016/j.fcr.2017.02.018

Chen L, Yang F, Xu J, Hu Y, Hu Q, Zhang Y, Pan G (2002) Determination of selenium concentration of rice in China and effect of fertilization of selenite and selenate on selenium content of rice. J Agric Food Chem 50:5128–5130. https://doi.org/10.1021/jf020137410.1021/jf0201374

Chilimba ADC, Young SD, Black CR, Meacham MC, Lammel J, Broadley MR (2012) Agronomic biofortification of maize with selenium (Se) in Malawi. Field Crop Res 125:118–128. https://doi.org/10.1016/j.fcr.2011.08.014

Combs GF Jr (2001) Selenium in global food systems. Brit J Nutr 85:517–547. https://doi.org/10.1079/BJN2000280

Dawe D, Pandey S, Nelson A (2010) Emerging trends and spatial patterns of rice production. In: Pandey S, Byerlee D, Dawe D, Dobermann A, Mohanty S, Rozelle S, Hardy B (eds) Rice in the Global economy: strategic research and policy issues for food security. Chapter 1.1. International Rice Research Institute, pp 15–35. ISBN 978-971-22-0258-2.

Djanaguiraman D, Devi DD, Shanker AK, Annie SJ, Bangarusamy U (2004) Impact of selenium spray on monocarpic senescence of soybean (*Glycine Max* L.). J Food Agric Environ 2:44–47. https://doi.org/10.1234/4.2004.162

Djanaguiraman M, Prasad P, Seppanen M (2010) Selenium protects sorghum leaves from oxidative damage under high temperature stress by enhancing antioxidant defense system. Plant Physiol Biochem 48:999–1007. https://doi.org/10.1016/j.plaphy.2010.09.009

Ducsay L, Ložek O (2006) Effect of selenium foliar application on its content in winter wheat grain. Plant Soil Environ 52:78–82. https://doi.org/10.17221/3349-PSE

Ellis DR, Salt DE (2003) Plants, selenium, and human health. Curr Opin Plant Biol 6:273–279. https://doi.org/10.1016/S1369-5266(03)00030-X

Fairweather-Tait SJ, Bao Y, Broadley MR, Collings R, Ford D, Hesketh JE, Hurst R (2011) Selenium in human health and disease. Antioxid Redox Signal 14:1337–1383. https://doi.org/10.1089/ars.2010.3275

FAO (Food and Agriculture Organization of the United Nations) (2009) How to feed the world in 2050. Synthesis report. Available online: http://www.fao.org/fileadmin/templates/wsfs/docs/expert_paper/How_to_Feed_the_World_in_2050.pdf. Accessed 15 Apr 2019

FAO (Food and Agriculture Organization of the United Nations) (2011) Looking ahead in world food and agriculture: perspectives to 2050. In: Conforti P (ed) FAO. ISBN 978-92-5-106903-5

FAO (Food and Agriculture Organization of the United Nations) (2014) Análise de incentivos e desincentivos de preço para o arroz em Moçambique, Série de Notas Técnicas, Loganemio DL, Rome, p 39

FAO (Food and Agriculture Organization of the United Nations) (2016) The State of food and Agriculture – Climate change, agriculture and food security. In Brief. Food and Agriculture Organization of the United Nations, Rome, p 24. ISBN 978-92-5-109374-0

FAO, IFAD, WFP (Food and Agriculture Organization of the United Nations, International Fund for Agricultural Development, World Food Program) (2015) The State of Food Insecurity in the World 2015. Meeting the 2015 International Hunger Targets: Taking Stock of Uneven Progress.. FAO, Rome. ISBN 978-92-5-108785-5

FAOSTAT (2017) FAO statistics division. Accessed Mar 2019. http://faostat.fao.org/site/567/DesktopDefault.aspx?PageID=567

FAOSTAT (2018) UN Food and Agriculture Organization, Corporate Statistical Database. Accessed Mar 2019. http://www.fao.org/faostat/en/#home

Feng R, Wei C, Tu S (2013) The roles of selenium in protecting plants against abiotic stresses. Environ Exp Bot 87:58–68. https://doi.org/10.1016/j.envexpbot.2012.09.002

Feng T, Chen S, Gao D, Liu G, Bai H, Li A, Peng L, Ren Z (2015) Selenium improves photosynthesis and protects photosystem II in pear (*Pyrus bretschneideri*), grape (*Vitis vinifera*), and peach (*Prunus persica*). Photosynthetica 53:609–612. https://doi.org/10.1007/s11099-015-0118-1

Fernandez-Martınez A, Charlet CL (2009) Selenium environmental cycling and bioavailability: a structural chemist point of view. Rev Environ Sci Biotechnol 8:81–110. https://doi.org/10.1007/s11157-009-9145-3

Filek M, Gzyl-Malcher B, Zembala M, Bednarska E, Laggner P, Kriechbaum M (2010) Effect of selenium on characteristics of rape chloroplasts modified by cadmium. J Plant Physiol 167:28–33. https://doi.org/10.1016/j.jplph.2009.07.003

Filippini T, Michalke B, Wise LA, Malagoli C, Malavolti M, Vescovi L, Salvia C, Bargellini A, Sieri S, Krogh V, Ferrante M, Vinceti M (2018) Diet composition and serum levels of selenium species: a cross-sectional study. Food Chem Toxicol 115:482–490. https://doi.org/10.1016/j.fct.2018.03.048

Fischer T, Byerlee D, Edmeades G (2014) Crop yields and global food security – Will yield increase continue to feed the world? The Australian Centre for International Agricultural Research, Grains Research & Development Corporation, Monograph 158, Canberra. XXIIi, pp 634. ISBN 978 1 925133 06 6

Fordyce FM (2013) Selenium deficiency and toxicity in the environment. In: Selinus O, Alloway B, Centeno JA, Finkelman R, Fuge R, Lindh U, Smedley PL (eds) Essentials of medical geology. Revised Edition. Chapter 16. Springer, Dordrecht, pp 375–416. https://doi.org/10.1007/978-94-007-4375-5_16

Forsman C (2014) Fortificação dos alimentos básicos em Moçambique. United States Agency for International Development, USAID/Mozambique, p 35

Freeman JL, Lindblom SD, Quinn CF, Fakra S, Marcus MA, Pilon-Smits EA (2007) Selenium accumulation protects plants from herbivory by *Orthoptera* via toxicity and deterrence. New Phytol 175:490–500. https://doi.org/10.1111/j.1469-8137.2007.02119.x

Fu DD, Duan ML, Liang DL, Wang SS, Wu XP (2011) Effects of selenite and selenate on growth and nutrient absorption of pak choi (*Brassica chinensis* L.) during successive planting conditions. Plant Nutr Fertil Sci 17:358–365. https://doi.org/10.1007/s11356-015-4344-7

Galeas ML, Zhang LH, Freeman JL, Wegner M, Pilon-Smits EA (2007) Seasonal fluctuations of selenium and sulfur accumulation in selenium hyper- accumulators and related nonaccumulators. New Phytol 173:517–525. https://doi.org/10.1111/j.1469-8137.2006.01943.x

Garcia-Bañuelos ML, Hermosillo-Cereceres MA, Sanchez E (2011) The importance of selenium biofortification in food crops. Curr Nutr Food Sci 7:181–190. https://doi.org/10.2174/157340111797264796

Garcia-Casal MN, Peña-Rosas JP, Giyose B, consultation working groups (2017) Staple crops biofortified with increased vitamins and minerals: considerations for a public health strategy. Ann N Y Acad Sci 1390:3–13. https://doi.org/10.1111/nyas.13293

Garg M, Sharma N, Sharma S, Kapoor P, Kumar A, Chunduri V, Arora P (2018) Biofortified crops generated by breeding, agronomy, and transgenic approaches are improving lives of millions of people around the world. Front Nutr 5:12. https://doi.org/10.3389/fnut.2018.00012

González-Morales S, Pérez-Labrada F, García-Enciso EL, Leija-Martínez P, Medrano-Macías J, Dávila-Rangel IE, Benavides-Mendoza A (2017) Selenium and sulfur to produce allium functional crops. Molecules 22:558. https://doi.org/10.3390/molecules22040558

Gupta M, Gupta S (2017) An overview of selenium uptake, metabolism. Toxicity Plants Front Plant Sci 7:2074. https://doi.org/10.3389/fpls.2016.02074

Hartikainen H (2005) Biogeochemistry of selenium and its impact on food chain quality and human health. J Trace Elem Med Biol 18:309–318. https://doi.org/10.1016/j.jtemb.2005.02.009

Harvest Plus (2014) Biofortified Rice. Available: http://www.harvestplus.org/content/zinc-rice. Accessed Mar 2019

Hatfield JL, Boote KJ, Kimball BA, Ziska LH, Izaurralde RC, Ort DR, Thomson AM, Wolfe D (2011) Climate impacts on agriculture: implications for crop production. Agron J 103:351–370. https://doi.org/10.2134/agronj2010.0303

Haug A, Graham RD, Christophersen OA, Lyons GH (2007) How to use the world's scarce selenium resources efficiently to increase the selenium concentration in food. Microb Ecol Health Dis 19:209–228. https://doi.org/10.1080/08910600701698986

Hawkesford MJ, Zhao FJ (2007) Strategies for increasing the selenium content of wheat. J Cereal Sci 46:282–292. https://doi.org/10.1016/j.jcs.2007.02.006

Hirschi KD (2009) Nutrient biofortification of food crops. Annu Rev Nutr 29:401–421. https://doi.org/10.1146/annurev-nutr-080508-141143

Houshmandfar A, Fitzgerald GJ, O'Leary G, Tausz-Posch S, Fletcher A, Tausz M (2018) The relationship between transpiration and nutrient uptake in wheat changes under elevated atmospheric CO_2. Physiol Plant 163:516–529. https://doi.org/10.1111/ppl.12676

Hu Q, Chen L, Xu J, Zhang Y, Pan G (2002) Determination of selenium concentration in rice and the effect of foliar application of Se- enriched fertilizer or sodium selenite on the selenium content of rice. J Sci Food Agric 82:869–872. https://doi.org/10.1002/jsfa.1115

Institute of Medicine (2000) Dietary reference intakes: vitamin C, vitamin E, selenium, and carotenoids. Food and Nutrition Board. National Academy Press, Washington, D.C. ISBN-10: 0-309-06935-1

IPCC (Intergovernmental Panel on Climate Change) (2014) Glossary, acronyms and chemical symbols. In: Climate change 2014: mitigation of climate change. Working Group III Contribution to the IPCC Fifth Assessment Report, Cambridge University Press, Cambridge. https://doi.org/10.1017/CBO9781107415416.023

Jablonska E, Vinceti M (2015) Selenium and human health: witnessing a copernican revolution? J Environ Sci Health 33:328–368. https://doi.org/10.1080/10590501.2015.1055163

Jiang C, Zu C, Shen J, Shao F, Li T (2015a) Effects of selenium on the growth and photosynthetic characteristics of flue-cured tobacco (*Nicotiana tabacum* L.). Acta Soc Bot Pol 84:71–77. https://doi.org/10.5586/asbp.2015.006

Jiang Y, Zeng Z, Bu Y, Ren C, Li J, Han J, Tao C, Zhang K, Wang X, Lu G (2015b) Effects of selenium fertilizer on grain yield, Se uptake and distribution in common buckwheat (*Fagopyrum esculentum* Moench). Plant Soil Environ 61:371–377. https://doi.org/10.17221/284/2015-PSE

Jones GD, Droz B, Greve P, Gottschalk P, Poffet D, McGrath SP, Seneviratne SI, Smith P, Winkel LHE (2017) Selenium deficiency risk predicted to increase under future climate change. Proc Natl Acad Sci U S A 114:2848–2853. https://doi.org/10.1073/pnas.1611576114

Kennedy G, Burlingame B (2003) Analysis of food composition data on rice from a plant genetic resources perspective. Food Chem 80:589–596. https://doi.org/10.1016/S0308-8146(02)00507-1

Kennedy G, Nantel G, Shetty P (2003) The scourge of "hidden hunger": Global dimensions of micronutrient deficiencies. Food Nutr Agric 32:8–16. ISSN 1014-806X

Kikkert J, Berkelaar E (2013) Plant uptake and translocation of inorganic and organic forms of selenium. Arch Environ Contam Toxicol 65:458–465. https://doi.org/10.1007/s00244-013-9926-0

Kuznetsov VV, Kholodova VP, Kuznetsov VV, Yagodin BA (2003) Selenium regulates the water status of plants exposed to drought. Dokl Biol Sci 390:266–268. https://doi.org/10.1023/A:1024426104894

Li HF, McGrath SP, Zhao FJ (2008) Selenium uptake, translocation and speciation in wheat supplied with selenate or selenite. New Phytol 178:92–102. https://doi.org/10.1111/j.1469-8137.2007.02343.x

Li Z, Liang D, Peng Q, Cui Z, Huang J, Lin Z (2017) Interaction between selenium and soil organic matter and its impact on soil selenium bioavailability: a review. Geoderma 295:69–79. https://doi.org/10.1016/j.geoderma.2017.02.019

Li X, Wu YS, Li BQ, Yang YH, Yang YP (2018) Selenium accumulation characteristics and biofortification potentiality in Turnip (*Brassica rapa* var. *rapa*) Supplied with selenite or selenate. Front Plant Sci 8:2207. https://doi.org/10.3389/fpls.2017.02207

Lidon FC, Almeida AS, Costa AR, Bagulho AS, Scotti-Campos P, Semedo JN, Maçãs B, Coutinho J, Pinheiro N, Gomes C, Leitão AE, Pais IP, Silva MM, Reboredo FH, Pessoa MF, Ramalho JC (2015) Sequential zinc and iron biofortification of bread-wheat grains: from controlled to uncontrolled environments. J Crop Past Sci 66:1097–1104. https://doi.org/10.1071/CP14270

Lidon FC, Oliveira K, Galhano C, Guerra M, Ribeiro MM, Pelica J, Pataco I, Ramalho JC, Leitão AE, Almeida AS, Scotti-Campos P, Ribeiro-Barros AI, Pais IP, Silva MM, Carvalho ML, Santos JP, Pessoa MF, Reboredo FH (2019) Selenium biofortification of rice through foliar application with selenite and selenate. Exp Agric. 55: 528-542. https://doi.org/10.1017/S0014479718000157

Lidon FC, Oliveira K, Ribeiro MM, Pataco I, Ramalho JC, Leitão AE, Almeida AS, Scotti-Campos P, Ribeiro-Barros AI, Pais IP, Silva MM, Pessoa MF, Reboredo FH (2018) Selenium biofortification of rice grains and implications on macronutrients quality. J Cereal Sci 81:22–29. https://doi.org/10.1016/j.jcs.2018.03.010

Lin CJ, Li CY, Lin SK, Yang FH, Huang JJ, Liu YH, Lur HS (2010) Influence of high temperature during grain filling on the accumulation of storage proteins and grain quality in rice (*Oryza sativa* L.). J Agric Food Chem 58:10545–10552. https://doi.org/10.1021/jf101575j

Loladze I (2014) Hidden shift of the ionome of plants exposed to elevated CO_2 depletes minerals at the base of human nutrition. eLIFE 3:e02245. https://doi.org/10.7554/eLife.02245

Longchamp M, Castrec-Rouelle M, Biron P, Bariac T (2015) Variations in the accumulation, localization and rate of metabolization of selenium in mature *Zea mays* plants supplied with selenite or selenate. Food Chem 182:128–135. https://doi.org/10.1016/j.foodchem.2015.02.137

Lyons G (2018) Biofortification of cereals with foliar selenium and iodine could reduce hypothyroidism. Front Plant Sci 9:730. https://doi.org/10.3389/fpls.2018.00730

Lyons GH, Lewis J, Lorimer MF, Holloway RE, Brace DM, Stangoulis JCR, Graham RD (2004) High-selenium wheat agronomic biofortification strategies to improve nutrition. Food Agric Environ 2:171–178

Lyons MP, Papazyan TT, Surai PF (2007) Selenium in food chain and animal nutrition: lessons from nature. Asian-Aust J Anim Sci 20:1135–1155. https://doi.org/10.5713/ajas.2007.1135

Lyons GH, Genc Y, Soole K, Stangoulis JCR, Liu F, Graham RD (2009) Selenium increases seed production in *Brassica*. Plant Soil 318:73–80. https://doi.org/10.1007/s11104-008-9818-7

Mackowiak C, Amacher M (2008) Soil sulfur amendments suppress selenium uptake by alfalfa and western wheatgrass. J Environ Qual 37:772–779. https://doi.org/10.2134/jeq2007.0157

Malagoli M, Schiavon M, Dall'acqua S, Pilon-Smith EAH (2015) Effects of selenium biofortification on crop nutritional quality. Front Plant Sci 6:280. https://doi.org/10.3389/fpls.2015.00280

Mangueze AV, Pessoa MFG, Silva MJ, Ndayiragije A, Magaia HE, Cossa VSI, Reboredo FH, Carvalho ML, Santos JP, Guerra M, Ribeiro-Barros AI, Lidon FC, Ramalho JC (2018) Simultaneous Zinc and selenium biofortification in rice. Accumulation, localization and implications on the overall mineral content of the flour. J Cereal Sci 82:34–41. https://doi.org/10.1016/j.jcs.2018.05.005

Manzanares W, Hardy G (2016) Can dietary selenium intake increase the risk of toxicity in healthy children? Nutrition 32:149–150. https://doi.org/10.1016/j.nut.2015.07.001

Mao H, Wang J, Wang Z, Zan Y, Lyons G, Zou C (2014) Using agronomic biofortification to boost zinc, selenium, and iodine concentrations of food crops grown on the loess plateau in China. J Soil Sci Plant Nutr 14:459–470. https://doi.org/10.4067/S0718-95162014005000036

Martins LD, Tomaz MA, Lidon FC, DaMatta FM, Ramalho JC (2014) Combined effects of elevated [CO_2] and high temperature on leaf mineral balance in *Coffea* spp. plants. Clim Chang 126:365–379. https://doi.org/10.1007/s10584-014-1236-7

Mayer JE, Pfeifer WH, Beyer P (2008) Biofortified crops to alleviate micronutrient malnutrition. Cur Op Plant Biol 11:166–179. https://doi.org/10.1016/j.pbi.2008.01.007

McGrath JM, Lobell DB (2013) Reduction of transpiration and altered nutrient allocation contribute to nutrient decline of crops grown in elevated CO_2 concentrations. Plant Cell Environ 36:697–705. https://doi.org/10.1111/pce.12007

Meenakshi JV, Johnson NL, Manyong VM, De Groote H, Javelosa J, Yanggen DR, Naher C, Gonzalez F, García J, Meng E (2010) How cost effective is biofortification in combating micronutrient malnutrition? An exante assessment. World Dev 38:64–75. https://doi.org/10.1016/j.worlddev.2009.03.014

Mehdi Y, Hornick J-L, Istasse L, Dufrasne I (2013) Selenium in the environment, metabolism and involvement in body functions. Molecules 18:3292–3311. https://doi.org/10.3390/molecules18033292

Mohanty S (2013) Trends in global rice consumption. Rice facts: Rice Today (International Rice Research Institute) 12:44–45

Mozafariyan M, Pessarakli M, Saghafi K (2017) Effects of selenium on some morphological and physiological traits of tomato plants grown under hydroponic condition. J Plant Nutr 40:139–144. https://doi.org/10.1080/01904167.2016.1201500

Mudema JA, Manjate G (2014) Rentabilidade da produção de arroz com utilização de um pacote tecnológico melhorado, no regadio de Chókwè, província de Gaza. Relatório Preliminar de Pesquisa. Instituto de Investigação Agrária de Moçambique, p. 28

Nakandalage N, Nicolas M, Norton RM, Hirotsu N, Milham PJ, Seneweera S (2016) Improving rice Zinc biofortification success rates through genetic and crop management approaches in a changing environment. Front Plant Sci 7:764. https://doi.org/10.3389/fpls.2016.00764

Natasha SM, Niazi NK, Khalid S, Murtaza B, Bibi I, Rashid MI (2018) A critical review of selenium biogeochemical behavior in soil-plant system with an inference to human health. Environ Pollut 234:915–934. https://doi.org/10.1016/j.envpol.2017.12.019

Navarro-Alarcon M, Cabrera-Vique C (2008) Selenium in food and the human body: a review. Sci Total Environ 400:115–141. https://doi.org/10.1016/j.scitotenv.2008.06.024

Oldfield JE (1999) Selenium World Atlas, Selenium-Tellurium Development Association, 71–83

Oliveira K, Pataco IM, Mourinho MP, Santos C, Pelica J, Ramalho JC, Leitão AE, Pais IP, Scotti-Campos P, Lidon FC, Reboredo FH, Pessoa MF (2015) Selenium biofortification in rice – A pragmatic perspective. Emir J Food Agric 27:231–241. https://doi.org/10.9755/ejfa.v27i3.19285

Oliveira VC, Faquin V, Guimarães KC, Andrade FR, Pereira J, Guilherme LRG (2018) Agronomic biofortification of carrot with selenium. Ciênc Agrotec 42:138–147. https://doi.org/10.1590/1413-70542018422031217

Padovani RM, Amaya-Farfan J, Colugnati FAB, Domene SMA (2006) Dietary reference intakes: application of tables in nutritional studies. Rev Nutr 19:741–760. https://doi.org/10.1590/S1415-52732006000600010

Pandey S, Bhandari H, Hardy B (2007) Economic costs of drought and rice farmers coping mechanisms: a cross country comparative analysis. International Rice Research Institute. p. 203

Papp LA, Lu J, Holmgren A, Khanna KK (2007) From Selenium to selenoproteins: synthesis, identity, and their role in human health. Antioxid Redox Signal 9:775–806. https://doi.org/10.1089/ars.2007.1528

Pennanen A, Xue T, Hartikainen H (2002) Protective role of selenium in plant subjected to severe UV irradiation stress. J Appl Bot 76:66–76

Pfeiffer WH, McClafferty B (2007) HarvestPlus: breeding crops for better nutrition. Crop Sci 47:88–105. https://doi.org/10.2135/cropsci2007.09.0020IPBS

Pilon-Smits EAH, Quinn CF (2010) Selenium metabolism in plants. In: Hell R, Mendel R-R (eds) Cell biology of metals and nutrients. Plant cell monographs 17. Springer Berlin Heidelberg, pp 225–241. https://doi.org/10.1007/978-3-642-10613-2_10

Pleijel H, Hogy P (2015) CO_2 dose-response functions for wheat grain, protein and mineral yield based on FACE and open-top chamber experiments. Environ Pollut 198:70–77. https://doi.org/10.1016/j.envpol.2014.12.030

Powell N, Ji X, Ravash R, Edlington J, Dolferus R (2012) Yield stability for cereals in a changing climate. Funct Plant Biol 39:539–552. https://doi.org/10.1071/FP12078

Ramalho JC, Pais IP, Leitão AE, Guerra M, Reboredo FH, Máguas C, Carvalho ML, Scotti-Campos P, Ribeiro-Barros AI, Lidon FC, DaMatta FM (2018) Can elevated air [CO_2] conditions mitigate the predicted warming impact on the quality of coffee bean? Front Plant Sci 9:287. https://doi.org/10.3389/fpls.2018.00287

Ramos SJ, Faquin V, Guilherme LRG, Castro EM, Avila FW, Carvalho GS, Bastos CEA, Oliveira C (2010) Selenium biofortification and antioxidant activity in lettuce plants fed with selenate and selenite. Plant Soil Env 56:584–588. https://doi.org/10.17221/113/2010-PSE

Rayman MP (2008) Food-chain selenium and human health: emphasis on intake. Brit. J Nutr 100:254–268. https://doi.org/10.1017/S0007114508939830

Rayman MP (2012) Selenium and human health. Lancet 379:1256–1268. https://doi.org/10.1016/S0140-6736(11)61452-9

Redfern SK, Azzu N, Binamira JS (2012) Rice in Southeast Asia: facing risks and vulnerabilities to respond to climate change. In: Meybeck A, Lankoski J, Redfern S, Azzu N, Gitz V (eds) Building resilience for adaptation to climate change in the agriculture, proceedings of a Joint FAO/OECD Workshop, Rome, 23–24 April. Food and Agriculture Organization of the United Nations Organisation for Economic co-Operation and Development, Rome. Italy, pp 295–314

Rehman HM, Cooper JW, Lam H-M, Yang SH (2018) Legume biofortification is an underexploited strategy for combatting hidden hunger. Plant Cell Environ 42:52–70. https://doi.org/10.1111/pce.13368

Rodrigo S, Santamaria O, Poblaciones MJ (2014) Selenium application timing: influence in wheat grain and flour selenium accumulation under mediterranean conditions. J Agric Sci 6:23–30. https://doi.org/10.5539/jas.v6n3p23

Rodrigues WP, Martins MQ, Fortunato AS, Rodrigues AP, Semedo JN, Simões-Costa MC, Pais IP, Leitão AE, Colwell F, Goulao L, Máguas C, Partelli FL, Campostrini E, Scotti-Campos P, Ribeiro-Barros AI, Lidon FC, DaMatta FM, Ramalho JC (2016) Long-term elevated air [CO_2] strengthens photosynthetic functioning and mitigates the impact of supra-optimal temperatures in tropical *Coffea arabica* and *C. canephora* species. Glob Chang Biol 22:415–431. https://doi.org/10.1111/gcb.13088

Ros G, van Rotterdam A, Bussink D, Bindraban P (2016) Selenium fertilization strategies for bio-fortification of food: an agro-ecosystem approach. Plant Soil 404:99–112. https://doi.org/10.1007/s11104-016-2830-4

Rosa V, Erkekoglu P, Forestier A, Favier A, Hincal F, Diamond AM, Douki T, Rachidi W (2012) Low doses of selenium specifically stimulate the repair of oxidative DNA damage in LNCaP prostate cancer cells. Free Radic Res 46:105–116. https://doi.org/10.3109/10715762.2011.647009

Schomburg L (2016) Dietary selenium and human health. Nutrients 9(22). https://doi.org/10.3390/nu9010022

Schwarz K, Foltz CM (1957) Selenium as an integral part of factor 3 against dietary necrotic liver degeneration. J Am Chem Soc 79:3292–3293. https://doi.org/10.1016/S0899-9007(98)00183-X

Seck PA, Diagne A, Mohanty S, Wopereis MCS (2012) Crops that feed the world 7: rice. Food Security 7(24). https://doi.org/10.1007/s12571-012-0168-1

Semedo JN, Rodrigues WP, Dubberstein D, Martins MQ, Martins LD, Pais IP, Rodrigues AP, Leitão AE, Partelli FL, Campostrini E, Tomaz MA, Reboredo FH, Scotti-Campos P, Ribeiro-Barros AI, Lidon FC, DaMatta FM, Ramalho JC (2018) Coffee responses to drought, warming and high [CO_2] in a context of future climate change scenarios. In: Alves F, Leal W, Azeiteiro U (eds) Theory and practice of climate adaptation, Chapter 26, Climate Change Management Series. Springer, pp 465–477, ISBN 978-3-319-72874-2. https://doi.org/10.1007/978-3-319-72874-2

Sors TG, Ellis DR, Salt DE (2005) Selenium uptake, translocation, assimilation and metabolic fate in plants. Photosynth Res 86:373–389. https://doi.org/10.1007/s11120-005-5222-9

Stein AJ (2015) The poor, malnutrition, biofortification, and biotechnology. In: Herring R (ed) The Oxford handbook of food, politics and society. Oxford University Press, New York, pp 1–26. https://doi.org/10.1093/oxfordhb/9780195397772.013.005

Sun GX, Liu X, Williams PN, Zhu YG (2010) Distribution and translocation of selenium from soil to grain and its speciation in paddy rice (*Oryza sativa* L.). Environ Sci Technol 44:6706–6711. https://doi.org/10.1021/es101843x

Tack J, Barkley A, Nalley LL (2015) Effect of warming temperatures on US wheat yields. Proc Natl Acad Sci U S A 112:6931–6936. https://doi.org/10.1073/pnas.1415181112

Taspinar MS, Agar G, Yildirim N, Sunar S, Aksakal O, Bozari S (2009) Evaluation of selenium effect on cadmium genotoxicity in *Vicia faba* using RAPD. J Food Agric Environ 7:857–860. https://doi.org/10.1234/4.2009.2811

Teixeira EI, Fischer G, van Velthuizen H, Walter C, Ewert F (2013) Global hot-spots of heat stress on agricultural crops due to climate change. Agric For Meteorol 170:206–215. https://doi.org/10.1016/j.agrformet.2011.09.002

Teng P, Caballero-Anthony M, Lassa J (2016) The future of rice security under climate change—NTS report. Centre for non-traditional security studies. Available on https://www.rsis.edu.sg/wp-content/uploads/2016/10/NTS-Report4-July2016-ClimateChangeAndRice.pdf

Terry N, Zayed A, Souza M, Tarun A (2000) Selenium in higher plants. Annu Rev Plant Biol 51:401–432. https://doi.org/10.1146/annurev.arplant.51.1.401

Thavarajah P, Vial E, Gebhardt M, Lacher C, Kumar S, Combs GF (2015) Will selenium increase lentil (*Lens culinaris* Medik) yield and seed quality? Front Plant Sci 6:1–9. https://doi.org/10.3389/fpls.2015.00356

Thiec DL, Dixon M, Loosveldt P, Garrec JP (1995) Seasonal and annual variations of phosphorus, calcium, potassium and manganese contents In different cross-sections of *Picea abies* (L.) Karst. needles and *Quercus rubra* L. leaves exposed to elevated CO_2. Trees 10:55–62. https://doi.org/10.1007/BF00192184

Thomson CD (2004) Assessment of requirements for selenium and adequacy of selenium status: a review. Eur J Clin Nutr 58:391–402. https://doi.org/10.1038/sj.ejcn.1601800

Tian H, Bao Z, Wei C, Zhang H, Wu S, Gu A, Xie S (2016) Improved selenium bioavailability of selenium enriched slate via calcination with a Ca based sorbent. J Geochem Explor 169:73–79. https://doi.org/10.1016/j.gexplo.2016.07.015

Turakainen M, Hartikainen H, Seppanen MM (2004) Effects of selenium treatments on potato (*Solanum tuberosum* L.) growth and concentrations of soluble sugars and starch. J Agric Food Chem 52:5378–5382. https://doi.org/10.1021/jf040077x

Uprety DC, Reddy VR (2016) Case histories: crops. In: Crop responses to global warming. Springer, Singapore, pp 41–116. isbn:978-981-10-2004-9

Wahid A, Gelani S, Ashraf M, Foolad MR (2007) Heat tolerance in plants: an overview. Environ Exp Bot 61:199–223. https://doi.org/10.1016/j.envexpbot.2007.05.011

Wallenberg M, Olm E, Hebert C, Björnstedt M, Fernandes AP (2010) Selenium compounds are substrates for glutaredoxins: a novel pathway for selenium metabolism and a potential mechanism for selenium-mediated cytotoxicity. Biochem J 429:85–93. https://doi.org/10.1042/BJ20100368

Wang J, Wang Z, Mao H, Zhao H, Huang D (2013) Increasing Se concentration in maize grain with soil- or foliar-applied selenite on the Loess Plateau in China. Field Crop Res 150:83–90. https://doi.org/10.1016/j.fcr.2013.06.010

Wang P, Zhang Z, Chen Y, Wei X, Feng B, Tao F (2016) How much yield loss has been caused by extreme temperature stress to the irrigated rice production in China? Clim Chang 134:635–650. https://doi.org/10.1007/s10584-015-1545-5

Wassmann R, Nelson GC, Peng SB, Sumfleth K, Jagadish SVK, Hosen Y, Rosegrant MW (2010) Rice and global climate change. In: Pandey S, Byerlee D, Dawe D, Dobermann A, Mohanty S, Rozelle S, Hardy B (eds) Rice in the global economy: strategic research and policy issues for food security, Chapter 4.1. International Rice Research Institute, Manila, pp 411–432. ISBN 978-971-22-0258-2

Whanger PD (2002) Selenocompounds in plants and animals and their biological significance. J Am Coll Nutr 21:223–232. https://doi.org/10.1080/07315724.2002.10719214

White PJ (2018) Selenium metabolism in plants. BBA General Subj 1862:2333–2342. https://doi.org/10.1016/j.bbagen.2018.05.006

White PJ, Broadley MR (2005) Biofortifying crops with essential mineral elements. Trends Plant Sci 10:586–593. https://doi.org/10.1016/j.tplants.2005.10.001

White PJ, Broadley MR (2009) Biofortification of crops with seven mineral elements often lacking in human diets iron, zinc, copper, calcium, magnesium, selenium and iodine. New Phytol 182:49–84. https://doi.org/10.1111/j.1469-8137.2008.02738.x

White PJ, Bowen HC, Parmaguru P, Fritz M, Spracklen WP, Spiby RE, Meacham MC, Mead A, Harriman M, Trueman LJ, Smith BM, Thomas B, Broadley MR (2004) Interactions between selenium and sulphur nutrition in *Arabidopsis thaliana*. J Exp Bot 55:1927–1937. https://doi.org/10.1093/jxb/erh192

WHO-World Health Organization (1987) Environmental Health Criterion 58-Selenium, World Health Organization, Geneva. Available at: http://whqlibdoc.who.int/publications/2009/9789241597494_eng.pdf. Accessed Mar 2019

WHO-World Health Organization. (1996) Trace Elements in Human Nutrition and Health, World Health Organization, Geneva. Available at: http://whqlibdoc.who.int/publications/2009/9789241597494_eng.pdf. Accessed Mar 2019

Williams PN, Lombi E, Sun G-X, Scheckel K, Zhu Y-G, Feng X, Zhu J, Carey A-M, Adomako E, Lawgali Y, Deacon C, Meharg AA (2009) Selenium characterization in the global supply chain of rice. Environ Sci Technol 43:6024–6030. https://doi.org/10.1021/es900671m

Winkel L, Vriens B, Jones G, Schneider L, Pilon-Smits E, Bañuelos G (2015) Selenium cycling across soil-plant-atmosphere interfaces: a critical review. Nutrients 7:4199–4239. https://doi.org/10.3390/nu7064199

Wise RR, Olson AJ, Schrader SM, Sharkey TD (2004) Electron transport is the functional limitation of photosynthesis in field-grown pima cotton plants at high temperature. Plant Cell Environ 27:717–724. https://doi.org/10.1111/j.1365-3040.2004.01171.x

Xiaoqin Y, Chu J, Wang G (2009a) Effects of Selenium on wheat seedlings under drought stress. Biol Trace Elem Res 130:283–290. https://doi.org/10.1007/s12011-009-8328-7

Xiaoqin Y, Jianzhou C, Guangyin W (2009b) Effects of drought stress and selenium supply on growth and physiological characteristics of wheat seedlings. Acta Physiol Plant 31:1031–1036. https://doi.org/10.1007/s11738-009-0322-3

Xue TL, Hartikainen H, Piironen V (2001) Antioxidative and growth-promoting effects of selenium on senescing lettuce. Plant Soil 237:55–61. https://doi.org/10.1023/A:1013369804867

Yang L, Zhao G-H, Yu F-F, Zhang R-Q, Guo X (2016) Selenium and iodine levels in subjects with Kashin-Beck disease: a meta-analysis. Biol Trace Elem Res 170:43–54. https://doi.org/10.1007/s12011-015-0463-8

Zayed AM, Lytle CM, Terry N (1998) Accumulation and volatilization of different chemical species of selenium by plants. Planta 206:284–289. https://doi.org/10.1007/s004250050402

Zembala M, Filek M, Walas S, Mrowiec H, Kornas A, Miszalski Z, Hartikainen H (2010) Effect of selenium on macro and microelement distribution and physiological parameters of rape and wheat seedlings exposed to cadmium stress. Plant Soil 329:457–468. https://doi.org/10.1007/s11104-009-0171-2

Zhang M, Tang S, Huang X, Zhang F, Pang Y, Huang Q, Yi Q (2014) Selenium uptake, dynamic changes in selenium content and its influence on photosynthesis and chlorophyll fluorescence in rice (*Oryza sativa* L.). Environ Exp Bot 107:39–45. https://doi.org/10.1016/j.envexpbot.2014.05.005

Zhang Y, Wang Y, Niu H (2017) Spatio-temporal variations in the areas suitable for the cultivation of rice and maize in China under future climate scenarios. Sci Total Environ 601:518–532. https://doi.org/10.1016/j.scitotenv.2017.05.232

Zhao C, Ren J, Xue C, Lin E (2005) Study on the relationship between soil selenium and plant selenium uptake. Plant Soil 277:197–206. https://doi.org/10.1007/s11104-005-7011-9

Zhao C, Liu B, Piao S, Wang X, Lobell DB, Huang Y, Huang M, Yao Y, Bassu S, Ciais P, Durand J-L, Elliott J, Ewert F, Janssens IA, Li T, Lin E, Liu Q, Martre P, Müller C, Peng S, Peñuelas J, Ruane AC, Wallach D, Wang T, Wu D, Liu Z, Zhu Y, Zhu Z, Asseng S (2017) Temperature increase reduces global yields of major crops in four independent estimates. Proc Natl Acad Sci U S A 114:9326–9331. https://doi.org/10.1073/pnas.1701762114

Zhu YG, Pilon-Smits EA, Zhao FJ, Williams PN, Meharg AA (2009) Selenium in higher plants: understanding mechanisms for biofortification and phytoremediation. Trends Plant Sci 14:436–442. https://doi.org/10.1016/j.tplants.2009.06.006

Zwolak I, Zaporowska H (2012) Selenium interactions and toxicity: A review. Cell Biol Toxicol 28:31–46. https://doi.org/10.1007/s10565-011-9203-9

Chapter 9
Rice: Prevention and Management of Type 2 Diabetes and Coronary Heart Disease

Christine J. Bergman

9.1 Rice Overview

Oryza sativa is the most commonly grown rice subspecies and is the staple food for more than one half of the world's population (IRRI 2013). There are two primary subtypes of this subspecies: indica types (cultivated in tropical climates and when cooked is relatively firm and not sticky) and japonica types (cultivated in temperate climates and when cooked is relatively soft and sticky). Other than *Oryza sativa* there is another cultivated subspecies of rice known as *Oryza glaberrima* (cultivated in West Africa) and 22 other related species (Jacquemin et al. 2013; Linares 2002). *Oryza glaberrima* is also grown for human food production while the related species are used in the study of rice evolution and the genetic control of traits and as a source of genes able to resist abiotic and biotic stressors.

Across the planet, rice from here on defined as *Oryza sativa* L. exists with various combinations of end-use quality traits. These include physical properties such as length-width ratio, grain weight, and bran color. Also, functional characteristics vary across varieties (Bao and Bergman 2018; Bergman 2018). Some of these traits include gelatinization temperature, rheological properties, and cooked rice texture. A great degree of variability also exists in terms of its nutritional and health-beneficial properties (Dipti et al. 2012); some of these properties are reviewed in this chapter.

Rice end-use quality desired by one region of the world may be completely unacceptable to another. Consumer's preference for rice lies largely in its cooked texture. Most rice does not enter the export market, but rather is consumed close to where it was produced. Consequently, professionals working in the food-processing industry and consumers are often exposed only to local cultivars and are unaware of the great

C. J. Bergman (✉)
Food & Beverage and Event Management Department, University of Nevada Las Vegas, Las Vegas, NV, USA
e-mail: Christine.Bergman@UNLV.EDU

A. Costa de Oliveira et al. (eds.), *The Future of Rice Demand: Quality Beyond Productivity*, https://doi.org/10.1007/978-3-030-37510-2_9

variation in the end-use quality of the world's rice germplasm. This can be appreciated by recognizing that there are more than 100,000 different genotypes of rice and its related species being maintained in germplasm collections across the world (Hay 2017). Within these valuable genetic resources, rice scientists can identify gene mutations associated with particular rice nutritional and health-beneficial properties, thus allowing rice breeders to bring desired traits into adapted germplasm suitable for use in developing cultivars with new food and industrial applications. This knowledge can also be used to create nutritionally enhanced cultivars using mutation breeding or gene-editing techniques.

Rice grains, after removing its outer husk layer, consist of three fractions: the endosperm (starch, protein, vitamins, and minerals), bran (protein, dietary fiber, lipids, vitamins, minerals, and phytochemicals), and germ (lipids, protein, vitamins, minerals, and phytochemicals). These grains are called whole grain rice, unpolished rice, unmilled rice, and brown rice, depending on the region of the world. The brown rice term will be used below to describe this form of rice. The name comes from the fact that most rice grown commercially have brown-colored bran. However, there are many genotypes with various other bran colors. The nutritional (i.e., macronutrient quality and quantity and micronutrient content) and health-beneficial properties (i.e., bioactive phytochemicals) of these genotypes will be covered in another chapter within this book.

Globally, most rice is polished/milled before cooking the intact grains. This involves a process that separates the endosperm from the bran and germ (Bao 2018). The resulting rice kernels are white colored to some degree. Today some of this rice is also ground into flour and some is extracted for its components, such as starch and protein. Historically, rice has been milled to increase its shelf life, speed the time of cooking, and reduce the need for chewing. These were originally positive things for humans when kilocalories were in short supply; however, by removing the bran and germ, rice loses most of its vitamins, minerals, dietary fiber, lipids, and phytochemicals, thus reducing its nutritional content. Also, there are many groups of consumers globally that eat rice as their primary staple and have significant micronutrient deficiencies, especially Fe, Zn, Vitamin A, and folate (Bailey et al. 2015). Consequently, some milled rice is fortified with vitamins and minerals and has been breed to have higher levels of some of these compounds, and others have been genetically modified to contain nutrients typically not found in rice, such as beta-carotene (Dipti et al. 2012; Birlat et al. 2017). Most rice that is consumed with elevated micronutrient levels are those that have been fortified postharvest (WHO 2018). The bioavailability, allergenicity, and acceptability of biofortified rice has received limited research attention and little commercialization of these varieties has been achieved (Birlat et al. 2017; Talsma 2017; Lockyer et al. 2018). In time, some scientists think these barriers to improving human health will be overcome and then the appropriate regulatory framework will need to be developed and adopted (Mejia 2017). Beyond micronutrient deficiency diseases, dietary intake is also known to influence human's risk for developing chronic disease.

The World Health Organization (WHO) statistics indicate that the top 10 causes of mortality account for 51.4% of all deaths globally and several of these are lifestyle-related (Karuga 2018). These chronic diseases include cardiovascular disease and type 2 diabetes (T2D). One of the risk factors associated with these chronic diseases is dietary makeup. Considering the large number of people that rely on rice as their staple food, the effects of consuming rice and its coproducts on chronic disease morbidity risk and management need to be understood by consumers, the rice industry, public health officials, and food service operators. The association between eating rice and type 2 diabetes and coronary heart disease is discussed below.

9.2 Type 2 Diabetes

9.2.1 Introduction

According to WHO statistics, type 2 diabetes is the seventh leading cause of death globally (Fig. 9.1). This chronic disease reportedly is associated with various risk factors such as obesity, lack of exercise, age, some ethnic groups, and poor diet. The global prevalence of T2D is increasing at a rapid rate, along with obesity. Research indicates that it can be delayed or even prevented by eating a healthy diet, maintaining a normal body mass index, and leading an active lifestyle (Gruss et al. 2019). However, what a healthy diet is remains controversial to some degree. Meta-analysis reviews have concluded that the Mediterranean diet score, the Healthy Eating Index, the Alternate Healthy Eating Index, and the DASH dietary score are associated with reduced risk of T2D (Schwingshack and Hoffmann 2015; Schwingshack et al. 2015; Schwingshack et al. 2017). In general, these styles of eating recommend consuming some combination of whole grains, vegetables, fruits, lean meats and fish, and limited

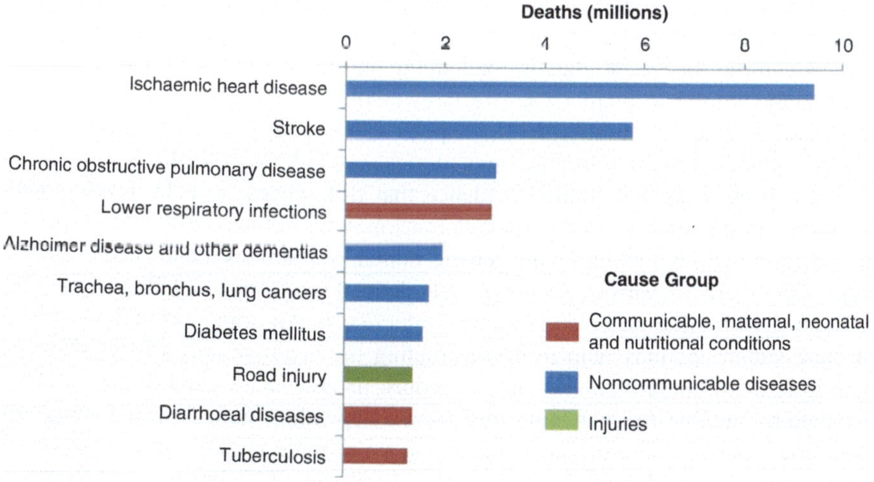

Fig. 9.1 Global health estimates 2016: deaths by cause, age, sex, by country, and by region, 2000–2016. (Geneva, World Health Organization 2018)

saturated fat or red meat and refined carbohydrates. A review of observational studies and randomized controlled trials found that dietary factors showing promise for reducing T2D risk are greater intakes of cereal grain fiber, unsaturated fatty acids, magnesium, and polyphenols, and a reduced dietary glycemic load, added sugars, and high-sugar beverages (Palacios et al. 2019). The latter three are encouraged to prevent people from persistently having plasma glucose levels above normal limits, which are known to be harmful to cells, tissues, and organs and are risk factors for developing T2D and cardiovascular disease (Priebe et al. 2018). None of the diet protocols mentioned above state do not eat rice, but they do all recommend that carbohydrates be consumed in large part as whole grains; thus rice eaters are encouraged by these diet plans to limit their milled rice consumption and focus on eating brown rice instead. Yet, many popular media and some health professionals recommend eating very low carbohydrate diets and no grain diets with the promise of preventing the development of T2D and as a means of managing this chronic disease, and some specifically state to not eat rice (Davis 2017). Additionally, multiple prospective cohort studies in Western and Asian populations have found positive associations between the intake of white rice and the risk of T2D (Hu et al. 2012).

There is no optimum way to evaluate the effects of eating rice on the risk and management of T2D due to the many possible correlated factors, large study expenses, and lack of subjects willing to be studied over a long time period. Thus, various methods are used to try and understand the related factors such as measurements of postprandial blood sugar and insulin levels via in vivo analysis after the consumption of planned meals or individual food intake. For example, the glycemic index is the ranking of carbohydrate-containing foods based on how quickly they raise blood sugar levels in comparison to glucose. Similarly, glycemic load is a measure of how quickly a source of carbohydrates raises blood sugar levels and how much glucose per serving it can deliver. But these methods have a shortcoming in that people don't typically eat one food, such as rice, at a time. We eat meals containing several foods. Whether one food or meals are evaluated, limiting extraneous variables during in vivo studies is challenging and these studies are expensive. In vitro methods used to predict the effects of eating carbohydrates on blood sugar levels are enzymatic digestibility (i.e., speed that digestive enzymes can change starch to glucose) and in vitro measures of glycemic index and load.

There is growing but limited evidence that differences in T2D development between individuals and between populations may be influenced by carbohydrate and fiber intake interactions with genetic human variants (Villegas et al. 2014). Reportedly there are various phenotypically defined subgroups of T2D that may be associated with these genetic variations (Udler 2019). An improved understanding of these subgroups may help in deconstructing the heterogeneity of this disease, both in terms of dietary choices able to reduce its risk factors and dietary recommendations suitable for managing this disease depending on a person's subgroup (McCarthy 2017). Therefore, it seems that some of the confusion in the effects of eating carbohydrate-rich foods on the risk of developing T2D has resulted from not controlling for differences in the study subject's genetic makeup.

In addition to the above discussion, some confusion may exist in the understanding of the influence of rice consumption on T2D risk factors and management because most epidemiological studies don't record and study the specific type of rice each subject typically eats, and some don't determine the quantity of rice eaten daily. Differences in a carbohydrate-rich food's rate of digestibility, percentage, and type of dietary fiber, quantity as a percentage of kilocalories, etc., may influence its effect on T2D risk factors and management of the disease. The many different genotypes of rice humans eat and the different types of processing this grain undergoes also likely influences its role in the development and management of T2D.

The question to be addressed in the sections below is which types of rice (i.e., chemical makeup), processing methods, and type of cooking fit in with our current understanding of a healthy diet? More specifically, which rice types and processing methods result in rice with carbohydrates that have a slower effect on postprandial plasma glucose levels and raise insulin levels to a lesser degree.

9.2.2 *Milled Rice*

Rice-cooked rice texture and processing properties are known to be associated with a cultivar's amylose content. Much of the discussion in this section discusses various traits associations with amylose content. But the reader should keep in mind that some of these results may not be 100% due to amylose per se because the iodine binding methods used to quantify this compound don't measure strictly amylose chains, but also extra-long amylopectin chains. Size exclusion chromatography, on the other hand, is a technique capable of measuring the actual proportion of amylose and amylopectin (Fitzgerald et al. 2009).

The GI of milled rice reportedly varies from 48 to 92 across different genotypes (Atkinson et al. 2008; Fitzgerald et al. 2011). This indicates that traditional breeding methods can be used to create adapted varieties with superior GI. The milled rice of varieties with higher levels of amylose and particular mutations within the waxy gene reportedly have slower rates of digestibility and GI (Chung, et al. 2011; Fitzgerald et al. 2011; Meera et al. 2019). However, Anacleto et al. (2019) reported that amylose content alone was not a good predictor of the rate at which rice grains digest ($R^2 = 0.28$). They studied a group of indica genotypes with amylose content ranging from 0% to 30%, thus indicating that other aspects of rice grains influence starch digestibility besides amylose content.

Butardo et al. (2017) reported identifying haplotypes in an indica-rice diversity panel that are associated with phenotypes with slower rice digestibility. These indica- types in terms of amylose content were in the categories of no amylose, medium-amylose, and high-amylose types. Their findings suggest that in addition to elevated amylose levels, differences in rice starch structure can also be selected for using marker-assisted breeding methods to create rice with eating qualities desired by some consumers and with improved digestibility.

Rice starch has very small (<4 μm), small (4–6 μm), medium (6–8 μm), and large (>8 μm) granules. Genotypes with greater amounts of the larger starch granules reportedly had lower in vitro starch digestibility (Dhital et al. 2015). The authors noted that smaller granules had larger pores and no fused granule agglomerates; thus, greater surface area for amylase activity may cause the higher digestibility rates of smaller granules. In this study, B-type granules were more resistant to enzymatic digestion than A-types, thus indicating that greater starch crystallinity also influences starch digestibility. Work by Alhambra et al. (2019) also found that longer chain amylopectin is associated with slower rates of starch digestion.

Starch that is resistant to enzymatic hydrolysis is influenced by many factors, specifically starch granule size and shape, size of pores in the starch granule, amylose/amylopectin ratio, the chain length of amylose, and degree of starch crystallinity (Leszczyñski 2004; Ma and Boye (2018). Known as resistant starch (RS), this type of starch is not digested in the small intestine, but in the large intestine where it is fermented into short-chain fatty acids. Consumption of RS has been associated with reduced risk of developing T2D and becoming obese. The exact mechanism is unknown but is thought to be associated with reduced kilocalorie intake and the modulation of the expression of genes involved in glucose and lipid metabolism, and insulin regulation (Meenu and Xu 2019).

Greater amounts of RS are reportedly in flour and starch from rice varieties with higher amounts of amylose (Zhu et al. 2011). This association was found across four cultivars containing from 1.7% to 55.4% amylose, the latter being a transgenic variety. Chen et al. (2017) studied 40 high amylose varieties and found that the amount of RS in cooked rice varied by 1.9-fold. High levels of RS were identified in high amylose varieties that are resistant to degradation during thermal processing, as is used in the canning industry, and in one variety that is expected to have slow retrogradation rates after cooking, thus having softer texture (Fig. 9.2).

Very-high-amylose rice types that are called amylose-extender (ae) mutants have been created using mutation breeding techniques. They have a mutation in the gene that codes for starch-branching enzyme IIb (BEIIb) which alters the structure of amylopectin in the endosperm by reducing short chains and increasing longer chains (Yano et al. 1985; Nishi et al. 2001). Consequently, the ae mutant has been used in traditional breeding efforts to create adapted varieties with higher levels of RS starch and thus slower rates of carbohydrate digestion (Wada et al. 2018). Figure 9.3 displays grains of 'Chikushi-kona 85' the most recent variety breed to contain the ae mutation and superior yield compared to previous similar mutants. 'Chou 2418' was developed by crossing a high-amylose rice cultivar Koshinomenjiman and the ae mutant EM10 (Noro et al. 2019). It contained 25 times more RS than that of a traditional Japanese rice cultivar after retrogradation. The greater level was attributed to not just the high level of amylose, but also long chains of amylopectin. The authors indicated that Chou 2418 is suitable as a flour for use in the production of processed foods, but doesn't have a suitable cooked texture for eating in its original form.

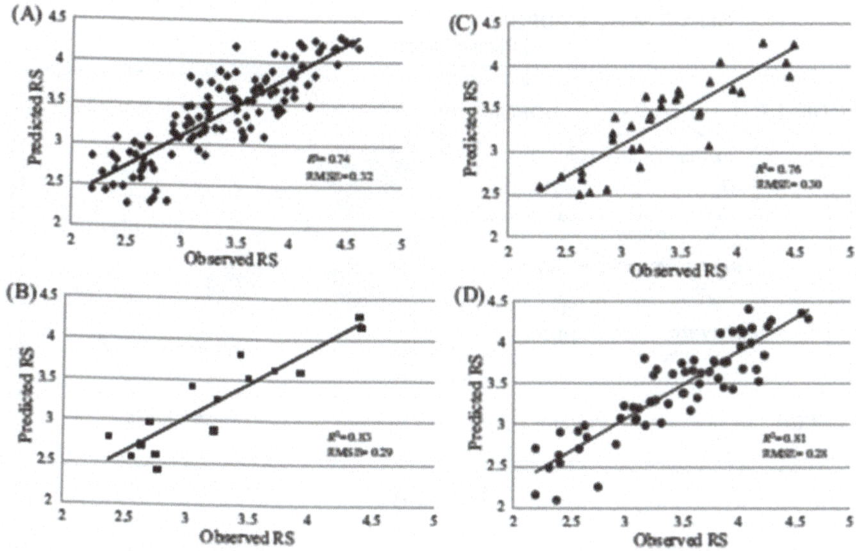

Fig. 9.2 Scatter plots of observed and predicted percent resistant starch (RS) in cooked rice. The predicted RS was calculated using a multiple linear regression backward stepwise selection model with (**a**) predictors of apparent amylose content (AA) and RVA parameters of peak viscosity, setback from peak viscosity, and pasting temperature for all variety group, and (**b**), (**c**), and (**d**) predictors of AA and RVA pasting temperature for varieties in haplotypes of SI, SL, and WL, where S and W indicate strong and weak paste viscosity and I and L stand for intermediate and low gelatinization temperature (Chen et al. 2017)

Fig. 9.3 Milled grains of 'Nishihomare' (conventional Japanese japonica type) and 'Chikushikona 85' (Japanese japonica line derived from a starch-branching enzyme IIb-deficient mutant). The opaque grains are suggestive of insufficiently filled starch granules in the endosperm (Wada et al. 2018)

Zhou et al. (2016) describing their research and the work of others have proposed a biosynthetic pathway that results in elevated RS (Fig. 9.4). A mutation that results in a deficiency in SSIIIa will decrease the production of amylopectin, thus resulting in greater carbon availability for use in amylose biosynthesis by GBSSI encoded by

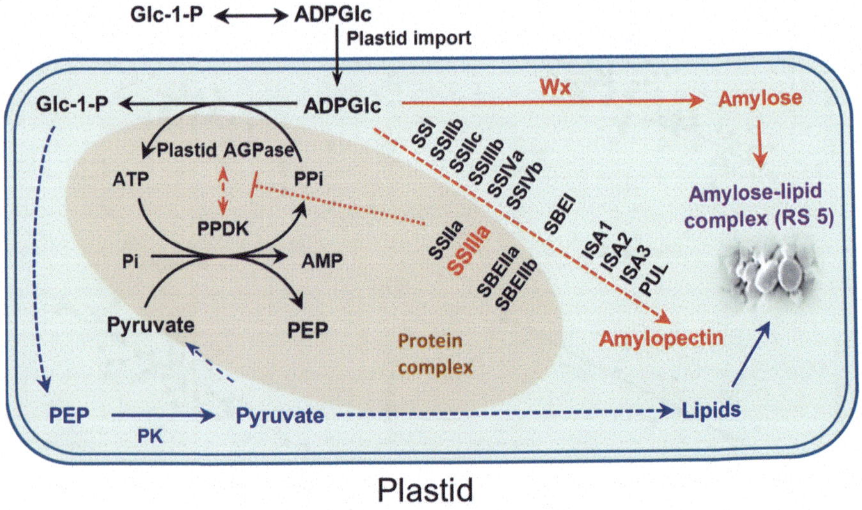

Fig. 9.4 Proposed RS biosynthetic pathway in the plastid. Biosynthesis of ADPGlc is brought about primarily by cytosolic AGPase, and ADPGlc is then imported into the plastid for starch biosynthesis. Several amyloplast enzymes exist in a large protein complex, which includes AGPase, PPDK, SSIIa, SSIIIa, SBEIIa, and SBEIIb, and this complex is thought to provide a means to control the partitioning of carbon between starch and lipids. The amyloplast contains a high level of pyrophosphatase, which keeps the concentration of PPi in the stroma very low. The presence of PPDK in a complex with AGPase may enable PPi to be channeled directly to AGPase for the conversion of ADPGlc to G-1-P and subsequently to lipid. The sequestering of PPDK in a protein complex may also prevent a futile cycle operating between PPDK and pyruvate kinase (PK). The starch biosynthetic enzymes in the complex are proposed to inhibit the activity of PPDK and AGPase (red dotted bar). In the absence of a functional SSIIIa protein, the complex is disrupted, and the influence on PPDK and AGPase is reduced, such that more carbon is directed from ADPGlc to Glc-1-P and into lipid. At the same time, the absence of SSIIIa means that relatively more ADPGlc can also be consumed by the Wx protein in the biosynthesis of amylose. This process leads to an increase in amount of amylose–lipid complex and hence RS 5. Dashed arrows indicate multiple steps. AGPase, ADP glucose pyrophosphorylase; ISA, isoamylase; PEP, phosphoenolpyruvate; PUL, pullulanase (Zhou et al. 2016)

the Wx gene. The increased levels of amylose plus lipids increase the content of amylose–lipid complexes, which are measured as a part of RS.

"Goami 2" formally called 'Suweon 464,' is a mutant with similar starch properties to those reported above with the ae mutation (Kim et al. 2005). However, a causal mutation wasn't found in BEIIb; thus, apparently there is more than one mutation able to create rice with a lower GI (Butardo et al. 2012). Additional phenotypic information about this mutant is known compared to the ae mutants. Specifically, it has smaller starch granules and higher levels of total, soluble, and insoluble dietary fiber than its parent 'Ilpumbyeo'. The authors indicated that there is a thick band of peripheral filaments that surround compound starch granules thought to be amylose–lipid complexes, which function as a structural barrier to the entrance of water into the granule, causing low swelling power, incomplete gelatinization, and texture not suitable for eating as intact grains. Consumption of this mutant was associated with weight loss and the reduction of triacylglycerol concentrations in obese subjects.

Although the ae mutants and Goami 2 have a lower GI, to date, all of these genotypes have cooking and eating quality that isn't acceptable to any rice-eating populations (Butardo et al. 2017). Specifically, they appear opaque, as with waxy genotypes, but are extremely firm in texture after cooking (Fig. 9.3). If ground into a flour, these mutants could be added to processed foods to decrease these foods' rate of digestion. Wada et al. (2018) reported evidence of this by making cookies using the ae mutant 'Chikushi-kona 85' and feeding them to test subjects. The increase in blood sugar levels after eating these cookies was significantly lower compared to that of a traditional cultivar Hinohikari.

9.2.3 *Processing Effects*

Various processing methods have been evaluated to determine its effects on the amount of RS in rice and its co-products. Chen et al. (2017) evaluated the effects of two cooking methods: fixed water-to-rice ratio/time and in excess-water/minimum-cook time using six rice varieties with high levels of naturally occurring RS; no differences were found. These two cooking methods approximate many of the cooking methods used in rice-consuming regions. No reports of the effect of different cooking methods on rice with lower amylose content were found.

A heat-moisture treatment using various combinations over time increased the RS contents of starch from a high-, medium- and low-amylose rice variety (Van Hung et al. 2016). Only the high-amylose variety changed enough to be commercially important for use in food tailored for the prevention or management of T2D. Another heat-moisture treatment is mentioned in a European Union patent for the production of a puffed rice product with a decrease in the rate of starch digestibility by 32–46% (TCI CO LTD [TW] 2016). The patent didn't describe the rice used to collect this data or what control was used.

Trying the reverse of an increased heat-moisture treatment, Alhambra et al. (2019) studied the effect of retrogradation conditions on rice starch. The method used to retrograde the starch was the storage of cooked rice at refrigerator temperature for 24 h. The effect of this treatment on starch digestibility was greater on rice with more long glucan chains. Amylose is known to undergo the formation of double helices during retrogradation conditions, thus forming crystallites that are less susceptible to enzymatic hydrolysis (Zhu et al. 2011). These glucan chains undergo this process at a faster rate than those of amylopectin (Trinidad et al. 2013). The effect of eating rice cultivars with high levels of amylose and RS ('Dixiebelle' and 'Rondo'), compared to commercial short-grain rice (Tamanishiki brand) with lower RS on glucose and insulin response, was studied by Zenel and Stewart (2015). This single-blind, randomized, crossover clinical trial, performed with 18 healthy subjects, found the high-amylose types attenuated postprandial blood glucose and insulin response compared to short-grain rice. Although the authors didn't point this out, they stored the rice after cooking for up to three days at refrigerator temperature. Thus, the rice was stored at a temperature that causes rice starch to at least partially retrograde, resulting in higher levels of RS being formed in the cooked rice. The authors should have concluded that high-amylose rice cooked and then allowed to retrograde decreased postprandial blood glucose and insulin response.

In addition to the type of rice one eats or how it was processed, there is evidence, although limited, that eating a meals primary carbohydrate at the end of a meal can significantly decrease postprandial glucose levels in individuals with prediabetes (Shukla et al. 2019). This study was performed using wheat bread as the carbohydrate source; thus, similar work is needed to determine if other sources of carbohydrates such as rice will result in a similar conclusion.

Hassan et al. (2017) reviewed the evidence that milled rice (and brown) throughout some global growing regions have elevated levels of inorganic arsenic (As). This compound is known to be toxic to pancreatic beta cells and has been associated with increased blood glucose and insulin levels (Bell 2015). Also, an association between As exposure and incidence of T2D has been reported. Elevated levels of As in rice are mostly due to the use of soils and water for production that has been contaminated by human activities such as metal mining and smelting, use of arsenic-containing pesticides and herbicides, wood preservatives, etc. (Mitra et al. 2017). A review by Kumarathilaka et al. (2019) indicates that milling, washing rice with low-As water (e.g., rainwater), and cooking in excess water with low levels of As lowers the levels of As in cooked rice. A QTL on chromosome 5 reportedly decreases grain As levels in cultivars by 10% in high-yielding cultivars from Bangladesh (Norton et al. 2019). Transgenic rice with vacuolar arsenic sequestration genes has been made that had a 70% reduction in As in its brown rice (Deng et al. 2018). However, more work is needed on this approach because these plants when grown in soil containing high levels of As, their high vacuolar sequestration of this mineral in nongrain tissues will create hyperaccumulation, leading to cell damage. Further details regarding As in rice are discussed in another chapter in this book.

9.2.4 Brown Rice

Whole grain rice (i.e., brown and its other various bran colors), its processed forms, and many of its fractions are used in traditional Asian remedies for many different human ailments including T2D (Ahuja et al. 2008). For example, a rice from India with red bran and its undermilled counterpart are marketed as being appropriate for people with T2D in Asian markets in the USA. As mentioned above most science-based diet plans today recommend eating rice as a whole grain or in products that include both its bran and germ, as well as the endosperm. However, there are commercial diet plans today, such as the paleo diet that does not include brown rice as part of the diet plan (Paleo 2019).

Significant in vitro work indicates that brown rice and parboiled brown rice have a lower GI than its milled rice counterparts (Wahengbam et al. 2019). Based on these findings, Pletsch and Hamaker (2017) provided support for this recommendation with their finding that human gastric emptying was slower for brown rice compared to white rice, regardless of the variation in amylose content and in vitro starch digestion rates. However, only six rice cultivars and ten adults were evaluated in this study.

A randomized crossover trial in India was performed using overweight adults to evaluate the difference between eating parboiled-brown rice and white rice on diabetes risk factors (Malik et al. 2019). The trial was for three months long with a two-week washout period. Biomarkers of glycemic control and lipid parameters were not different between the groups. However, a significant decrease in HbA1c was found in the brown rice group compared to the white rice group in subjects with the metabolic syndrome. The brown rice group reported having greater satiety after meals compared to those that ate white rice, but this was only significant during the second month.

Patients ($n = 160$) with T2D that stayed in a South Korean hospital and required a drug intervention to maintain regular blood glucose levels were placed on a vegetarian diet (~1800 kcal) consisting of brown rice at each meal along with traditional Korean side dishes, such as kimchi and fermented soybean products (Jung et al. 2018). This was a retrospective study using patient charts; however, no length of hospital stay was reported. Although patients did not limit their sugar or salt intake, approximately 86% of the patients were able to stop using pharmaceutical agents to control T2D and lost weight. The authors concluded that dietary interventions with brown rice as the staple can control T2D just as effectively as medications.

Germinated and pre-germinated brown rice is produced generally by soaking the brown rice grain in water, germinating in air, to induce a degree of germination and then gentle drying to approximately 12% moisture. Germinated rice generally has a radicle of 2–5 mm and the radical of pre-germinated is approximately 0.5–1 mm. The specific time length of germination is sometimes controlled by measuring the alpha-amylase of the grains. Abubakar et al. (2018) studied germinated Malaysian rice varieties with high-amylose content. Using Sprague–Dawley rats and in vitro assays, the authors concluded that the high-amylose rice that was germinated had a lower GI and load compared to their white and low-amylose counterparts. As discussed above, varieties with very high-amylose levels are not suitable for table rice because of its objectionable texture. However, EM10 after pre-germination and cooking was found to be suitable as a blending material to increase the RS levels of leavened bread and noodles by 12% (Nakamura et al. 2010). These products according to the authors had acceptable taste and texture, although this determination appears to have based on their subjective evaluation.

9.3 Coronary Heart Disease

9.3.1 Introduction

According to the WHO statistics, coronary heart disease (CHD), also known as ischemic heart disease, is the leading cause of death globally (Fig. 9.1). As described by Ross (1999), this disease develops when accumulation of lipoprotein droplets in the intima of the coronary vessels occurs and then a cascade of biochemical events leads to plaque buildup in the walls of the arteries. The final atherosclerotic plaque formed narrows the arteries, making it harder for blood to flow through. If a blood clot forms, it can block the blood flow; this can result in a heart attack or stroke.

Environmental and genetic factors reportedly interact with one another to determine the clinical and metabolic phenotype of cardiovascular diseases (Girelli et al. 2009). The typical phenotype that is watched for today includes the following: T2D, hypertension, smoking, hyperlipidemia, obesity, homocystinuria, and psychosocial stress (Malakar et al. 2019). The environmental effects on developing CHD include a person's diet.

9.3.2 Milled Rice

The American Heart Association's dietary guidelines to promote cardiovascular health emphasized that adults increase high-fiber foods and decrease high-saturated-fat foods. It seems that consumers, not fully understanding this message, increased their consumption of carbohydrates, not those with significant levels of dietary fiber, thinking they were increasing their fiber intake and reducing saturated fats. This is thought to have resulted in people chronically consuming high glycemic foods and diets with a high glycemic load (HGL). Both of these have been reported to have increased the risk for developing CHD (Mirrahimi et al. 2012).

Mathews et al. (2015) investigated the proposed interconnectivity of the effects of HGL diets with CHD disease-causing pathways. Specifically, they conducted individual meta-analyses of individual biomarkers and lifestyle effects associated with CHD. They determined that HGL diets do not only influence the traditional lipid (e.g., low-density lipoprotein) and metabolic biomarkers (e.g., plasma nitrate) used to evaluate CHD risk, but also inflammation, coagulation, and vascular function biomarkers. There is also growing evidence that there are genetic human variants and epigenetic factors, specifically associated with CHD risk, and common sets of molecular networks that drive the pathogenesis of both CHD and T2D across ethnicities (Wolters et al. 2019; Shu et al. 2017).

Muraki et al. (2015) studied the dietary intake of three large cohorts (prospectively) of Caucasian adults living in the USA that didn't have CHD. The authors concluded that eating white and brown rice wasn't associated with increased risk of developing CHD, at rates at which these people eat rice. The rates of rice consumption in this population are known to be relatively low compared to the people of Asia and much of Latin America. Research is needed to clarify the effects of eating low, medium, or high levels of milled rice on CHD risk factors.

It is this author's opinion that the discussion above related to rice variability and association with T2D risk factors can also be considered similar to the influences of eating rice as a large portion of one's kilocalories on the risk factors associated with CHD. That is, when eating rice, whole grain (brown) should be consumed over white rice, high-amylose types should be preferred over low, and rice with low levels of As are encouraged. If processed foods are eaten made with rice, those types made with the low-glycemic rice flour should be chosen.

9.3.3 *Brown Rice*

In the USA, brown rice is approved to carry the following health claim: "Diets rich in whole grain foods and other plant foods and low in total fat, saturated fat and cholesterol may reduce the risk of heart disease and some cancers" (Health Claim 2008). Significant supportive research is required for food companies to be allowed to market food using a health claim. Various nutritional and bioactive components are thought to be associated with the ability of brown rice consumption to decrease the risk of developing CVD. For example, dietary fiber, phytosterols, phenolic compounds, and gamma-aminobutyric acid (Burlando and Cornara 2014). Figure 9.5 displays the relative number of studies that have been reported related to rice components and extracts influence on CHD compared to other chronic disease. These authors also discussed how these compounds and fractions may reduce the risk of developing CHD.

A few human-based studies examining the effects of eating brown rice on CHD risk factors have been published since the health claim for brown rice was approved. The study by Jung et al. (2018), also described above, reported that approximately 86% of patients with CHD risk factors were able to reduce or stop taking their pharmaceutical agents for controlling hypertension and hypercholesterolemia by eating

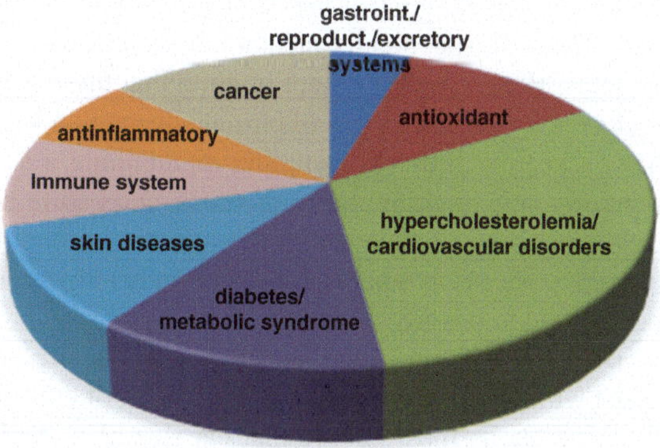

Fig. 9.5 Frequency of citations of different syndromes in the survey of scientific literature conducted concerning the therapeutic properties of rice and its derivatives (Burlandoa and Cornara 2014)

brown rice with traditional Korean side dishes, for three meals a day. Another study related to hypertension and rice consumption was performed by Kondo et al. (2017). This was a randomized, open-labeled, parallel-controlled trial to investigate the effects of eating brown rice ($n = 14$) versus white rice ($n = 14$) on endothelial function in Japanese adults with T2D. Endothelial dysfunction is documented to proceed to the development of CHD. This intervention which lasted for eight weeks resulted in significantly improved endothelial function. No significant differences in serum amounts of triglyceride, total cholesterol, HDL, or LDL cholesterol were found. This study had relatively few subjects and thus will need to be replicated. This will be difficult because the rice was purchased from a company and was blended with amaranth and barley. It can be assumed, however, that the rice was a traditional Japanese short-grain cultivar with low amylose content.

A review by Cho and Lim (2016) summarized research using in vitro and animal models that were performed to study the possible effects of eating germinated rice (described above) on chronic disease risk factors. They found that most of the studies reported to date concluded that germinated brown rice has the potential to improve blood pressure regulation, reduce blood cholesterol levels, and assist with obesity avoidance.

9.4 Conclusion

Commercially available milled rice is needed that has a reduced glycemic index and glycemic load as a carbohydrate source for humans in order to decrease the risk of and management of T2D and CHD. Some genotypes have been identified and others developed that have relatively low levels of these traits; however, very little of these rice varieties are grown commercially. These rice varieties in general are more difficult to cook and are firmer in texture than most consumers will tolerate; thus they are only suitable for use in food products. It will take some creative science to produce varieties that have acceptable cooked rice sensory properties and can limit blood glucose "spiking" after a meal. This will require a more complete understanding of the needed rice biochemical structural properties and related metabolic and genetic control mechanisms. Scientific tools are available that can help with this goal, specifically metabolomics, phenomics, and the various gene-editing systems. But it should be kept in mind that these new desired rice types don't need to decrease rice eater's risk of chronic disease on its own. These people also need to have a healthy lifestyle including maintaining a suitable body weight, exercising regularly, not smoking, limiting stress, and eating the new rice along with what is currently known to be a healthy diet. Some of these lifestyle factors, however, such as having access to what is considered a healthy diet are difficult for many of these people due to poverty, etc.

Brown rice compared to its white rice counterpart, today, has greater potential to assist with the prevention and management of chronic disease. The difficulty is that for generations most people that eat rice as a staple eat it in its white rice form.

Brown rice types to many of these people are considered, in terms of sensory properties, to be undesirable. There are processing methods such as undermilling, pregermination, and perhaps others to be developed in the future that have the potential to improve the consumer acceptability of brown rice. Also, the rice genotypes that have bran that is colored white, if undermilled, may meet consumers' visual demand and have sensory properties more closely aligning with their desires.

Public health professionals, dietitians, nurses, and physicians have a role to play in educating consumers on the types of rice and rice-based products to eat to prevent and manage chronic diseases. However, to assist them in making science-based recommendations, more media resources are needed that translate scientific findings related to the health-beneficial properties of certain rice into language that these professionals will fully appreciate.

As rice science advances, varieties with enhanced health-beneficial traits will hopefully become available. Strong research evidence, however, documenting the efficacy of these varieties will be needed for health claims to be approved by government agencies. Also, if compounds not currently in rice are added or created in rice, the safety of these types must be determined to alleviate the concerns of consumers. Food companies will also need to identify consumers willing to pay for these types of rice and related products. Some of the biggest challenges in this "plot to plate" path is how to encourage farmers to take a risk and grow the new cultivars. Agronomists and agriculture extension agents must demonstrate that these new cultivars have enhanced agronomic properties for farmers to think growing them will result in good yields, as is needed for them to remain competitive and meet the food needs of a constantly growing human population. Once harvested the new rice types will need to be segregated from traditional varieties. This is especially challenging in the Western world where rice is grown as a commodity; specifically, most long grain types and short grain-types are comingled with all varieties of the same grain type. These challenges can and should be overcome to improve the well-being of people that eat rice regularly.

References

Abubakar B, Yakasai HM, Zawawi N, Ismail M (2018) Compositional analyses of white, brown and germinated forms of popular Malaysian rice to offer insight into the growing diet-related diseases. J Food Drug Anal 26:706–715

Ahuja U, Ahuja SC, Thakrar R, Singh RK (2008) Rice a nutraceutical. Asian Agri-Hist 12:93e–e108

Alhambra C, De Guzman M, Dhital S, Bonto A, Dizon E, Israel K, Hurtada WA, Butardo VM, Sreenivasulu N (2019) Long glucan chains reduce in vitro starch digestibility of freshly cooked and retrograded milled rice. J Cereal Sci 86:108–116

Anacleto R, Badoni S, Parween S, Butardo VM, Misra G, Cuevas RP, Kuhlmann M, Trinidad TP, Mallillin AC, Acuin C, Bird AR, Morell MK, Sreenivasulu N (2019) Integrating a genome-wide association study with a large-scale transcriptome analysis to predict genetic regions influencing the glycaemic index and texture in rice. Plant Biotechnol 17:1261–1275

Atkinson FS, Foster-Powell K, Brand-Miller JC (2008) International tables of glycemic index and glycemic load values: 2008. Diabetes Care 31:2281–2283

Bailey RL, West KP Jr, Black RE (2015) The epidemiology of global micronutrient deficiencies. Ann Nutr 66:22–33

Bao J (2018) Rice milling quality. In: Bao J (ed) Rice, 4th edn. AACC International. Published by Woodland Publishing Series in Food Science, Technology and Nutrition, Oxford UK, pg. 339–369

Bao J, Bergman CJ (2018) Rice flour and starch functionality. In: Sjöö M, Nilsson L (eds) Starch in food: structure, function and applications, 2nd edition, Woodland Publishing Series in Food Science, Technology and Nutrition, Oxford UK pp 373–419

Bell D (2015) Riceabetes: is the association of type 2 diabetes with rice intake due to a high carbohydrate intake or due to exposure to excess inorganic arsenic? Postgrad Med 127:781–782

Bergman CJ (2018) Rice end-use quality analysis. In: Bao J (ed) Rice chemistry and technology, 4th edn. AACC International. Published by Elsevier Inc. pp 273–337.

Birlat DS, Malik K, Sainger M, Chaudhary D, Jaiwal R, Jaiwal PK (2017) Progress and challenges in improving the nutritional quality of rice (Oryza sativa L.). Crit Rev Food Sci Nutr 57:2455–2481

Burlando B, Cornara L (2014) Therapeutic properties of rice constituents and derivatives (Oryza sativa L.): a review update. Trends Food Sci Technol 40:82–98

Butardo VM Jr, Anacleto R, Parween S, Samson I, de Guzman K, Alhambra CM, Misra G, Sreenivasulu N (2017) Systems genetics identifies a novel regulatory domain of amylose synthesis. Plant Physiol 173:887–906

Butardo VM, Venea DD, Colgrave ML, Campbell PM, Resurreccion A, Cuevas RP, Jobling SA, Tetlow I, Rahman S, Morell M, Fitzgerald M (2012) Biomolecular analyses of starch and starch granule proteins in the high-amylose rice mutant Goami 2. J Agric Food Chem 60:11576–11585

Chen M-H, Bergman CJ, McClung AM, Everette JD, Tabien R (2017) Resistant starch: variation among high amylose rice varieties and its relationship with apparent amylose content, pasting properties and cooking methods. Food Chem 234:180–189

Cho D, Lim S (2016) Germinated brown rice and its bio-functional compounds. Food Chem 196:259–271

Chung H-J, Liu Q, Lee L, Wei D (2011) Relationship between the structure, physicochemical properties and in vitro digestibility of rice starches with different amylose contents. Food Hydrocoll 25:968–975

Davis (2017) Don't eat anything white. https://www.wheatbellyblog.com/membership-account/membership-levels/

Deng F, Yamaji N, Ma JF, Lee S-K, Jeon J-S, Martinoia E, Lee Y, Song W-U (2018) Engineering Rice with lower grain arsenic. Plant Biotechnol J 16:1691–1699

Dhital S, Butardo VM, Jobling SA, Gidley MJ (2015) Rice starch granule amylolysis – differentiating effects of particle size, morphology, thermal properties and crystalline polymorph. Carbohydr Polym 115:305–316

Dipti S, Bergman C, Indrasari S, Herath T, Hall R, Lee H, Bassinello H, Hibbi F, Bassinello P, Graterol E, Ferraz J, Fitzgerald M (2012) The potential of rice to offer solutions for malnutrition and chronic diseases. Rice 5:1–18

Fitzgerald MA, Bergman CJ, Resurreccion AP, Möller J, Jimenez R, Reinke RF, Martin M, Blanco P, Molina F, Chen M-H, Kuri V, Romero MV, Habibi F, Umemoto T, Jongdee S, Graterol E, Reddy KR, Bassinello PZ, Sivakami R, Rani NS, Das S, Wang YJ, Indrasari SD, Ramli A, Ahmad R, Dipti SS, Xie L, Lang NT, Singh P, Toro DC, Tavasoli F, Mestres C (2009) Addressing the dilemmas of measuring amylose in rice. Cereal Chem 86:492–498

Fitzgerald MA, Rahman S, Resurreccion AP, Concepcion J, Daygon VD, Dipti SS, Kabir KA, Klingner B, Morell MK, Bird AR (2011) Identification of a major genetic determinant of glycaemic index in rice. Rice 4:66–74

Girelli D, Martinelli N, Peyvandi F, Olivieri O (2009) Genetic architecture of coronary artery disease in the genome-wide era: implications for the emerging "golden dozen" loci. Semin Thromb Hemost 35:671–682
Gruss SM, Nhim K, Gregg E, Bell M, Luman E, Albright A (2019) Public health approaches to type 2 diabetes prevention: the US National Diabetes Prevention Program and beyond. Curr Diabetes Rep 19:78–98
Hassan FI, Niaz K, Khan F, Maqbool F, Abdollahi M (2017) The relation between rice consumption, arsenic contamination, and prevalence of diabetes in South Asia. EXCLI J 16:1132–1143
Hay FR (2017) Seed banks. In: Encyclopedia of applied plant sciences, vol 2, 2nd edn. Academic, Waltham, pp 327–333
Health Claim (2008) Health claim extended to brown rice. Tufts Univ Health Nutr Lett 26:3. http://search.ebscohost.com/login.aspx?direct=true&db=aph&AN=33228427&site=ehost-live
Hu EA, Pan A, Malik V, RamyaBai M, Vijayalakshmi P, Lakshmipriya N, Gayathri R, Kokila A, Jones C, Hong B, Li R, Krishnaswamy K, Anjana RM, Spiegelman D, Willett WC, Hu FB, Mohan V (2012) White rice consumption and risk of type 2 diabetes: meta-analysis and systematic review. BMJ 344:e1454
International Rice Research Institute (2013) World rice statistics 2013. The Philippines: IRRI, Los Baños. June 29, 2013
Jacquemin J, Bhatia D, Singh K, Wing RA (2013) The international Oryza map alignment project: development of a genus-wide comparative genomics platform to help solve the 9 billion-people question. Curr Opin Plant Biol 16:147–156
Jung SJ, Cho BW, Chae SW (2018) Does Korean diet based on brown rice really have the effect on treating chronic diseases and on suspending drug use? J Ethnic Foods 5:231–238
Karuga J (2018) Top ten leading causes of death in the world. WorldAtlas. https://www.worldatlas.com/articles/top-ten-leading-causes-of-death-in-the-world.html
Kim KS, Hwang HG, Kang HT, Hwang IK, Lee YT, Choi HC (2005) Ultrastructure of individual and compound starch granules in isolation preparation from a high-quality, low-amylose rice, ilpumbyeo, and its mutant, G2, a high-dietary fiber, high-amylose rice. J Agric Food Chem 23:8745–8751
Kondo K, Morino K, Nishio Y, Ishikado A, Arima H, Nakao K, Nakagawa F, Nikami F, Sekine O, Nemoto K, Suwa M, Matsumoto M, Miura K, Makino T, Ugi S, Maegawa H (2017) Fiber-rich diet with brown rice improves endothelial function in type 2 diabetes mellitus: a randomized controlled trial. PLoS One 12:1–16
Kumarathilaka P, Seneweera S, Ok YS, Meharg A, Bundschuh J (2019) Arsenic in cooked rice foods: assessing health risks and mitigation options. Environ Int 127:584–591
Leszczyński W (2004) Resistant starch–classification, structure, production. Pol J Food Nutr Sci 13:37–50
Linares OF (2002) African rice (Oryza glaberrima): history and future potential. Proc Natl Acad Sci 99:16360–16365
Lockyer S, White A, Buttriss JL (2018) Biofortified crops for tackling micronutrient deficiencies. What impact are these having in developing countries and could they be of relevance within Europe? Nutr Bull 43:319–357
Ma Z, Boye J (2018) Research advances on structural characterization of resistant starch and its structure-physiological function relationship: a review. Crit Rev Food Sci Nutr 58:1059–1083
Malakar AK, Choudhury D, Halder B, Paul P, Uddin A, Chakraborty S (2019) A review on coronary artery disease, its risk factors, and therapeutics. J Cell Physiol 234:16812–16823
Malik VS, Sudha V, Wedick NM, Ramy BM, Vijayalakshmi P, Lakshmipriya N, Gayathri R, Kokila A, Jones C, Hong B, Li R, Krishnaswamy K, Anjana RM, Spiegelman D, Willett WC, Hu FB, Mohan V (2019) Substituting brown rice for white rice on diabetes risk factors in India: a randomised controlled trial. Br J Nutr 121:1389–1397
Mathews MJ, Liebenberg L, Mathews EH (2015) How do high glycemic load diets influence coronary heart disease? Nutr Metab 12:1–15

McCarthy MI (2017) Painting a new picture of personalized medicine for diabetes. Diabetologia 60:793–799

Mejia LA, Omar D, Boukerdenna H (2017) Global regulatory framework for production and marketing of crops biofortified with vitamins and minerals. Ann N Y Acad Sci 1390:47–58

Meera K, Smita M, Haripriya S, Sen S (2019) Varietal influence on antioxidant properties and glycemic index of pigmented and non-pigmented rice. J Cereal Sci 87:202–208

Meenu M, Xu B (2019) A critical review on anti-diabetic and anti-obesity effects of dietary resistant starch. Crit Rev Food Sci Nutr 59:3019–3031

Mirrahimi A, de Souza RJ, Chiavaroli L, Sievenpiper JL, Beyene J, Hanley AJ, Augustin LSA, Kendall CWC, Jenkins DJA (2012) Associations of glycemic index and load with coronary heart disease events: a systematic review and meta-analysis of prospective cohorts. J Am Heart Assoc 1:e000752. https://doi.org/10.1161/JAHA.112.000752

Mitra A, Chatterjee S, Moogouei R, Gupta DK (2017) Arsenic accumulation in rice and probable mitigation approaches: a review. Agronomy 7:67–88

Muraki I, Wu H, Imamura F, Laden F, Rimm EB, Hu FB, Willett WC, Sun Q (2015) Rice consumption and risk of cardiovascular disease: results from a pooled analysis of 3 US cohorts 1234. Am J Clin Nutr 101:164–172

Nakamura S, Satoh H, Ohtsubo K (2010) Palatable and bio-functional wheat/rice products developed from pre-germinated brown rice of super-hard cultivar EM10. Biosci Biotechnol Biochem 74:1164–1172

Nishi A, Nakamura Y, Tanaka N, Satoh H (2001) Biochemical and genetic analysis of the effects of amylose-extender mutation in rice endosperm. Plant Physiol 127:459–472

Noro W, Itayagoshi S, Hosono R, Matsui T, Takahashi M, Nakamura S, Ishizaki K, Watanabe S, Ohtsubo K (2019) Characterization of the new amylopectin long-chain rice cultivar Chou 2418 and its boiled rice grains. Food Sci Technol Res 25:227–235

Norton GJ, Travis AJ, Talukdar P, Hossain M, Islam MR, Douglas A, Price AH (2019) Genetic loci regulating arsenic content in rice grains when grown flooded or under alternative wetting and drying irrigation. Rice 12:54

Palacios OM, Kramer M, Maki KC (2019) Diet and prevention of type 2 diabetes mellitus: beyond weight loss and exercise. Expert Rev Endocrinol Metab 14:1–12

Paleo (2019) All about rice. https://paleoleap.com/about-rice/

Pletsch EA, Hamaker BR (2017) Brown rice compared to white rice slows gastric emptying in humans. Eur J Clin Nutr 72:367–373

Priebe MG, Eelderink C, Wachters-Hagedoorn RE, and Vonk R (2018) Starch digestion and applications of slowly available starch. In: Malin Sjöö and Lars Nilsson (ed) Starch in food, 2nd edn. Woodhead Publishing Series in Food Science, Technology and Nutrition, Oxford UK pg 805–826

Resurreccion J, Muller R, Jimenez RF, Reinke M, Martin P, Blanco F, Molina MH, Chen V, Kuri MV, Romero F, Habibi T, Umemoto S, Jongdee E, Graterol KR, Reddy PZ, Bassinello R, Sivakami NS, Rani S, Das YJ, Wang SD, Indrasari A, Ramli R, Ahmad SS, Dipti L, Xie NT, Lang P, Singh DC, Toro F, Tavasoli, Mestres C (2009) Addressing the dilemmas of measuring amylose in rice. Cereal Chem 86(5):492–498

Ross R (1999) Atherosclerosis – an inflammatory disease. N Engl J Med 340:115–126

Schwingshack L, Hoffmann G (2015) Diet quality as assessed by the healthy eating index, the alternate healthy eating index, the dietary approaches to stop hypertension score, and health outcomes: a systematic review and meta-analysis of cohort studies. J Acad Nutr Diet 115(5):780e5–800e5

Schwingshack L, Missbach B, Konig J, Hoffmann G (2015) Adherence to a Mediterranean diet and risk of diabetes: a systematic review and meta-analysis. Public Health Nutr 18:1292–1299

Schwingshack L, Hoffmann G, Lampousi AM, Knüppel S, Iqbal K, Schwedhelm C, Bechthold A, Schlesinger S, Boeing H (2017) Food groups and risk of type 2 diabetes mellitus: a systematic review and meta-analysis of prospective studies. Eur J Epidemiol 32:363

Shu L, Zhang G, Kurt Z, Zhao Y, Yang X, Luo X, Lusis AJ, Liu S, Chan KHK, Huan T, Levy D, Codoni V, Trégouët D-A, Yang J, Wilson JG (2017) Shared genetic regulatory networks for cardiovascular disease and type 2 diabetes in multiple populations of diverse ethnicities in the United States. PLoS Genet 13:1–25

Shukla AP, Dickison M, Coughlin N, Karan A, Mauer E, Truong W, Casper A, Emiliano AB, Kumar RB, Saunders KH, Igel LI, Aronne LJ (2019) The impact of food order on postprandial glycaemic excursions in prediabetes. Diabetes Obes Metab 21:377–381

Talsma EF (2017) Acceptance and adoption of biofortified crops in low- and middle-income countries: a systematic review. Nutr Rev 75:798–829

TCI Co. Ltd [TW] (2016) Method of preparing low glucose rice products and application thereof Lin, Y.-H., Su, H.-L. and Yu, C.-H. (inventors). European Patent TW2016117. https://worldwide.espacenet.com/publicationDetails/biblio?FT=D&date=20160401&DB=EPODOC&locale=&CC=TW&NR=201611733A

Trinidad T, Mallillin A, Encabo R, Sagum R, Felix A, Juliano B (2013) The effect of apparent amylose content and dietary fibre on the glycemic response of different varieties of cooked milled and brown rice. Int J Food Sci Nutr 64:89–93

Udler MS (2019) Type 2 diabetes: multiple genes, multiple diseases. Curr Diab Rep 19:1–9

Van Hung P, Vien N, Phi N (2016) Resistant starch improvement of rice starches under a combination of acid and heat-moisture treatments. Food Chem 191:6773

Villegas R, Goodloe RJ, McClellan BE Jr, Boston J, Crawford DC (2014) Gene-carbohydrate and gene-fiber interactions and type 2 diabetes in diverse populations from the National Health and nutrition examination surveys (NHANES) as part of the epidemiologic architecture for genes linked to environment (EAGLE) study. BMC Genet 15:1–16

Wada T, Yamaguchi O, Miyazaki M, Miyahara K, Ishibashi M, Aihara T, Shibuta T, Inoue T, Tsubone M, Toyosawa Y, Satoh H, Akaishi R, Yoshii Y, Ohtsubo K (2018) Development and characterization of a new rice cultivar, 'Chikushi-Kona 85', derived from a starch-branching enzyme IIb-deficient mutant line. Breed Sci 68:278–283

Wahengbam ED, Das AJ, Green BD, Shooter J, Hazarika MK (2019) Effect of iron and folic acid fortification on in vitro bioavailability and starch hydrolysis in ready-to-eat parboiled rice. Food Chem 292:39–46

WHO (World Health Organization) (2018) Guideline: fortification of rice with vitamins and minerals as a public health strategy. Available at: www.apps.who.int/iris/bitstream/handle/10665/272535/9789241550291-eng.pdf?ua=1

Wolters FJ, Yang Q, Biggs ML, Jakobsdottir J, Li S, Evans DS, Bis JC, Harris TB, Vasan RS, Zilhao NR, Ghanbari M, Ikram MA, Launer L, Psaty BM, Tranah GJ, Kulminski AM, Gudnason V, Seshadri S (2019) The impact of APOE genotype on survival: results of 38,537 participants from six population-based cohorts (E2-CHARGE). PLoS One 14:e0219668

Yano M, Okuno K, Kawakami J, Satoh H, Omura T (1985) High amylose mutants of rice Oryza sativa L. Theor Appl Genet 69:253–257

Zenel AM, Stewart ML (2015) High amylose white rice reduces post-prandial glycemic response but no appetite in humans. Nutrients 7:5362–5374

Zhou H, Lijun W, Guifu L, Xiangbing M, Yanhui J, Shu S, Kong X, Sun J, Yu H, Smith SM, Wu D, Li J (2016) Critical roles of soluble starch synthase SSIIIa and granule-bound starch synthase waxy in synthesizing resistant starch in rice. Proc Natl Acad Sci U S A 113:12844–12849

Zhu L-J, Liu Q-Q, Wilson JD, Gu M-H, Shi Y-C (2011) Digestibility and physicochemical properties of rice (Oryza sativa L.) flours and starches differing in amylose content. Carbohydr Polym 86:1751–1759

Part IV
Special Rice

Chapter 10
Black Rice (*Oryza sativa* L.) Breeding

Ujjawal Kumar Singh Kushwaha, Indra Deo, Narendra Kumar Singh, and Surendra Nath Tripathi

10.1 Introduction

Rice (*Oryza sativa* L.) is the staple food for about half of the world population. It is an annual crop grows in diverse agroecological regions. Nearly 90% of rice is cultivated in Asian countries (Ricepedia). China leads rice production, followed by India. Rice is of different types based on early domestication, i.e., *indica*, *japonica*, and *javanica*. Rice can be classified on different basis such as maturity days (long maturity days, medium maturity days, and short/early maturity days); adaptation and cultivation practices, i.e., cultivated rice and wild rice; aroma, i.e., aromatic rice (basmati rice) and non-aromatic rice; and endosperm color: black, brown, red, and white rice. Each rice type (based on color) has special health importance. Mostly white rice is cultivated and consumed worldwide. Surprisingly, white rice is considered inferior than the black, brown, and red rice as most nutritious part of white rice is milled during milling and polishing process (International Rice Research Institute).

Black rice is black in color due to presence of a pigment called anthocyanin, an antioxidant. It is considered as twenty-first-century modern superfood because of its high nutrition profile. It is glutinous, packed with high level of nutrients, and mainly cultivated in Asia. This rice is mostly grown in China, Korea, Japan, India, Thailand,

U. K. S. Kushwaha (✉)
Department of Genetics and Plant Breeding, College of Agriculture, G. B. Pant University of Agriculture and Technology, India/Agriculture Botany Division, Nepal Agricultural Research Council, Khumaltar, Lalitpur, Nepal
e-mail: kushujjwal@gmail.com

I. Deo · N. K. Singh
Department of Genetics and Plant Breeding, College of Agriculture, G. B. Pant University of Agriculture and Technology, Pantnagar, India

S. N. Tripathi
Department of Entomology, College of Agriculture, G. B. Pant University of Agriculture and Technology, Pantnagar, India

A. Costa de Oliveira et al. (eds.), *The Future of Rice Demand: Quality Beyond Productivity*, https://doi.org/10.1007/978-3-030-37510-2_10

Indonesia, and Philippines (Kong et al. 2008). The remarkable health benefits of black rice have rated it high in the market. As compared to other rice types, black rice achieves highest price. The cultivation practices of black rice are quite similar to other rice types with recommended agronomic management practices.

The word "black rice" refers to a type of rice from the species *Oryza sativa*, distinctive with the color of grain, rather than other qualities (Kushwaha 2016a, b). Black rice could be seen in a number of short grain, long grain, and glutinous varieties similar to brown rice. It contains large amount of antioxidants called anthocyanin, which is more than in blueberries. This rice is also known as long-life rice because its consumption is supposed to extend the consumer's life span. Modern molecular research also supports the evidences on black rice as a superfood (Kushwaha 2016a, b). This is only a natural rice variety that could be safely taken by diabetic patients since it contains low glycemic index (Jang et al. 2015).

Black rice has multiple applications. Black rice extract is a rich source of antioxidants and is regarded a healthy coloring dye. Black rice powder extracted from the rice bran could be used as a food coloring dye in sodas and other products. It can also be utilized for developing functional foods and cosmetics and in nutraceutical, pharmaceutical, and other health products. Black and red rice bran could be used as an alternate raw material in the industry for manufacturing the food with high antioxidant activity (Jang et al. 2012). Black Rice Bran (BRB) is widely used as an excellent ingredient to increase the nutritional value and antioxidant properties of noodles (Kong et al. 2012).

This chapter presents detailed introduction of black rice, its importance in modern food system for benefit of the consumers, and breeding procedures, health importance, special features of black rice, genetic resources characteristics, cultivation practices, traditional breeding approaches and research progress of black rice on tissue culture, mutation breeding, genetic engineering, and molecular breeding.

10.1.1 History

Black rice is popularly known as forbidden rice because it was forbidden for the common masses and was only meant for the Chinese emperors and their family. Emperors believed that this rice could lengthen their lives; therefore they could rule over long time. Consumption to others was completely forbidden and if one found guilty was punished. Black rice was free to cultivate by local farmers only after end of Chinese dynasty (Fig. 10.1).

There is no debate that black rice is originated from Asian countries, including China (Mingwei et al. 1995; Hoahua et al. 1996), India (Sastry 1978), Japan (Natsumi and Noriko 1994), and Vietnam (Quan 1999). Modern archeological evidences show that black rice was cultivated near Yangzi river of China since prehistory. Farmers still cultivate different varieties of this rice types. More than 200 types of local landraces are available in the world and only China is responsible for 62% of global production of black rice and has developed more than 54 improved

Fig. 10.1 Worldwide map showing black rice–growing countries

varieties. China dominates black rice cultivation followed by Sri Lanka, Indonesia, India, Philippines, etc. Thailand occupies the ninth position to black rice cultivation (Ichikawa et al. 2001; Sompong et al. 2011).

10.1.2 Economic Importance

Black rice is mostly used as a coloring dye in food industries. Food items like pasta, ice cream, donuts, etc., are colored by natural dying materials; black rice serves as the best organic dye materials. The outer part, endosperm, of black rice is black in color which is extracted for coloring purposes and is highly used in different food products (Kushwaha 2016a, b).

Since black rice is a rich source of calcium, iron, and other micronutrients, it is very much useful food ingredients for human health improvement (Kushwaha 2016a, b). High nutrition profile in black rice make it highly priced rice in the market. A farmer who grows black rice might harvest more benefits than the normal white rice.

10.1.3 Nutritional and Medicinal Importance

Black rice contains anthocyanin which is mainly composed of cyanidin 3-O-glucoside and peonidin 3-O-glucoside (Lee et al. 2014). Black rice is rich with nutritional components like dietary fibers, phytic acid, vitamin E, and vitamin B as compared to the ordinary milled rice (Table 10.1) (Kushwaha 2016b). It is rich in vitamin B

Table 10.1 Comparison between chemical analysis and phenolic compounds of black rice (IAC 600) and whole white rice in dry matter

S. no	Analysis	Black rice	Whole rice
1	Humidity (%)	87.95	73.09
2	Lipids (%)	3.87	0.9
3	Ash (%)	1.98	0.46
4	Protein (%)	11.9	2.58
5	Dietary fiber (%)	5.67	1.8
6	Carbohydrates (%)	63.45	22.96
7	Phenolic compounds (mg = g)	23.78	2.45

Source: Salgado et al. (2010)

complex, riboflavin, thiamine, and minerals like iron, calcium, and magnesium. Salgado et al. (2010) reported that anthocyanin from black rice has higher antioxidant activity than red rice. Phenolics have been reported as major hydrophilic antioxidants in black rice whereas carotenoids, tocopherol, and gamma oryzanols are lipophilic antioxidative substitutes (Min et al. 2011). These substances are associated with prevention of cardiovascular disease and ageing (Shen et al. 2009). It is reported that colored rice cause decrease of oxidative stress and simultaneous increase of antioxidant capacity (Hu et al. 2003). Black rice provides health benefits by reducing risk of chronic disease such as cardiovascular problems, cancers (Xia et al. 2006), diabetics (Walter and Marchesan 2011), iron-deficiency anemia (Wang and Guo 2007). Ichikawa et al. (2001) reported that black rice is two fold stronger with respect to antioxidant activities of blueberries.

A series of reports reveals the facts that black rice has several health benefits. An antioxidant present in black rice cyanidin-3-O-glycoside has antioxidant activity 3.5 times more than Trolox, a vitamin E analog (Zuo et al. 2012). Black rice extract consumption might induce superoxide dismutase (SOD), catalase (CAT) (Chiang et al. 2006), and glutathione peroxidase activities (Auger et al. 2010), suppresses reactive oxygen species (ROS) and nitric oxide (NO) radicals (Hu et al. 2003), and inhibits xanthine oxidase, one of the generators of superoxide anions (Zuo et al. 2012). Regular consumption of this rice may reduce many malefic health problems.

Zuo et al. (2012) research reveals the fact that black rice extract might enhance the mean life span of fruit flies by 14%. This effect was accompanied by an upregulation of CuZnSOD (SOD1), MnSOD (SOD2), and CAT. It is also found that anthocyanin and tocopherol in black rice bran significantly prevent lipid oxidation. The capability of anthocyanin is greater than the tocopherol in retarding cholesterol oxidation. The prevention of lipid oxidation could reduce the inflammation associated with chronic diseases. Both the hydrophilic and lipophilic antioxidants contribute to the health-promoting function (Zhang et al. 2013). In the same way, dietary intake of anthocyanin extract from black rice (AEBR) decreases platelet hyperactivity, hypertriglyceridemia, and body weight gain and facilitates in the maintenance of optimal platelet function in dyslipidemic rats induced by high-fat diets (Yang et al. 2011).

Consumption of black rice provides multiple benefits; for example, it enhances health and increases longevity, protects heart health, reduces atherosclerosis, controls hypertension, reduces stroke risk level in women, improves digestive health, provides anti-inflammatory action, reduces allergy, detoxifies the body, and improves lipid profile. Furthermore, it reduces risk of diabetics, improves eye vision, helps in satiation and weight management, reduces cancer growth, increases life span, protects from osteoporosis, enhances hair growth, reduces asthma risk, facilitates antioxidative effects (anti-aging), prevents Alzheimer's disease, acts as y-amino butyric acid (GABA), works as nature great superfood, and works as panacea (Kushwaha 2016a, b; Choi et al. 2010).

Black rice has more medicinal value. It is considered as fatigue fighter. In China, black rice is recommended for elderly people as it increases appetite and is a good treatment for digestive disorders. Since it is rich in iron, fiber, and antioxidants (vitamin E) and lower in sugar, therefore, it is considered alleviator of chronic inflammation, allergies, high cholesterol, high blood pressure, diabetics, Alzheimer's disease, heart disease, certain cancers, and even aging (Kushwaha 2016a, b; Yao et al. 2013).

10.1.4 Special Features of Black Rice

Black rice is an heirloom rice, i.e., it was widely grown earlier but occasionally grown in present context. Black rice is available in both type: *japonica* and *indica* rice. Typically, black rice (endosperm) is black in color due to anthocyanin. When cooked, its color become purple like deep dark vermillion hue.

Black glutinous rice has small slender culm, narrow short leaves, purple leaf margins, purple leaf tips, and purple stripes on leaf blades and sheathes (Fig. 11.2) (Bounphanousay 2008). The newly developed Thai black rice has longer and more translucent grains and retains Jasmine cooking qualities of softness, better taste, and cherished smell. Some new black rice varieties have advantage of fast growing which can be harvested with three crop per year. Similarly, a short-grained black rice which is sticky, sweet, and glutinous is especially sticky when cooked. It is called glutinous in the sense of being glue-like (hence, the name glutinous) or sticky, but do not contain gluten. Based on waxiness, black rice may be black nonwaxy rice and black waxy rice.

Black rice is available in different color based on anthocyanin content and other morphological characteristics (Fig. 10.2). Black glutinous rice is natural rice type with grains unevenly colored which look like wild rice when dry. Its rich nutty flavor is distinctly different from more subtle delicateness of white glutinous rice. Black rice variety studies have been mostly done in China (Zuo et al. 2012), Thailand (Sangkitikomol et al. 2010), and Korea (Seo et al. 2013). It has been reported that local black rice varieties have low threshing quality and give farmers a hard time during threshing. Thus, black rice varieties have been the subject of exploratory research for its potential biomedical applications (Sangkitikomol et al. 2010).

Fig. 10.2 Black rice leaf with purple pigmentation on leaf surface (**a**). Black rice endosperm after removal of hull (**b**)

10.2 Genetic Resources and Conservation

10.2.1 Genetic Resources Characteristics

Black rice is rich with local landraces. Many landraces have been diminished due to timely collection and their preservation in gene banks. Farmers have saved many rice landraces in different parts of the world like China, Korea, Japan, India, Thailand, Indonesia, and Philippines (Table 10.2). Only China has more than 200 types of local landraces and many improved varieties. Most local landraces have bold- and coarse-type grain (Fig. 10.3). The rice after cooking looks purple and becomes glutinous. Sticky black rice is mostly grown in different Asian countries.

Recently many attempts have been made to develop better rice varieties with high yield packed with high level of antioxidants/bioactive compounds. The bioactive compounds, like polyphenol, might be influenced by genotype, environment, and genotype-environment interaction (Goffman and Bergman 2004). For example, C3GHi is a new black rice variety developed through conventional breeding has high level of bioactive compound cyanidin-3-glucoside and antioxidants than the normal local black rice landraces (Kim et al. 2010). Black scented rice (Chakhao) is more tolerant to pest and diseases as the plant contains distributed anthocyanin and phenolic compounds to its different parts, i.e., root to leaf. The rice grains are small in size, 100 grain weight have higher phenolic and flavonoid content, and have antioxidant capacity than grain with normal or large size. Phenolic content and antioxidant capacity of rice are correlated with each other. The phenolic content could be indirectly predicted by grain length and 100 grain weight. Thus, new rice varieties with high antioxidant accumulation could be achieved by breeding for extremely small grain size (Yafang et al. 2011). Kushwaha (2016a, b) also reported that black

Table 10.2 Name of black rice varieties/landraces found in different parts of the world

Black rice variety name	Seed color	Origin
China black rice (CNB)	Black	China
Hong XieNuo	Black	China
Toketsumochi	Black	China
YunanShiping	Black	China
Xiang XieNuo	Black	China
Zhejiang	Black	China
Heimi	Black	China
Jing Nian	Black	China
Thai glutinous black rice (Kaoniow Dam)	Black	Thailand
Homnin black rice (HN)	Black	Thailand
Munpu red rice (MP)	Black	Thailand
KhaohomMaephayatongdam	Black	Thailand
Ballatinao black rice	Black	Thailand, Philippines
Thai Jasmine rice	Black	Thailand
WojaLaka	Black	Thailand
Laka	Black	Thailand
Kamklaing	Black	Thailand
Nang Dum	Black	Thailand
Br no 3, 19, 20, 26–29, 32, 42, 44, 46, 50, 52, 54, 64, 65,68, 70, 71	Black	Thailand
Br no. 16, 53, 55, 56, 58	Brown	Thailand
Br no. 18, 30, 59, 63	Straw	Thailand
PadiHitamMagelang(hairy)	Black	Central Java
PadiHitamMagelang (hairless)	Black	Central Java
PadiHitamSragen	Black	Central Java
PadiHitamWonosobo	Black	Wonosobo-Central Java
PadiHitamBanjarnegara	Black	Central Java
Indonesian black rice	Black	Indonesia
PuluikArang	Black	Indonesia
Cempo Ireng	Black	Indonesia
Ladang	Black	Indonesia
Mitak	Black	Indonesia
KetanHitam	Black	Indonesia
Kurogome	Black	Japan
Yayoimurasaki	Black	Japan
Akamai	Black	Japan
Kuromai	Black	Japan
Asamurasaki	Black	Japan
Shiho	Black	Japan
Xiang Xie Nuo	Black	Japan
Hong Xie Nuo	Black	Japan

(continued)

Table 10.2 (continued)

Black rice variety name	Seed color	Origin
Toketsumochi	Black	Japan
Yunan Shiping	Black	Japan
KH Kam	Black	Laos
Khao Kam (JP#80521)	Black	Laos
Khao Kam (JP#80522)	Black	Laos
KH Chepheum	Black	Laos
Alus	Black	Malaysia
HitamPulut	Black	Malaysia
Artemide	Black	Piedmont
Venere	Black	Piedmont
Nerone	Black	Piedmont
Galo	Black	Philippines
KinangdangItim	Black	Philippines
Chakhao (Scented black rice)	Black	India
Poreiton Chakhao	Black	Manipur, India
Chakhao Amubi	Black	Manipur, India
Burma Black Rice	Black	Manipur, India
Galsaekchalmi	Black	South Korea
Jeoktomi	Black	South Korea
Hongchalmi	Black	South Korea
Nogwonmi	Black	South Korea
IC 60	Black	South Korea
Dragon eyeball 100	Black	Korea
Heukjinjubyeo	Black	Korea
Heukgwangbyeo	Black	Korea
Heuknambyeo	Black	Korea
Josaengheukchal	Black	Korea
Melik	Black	Kedon, Ganjuran, Bantul,Yogyakarta
Jlitheng	Black	Sleman, Yogyakarta
CempoIreng	Black	Seyegan, Sleman, Yogyakarta
PariIreng	Black	Padasan, Pakembinangun, Sleman, Yogyakarta
PadiHitam NTT	Black	Alor, NTT
PadihitamBantul	Black	Njayan, Imogiri, Bantul

Source: Kushwaha (2016a, b), Oikawa et al. (2015), Phonsakhan and Kong-Ngern (2015)

rice anthocyanin content provided with sufficient soil moisture and optimum sunshine during grain-filling period increases. Black rice anthocyanin content increases when the altitude of its plantation is increased. It means that if a rice variety is grown in a cooler climate, the anthocyanin content is higher than those growing in the warm environment provided better agronomic management (Kushwaha 2016a).

Fig. 10.3 Black rice local landraces. Black rice Chakhao from India, medium bold type (**a**); black rice from China, fine type (**b**); black rice from China bold and coarse type (**c**)

The antioxidant capacity of some rice varieties including black rice, Galsaekchalmi, Jeoktomi, Hongchalmi, and Nogwonmi was analyzed in South Korea. Phytic acid content was found highest in Hongchalmi cultivar. The inhibition of lipid peroxidation was higher in Jeoktomi cultivar than other rice varieties. Similarly, Nogwonmi showed lowest antioxidant capacity than the other rice varieties. This finding provide valuable information on antioxidant potential of newly developed pigmented rice varieties and may assist breeders in the selection of cultivar for the development of new lines with enhanced functional quality (Kang et al. 2013).

10.2.2 Genetic Resources Conservation

Once black rice was rich with its genetic resources, but most rice landraces are being depleted now. Only rice landraces which had higher yield and better quality were preserved by the farmers in their fields. Many countries have conserved black rice landraces in their in situ and ex situ gene banks; for example, China has access over 359 black rice landraces in its gene banks. Similarly, there are 50 in Sri Lanka, 42 in Indonesia, 30 in India, 25 in the Philippines, and 24 in Bangladesh gene bank (Chaudhary 2003). The gene responsible for origin of black rice is *kala4* (Oikawa et al. 2015).

Black rice genetic resources were depleted mainly during Chinese emperor's dynasty. There was no access of black rice seed for the farmers; therefore, many landraces were disappeared (Kushwaha 2016a, b). Recently, scientific evidences depict the importance of black rice; hence, both government and scientists are highly motivated to conserve black rice genetic resources.

(a) *In situ conservation*: In situ conservation involves conservation of a living population, species, or evolutionary process in natural state. In situ conservation is the way of protecting the natural habitat of a living body. For example, large ecosystems are national parks and wildlife conservation (https://www.nap.edu/read/2116/chapter/7). Farmers are still working as volunteer for conserving many local crop landrace including black rice worldwide.

(b) *Ex situ conservation*: Ex situ conservation is the conservation of living organisms, germ cell, somatic cell, parts of plants, etc., out of natural habitat where they occur. Ex situ includes green house, gene banks, botanical gardens, etc.

(c) *In vitro conservation*: In vitro conservation is a part of gene bank conservation system. In this conservation system, a part of plant, tissue, or cell is conserved in sterile media, flask/tubes, or in controlled environment. The prerequisite of successful in vitro conservation is that parts of plants are able to regenerate into the whole plants when provided suitable environments (https://cropgenebank.sgrp.cgiar.org/index.php/).

(d) *Cryopreservation*: Cryopreservation is the process of preserving cell, tissue, or minute part of living organelles in a very low temperature so as to protect them from any biological constructs susceptible to damage caused by unregulated chemical kinetics. This process involves liquid nitrogen at −196 °C or solid carbon dioxide at −80 °C as a cooling agent.

(e) *Gene banks*: Gene banks mainly involve ex situ conservation methods. Mainly plants propagules, seeds, or parts of plants are conserved in gene banks. In situ gene bank involves maintenance of rare plants or germplasm in field gene banks either through seed or by plant propagates. Gene banks are established worldwide to protect thousands of plants from being extinct. Many black rice landraces have already been conserved in gene banks; for example, two landraces are conserved in National Agricultural Genetic Resource Center, Khumaltar Lalitpur, Nepal (Kushwaha 2016a, b).

10.3 Cultivation Challenges and Breeding Objectives

Black rice is cultivated similar to other rice types. The cultivation practice mainly involves land preparation, seed sowing, seedlings transplanting, crop agronomic care, and management practices like timely weeding, recommended fertilization, adequate irrigation, and protection from diseases and pests. Similarly, harvesting involves cutting fully ripened crop, drying, threshing, winnowing, and packaging.

Since black rice local landraces have short stature and thick culm, it makes them lodging resistance. These rice varieties also have resistance capacity against many disease and pests. Chinese black rice is resistant to moderately resistant on Bacterial leaf blight (*Xanthomonas axonopodis* pv. Oryzae) and rice blast (*Magnaporthe grisea*). The plant's whole parts like leaf, stem and grain contain anthocyanin and many other phenolic compounds which make the crop repellent from pests and resistance from many diseases (Kushwaha 2016a, b).

10.3.1 Cultivation Challenges

Farmers grow black rice similar to other rice types. The challenges and constraints faced for rice cultivation apply here in black rice too. The most challenging issues are seed quality, land preparation, agronomic management, irrigation facilities, fertilizer dose, and harvesting at appropriate time. A well-suited farmland produces higher yield, only when all the facilities are available during cropping period (Villalobos and Fereres 2016).

Black rice varieties are also highly affected by climatic constraints like biotic and abiotic stresses (Oort and Zwart 2017). Biotic stress like diseases and pests cause severe loss to crop yield. The uncertainty caused by the climate possesses threat to the crop, making favorable environment for disease outbreak. High temperature followed by higher humidity might favor bacterial blight, blast, and many more diseases. Similarly, rise in temperature is also responsible for increase in reproduction rate of many insects that might affect crop yield irreversibly. Since temperature is one of the deciding factors of yield, a rise in temperature might reduce the photosynthetic yield and ultimately decrease the yield (Stuecker et al. 2018). The rainfall pattern is also irregular that affects crop and its growth negatively. Thus, climate resilient black rice variety is a need to address climate change.

Black rice possesses an antioxidant, anthocyanin, and other phenolic products in its whole plant parts, i.e., from leaf to stem, root, panicles, and grains. These phenolic compounds work as insect repellent and provide resistance to plants from many diseases. Black rice shows resistance to many diseases like bacterial leaf blight (BLB) and blast and sheath blight and works as donor for disease-resistance genes (Kuswaha 2016).

10.3.2 Breeding Objectives

The overall breeding objectives are to develop superior-quality improved black rice variety with higher anthocyanin content. The variety should have wide range of adaptability and resilient with climatic effects. The variety must possess desirable grain qualities like higher protein content (>7%), higher anthocyanin, adequate flavonoids and phenolic compounds, high in fiber, and rich with Fe, Ca, Zn, and Mg. The black rice grains should have high digestibility and should have short cooking time (Kushwaha 2016a, b).

10.4 Traditional Breeding Approaches

Rice breeding is the most important component for development of new rice varieties. Since rice being a self-pollinated crop, it follows breeding approaches which are used in self-pollinating crop. The appropriate crop breeding methods depend on the

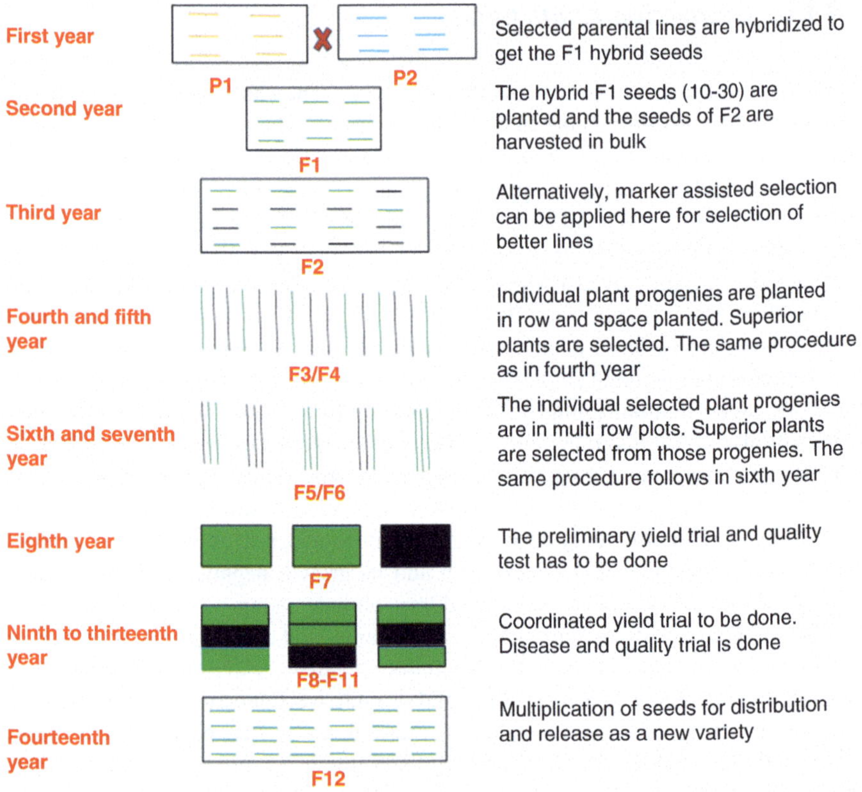

Fig. 10.4 Schematic representation of pedigree breeding method. In this schematic diagram, green color represents elite lines and black color represents inferior lines. Only green lines and green box proceeded for varietal selection

objective of the breeder, type of crop chosen for breeding, country's breeding scheme, and available crop management facilities (Singh 2013).

Black rice is available in both *indica* and *japonica* type. The conventional rice breeding method is mostly applied in different parts of the world followed by marker-assisted selection. The most appropriate method of black rice breeding is pedigree breeding (Fig. 10.4). In pedigree method, individual plants are selected from F2 and subsequent generation and their progenies are tested. This method is most commonly used for selection of segregating generations. It is also useful in the selection of new superior genotypes. This method is also suitable for improving specific characteristics, i.e., for improving disease resistance. Conventional breeding methods take nearly 12–14 years to develop a new variety whereas incorporation of marker-assisted breeding involves only 4–6 years. The detailed procedure of conventional pedigree method is presented below (Singh 2013).

10.4.1 Selection of Appropriate Parents

Conventional rice breeding starts with the selection of most appropriate parents. Parent selection is a difficult task for a breeder when the pedigree history is unknown. Parents must show contrasting characters. The parents, used for crossing, should have high general and specific combining ability so that the F1 and its proceeding generations would have desirable characteristics like high yield and resistance to disease and pests.

The suitable crossing design is made before hybridization. The appropriate design depends on the number of parents available and breeder choices. The design may be line X tester design, diallel cross, partial diallel cross, half diallel cross, etc. With appropriate design, sow the parental seeds in the field in such a way that both genotypes flowering synchronize at the same time. Alternate sowing could be done to synchronize flowering time. For black rice, one parental rice cultivar should be black while the other should be disease resistant or should possess any quality traits like high anthocyanin content.

10.4.2 Crossing the Parents

Plant hybridization is done at the time of perfect flower synchronization. Flower synchronization is a technique of designing flowering time of male and female parents in approximate same date. With alternate sowing, synchronization is possible. Check the pollen grain for its viability before starting crossing. A well-trained person can perform crossing activities well. First, to start emasculation (the removing of immature pollen grain from the floral organ), choose a plant that possesses higher yield but disease susceptible. Wash the pollen grain with a sharp knife and brush. The washing is done when pollen grain starts maturing. Now bag with a butter paper so that there would be no chance of pollination from other rice genotypes. Perform this action in mid evening nearly 4–5 PM. Next day in the morning, collect pollen grains from the source parents (possess disease resistant genes) and pollinate by hand.

10.4.3 Raising Generation Lines

Plant the F1 (obtained from crossing between two parents) in normal growing season to obtain abundant seeds. Sow the seeds in next year. At least 2000–3000 plants must be maintained in F2 to select the better genotypes. In F2 generation, plant segregates and there is a better chance to choose best lines for the F3 generation.

The lines from F1 to F9 generation are called generation lines (Singh and Singh 2015). Plant selection starts with F2 generation and so on. Best genotypes are chosen and maintained up to F7/F8 generation, and then the selected best F8 generation lines are maintained for observation nursery. Generation lines are maintained up to F7/F8 generations because it is believed that there would be no segregation onwards.

The observation nursery seed material is also used for disease and pest-testing trial sidewise. If the genotypes are found better with higher yield and resistant to disease and pests, such genotypes are selected and maintained for initial yield trial (IYT).

10.4.4 Comparing Genotypes at Trial Level

Maintain the best genotypes at initial yield/preliminary yield trial, obtained from observation nursery trial. In this trial, only few genotypes (<30) are maintained and compared with popular standard check or with popular local variety. The best genotypes screened from initial yield trial are again maintained for Coordinated Varietal Trial (CVT) next year. The same routine (similar to Initial Evaluation Trial – IET) is also followed in CVT trial. In CVT, again genotypes are compared with the best check variety and most suitable genotypes are selected for Farmer's Field Trial (FFT). The IET and CVT may be maintained for 2 years but normally considered for 1 year. IET, CVT, and FFT are normally set up in randomized complete block design with three replications (Kushwaha and Pandey 2017). FFT is maintained for at least 2 years at different locations, i.e., multilocation trial. The best and suitable genotypes found in FFT are selected for release as variety. It takes nearly 12–14 years (Singh 2013).

10.4.5 Trait Improvement

Black rice cultivar Matesih has short heading time and harvest time than other accessions. Matesih has highest amylose content, lower protein content than klaten accessions, and lower content of anthocyanin than toraja accessions (Nandariyah et al. 2018).

Kang et al. (2011) studied physicochemical properties of endosperm of five black rice (Dragon eyeball 100, Heukjinjubyeo, Heukgwangbyeo, Heuknambyeo, and Josaengheukchal) and a white rice (Hwayoungbyeo) varieties. These black rice varieties showed elevated amount of minerals and rapid hydrolysis rate. High amino acid and sugar contents were reported in Heuknambyeo and Heukgwangbyeo varieties, respectively. Lower oleic acid, but higher linoleic acid, was reported in Dragon eyeball 100 and Heukjinjubyeo compared with the other samples. Likewise, the pasting and viscosity values were generally large in quantity in Heuknambyeo and Heukgwangbyeo, while lowest in Josaengheukchal. This study thus illustrates the wide variation in the physicochemical properties of the black rice varieties analyzed.

The results could serve as baseline information for food processors in evaluating the quality of black rice suitable for specialty food processing. Lee (2010) reported that the chemical contents of grains vary per varieties. Anthocyanin showed significant differences and cyanidin-3-O-glucoside (52.1 ± 6.3–1601.0 ± 8.5 μg/g) had a markedly higher content than peonidin-3-O-glucoside (ND-82.6 ± 1.2 μg/g). Among varieties, Heugjinjubyeo showed the highest anthocyanin, whereas Heughyang had the lowest.

Farmyard manure 10 tons +75–40-40 N-P-K kg per hectare performed significantly higher with higher number of effective tillers/hill, panicle length, number of spikelets/panicle, number of filled spikelets/panicle, test weight, grain protein content, and consequently grain yield and straw yield. This research depicts that integrated use of organic and inorganic sources of nutrients to be the best method for increasing productivity of black rice (Gangmei and George 2017).

10.5 Biotechnology Approaches

Black rice is a highly demanded type of rice because of its nutrition profile. Modern research profiles show that biotechnological approaches also apply to study the black rice as well. Marker-assisted breeding applies to develop a high-yielding black rice variety in a short period.

10.5.1 Tissue Culture Methodology

Rice research uses in vitro technique like tissue culture technique for research purpose only. No reports have been observed that applied tissue culture technique commercially used for rice seedling production. Rice parts like leaf, shoot, culm, anther, and seed are considered as explants for tissue culture technique (Fig. 10.5). This technique might be widely applied for large amount of healthy seedling production (Yoshida and Kato 1996). For callus induction, a variety of nutrient media are available in the market. Niroula et al. (2005) reported that callus induction was best obtained with a nutrient media containing (N6 mineral salts and N6 vitamins, 2 mg/l each + myoinositol, 100 mg/l + 2,4-D, 2.5 mg/l + kinetin, 0.5 mg/l + $AgNO_3$, 10 mg/l + maltose, 50 g/l).

10.5.2 Mutation Breeding

Mutation is highly used techniques to create variation among black rice germplasm. The agent which stimulates mutation is mutagen. Mutagen might be a chemical substance or x-rays/gamma rays. Riyatun et al. (2018) conducted a study on third

Fig. 10.5 Tissue culture technique in rice

generation of gamma irradiated black rice (*Oryza sativa* L. cv. Cempo Ireng) and found there was no significant changes of moisture, lipids, carbohydrates, and fiber contents compared with control. Hartanti et al. (2017) reported that black rice irradiated by 200 Gy gamma rays showed significant influence on the length of the stem, number of internodes, and length of leaves whereas with dose of 300 Gy showed significant influence of the decreased value of diameter of third internode, branch number, and leaves width. Flowering time was getting faster with increase in dose of irradiation. However, there were no differences in the esterase banding pattern between radiation dose of 200 and 300 Gy than the control plants, indicated that randomly mutation had occurred. Warman et al. (2015) also irradiated 200 Gy dose of gamma rays to black rice seeds and studied M2 population to select semidwarf and dwarf mutated plants for genetic improvement of West Sumatra local rice. They got 31 candidate dwarf/semidwarf mutants with frequency of 0.062%.

Dewi et al. (2016) conducted an experiment to improve agronomic traits of black rice Cempo Ireng by artificial induction of mutagen sodium azide (NaN_3); the black rice seeds soaked in sodium azide for 18 h showed slow germination, reduced percentage of seed germination, and reduced plant height. But plant treated with 2.5 mM of sodium azide combined with cytokinin of 25 ppm or 50 ppm showed a better vegetative growth and a slight increase in yield compared to control. Yuwono and Sutoyo (2017) found that black rice local landraces Cempo Ireng, Cempo Melik, and Melik when irradiated with gamma rays with dose of 0 Gy, 100 Gy, 200 Gy, and 300 Gy shortened the seedling's height, plant height, and accelerated the days of flowering. It indicated there were mutant lines of M1.

Tada et al. (2015) mitigated seed-shattering trait by gamma rays, a mutant breeding approach. In their research, several lines were selected with reduced seed shattering from exhibited various phenotypes for seed shattering, fertility, and panicle morphology. These lines were good resources for the improvement of Yayoi-Murasaki and could be useful to understand the molecular mechanism of seed-shattering trait in rice.

10.5.3 Genetic Engineering

Oikawa et al. (2015) used multiple techniques such as molecular breeding, genetic engineering, tissue culture, mutation methods, and classical breeding approach to confirm the origin of black rice, and the gene responsible for black rice was *kala4*. They constructed a vector for rice transformation; suppressor mutants were also isolated by screening. Oh et al. (2015) used transcriptome data from black rice varieties to identify anthocyanin-related genes. They used 135 K *Oryza sativa* microarray and found 3728 genes were related with anthocyanin synthesis. Enriched pathway analysis confirmed 53 candidates gene for anthocyanin biosynthesis.

Chen et al. (2016) used Isobaric tags for relative and absolute quantification (iTRAQ) MS/MS to determine the difference in the metabolic pathways which are involved in the dynamic grain proteome in the development of black rice *indica* cultivar (*Oryza sativa* L. *indica* var. SSP). They identified eight targeted genes encoding different metabolic proteins via quantitative real-time polymerase chain reaction (qRT-PCR) analysis; these genes had difference in transcriptional and translational expression during grain development. Purwestri et al. (2015) also conducted an experiment to transform rolC::Hd3a-GFP by *Agrobacterium tumefaciens* on black rice (*Oryza sativa* L. cv. Cempo Ireng) to promote early flowering. Chen et al. (2017) reported that the proteomics of relation between leaf and grain in anthocyanin biosynthesis might be regulated by signal transduction and sugar metabolism in black rice leaf.

10.5.4 Molecular Breeding

Varietal developmental is a time-taking and long process that nearly takes 12–14 years through conventional breeding approach. Nowadays conventional breeding is integrated with molecular breeding for enhancing varietal developmental activities. Marker-assisted selection (MAS), a molecular breeding technique, applies suitable markers to identify candidate genes of favorable traits in a crop

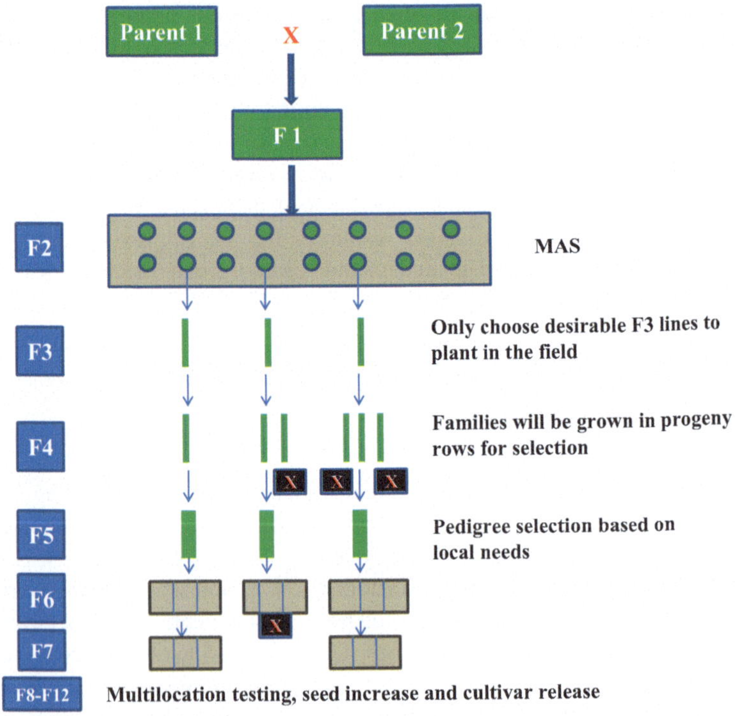

Fig. 10.6 Molecular breeding method used for black rice breeding

(Fig. 10.6). The target traits are introgressed into the desired genotype and released as a variety that might take 5–6 years only (Singh and Singh 2015).

All the procedure of marker-assisted selection is similar to conventional pedigree breeding method, but at the same time, different markers (SSR, RFLP, RAPD, SNPs) are used at F2 generations. The lines or genotypes which match the desirable characteristics are selected for the next generation. Therefore, those fixed lines are only selected that possess the targeted genes/QTLs. Marker-assisted selection is a rapid method of variety development which takes normally 5–6 years to develop a variety (Singh 2013).

Generally, black rice has lower yield than white rice. Therefore, the novel rice varieties having maximum yield and better quality are expected to be bred across the hybrid between the white and black rice. Recently, 15 breeding lines with high yields have been bred through hybrid between the white (II32 B) and black rice (Yunanheixiannuo) as the parent. Their phenotypes in rice bran layer are easy to identify because of the difference in color shades, including black, light purple, and white (Zhang et al. 2015). Zhang et al. (2015) made advanced lines from the cross between white and black rice as parents to evaluate phenolic levels and antioxidant properties. The breeding line YF53 has highest total phenolic and anthocyanin content and antioxidant activities. The experimental results reveal that advanced breeding lines could be developed for improvement of phenolic profiles and antioxidant activities with high yield. Shen et al. (2009) also studied the nutritional quality of red, white, and black rice so that they could turn to develop a new rice variety with enhanced antioxidant activity.

Bounphanousay et al. (2008) characterized 53 accessions of black glutinous rice from Lao PDR along with three white rice varieties to evaluate variations in 2-acetyl-1 pyrroline. Thirty-six genotypes showed aromatic properties which could be used for further breeding program.

10.6 Conclusions and Prospects

Black rice is most precious food material with its distinct taste and color. It is known as long-life rice as its consumption increases the longevity. Its ever-increasing demand in the market shows its high value among the consumers. Now Asian countries are interested to promote black rice and scientists are also involved in developing climate-resilient varieties.

Recent advancement in molecular breeding and high-throughput genotyping makes it possible to develop high-yielding new variety within a short time period. Modern breeding techniques, for example, mutation breeding, are the most suitable approach for creating variation which could be highly applied in case of black rice since black rice landraces are limited. Tissue culture technique can also be applied to generate new healthy seedlings, but still it has no commercial application in rice program. Genetic engineering is widely used to study black rice landraces with varied research purposes. Molecular breeding techniques, like MAS, are widely used for the development of new black rice varieties globally. Modern breeding approaches facilitated black rice research efficiently, and hence abundant knowledge is gained within a short time. Many black rice varieties have been developed that help to address the global demand.

Nowadays many high-yielding black rice varieties have been developed, but they are challenged in different agroclimatic zones, which might be due to climatic effects. Since global climate is a serious threat to our planet, scientists must think and accelerate breeding approaches that must develop climate-resilient high-yielding varieties to secure our future at appropriate time. The increased environmental temperature due to climate change is a great risk for both black and Basmati type rice. As temperature increases, the quality of black rice might reduce with decrease in anthocyanin, mineral, and nutrient content of the grain. Black rice cultivation might be favorable in mountainous regions with high altitude where environmental temperature is most suitable for securing rice grain quality.

Gene editing, a molecular tool to edit gene of desirable trait, is highly appreciated globally. This tool could also be applied here in black rice to develop high-yielding varieties packed with high level of nutrition. Since black rice is natural product with therapeutic value, gene editing techniques should be used for research purposes only that could maintain the beauty of natural products.

Many black rice varieties have been developed, and still its demand is high. People become more aware with the health benefits of this rice that make it widely popular among the consumers. Thus, abundant research is needed that could propel to understand the full functions of black-type rice grain on human health.

Appendix I: Research Institutes Relevant to Black Rice

Institution	Specialization and research activities	Contact information and website
Nepal Agricultural Research Council	Research on food and fodder crop, pulses, and animals	Nepal Agricultural Research Council, Singh Darbar Plaza, Kathmandu, Nepal www.narc.gov.np Phone: 977-1-5523041
Dharwad University of Agricultural Science	Research on agricultural crops including rice	Department of Food Science and Nutrition, College of Rural Home Science, Dharwad University of Agricultural Science, Dharwad-580,005, India Website: www.uasd.edu
State University of Ponta Grossa, Av. Carlos Cavalcanti	Research activities on food science	Graduate Program in Food Science and Technology, State University of Ponta Grossa, Av. Carlos Cavalcanti, 4748, CEP 84030-900 Ponta Grossa, PR, Brazil Website: https://portal.uepg.br/
Khon Kaen University	Research activities in biological science	Department of Biology, Faculty of Science, Khon Kaen University, Khonkaen 40002, Thailand Website: https://www.kku.ac.th/?l=en
Universitas Gadjah Mada	Research activities on biological science	Faculty of Biology, Universitas Gadjah Mada, Jl. Teknika Selatan, Sekip Utara, Sinduadi, Mla, Sleman, Daerah Ismewa Yogyakarta 55281,Indonesia Website: https://ugm.ac.id/en/
Kyungpook National University	Research related with food science	Department of Food Science and Nutrition, Kyungpook National University, Daegu, Korea Website: http://en.knu.ac.kr/main/main.htm
Ajou University	Research related with biological science	Department of Biological Science, Ajou University, Suwon, Korea Website: https://www.ajou.ac.kr/en/
Fujian Agriculture and Forestry University	Research related with crop and forestry science	College of Life Sciences, Fujian Agriculture and Forestry University, Fuzhou 350002, China Website: www.english.fafu.edu.cn
Hefei University of Technology	Research of food and food engineering	School of Food and Biological Engineering, Hefei University of Technology, Hefei, Anhui, PR China Website: http://spyswe.hfut.edu.cn/xueyuangaikuo/

Institution	Specialization and research activities	Contact information and website
Universitas Sebelas Maret (UNS) Surakarta	Research on agricultural science	Faculty of Agriculture, Universitas Sebelas Maret (UNS) Surakarta Jl. Ir. Sutami 36A, Surakarta, Central Java, Indonesia 57126 Website: www.en.hfut.edu.cn/index.php
Indian Agricultural Research Institute	Research on agricultural crops	Indian Agricultural Research Institute, Hillside road, Pusa, New Delhi-110012, India Website: www.iari.res.in

Appendix II: Black Rice Genetic Resources

Cultivar	Important traits	Cultivation location
Yayoi -murasaki	Late heading and easy seed shattering	Tokushima, Japan
Chakhao Amubi	High grain protein content	Allahabad, India
Br no. 3	Black brown pericarp	Khon Kaen, Thailand
Br no. 18	Straw color grain	Khon Kaen, Thailand
Br no. 53	Brown color grain	Khon Kaen, Thailand
Br no. 19	Purple color leaf and high anthocyanin content (2.53>)	Khon Kaen, Thailand
Br no 16	Green color leaf	Khon Kaen University, Khon Kaen, Thailand
Cempo Ireng	Black rice bran contains cyanidin 3-O-glucoside	Sayegan, Sleman, Yogyakarta, Indonesia
Heukjinjubyeo	High Ca content	Korea
Heukgwangbyeo	High K content	Korea
Heuknambyeo	High Mg content	Korea
Matesih	Shorter maturity period	Matesih area, Karanganyar, Indonesia
Klaten	High production	Matesih area, Karanganyar, Indonesia
Poreiton Chakhao	Aromatic	Manipur, India
Chakhao Amubi	Aromatic	Manipur, India
Burma Black Rice	Non aromatic	Manipur, India

Source: Phonsakhan and Kong-Ngern (2015), Kushwaha (2016a, b)

References

Auger C, Caporiccio B, Landrault N et al (2010) Increasing the bioactive compounds contents by optimizing the germination conditions of oSuthern Thai brown rice. Songklanakarin J Sci Technol 32(3):219–230

Bounphanousay C (2008) Chemical and molecular characterization of fragrance in black glutinous rice from Lao PDR. Asian J Plant Sci 7(1):1–7

Chaudhary RC (2003) Speciality rices of the world: effect of WTO and IPR on its production trend and marketing. J Food Agric Environ 1(2):34–41

Chen L, Huang Y, Xu M et al (2016) iTRAQ-based quantitative proteomics analysis of black rice grain development reveals metabolic pathways associated with anthocyanin biosynthesis. PLoS One 11(7):e0159238. https://doi.org/10.1371/journal.pone.0159238

Chen L, Huang Y, Xu M et al (2017) Proteomic analysis reveals coordinated regulation of anthocyanin biosynthesis through signal transduction and sugar metabolism in black rice leaf. Int J Mol Sci 18:2722. https://doi.org/10.3390/ijms18122722

Chiang AN, Wu HL, Yeh HI et al (2006) Antioxidant effects of black rice extract through the induction of superoxide dismutase and catalase activities. Lipid 41:797–803

Choi SP, Kim SP, Kang MY et al (2010) Protective effects of black rice bran against chemically-induced inflammation of mouse skin. J Agric Food Chem 58(18):10007–10015. https://doi.org/10.1021/jf102224b

Dewi K, Meidiana G, Sudgino S (2016) Effects of sodium azide (NaN3) and cytokinin on vegetative growth and yield of black rice plant (Oryza sativa L. Cempo Ireng). AIP Conf Proc 1755:130005. https://doi.org/10.1063/1.4958549

Gangmei Tigangam P, George PJ (2017) Black rice CV. 'Chakhao Amubi' (Oryza sativa L.) response to organic and inorganic sources of nutrients on growth, yield and grain protein content. J Pharmacog Phytochem 6(4):550–555

Goffman FD, Bergman CJ (2004) Rice kernel phenolic content and its relationship with antiradical efficiency. J Sci Food Agric 84:1235–1240

Hartanti RS et al (2017) IOP Conf Ser. Mater Sci Eng 193:012038

Hoahua HE, Pan X, Zao Z et al (1996) Properties of the pigment in black rice. Chin Rice Res News 4(2):11–12

Hu C, Zawistowski J, Ling W, Kitts DD (2003) Black rice (Oryza sativa L. indica) pigmented fraction suppresses both reactive oxygen species and nitric oxide in chemical and biological model systems. J Agric Food Chem 51:5271–5277

Ichikawa H, Ichiyanagi T, Xu B et al (2001) Antioxidant activity of anthocyanin extract from purple black rice. J Med Food 4(4):211–218

Jang HH, Park MY, Kim HW et al (2012) Black rice (Oryza sativa L.) extract attenuates hepatic steatosis in C57BL/6 J mice fed a high-fat diet via fatty acid oxidation. Nutr Metab 9:27

Jang WS, Seo CR, Jang HH et al (2015) Black rice (Oryza sativa L.) extracts induce osteoblast differentiation and protect against bone loss in ovariectomized rats. Food Funct 6(1):265–275

Kang M-Y, Kim J-H, Rico CW et al (2011) A comparative study on the physicochemical characteristics of black rice varieties. Int J Food Proper 14(6):1241–1254

Kang MY, Rico CW, Bae HJ et al (2013) Antioxidant capacity of newly developed pigmented rice cultivars in Korea. Cereal Chem 90(5):497–500

Kim C, Kikuchi S, Kim YK et al (2010) Computational identification of seed-specific transcription factors involved in anthocyanin production in black rice. Biochip J 4(3):247–255

Kong L, Wang Y, Cao Y (2008) Determination of myoinositol and D-chiroinositol in black rice bran by capillary electrophoresis with electrochemical detection. J Food Compos 7(1):1–8

Kong S, Kim DJ, Oh SK et al (2012) Black rice bran as an ingredient in noodles: chemical and functional evaluation. J Food Sci 77(3):303–307

Kushwaha UKS (2016a) Black rice anthocyanin content increases with increase in altitude of its plantation. Adv Plants Agric Res 5(1):00170. https://doi.org/10.15406/apar.2016.05.00170

Kushwaha UKS (2016b) Black rice: research, history, and development. Springer International Publishing, Switzerland. https://doi.org/10.1007/978-3-319-30153-2

Kushwaha UKS and Pandey ID (2017) Plant height enhances fertile grain per panicle in rice. Second all India seminar on advances in engineering and technology for sustainable development, 25–26 November 2017, proceedings, GBPUAT, Pantnagar, pp 293–301. ISBN: 978-81-935237-6-6

Lee JH (2010) Identification and quantifications of anthocyanins from the grains of black rice (Oryza sativa L.) varieties. Food Sci Biotechnol 19(2):391–397

Lee SM, Choi Y, Sung J, Lee J (2014) Protective effects of black rice extracts on oxidative stress in mice. J Nutr 136:2220–2225

Min B, McClung AM, Chen MH (2011) Phytochemicals and antioxidant capacities in rice brans of different color. J Food Sci 76:117–126

Mingwei Z, Peng Z, Xu Y (1995) Genetic effect analysis on pigment content in pericarp of black rice grain. Chin J Rice Sci 9(3):149–155

Nandariyah et al (2018) IOP Conf Ser Earth Environ Sci 142:012052

Natsumi T, Noriko O (1994) Physicochemical properties of kurogome, a Japanese native black rice. Part 1 Bull Gifu Women's Coll 23:105–113

Niroula RK, Sah BP, Bimb HP et al (2005) Effect of genotype and culture media on callus induction and plant regeneration from matured rice grain culture. J Inst Agric Anim Sci 26:21–26

Oh JH, Won SY, Lee JH et al (2015) Computational identification of anthocyanin-related genes using transcriptome data from black rice plants. Cell Dev Biol 4:157. https://doi.org/10.4172/2168-9296.1000157

Oikawa T, Maeda H, Oguchi T et al (2015) The birth of a black rice gene and its local spread by introgression. Plant. https://doi.org/10.1105/tpc.15.00310

Phonsakhan W, Kong-Ngern K (2015) A comparative proteomic study of white and black glutinous rice leaves. Electron J Biotechnol 18:29–34

Purwestri YA, Sari RDK, Anggraeni LN et al (2015) Agrobacterium tumefaciens mediated transformation of rolC: Hd3a-GFP in black rice (Oryza sativa L. cv. Cempo Ireng) to promote early flowering. Procedia Chem 14:469–473

Quan LH (1999) Selection of yeast for beverage production from black rice. Nong Nghiep Cong NghiepThue Pham 8:375–376

Riyatun et al (2018) IOP Conf Ser: Mater Sci Eng 333:012073

Salgado JM, Anderson GCO, Débora NM et al (2010) The role of black rice (Oryza sativa L.) in the control of hypercholesterolemia in rats. J Med Food 13(6):1355–1362

Sangkitikomol W, Tencomnao T, Roccjanasaroj A (2010) Antioxidant effects of anthocyaninsrich extract from black sticky rice on human erythrocytes and mononuclear leukocytes. African J Biotechnol 9(48):8222–8229

Sastry SVS (1978) Inheritance of genes controlling glume size, pericarp colour, and their interrelationships in indica rice. Oryza 15:177–179

Seo WD, Kim JY, Song YC et al (2013) Comparative analysis of physicochemicals and antioxidative properties in new red rice (Oryza sativa L. cv. Gunganghongmi). J Crop Sci Biotechnol 16:63–68

Shen Y, Jin L, Xiao P et al (2009) Total phenolics, flavonoids, antioxidant capacity in rice grain and their relations to grain color, size and weight. J Cereal Sci 49(1):106–111

Singh BD (2013) Plant breeding: principles and methods. Kalyani Publishers, New Delhi

Singh BD, Singh AK (2015) Marker assisted breeding: principles and practices. Springer, New Delhi. https://doi.org/10.1007/978-81-322-2316-0

Sompong R, Siebenhandl ES, Linsberger MG et al (2011) Physicochemical and antioxidative properties of red and black rice varieties from Thailand, China and Sri Lanka. Food Chem 124(1):132–140

Stuecker MF, Tigchelaar M, Kantar MB (2018) Climate variability impacts on rice production in the Philippines. PLoS One 13(8):e0201426. https://doi.org/10.1371/journal.pone.0201426

Tada N, Nii K, Sugita SK (2015) Mutant breeding of a Japanese traditional black rice cultivar Yayoi-murasaki to improve seed shattering trait. Nucleus 58(3):217–223

Van Oort Pepijn AJ, Zwart Sander J (2017) Impacts of climate change on rice production in Africa and causes of simulated yield changes. Glob Chang Biol 24(3):1029–1045

Villalobos FJ, Fereres E (eds) (2016) Principles of agronomy for sustainable agriculture. Springer, New York. https://doi.org/10.1007/978-3-319-46116-8

Walter M, Marchesan E (2011) Phenolic compounds and antioxidant activity of rice. Braz Arch

Wang JT, Guo L (2007) Research advance of black rice pigment and its biological effect. J Anhui Agric Sci 35(32):10199–10200

Warman B, Suliansyah I, Swasti E et al (2015) Selection and semi dwarf allele mutants segregation pattern as the result of gamma rays irradiation of West Sumatera black rice. Intr J Adv Sci Eng Inf Tech 5(5):1–10. ISSN: 2088–5334

Xia X, Ling W, Ma J et al (2006) An anthocyanin-rich extract from black rice enhances atherosclerotic plaque stabilization in apolipoprotein E-deficient in mice. J Nutr 136:2220–2225

Yafang S, Gan Z, Jinsong B (2011) Total phenolic content and antioxidant capacity of rice grains with extremely small size. Afr J Agric Res 6(10):2289–2293

Yang Y, Marc CA, Yan H et al (2011) Anthocyanin extract from black rice significantly ameliorates platelet hyperactivity and hypertriglyceridemia in dyslipidemic rats induced by high fat diets. J Agric Food Chem 59(12):6759–6764

Yao SL, Xu Y, Zhang YY, Lu YH (2013) Black rice and anthocyanins induce inhibition of cholesterol absorption in vitro. Food Funct 4(11):1602–1608

Yoshida T, Kato H (1996) In vitro propagation of hybrid rice (Oryza sativa L.). JARQ 30:9–14

Yuwono SS, Sutoyo (2017) Early growth performance some varieties of black rice (Oryza sative L.) irradiated using gamma ray. IntJ Adv Engin Technol 10(2):145–153

Zhang X, Shen Y, Prinyawiwatkul W et al (2013) Comparison of the activities of hydrophilic anthocyanins and lipophilic tocols in black rice bran against lipid oxidation. Food Chem 141(1):111–116

Zhang H, Shao Y, Bao J et al (2015) Phenolic compounds and antioxidant properties of breeding lines between the white and black rice. Food Chem 172(1):630–639

Zuo Y, Peng C, Liang Y et al (2012) Black rice extract extends the lifespan of fruit flies. Food Funct 3(12):1271–1279

Chapter 11
Aromatic Rice

Somnath Roy, Amrita Banerjee, N. Basak, J. Kumar, and N. P. Mandal

11.1 Introduction

The aromatic or fragrant rice derives its name due to the presence of characteristic aroma in the grains and is generally identified by three main factors: aroma, appearance, and taste (Chaudhury et al. 2003). Aromatic rice is considered premium because of its pleasant and subtle aroma, unique taste, and excellent grain quality such as superfine kernels, high kernel elongation after cooking, and good texture of cooked rice. The tropical Asian countries have their own set of high-quality aromatic rice varieties having immense cultural significance (Singh et al. 2000a). Within the Asian cultivated rice (*Oryza sativa* L.) genepool, aromatic rice varieties constitute a small but special group which fetches a premium in global markets. Among the different aromatic rice cultivar types, the Basmati cultivars from north and northwestern India and Pakistan and Jasmine rice such as Khao Dwak Mali (KDML) 105 from north and northeastern Thailand are considered to be the most valued rices. Other important aromatic varieties in the world market are Siamati (Thailand), Bahra (Afghanistan), Sadri (Iran), and Della, Texamati, and Kasmati (USA) (Singh et al. 2000a). The Basmati and Jasmine rice continue to be the most premium rice varieties in the present rice trading. Fragrant rice is priced on the trade market at $1023–1167/T (tonne), whereas the nonfragrant rice is traded for $360–888/T in 2018. The global trading prices of different coarse rice categories have declined sharply during recent years after record highs in 2007–2008 (Childs and Kiawu 2009). However, the prices of fragrant rices only declined slightly and

S. Roy (✉) · A. Banerjee · J. Kumar · N. P. Mandal
Central Rainfed Upland Rice Research Station, ICAR-National Rice Research Institute, Hazaribag, Jharkhand, India
e-mail: Somnath.Roy@icar.gov.in

N. Basak
Crop Physiology and Biochemistry Division, ICAR-National Rice Research Institute, Cuttack, Odisha, India

A. Costa de Oliveira et al. (eds.), *The Future of Rice Demand: Quality Beyond Productivity*, https://doi.org/10.1007/978-3-030-37510-2_11

they were always the highest (Fig. 11.1). The price premium of aromatic rice attracts many stakeholders and raises the competition between domestic and trade markets; as a result, often the domestic aromatic rice prices are closer to the world prices in India and Thailand (Dawe 2007). Globally, rice trading is low and accounts for only 7.13% of the total rice production. The aromatic rice (Basmati and Jasmine) accounts for around 15–18% of worldwide rice trade (Giraud 2013).

The unique aroma and its strength in scented rice are considered to play a vital role in determining the marketability and consumer acceptability. The flavor chemistry of aromatic rice grains revealed the existence of more than 200 volatile compounds such as alcohols, aldehydes, acids, esters, hydrocarbons, ketones, phenols, pyridines, and other compounds. However, the relationships among the volatile compositions and rice aroma have not been established, except for 2-acetyl 1 pyrroline (2-AP). Comparative analysis of the volatile compounds of aromatic and nonaromatic rices revealed that 2-AP, which imparts specific "popcorn" flavor in aromatic rice and has comparably lower odor threshold among rice volatiles, occurs at significantly higher levels in aromatic rice than the nonaromatic rice varieties (Buttery et al. 1983). Several studies reported that 2AP is the only volatile compound for which the relationship between its concentration in rice and sensory intensity could be established (Paule and Powers 1989). Yet the distinct aroma of different aromatic varieties indicated that rice aroma chemistry is a complex interplay between numerous volatile aroma compounds where 2-AP is the most important compound (Maga 1984). Even though 2-AP was identified as the principal aroma compound, the biochemical pathways of its synthesis leading to fragrance remained inadequately understood.

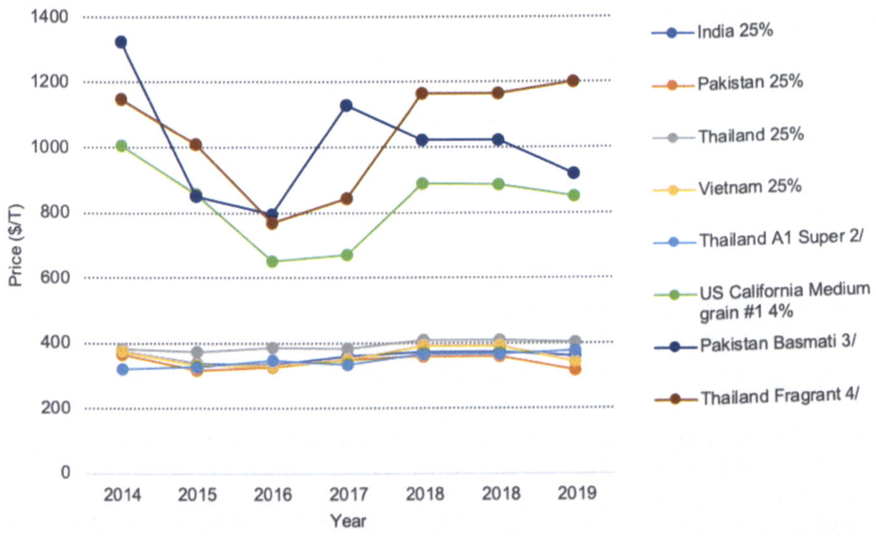

Fig. 11.1 Rice export price, $/T, F.o.b., 2014- January 2019. (Source: FAO rice price update; http://www.fao.org. %, broken grain percentage; 2/, white broken rice; 3/, Basmati ordinary up to May 2011. Super kernel white basmati 2% from June 2011 onwards; 4/, Hom Mali rice, grade A)

The aroma in rice grains was reported to be governed by a single recessive locus, *fgr*, on rice chromosome 8 (Ahn et al. 1992). Subsequently, a nonfunctional allele of the gene encoding putative *Betaine aldehyde dehydrogenase 2* (*BADH2*) was identified in the *fgr* region as the best aroma candidate locus (Bradbury et al. 2005a). The *badh2* allele in aromatic cultivars carried an eight base-pair (8-bp) deletion in the exon 7 leading to premature stop codon and ultimately loss of function of the BADH2 enzyme. This functional mutation is found in most of the aromatic varieties around the world. Molecular markers developed based on this mutation in *BADH2* have been used extensively for selecting the aroma trait (Bradbury et al. 2005b; Amarawathi et al. 2008).

11.2 Origin and Distribution of Aromatic Rice

Oryza sativa is one of the oldest crops and Asian civilization was built on it. Around 9000 years ago, humans started domesticating perennial Asian wild grass, *Oryza rufipogon*, in different parts of Asia. It was proposed that the domestication of rice in Asia occurred independently and concurrently at several sites including the Ganges valley in India, west China, the Yangtze river area in China, coastal swamp habitats in southeast Asia, the "belt-region" with its great diversity in *Oryza* species along the southern slopes of the Himalayas, and several other places (Solheim 1972; Chang 1976; Yan 1989).

The ecogeographical and ethno-cultural diversity of this vast rice growing areas has resulted in the evolution of different cultivar groups with rich phenotypic diversity. Two varietal groups – *Hsien* and *Keng* – have been recognized in the ancient China since Han dynasty dating back more than 2000 years ago (Oka 1988). The *Hsien* and *Keng* correspond to *indica* and *japonica,* respectively. Studies involving seven wild and domesticated strains suggested that *indica* and *japonica* have been originated from different genetic stocks of wild rice through two separate domestication events: *japonica* in China and *indica* in India (Gross and Zhao 2014). However, studies considering domestication-related genes, such as *sh4* (non-shattering of grains) and *Rc* (gene for red pericarp; domestication mutated *Rc* which caused white pericarps), proposed an overlapping in *japonica–indica* domestication: the mutation in these genes arose in *japonica* first and then spread to *indica* (Kovach et al. 2007; Sang and Ge 2007). Recent studies using both domestication and neutral genes, and whole genome sequences from *O. rufipogon, indica,* and *japonica* accessions, generated contrasting views on evolution of *indica* and *japonica* from single or separate domestication events (He et al. 2011; Molina et al. 2011). The 3K Rice Genomes Project (The 3000 Rice Genomes Project 2014) offered supporting views on independent domestication of some of the *indica* pool by analyzing haplotype plots of nine important domestication genes: *Rc, Bh4, PROG1, OsC1, Sh4, Wx, GS3, qSH1,* and *qSW5* (Wang et al. 2018). As it currently stands, the origin of *indica* and *japonica* is still an open end.

Indica and *japonica* cultivars differ in many phenotypic attributes, including their grain shape (*indica* – long grained and *japonica* – small grained) and grain amylose content (*indica* has higher amylose content than *japonica* and thus less sticky after cooking). Earlier, *indica* and *japonica* types are each characterized by certain diagnostic characters, such as $KClO_3$ resistance, cold tolerance, apiculus hair length, and phenol reaction (Oka 1958). Genetically, these two varietal groups are well differentiated; however, few additional groups can also be recognized (Glaszmann 1987; Garris et al. 2005; Wang et al. 2014). In his landmark study, Glaszmann (1987) reported six cultivar groups (Group I–Group VI) within *O. sativa* by examining 1688 rice cultivars from different countries for allelic variations at 15 polymorphic loci of eight isozymes. The *indica* and *japonica* groups were represented by Group I and Group VI, respectively. Group II corresponds to the *aus* groups representing early maturing and drought-tolerant upland rice varieties grown in eastern India and Bangladesh. Group III and Group IV correspond to the floating rices of Bangladesh and India, popularly known as *ashina* and *rayada*. Group V includes aromatic rices of the Indian subcontinent, including Basmati.

This grouping was further supported by studies with different molecular markers involving diverse set of rice germplasm. Five distinct genetic groups – *indica, aus, aromatic, tropical japonica, and temperate japonica* – were reported within 234 samples of *O. sativa* using 169 simple sequence repeat (SSR) markers and two chloroplast loci (Garris et al. 2005). This grouping corresponds to Glaszmann's (Glaszmann 1987) groups I, II, V, and VI (includes both *temperate* and *tropical japonica*). This classification of *O. sativa* into five groups was supported by analyses of genome-wide single nucleotide polymorphism (SNP) data (Caicedo et al. 2007; Zhao et al. 2010; Huang et al. 2012). A few studies also reported additional subdivisions within *indica* and *japonica* (de Oliveira Borba et al. 2009; Zhang et al. 2009; Ebana et al. 2010; Li et al. 2010; Wang et al. 2014). Under 3K Rice Genomes Project (The 3000 Rice Genomes Project 2014), whole genome sequencing of 3010 *O. sativa* accessions provided scope for understanding the genetic structure within cultivated rice genepool at whole genome scale. Nine subpopulations were reported by analyzing approximately 17 million SNPs which largely correspond to the geographic origins of the accessions (Wang et al. 2018). Four groups were identified within *indica* and three within *japonica* clusters. Two other subpopulations represented the circum-aus (including *aus, boro,* and *rayada* ecotypes) and circum-aromatic (Basmati and Sadri aromatic rices) (Fig. 11.2). In all the genetic diversity studies, *aromatic* group showed genetic proximity with the *japonica*.

In Glaszmann's classification, the Group V represented the aromatic rices from Iran, Afghanistan, Pakistan, India, Bangladesh, Myanmar, and China. This group thus includes Basmati and Sadri cultivars along with long- and medium-grain traditional aromatic varieties from different countries (Glaszmann 1987). Moreover, aromatic accessions can also be identified within Group I (*indica*) and Group VI (*japonica*). Many aromatic varieties of Southeast Asia and China belong to *indica*, including the Jasmine rice cultivars from Thailand. The aromatic *chakhao* cultivars from the Indian state of Manipur also categorized as *indica* (Roy et al. 2015). A few aromatic varieties from Indonesia, Philippines, and China have been classified as

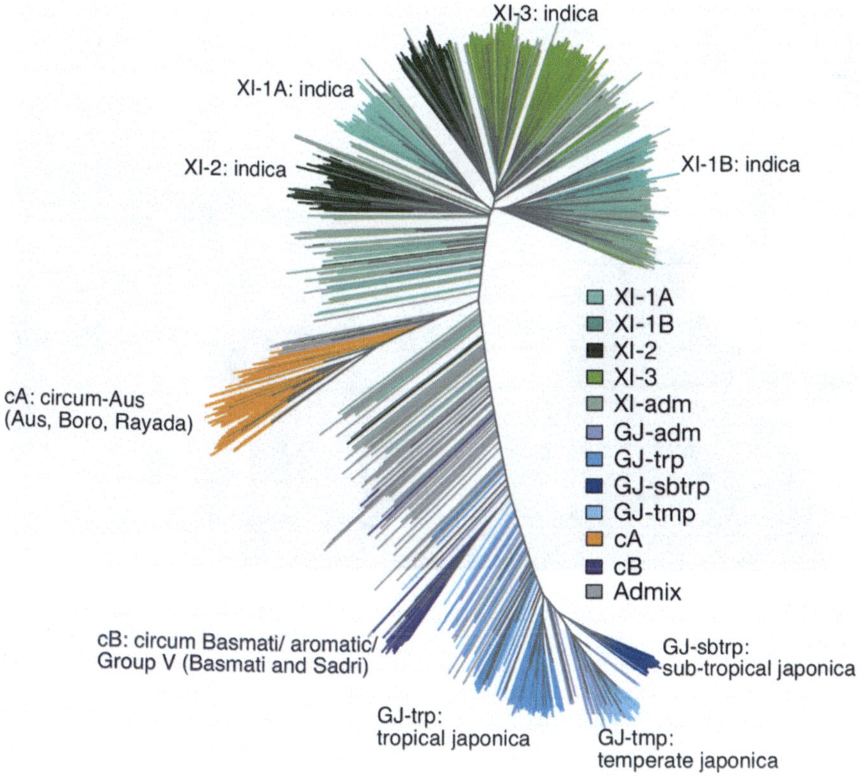

Fig. 11.2 Unweighted neighbor-joining tree showing subpopulation structure in 3010 samples of *Oryza sativa*. The tree was computed on a simple matching distance matrix for approximately 17 million SNPs. Rice accessions are colored by their assignment to $k = 9$ subpopulations from ADMIXTURE. The Basmati and sadri aromatic rices falling under Group V according to Glaszmann (1987) were grouped under circum-Basmati (cB). However aromatic accessions also found in *indica* and *japonica* clade. (Reproduced and modified from Wang et al. 2018)

japonica. Due to genetic affinity of *aromatic* and *japonica* rice, it is assumed that aromatic rice is a subtype of *japonica*. However, occurrence of aroma trait within different genetic subpopulation of rice obscures our understanding of aromatic rice evolution.

It is widely accepted that the center of diversity of aromatic rice of Group V (Glaszmann 1987) is the foothills of Himalayas covering the Indian states of Uttarakhand, Uttar Pradesh, Bihar, and *Tarai* region of Nepal, where a large number of aromatic rice cultivars were being cultivated (Fig. 11.3) (Khush 2000). Over the period, aromatic rices spread to adjoining areas of India, Pakistan, Afghanistan, Iran, Iraq, Bangladesh, and Thailand. Aromatic rice varieties found in Italy, France, and other European countries are believed to have been introduced from Asia. The geographical limit of distribution of *japonica* (east and Southeast Asia) and *aromatic* rice indicates a complex evolutionary history of aromatic rice. There are limited studies targeting unravelling the origin of *aromatic* rice.

Fig. 11.3 Center of diversity of aromatic rice. Aromatic rice is believed to be originated in the foothills of Himalaya covering the Indian states of Uttrakhand, Uttar Pradesh, Bihar, and the *Tarai* region of Nepal (Khush 2000). From this center of diversity aromatic rice distributed to different parts of India and the globe as showed by arrows. The cultivation areas of Basmati, Jasmine, and Sadri aromatic rices are shown in red circles. Aromatic rice is grown in many countries around the world

A study on the origin and evolution of *badh2* in a diverse set of 242 *O. sativa* accessions sampled from 38 Asian countries was carried out to analyze the haplotypic diversity across the *BADH2* region (Kovach et al. 2009). It was shown that out of eight putative nonfunctional alleles of *badh2*, the *badh2.1* allele (carrying 8-bp deletion in exon 7) is predominant in all Group V (*aromatic*) cultivars. However, the *badh2.1* allele also existed in 6% *indica* and 15% *japonica* accessions. This study highlighted the evolution of fragrance trait in rice in the light of *BADH2* haplotype diversity only. However, a whole-genome-based evolutionary analysis will shed light on genetic distinctness of the *aromatic* group. The 3K Rice Genomes Project has opened many possibilities to fine tune the molecular basis of *O. sativa* evolution by analyzing genome-wide variation. On this direction, a recent study showed that *aromatic rice* originated in the Indian subcontinent from hybridization between *japonica* and a wild rice (*O. rufipogon*) populations, which was also ancestral to *aus*, in the South of the Himalayas (Civàñ et al. 2019). Earlier it was postulated that the fragrance trait was first appeared in the wild rice and that aromatic trait later appeared in some cultivars as preferentially selected by the local farmers during the process of domestication (Prathepha 2008). Diversity of aromatic rice cultivars has evolved over thousands of years as local farmers selected different types to suit their needs and also due to adaptation in a wide range of agroecological conditions. Most of the aromatic rice-growing countries around the world have their own traditional aromatic rices.

11.3 Aromatic Rice Varieties

11.3.1 Basmati Rice

Basmati (meaning “the fragrant one”) is a special quality long-grain aromatic rice cultivated in a particular geographical region of the Indian subcontinent with a recorded history of over 200 years. Basmati rice is now a registered geographical indication (GI) product since February 2016. The GI for Basmati has been given to seven stated Indian states (Punjab, Haryana, Uttarakhand, Himachal Pradesh, Delhi, Jammu & Kashmir, and parts of Uttar Pradesh) in the Indo-Gangetic plains on the foothills of Himalaya. As compared to non-Basmati rices, it has characteristics long-slender grains with pleasant aroma. The quality of Basmati rice is estimated by a harmonious combination of grain quality traits such as minimum kernel dimension, intensity of aroma, texture of cooked kernels, high volume expansion due to linear kernel elongation with minimum breadth-wise swelling, and a medium glycemic index (around 60), fluffiness, palatability, and longer shelf life (Table 11.1) (Singh 2000).

These unique features of traditional Basmati are the product of centuries of selection and cultivation by farmers and are preserved in their purest form in the traditional cultivars. However, the traditional cultivars have a number of undesirable

Table 11.1 Unique quality parameter of Basmati rice

Quality parameters	Details
Milled rice length	>6.6 mm
Milled rice breadth	<2.0 mm
Length/breadth ratio	>3.0
Kernel shape and size	Long, slender
Kernel appearance	Creamy white, translucent
Head rice recovery	>40%
Cooked kernel texture	Cooked kernel to be firm, tender, and nonsticky without splitting and bursting. Amylose content 20–22%
Kernel elongation ratio	Unique feature of Basmati grains. Upon cooking kernels elongate almost twice, but do not flatten much. Elongation ratio >1.80
Flavor and taste	The Basmati cultivars have a distinctive fragrance which differentiated them from other aromatic cultivars. Aroma is categorized as appetizing and strong. Taste is sweet and appealing

Source: Singh (2000)

traits such as tall stature (prone to lodging), low yield, photoperiod, and temperature sensitivity; as a result their production and productivity are less.

Improvement work on Basmati had led to development of many improved "elite" Basmati varieties. But combining all the unique features of Traditional Basmati into improved Basmati varieties has been a challenge. Initial research efforts during 1930 on Basmati rice improvement based on germplasm collection, evaluation, and selection led to the identification and release of "Basmati 370" for cultivation in 1933 which dominated Basmati rice export until the late 1980s (Ramaiah and Rao 1953).

Systematic research on Basmati rice improvement in India was initiated during late 1960s and significant research work was carried out at the ICAR-Indian Agricultural Research Institute (ICAR-IARI), New Delhi, during 1970s on standardization of protocols for estimating various Basmati quality parameters and analyzing their inheritance pattern (Singh et al. 2004). The first semidwarf, photoperiod-insensitive, and high-yielding Basmati rice variety Pusa Basmati 1 was released for cultivation in 1989. A popular traditional Basmati variety, Taraori Basmati, was released in 1996 with premium grain and cooking quality; later this variety has been used as donor for grain quality at International Rice Research Institute (IRRI), Philippines (Singh and Singh 2010). This was a selection (also known as Karnal Local) from the traditional Basmati rice collection, Haryana Basmati Collection (HBC19) from Karnal district of Haryana (Singh et al. 2004). Later ICAR-IARI developed Pusa Basmati 1121 in 2003 with pleasant aroma, extra slender grains (9.00 mm), and exceptional kernel elongation (elongation ration 2.5) and volume expansion (more than four times) after cooking (Fig. 11.4a; Table 11.2) (Singh et al. 2018).

The traditional varieties command a considerable price advantage over many of the improved Basmati varieties due to their superior grain quality and authenticity. A list of traditional and improved Basmati rice is given in Table 11.3. The Punjab province of Pakistan also produces export quality Basmati rice. In terms of exporting Basmati rice, India and Pakistan have monopoly in the world markets. India exported 2,294,787 tonnes of Basmati rice around the world in 2018–2019, and the value is considerably higher than that exported in 2000 (638,380 tonnes).

The genetic diversity within Basmati rice germplasm is well documented. A set of 141 Basmati rice accessions, including traditional and improved varieties, collected from northwestern states of India was classified using trait-linked SSR markers into four clusters which largely correlate with their geographical distribution (Salgotra et al. 2015). Traditional Basmati cultivars were effectively differentiated

Fig. 11.4 (continued) cultivars from southern states of India: (A) Ambemohar Ajra, (B) Kothimbiri, (C) Raibhog, (D) Ambemohar-Tambda, (E) Basumati, (F) Kate chinoor, (G) Gham, (H) Basmati 370, (I) Ghansal, (J) Chimansal, (K) Parabhani chinoor, (L) Kalsal, (M) Dubraj, (N) Kalanamak. (The photo in Panel **a** was reproduced with permission from Sing et al. (2018). Creative Commons Attribution 4.0 International License (http://creativecommons.org/licenses/by/4.0/). The photo of KDML 105 (Panel **b**) was kindly provided by Dr. A. Vanavichit, Professor & Director, Rice Gene Discovery/Rice Science Center, Kasetsart University, Thailand. The Photo of rice cultivars in Panel **c** was reproduced with permission Nadaf et al. 2016)

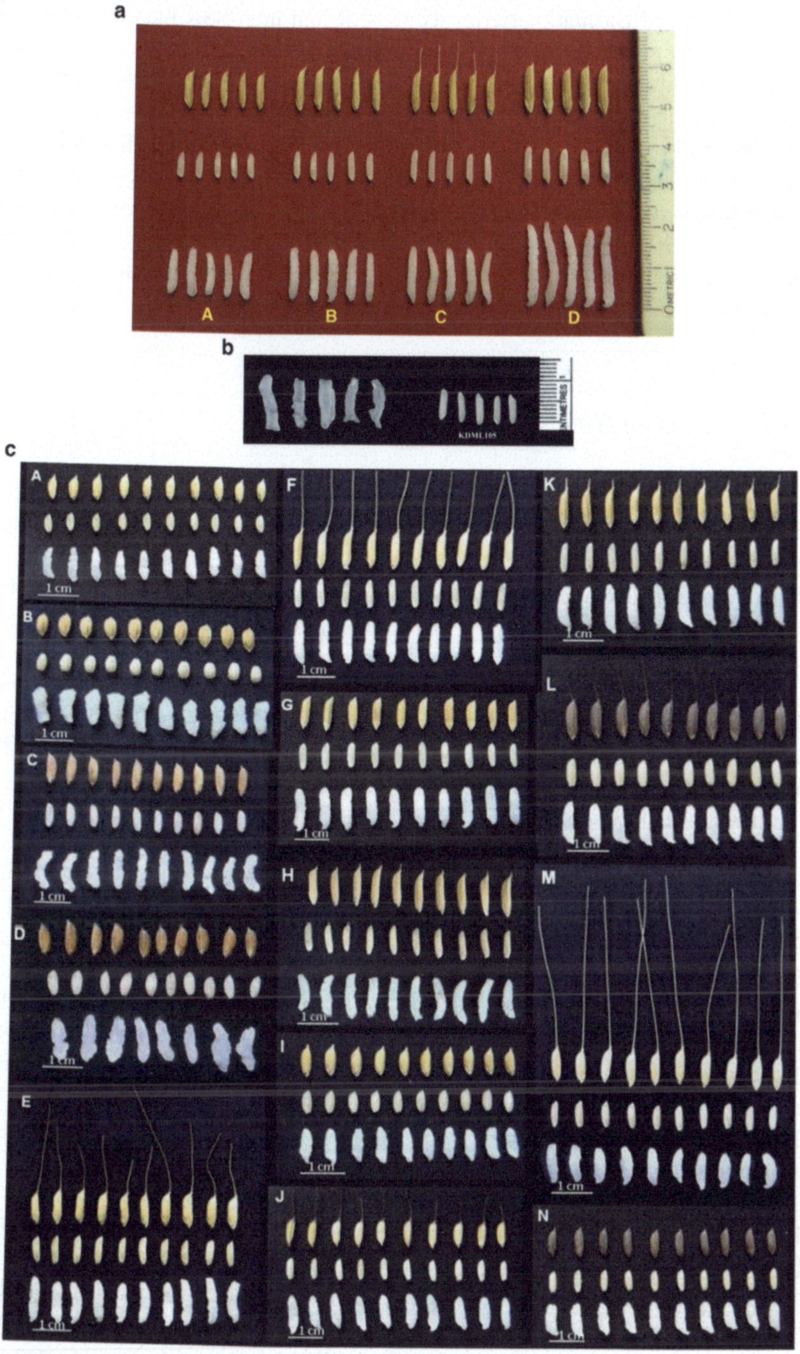

Fig. 11.4 Grain morphological diversity in aromatic rice. (**a**): Grain and cooking quality of traditional and improved Basmati rice varieties: (A) Basmati 370, (B) Taraori Basmati, (C) Pusa Basmati 1, and (D) Pusa Basmati 1121; (**b**): Kernel elongation in Jasmine rice variety KDML105 of Thailand; (**c**): Grain and cooked kernel elongation of some medium- and short-grained rice

from the improved Basmati varieties using SSR and inter-SSR markers (Nagaraju et al. 2002). The genetic diversity within traditional cultivars was lower than that of the improved cultivars. However, this study included only six traditional Basmati varieties. Genetic diversity studies have showed that most of the Basmati cultivars were genetically distinct from the traditional non-Basmati aromatic cultivars of India (Aggarwal et al. 2002; Nagaraju et al. 2002; Jain et al. 2004). A genetic diversity analysis of 31 Basmati cultivars using 30 SSR markers showed a similarity coefficient of ~90% and grouped in close association with *japonica*. On the other hand, the traditional non-Basmati cultivars were found to be closer to *indica* (Jain et al. 2004).

11.3.2 Jasmine Rice

Jasmine rice (Thai: *Khao Hom Mali*) is a long and fine-grain aromatic rice produced primarily in Thailand and in other neighboring countries such as Cambodia, Laos, and Vietnam. Its fragrance has the resemblance with jasmine (*Pandanus amaryllifolius*) and popcorn. The name "jasmine" sometimes refers to the color of the kernels, which is as white as jasmine flower. In Thailand more than 600 aromatic rice landraces are grown in diverse ecologies ranging from deep-water to uplands. The official Thai jasmine rice *Hom Mali* varieties are widely distributed throughout Thailand under a variety of names: *Khao Hom, Hom Mali* and *Khao Mali* (Table 11.2). The famous jasmine rice variety, *Khao Dawk Mali* 105 (KDML105), is a pure line selection from traditional *Hom Mali* rice in Chonburi province of Thailand and released for cultivation in 1959 (Vanavichit 2018). The Jasmine rice varieties (KDML105) are soft texture, aromatic, long-slender white rice with low amylose content (18%) and a low degree of gelatinization temperature (Fig. 11.4b; Table 11.3). When grown in its optimal environment, the physical appearance of polished kernels is pearly, clear, very long, and slender.

Several improved jasmine varieties were developed from KDML105 through mutation breeding (RD6 and RD15), conventional breeding (RD33), and marker-assisted backcross breeding (RD51, RD73, HM80, HM811, HM812, HM821, HM822, and HM84). In Thailand, KDML 105 rice is considered a vital crop for domestic consumption and a primary export commodity (Chitrakorn 2003). The prominent jasmine rice varieties from Cambodia include *Phka rumduol, phka romeat*, and *phka rumdeng*.

The level of genetic diversity within jasmine rice is less explored. Earlier study with isozyme markers categorized jasmine rice varieties from Thailand, Cambodia, and Vietnam as *indica* (Group I) (Glaszmann 1987). Inherently, jasmine rice varieties including KDML105 was tall, droopy, photoperiod sensitive, and susceptible to insects and diseases. Over the last 20 years, concerted breeding programs were undertaken to develop many improved high-yielding, high-quality jasmine rice varieties with multiple stress tolerance (Vanavichit 2018).

Table 11.2 Popular varieties of Basmati, Jasmine, and Sadri aromatic rices

Aromatic rice groups	Popular varieties	Reference
Basmati (India and Pakistan)	Traditional: Basmati 217, Basmati 370, Taraori Basmati, Type-3 (Dehradun Basmati), Ranbir Basmati and Pakistani Basmati Improved: Pusa Basmati 1, Punjab Basmati 1, Haryana Basmati 1, Mahi Sugandha, Pusa Sugandha 3, Kasturi and Pusa Basmati 1121	Nagaraju et al. (2002), Pachauri et al. (2010)
Jasmine (Thailand)	Traditional: Hom Mali, Hom Klong Luang, Hom Supanburi, Niaw San Pahtawng, Phka rumduol, phka romeat and phka rumdeng Improved: Khao Dawk Mali 105 (KDML 105), RD6, RD15, RD33, RD51, RD73, HM80, HM84, HM811, HM812, and HM821	Sarkarung et al. (2000), Vanavichit et al. (2018)
Sadri (Iran)	Domsiah, Gerdeh, Hassani, Hassan Sarai, Mehr, Mirza, Anbar-boo, Mosa Tarom, Sadri, Salari	Shobha Rani (1998)

Table 11.3 Grain and cooking quality parameters of Basmati and other aromatic rice varieties

Variety	Milling %	Head rice recovery %	Kernel length (mm)	Kernel breadth (mm)	Cooked kernel length	Kernel elongation ratio after cooking
Traditional Basmati[a]						
Basmati 370	72.5	53.0	6.89	1.85	13.40	1.94
Taraori Basmati	69.0	49.9	7.15	1.78	13.97	1.95
Improved Basmati[a]						
Pusa Basmati 1	67.0	48.5	7.38	1.80	14.75	2.00
Pusa Basmati 1121	70.5	54.5	8.00	1.90	21.50	2.69
Non-Basmati aromatic						
Jasmine (KDML105)[b,c,d]	–	57.84	7.50	1.98	9.91	1.39
Sadri (Domsiah)[b]	–	63.0	6.8	1.90	14.20	2.10
Della AR[e]	–	–	7.11	2.08	9.68	1.36
Short- and medium-grain aromatic						
Kataribhog[f]	70.4	–	3.79	1.75	2.99	1.71
Randhuni Pagol[f]	70.0	–	4.39	1.70	2.57	1.51
Badshabhog[b]	72.0	67.0	4.53	1.74	7.4	1.86
Kalanamak[b]	73.0	64.0	5.47	2.10	11.3	2.00
Ambemohar ajra[f]	–	–	4.09	1.78	7.40	1.81
Kate Chinoor[f]	–	–	6.40	1.60	10.76	1.67
Velchi[f]	–	–	5.63	2.22	8.42	1.50

[abcdeff]Source: Singh et al. (2018), Singh et al. (2000a), Vanavichit et al. (2018), Yadav and Jindal (2007), Sharp (1986), Shahidullah et al. (2009), Nadaf et al. (2016)

11.3.3 Sadri Rice

The *sadri* rices of Iran look like Basmati and are genetically categorized in the same group as Basmati (Glaszmann 1987; Garris et al. 2005). These rices have a similar kernel elongation after cooking but were not as aromatic as Basmati. Overall, *sadri* varieties are categorized by tall stature (130–135 cm), droopy leaves, long-slender grains (kernel length >7 mm), mild to strong aroma, intermediate amylose content (~22%), good kernel elongation ratio, and overall superior cooking and eating qualities (Table 11.2) (Shobha Rani 1998; Nematzadeh et al. 2000). These varieties are poor yielder (2.5–4.0 tones/ha) and susceptible to pests and diseases (Nematzadeh et al. 2000). In Iranian markets, *sadri* varieties fetch premium price due to high-quality aromatic grains. The notable cultivars of *sadri* group are mentioned in Table 11.3.

11.3.4 Short- and Medium-Grain Aromatic Rice of India

Other than Basmati, a large number of medium- and short-grain aromatic rice cultivars suited to local agroclimatic conditions and cultural preferences are grown in different parts of India. These cultivars are maintained under different names often derived from the main feature of the cultivar, quality, and originating locality/ethnic group (Roy et al. 2014).

Almost every Indian state has its own set of aromatic cultivars which are cultivated for their special quality and use in social ceremonies. The examples of various traditional non-Basmati cultivars grown in India are listed in different literatures (Singh et al. 2000b; Pachauri et al. 2010; Nadaf et al. 2016; Verma et al. 2019). Cultivation of these traditional cultivars is decreasing day by day due to farmers' preference for high-yielding improved varieties, inherent low-yield potential, lodging, and disease and pest susceptibility.

The non-Basmati traditional aromatic rices have one or more of the Basmati characteristics, for example, similar grain shape (length/breadth ration ≥3.0) or sometimes similar cooked kernel elongation rate, but not all of them, especially they have small and medium kernels. These non-Basmati aromatic cultivars have unique aroma, and cooking and eating qualities are consumed locally as a delicacy. Distinction between two groups is also exists in terms of their specific area of adaptation.

As compared to Basmati, the available diversity of non-Basmati aromatic rices is underexploited. The diversity in aromatic rice germplasm in northeastern parts of India was earlier reported by Roy et al. (2015). The traditional aromatic rice cultivars of Assam are locally known as *joha*. It is a special class under *sali* or winter rice. The *joha* rice cultivars are popular for its unique aroma, good cooking qualities, and excellent palatability (Das et al. 2010). The aromatic rice cultivars are locally known as *chakhao* (*chak* = rice and *hao* = delicious) in the state of Manipur.

Considerable diversity exists within the *chakhao* germplasm (Roy et al. 2014). The aromatic rice in Mizoram is known as *tai* rice and is mostly cultivated under wetland conditions.

A substantial number of aromatic rice cultivars, namely, Krishna Bhog, Brimphul, Kalanunia, and local Basmati, are grown in the states of Sikkim and Tripura. Population genetics study in a collection of 107 traditional aromatic rice landraces from the states like Arunachal Pradesh, Assam, Manipur, Mizoram, Nagaland, and Sikkim showed the existence of three genetically distinct clusters. The first cluster included accessions of *joha* rice from Assam, *tai* rice from Mizoram, and aromatic landraces from the state of Sikkim. The *chakhao* landraces of Manipur formed a second cluster and those from Nagaland formed the third one (Roy et al. 2015).

In the Southern states of India such as Maharashtra and Karnataka, considerable number of traditional aromatic rice cultivars are grown. Systematic collection and characterization these aromatic rice cultivars have been done recently (Mathure et al. 2011). Notable aromatic landraces from this region are Ambemohar, Belgaum basmati, Chimansal, Chinoor, Girga, Kala bhat, Kali kumud, Kothmirsal, Jiri, Kalsal, and Velchi. Based on a study of grain size, most of the non-Basmati aromatic cultivars had short to medium grain size and medium grain shape (Fig. 11.4c). Recently, Nadaf et al. (2016) published detailed report on the evaluation of these rice germplasm for various grain quality parameters. Some of the short-grain aromatic varieties are known for their superior taste, aroma, special quality attributes, and adaptation to stress-prone ecologies where long-grain aromatic rice could not be grown. However, the market potentiality of these rice cultivars is very less explored. Efforts are needed to identify, improve, and promote cultivation of such short-grained aromatic rice varieties.

11.4 Biochemistry of Aroma in Rice

11.4.1 Rice Aroma Chemistry

A large number of volatile and semi-volatile compounds contribute to the aroma and flavor of rice. More than 200 volatile compounds were observed in rice; however, only a few of them have been identified as affecting the aroma and flavor of cooked rice. Determining which volatile compounds are responsible for the perceived aroma of different types of aromatic rice is difficult. The perceived aroma/flavor is not only strictly additive but may result from interactions of several compounds (Buttery et al. 1988; Jezussek et al. 2002).

Among the numerous volatiles, 2-acetyl-1-pyrroline (2-AP), responsible for characteristic "popcorn" aroma, has been identified as the principal compound that imparts natural fragrance of rice (Buttery et al. 1982). Several researchers have calculated and compared the aroma values (AV) of different rice cultivars to determine which of the numerous volatile compounds are candidates as important contributors to aroma and flavor. Based on this approach, Buttery et al. (1988) reported that aldehydes (E)-2-nonenal and (E,E)-2,4-decadienal had the lowest odor threshold and were considered to likely contribute to the rice aroma considering their amounts in rice grains. Other aldehydes with relatively low thresholds that are also likely to contribute were (E)-2-decenal, octanal, nonanal, and decanal. Based on AV, hexanal (grassy flavor) and 2-pentylfuran (beany flavor) were identified as principal compounds contributing more to flavor change in milled rice early in storage, while 2-nonenal (rancid flavor) and octanal (fatty flavor) contributed more to the overall flavor of milled rice during long-term storage (Lam and Proctor 2003).

Comparative profiling of volatile compounds in aromatic and nonaromatic rice cultivars has been done by many researchers to understand the role of different compounds in producing perceived aroma/flavor. Higher concentration of pentadecan-2-one, hexanol, and 2-pentylfuran was recorded in Basmati rice than nonaromatic cultivar (Hussain et al. 1987). Petrov et al. (1996) reported nine compounds to discriminate aromatic and nonaromatic rice: pentanol, hexanol, 2-AP, (E)-hept-2-enal, benzaldehyde, octanal, pentadecan-2-one, 6, 10, 14-trimethylpentadecan- 2-one, and hexadecanol. More than 70 volatiles were identified to discriminating fragrant and nonfragrant rice where alkanals, alk-2-enals, alka(E)- 2,4-dienals, 2-AP, 2-pentylfuran, and 2-phenylethanol were the major compounds (Widjaja et al. 1996). Tava and Bochi (1999) also reported that the only differences between Basmati and Italian aromatic cultivar B5-3 were in concentration of 2-AP and lipid oxidation products. To note that, these studies were conducted on a small set of germplasm, and the difference in volatile constituents was influenced by growing/handling differences. For a stable result, aroma comparison studies need to be conducted on aromatic and nonaromatic germplasm grown under identical environments and similarly processed during postharvest. The list of olfactory-active volatile compounds that may affect aroma and flavor of rice in rice is given in Table 11.4 (Champagne 2008).

A total of 41 odor-active compounds were identified from cooked brown rice of three varieties Improved Malagkit Sungsong, Basmati 370, and Khaskhani, and an *indica* variety, out of which 2-AP was identified as the key aroma component (Jezussek et al. 2002). 2-Amino acetophenone (medicinal, phenolic) and 3-hydroxy-4,5-dimethyl-2(5H)-furanone (Sotolon; seasoning-like) were found in higher concentrations in three varieties except the *indica* one.

The aroma chemistry of six distinctly different rice flavor types (basmati, jasmine, two Korean *japonica* cultivars, black rice, and a nonaromatic rice) was analyzed, and a total of 36 odorants from cooked samples were characterized (Yang et al. 2008). Thirteen odor-active compounds [2-AP, hexanal, (E)-2-nonenal, octanal, heptanal, nonanal, 1-octen-3-ol, (E)-2-octenal, (E,E)-2,4-nonadienal, 2-heptanone, (E,E)-2,4-decadienal, decanal, and guaiacol] among the six flavor types were the primary compounds explaining the differences in aroma. 2-AP had the lowest odor threshold (0.02 ng/L) followed by 11 aldehydes (0.09–3.1 ng/L), guaiacol (1.5 ng/L), and 1-octen-3-ol (2.7 ng/L). This study used multivariate analysis to report that six flavor types could be separated and characterized using these compounds.

Table 11.4 List of olfactory-active compounds identified from brown and/or milled rice

Volatile compound	Aroma type
Butan-2,3-dione	Buttery
n-Hexanal	Green, grassy
(Z)-Hex-3-enal	Green, leaf-like
Octanal	Fatty, citrus-like
Oct-1-en-3-one	Mushroom-like
2-Methyl-3-furanthiol	Meaty, sulfurous
2-Acetyl-1-pyrroline	Popcorn
Non-1-en-3-one	Mushroom-like
2-Metthoxy-3,5-dimethylpyrazine	Earthy
Acetic acid	Sour
Methional	Cooked potato
Decanal	Soapy
(Z)-Non-2-enal	Fatty, green
2-Isobutyl-3-methoxypyrazine	Earthy, green bell pepper
(E)-Non-2-enal	Fatty, tallow
(E,Z)-Nona-2,6-dienal	Cucumber-like
(Z)-Dec-2-enal	Fatty, green
Butanoic acid	Sweaty, rancid
Phenylacetaldehyde	Honey-like
2-and 3-Methylbutanoic acid	Cheese-like, sweaty
(E,E)-Nona-2,4-dienal	Fatty
Pentanoic acid	Sweaty
(E,Z)-Deca-2,4-dienal	Fatty, green
(E,E)-Deca-2,4-dienal	Fatty
Hexanoic acid	Sweaty
2-Methoxyphenol	Smoky
2-Phenylethanol	Honey-like
γ-Octalactone	Coconut-like
β-Ionone	Violet-like
4,5-Epoxy-(E)-dec-2-enal	Metallic
4-Hydroxy-2,5-dimethyl-3(2H)-furanone	Caramel-like
γ-Nonalacton	Coconut-like
4-Methylphenol	Phenolic
3-Methylphenol	Spicy
Bis-(2-methyl-3-furyl)-disulfide	Meaty
3-Hydroxy 4,5-dimethyl-2(5H)-furanone	Seasoning-like
4-Vinyl-2-methoxyphenol	Spicy, clove-like
2-Amino acetophenone	Medicinal, phenolic
3a,4,5,7a-Tetrahydro-3,6-dimethyl-2(3H)-benzofuranone	Sweaty, spicy
5-Ethyl-3-hydroxy-4-methyl-2(5H)-furanone	Seasoning-like
4-Vinylphenol	Phenolic
Indole	Sweet, burnt
3-Methylindole	Mothball-like
Phenylacetic acid	Honey-like
Vanillin	Vanilla-like
2-Pentylfuran	Beany

Source: Champagne (2008)

Laguerre et al. (2007) estimated the volatile compounds differentiating aroma of 29 aromatic and 32 nonaromatic rice cultivars using mass spectrometry with no chromatography. Only 2-AP was identified as the differentiating compound between aromatic and nonaromatic rices. No other compounds had odor threshold low enough to contribute to rice aroma in these 61 rice cultivars. A review of the published reports indicating that the role of different volatile compounds, other than 2-AP, in creating unique aroma/flavor types in diverse rice varieties is still to resolve.

11.4.2 The Aroma Compound 2-AP

Fragrant rice varieties have a pleasant aroma mainly due to the accumulation of the key odor-active component 2-acetyl-1-pyrroline (2-AP). 2-AP [IUPAC: 5-acetyl-3,4-dihydro-2H-pyrrole and 1-(3,4-dihydro-2H-pyrrol-5-yl)-ethanone] is regarded as the basic aroma compound in rice because it was isolated and characterized from the basic fraction of a steam distillation extract of aromatic rice (Buttery et al. 1982). 2-AP is reported in all parts of the rice plant, except the roots (Yoshihashi et al. 2002). It is also found in nonaromatic varieties, but in much lower concentration. 2-AP has been reported as the aroma compound in several other species: *Pandanus amaryllifolius* (Thimmaraju et al. 2005), popcorn (Schieberle 1995), mung bean (Brahmachary and Ghosh 2002), soybean, sorghum, and *Bacillus cereus* (Romanczyk et al. 1995).

The biosynthetic pathway of 2-AP is less explored; however, amino acids including proline, ornithine, and glutamate are the most probable precursors for 2-AP biosynthesis. For example, in *Bacillus cereus* strain 35 cultured on agar containing proline, ornithine, and glutamate produced significant amount of 2-AP (Romanczyk et al. 1995). L-proline was found to enhance the in vitro synthesis of 2-AP in callus culture of *Pandanus amaryllifolius* (Thimmaraju et al. 2005). Proline, glutamic acid, and ornithine were identified as the precursors of 2-AP in the seedlings and callus of aromatic cultivar KDML105 (Yoshihashi et al. 2002). Vanavichit et al. (2005) proposed that the 2-AP in rice is synthesized via the polyamine pathway, and the immediate precursor of 2-AP was 1-pyrroline, which is formed from 4-aminobutyraldehyde (AB-ald), an immediate precursor of γ-aminobutyric acid (GABA).

In higher plants, ornithine, proline, and glutamate can be converted to a common metabolite, Δ^1-pyrroline-5-carboxylic acid, via three distinct enzymes: ornithine aminotransferase (OAT), proline dehydrogenase (PRODH), and Δ^1-pyrolline-5-carboxylic acid synthetase (P5CS). Based on a model system using *Bacillus subtilis* ssp. *natto* to investigate the formation of 2-acetyl-1-pyrroline via methylglyoxal and Δ^1-pyrolline-5-carboxylic acid, Huang et al. (2007) reported that Δ^1-pyrolline-5-carboxylic acid could react with methylglyoxal to form 2-acetyl-1-pyrroline. Huang et al. (2008) studied the formation of putative 2-AP precursors, Δ^1-pyrroline-5-carboxylic acid and methylglyoxal, in aromatic and nonaromatic rice cultivars. They reported the formation of 2-AP from Δ^1-pyrroline-5-carboxylic acid by upregulation of *P5CS1* and *P5CS2* genes. This study also showed that the nitrogen atom of 2-AP was derived from glutamic acid based on a tracer experiment with ^{15}N-labeled glutamic acid. Similarly, Yoshihashi (2002) used ^{15}N-labelled proline to conclude that the pyrroline ring of 2-AP originated from proline. Moreover, this study also showed that the acetyl group of 2-AP did not originate from proline.

Other model studies with isotopically labelled proposed that the acetyl group of 2-AP may be contributed by 2-oxopropanal under thermal conditions (Schieberle 1995; Hofmann and Schieberle 1998). The compound 2-oxopropanal could react with ornithine and proline under thermal conditions and produce 2-AP and 6-ATHP (6-acetyl-1,2,3,4-tetrahydropyridine), respectively. 6-ATHP is a tautomer of 2-AP and both have organoleptic properties (Yoshihashi 2002). However, 6-ATHP was not reported in aromatic rice flavor. Huang et al. (2008) also proposed 2-oxopropanal as the precursor of the acetyl group in aromatic rice, even in nonthermal conditions.

In organisms, 2-oxopropanal is common because it is the intermediate of glycolysis; the reaction of AB-ald always produces 6-ATHP. 2-oxopropanal is also known for its high cytotoxicity and high reactivity and also as a potent glycation agent to DNA and proteins (Vanavichit and Yoshihashi 2010). 2-Oxopropanal is known for its high cytotoxicity and high reactivity and also as the most important glycation agent to DNA and proteins. The pathway of 2-AP biosynthesis in rice under genetic control of *BADH2* is discussed in later section.

11.5 Genetic and Molecular Basis of Aroma in Rice

11.5.1 Gene for Aroma in Rice

The aroma trait in rice is believed to have arisen in domesticated rice varieties from gene mutation(s) during evolution or as outcome of domestication process because this trait is not found in wild gene pool. The genetic control of aroma in rice is complex and many studies have reported that inheritance of aroma is controlled by one or two or three dominant or recessive genes or by QTLs (Table 11.5).

The conflicting reports on the genetics of aroma in rice might have resulted due to (i) different phenotyping methods used and detection of aroma from both leaves and grains, (ii) segregation distortion (Lorieux et al. 1996), and (iii) use of diverse genotypes used under study as the nature of inheritance of aroma is cross or genotype specific. Mapping of grain aroma was first reported in 1992 by Ahn et al. (1992) by using restriction fragment length polymorphism (RFLP) markers in a pair of near isogenic lines (NILs) of rice cv. Lemont, one carrying the scent allele from rice cv. Della (Jasmine-derived aromatic variety). This study reported that the gene for fragrance (*fgr*) is linked to RFLP clone RG28 on the long arm of chromosome 8 at a distance of 4.5 cM.

Subsequently many studies on molecular mapping of rice grain aroma have been reported (Table 11.6), and most of the studies have supported the results of Ahn et al. (1992). Lorieux et al. (1996) tagged *fgr* as a major and recessive QTL in the

Table 11.5 A summary of studies on the genetics of aroma in rice

Cross used	No of genes/ QTLs	Gene action	Method of phenotyping	Reference
–	3	Dominant, complementary	Leaf analysis in water	Nagaraju et al. (1975)
Pokura (aromatic) x Pokura (nonaromatic)	1	Recessive	Leaf analysis in KOH	Reddy and Reddy (1987)
T-141 × K-44–1	4	Dominant, complementary	Chewing of grains	Dhulappanavar (1976)
–	1	Recessive	Leaf analysis in KOH	Sood and Siddiq (1978)
Pankaj × Kalabhat (aromatic)	2	Dominant, complementary	Leaf analysis	Tripathi and Rao (1979)
Sons × basmati 370 (aromatic); Jaya × HR 47 (aromatic); Jaya × HR 59 (aromatic); Jaya × HR 22 (aromatic)	3	Dominant, complementary	Leaf analysis	Reddy and Sathyanarayanaiah (1980)
Della (aromatic) × Dawn	1	Recessive	Leaf and seed analysis in KOH, chewing	Berner and Hoff (1986)
Brimphul (aromatic) × Koshihikari; Brimphul (aromatic) × Nipponbare	1	Recessive, inhibitory	Chewing	Tsuzuki and Shimokawa (1990)
B8462T3–710 × Lemont NIL (aromatic)	1	Recessive	Chewing and leaf analysis	Ahn et al. (1992)
–	1	Recessive	Leaf analysis in KOH	Ali et al. (1993)
Toro2 × Jasmine 85; Rico1 × Jasmine 85; Gulfmont × PI457917; Amber × Lemont; Rosemont × dragon eyeball 100	1/2	Recessive	Chewing	Pinson (1994)
ADT37 × Pusa Basmati 1 and IR50 × Pusa Basmati 1	2	Recessive	Leaf analysis in KOH	Vivekanandan and Giridharan (1994)
KDML105 (aromatic) × CT9993-5-10-1-M	1	Recessive	Chewing	Tragoonrung et al. (1996)
IR64 × Azucena (aromatic)	3	Polygenic	Chewing and leaf analysis	Lorieux et al. (1996)
Kyeema (aromatic) × Doongara	1	Recessive	Chewing	Garland et al. (2000)
Kyeema (aromatic) × Doongara	1	Recessive	Chewing	Cordeiro et al. (2002)

(continued)

Table 11.5 (continued)

Cross used	No of genes/ QTLs	Gene action	Method of phenotyping	Reference
Basmati 370 (aromatic) × IR36	No clear-cut single gene segregation detected	–	Seed in KOH	Nematzadeh et al. (2004)
IRRI 38 × Jeerigesanna (aromatic); BI 33 × Jeerigesanna (aromatic); Azucena (aromatic) × Moro mutant	1	Recessive	Leaf analysis	Patil et al. (2012)

Modified after Sakthivel et al. (2009)

Table 11.6 Genetic mapping of grain aroma locus in rice

Gene(s)/ QTL(s)	Chromosome	Marker type	Reference
1	8	RFLP	Ahn et al. (1992)
1	8	RFLP	Yano et al. (1992)
1	–	RAPD	Tragoonrung et al. (1996)
1 major and 2 minor	4, 8 (major), 12	RFLP and STS	Lorieux et al. (1996)
1	8	SSR	Garland et al. (2000)
1	8	SSR	Cordeiro et al. (2002)
1	8	SNP	Jin et al. (2003)
			Nematzadeh et al. (2004)
1	8	SSR	Bradbury et al. (2005a)
1	8	SSR, EST	Wanchana et al. (2005)
1	8	SSR	Chen et al. (2006)
1	8	SSR	Li et al. (2006)
3	3, 4, and 8	SSR	Singh et al. (2007)
3	3, 4, and 8	SSR	Amarawathi et al. (2008)
1	8	SSR	Sun et al. (2008)

Modified after Sakthivel et al. (2009)

same region flanked by RG1 and RG28 with the estimated genetic distance of 12 cM in a (IR64 x Azucena) DH (doubled haploid) population where the traditional upland variety Azucena was the donor of aroma. This study demonstrated that 2-AP detected by sensitive methods and gas chromatography (GC) was perfectly correlated and that large quantitative variations were observed among aromatic DH lines. Two minor QTLs were also identified in this study which may affect the strength of aroma in rice grains (Lorieux et al. 1996). Later, development of molecular markers linked to *fgr* locus was undertaken to facilitate selection of the aroma trait in rice breeding (Garland et al. 2000; Cordeiro et al. 2002).

The availability of genome sequences and high-density molecular maps in rice have allowed fine mapping and positional cloning of aroma gene in rice. By using a F_2 mapping population of Kyeema (also known as Della, an aromatic long-grain

Australian variety) x Gulfmont (nonaromatic), Bradbury et al. (2005a) mapped grain aroma on chromosome 8 between SSR markers RM515 and SSRJ07. By sequencing 17 genes in the *fgr* region, they suggested that a gene encoding a putative *Betaine aldehyde dehydrogenase homologue 2* (*BADH2*) is most likely to be the *fgr* gene, due to its sequence divergence between fragrant and nonfragrant rice varieties. It was shown that the *BADH2* allele in fragrant genotypes has a mutation, eight base-pair (8-bp) deletion in exon 7, introducing a stop codon upstream of key amino acid sequences conserved in other BADHs (Fig. 11.5).

During screening of 14 fragrant and 64 nonfragrant rice varieties, the 8-bp mutation on exon 7 was perfectly matched with aroma. Therefore, the loss of function of the *fgr* gene product resulted in the development of aroma which corresponded well with the recessive nature of rice aroma.

Fine mapping and map-based cloning of aroma gene (named as *Os2AP*) were also reported in a parallel study with advanced mapping population of the jasmine rice (KDML105) cross which also identified same mutations (Vanavichit et al. 2004). In silico physical mapping within the rice grain aroma QTL located on a 4.5-cM region of the rice chromosome 8 was undertaken by comparing genomic BAC-end sequences of Nipponbare in an advanced mapping population of KDML105 x CT9993. The fine location of aroma gene was narrowed down to 170 kb by the in silico approach (Wanchana et al. 2005).

The *fgr* locus was fine-mapped and further narrowed down to 69 kb using 2891 individuals from F_2 and BC_1 generations of four different aromatic x nonaromatic crosses (Chen et al. 2006). Sequence analysis of the *fgr* region revealed three candidate genes encoding respective eukaryotic-type carbonic anhydrase, 3-methylcrotonyl-CoA carboxylase beta chain, and betaine aldehyde dehydrogenase (BADH2). *BADH2* also identified as the candidate locus for aroma in rice cv. Azucena (Bourgis et al. 2008). Therefore, the functional mutation including 8-bp deletion and 3 single nucleotide polymorphisms (SNPs) in the seventh exon of *Badh2* have led to a premature stop codon and putatively truncated BADH2 protein (252 amino acids in aromatic rice) that result in abrogation of the function of BADH2 enzyme and consequently accumulate substrate 2-AP in aromatic genotypes. In contrast, the functional *BADH2* codes for 503 amino acids mature protein in nonaromatic rices. Surveys of diverse fragrant germplasm supported the association of *badh2* with fragrance (Bourgis et al. 2008; Fitzgerald et al. 2008; Shi et al. 2008). In addition, transformation of a fragrant variety with the dominant nonfragrant *BADH2* allele has been shown to abolish fragrance, confirming that *BADH2* is the major genetic determinant of fragrance in rice (Chen et al. 2008).

11.5.2 Evolution of badh2 *and Its Haplotypic Variations in Rice*

After the identification of the gene for grain aroma in rice, major emphasis was given to understand the allelic variations and origin of *badh2* by utilizing diverse rice genepool. Kovach et al. (2009) reported the origin of *badh2* in *O. sativa* by

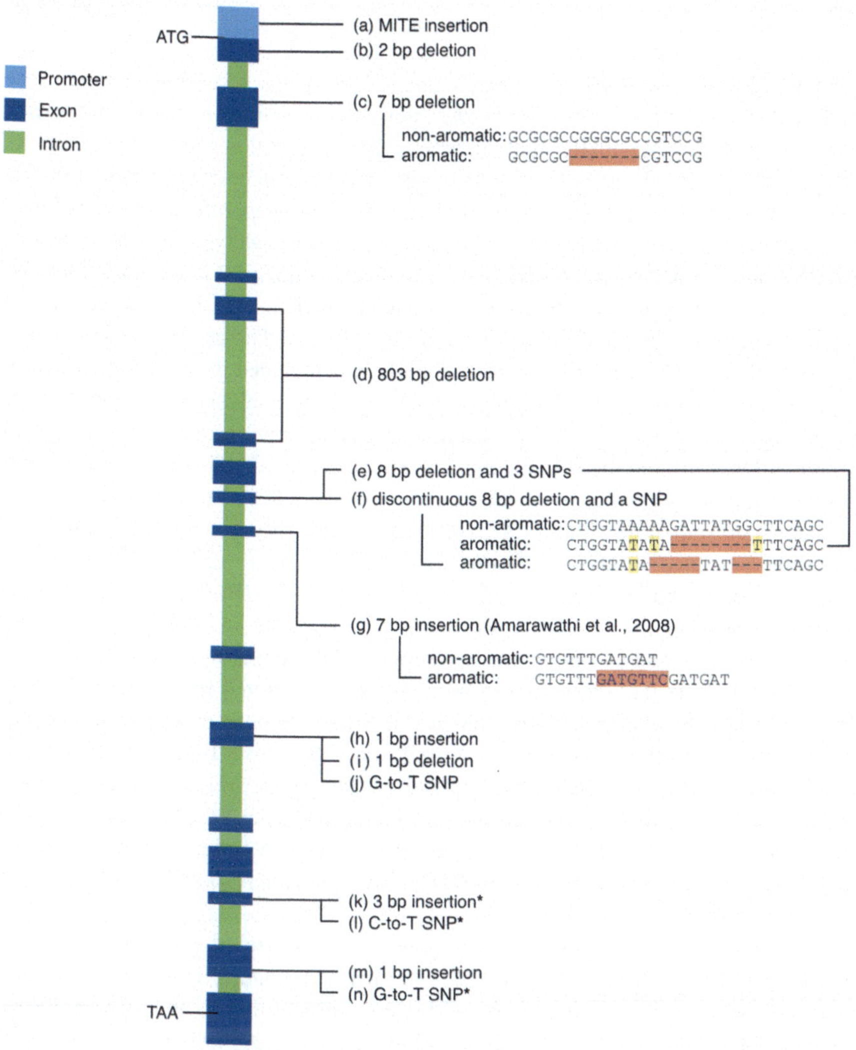

Fig. 11.5 The structure of *Betaine aldehyde dehydrogenase 2* (*Badh2*) locus. The functional *Badh2* gene consists of 15 exons (blue boxes) with its start codon in exon 1 and stop codon in exon 15. Multiple null *badh2* alleles in aromatic rices have been reported possessing various mutations in exons and introns. Some of mutations in different exons result in premature stop codon and putatively truncated BADH2 protein which leads to loss of function of BADH2 enzyme and consequently accumulation of substrate 2-AP in aromatic cultivars. In the figure only exonic mutations are shown: (a) a miniature interspersed transposable element (MITE) insertion in promoter region (Bourgis et al. 2008), (b) 2-bp deletion in exon 1 (Kovach et al. 2009), (c) a 7-bp deletion in the second exon (Shi et al. 2008), (d) 803-bp deletion between the exons 4 and 5 (Shao et al. 2011), (e) an 8-bp deletion (highlighted in red) and 3 SNPs (highlighted in yellow) in exon 7 of fragrant allele (Bradbury et al. 2005a), (f) a discontinuous 8-bp deletion (highlighted in red) and a SNP (highlighted in yellow) of fragrant allele (Amarawathi et al. 2008), (g) an 7-bp insertion (highlighted in red) in exon 8 of fragrant allele (Amarawathi et al. 2008), (h) T insertion in exon 10 (Kovach et al. 2009), (i) T deletion in exon 10 (Kovach et al. 2009), (j) G-to-T substitution in exon 10 (Kovach et al. 2009), (k) 3-bp (TAT) insertion (Kovach et al. 2009) and (l) C-to-T substitution in exon 13 (Kovach et al. 2009), (m) G insertion in exon 14 (Kovach et al. 2009), and (n) G-to-T substitution in exon 14 (Kovach et al. 2009). ∗The polymorphisms do not alter the function of the gene

examining 280 accessions of wild rice and 176 *O. sativa* accessions belonging to different rice genetic groups. The *badh2* allele was absent from all wild genotypes except one accession with the allele in heterozygous condition suggesting introgression from fragrant *O. sativa*. Within the cultivated rice, highest frequency of *badh2* was observed in Group V (*aromatic*) and the lowest in *temperate japonica* and *aus*.

To trace the origin of *badh2* in *O. sativa,* approximately 5 kb region across *BADH2* and 5.3 Mb region flanking the gene were sequenced from more than 240 accessions. Eight gene haplotypes were obtained which largely differentiated in two distinct clusters: *japonica* and *indica* gene haplotypes. The accessions (including *indica* and Group V) carrying the *badh2* allele was included in the *japonica* haplotype. The extended (*badh2* flanking region) haplotype analysis also supported the gene haplotypes, with all *badh2*-containing accessions having an extended region of *japonica* DNA around the *BADH2* gene. It was proposed that the *badh2* allele has a single origin in a *japonica*-like background and the allele was transferred via introgression into *indica* (Kovach et al. 2009). A similar theory was earlier proposed based on sequence variation in *badh2* locus of 81 varieties extracted from a core collection representative of Asian rice isozyme diversity (Bourgis et al. 2008). A separate study on the haplotype diversity across *badh2* in 144 rice accessions (*indica, japonica,* and *O. rufipogon*) also supported these findings (Shao et al. 2013).

Many aromatic rice cultivars did not carry the functional mutation (8-bp deletion) in exon 7, for example, traditional non-Basmati aromatic cultivars, Tarunbhog, Ganjeikalli, Bishnubhog, Bansphool A, and Adamchini from India (Sakthivel et al. 2009), and mildly aromatic cultivar 'Masino' (Roy et al. 2012). Fifteen aromatic varieties from nine different countries were identified as devoid of the *fgr* allele carrying the functional mutation (Fitzgerald et al. 2008). All these reports indicated that there are additional mutations in *BADH2* that may be responsible for enhanced 2-AP synthesis.

Availability of diverse aromatic rice gene pool has offered excellent opportunities for exploring allelic variants at *BADH2* locus. The *badh2* gene was sequenced in 26 aromatic rices originated from wide geographical region and lacking the *fgr* allele to detect additional functional mutations within the gene (Kovach et al. 2009). Their analysis revealed eight nonsynonymous polymorphisms, four of which (7-bp deletion in exon 2 in Chinese cultivar Hsiang-Keng-Nuo, 2-bp deletion in first exon in a cultivar from Madagascar, T insertion in tenth exon in cultivar Pare Baine Pulut from Indonesia, G-to-T substitution in tenth exon of Indian rice cultivar Vishunparag, and T deletion in 10 exon in a cultivar from Malaysia) were frameshift-inducing indels and one (G insertion in 14th exon six cultivars from the Indian subcontinent) of which was a SNP creating a premature stop codon, all putatively resulting in a truncated BADH2 protein (Fig. 11.5). These mutations resulted in truncation of the BADH2 protein before critical residues that form the catalytic and/or substrate binding domains and predicted to eliminate an oligomerization domain of the protein. Furthermore, the *BADH2* sequence polymorphisms have strong geographic associations (Kovach et al. 2009). Different authors have also reported functional variations within the *BADH2* gene (Fig. 11.5). The 7-bp deletion in exon 2 was earlier detected as a frameshift mutation in 12 Chinese aromatic accessions which did not possess the 8-bp deletion in exon 7 (Shi et al. 2008).

Absence of a transposable element, classified as a MITE (miniature interspersed transposable element), in the promotor region of the *BADH2* in aromatic cultivars from different rice subpopulations (Basmati, *japonica*, and *indica* KDM105) was shown to be associated with aroma (Bourgis et al. 2008). Several other nucleotide polymorphisms detected in exons and introns of *BADH2* have not yet been functionally validated.

11.5.3 Involvement of Other Genes in the Production of Rice Aroma

Although the *BADH2* gene on rice chromosome 8 has been identified as the gene for aroma in majority of fragrant rice cultivars and its role in 2-AP biosynthesis has been elucidated, it is likely that several other genes may influence the aroma levels in rice (Fitzgerald et al. 2008). There are other QTLs for aroma mapped on rice chromosomes 3, 4, and 12 (Table 11.6). The QTL (*aro4-1*) on chromosome 4 of Basmati rice reported in Amarawathi et al. (2008) was same QTL detected in Lorieux et al. (1996). However, this QTL had low LOD score in both QTL studies. Interestingly, this minor aroma QTL on chromosome 4 is colocalized with the *betaine aldehyde dehydrogenase 1* (*BADH1*) gene. This gene has similar molecular function as *BADH2* on rice chromosome 8. It was suggested that mutations that lead to a nonfunctional *BADH1* or a reduction in expressions levels may cause an increase in 2-AP accumulation.

Bradbury et al. (2008) suggested that despite the low affinity of BADH1 enzyme to AB-ald, AB-ald may still be a substrate for BADH1 in vivo and that the presence of BADH1 may diminish the pool of AB ald available to form 2-AP. The similarity of preferred substrates of BADH1 and BADH2 makes it unclear as to why a nonfunctional BADH2 has such a major impact on the fragrance phenotype of rice.

In higher plants, betaine is known as a protective cytoplasmic osmolyte, allowing normal growth of plants under saline and moisture stress environments. Betaine is synthesized via a two-step oxidation of choline, and the second step (from Bet-ald to betaine) is catalyzed by BADH. Two BADH isozymes (BBD1 and BBD2) were reported to induce to higher levels by salt, drought, and abscisic acid treatments in barley (Nakamura et al. 2001). The rice BADH2 protein shows high betaine aldehyde dehydrogenase activity, suggesting that *BADH2* also may play a key role in osmoregulation in nonaromatic rice. However, the loss of function of *BADH2* in aromatic cultivars is supposed to be compensated by other *BADH* genes to allow normal plant grown under stress environments. *BADH1* accounts for ~30% of the total *BADH* gene expression within nonaromatic rice leaves and the expression is increased under salt stress (Fitzgerald et al. 2008).

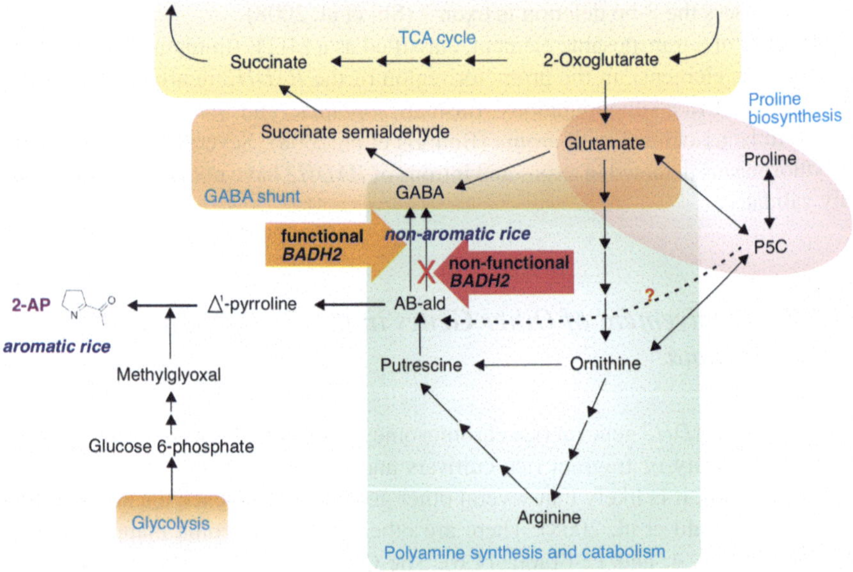

Fig. 11.6 2-AP biosynthesis pathways proposed for aromatic rice. The nonfunctional BADH2 enzyme encoded by the *badh2* allele in aromatic rice could not convert 4-aminobutyric acid (GABA) into AB-ald. As a result, accumulation of AB-ald activates 2-AP biosynthesis in aromatic rice (Bradbury et al. 2008; Chen et al. 2008). The synthesis pathway of 4-aminobutyraldehyde (AB-ald) from Δ^1-pyrolline-5- carboxylic acid is unknown. P5C could react with methylglyoxal to form 2-acetyl-1-pyrroline (Huang et al. 2007). Formation of 2-AP from Δ^1-pyrroline-5-carboxylic acid by upregulation of *P5CS1* and *P5CS2* genes in aromatic rice has been reported (Huang et al. 2008). This pathway is *badh2* independent

The gene haplotypic diversity of *BADH1* in 92 diverse rice accessions has been reported (Singh et al. 2010). Out of three exonic SNPs (all causing non-synonymous substitutions), two were sowed to be associated with aromatic varieties. These two SNPs were included in BADH1 protein haplotype 2 or PH2 and were predicted to cause a reduction in substrate binding capacity of BADH1 enzyme toward AB-ald. It was suggested that PH2 haplotype could be loss-of-function allele of *BADH1* gene with implications in rice aroma similar to the null-*BADH2*. However, further research is required to understand the functions of different *BADH* genes in both tolerance to abiotic stresses and the regulation of 2-AP biosynthesis.

11.5.4 Genetic Control of Biosynthetic Pathways of 2-AP in Rice

The 2-AP biosynthetic pathway in rice became unraveled after the discovery of candidate gene for aroma, and many conceivable pathways have been proposed (Fig. 11.6). Chen et al. (2008) used transgenic approach to determine the molecular basis of 2-AP synthesis in rice. They transformed aromatic rice cv. rice recipient Wuxiangjing (*japonica*) with pCAM-*Badh2* (carrying the wild-type *BADH2*) and two other gene constructs. The 2-AP level was significantly reduced in *Badh2* transgenic lines (1.99 ± 1.03 ng/g) in comparison to the non-transgenic lines (aromatic recipient) (2-AP level: 21.33 ± 7.37 ng/g). It was also shown that the intact full-length BADH2 protein (503 residues) appeared exclusively in nonfragrant transgenic lines. These results indicate that the full-length BADH2 protein encoded by *BADH2* renders rice nonfragrant by inhibiting 2-AP biosynthesis. Compared with the functional *BADH2* allele, the nonfunctional alleles (carrying mutations in exon7 and exon2) had very low transcription levels, as revealed by real-time RT-PCR and RNA gel blot analysis. Therefore, the loss-of-function mutations in *BADH2* severely suppress mRNA transcription.

BADH2 protein was localized immunologically in the cytoplasm with the C-terminal serine-lysine-leucine (SKL) signal peptide specific for targeting to the peroxisomes. It is predicted to catalyze the oxidization of betaine aldehyde, 4-aminobutyraldehyde (AB-ald), and 3-aminopropionaldehyde. It is suggested that *BADH2* is not directly involved in 2-AP biosynthesis, as the absence of BADH2 protein results in aroma. BADH2 enzyme is predicted to involve in a competing pathway in which one of the 2-AP precursors serves as a BADH2 substrate (Bradbury et al. 2005a) or that BADH2 participates in 2-AP catabolism. BADH protein reported to catalyze the oxidation of Bet-ald and the substrates structurally similar to Bet-ald, such as 3-dimethylsul- foniopropionaldehyde, AP-ald, and AB-ald in sugar beet (Trossat et al. 1997). AB-ald is known to be maintained in an equimolar ratio with Δ^1-pyrroline, an immediate 2-AP precursor. Therefore, AB-ald levels are important in regulating the rate of 2-AP biosynthesis. On the other hand, AB-ald can be converted into 4-aminobutyric acid (GABA) by the BADH2 enzyme, thus inhibiting 2-AP biosynthesis. In fragrant rice lacking intact *BADH2*, failure to convert AB-ald into GABA due to the absence of BADH2 enzymatic activity results in AB-ald accumulation, which activates 2-AP biosynthesis.

A similar biosynthetic pathway for 2-AP was suggested in Bradbury et al. (2008). They proposed that the biochemical pathway of 2-AP production in rice starts with proline being catabolized via putrescine into AB-ald. BADH2 showed greatest affinity toward AB-ald suggesting AB-ald is an effective substrate for BADH2. If BADH2 is present and functional, it is able to convert the majority of AB-ald to γ-aminobutyric acid (GABA), but if BADH2 is absent or nonfunctional, the majority of AB-ald are accumulated and form Δ^1-pyrroline, which then acetylated to form 2-AP.

Niu et al. (2008) also showed that RNAi-mediated disrupting of *BADH2* in non-aromatic varieties leads to significant increase in 2-AP production. As described in earlier section, formation of 2-AP from Δ^1-pyrroline-5-carboxylic acid (P5C) by upregulation of P5CS1 (Δ^1-pyrroline-5-carboxylate synthetase1) and P5CS2 genes in fragrant rice varieties was reported (Huang et al. 2008). This pathway did not indicate any direct role proposed for *badh2*. Moreover, the authors mentioned 1-pyrroline as the intermediate to 2-AP formation although the detailed formation pathway from P5C to 1-pyrroline was not then described.

11.6 Perspective and Future Thrust

Among different quality traits, aroma is considered most important. The preference for rice aroma has resulted in strong human selection for this trait during domestication. The 2-acetyle-l-pyrroline (2-AP) has been reported as an important flavor component giving "popcorn"-like aroma of several aromatic varieties. However, the aroma we perceive from cooked or uncooked aromatic rice cultivars is a result of a more than 200 volatile compounds present in specific proportion. A unique aroma specific to certain cultivars might be due to the differential combination of the aroma compounds. However, we have little idea how these compounds differ from one variety to other. Understanding the genetic basis and phenotypic expression of rice aroma is an important objective of the rice breeders.

Rice aroma has been appeared as a complex quantitative trait, even though a major gene controlling the level of principal aroma compound 2-AP has been identified, and its expression is modulated by environmental factors, as well as pre- and postharvest handling of the crop (Champagne 2008). Soil type and growth environments play important roles in the expression of rice aroma gene(s) and concentration of 2-AP in the rice grains. Same rice variety grown in different environments, other than their native area of cultivation, exhibits significant difference in its aroma level and other grain quality parameters. For this reason, GI tag is enforced for Basmati. Similar observations have been recorded in case of black aromatic *chakhao* rice cultivars of Manipur. Hence, in-depth investigations are required to identify all the factors involved in the regulation of expression of rice aroma.

The cultivation of low-yielding high-quality traditional aromatic rices has been reduced significantly after the green revolution. A large number of small- and medium-grained aromatic rices have already been lost and many are at the verge of extinction. Some of the small- and medium-grained aromatic rices possess excellent aroma and other quality traits like taste, kernel elongation after cooking, special uses, etc. The gene banks around the world have conserved sizable aromatic germplasm collections which could be excellent sources for improving quality in high-yielding varieties.

The *badh2* gene seems to explain only the 2-AP accumulation in most of the aromatic cultivars. There is a lack of information regarding the genetic relationship of various other aroma compounds in determining specific fragrance of some aromatic rices. Identification and characterization of other gene(s) in aroma development is required for developing further insights into molecular basis of rice aroma. A further understanding on the aroma of short-grained aromatic rice varieties that have strong aroma even when grown outside the geographical indication of the Basmati varieties is required. Molecular mapping of genes conditioning fragrance in these non-Basmati cultivars will enhance our knowledge on the inheritance of rice aroma.

Molecular breeding strategies have to be developed for more precise and effective transfer of genes aroma and superior grain quality in high-yielding rice varieties. To achieve this, we have to develop precise phenotyping protocols/quantitative assays for the aroma and grain quality parameters. Overall, future research on aromatic rice needs to be targeted on elucidating the molecular basis of the evolution of aroma in rice, determination of additional genetic and biochemical control of aroma in rice grains, and utilization of high-quality traditional small- and medium-grained aromatic landraces in rice breeding.

References

Aggarwal RK, Shenoy VV, Ramadevi J, Rajkumar R, Singh L (2002) Molecular characterization of some Indian Basmati and other elite rice genotypes using fluorescent-AFLP. Theor Appl Genet 105:680–690

Ahn SN, Bollich CN, Tanksley SD (1992) RFLP tagging of a gene for aroma in rice. Theor Appl Genet 84:825–828

Ali SS, Jafri SJH, Khan MG, Butt MA (1993) Inheritance studies for aroma in two aromatic cultivars of Pakistan. Intl Rice Res Notes 18:6

Amarawathi Y, Singh R, Singh AK, Singh VP, Mohapatra T, Sharma TR, Singh NK (2008) Mapping of quantitative trait loci for basmati quality traits in rice (Oryza sativa L.). Mol Breed 21:49–65

Berner DK, Hoff BJ (1986) Inheritance of scent in American long grain rice. Crop Sci 26:876–878

Bourgis F, Guyot R, Gherbi H, Tailliez E, Amabile I, Salse J et al (2008) Characterization of the major fragrance gene from an aromatic japonica rice and analysis of its diversity in Asian cultivated rice. Theore Appl Genet 117:353–368

Bradbury LM, Fitzgerald TL, Henry RJ, Jin Q, Waters DL (2005a) The gene for fragrance in rice. Plant Biotechn J 3:363–370

Bradbury LM, Henry RJ, Jin Q, Reinke RF, Waters DL (2005b) A perfect marker for fragrance genotyping in rice. Mol Breed 16:279–283

Bradbury LM, Gillies SA, Brushett DJ, Waters DL, Henry RJ (2008) Inactivation of an aminoaldehyde dehydrogenase is responsible for fragrance in rice. Plant Mol Biol 68:439–449

Brahmachary RL, Ghosh M (2002) Vaginal pheromone and other compounds in mung bean aroma. J Sci Indus Res 61:625–629

Buttery RG, Ling LC, Juliano BO (1982) 2-acetyl-1-pyrroline: an important aroma compound in cooked rice. Chem Indus 12:958–959

Buttery RG, Ling LC, Juliano BO, Turnbauhg JG (1983) Cooked rice aroma and 2-acetyl 1 pyrroline. J Agric Food Chem 31:823–826

Buttery R, Turnbaugh J, Ling L (1988) Contributions of volatiles to rice aroma. J Agric Food Chem 36:1006–1009

Caicedo AL, Williamson SH, Hernandez RD, Boyko A, Fledel-Alon A, York TL et al (2007) Genome-wide patterns of nucleotide polymorphism in domesticated rice. PLoS Genet 3:1745–1756

Champagne ET (2008) Rice aroma and flavor: a literature review. Cereal Chem 85:445–454

Chang TT (1976) The origin, evolution, cultivation, dissemination and diversification of Asian and African rices. Euphytica 25:435–444

Chaudhury D, Tran DV, Duffy R (2003) Speciality rices of the world: breeding, production and marketing. FAO books, Roma

Chen S, Wu J, Yang Y, Shi W, Xu M (2006) The fgr gene responsible for rice fragrance was restricted within 69 kb. Plant Sci 171:505–514

Chen S, Yang Y, Shi W, Ji Q, He F, Zhang Z et al (2008) Badh2, encoding betaine aldehyde dehydrogenase, inhibits the biosynthesis of 2-acetyl-1-pyrroline, a major component in rice fragrance. Plant Cell 20:1850–1861

Childs N, Kiawu J (2009) Factors behind rise in global rice prices in 2008. RCS-09D-01, ERS USA

Chitrakorn S (2003) Rice and Thai way of life. In: Lorlowhakarn S (ed) Science and technology with Thai rice. Public Information Department Thailand's National Science and Technology Development Agency, Bangkok

Civàñ P, Ali S, Batista-Navarro R, Drosou K, Ihejieto C, Chakraborty D et al (2019) Origin of the aromatic group of cultivated rice (Oryza sativa L.) traced to the Indian subcontinent. Genome Biol Evol 11:832–843

Cordeiro GM, Christopher MJ, Henry RJ, Reineke RF (2002) Identification of microsatellite markers for fragrance in rice by analysis of the rice genome sequence. Mol Breed 9:245–250

Das A, Kesari TV, Rangan L (2010) Aromatic joha rice of Assam-a review. Agric Rev 31:1–10

Dawe D (2007) Have recent increases in international cereal prices been transmitted to domestic economies? The experience in seven large Asian countries. FAO ESA working paper (08–03)

de Oliveira Borba TC, Brondani RP, Rangel PH, Brondani C (2009) Microsatellite marker- mediated analysis of the EMBRAPA Rice Core collection genetic diversity. Genetica 137:293–304

Dhulappanavar CV (1976) Inheritance of scent in rice. Euphytica 25:659–662

Ebana K, Yonemaru J, Fukuoka S, Iwata H, Kanamori H, Namiki N et al (2010) Genetic structure revealed by a whole-genome single-nucleotide polymorphism survey of diverse accessions of cultivated Asian rice (Oryza sativa L.). Breed Sci 60:390–397

Fitzgerald MA, Hamilton NR, Calingacion MN, Verhoeven HA, Butardo VM (2008) Is there a second fragrance gene in rice? Plant Biotechnol J 6:416–423

Garland S, Lewin L, Blakeney A, Reineke R, Henry R (2000) PCR- based molecular markers for the fragrance gene in rice (Oryza sativa L.). Theor Appl Genet 101:364–371

Garris AJ, Tai TH, Coburn J, Kresovich S, McCouch S (2005) Genetic structure and diversity in Oryza sativa L. Genetics 169:1631–1638

Giraud G (2013) The world market of fragrant rice, main issues and perspectives. Int Food Agribus Man 16:1–20

Glaszmann JC (1987) Isozymes and classification of Asian rice varieties. Theor Appl Genet 74:21–30

Gross BL, Zhao Z (2014) Archaeological and genetic insights into the origins of domesticated rice. Proc Natl Acad Sci U S A 111:6190–6197

He Z, Zhai W, Wen H, Tang T, Wang Y, Lu X et al (2011) Two evolutionary histories in the genome of rice: the roles of domestication genes. PLoS Genet 7:e1002100

Hofmann T, Schieberle P (1998) 2-Oxopropanal, hydroxy-2-propanone, and 1-pyrrolines important intermediates in the generation of the roast- smelling food flavor compounds 2-acetyl-1-pyrroline and 2-acetyltetrahydropyridine. J Agric Food Chem 46:2270–2277

Huang TC, Huang YW, Hung HJ, Ho CT, Wu ML (2007) Δ1-Pyrroline-5-carboxylic acid formed by proline dehydrogenase from the Bacillus subtilis ssp. natto expressed in Escherichia coli as a precursor for 2-acetyl-1-pyrroline. J Agric Food Chem 55:5097–5102

Huang T-C, Teng C-S, Chang J-L, Chuang H-S, Ho C-T, Wu M-L (2008) Biosynthetic mechanism of 2-Acetyl-1-pyrroline and its relationship with Δ1-Pyrroline-5-carboxylic acid and methylglyoxal in aromatic Rice (Oryza sativa L.) callus. J Agric Food Chem 56:7399–7404

Huang XH, Zhao Y, Wei XH, Li CY, Wang A, Zhao Q et al (2012) Genome-wide association study of flowering time and grain yield traits in a worldwide collection of rice germplasm. Nat Genet 44:32–43

Hussain A, Mujtaba-Naqvi SH, Hammerschmidt FJ (1987) The isolation and identification of volatile components from basmati rice (Oryza sativa L.). In: Martens M, Dalen GA, Russwurm H Jr (eds) Flavor science and technology, proceedings of the 5th Weurman flavour research symposium. Wiley, New York

Jain S, Jain KR, McCouch SR (2004) Genetic analysis of Indian aromatic and quality rice (Oryza sativa L.) germplasm using panels of fluorescently-labeled microsatellite markers. Theor Appl Genet 109:965–977

Jezussek M, Juliano BO, Schieberle P (2002) Comparison of key aroma compounds in cooked brown rice varieties based on aroma extract dilution analyses. J Agric Food Chem 50:1101–1105

Jin Q, Waters D, Cordeiro GM, Henry RJ, Reinke RF (2003) A single nucleotide polymorphism (SNP) marker lined to the fragrance gene in rice (Oryza sativaL.). Plant Sci 165:359–364

Khush GS (2000) Taxonomy and origin of rice. In: Singh RK, Singh US, Khush GS (eds) Aromatic rices. Oxford & IBH Publishing Co, New Delhi, pp 5–13

Kovach MJ, Sweeney MT, McCouch SR (2007) New insights into the history of rice domestication. Trends Genet 11:578–587

Kovach MJ, Calingacion MN, Fitzgerald MA, McCouch SR (2009) The origin and evolution of fragrance in rice (Oryza sativa L.). Proc Natl Acad Sci U S A 106:14444–14449

Laguerre M, Mestres C, Davrieux F, Ringuet J, Boulanger R (2007) Rapid discrimination of fragrant rice by solid-phase microextraction, mass spectrometry, and multivariate analysis used as a mass sensor. J Agric Food Chem 55:1077–1083

Lam HS, Proctor A (2003) Milled rice oxidation volatiles and odor development. J Food Sci 68:2676–2681

Li JH, Wang F, Liu WG, Jin SJ, Liu YB (2006) Genetic analysis and mapping by SSR marker for fragrance gene in rice Yuefeng B. Mol Plant Breed 4:54–58

Li X, Yan W, Agrama H, Hu B, Jia L, Jia M et al (2010) Genotypic and phenotypic characterization of genetic differentiation and diversity in the USDA rice mini-core collection. Genetica 138:1221–1230

Lorieux M, Petrov M, Huang N, Guiderdoni E, Ghesquiere A (1996) Aroma in rice: genetic analysis of a quantitative trait. Theor Appl Genet 93:1145–1151

Maga JA (1984) Rice product volatile: a review. J Agric Food Chem 32:924–970

Mathure S, Shaikh A, Renuka N, Wakte K, Jawali N, Thengane R, Nadaf A (2011) Characterisation of aromatic rice (Oryza sativa L.) germplasm and correlation between their agronomic and quality traits. Euphytica 179:237–246

Molina J, Sikora MG, Flowers JM, Rubinstein S, Reynolds A, Huang P et al (2011) Molecular evidence for a single evolutionary origin of domesticated rice. Proc Natl Acad Sci U S A 108:8351–8356

Nadaf A, Mathure S, Jawali N (2016) Scented rice (Oryza sativa L.) cultivars of India: a perspective on quality and diversity. Springer (India) pvt Ltd, New Delhi

Nagaraju M, Chaudhary D, Rao MJB (1975) A simple technique to identify scent in rice and inheritance pattern of scent. Curr Sci 44:599

Nagaraju J, Kathirvel M, Kumar RR, Siddiq EA, Hasnain SE (2002) Genetic analysis of traditional and evolved basmati and non-basmati rice varieties by using fluorescence-based ISSR-PCR and SSR markers. Proc Natl Acad Sci U S A 99:5836–5841

Nakamura T, Nomura M, Mori H, Jagendorf AT, Ueda A, Takabe T (2001) An isozyme of betaine aldehyde dehydrogenase in barley. Plant Cell Physiol 42:1088–1092

Nematzadeh GA, Karbalaie MT, Farrokhzad F, Gharcyazic B (2000) Aromatic rices of Iran. In: Singh RK, Singh US, Khush GS (eds) Aromatic rices. Oxford & IBH Publishing Co, New Delhi, pp 191–200

Nematzadeh GA, Huang N, Khush GS (2004) Mapping the gene for aroma in rice (Oryza sativa L.) by bulk segregant analysis via RAPD markers. J Agric Sci Tech 6:129–137

Niu X, Tang W, Huang W, Ren G, Wang Q, Luo D et al (2008) RNAi-directed downregulation of OsBADH2 results in aroma (2-acetyl-1-pyrroline) production in rice (Oryza sativa L.). BMC Plant Biol 8:100

Oka HI (1958) Intervarietal variation and classification of cultivated rice. Indian J Genet Plant Breed 18:79–89

Oka HI (1988) Origin of cultivate rice. Japan Scientific Societies Press, Tokyo

Pachauri V, Singh MK, Singh AK, Singh S, Shakeel NA, Singh VP, Singh NK (2010) Origin and genetic diversity of aromatic rice varieties, molecular breeding and chemical and genetic basis of Rice aroma. J Plant Biochem Biotechnol 19:127–143

Patil KG, Nagesh Patil VG, Kulkarni RS, Sashidhar HE (2012) Inheritance of aroma in aromatic rice (Oryza sativa L.) genotypes. Ann Biol Res 3:5472–5474

Paule CM, Powers JJ (1989) Sensory and chemical examination of aromatic and nonaromatic Rices. J Food Sci 54:343–346

Petrov M, Danzart M, Giampaoli P, Faure J, Richard H (1996) Rice aroma analysis: discrimination between a scented and a non- scented rice. Sci Aliments 16:347–360

Pinson SR (1994) Inheritance of aroma in six rice cultivars. Crop Sci 34:1151–1157
Prathepha P (2008) The fragrance (fgr) gene in natural populations of wild rice (Oryza rufipogon Griff.). Genet Resour Crop Evol 56:13–18
Ramaiah K, Rao MV (1953) Rice breeding and genetics. ICAR science monograph 19. Indian Council of Agricultural Research, New Delhi
Reddy VD, Reddy GM (1987) Genetic and biochemical basis of scent in rice (Oryza sativa L.). Theor Appl Genet 73:699–700
Reddy PR, Sathyanarayanaiah K (1980) Inheritance of scent in rice. Indian J Genet Plant Breed 40:327–329
Romanczyk LJ, McClell CA, Post LS, Aitken WM (1995) Formation of 2-acetyl-1-pyrroline by several Bacillus cereus strains isolated from cocoa fermentation boxes. J Agric Food Chem 43:469–475
Roy S, Banerjee A, Senapati BK, Sarkar G (2012) Comparative analysis of agro-morphology, grain quality and aroma traits of traditional and basmati-type genotypes of rice, *Oryza sativa* L. Plant Breed 131:486–492
Roy S, Banerjee A, Pattanayak A, Roy SS, Rathi RS, Misra AK et al (2014) Chakhao (delicious) rice landraces (Oryza sativa L.) of north-East India: collection, conservation and characterization of genetic diversity. Plant Genet Resour: Charact Util 12:264–272
Roy S, Banerjee A, Mawkhlieng B, Misra AK, Pattanayak A, Harish GD et al (2015) Genetic diversity and population structure in aromatic and quality rice (Oryza sativa L.) landraces from north-eastern India. PLoS One 10:e0129607
Sakthivel K, Shobha Rani N, Pandey MK, Sivaranjani AK, Neeraja CN, Balachandran SM et al (2009) Development of a simple functional marker for fragrance in rice and its validation in Indian basmati and non-basmati fragrant rice varieties. Mol Breed 24:185–190
Salgotra RK, Gupta BB, Bhat JA, Sharma S (2015) Genetic diversity and population structure of basmati rice (Oryza sativa L.) germplasm collected from North Western Himalayas using trait linked SSR markers. PLoS One 10:e0131858
Sang T, Ge S (2007) Genetics and phylogenetics of rice domestication. Curr Opin Genet Dev 17:533–538
Sarkarung S, Somrith B, Chitrakorn S (2000) Aromatic rices of Thailand-aromatic rices of other countries. In: Singh RK, Singh US, Khush GS (eds) Aromatic rices. Oxford & IBH Publishing Co Pvt Ltd, New Delhi, pp 180–183
Schieberle P (1995) Quantification of important roast-smelling odorants in popcorn by stable-isotope dilution assay and model studies on flavor formation during popping. J Agric Food Chem 50:2442–2448
Shahidullah SM, Hanafi MM, Ashrafuzzaman M, Ismail MR, Khair A (2009) Genetic diversity in grain quality and nutrition of aromatic Rices. Afr J Biotechnol 8:1238–1246
Shao GN, Tang A, Tang SQ, Luo J, Jiao GA, Wu JL, Hu PS (2011) A new deletion mutation of fragrant gene and the development of three molecular markers for fragrance in rice. Plant Breed 130:172–176
Shao G, Tang S, Chen M, Wei X, He J, Luo J et al (2013) Haplotype variation at Badh2, the gene determining fragrance in rice. Genomics 101:157–162
Sharp RN (1986) Quality evaluation of milled aromatic rice from India, Thailand and United States. J Food Sci 51:634–636
Shi W, Yang Y, Chen S, Xu M (2008) Discovery of a new fragrance allele and the development of functional markers for the breeding of fragrant rice varieties. Mol Breed 22:185–192
Shobha Rani N (1998) The rice situation in Iran. Intl Rice Comm Newsl 47:36
Singh VP (2000) Basmati rice of India. In: Singh RK, Singh US, Khush GS (eds) Aromatic rices. Oxford & IBH Publishing Co, New Delhi, pp 135–154
Singh VP, Singh AK (2010) Role of Indian Agricultural Research Institute in collection, acquisition, evaluation, enhancement, utilization and conservation of rice germplasm. In: Sharma SD, Rap P (eds) Genetic resources of rice in India. Today and Tomorrow's Printers and Publications, New Delhi, pp 135–150

Singh RK, Gautam PL, Saxena S, Singh S (2000a) Scented rice germplasm: conservation, evaluation and utilization. In: Singh RK, Singh US, Khush GS (eds) Aromatic rices. Oxford & IBH Publishing Co Pvt Ltd, New Delhi, pp 107–134

Singh RK, Singh US, Khush GS, Rohilla R, Singh JP, Singh S, Shekhar KS (2000b) Small and medium grained aromatic rices of India. In: Singh RK, Singh US, Khush GS (eds) Aromatic rices. Oxford & IBH Publishing Co Pvt Ltd, New Delhi, pp 155–178

Singh VP, Pratik S, Gopala Krishnan S, Singh AK (2004). Role of Indian Agricultural Research Institute in genetic improvement of rice varieties in India. In: Sharma SD, Rao P (ads) Genetic improvement of rice varieties in India, Today and Tomorrow's Printers and Publications, New Delhi, p 141–187

Singh R, Singh AK, Sharma TR, Singh A, Singh NK (2007) Fine mapping of aroma QTLs in basmati rice (Oryza sativa L) on chromosomes 3, 4 and 8. J Plant Biochem Biotechnol 16:75–82

Singh A, Singh PK, Singh R, Pandit A, Mahato AK, Gupta DK et al (2010) SNP haplotypes of the BADH1 gene and their association with aroma in rice (Oryza sativa L.). Mol Breed 26:325–338

Singh V, Singh AK, Mohapatra T, Gopala Krishnan S, Ellur RK (2018) Pusa basmati 1121 – a rice variety with exceptional kernel elongation and volume expansion after cooking. Rice 11:19

Solheim WG (1972) An earlier agricultural revolution. Sci Am 226:34–41

Sood BG, Siddiq EA (1978) A rapid technique for scent determination in rice. Indian J Genet Plant Breed 38:268–271

Sun SH, Gao FY, Lu XJ, Wu XJ, Wang XD, Ren GJ, Luo H (2008) Genetic analysis and gene fine mapping of aroma in rice (Oryza sativa L. Cyperales, Poaceae). Genet Mol Biol 31:532–538

Tava A, Bocchi S (1999) Aroma of cooked rice (Oryza sativa): comparison between commercial basmati and Italian line B5-3. Cereal Chem 76:526–529

The 3000 Rice Genomes Project (2014) The 3,000 rice genomes project. GigaScience 3:7

Thimmaraju R, Bhagyalakshmi N, Narayan MS, Venkatachalam L, Ravishankar GA (2005) In vitro culture of Pandanus amaryllifolius and enhancement of 2-acetyl-1-pyrroline, the major flavouring compound of aromatic rice, by precursor feeding of L-proline. J Sci Food Agric 85:2527–2534

Tragoonrung S, Sheng JQ, Vanavichit A (1996) Tagging an aromatic gene in lowland rice using bulk segregant analysis. In: Rice genetics III. International Rice Research Institute, Manila, pp 613–618

Tripathi RS, Rao MJ (1979) Inheritance and linkage relationship of scent in rice. Euphytica 28:319–323

Trossat C, Rathinasabapathi B, Hanson AD (1997) Transgenically expressed betaine aldehyde dehydrogenase efficiently catalyzes oxidation of dimethylsulfoniopropionaldehyde and omega-aminoaldehydes. Plant Physiol 113:1457–1461

Tsuzuki E, Shimokawa E (1990) Inheritance of aroma in rice. Euphytica 46:157–159

Vanavichit A, Kamolsukyeunyong W, Siangliw M, Siangliw JL, Traprab S, Ruengphayak S et al (2018) Thai Hom Mali Rice: Origin and Breeding for Subsistence Rainfed Lowland Rice System. Rice 11:20

Vanavichit A, Kamolsukyurnyong W, Wanchana S, Wongpornchai S, Ruengphayak S, Toojinda T, Tragoonrung S (2004) Discovering genes for rice grain aroma. In: Proceedings of the 1st international conference on rice for the future. Kasetsart University, Bangkok, pp 71–80

Vanavichit A, Yoshihashi T (2010) Molecular aspects of fragrance and aroma in rice. Adv Bot Res 56:49–73

Vanavichit A, Yoshihashi T, Wanchana S, Areekit S, Saengsraku D, Kamolsukyunyong W (2005) Cloning of Os2-AP, the aromatic gene controlling the biosynthetic switch of 2- acetyl-1-pyrroline and gamma-aminobutyric acid (GABA) in rice. In: 5th international rice genetics symposium. IRRI, Philippines, p 44

Verma DK, Srivastav PP, Nadaf A (2019) Aromatic rice from different countries: an overview. In: Verma DK, Srivastav PP (eds) Science and technology of aroma, flavor, and fragrance in rice. Apple Academic Press, Inc, Oakville, pp 35–58

Vivekanandan P, Giridharan P (1994) Inheritance of aroma and breadthwise grain expansion in basmati and non-basmati rices. Intl Rice Res Notes 19:4–5

Wanchana S, Kamolsukyunyong W, Ruengphayak S, Toojinda T, Tragoonrung S, Vanavichit A (2005) A rapid construction of a physical contig across a 4.5 cM region for rice grain aroma facilitates marker enrichment for positional cloning. Sci Asia 31:299–330

Wang C-H, Zheng X-M, Xu Q, Yuan X-P, Huang L, Zhou H-F et al (2014) Genetic diversity and classification of *Oryza sativa* with emphasis on Chinese rice germplasm. Heredity 112:489–496

Wang W, Mauleon R, Hu Z, Chebotarov D, Tai S, Wu Z et al (2018) Genomic variation in 3,010 diverse accessions of Asian cultivated rice. Nature 557:43–49

Widjaja R, Craske JD, Wootton M (1996) Changes in volatile components of paddy, brown, and white fragrant rice during storage. J Sci Food Agric 71:218–224

Yadav BK, Jindal VK (2007) Modeling changes in milled rice (Oryza sativa L.) kernel dimensions during soaking by image analysis. J Food Eng 80:359–369

Yan W (1989) Rethinking the origin of rice agriculture in China. Nongye Kaogu 89:74–83

Yang DS, Shewfelt RL, Lee KS, Kays SJ (2008) Comparison of odor-active compounds from six distinctly different rice flavor types. Jo J Agric Food Chem 56:2780–2787

Yano M, Shimosaka E, Saito A, Nakagahra M (1992) Linkage analysis of a gene for scent in indica rice variety, Surjamkhi, using restriction fragment length polymorphism markers. Jpn J Breed 41:338–339

Yoshihashi T (2002) Quantitative analysis on 2-acetyl-1-pyrroline of aromatic rice by stable isotope dilution method and model studies on its formation during cooking. J Food Sci 67:619–622

Yoshihashi TN, Huong TT, Inatomi H (2002) Precursors of 2-acetyl-1-pyrroline, a potent flavor compound of an aromatic rice variety. J Agric Food Chem 50:2001–2004

Zhang DL, Zhang HL, Wang MX, Sun JL, Qi YW, Wang F et al (2009) Genetic structure and differentiation of Oryza sativa L. in China revealed by microsatellites. Theor Appl Genet 119:1105–1117

Zhao KY, Wright M, Kimball J, Eizenga G, McClung A, Kovach M et al (2010) Genomic diversity and introgression in *O. sativa* reveal the impact of domestication and breeding on the rice genome. PLoS One 5:e10780

Chapter 12
Red Rice

José Manoel Colombari Filho, Aluana Gonçalves de Abreu, and José Almeida Pereira

12.1 Introduction

The term "red rice" refers to the color of rice kernel which is determined by deposits of natural reddish pigments in different layers of the pericarp during rice seed development, covered with dark- or light-colored husk. Although it is often associated with weedy rice, and sometimes used as a synonym, "red rice" can also be wild or cultivated rice (Ahuja et al. 2007). Red rice has been grown for a long time in Asia and consumed as unpolished whole grain because the milling process strips the color along with the bran layer, making it similar to polished white rice. Brown rice, also called whole grain rice, represents the unpolished version of the more common white (noncolored) rice including bran, germ, and endosperm (Fig. 12.1).

Red pericarp coloration in rice grains is inherited by two independent genes with epistatic interaction, *Rc* and *Rd*. *Rc* is located on chromosome 7 and *Rd* is on chromosome 1, and only when they are together do grains have red pericarp. The *Rc* gene encodes a basic helix–loop–helix (bHLH) regulatory protein in the red pigment synthesis pathway. One mutation generated the new *rc* allele, which is characterized by a 14-bp frameshift deletion in exon 7 and has a nonfunctional gene product present in the majority of the noncolored cultivars (Sweeney et al. 2006; Furukawa et al. 2007). Thus, *Rc* in the absence of *Rd* produces brown grains, whereas *Rd* alone has no phenotype. The genes *Rc* and *Rd* segregate in the ratio 9:3:4, in recessive epistasis, where the genotype *Rc_RdRd* produces red-colored grains; *Rc_rdrd* produces brown-colored grains; and *rcrcRd_* or *rcrcrdrd* produce light brown- or white-colored grains.

J. M. Colombari Filho (✉) · A. G. de Abreu
Embrapa Arroz e Feijão, Santo Antônio de Goiás, Goiás, Brazil
e-mail: jose.colombari@embrapa.br

J. A. Pereira
Embrapa Meio Norte, Teresina, Piauí, Brazil

A. Costa de Oliveira et al. (eds.), *The Future of Rice Demand: Quality Beyond Productivity*, https://doi.org/10.1007/978-3-030-37510-2_12

Fig. 12.1 Paddy, unpolished and polished grains, respectively, from left to right. Noncolored rice (**a**) and red rice (**b**)

Fig. 12.2 Weedy rice plant in a field of white rice production in the Southern Region of Brazil (**a**) and BRS 902: Brazilian red rice cultivar (**b**)

Weedy rice are de-domesticated plants from cultivated rice (Fig. 12.2), mainly *Oryza sativa*, that infest and compete with this crop culture and cause reduction in yield (Delouche et al. 2007; Ziska et al. 2015). The plant de-domestication is a distinct evolutionary process involving a loss of traits aggregated under domestication, during which domesticated crops are turned into self-sustainable "wild-like" plants mainly driven by natural selection (Gressel 2005; Ellstrand et al. 2010; Gamborg et al. 2010). Major genes associated with de-domestication are *Sh4* (Li et al. 2006), *Rc* (Sweeney et al. 2006), *PROG1* (Tan et al. 2008), and *LABA1* (Hua et al. 2015), whose recessive alleles are associated with non-shattering rachis, white pericarp, erect growth, and barbless awns, respectively. So, the weedy rice have high degree of seed shedding and dormancy and occurrence of high rate of outcrossing, which have been contributed to these rice coevolving through hybridization among weedy, wild, and cultivated rice. They are also hardy, more prolific in nature, and carry genes for tolerance to various adverse conditions. Weedy rice, especially types with a red pericarp, are globally well known as a problem in the rice industry.

The economic and environmental problems they pose include those related to rice crop production, rice milling for commerce, quarantine regulations, and seed trade (Delouche et al. 2007; Ziska et al. 2015).

Wild rice are any *Oryza* species other than cultivated ones: *Oryza sativa* and *Oryza glaberrima*. Red pericarp is present in most wild rice species, and the cultivated rice pericarp was originally red. Although *Rc* allele is referred to as a mutant, its action is dominant over white pericarp (*rc* allele), suggesting that domestication trait arose through human selection for loss-of-function mutations at the *Rc* ancestral locus (red) from *O. rufipogon* (Sweeney et al. 2006). The change in pericarp color from red to white was an important hallmark of rice domestication. Most rice cultivars grown and consumed throughout the world today, mainly *O. sativa*, have a white pericarp, while all known accessions of the wild ancestor, *O. rufipogon*, invariably have grains with a red pericarp. Some presumed reasons for selection against pigmentation in rice are the following: insects and pathogens could be more easily detected and eliminated against a light than a darker red background, and colored rice, due to its harder seed coat, would take more time to cook and would require more time to find fuel with which to cook and more work to remove the bran layer (Sweeney et al. 2007).

Although white rice is one of the most important cereal crops in the world, cultivated red varieties are still grown in restricted regions, where they have a cultural, ceremonial, or medicinal value (Sweeney et al. 2006). Despite its low diffusion, in the recent years, the global food trends have shown an increase in interest in consuming colored rice as people seek the well-being and the alimentary health or that value the sensoriality and pleasure of new gastronomic experiences. Many health-promoting properties have been associated to the consumption of colored grain rice, which contains high levels of several antioxidant compounds associated with reduced risk of chronic diseases (Liu 2004; Ciulu et al. 2018). Also, only colored rice can add a fifth attribute to grain quality, the color, besides different shapes, flavors, fragrances, and textures, which may also be present in noncolored rice. As in white rice, great diversity exists among red rice. They can be short- or long grained, scented or non-scented, glutinous or non-glutinous, late- or early maturing, short or tall plant height, and traditional or modern plant architecture, among other traits. Before the advent of high-yielding varieties, which are mostly white, red rice formed an important group in almost all the rice-growing Asian countries such as Sri Lanka, the Philippines, Korea, China, Japan, and India (Ahuja et al. 2007). Flavored, black rice were the favorites of the royals of China, while red rice were preferred by people in many parts of India, Sri Lanka, and Bhutan (Ahuja et al. 2007).

12.2 Cultural Habits

In India, red rice have occupied a special position since time immemorial (Ahuja et al. 2007). From the ancient Ayurveda to the Buddhist scripture, red rice is mentioned for its nutritive and medicinal value. Glutinous type is useful for special preparations like "puttu" in South India. In Himachal Pradesh, Jatu red rice is prized for its aroma and taste, Matali and Lal dhan are used for curing blood pressure and fever, and Kafalya is used for treating leucorrhea and abortion complications (Ahuja et al. 2007).

The Chinese use red rice for preparing vinegar, tart, cosmetics, red kojic, and red rice yeast, which is used for medicinal purposes. Red rice yeast is prepared by fermenting yeast *Monascus purpureus* over red rice. It is said to promote blood circulation and is used in treating upset stomach, indigestion, bruised muscles, and hangovers; and it is a cholesterol-lowering product that is commercially marketed the world over (Chaudhary and Tran 2001).

Red rice in Japan is used for preparation of red sake, colored noodles, and cakes for ceremonial occasions. In Sri Lanka, red rice are a favorite as food and some are used as medicines (Ahuja et al. 2007).

Bhutanese red rice is a staple food in this country and is cultivated at subsistence levels, while as little as 1% of rice is marketed commercially. It is a medium-grain rice that has been grown in Bhutan for thousands of years, at nearly 9000 ft. above sea level and irrigated with mineral-rich glacial waters. Bhutan red rice is considered to be of superior quality with a distinctive earthy taste; a complex, nutty, soft texture; and a russet color and has a shorter cooking time of only 20 min. As red rice is currently not being farmed by a majority of Bhutanese farmers, it has become more expensive, and most people are not able to afford to eat it as part of a daily meal. It is now more frequently eaten on special occasions such as weddings or national and religious holidays (Slow Food Foundation for Biodiversity 2019).

In the Americas, a similar tradition occurs in Brazil where the red rice has been cultivated in the Northeast Region since the sixteenth century, called "arroz da terra" ("land rice," in English), brought by Portuguese colonizers (Pereira 2002). After the introduction of white rice, in the eighteenth century, more productive and preferred by the Portuguese consumers, the cultivation of red rice declined (Pereira 2002). Red rice production continued in less favorable areas, using low-input agricultural practices, where its consumption became a tradition, as in the Paraiba state, on the valleys of Piancó and Rio do Peixe, and in the Rio Grande do Norte state, on Apodi Valley (Pereira 2004). In addition, most of the varieties cultivated nowadays in these sites come from the ones introduced centuries ago, and therefore, have a very narrow genetic base. In these states, red rice varieties present grains with rounded shape and are traditionally consumed as semi-polished and cooked with milk, salt, and spices (Fig. 12.3) (Pereira et al. 2009). However, in modern times, this Brazilian red rice also started to be commercialized and used in gourmet food products in other regions, as South and Southeast, where people have greater purchasing power. These consumers recognize that red rice is a higher value-added product, due to its functional properties and its unique flavor, color, and texture. In Brazil, rice for specific culinary, or which have novel cooking, appearance, aroma, and flavor characteristics, have been called "special rice." This niche market is growing rapidly and usually practices a premium market price. The added value is benefiting mainly small-scale farmers, as they can achieve higher profitability than with polished white rice (Colombari-Filho et al. 2018).

Fig. 12.3 Appearance of the cooked grains of the traditional Brazilian red rice ("arroz da terra"): semi-polished grains cooked with milk (**a**) and whole grains (**b**). (Photos by Maria Eugênia Ribeiro, Embrapa Meio-Norte)

12.3 Functional and Nutraceutical Properties

Red rice is consumed as whole grain, which consists of 6–7% bran, 2–3% embryo, and 90% endosperm by weight, in general (Chen et al. 1998). The whole grain is rich in proteins, lipids, dietary fibers, minerals, vitamins, and phytochemicals including several phenolic compounds (Shao and Bao 2015). Besides this high functional food value, the pericarp of red rice has a high measured content for total soluble phenolic compounds (81%) and antioxidant properties (de Mira et al. 2009).

Phenolic compounds are derived from the secondary metabolism (also known as specialized metabolism) that are not absolutely required for the survival of the plants, as for growth and reproduction, but are generally involved in defense against ultraviolet radiation, aggression by pathogens or pests, and pollinator attraction. These compounds possess one or more aromatic rings with one or more hydroxyl groups and generally are categorized as phenolic acids, flavonoids, stilbenes, coumarins, and tannins (Manach et al. 2004; Liu 2007).

Over the past years, phenolic compounds have been studied due to functional and nutraceutical properties from their antioxidant activities and their probable role in the prevention of various diseases associated with oxidative stress, such as cancer and cardiovascular and neurodegenerative diseases (Manach et al. 2004). Recent research results have shown that the phytochemical content and antioxidant activity of whole grain rice have been previously underestimated, as the literature only considered free phenolics and the major portion of phenolics in rice (62%) is composed of the bound form (Adom and Liu 2002). In addition, studies have shown that there are beneficial effects associated with whole grain consumption due to the existence of the unique phytochemicals that complement those in fruits and vegetables when consumed together (Liu 2007). However, these benefits are found only in whole grain, because polishing significantly reduces the concentration of total soluble phenolic compounds of rice grains with any pericarp color (Walter et al. 2013). Thus, these are the main reasons for the increase in whole grain consumption, both for

Hydroxybenzoic acids	Name	R1	R2	R3	R4
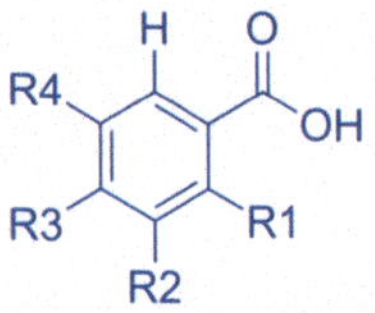	Gallic acid	H	OH	OH	H
	Protocatechuic acid	H	OCH_3	OH	OCH_3
	p-Hydroxybenzoic acid	H	H	OH	H
	Vanillic acid	H	OH	OH	OH
	Syringic acid	OH	H	H	OH

Hydroxycinnamic acids	Name	R1	R2	R3	R4
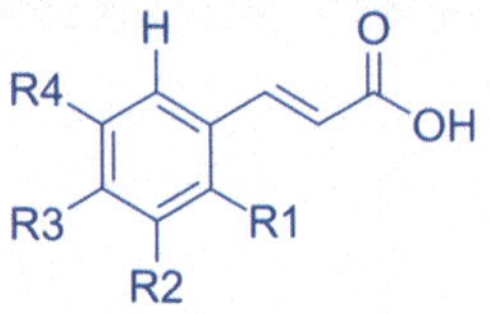	Ferulic acid	H	OCH_3	OH	H
	p-Coumaric	H	H	OH	H
	Sinapic acid	H	OCH_3	OH	OCH_3

Fig. 12.4 Chemical structure of phenolic acids present in whole rice grains. (Adapted from Goleniowski et al. 2013)

colored and noncolored rice, by people seeking well-being and alimentary health (Liu 2004).

The main phenolic compounds in noncolored rice are classified as non-flavonoids (phenolic acids), while in red rice, the main phenolics are classified as flavonoids (polyphenols) (Zhou et al. 2014). Phenolic acids are simple compounds and are present at the beginning of the synthesis pathway of specialized metabolites, while flavonoids are more complex and are located at the end of the metabolic pathway.

Phenolic acids contain a phenolic ring and at least one organic carboxylic acid function. They constitute an important class of the phenolic fraction of whole grain rice, including both hydroxybenzoic and hydroxycinnamic acids, and can be divided according to their structure (Goleniowski et al. 2013; Ciulu et al. 2018). These acids have been reported in higher concentrations in colored rice than in noncolored rice (brown rice) (Goufo and Trindade 2014), and the main ones are (Fig. 12.4) ferulic acid (from 56% to 77%), p-coumaric acid (from 8% to 24%), sinapic acid (from 2% to 12%), gallic acid (from 1% to 6%), protocatechuic acid (from 1% to 4%), p-hydroxybenzoic acid (from 1% to 2%), vanillic acid (1%), and syringic acid (1%) (Goufo and Trindade 2014).

Flavonoids contain a basic structure constituted of 15 carbon atoms arranged in 3 rings (C6-C3-C6), which consist of 2 phenyl rings and a heterocyclic ring capable of performing free radical scavenging activity. These compounds may themselves be divided into six subclasses depending on the type of heterocycle involved: flavonols, flavones, isoflavones, flavanones, anthocyanidins, and flavanols (Fig. 12.5). Also, they may be associated with various carbohydrates and organic acids and with each other (Manach et al. 2004). The flavonoids present in red rice are the proanthocyanidins, which belong to the class of flavanols (Min et al. 2011).

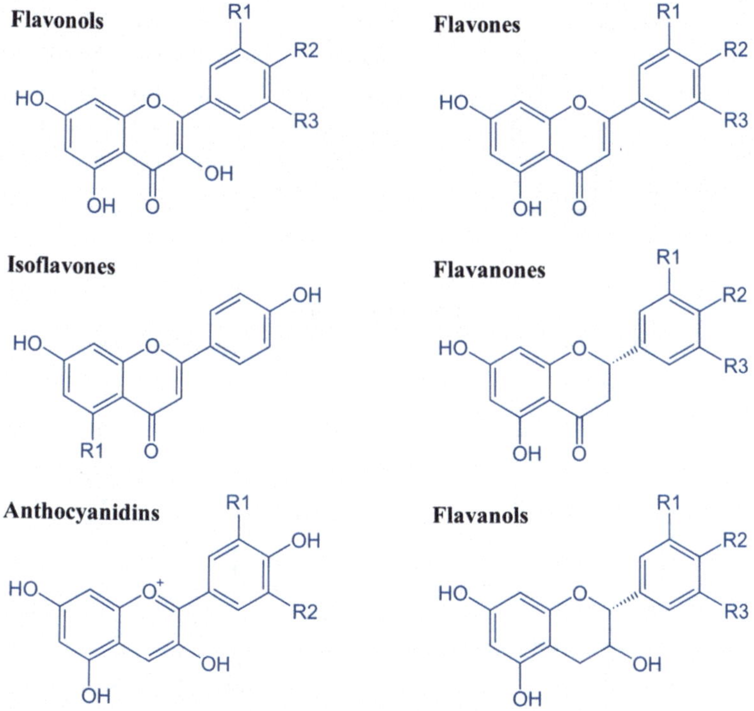

Fig. 12.5 Chemical structures of flavonoids. (Adapted from Manach et al. 2004)

The proanthocyanidins, or condensed tannins, are the most abundant polyphenols in the plant kingdom. They are oligomers of flavan-3-ol and/or flavan-3,4-diol (Smeriglio et al. 2017), and, probably, they are so named because of red pigments from the classes of anthocyanidins, cyanidin, and delphinidin (de Jesus et a. 2012). They can be classified into several classes depending on the hydroxylation of the constitutive units and the linkages between them. The proanthocyanidins present in red rice can be procyanidins (bond between units of catechin and/or epicatechin) and prodelphinidin (bond between units of epigallocatechin and/or gallocatechin), as the chemical structures shown in Fig. 12.6 (Bordiga et al. 2014; Pintha et al. 2015).

In the last years, a remarkable attention has been paid to proanthocyanidins due to the human health-promoting benefits, which are beneficial not only as an antioxidant and anti-inflammatory but have also been included immunomodulatory, anticancer, cardioprotective, and antithrombotic properties (Tsuda et al. 2002; Hyun and Chung 2004; Zhao et al. 2004; Nile and Park 2014; Smeriglio et al. 2014; Sieniawska 2015; Huang and Lai 2016; Smeriglio et al. 2017). Mainly due to the sharp increase in lifestyle-related health issues and diseases – such as diabetes, cancer, and heart problems – scientists are looking at quality traits other than carbohydrates, protein, and fat in foodstuffs. Evaluation of foodstuffs now places greater focus on their antioxidant properties, glycemic index, and mineral content (Ahuja

Flavan-3-ol	Name	R1	R2	R3	R4	R5
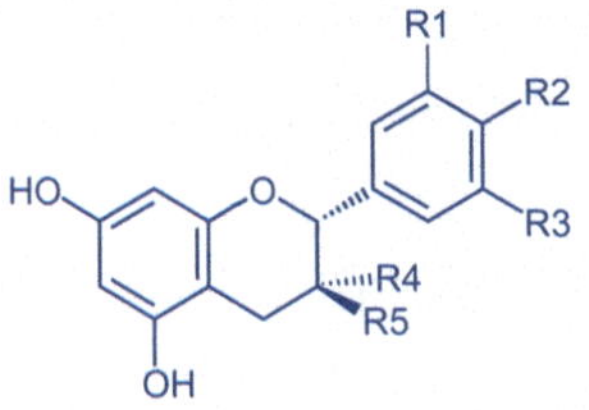	Catechin	H	OH	OH	H	OH
	Epicatechin	H	OH	OH	OH	H
	Gallocatechin	OH	OH	H	H	OH
	Epigallocatechin	OH	OH	H	OH	H

Proanthocyanidins

Procyanidin

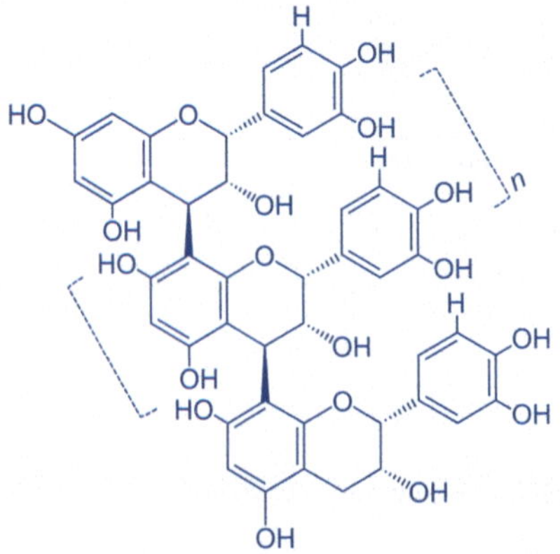

Prodelphinidin

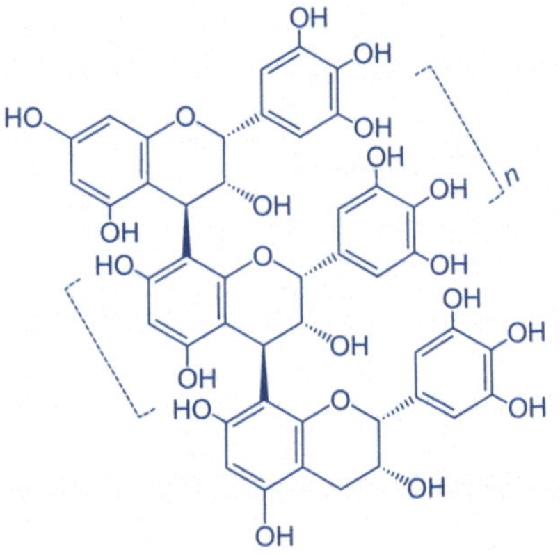

Fig. 12.6 Chemical structures of flavan-3-ols and proanthocyanidins (procyanidin and prodelphinidin). (Adapted from Smeriglio et al. 2017)

et al. 2007). Red rice may also be helpful for the prevention or cure of diseases caused by vitamin A and B deficiencies (Kristamtini et al. 2012), and the zinc and iron content of red rice is from two to three times higher than that of white rice (Ramaiah and Rao 1953).

Scientists are looking at the colored rice as more than just starchy food and are analyzing their antioxidant properties and mineral content. Red rice have been found to have greater radical scavenging activity than black and brown rice by the 2,2-Azinobis (3-ethyl-benzothiazoline-6-sulfonate) (ABTS) and 1,1-diphenyl-2-picrylhydrazyl (DPPH) methods, and this activity is correlated with polyphenols and proanthocyanidin content (Oki et al. 2005; Colombari-Filho et al. 2018). With this rediscovery of their nutritive and medicinal value, red rice have begun to regain their old position and prestige (Ahuja et al. 2007; Colombari-Filho et al. 2018). In Japan, for example, there is a revival of red rice, what has resulted in the emergence of many kinds of secondary products like colored noodles, cakes, alcoholic beverages, etc. (Itani and Ogawa 2004).

The cosmetic industry is also interested in red rice potential as an antiaging ingredient. Rice by itself is a source of phytochemicals such as γ-oryzanol, vitamin E homologues, and phenolic acids (Finocchiaro et al. 2007; Yawadio et al. 2007). Red rice aggregates the proanthocyanidins, which antioxidant properties were discussed above. However, one constraint to use rice as an active ingredient in the cosmetic industry is still the availability of the material, but plant cell culture could be an alternative to obtain enough material for extracting active compounds. Vichit et al. (2018) demonstrated the effective antiaging properties of Thai red rice stem cell extract, which can be used as an effective ingredient for cosmetic products functionalized in skin lightening, hydration, and elasticity.

12.4 Other Features

In addition of being nutritive and having medicinal and cosmetic value, red rice possess many other special features. The cytoplasmic male sterile (CMS-WA) was discovered in a wild rice (*O. rufipogon*), and this cytoplasm was transferred by backcrossing into *O. sativa* ssp. *indica* (cultivated rice). This feature was fundamental to trigger the hybrid rice breeding by the three lines system (CMS, maintainer, and restorer) in China (Lin and Yuan 1980).

Red rice have proven to be suitable for various edaphoclimatic conditions and adverse situations as they can possess tolerance to drought, flood, submergence, alkalinity, salinity, and resistance to pests and diseases in the field (Chaudhary and Tran 2001; Ahuja et al. 2007). Also, they are comparatively more resistant to storage insect pests than brown rice (Ahuja et al. 2007).

Indian red landraces, such as Ptb 18 and Ptb 19, have been used for broad-spectrum resistance to the dreaded brown plant hopper, in breeding programs all over the world (Ahuja et al. 2007). Red rice Niver and Zag cultivars were often cultivated in high-altitude areas as these varieties were hardier than white rice and withstood the chill of cold weather and were more tolerant to cold water from mountains entering first into the fields there. Besides, spiky glumes of red rice gave some protection to the crop against some wild beasts prevalent in these areas (Sultan and Subba Rao 2013).

In Brazil, red rice have been mainly used to impart resistance against many diseases. A Brazilian red landrace, called Três Marias, was rated as resistant to *Magnaporthe oryzae* in the 1982 International Rice Blast Nursery (Prabhu et al. 1982; Chaudhary 1996). Since then, it has been used as a source for blast resistance in the rice breeding program at the Brazilian Agricultural Research Corporation (Embrapa), mainly as parent in recurrent selection populations (Morais-Júnior et al. 2015). This accession is already part of the genealogy of a modern upland cultivar, BRS Sertaneja (Breseghello et al. 2006).

12.5 Production Limitations

Despite its high medicinal and nutraceutical value, the low yield of red rice makes farmers less interested in cultivating it. However, red rice is still widely cultivated because it can be grown under low inputs and under the seasonal environmental conditions to which modern cultivars are not adapted (Islam et al. 2018). In Bangladesh, more than 30% of *aus* rice are a traditional red variety, which characteristically has limited grain yield (Islam et al. 2018). In Brazil, until a few years ago, there was also no red rice cultivar, only landraces, because rice breeding programs have prioritized only the development of white rice cultivars (Pereira 2002).

Typically, the traditional red rice varieties have low tillering ability, tall and overgrowth plants, long droopy leaves that turn yellow during grain development, weak stems, and become logged at maturity (Pereira et al. 2008; Mau et al. 2017; Islam et al. 2018). Therefore, commercial availability of red rice in markets is limited by a lack of improved cultivars at the farmer level. With the advent of genome editing, the rapid conversion of elite white pericarp varieties into red ones is possible, accelerating the breeding of new red rice varieties with elite agronomic traits (Zhut et al. 2019). Modern breeding techniques could also improve the yield of red rice to overcome this problem. In Brazil, two research institutions, the Brazilian Agricultural Research Corporation (Embrapa) and the Agricultural Research and Rural Extension Company of Santa Catarina (Epagri), started, in partnership, to develop rice breeding lines other than white grain, for the several international cuisines (or simply "Brazilian special rice") (Colombari-Filho et al. 2018). Recently, these programs have already released modern red rice cultivars: BRS 901 and BRS 902, released by Embrapa (Fig. 12.7); and SCS119 Rubi, released by Epagri (Wickert et al. 2014).

SCS119 Rubi was developed from populations of weedy rice accessions. These accessions were collected by Epagri's researchers from different cultivated white rice fields of Santa Catarina state. The selection of plants within these populations

Fig. 12.7 Red rice cultivars: grains of BRS 901 (**a**). (Photo by Luciana Pereira dos Santos Fernandes, Embrapa Meio-Norte) and panicle of BRS 902 (**b**). (Photo by Maria Eugênia Ribeiro, Embrapa Meio-Norte)

focused mainly on red long-shaped grains with low grain shattering, modern plant architecture, resistance to diseases (mainly blast), iron toxicity tolerance, and high grain yield (Wickert et al. 2014). BRS 901 and BRS 902, on the other hand, were developed from artificial crossing between traditional varieties collected by researchers at Embrapa. These red rice varieties were PB 01 (female parent) and PB 05 (male parent), which present traditional and modern plant architecture, respectively. The selections for the development of these two new red rice cultivars focused mainly on round-shaped grains (like the "arroz da terra," Fig. 12.7) with high percentage of head whole grains (greater than 65%), resistance to diseases (mainly blast), high grain yield, and high levels of phenolic compounds. BRS 901 and BRS 902 cultivars have an important impact socioeconomically in the Northeast Region of Brazil, because their grain yield is 30% higher than the traditional red rice varieties. In addition, among the Brazilian cultivars of colored rice (black or red), the BRS 902 presents high levels of free phenolic compounds and the highest antioxidant activity by the DPPH and ABTS methods (Colombari-Filho et al. 2018).

References

Adom KK, Liu RH (2002) Antioxidant activity of grains. J Agric Food Chem 50:6182–6187. https://doi.org/10.1021/jf0205099

Ahuja U, Ahuja SC, Chaudhary N, Thakrar R (2007) Red rices – past, present, and future. Asian Agri-Hist 11:291–304. http://asianagrihistory.org/articles/Red-Rices-Uma-Ahuja.pdf

Bordiga M, Gomez-Alonso S, Locatelli M, Travaglia F, Coïsson JD, Hermosin-Gutierrez I, Arlorio M (2014) Phenolics characterization and antioxidant activity of six different pigmented *Oryza sativa* L. cultivars grown in Piedmont (Italy). Food Res Int 65:282–290. https://doi.org/10.1016/j.foodres.2014.03.007

Breseghello F, Morais OP, Castro EM, Pereira JA, Utumi MM, Lopes IM, Cordeiro ACC, Bassinello PZ, Fonseca JR, Prabhu AS, Peters V, Soares AA (2006) BRS Sertaneja: cultivar precoce de arroz de terras altas. Comunicado Técnico 133. Santo Antônio de Goiás, Embrapa Arroz e Feijão, 4p. https://www.embrapa.br/busca-de-publicacoes/-/publicacao/214893/brs-sertaneja-cultivar-precoce-de-arroz-de-terras-altas

Chaudhary RC (1996) Internationalization of elite germplasm for farmers: collaborative mechanisms to enhance evaluation of rice genetic resources. In: Plant genetic resources: characterization and evaluation. New approaches for improved use of plant genetic resources. National Institute of Agrobiological Resources, Tsukuba, pp 221–243

Chaudhary RC, Tran DV (2001) Specialty rices of the world – a prologue. In: Chaudhary RC, Tran DV, Dutty R (eds) Specialty rices of the world: breeding, production, and marketing. FAO, Rome, pp 3–14

Chen H, Siebenmorgen TJ, Griffin K (1998) Quality characteristics of long-grain rice milled in two commercial systems. Cereal Chem 75:560–565. https://doi.org/10.1094/CCHEM.1998.75.4.560

Ciulu M, Cádiz-Gurrea ML, Segura-Carretero A (2018) Extraction and analysis of phenolic compounds in rice: a review. Molecules 23:2890. https://doi.org/10.3390/molecules23112890

Colombari-Filho JM, Pereira JA, Vanier NL (2018) Mercado brasileiro para tipos especiais de arroz: pigmentados, aromáticos e para culinárias japonesa e italiana. In: Arroz: do campo à mesa. Informativo Agropecuário, vol 39. EPAMIG, Belo Horizonte, pp 86–95

De Jesus NZT, Falcão HS, Gomes IF, Leite TJA, Lima GRM, Barbosa-Filho JM, Tavares JF, Silva MS, Athayde-Filho PF, Batista LM (2012) Tannins, peptic ulcers and related mechanisms. Int J Mol Sci 13:3203–3228. https://doi.org/10.3390/ijms13033203

de Mira NVM, Massaretto IL, Pascual CSCI, Marquez UML (2009) Comparative study of phenolic compounds in different Brazilian rice (Oryza sativa L.) genotypes. J Food Compos Anal 22:405–409. https://doi.org/10.1016/j.jfca.2008.06.012

Delouche JC, Burgos NR, Gealy DR, de San Martin GZ, Labrada R, Larinde M, Rosell C (2007) Weedy rices: origin, biology, ecology and control, vol 188. FAO Plant Production and Protection Paper, Rome, p 144

Ellstrand NC, Heredia SM, Leak-Garcia JA, Heraty JM, Burger JC, Yao L, Nohzadeh-Malakshah S, Ridley CE (2010) Crops gone wild: evolution of weeds and invasives from domesticated ancestors. Evol Appl 3:494–504. https://doi.org/10.1111/j.1752-4571.2010.00140.x

Finocchiaro F, Ferrari B, Gianinetti A, Dall'asta C, Galaverna G, Scazzina F, Pellegrini N (2007) Characterization of antioxidant compounds of red and white rice and changes in total antioxidant capacity during processing. Mol Nutr Food Res 5:1006–1019. https://doi.org/10.1002/mnfr.200700011

Furukawa T, Maekawa M, Oki T, Suda I, Iida S, Shimada H, Takamure I, Kadowaki K (2007) The Rc and Rd genes are involved in proanthocyanidin synthesis in rice pericarp. Plant J 49:91–102. https://doi.org/10.1111/j.1365-313X.2006.02958.x

Gamborg C, Gremmen B, Christiansen SB, Sandoe P (2010) De-domestication: ethics at the intersection of landscape restoration and animal welfare. Environ Values 19:57–78. https://doi.org/10.3197/096327110X485383

Goleniowski M, Bonfill M, Cusido RM, Palazón J (2013) Phenolic acids. In: Ramawat KG, Mérillon JM (eds) Natural products. Springer-Verlag, Berlin, pp 1951–1973

Goufo P, Trindade H (2014) Rice antioxidants: phenolic acids, flavonoids, anthocyanins, proanthocyanidins, tocopherols, tocotrienols, γ -oryzanol, and phytic acid. Food Sci Nutr 2:75–104. https://doi.org/10.1002/fsn3.86

Gressel J (2005) Crop ferality and volunteerism. CRC Press, Boca Raton, p 422

Hua L, Wang DR, Tan L, Fu Y, Liu F, Xiao L, Zhu Z, Fu Q, Sun X, Gu P, Cai H, McCouch SR, Sun C (2015) LABA1, a domestication gene associated with long, barbed awns in wild rice. Plant Cell 27:1875–1888. https://doi.org/10.1105/tpc.15.00260

Huang Y-P, Lai H-M (2016) Bioactive compounds and antioxidative activity of colored rice bran. J Food Drug Anal 24:564–574. https://doi.org/10.1016/j.jfda.2016.01.004

Hyun JW, Chung HS (2004) Cyanidin and malvidin from Oryza sativa cv. Heugjinjubyeo mediate cytotoxicity against human monocytic leukemia cells by arrest of G2/M phase and induction of apoptosis. J Agric Food Chem 52:2213–2217. https://doi.org/10.1021/jf030370h

Islam MZ, Khalequzzaman M, Prince MFRK, Siddique MA, Rashid ESMH, Ahmed MSU, Pittendrigh BR, Ali MP (2018) Diversity and population structure of red rice germplasm in Bangladesh. PLoS One 13:e0196096. https://doi.org/10.1371/journal.pone.0196096

Itani T, Ogawa M (2004) History and recent trends of red rice in Japan. Jpn J Crop Sci 73:137–147. https://doi.org/10.1626/jcs.73.137

Kristamtini T, Basunanda P, Murti RH, Supriyanta SW (2012) Morphological of genetic relationships among black rice landraces from Yogyakarta and surrounding areas. ARPN J Agric Biol Sci 7:982–989. http://www.arpnjournals.com/jabs/research_papers/rp_2012/jabs_1212_494.pdf

Li C, Zhou A, Sang T (2006) Rice domestication by reducing shattering. Science 311:1936–1939. https://doi.org/10.1126/science.1123604

Lin S, Yuan L (1980) Hybrid rice breeding in China. In: Innovative approaches to rice breeding. International Rice Research Institute, Los Banos, pp 35–51

Liu RH (2004) Whole grain phytochemicals and health. J Cereal Sci 46:207–219. https://doi.org/10.1016/j.jcs.2007.06.010

Liu RH (2007) Whole grain phytochemicals and health. J Cereal Sci 46:207–219. https://doi.org/10.1016/j.jcs.2007.06.010

Manach C, Scalbert A, Morand C, Remesy C, Jimenez L (2004) Polyphenols: food sources and bioavailability. Am J Clin Nutr 79:727–747. https://doi.org/10.1093/ajcn/79.5.727

Mau YS, Markus JER, Shirly S, Oematan S, Ndiwa ASS, Handoko DD, Nasution A, Makbul K (2017) Genetic diversity of red and black upland rice accessions from East Nusa Tenggara, Indonesia as revealed by agro-morphological characters. Biodiversitas 18:197–211. https://doi.org/10.13057/biodiv/d180126

Min B, Mcclung AM, Chen MH (2011) Phytochemicals and antioxidant capacities in rice brans of different color. J Food Sci 76:C117–C126. https://doi.org/10.1111/j.1750-3841.2010.01929.x

Morais-Júnior OP, Melo PGS, Morais OP, Castro AP, Breseghello F, Utumi MM, Pereira JA, Wruck FJ, Colombari-Filho JM (2015) Genetic progress after cycles of upland rice recurrent selection. Sci Agric 72:297–305. https://doi.org/10.1590/0103-9016-2014-0137

Nile SH, Park SW (2014) Edible berries: bioactive components and their effect on human health. Nutrition 30:134–144. https://doi.org/10.1016/j.nut.2013.04.007

Oki T, Masuda M, Nagai S, Take'ichi M, Kobayashi M, Nishiba Y, Sugawara T, Suda I, Sato T (2005) Radical-scavenging activity of red and black rice. In: Toriyama K, Heong KL, Hardy B (eds) Rice is life: scientific perspectives for the 21st century. Proceedings of the world rice research conference, 4–7 November 2004, Tokyo and Tsukuba, Japan. International Rice Research Institute, Los Baños, Philippines; and Japan International Research Center for Agricultural Sciences, Tsukuba, pp 256–259

Pereira JA (2002) Cultura do arroz no Brasil: subsídios para a sua história. Embrapa Meio-Norte, Teresina, p 226

Pereira JA (2004) O arroz-vermelho cultivado no Brasil. Embrapa Meio-Norte, Teresina, p 90

Pereira JA, Morais OP, Breseghello F (2008) Análise da heterose de cruzamentos entre variedades de arroz-vermelho. Pesq Agropec Bras 43:1135–1142. https://doi.org/10.1590/S0100-204X2008000900006

Pereira JA, Bassinello PZ, Cutrim VA, Ribeiro VQ (2009) Comparação entre características agronômicas, culinárias e nutricionais em variedades de arroz branco e vermelho. Rev Caatinga 22:243–248. https://ainfo.cnptia.embrapa.br/digital/bitstream/item/33406/1/CAATINGAPZB.pdf

Pintha K, Yodkeeree S, Limtrakul P (2015) Proanthocyanidin in red rice inhibits MDA-MB-231 breast cancer cell invasion via the expression control of invasive proteins. Biol Pharm Bull 38:571–581. https://doi.org/10.1248/bpb.b14-00719

Prabhu AS, Bebendo IP, Faria JC, Souza DM, Soave J, Amaral JRM (1982) Fontes de resistência vertical a Pyricularia oryzae em arroz. Summa Phytopathol 8:78–90. https://ainfo.cnptia.embrapa.br/digital/bitstream/item/83056/1/prabhu.pdf

Ramaiah K, Rao MVBN (1953) Rice breeding and genetics. ICAR science monograph 19. Indian Council of Agricultural Research, New Delhi

Shao Y, Bao J (2015) Polyphenols in whole rice grain: genetic diversity and health benefits. Food Chem 180:86–97. https://doi.org/10.1016/j.foodchem.2015.02.027

Sieniawska E (2015) Activities of tannins from in vitro studies to clinical trials. Nat Prod Commun 10:1877–1884. https://doi.org/10.1177/1934578X1501001118

Slow Food Foundation for Biodiversity (2019) Bhutanese red rice. https://www.fondazioneslowfood.com/en/ark-of-taste-slow-food/bhutanese-red-rice/. Accessed on 1 Apr 2019

Smeriglio A, Monteleone D, Trombetta D (2014) Health effects of Vaccinium myrtillus L.: evaluation of efficacy and technological strategies for preservation of active ingredients. Mini Rev Med Chem 14:567–584. https://doi.org/10.2174/1389557514666140722083034

Smeriglio A, Barreca D, Bellocco E, Trombetta D (2017) Proanthocyanidins and hydrolysable tannins: occurrence, dietary intake and pharmacological effects. Br J Pharmacol 174:1244–1262. https://doi.org/10.1111/bph.13630

Sultan SM, Subba Rao LV (2013) Germplasm collection from last remnants of rice Landrace genetic diversity in high altitude areas of Kashmir Himalayas. Int J Conserv Sci 4:467–476. http://www.ijcs.uaic.ro/public/IJCS-13-44-Sultan.pdf

Sweeney MT, Thomson MJ, Pfeil BE, McCouch S (2006) Caught red-handed: Rc encodes a basic helix-loop-helix protein conditioning red pericarp in rice. Plant Cell 18:283–294. https://doi.org/10.1105/tpc.105.038430

Sweeney MT, Thomson MJ, Cho YG, Park YJ, Williamson SH, Bustamante CD, McCouch SR (2007) Global dissemination of a single mutation conferring white pericarp in rice. PLoS Genet 3:e133. https://doi.org/10.1371/journal.pgen.0030133

Tan L, Li X, Liu F, Sun X, Li C, Zhu Z, Fu Y, Cai H, Wang X, Xie D, Sun C (2008) Control of a key transition from prostrate to erect growth in rice domestication. Nat Genet 40:1360–1364. https://doi.org/10.1038/ng.197

Tsuda T, Horio F, Osawa T (2002) Cyanidin 3-O-β-D-glucoside suppresses nitric oxide production during a zymosan treatment in rats. J Nutr Sci Vitaminol 48:305–310. https://doi.org/10.3177/jnsv.48.305

Vichit W, Saewan N, Prinyarux T (2018) Anti-aging efficacy of Thai red rice callus cosmetic product. J Appl Sci 17:S63–S72. https://doi.org/10.14416/j.appsci.2018.02.S08

Walter M, Marchesan E, Massoni PFS, da Silva LP, Sartori GMS, Ferreira RB (2013) Antioxidant properties of rice grains with light brown, red and black pericarp colors and the effect of processing. Food Res Int 50:698–703. https://doi.org/10.1016/j.foodres.2011.09.002

Wickert E, Schiocchet M, Noldin J, Raimondi J, Andrade A, Scheuermann K, Marschalek R, Martins G, Hickel E, Eberhardt D, Knoblauch R (2014) Exploring variability: new Brazilian varieties SCS119 Rubi and SCS120 Onix for the specialty rices market. Open J Genet 4:157–165. https://doi.org/10.4236/ojgen.2014.42016

Yawadio R, Tanimori S, Morita N (2007) Identification of phenolic compounds isolated from pigmented rices and their aldose reductase inhibitory activities. Food Chem 101:1616–1625. https://doi.org/10.1016/j.foodchem.2006.04.016

Zhao C, Giusti MM, Malik M, Moyer MP, Magnuson BA (2004) Effects of commercial anthocyanin-rich extracts on colonic cancer and nontumorigenic colonic cell growth. J Agric Food Chem 52:6122–6128. https://doi.org/10.1021/jf049517a

Zhou Z, Chen X, Zhang M, Blanchard C (2014) Phenolics, flavonoids, proanthocyanidin and antioxidant activity of brown rice with different pericarp colors following storage. J Stored Prod Res 59:120–125. https://doi.org/10.1016/j.jspr.2014.06.009

Ziska LH, Gealy DR, Burgos N, Caicedo AL, Gressel J, Lawton-Rauh AL, Avila LA, Theisen G, Norsworthy J, Ferrero A, Vidotto F, Johnson DE, Ferreira FG, Marchesan E, Menezes V, Cohn MA, Linscombe S, Carmona L, Tang R, Merotto A (2015) Chapter three-weedy (red) rice: an emerging constraint to global rice production. Adv Agron 129:181–228. https://doi.org/10.1016/bs.agron.2014.09.003

Zhu Y, Lin Y, Chen S, Liu H, Chen Z, Fan M, Hu T, Mei F, Chen J, Chen L, Wang F (2019) CRISPR/Cas9-mediated functional recovery of the recessive rc allele to develop red rice. Plant Biotechnol J 17:2096-2105. https://doi.org/10.1111/pbi.13125

Part V
Oryza Species and Rice Grain Quality

Chapter 13
Wild *Oryza* for Quality Improvement

Sharmin Hasan and Robert J. Henry

13.1 Introduction

The genus *Oryza* evolved around 130 million years ago in Gondwana and was later found in the current distributions of this supercontinent: Africa, Asia, the Americas, and Australia (Khush 1997; Brar and Singh 2011; Brar and Khush 2018). Modern *Oryza* species have experienced varying periods of a few thousands to 15 million years of independent evolutionary processes that continue today (Stein et al. 2018). Evidence from archeological data and molecular clock analysis of DNA sequences indicates that rice domestication probably commenced in the Yangtze Valley of China dating back around 8200–13,500 years (Molina et al. 2011). The genus comprises a total of 11 genomes that are found in 2 cultivated (A-genome) and 24 wild rice species (AA-HHJJ) (Brar and Khush 2018). The discriminating taxonomic attributes of *Oryza* species are associated with varying factors, including geographical distribution, annual or perennial habitats, morphological and physiological traits, genomic composition, and cultivated or wild habitats (Vaughan et al. 2008a). Depending on all these aspects, the genus has been grouped into five species complexes: *O. sativa*, *O. officinalis*, *O. meyeriana*, *O. ridleyi*, and Unclassified (Brar and Khush 2018) (Table 13.1).

Wild *Oryza* species like the wild relatives of other grass crop species display phenotypically generally poorer agronomic characteristics, including prosthetic

S. Hasan
Queensland Alliance for Agriculture and Food Innovation, The University of Queensland, Brisbane 4072, QLD, Australia

Department of Botany, Jagannath University, Dhaka 1100, Bangladesh
e-mail: s.hasan@uq.net.au

R. J. Henry (✉)
Queensland Alliance for Agriculture and Food Innovation, The University of Queensland, Brisbane 4072, QLD, Australia
e-mail: robert.henry@uq.edu.au

A. Costa de Oliveira et al. (eds.), *The Future of Rice Demand: Quality Beyond Productivity*, https://doi.org/10.1007/978-3-030-37510-2_13

Table 13.1 A brief overview of the genus *Oryza*: chromosome number, genomic composition, distribution, and their useful traits

Species	2n	Genome	No. of accessions[a]	Distribution	Useful traits
Oryza sativa complex: 02 cultivar and 06 wild species					
O. sativa L. (Cultivar)	24	AA	116,751	Worldwide	Cultigen, highly productive
O. glaberrima Steud. (Cultivar)	24	A^gA^g	1655	West Africa	Cultigen; low yielding but has tolerance to drought, acidity, P deficiency; iron toxicity; resistance to BB and blast, RYMV, African gall midge, nematodes; weed competitiveness
O. nivara Sharma et Shastry	24	AA	1503	Tropical and subtropical Asia	Resistance to grassy stunt virus, BB, blast, BPH, yield, and grain quality-enhancing loci (QTLs); grain micronutrient traits loci
O. rufipogon Griff.	24	AA	1034	Tropical and subtropical Asia, tropical Australia	Resistance to BB, blast, BPH, tungro virus; moderately tolerant to Shb, tolerance to aluminum and soil acidity; source of CMS, yield, and grain quality-enhancing QTLs; grain micronutrient traits loci
O. breviligulata A. Chev.et Roehr	24	A^gA^g	218	Africa	Resistance to GLH, BB; drought avoidance
O. longistaminata A. Chev. et Roehr	24	A^lA^l	216	Africa	Resistance to BB, nematodes, stem borer, drought avoidance
O. meridionalis Ng	24	A^mA^m	56	Tropical Australia	Elongation ability; drought avoidance; grain quality-enhancing loci
O. glumaepatula Steud.	24	$A^{gp}A^{gp}$	54	South and Central America	Resistance to blast, elongation ability; source of CMS
O. officinalis complex: 12 wild species					
O. punctata Kotschy ex Steud	24	BB	71	Africa	Resistance to BPH, zigzag leafhopper
O. schweinfurthiana Prodoehl	48	BBCC	35	Africa	Resistance to BPH, zigzag leafhopper
Species	2n	Genome	No. of accessions[a]	Distribution	Useful traits
O. minuta J.S. Presl. ex C.B. Presl.	48	BBCC	63	The Philippines and Papua New Guinea	Resistance to BB, blast, BPH, GLH, grain quality-enhancing loci
O. malampuzhaensis Krishnasw. and Chandrasekh	48	BBCC	13	India	Resistance to thrips, BPH, GLH, WBPH, BB, stem rot

(continued)

Table 13.1 (continued)

Species	2n	Genome	No. of accessions[a]	Distribution	Useful traits
O. officinalis Wall ex Watt	24	CC	275	Tropical and subtropical Asia, tropical Australia	Resistance to thrips, BPH, GLH, WBPH, BB, stem rot; tolerance to heat, grain quality-enhancing loci
O. rhizomatis Vaughan	24	CC	20	Sri Lanka	Drought avoidance, resistance to blast; tolerance to heat
O. eichingeri A. Peter	24	CC	30	South Asia and East Africa	Resistance to BPH, WBPH, GLH; grain micronutrient traits loci
O. latifolia Desv.	48	CCDD	58	South and Central America	Resistance to BPH, BB, high biomass production; grain quality-enhancing loci
O. alta Swallen	48	CCDD	12	South and Central America	Resistance to striped stemborer; high biomass production; grain quality-enhancing loci
O. grandiglumis (Doell) Prod.	48	CCDD	10	South and Central America	High biomass production
O. australiensis Domin	24	EE	36	Tropical Australia	Resistance to BPH, BB, blast; drought avoidance; tolerance to heat and drought; grain quality-enhancing loci
O. brachyantha A. Chev. et Roehr.	24	FF	19	Africa	Resistance to BB, yellow stemborer, leaf folder, whorl maggot; tolerance to laterite soil
O. meyeriana complex: 03 wild species					
O. granulata Nees et Arn. ex Watt	24	GG	24	South and Southeast Asia	Shade tolerance; adaptation to aerobic soil
Species	2n	Genome	No. of accession[a]	Distribution	Useful traits
O. meyeriana (Zoll. et (Mor. ex Steud.) Baill.)	24	GG	11	Southeast Asia	Shade tolerance; adaptation to aerobic soil; grain quality-enhancing loci
O. neocaledonica Morat	24	GG	1	New Caledonia	
O. ridleyi complex: 02 wild species					
O. longiglumis Jansen	48	HHJJ	6	Irian Jaya, Indonesia, and Papua New Guinea	Resistance to blast, BB; grain quality-enhancing loci

(continued)

Table 13.1 (continued)

Species	2n	Genome	No. of accessions[a]	Distribution	Useful traits
O. ridleyi Hook. F.	48	HHJJ	15	South Asia	Resistance to blast, BB, blast, tungro virus, stem borer, whorl maggot
Unclassified: 02 wild species					
O. schlechteri Pilger	48	HHKK	1	Papua New Guinea	Stoloniferous
O. coarctata Roxb	48	KKLL	1	South Asia	Salt tolerance

Modified from Brar and Khush (2018); Sanchez et al. (2013)
BPH brown plant hopper, *GLH* green leafhopper, *WBPH* white-backed plant hopper, *BB* bacterial blight, *Shb* sheath blight, *CMS* cytoplasmic male sterility, *RYMV* rice yellow mottle virus; a: Accessions maintained in rice genebank at IRRI, the Philippines
Comment: *O. nivara* may be an annual ecotype of *O. rufipogon*; *O. breviligulata* is now *O. barthii*; *Porteresia coarctata* is now *O. coarctata*; tetraploid *O. punctata* is now *O. schweinfurthiana*; tetraploid *O. officinalis* is *O. malampuzhaensis*

plant habit, less grain yield, inferior grain quality, and shattering nature in seed (Brar and Khush 2018). Intriguingly, these species contain many beneficial genes of interest for yield- and quality-related traits. However, an enormous number of these genes still remain unexploited because of the difficulty of transferring these genes into cultivated rice. Over time, wild species have been under selection for a collection of phenotypic traits (i.e., domestication syndrome) such as non-shattering spikelet, low seed dormancy, synchronized flowering and seed development, few tiller and branches, and numerous inflorescences and seeds. This has led them to become domesticated as their survival and dispersal depend on humans (Molina et al. 2011). Domestication is an intricate evolutionary process wherein wild species experience a genetic-based morphological and physiological modification, leading to improved productivity and quality traits (Tan et al. 2008; Vaughan et al. 2008a; Purugganan and Fuller 2009; Molina et al. 2011). In this process, a series of genomic modifications, including selection of nucleotide polymorphisms resulting from spontaneous mutation, recombination, and fixation of alleles, altered wild rice into domesticated rice (Vaughan et al. 2008a; Molina et al. 2011; He et al. 2011). Artificial selection pressure for many agronomically important traits facilitated this domestication process primarily influenced by cultural preferences for distinct food and a wide range of ecological conditions (Vaughan et al. 2008a). Such agronomic features of rice include the yield, grain size and shape, grain color, aroma, and amylose content (Kovach et al. 2007). The domestication process provides insights into evolutionary mechanism such as founder events, selection, introgression, and genetic bottlenecks as well as the cultural context for the human selection of cultivated rice (Vaughan et al. 2008a, b; Molina et al. 2011). As domesticated rice species originated from a founder population of wild rice species, genetic bottlenecks and artificial selection together resulted in a lower genetic diversity in cultivated rice. Two cultivated rice species, namely, *Oryza sativa* L. and *Oryza glaberrima* Steud., were domesticated independently from the AA-genome of wild *Oryza*, known as Asian rice and African rice, respectively (Vaughan et al. 2008a; Londo et al. 2006) (Fig. 13.1). Asian rice is more prevalent than African rice, which is largely confined to West Africa. African rice exhibits annual habit with limited secondary panicle branches, red color grain pericarp, and short ligule relative to Asian rice (Vaughan et al. 2008a).

O. sativa is domesticated from two wild progenitors: the perennial *Oryza rufipogon* Griff. and the annual *Oryza nivara* Sharma et Shastry. The perennial strains are widely found in Australia and Asia, while the annual ecotypes are confined to South and Southeast Asia (Vaughan et al. 2008a; Sang and Ge 2013). However, recent evidence suggests that the Australian populations are a distinct taxon (Wambugu et al. 2015). Lately researchers have suggested that *O. nivara* is an annual ecological ecotype of *O. rufipogon* (Molina et al. 2011). The annual and perennial ecotypes are different from their life history, mating system, and habitat preferences (Chen et al. 2004; Li et al. 2006a; Vaughan et al. 2008a). Perennial ecotypes are found in wet areas such as lakes and waterways, while annual ecotypes prefer habitats in seasonally dry areas (Vaughan et al. 2008a). In perennial *O. rufipogon*, an outcrossing mating system is predominant, resulting in high genetic variability that would support evolutionary diversification. In contrast, selfing is the mating system found in *O. nivara* that exhibits high seed productivity and seed dormancy concordant with cultivated *O. sativa* (Cheng et al. 2003; Li et al. 2006a; Vaughan et al. 2008a; Sang and Ge 2013). Six groups of Asian rice varieties, namely, *indica*, *japonica*, *aus*, *aromatic*, *rayada*, and *ashina*, were reported based on population structure analysis with 15 isozyme loci (Glaszmann 1987). Another study using 234 accessions of rice with 2 chloroplast loci and 169 simple sequence repeats (SSRs) distinguished 5 distinct classes: *indica*, *aromatic*, *aus*, *temperate japonica*, and *tropical japonica* (Garris et al. 2005). This research also postulated that both *indica* and *aus* had close evolutionary relationship and the groupings emerged due to their differential demographic histories. The *aus*, *rayada*, and *ashina* varieties are distributed in the Himalayan foothills. *aus* and *rayada* varieties are especially grown in Bangladesh and *ashina* in India. *Aromatic* rice varieties are grown in Pakistan, Nepal, and India. Two varieties, namely, *indica* and *japonica*, are genetically diverse

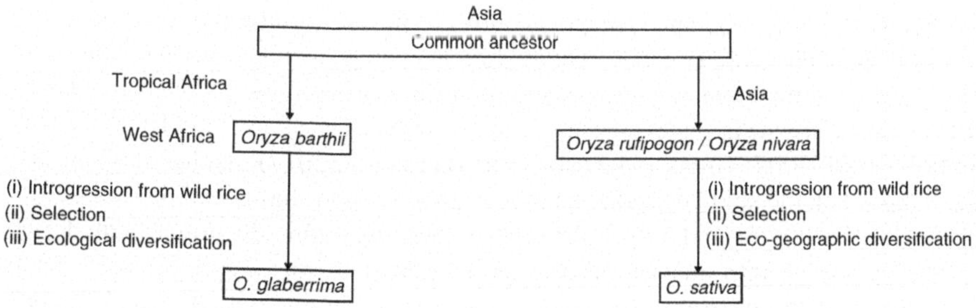

Fig. 13.1 Schematic diagram for the domestication events of cultivated rice. (Adapted from Vaughan et al. 2008a)

subspecies of Asian rice; however, the evolution history of these two subspecies is still convoluted (Chen et al. 2004; Vaughan et al. 2008a; Molina et al. 2011; He et al. 2011).

Despite having evidence from genomic and archeological studies, there is still an unresolved debate about the geographic location of the origin of domesticated rice, the contributing strains of *O. rufipogon* as an ancestor, and single (monophyletic) or multiple (polyphyletic) domestication events that were associated with the origin of the two subspecies of *O. sativa* (Gross and Zhao 2014; Huang et al. 2012). The hypothesis of a single origin or snowballing model suggests that a single domestication event resulted in an early rice cultivar *O. sativa*, and then subsequently *indica* and *japonica* originated from post-domestication introgression–hybridization with different wild progenitors (Vaughan et al. 2008a, b; Molina et al. 2011; Huang et al. 2012). In one version of this, *indica* rice evolved first from *O. rufipogon* in South and Southeast Asia, and then *japonica* originated from *indica* subspecies when they moved and adapted to the higher latitudes and elevations (Lu et al. 2002; Gao and Innan 2008). In contrast, a different hypothesis for the snowballing model or single-origin model came from Huang et al. (2012) suggesting that at first *japonica* evolved in southern China and then *indica* was domesticated from wild populations following the successive introgression of domesticated alleles of *japonica* subspecies. One of the most convincing lines of evidence that supports this model based on recent molecular studies is that both subspecies share closely related genes such as *sh4* and *prog1* related to two key domesticated traits: seed non-shattering (Li et al. 2006b) and erect habit (Tan et al. 2008), respectively. This suggests that the mutation occurred in one genotype that was successively transmitted into these two subspecies (Huang et al. 2012; Choi and Purugganan 2018). Furthermore, Gao and Innan (2008) conducted a Bayesian demographic analysis of 60 multilocus microsatellite loci using 92 *O. rufipogon* accessions and 2 subspecies of *O. sativa* and claimed the *japonica* subspecies exhibited substantially a reduced level of genetic variability relative to the *indica* subspecies due to a rigorous bottleneck effect that resulted from demography and selection pressure during their domestication. Furthermore, they advocated for nonindependent domestication of the two subspecies because somewhat they had common founder populations and the ample genetic exchange occurred between them through gene flow.

Unlike a single domestication model, *indica* and *japonica* subspecies are assumed to have originated independently from pre-differential progenitor of *O. rufipogon* following two different evolutionary trajectories: *japonica* from wild *O. rufipogon* in the Yangtze Valley of southern China as early as 8000 BP and *indica* from wild O. *nivara* in the Ganges of northern India as early as 5000 BP (Sweeney and McCouch 2007; Gross and Zhao 2014; Bates et al. 2017). Archaeological evidence based on the study of spikelet bases and weeds from Indus Civilization settlements in North-west India speculated that there was already domesticated rice cultivation before introduction of fully domesticated *O. sativa* spp. *japonica* from China to India around 2000 BCE (Bates et al. 2017). Phylogeographic analysis further revealed that *indica*-like and *japonica*-like haplotypes were related to wild rice that were distributed in South and Southeast Asia, and in China, respectively (Londo et al. 2006; Vaughan et al. 2008a). Another study based on phylogenetic data using all the DNA sequences explained that *O. sativa* formed two distinct clades for *indica* and *japonica* and they were closely associated with different *O. rufipogon* strains (Rakshit et al. 2007; Vaughan et al. 2008a; Molina et al. 2011).

A piece of far more clear evidence from molecular clock studies conferred that genetic divergences in AA genome wild rice occurred prior to the origin of *O. sativa* (Vaughan et al. 2008b). A very recent investigation provided a contrasting revelation about the domestication of Asian rice and raised a question of whether both models can act in the domestication process. This view supports multiple origins of cultivated rice; however, de novo domestication happened once, and later the introgression events facilitated the transfer of domestication alleles between *japonica* and *indica* subpopulations (Choi and Purugganan 2018). The evolution of African-cultivated rice took place more than 3000 years ago after the origin of *O. sativa* (Wang et al. 2014). *O. barthii* A. Chev.et Roehr is assumed to be an ancestor of African cultivated rice, which followed dual evolutionary trajectories: firstly, the divergence occurred in AA-genome Asian *Oryza*, which was introduced in Africa and assumed to be an ancestor of *O. barthii*, and secondly, introgression, selection, and ecological diversification of African wild progenitor led to evolving African cultivated rice (Vaughan et al. 2008a). A strong claim to strengthen the hypothesis of the domestication of this rice cultivar near the Niger River was made by Wang et al. (2014) through population genomic analyses of *O. glaberrima* and *O. barthii* accessions. They observed that these two species shared a set of common genes which probably resulted from the domestication process. Despite the debate on the genetic divergence and domesticated gene uniformity, it is unequivocally viewed that hybridization or introgression had a significant role in shaping the genomic structure of the present rice cultivars and artificial selection balanced the fixation of key domesticated alleles and genetic diversity across the cultivar gene pools (Sang and Ge 2013).

13.2 Challenges for Crop Improvement

Rice has long been regarded as a major dietary source for over half the world's population. Rice meets over 35–60% of the caloric needs of over three billion people (Zhang et al. 2018). The global population is predicted to rise by at least 25% by the year 2050. As a result, global rice demands will increase by 40 million tons from 2015 to 2020, with a further rise to 852 million tons by 2035 (Brar and Khush 2018). Rice production has taken a leap since the development of the Green Revolution technology introduced by the International Rice Research Institute (IRRI) in the 1960s (Khush 1997; Jena and Mackill 2008). The improvement of grain quality of rice has not been prioritized in comparison with the improvement of yield to date. Rice is cultivated on around 154 million hectares annually in more than 100 countries across the world. In 2017, it is estimated that around 770 million tons of rice are being produced in around 167-million-hectare area (FAOSTAT

2017). There is a continuous growth in rice production; however, it has kept a slow pace relative to that required meeting the increased global demands of the ever-increasing population. It is calculated that the growth in annual rice production is at below 1%, while the population grows at a rate of 1.5% per year.

The major underlying challenges of increasing rice production of the required 25–30% in the next 30 years include irresistible population growth, loss of farming land, soil degradation, fertilization cost for higher labor and high energy, biofuel policy, declining rice yield, and abiotic and biotic stress because of climate change and other resources required for cultivation (Fitzgerald et al. 2009). Climatic changes, especially rising temperature, an intensified hydrological cycle, and increasing atmospheric carbon dioxide (CO_2) and tropospheric ozone (O_3) impose a cascade of challenges to agriculture (Lobell and Gourdji 2012). Global warming has a direct negative impact on crop production. Peng et al. 2004 suggested that an increment of only 1 °C in temperature resulted in the reduction of 10% grain yield in a dry season. Rice cultivars need to be improved so that they have an increased ability to adapt to a new and stressful environment as well as faster growth with increased yield, superior nutritional quality, and resistance to abiotic and biotic stresses. Improved cultivars are required to combat these threats enabling enhanced yield to feed the ever-increasing global population. Wild rice species renders an enormous source of valuable genes that can withstand many existing and emerging stresses. Numerous genes, especially for biotic and abiotic stresses from wild rice, have been incorporated into cultivated rice to produce new elite varieties. These varieties show resistance to brown plant hopper, grassy stunt, tungro virus, blast, bacterial blight, and tolerance to salinity, drought, and acid sulfate (Brar and Singh 2011). In this way, food security can be achieved by searching for variation in wild rice species, identifying new alleles and genes required for developing improved rice varieties. Breeding efforts need to monitor the exact impacts of climate change on world crop production in association with climatologists and adopt those traits that are currently being measured and modeled in an integrated approach combining genome-based technologies, data mining, and phenotyping methods to develop climate-resilient crops (Varshney et al. 2018).

Even though rice productivity and consumption increase over time, global population health is deteriorating for lack of nutrition because some micronutrients like iron, zinc, and vitamins A, B, C, and D, essential for human health, are insufficient in milled rice. The world, especially developing countries, faces major health-related problems, including malnutrition due to lack of micronutrients and vitamins and chronic diseases (obesity and type II diabetes) (Dipti et al. 2012). Over 3 billion people are suffering from at least one or more vitamin and/or mineral deficiencies and 3.1 million children decease every year due to malnutrition (Mahender et al. 2016). Developing countries usually treat undernutrition-related diseases by adopting pharmaceutical supplements and fortification programs; however, those options are expensive for the poor. The debate continues on whether the policy to increase the yield of lower cost and energy-dense crops, e.g., rice, wheat, and corn, which are widely consumed by the poor can solve this problem or amelioration of crop nutritional status can combat micronutrients deficiency-related health problems

(Underwood 2000). Keeping this in mind, nutritionists and agriculturists need to work hand in hand and focus on fortification research, with an aim to increase mineral contents in crops via selective breeding or genetic modification. Recent medical research demands a shift in focus to enhancing the quality of crops for human health (Fernie et al. 2006). Micronutrients have a vital role in physiological functions in the human body that must be provided in foods or as supplements because they cannot be synthesized in the body in a sufficient amount to meet human nutritional need (Underwood 2000). Foods like rice that we intake in our daily life can meet our mineral need as the components and content of minerals in rice grains directly or indirectly influence human metabolism (Zhang et al. 2018). As a proximate solution, biofortification – a process to enrich nutritional contents in stable crops via plant breeding – first began in the early 1900s and still continues to increase the amount of iron (Fe), zinc (Zn), and beta-carotene in staple crops. This can be done in a crop through plant breeding, transgenic techniques, and agronomic practices. For instance, through transgenic technology, a nicotianamine synthase (NAS) gene was incorporated into the rice endosperm from barley that resulted in an elevated Fe content in rice (Lee et al. 2009). Another striking example of biofortification is genetically engineered golden rice, which is a source of provitamin A containing 1.6–35 $\mu g\ g^{-1}$ β-carotene of rice (Tang et al. 2009). The protein content was increased in endosperm of rice by inserting a soybean glycinin gene through the application of biotechnology (Momma et al. 1999). Biofortification programs cannot be successful in places where producers or consumers show a reluctance to accept biofortified crops.

13.3 Rice Quality Criteria Used by Consumers

Grain quality traits of crops regulate the market values of a new variety and are governed by a consumer's preferences. Rice grain quality includes physical appearance, cooking, eating, and sensory properties (Fitzgerald et al. 2009). A large proportion of consumers prefer to consume polished milled rice. Consumers, millers, whole sellers, and retailers focus on the appearance of rice grain (Fitzgerald et al. 2009). The attributes of grain appearance include the shape and size of kernels, chalkiness, and translucency of grains (Cruz and Khush 2000). Rice varieties are mainly classified on the basis of two physical parameters: length and shape. Based on kernel dimensions, grain length is divided into four groups: extra-long (>7.50 mm), long (6.61–7.50 mm), medium (5.51–6.60 mm), and short (≥ 5.50 mm). Similarly, on the basis of grain length and width ratio, grain shape is categorized into three groups: bold (<2), medium (2.1–3), and slender (>3) (Cruz and Khush 2000; Calingacion et al. 2014). The global patterns of rice consumption based upon grain appearances, cooking behavior, taste, tenderness, and aroma vary from region to region depending on historical and cultural diversity and environmental conditions for growth in a given area (Cruz and Khush 2000) (Fig. 13.2). For instance, people from parts of Southeast Asia, Thailand, Lao PDR, Cambodia, Malaysia, and

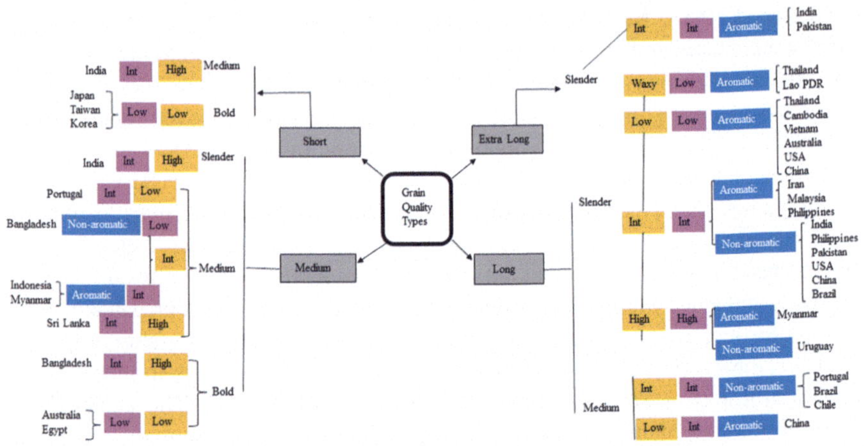

Fig. 13.2 Different combinations of grain quality types in each length class based on few quality parameters: grain length (grey), shape, amylose content (yellow), gelatinization temperature (pink), and fragrance (blue). Int: Intermediate. (Adapted from Calingacion et al. 2014)

the Philippines consume long and slender grains. In Bangladesh and Indonesia, consumers favor slender and medium grain with intermediate amylose content. The Japanese, the Taiwanese, and the South Koreans prefer short and bold grains. People from parts of India, Sri Lanka, and Myanmar consumed bold and short grains. Extra-long grains are mainly consumed by most parts of Pakistan, Punjab, and Haryana states in India and Iran (Calingacion et al. 2014). The whiteness or translucency of polished rice is also an attraction to the rice consumers who pay a high price for these traits, although this rice contains an insignificant amount of the necessary vitamins and minerals compared to brown rice with a hull and bran, which have more Fe, minerals, and phytochemicals (Dipti et al. 2012). Rice with chalky endosperm has a poor appearance and low market value, even though the opacity goes away after cooking without altering eating quality (Cruz and Khush 2000).

Cooking and eating qualities that are due to the amount of starch, types of starch (i.e., amylose content), aroma, texture, and flavor are important criteria for discerning consumers (Calingacion et al. 2014). Consumer preferences not only vary from different countries across the world but also across the regions of a country. For instance, a smooth-textured rice with intermediate amylose content is popular to the Javanese rice consumers, whereas North and West Sumatrans prefer rice with hard texture and high amylose content (Septiningsih et al. 2003). Waxy (glutinous) or sticky rice, which does not expand in volume, is popular to the consumer from Lao PDR and the Isan province of Thailand (Cruz and Khush 2000). Low amylose-containing rice is popular with people from Japan, Taiwan, Cambodia, Thailand, Egypt, Australia, the northern and southwestern province of China and Southern China, and the areas of Lao PDR. Intermediate amylose rice is preferable to Malaysia, Pakistan, the Philippines, India, Vietnam, Uruguay, and Indonesia (Cruz and Khush 2000; Calingacion et al. 2014). Consumers in Ghana, Senegal, Surinam,

Colombia, and parts from Uruguay prefer high-amylose rice varieties (Calingacion et al. 2014). Aromatic rice characterized by the presence of 2-acetyl-1-pyrroline (2AP) has a great influence on both the market value and human preferences. This rice appears to be very popular with the Greater Mekong Subregion (GMS), Pakistan, Malaysia, Iran, and Punjab and Haryana states in India, and in some provinces of Indonesia. In GMS, the fragrant rice is an *indica* jasmine type (Calingacion et al. 2014), while consumers from the Middle East prefer the strong aroma. In contrast, European consumers have a negative approach to aromatic rice, as they feel that fragrance may arise from contamination (Cruz and Khush 2000). Two combinations of rice traits are most preferred by the consumers across the world: one type combines aromatic slender and long grain with low amylose content and gelatinization temperature, and the other type consists of nonaromatic long and slender grain size with intermediate gelatinization temperature and amylose content (Calingacion et al. 2014). Hence, these traits need to be improved to fulfil the growing consumer need for quality rice varieties.

13.4 Potential of Wild *Oryza* for Quality Improvement

Due to the dwindling genetic diversity of cultivated rice, many efforts have failed to find sufficient natural variation in cultivated rice germplasm to increase the productivity, adaptation, nutritional values, and grain quality of rice. Wild species are considered as an inherent source of genetic variation for crop improvement as this genetic resource is almost untapped or much less explored. Some other attributes of wild rice, including tolerance to unfavorable soil, climate, weathering conditions, and pests, suggest a potential role in improving cultivars The use of wild *Oryza* provides access to a relatively unexplored reservoir of useful quantitative trait loci (QTLs) and genes that exhibits a gene pool with allelic diversity required for rice improvement. Grain appearance, cooking, eating and sensory, and nutritional and processing quality are the main quality attributes of rice that are assessed by their physical and chemical properties as well as social and cultural context (Septiningsih et al. 2003; Aluko et al. 2004). In this section, we will document the progress in understanding the genetic background that underlies rice quality traits as well as the nascent potential of wild rice germplasm to contribute new variation for these traits.

13.5 Grain Appearance Quality

Grain appearance is evaluated by its size and shape, length and width, length and width percentage (L/W), translucency, and chalkiness of the endosperm (Tan et al. 2000), and this is important, as it is what consumers first see. Grain size and shape are regarded as domesticated syndrome that has been under selection during crop domestication as these traits are very crucial to human (Wu et al. 2017). Rice grain development differs from that of the other cereal crops because the grain weight of cereal crops generally depends on the supply of carbohydrate by photosynthetic leaves (source capacity) and the capacity to accumulate carbohydrate by the developing grain (sink capacity). Whereas rice grain development is largely dependent on the sink capacity, which in turn depends on the number of grains and their potential size meaning larger grain size corresponds to greater carbohydrate accumulation (Chen et al. 1994; Tikapunya et al. 2017a).

Many QTLs contributed to rice grain shape, size, and weight have been mapped in a wide range of populations over the last decade (Fitzgerald et al. 2009). So far, the cloning and characterization of these QTLs, including *GS3*, *GS5*, *GW2*, *GL3.1* (*qGL3*), *GL4*, *GL7*, *GLW7*, *GW5* (*qSW5*), *GW8*, and *GIF7*, have revealed that these loci regulate grain shape and width through triggering the cell cycle mechanism to accelerate cell division in cultivated rice (Tan et al. 2000; Wan et al. 2005; Fan et al. 2006; Song et al. 2007; Weng et al. 2008; Shomura et al. 2008; Mao et al. 2010; Li et al. 2011; Wang et al. 2012, 2015a; Si et al. 2016; Wu et al. 2017). Grain weight, which is a determinant of grain size, can be measured by grain length, width, and thickness (Si et al. 2016). The genetic study of grain shape reveals that grain width and length are regulated by both major and minor QTLs, wherein few major genes have large effects to a particular trait, while many minor genes confer small effects. Fan et al. (2006) derived near-isogenic lines (NILs) from a backcross between Chuan 7 (small grain) and Minghui 63 (large grain) and determined that a QTL (*GS3*), which was around 7.9 kb in length, mapped to chromosome 3 contributed to grain length and weight. This locus acted as both major and minor QTL for grain length and weight and grain thickness and width, respectively. Furthermore, another study confirmed that grain size was regulated by at least four different alleles, *GS3–1*, *GS3–2*, *GS3–3*, and *GS3–4*, of which *GS3–1* was perhaps the prototype and resembled the genotype of medium grain length that is found in widely grown *indica* type varieties, while *GS3–2* was associated with medium-sized grain of the *japonica* variety Nipponbare, and the other two alleles *GS3–3* and *GS3–4* contributed to long and short grain for the Minghui 63 and Chuan 7 varieties, respectively (Mao et al. 2010).

From a double-haploid (DH) population developed by crossing between two *indica* accessions, namely, H94 and Zhenshan 97, Li et al. (2011) detected a *GS5* locus that translates a putative serine carboxypeptidase, which positively contributed to the development of grain size. A major QTL (*GL7*) mapped to chromosome 7 for grain length was identified in an F_2 population developed by crossing between the *indica* variety, P13 (superior grain length and appearance quality), and *O. sativa* spp. *japonica* cv. Nipponbare that affected grain L/W ratio and the development of starch particles in the endosperm resulting in improvement of grain appearance quality (Wang et al. 2015a). In a study of the grain size of African cultivated rice, Wu et al. (2017) derived an F_2 population with 186 individuals from a cross between several introgression lines developed by a cross between a donor *O. barthii* accession (IRGC101248) and a recipient *O. glaberrima* accession (IRGC102305) and IRGC102305 and identified a locus (*GL4*) for grain length mapped to chromosome 4 between the marker RM3335 and RM5608 that regulated longitudinal cell division of both inner and outer glumes. The same study also postulated that during

African rice domestication, an *O. barthii* allele contributed to the enlarged grain length and an early stop codon in *GL4* locus due to a single-nucleotide polymorphism (SNP) contributed to small seeds but non-shattering in nature. Several QTLs corresponding to grain weight have been documented in the literature. Grain weight and width-related QTL on chromosome 2 (GW2) identified in a NIL population developed by crossing between *japonica* variety Waiyin- 3 (WY3) and *indica* variety Fengaizhan-1 (FAZ1) (Song et al. 2007). This study determined that this locus encoded a new RING-type protein, which exhibited intrinsic E3 ubiquitin ligase activity in the cytoplasm and developed expression in several tissues. Its reduced expression increased grain size and grain weight, while overexpression decreased grain weight and size. The WY3 alleles encode a truncated protein, which enlarged spikelet hull size and enhanced grain milk fillings, subsequently increasing grain yield and weight.

Another study also postulated that a 1212-bp deletion from *GW5* loci located on chromosome 5 resulted in the accelerated grain width in Asominori rice cultivar relative to rice cultivar IR24 with slender grain (Weng et al. 2008). Complementary research using an F_2 population derived from Nipponbare cultivars (*japonica*) × Kasalath cultivars (*indica*) identified a new QTL on chromosome 5 for seed width (*qSW5*) (Shomura et al. 2008). Also, a weight-related QTL (*GW8*) homologous to *OsSPL16* encodes a protein that increases cell division and consequently enhancing grain width and yield in rice (Wang et al. 2012). A major QTL (*GLW7*) detected for grain length and width from genome-wide association analysis, encoding a plant-specific transcription factor *OsSPL13*, was associated with the enlargement of grain hull and subsequently led to an increased grain yield and length. This study also suggests that in *tropical japonica* rice, the overexpression of *OsSPL13* contributed to produce large grains (Si et al. 2016). Grain filling is another important trait that is used to determine grain weight and controlled by the *GIF1* gene, which produces cell wall invertase enzymes controlling carbon partitioning at the initial grain filling stage. In cultivated rice, a mutation occurred in the regulatory sequence of a gene that resulted in restricted expression during grain filling, subsequently increasing grain weight in comparison with wild rice (Wang et al. 2008).

In comparison with domesticated rice, there are many grain appearance-related QTLs from wild *Oryza* species that have been documented in the literature. Recent research based on single point analysis by Rahman et al. (2007) revealed four QTL alleles (*gl3*, *gl6*, *gl7*, and *gl9*) for grain length mapped to chromosome 3, 6, 7, and 9, respectively, in an $F_{2:3}$ population developed by crossing between an introgression line IR71033 (an introgression line derived from *O. minuta* J.S. Presl. ex C.B. Presl.) and Junambyeo (a Korean *japonica* cultivar) promoting increased grain length. In addition, this study revealed one QTL (*glw2*) for grain length to weight ratio on chromosome 2 and two QTLs for 1000-grain weight (*tgw7* and *tgw11*) that were introgressed from the wild species and noticeably contributed to increase L/W ratio of grains and grain weight, respectively, in cultivars. Two QTLs for thousand-grain weight (*gw1.1*, *gw2.1*) were detected in *O. nivara* when backcross QTL analysis was conducted in two BC_2F_2 populations by Swamy et al. (2014). Furthermore, one QTL (lwr12.1) for grain L/W ratio was detected on chromosome 12 in *O. rufipogon*

(Swamy et al. 2012). A research group developed more than 2000 introgression lines, which accumulated genes from 45 AA genome wild species, namely, *O. rufipogon*, *O. nivara*, *O. longistaminata* A. Chev. et Roehr, *O. barthii*, *O. meridionalis* Ng, *O. glumaepatula* Steud., and one cultivar *O. glaberrima* to improve the yield of two *indica* varieties (Pusa44 and PR114). Of these, 318 were used to develop hybrids to assess the effects of introgression and detected grain weight-related two QTLs (*gw10*, *gw11*) on chromosomes 10 and 11, respectively, in an IL951 population introgressed from *O. nivara*, while two QTLs (*gw3.1*, *hgw11*) mapped to chromosomes 3 and 11, respectively, in an IL299 population were from *O. rufipogon* (Gaikwad et al. 2014).

Apart from QTLs analysis, it is also assumed that there is a positive correlation between grain length and grain weight from the results of field experiments. This is the case for extra-long grain size *indica* variety Minghui 63 grain usually resulted in the maximum 1,000-grain weight (26-28 g) compared to the medium grain size variety Zhenshan 97 (grain weight 22–24 g) and short grain variety Chuan 7 (grain weight 12.5 g) (Mao et al. 2010). In addition to the genetic perspective, environmental factors, for example, air temperature, in the grain filling stage affect grain quality. For instance, high temperatures resulted in decreased yield, milling quality, and amylose content but enhanced chalkiness in rice cultivars (Septiningsih et al. 2003). Zhang et al. (2016) examined the effects of different temperature on the grain quality of two *japonica* rice cultivars (Nanjing 46 and Jia 58) in both hot and normal seasons and found that high temperatures resulted in the reduction of grain width, grain L/W ratio, and grain weight in Nanjing 46 compared to Jia 58, indicating that Nanjing 46 was more sensitive to temperature. Nevertheless, chalkiness in the former variety was higher in a dry growing season. All these findings are consistent with the notion that high temperature severely affects the grain-filling stages. Like cultivated rice, wild rice grain varies greatly in size, shape, and weight. During the seed development stage, the expansion of grain size appears to be limited due to the hardness of the grain hull, subsequently affecting its size, shape, and weight. Such rigidity of the hull has been observed in wild rice, and, therefore, a slender grain shape resulted in the lowering of grain weight. Recent research on Australian wild rice provides some important insights into the physical appearances that are noteworthy to mention in this context. A study by Tikapunya et al. (2017a) revealed that two Australian wild rice species exhibited an extra-long paddy rice (>7.5 mm) where Taxa A (like *O. rufipogon*) produced medium-sized kernel and the kernels of Taxa B (like *O. meridionalis*) wer long. However, *O. australiensis* Domin. had short grains (4.90–5.00 mm), which resembled the grain size of cultivated *O. sativa* spp. *japonica* cv. Nipponbare (5.30 mm). They also deduced that generally the grain shape of cultivated rice is bold; however, Taxa A and Taxa B were slender-shaped kernels, and *O. australiensis* was medium. This indicates that these dimensions of Australian wild rice may meet the demands of Chinese consumers, as they prefer long grain with different grain shapes (medium to slender) based on the consumers' preference from different provinces.

13.6 Nutritional Quality

The nutrition of crop plants is determined by their chemical structure, i.e., the content of carbohydrates, proteins, lipids, minerals, vitamins, and other micronutrients. Most of the carbohydrate in the rice grain is starch, constituting about 90% of dry grain weight (Kasem et al. 2012; Kasem et al. 2014). Rice bran is another significant source of nutrients, including vitamins, protein, minerals, antioxidants, and phytosterols, but is not consumed when rice is eaten following milling. Starch is composed of the linear and helical α-polyglucan amylose and the branched α-polyglucan amylopectin where glucose units are connected linearly with α (1 → 4) glycosidic bonds in amylose, while branching in amylopectin occurs in every 24–30 glucose units connected by α (1 → 6) glycosidic bonds. The degree of polymerization (DP) of glucose chains results in a variation of amylopectin branch length distribution with $DP \leq 12$ (A chain), $13 \leq DP \leq 24$ (B1 chain), $25 \leq DP \leq 36$ (B2 chain), and $DP \geq 37$ (B3 or longer chain) (Nishi et al. 2001). The main differences in the composition of rice starch result from the relative proportion of amylose and amylopectin molecules in starch (Zhou et al. 2002).

Starch biosynthesis genes contribute to the variation in starch structures. Wild *Oryza* represents a source of useful variation in these genes that has potential for use in breeding rice with desirable starch attributes. Starch is synthesized in rice endosperm through a biochemical pathway using six different groups of enzymes, i.e., ADP glucose pyrophosphorylase (AGPase), soluble starch synthase (SS), starch branching enzyme (SBE), starch debranching enzyme (SDBE), starch phosphorylase, and disproportioning enzyme (DPE) (Ohdan et al. 2005). Thus far, there are 27 isoforms of 6 different classes of starch biosynthesis enzyme that have been identified, and each isoform has a distinct function in amylose and amylopectin synthesis. Amylose synthesis involves mainly AGPase and the granule-bound starch synthase (*GBSS*), whereas multiple enzymes such as several AGPase, SS, SBE, and DBE isoforms are involved in amylopectin synthesis (Nakamura et al. 2002; Ohdan et al. 2005; Wang et al. 2015b). *GBSSI* is controlled by a *Waxy* gene (*Wx*) due to the fact that null alleles (*wx*) give the grain a waxlike appearance. Since SNPs in the structural genes for *GBSSI* and their corresponding amino acid substitutions have been identified, it is evident that *GBSSI* varies between cultivated and wild rice, leading to variation in starch structure. Variation in *GBSSI* exon 6 and 10 and *SBIII* exons 11, 13, 14, and 16 impacts on starch properties in cultivated rice (Larkin et al. 2003; Chen et al. 2008; Kasem et al. 2012). Two SNPs in *GBSSI* exon 6 (A/C) and exon 8 (C/T) at bases 2294 and 3286 resulted in amino acid replacement from serine to tyrosine and from serine to proline, respectively, differentiating *O. sativa japonica* from *O. sativa indica* varieties (Larkin et al. 2003; Kasem et al. 2012). These SNPs were unique to the domesticated O. *sativa* cultivars and may have been selected by human in the domestication process.

In African cultivated rice, Wambugu et al. (2018) found novel polymorphisms affecting *GBSSI* activity. This was detected while conducting the whole genome-based bulk segregant analysis of progenies derived from *O. sativa* spp. *japonica*

(WAB56104) × *O. glaberrima* (CG14) to identify polymorphisms linked with amylose content. Kasem et al. (2012) conducted a study of 13 wild *Oryza* species (AA to HHJJ genome), 1 interspecific hybrid (*O. sativa* × *O. nivara*) and 2 Asian rice cultivars (Nipponbare and Teqing) to compare the main rice starch biosynthesis-related genes in rice. They found a nonconservative amino acid substitution in both *GBSSI* exon 6 and 12 and SSIIa exon 8 in wild species that brought about changes in the amino acid sequence. For instance, an A/C SNP in *O. longiglumis* Jansen and *O. meyeriana* Zoll. et (Mor. ex Steud.) Baill. and an A/G SNP in *O. australiensis* and *O. latifolia* Desv. were detected in *GBSSI* exon 6 at bases 2286, and this produced a nonconservative change of lysine to serine (Kasem et al. 2012). An A/C SNP at site 2290 was found altering threonine to proline in *O. longiglumis* and *O. meyeriana*. In *GBSSI* exon 10, a C/A SNP at site 3322 produced an amino acid substitution from glutamine to lysine in *O. longiglumis*. Noteworthy, a G/A SNP in exon 10 at bases 3328 played a significant role in dividing all rice species into two discrete groups: non-AA genome species with asparagine amino acid, except for *O. meyeriana*, and AA-genome species containing aspartic acid (Kasem et al. 2012). Hence, these nonconservative amino acid substitutions in non-AA genome species could contribute to change *GBSSI* activity and influence the amylose biosynthesis pathway providing a possible explanation for starch structural differences that are found.

The SSIIa enzyme (i.e., an isoform of SS) encoded by *Alk* gene on chromosome 6 controls the formation of A and B1 chains and amylopectin branch length distribution (Nakamura et al. 2002; Chen et al. 2008; Kasem et al. 2012). A total of 24 nonconservative amino acid constitutions were identified in *O. australiensis* and *O. longiglumis* due to SNPs in exon 8; however, the remaining amino acid substitutions were conservative in the wild rice (Kasem et al. 2012). Two Australian wild rice species, namely, Taxa A and Taxa B, showed morphological similarity with *O. rufipogon* and *O. meridionalis*, respectively; however, Taxa A is genetically distinct and may be a different species (Brozynska et al. 2014). The molecular structure of the SSIIa enzyme of Taxa A and Taxa B is different from that of cultivated rice, leading to insights into new starch properties (Tikapunya et al. 2017b). Modification in starch biosynthesis genes due to the presence of high rates of non-synonymous substitutions in exons of *GBSSI* and SSIIa in wild rice may bring about novel variation in starch structure. Therefore, the genetic resources of these wild species could be exploited in rice starch structure manipulation and could help develop rice varieties with potential health benefits due to altered starch digestibility.

Protein content (PC) is a crucial determinant for the improvement of nutritional value of rice. Rice protein exhibits good quality due to lysine (3.8%). Rice bran protein has been reported to have nutraceutical properties, which have a huge prospective in the food industry (Mahender et al. 2016). Clinical evidence revealed that type 2 diabetes patients who were provided with stabilized rice bran as a supplementary diet had lower level of adiponectin and glycated hemoglobin (HbA1c) than the control group that may help to improve glycemic and lipid anomalies in type 2 diabetic patients (Cheng et al. 2010). The amount of protein in rice is lower in cultivated rice (8.5%) than other cereals, including wheat (12.3%), millet (13.4%), and barley (12.8%) (Mahender et al. 2016). Five QTLs for protein content were mapped to chromosomes 3, 4, 5, 6, and 10, respectively, of which *qPc-6* located near the *Waxy* gene region showed a major effect on the regulation of protein content in an *indica* type variety (Yu et al. 2009). They also observed that these alleles were introgressed from a restorer line MY46. In recent times, three QTLs such as *qPro-8*, *qPro-9*, and *qPro-10* on chromosomes 8, 9, and 10, respectively, were identified from 120 DH populations derived from a crossing between Cheongcheong (indica type cultivars) and Nagdong (japonica type cultivars) (Yun et al. 2014). Protein content was reported to be higher in African rice and its wild progenitor than Asian rice (Chen et al. 2008). When the protein content of wild rice was examined, the protein content of the wild species was higher than that of cultivars ranging from 12% to 16%, of which *O. australiensis* contained the highest amount of 16.2% compared to the two cultivated rice cultivars Nipponbare (12.8%) and Teqing (14.3%) (Kasem et al. 2014). The high protein content of wild rice could be exploited to meet the human protein need either directly or through introgression into domesticated cultivars.

Rice is usually consumed as a polished grain having the husk or bran layer removed. Abrasive milling to eliminate seed coat, pericarp, aleurone layer, testa, and embryo causes the loss in protein, fat, vitamin B, Fe, ash, thiamin, riboflavin, niacin, and a-tocopherol. As a result, nutritional properties are either absent or less in milled rice grain (Fitzgerald et al. 2009). Rice pericarp color is associated with the presence of antioxidant, red and brown pericarp of seed in particular. In *O. sativa*, both *Rc* and *Rd* loci control the coloration of rice bran. *Rc* on chromosome 7 results in brown seed coat and pericarp; however, it also produces a red seed coat and pericarp together with *Rd* locus located on chromosome 1 as *Rd* alone cannot express any phenotype (Sweeney et al. 2006). A basic helix–loop–helix (bHLH) protein is encoded in *Rc* locus that regulates the formation of proanthocyanidin (Sweeney et al. 2006; Mahender et al. 2016).

Genome-wide association mapping using 416 rice accessions with varying nutri tional quality and grain color revealed that the purple pericarp producing *Ra* and *Rc* genes were the main loci that had a great effect on pericarp color and nutritional quality (Shao et al. 2011). A total of 41 markers had been detected of which 4 loci, including *RM228, RM339, fgr, and RM316*, were mainly important that have linked with nutrition and pericarp color traits. These six pericarp color-related loci also contribute to the genetic changes of flavonoid and phenolic content, and antioxidant properties (Shao et al. 2011), which lessen the risk of type 2 diabetes, cardiovascular disease, oxidative stress, and cancer (Mahender et al. 2016). The bran layer of nonpigmented rice grains contained 596.3 mg gallic acid equivalent (GAE) 100 g^{-1}), which was higher than that of the rice endosperm (56.9 mg GAE 100 g^{-1}) and the whole grain (263.0 mg GAE 100 g^{-1}) (Goufo and Trindade 2014; Mahender et al. 2016), Whereas in pigmented red rice and black rice, a total of phenolic compounds ranged from 165.8 to 731.8 mg (GAE) 100 g^{-1} and 841.0 to 1244.9 mg (GAE) 100 g^{-1} respectively (Shen et al. 2009; Mahender et al. 2016). In the context of plant breeding, *japonica* rice varieties are rich in antioxidant compounds relative to *indica* rice varieties (Goufo and Trindade 2014). Fasahat et al. (2012) developed two red pericarp rice variants (G33 and G37) by crossing between *O. rufipogon* and *indica* cultivar (MR219) and suggested that these two variants produced more total phenolic content (0.49 mgGAE/g by G33 and 0.51 mgGAE/g by G37) in comparison with the white pericarp *indica* cultivar (0.32 mgGAE/g). Ferric-reducing ability power (FRAP) was also higher in both rice variants (G33 and G37) than in *indica* cultivar MR219, and the content of total E vitamin was high in G37. In this connection, since Australian wild rice pericarp color ranges from dark-brown (Taxa A), red-brown (Taxa B), to light red-brown (*O. australiensis*), they may contain potential antioxidant properties (Kasem et al. 2014).

Wild rice is rich in minerals, including macro- and micronutrients compared to other cereal crops like wheat, oat, and corn. Studies on one non-AA genome wild rice accession of *O. punctata* Kotschy ex Steud and four accessions of *O. eichingeri* A. Peter revealed that the minerals, including phosphorous (P), potassium (K), calcium (Ca), magnesium (Mg), sodium (Na), zinc (Zn), iron (Fe), manganese (Mn), selenium (Se), and copper (Cu), were higher in the wild rice than in the cultivated variety Zhou 903 (Jiang et al. 2009). Hu et al. (2016) detected numerous QTLs controlling the content of six minerals, including Mg, Fe, Zn, Mn, Se, and Cu, in rice grain through QTL analysis using backcross inbreed populations (BIL) that were derived from a cross between *O. sativa* and *O. rufipogon*. From their research, 24 QTLs related to minerals were detected of which 5 QTLs (*qMg2*, *qMg3*, *qMg4*, *qMg9*, and *qMg12*) for Mg, 6 QTLs (*qZn3*, *qZn4*, *qZn6*, *qZn7*, *qZn10*, and *qZn12*) for Zn, 3 QTLs (*qFe3*, *qFe6*, and *qFe9*) for Fe content, 5 QTLs (*qMn1*, *qMn2*, *qMn3*, *qMn6*, and *qMn9*) for Mn, 4 QTLs (*qCu4*, *qCu6*, *qCu9*, and *qCu12*) for Cu, and 1 QTL (*qSe7*) for Se were identified. Besides *O. rufipogon*, QLTs for Zn (*qZn3.1*, *qZn5.1*, *qZn12.1*) and Fe (*qFe1.2*, *qFe3.2*, *qFe4.1*, *qFe11.1*, *qFe11.2*) were detected in *O. nivara* (Swamy et al. 2011; Haritha et al. 2018). These suggest that these nutrient-efficient QTLs derived from wild *Oryza* could facilitate genetic improvement of grain mineral nutrients in modern rice through their introgression.

13.7 Functional Quality

The molecular properties of amylose and amylopectin, in terms of branching ratio, molecular weight, and molecular branch chain length, influence physicochemical properties of a starch such as rice pasting viscosity, breakdown and setback viscosity, enthalpy changes, gel texture gelatinization temperature (GT), and retrogradation (Chen et al. 2008). Most starches comprise 20–30% amylose and 70–80% amylopectin; however, ratios of amylose and amylopectin vary with the botanical sources of the starch (Jane et al. 1999). Amylose content in *O. sativa* consists of a few long-chain branches with lower molecular mass, while African domesticated rice and its wild ancestor *O. barthii* both possess larger amounts of amylose chains with a higher amount of amylose (Wang et al. 2015b). The length of A and B1 chains in amylopectin affects the crystalline configuration of starch where B1 chains are to be found in the lamellae and A chains reside in the crystal lamellae (Nakamura et al. 2002). Rice pasting viscosity characteristics are measured by mixing starch granules with a small amount of water, as the temperature is steadily increasing to approximate cooking conditions (Chen et al. 2008). Rapid visco analyzer (RVA) is used to determine pasting viscosity.

Several studies have been documented that the largest QTL for pasting viscosity characteristics was detected on chromosome 6 of *Waxy* locus (Larkin et al. 2003; Chen et al. 2008). Genetic association studies of pasting characteristics using 171 rice genotypes from different geographic origin detected 3 SNPs in intron 1, exon 6, and exon 10 of *Waxy* gene and a total of 4 *Waxy* SNP-haplotypes (*Waxy-L*, *Waxy-I*, *Waxy-H*, and *Waxy-HH*), which resulted in a significant variation in RVA pasting properties. Three of four haplotypes containing exon 10 adenine nucleotides contributed to the strong correlation between AAC and pasting properties, including peak hot paste, final peak, breakdown, and setback viscosity (Chen et al. 2008). Recent study postulated a non-synonymous T/G SNP at 1630 position, which is an intron1/exon1 junction site of *GBSSI* impacted the physicochemical properties of rice starch (Kharabian-Masouleh et al. 2012).

There are 17 genes related to starch synthesis, and only 10 genes contributed to RVA profiles (Yan et al. 2011). Some other experiments confirmed that AAC is the major factor, which has a positive correlation with thermal, pasting, and textural properties (Kong et al. 2015). Therefore, rice varieties with higher amylose content correspond to higher pasting temperature and setback viscosity and lower peak viscosity. In contrast, rice varieties containing lower amylose content have lower retrogradation but higher swelling power (Jane et al. 1999; Larkin et al. 2003; Kong et al. 2015). The molecular structure of amylopectin significantly affects the thermal properties of rice starches. The high amylopectin chain ratio (ACR) in Asian cultivars results in lower-onset gelatinization temperature relative to those with starch with a lower ACR (Nakamura et al. 2002). In contrast, African domesticated rice and its wild ancestor display higher onset, peak, and conclusion gelatinization temperature in comparison with *O. sativa* (Wang et al. 2015b). Retrogradation is another physicochemical property of starch that is positively correlated with peak viscosity. GT and crystalline structure determines the properties of retrogradation (Kong et al. 2015). Considerable evidence suggests that variation in the amylopectin chain length enables retrogradation to form different types of crystalline structures (Gudmundsson 1994). Starch retrogradation begins faster when amylopectin with long A and B1 chain start to form double helices (Umemoto et al. 2004). However, the amylose content influences short-term changes in retrogradation. Studies revealed that low-amylose starch led to having low degree of retrogradation (% R) and the setback value usually reflected the degree of retrogradation of a starch (Varavinit et al. 2003).

In Asian cultivated rice, amylopectin types (L or S) are identified on the basis of amylopectin structure. L-type amylopectin contains the highest proportion of B1 chains (DP 12–24) and low proportion of short chain (DP $\leq$ 10) compared to S-type; however, there are no significant differences in the percentage of long B2 chains and longer chain (DP $\leq$ 25) for both amylopectin types (Nakamura et al. 2002). Starch *synthase IIa (SSIIa)* and starch branching enzyme (SBE) regulate the branching and elongation reactions during the synthesis of amylopectin that take place in rice endosperm. When the SSIIa gene is defective, it synthesizes more short A1 chains (DP 6–10), which has no branches and less branch containing long B1 chains, resulting S-type amylopectin relative to L-type amylopectin, whereas active SSIIa synthesizes more long B1 chains in the amylopectin cluster (Nakamura et al. 2005). *SBEIIb* isoforms play a crucial role in transferring short chains to B1 chain during the synthesis of amylopectin. Therefore, S-type amylopectin is formed when SS activity is defective and *SBEIIb* is active; in contrast, the superior activity of SSIIa and restricted function of SBE result in L-type amylopectin (Nakamura et al. 2002).

African domesticated rice and its wild progenitor were likely to have similar enzymatic activities of *SBEIIb* and SSIIa where SSIIa showed higher activity over *SBEIIb* resulting higher proportion of intermediate and long amylopectin chains (L-type amylopectin) (Wang et al. 2015b). L-type amylopectin corresponds to *indica* type cultivars with high GT, whereas S-type amylopectin is a feature of *japonica* type cultivars associated with low GT, leading to them becoming susceptible to alkali or urea. The *japonica*-type amylopectin has a large number of short A chain ($DP \leq 11$) with a depleted number of intermediate B1 chain ($12 \leq DP \leq 24$) compared to the *indica*-type amylopectin (Nakamura et al. 2002; Umemoto et al. 2002; Kasem et al. 2014). Like cultivated rice, wild *Oryza* species may be exploited to investigate and utilize their physicochemical starch properties, with an aim to improve cultivars. In *O. nivara*, two QTLs were identified on chromosome 2 (*ac2.1*) and chromosome 5 (*ac5.1*) that enhanced amylose content (Swamy et al. 2012). Also, a QTL controlling amylose content was found in an IR64/*O. rufipogon* BC_2F_2 population where this allele was introgressed from *O. rufipogon* (Septiningsih et al. 2003). Furthermore, an intensive research conducted on wild *Oryza* species, namely, *O. alta* Swallen, *O. australiensis*, *O. meridionalis*, *O. latifolia*, *O. nivara*, *O. officinalis* Wall ex Watt, and *O. rufipogon*, by Kasem et al. (2014) found that all wild species contained few A chains, but higher amount of B1 chains was found in amylopectin along with a high apparent amylose content and GT. They also concluded that wild rice species had low peak and final viscosity compared to both *indica* and *japonica* cultivars. These results suggested that the chain length distribution in amylopectin and GT of wild rice were similar to that of *indica*-type cultivars. Therefore, wild *Oryza* species could be a source of useful traits for enhancing the functional properties of rice.

13.8 Cooking, Eating, and Sensory Quality

The cooking, eating, and sensory qualities of cooked rice that affect the palatability and consumers' choice have become a prime concern in rice-producing countries. The cooking qualities of rice are evaluated by their cooking time, grain elongation during cooking, and length–breadth ratio after cooking, while eating quality is related to the texture of cooked rice. These parameters of rice are highly dependent upon apparent amylose content, gel consistency, and gelatinization temperature (Septiningsih et al. 2003; Larkin et al. 2003; Zhang et al. 2016). *Waxy* locus with two different alleles, W^{xa} and W^{xb}, in non-waxy varieties regulates the synthesis of amylose content (He et al. 1999). The former allele is found in the *indica*-type rice, whereas the latter one is found in *japonica*-type rice (He et al. 1999; Larkin et al. 2003; Nishi et al. 2001; Nakamura et al. 2002). W^{xa} promotes substantially enhanced levels of *Wx* protein as well as ACC relative to W^{xb}. Differences between the *indica* W^{xa} and *japonica* W^{xb} alleles may reflect a subtle difference in protein structure that could affect the synthesis of long-chain amylopectin (He et al. 1999; Larkin et al. 2003).

Based on amylose content, cultivated rice varieties are usually classified as waxy rice (1–2%) and non-waxy rice (>2%). Furthermore, non-waxy rice is classified into four groups: very low (2–10%), low (10–20%), intermediate (20–25%), and high (25–33%) (Kasem et al. 2014). The high-amylose *indica* cultivars are noted for their firm, slightly cohesive, and dry texture, whereas low-amylose *japonica* cultivars have a soft and sticky texture (Jin et al. 2010; Kasem et al. 2014). African rice generally has higher gelatinization temperature than the Asian cultivar (Wang et al. 2015b). The *tropical japonica* or *japonica*-type varieties grown in the southern USA have a long grain shape with intermediate apparent amylose contents, and cooked rice texture is similar to that of *indica* varieties (Larkin et al. 2003). This suggests that the most preferred cultivars are intermediate- to high-amylose cultivars for their soft but not sticky texture, while waxy rice with nearly zero amylose is mainly used as an ingredient for special foods like desserts and snack (Jin et al. 2010). As far as the prospect of wild *Oryza* is concerned in this context, the recent study of wild *Oryza* species, including *O. alta*, *O. australiensis*, *O. latifolia*, *O. meridionalis*, *O. nivara*, and *O. rufipogon*, revealed that these species contained high amylose contents that would be more preferable to the consumers who are accustomed to eating grain with high amylose content (Kasem et al. 2014). It is evident that varieties with high AAC are digested slowly like resistant starch and thus contribute to health benefits (Wang et al. 2015b).

Apart from amylose content, GT plays a crucial role in cooking time as well as the texture of cooked rice. Low gelatinization temperature facilitates complete starch breakdown, while an intermediate gelatinization temperature displays an incomplete breakdown of starch during cooking. GT in different rice varieties can be categorized into three groups: low (55–69 °C), intermediate (70–74 °C), and high (>74 °C) (Cruz and Khush 2000). The alkali-spreading values determine the gelatinization temperature. GT can also be determined by differential scanning calorimeter (DSC) to elucidate the cooking and eating qualities of rice (Kong et al. 2015). Many analyses have implied that the *Alk* locus on chromosome 6 was linked to an alkali degeneration gene for an alkali-spreading score, which indicates a gelatinization temperature (He et al. 1999; Aluko et al. 2004). *Synthase IIa (SSIIa)* gene regulates the gelatinization temperature of rice starch (Umemoto and Aoki 2005; Zhou et al. 2016). Natural variation in SSIIa of rice results in variation in GT. Umemoto et al. (2004) postulated that two SNPs (SNP2 and SNP3) in exon 8 resulted in base substitution from G to A that made SSIIa defective, leading to a change in chain length distribution, gelatinization, retrogradation, and alkali tolerance; however, SNP1 alone did not cause any significant difference in SSIIa properties in rice. In 2005, Umemoto and Aoki (2005) further revealed that SNP2 alone did not affect SSIIa function; rather, two single-nucleotide polymorphism in the SSIIa gene of *O. sativa*, SNP3 (A2209G) and SNP5 (C2341T), were associated with the modification of an amino acid near the C-terminus and changed the enzyme activity and starch granule binding ability and, subsequently, starch gelatinization function. Waters et al. (2006) analyzed sequence data of 70 genotypes and identified polymorphisms in exon 8 wherein 1 SNP (SNP3) at position 2412 (A/G) coded for methionine (A) or valine (G) and 2 SNPs (SNP4) at positions 2543 and 2544 represented leucine (GC) or phenylalanine (TT), respectively, that contributed to the variation of GT. L-type rice starch (*indica* rice) has high GT due to valine and leucine at SNP3 and SNP4, respectively, while methionine and phenylalanine together contribute to the low GT in S-type rice starch (*japonica* rice) (Waters et al. 2006).

The study of Nakamura et al. (2005) revealed *japonica* SSIIa differed from *indica* SSIIa for four amino acid replacements. That said, Glu-88, Val-737, Gly-604, and Leu-781 in *indica* rice were substituted by Asp-88, Met-737 Ser-604, and Phe-781 in *japonica* rice, respectively. These replacements significantly reduced SSIIa activity in *japonica* cultivar resulting in much lower GT compared to *indica* cultivars. Gel consistency (GC) determines an affinity of cooked rice to be hard when it cools. GC can be classified into three types: soft (61–100 mm), medium (41–60 mm), and hard (26–40 mm). The high amount of amylose and long-chain amylopectin containing rice exhibits high GC and an affinity to iodine, while rice with soft GC contains a high amount of short-chain amylopectin. QTLs for GC are concomitantly associated with the *Wx* locus. Two QTLs such as *qGC-2* and *qGC-7* were detected for GC that showed relatively major effects (He et al. 1999; Septiningsih et al. 2003). The GC has a negative correlation with AC, therefore rice with high AC results in low GC and hard gel texture (Septiningsih et al. 2003). Wild species could be acceptable to consumers who are accustomed to eating *indica*-type rice, as these species contain a high amount of amylose in grain endosperm and high GT. High amylose corresponds to high GT, ultimately *indica* cultivars require more cooking time than *japonica* type cultivars. The elongation ratio is considered as a parameter to assess the cooking quality of rice. Only one QTL was identified for rice grain elongation from wild *O. nivara* species (Swamy et al. 2012). As a result, this species possessed an elevated elongation ratio of 1.8, which was close to *O. sativa* cv. Nipponbare (1.9) and *O. sativa* cv. Teqing (1.7), as described by Kasem et al. (2014). Nonetheless, the hardness of cooked wild species makes them undesirable to consumers. This information also suggests that soft cooking traits were selected over harder texture during domestication.

Generally, rice protein content and amylose content determine the palatability of rice. Rice that exhibits a relatively good taste has a protein content below 7%. The low-protein content rice tends to absorb more water and, thereby, increase in volume resulting in higher gel consistency and softer taste (Martin and Fitzgerald 2002). In contrast, rice varieties, which contain high protein content exhibit lower elasticity and viscosity but becomes harder (Yun et al. 2014) Therefore, a low level of protein may have positive effects on eating quality. Wild species *O. latifolia* contain low protein that could be a positive cooked texture attribute, ultimately this could be used in the manipulation of low protein in cultivars (Kasem et al. 2014). Research conducted on the sensory and cooking properties of Australian wild rice (Taxa B) revealed that this species had mild smell and flavor like red rice and red basmati rice cultivars (Tikapunya et al. 2018). Moreover, its texture is firmer, chunky, and chewy, and thus this species has potential commercial value in the unpolished rice market.

13.9 Milling Quality

Milling quality depends on both the yield and the rate of broken kernels of milled rice. Three parameters including milled rice rate (MRR), brown rice rate (BRR), and head milled rice rate (HMRR) are the determinant of milling quality of rice (Qiu et al. 2015). The milling quality is regarded as rice grain's ability to withstand milling and polishing processes without excessive breakage (Cruz and Khush 2000). The head recovery is lower for African-cultivated rice *O. glaberrima* as they have lower milling quality compared to *O. sativa* (Ndjiondjop et al. 2018). Several factors such as rice genotype, the quality of the endosperm, and postharvest handling process impacts the milling quality. One QTL for milling percentage (*mp1.2*) identified between RM24 and RM595 markers on the chromosome 1 in a hybrid population (*O. sativa* cv Swarna × *O. nivara*) in which this QTL introgressed from *O. nivara*, and thereby this hybrid showed maximum milling percentage in terms of grain shape, size and chalkiness (Swamy et al. 2012).

Chalkiness is a crucial factor for milling quality, especially for HMRR. In chalky rice, starch granules are loosely attached in endosperm relative to the translucent endosperm. The decreased grain density ultimately makes the grain prone to break age during milling. Researchers found that the relationship between chalkiness and head rice yield is inverse, indicating that an increase in chalkiness results in a decrease in head rice yield (Zhao et al. 2016). During grain filling stage, an elevated daytime temperature, nighttime temperature, and humidity affect milling quality of rice, in terms of milled rice yield, head rice yield, and chalkiness (Peng et al. 2004; Siebenmorgen et al. 2013; Zhao et al. 2016). This was the case for two *japonica* cultivars, namely, Nanjing 46 and Jia 58, as they contained 42.6% and 19.8% areas of chalkiness during the high hot season compared to the percentage areas of chalkiness determined for the normal season, representing 12.3% and 12.5%, respectively (Septiningsih et al. 2003). It is also evident that not all chalky grains break during milling as well as many varieties develop chalkiness regardless of the environmental condition, rendering a mean to the genetic contribution (Zhao et al. 2016). A major QTL that affects grain chalkiness was detected on chromosome 5 (*Chalk 5*) using a doubled-haploid population developed by crossing between H95 (an *indica* variety with low chalkiness) and Zhenshan 97 (an *indica* variety with high chalkiness) (Li et al. 2014). Using another doubled haploid line from cultivar accession

IR64 × wild *O. rufipogon*, Septiningsih et al. (2003) identified that two QTLs on chromosomes 7 and 12 contributed to green rice grains and these alleles were introgressed from wild parents. They also postulated that green-colored grain had a positive correlation with chalkiness, consequently breakage. From 66 chromosome segment substitution lines (CSSLs) populations, three QTLs such as *qACE-8*, *qPGWC-8*, and *qDEC-8* for percentage of grains with area of chalky endosperm (ACE), chalkiness (PGWC), and degree of endosperm chalkiness (DEC) were found between G1149 and R727 markers on chromosome 8, respectively, and one more QTL for PGWC (*qPGWC-9*) was recognized in on chromosome 9 (Wan et al. 2005). Therefore, these QTLs represent useful genetic resources in the perspective of rice yield and quality improvement through a design-breeding approach.

13.10 Conclusion and Future Prospective

Consumer demand for high-quality rice in terms of grain appearance, nutritional values, and cooking and eating quality is increasing. The global rice market compels a shift in research focus from productivity to grain quality. The genes associated with grain quality have been widely studied in domesticated and wild rice. The lack of genetic variability in cultivated rice often restricts plant breeding. In contrast, germplasm of wild *Oryza* has already been extensively exploited to combat biotic and abiotic stresses to increase rice yield. A considerable amount of research reveals that wild *Oryza* harbors many genes related to grain size and shape with potential for use to improve grain appearance. Furthermore, natural variations in wild genes that regulate starch properties, functional properties, cooking, eating, and sensory properties have been identified. These genes from wild *Oryza* can readily be introgressed into domesticated rice to produce elite varieties with improved grain quality.

References

Aluko G, Martinez C, Tohme J, Castano C, Bergman C, Oard JH (2004) QTL mapping of grain quality traits from the interspecific cross *Oryza sativa* x *O. glaberrima*. Theor Appl Genet 109(3):630–639. https://doi.org/10.1007/s00122-004-1668-y

Bates J, Petrie CA, Singh RN (2017) Approaching rice domestication in South Asia: new evidence from Indus settlements in northern India. J Archaeol Sci 78:193–201. https://doi.org/10.1016/j.jas.2016.04.018

Brar DS, Khush GS (2018) Wild relatives of rice: a valuable genetic resource for genomics and breeding research. In: Mondal TK, Henry RJ (eds) The wild Oryza genomes. Springer International Publishing, Cham, pp 1–25. https://doi.org/10.1007/978-3-319-71997-9_1

Brar DS, Singh K (2011) Oryza. In: Kole C (ed) Wild crop relatives: genomic and breeding resources: cereals. Springer, Berlin, pp 321–365. https://doi.org/10.1007/978-3-642-14228-4_7

Brozynska M, Omar E, Furtado A, Crayn D, Simon B, Ishikawa R, Henry R (2014) Chloroplast genome of novel rice germplasm identified in Northern Australia. An International Journal devoted to original research in tropical plants. Trop Plant Biol 7(3):111–120. https://doi.org/10.1007/s12042-014-9142-8

Calingacion M, Laborte A, Nelson A, Resurreccion A, Concepcion JC, Daygon VD, Mumm R, Reinke R, Dipti S, Bassinello PZ, Manful J, Sophany S, Lara KC, Bao J, Xie L, Loaiza K, El-hissewy A, Gayin J, Sharma N, Rajeswari S, Manonmani S, Rani NS, Kota S, Indrasari SD, Habibi F, Hosseini M, Tavasoli F, Suzuki K, Umemoto T, Boualaphanh C, Lee HH, Hung YP, Ramli A, Aung PP, Ahmad R, Wattoo JI, Bandonill E, Romero M, Brites CM, Hafeel R, Lur HS, Cheaupun K, Jongdee S, Blanco P, Bryant R, Thi Lang N, Hall RD, Fitzgerald M (2014) Diversity of global rice markets and the science required for consumer-targeted rice breeding. PLoS One 9(1):e85106. https://doi.org/10.1371/journal.pone.0085106

Chen CL, Li CC, Sung JM (1994) Carbohydrate metabolism enzymes in CO2-enriched developing rice grains of cultivars varying in grain size. Physiol Plant 90(1):79–85

Chen LJ, Lee DS, Song ZP, Suh HS, Lu BR (2004) Gene flow from cultivated rice (Oryza sativa) to its weedy and wild relatives. Ann Bot 93(1):67–73. https://doi.org/10.1093/aob/mch006

Chen MH, Bergman CJ, Pinson SRM, Fjellstrom RG (2008) Waxy gene haplotypes: associations with pasting properties in an international rice germplasm collection. J Cereal Sci 48(3):781–788. https://doi.org/10.1016/j.jcs.2008.05.004

Cheng C, Motohashi R, Tsuchimoto S, Fukuta Y, Ohtsubo H, Ohtsubo E (2003) Polyphyletic origin of cultivated rice: based on the interspersion pattern of SINEs. Mol Biol Evol 20(1):67–75. https://doi.org/10.1093/molbev/msg004

Cheng HH, Huang HY, Chen YY, Huang CL, Chang CJ, Chen HL, Lai MH (2010) Ameliorative effects of stabilized rice bran on type 2 diabetes patients. Ann Nutr Metab 56(1):45–51. https://doi.org/10.1159/000265850

Choi JY, Purugganan MD (2018) Multiple origin but single domestication led to Oryza sativa. G3 (Bethesda) 8(3):797–803. https://doi.org/10.1534/g3.117.300334

Cruz ND, Khush GS (2000) Rice grain quality evaluation procedures. In: Singh US, Singh RK, Khush GS (eds) Aromatic rices. Oxford & IBH Publishing Co. Pvt. Ltd., New Delhi, pp 15–28

Dipti SS, Bergman C, Indrasari SD, Herath T, Hall R, Lee H, Habibi F, Bassinello PZ, Graterol E, Ferraz JP, Fitzgerald M (2012) The potential of rice to offer solutions for malnutrition and chronic diseases. Rice (N Y) 5(1):16. https://doi.org/10.1186/1939-8433-5-16

Fan C, Xing Y, Mao H, Lu T, Han B, Xu C, Li X, Zhang Q (2006) GS3, a major QTL for grain length and weight and minor QTL for grain width and thickness in rice, encodes a putative transmembrane protein. Theor Appl Genet 112(6):1164–1171. https://doi.org/10.1007/s00122-006-0218-1

FAOSTAT (2017). http://www.fao.org/faostat/en/#data/QC. Accessed 23 Mar 2019

Fasahat P, Abdullah A, Muhammad K, Karupaiah T, Ratnam W (2012) Red pericarp advanced breeding lines derived from Oryza rufipogon× Oryza sativa: physicochemical properties, total antioxidant activity, phenolic compounds and vitamin E content. Adv J Food Sci 4(3):155–165

Fernie AR, Tadmor Y, Zamir D (2006) Natural genetic variation for improving crop quality. Curr Opin Plant Biol 9(2):196–202. https://doi.org/10.1016/j.pbi.2006.01.010

Fitzgerald M (2017) Advances in understanding the role of rice nutrition. In: Sasaki T, Sage PRF, Spindel J et al (eds) Achieving sustainable cultivation of rice 1: breeding for higher yield and quality. Burleigh Dodds Science Publishing Limited, Cambridge, UK, pp 167–183

Fitzgerald MA, McCouch SR, Hall RD (2009) Not just a grain of rice: the quest for quality. Trends Plant Sci 14(3):133–139. https://doi.org/10.1016/j.tplants.2008.12.004

Gaikwad KB, Singh N, Bhatia D, Kaur R, Bains NS, Bharaj TS, Singh K (2014) Yield-enhancing heterotic QTL transferred from wild species to cultivated rice Oryza sativa L. PLoS One 9(6):e96939. https://doi.org/10.1371/journal.pone.0096939

Gao LZ, Innan H (2008) Nonindependent domestication of the two rice subspecies, Oryza sativa ssp. indica and ssp. japonica, demonstrated by multilocus microsatellites. Genetics 179(2):965–976. https://doi.org/10.1534/genetics.106.068072

Garris AJ, Tai TH, Coburn J, Kresovich S, McCouch S (2005) Genetic structure and diversity in Oryza sativa L. Genetics 169(3):1631–1638. https://doi.org/10.1534/genetics.104.035642

Glaszmann JC (1987) Isozymes and classification of Asian rice varieties. Theor Appl Genet 74(1):21–30. https://doi.org/10.1007/BF00290078

Goufo P, Trindade H (2014) Rice antioxidants: phenolic acids, flavonoids, anthocyanins, proanthocyanidins, tocopherols, tocotrienols, gamma-oryzanol, and phytic acid. Food Sci Nutr 2(2):75–104. https://doi.org/10.1002/fsn3.86

Gross BL, Zhao Z (2014) Archaeological and genetic insights into the origins of domesticated rice. Proc Natl Acad Sci U S A 111(17):6190–6197. https://doi.org/10.1073/pnas.1308942110

Gudmundsson M (1994) Retrogradation of starch and the role of its components. Thermochim Acta 246:329–341

Haritha G, Malathi S, Divya B, Swamy BPM, Mangrauthia SK, Sarla N (2018) Oryza nivara Sharma et Shastry. In: Mondal TK, Henry RJ (eds) The wild Oryza genomes. Springer International Publishing, Cham, pp 207–238. https://doi.org/10.1007/978-3-319-71997-9_20

He P, Li SG, Qian Q, Ma YQ, Li JZ, Wang WM, Chen Y, Zhu LH (1999) Genetic analysis of rice grain quality. Theor Appl Genet 98(3–4):502–508. https://doi.org/10.1007/s001220051098

He Z, Zhai W, Wen H, Tang T, Wang Y, Lu X, Greenberg AJ, Hudson RR, Wu CI, Shi S (2011) Two evolutionary histories in the genome of rice: the roles of domestication genes. PLoS Genet 7(6):e1002100. https://doi.org/10.1371/journal.pgen.1002100

Hu BL, Huang DR, Xiao YQ, Fan YY, Chen DZ, Zhuang JY (2016) Mapping QTLs for mineral element contents in brown and milled rice using an Oryza sativa× O. rufipogon backcross inbred line population. Cereal Res Commun 44(1):57–68. https://doi.org/10.1556/0806.43.2015.044

Huang X, Kurata N, Wei X, Wang ZX, Wang A, Zhao Q, Zhao Y, Liu K, Lu H, Li W, Guo Y, Lu Y, Zhou C, Fan D, Weng Q, Zhu C, Huang T, Zhang L, Wang Y, Feng L, Furuumi H, Kubo T, Miyabayashi T, Yuan X, Xu Q, Dong G, Zhan Q, Li C, Fujiyama A, Toyoda A, Lu T, Feng Q, Qian Q, Li J, Han B (2012) A map of rice genome variation reveals the origin of cultivated rice. Nature 490(7421):497–501. https://doi.org/10.1038/nature11532

Jane J, Chen YY, Lee LF, McPherson AE, Wong KS, Radosavljevic M, Kasemsuwan T (1999) Effects of amylopectin branch chain length and amylose content on the gelatinization and pasting properties of starch. Cereal Chem 76(5):629–637. https://doi.org/10.1094/Cchem.1999.76.5.629

Jena KK, Mackill DJ (2008) Molecular markers and their use in marker-assisted selection in rice. Crop Sci 48(4):1266–1276. https://doi.org/10.2135/cropsci2008.02.0082

Jiang S, Shi C, Wu J (2009) Studies on mineral nutrition and safety of wild rice (Oryza L.). Int J Food Sci Nutr 60(S1):139–147. https://doi.org/10.1080/09637480802498838

Jin L, Lu Y, Shao YF, Zhang G, Xiao P, Shen SQ, Corke H, Bao JS (2010) Molecular marker assisted selection for improvement of the eating, cooking and sensory quality of rice (Oryza sativa L.). J Cereal Sci 51(1):159–164. https://doi.org/10.1016/j.jcs.2009.11.007

Kasem S, Waters DLE, Henry RJ (2012) Analysis of starch gene diversity in the wild relatives of Oryza sativa. Trop Plant Biol 5(4):286–308. https://doi.org/10.1007/s12042-012-9111-z

Kasem S, Waters DLE, Ward RM, Rice NF, Henry RJ (2014) Wild Oryza grain physico-chemical properties. Trop Plant Biol 7(1):13–18. https://doi.org/10.1007/s12042-013-9133-1

Kharabian-Masouleh A, Waters DL, Reinke RF, Ward R, Henry RJ (2012) SNP in starch biosynthesis genes associated with nutritional and functional properties of rice. Sci Rep 2:557. https://doi.org/10.1038/srep00557

Khush GS (1997) Origin, dispersal, cultivation and variation of rice. Plant Mol Biol 35:25–34. https://doi.org/10.1023/A:1005810616885

Kong X, Zhu P, Sui Z, Bao J (2015) Physicochemical properties of starches from diverse rice cultivars varying in apparent amylose content and gelatinisation temperature combinations. Food Chem 172(C):433–440. https://doi.org/10.1016/j.foodchem.2014.09.085

Kovach MJ, Sweeney MT, McCouch SR (2007) New insights into the history of rice domestication. Trends Genet 23(11):578–587. https://doi.org/10.1016/j.tig.2007.08.012

Larkin PD, McClung AM, Ayres NM, Park WD (2003) The effect of the Waxy locus (Granule Bound Starch Synthase) on pasting curve characteristics in specialty rices (Oryza sativa L.). Euphytica 131(2):243–253. https://doi.org/10.1023/A:1023962406605

Lee S, Jeon US, Lee SJ, Kim YK, Persson DP, Husted S, Schjorring JK, Kakei Y, Masuda H, Nishizawa NK, An G (2009) Iron fortification of rice seeds through activation of the nicotianamine synthase gene. Proc Natl Acad Sci U S A 106(51):22014–22019. https://doi.org/10.1073/pnas.0910950106

Li C, Zhou A, Sang T (2006a) Genetic analysis of rice domestication syndrome with the wild annual species, Oryza nivara. New Phytol 170(1):185–193. https://doi.org/10.1111/j.1469-8137.2005.01647.x

Li C, Zhou A, Sang T (2006b) Rice domestication by reducing shatterin. Science 311:1936–1939. https://doi.org/10.1126/science.1123604

Li Y, Fan C, Xing Y, Jiang Y, Luo L, Sun L, Shao D, Xu C, Li X, Xiao J, He Y, Zhang Q (2011) Natural variation in GS5 plays an important role in regulating grain size and yield in rice. Nat Genet 43(12):1266–1269. https://doi.org/10.1038/ng.977

Li Y, Fan C, Xing Y, Yun P, Luo L, Yan B, Peng B, Xie W, Wang G, Li X, Xiao J, Xu C, He Y (2014) Chalk5 encodes a vacuolar H(+)-translocating pyrophosphatase influencing grain chalkiness in rice. Nat Genet 46(4):398–404. https://doi.org/10.1038/ng.2923

Lobell DB, Gourdji SM (2012) The influence of climate change on global crop productivity. Plant Physiol 160(4):1686–1697. https://doi.org/10.1104/pp.112.208298

Londo JP, Chiang YC, Hung KH, Chiang TY, Schaal BA (2006) Phylogeography of Asian wild rice, Oryza rufipogon, reveals multiple independent domestications of cultivated rice, Oryza sativa. Proc Natl Acad Sci U S A 103(25):9578–9583. https://doi.org/10.1073/pnas.0603152103

Lu BR, Zheng KL, Qian HR, Zhuang JY (2002) Genetic differentiation of wild relatives of rice as assessed by RFLP analysis. Theor Appl Genet 106(1):101–106. https://doi.org/10.1007/s00122-002-1013-2

Mahender A, Anandan A, Pradhan SK, Pandit E (2016) Rice grain nutritional traits and their enhancement using relevant genes and QTLs through advanced approaches. Springerplus 5(1):2086. https://doi.org/10.1186/s40064-016-3744-6

Mao H, Sun S, Yao J, Wang C, Yu S, Xu C, Li X, Zhang Q (2010) Linking differential domain functions of the GS3 protein to natural variation of grain size in rice. Proc Natl Acad Sci U S A 107(45):19579–19584. https://doi.org/10.1073/pnas.1014419107

Martin M, Fitzgerald MA (2002) Proteins in rice grains influence cooking properties. J Cereal Sci 36(3):285–294. https://doi.org/10.1006/jcrs.2001.0465

Molina J, Sikora M, Garud N, Flowers JM, Rubinstein S, Reynolds A, Huang P, Jackson S, Schaal BA, Bustamante CD, Boyko AR, Purugganan MD (2011) Molecular evidence for a single evolutionary origin of domesticated rice. Proc Natl Acad Sci U S A 108(20):8351–8356. https://doi.org/10.1073/pnas.1104686108

Momma K, Hashimoto W, Ozawa S, Kawai S, Katsube T, Takaiwa F, Kito M, Utsumi S, Murata K (1999) Quality and safety evaluation of genetically engineered rice with soybean glycinin: analyses of the grain composition and digestibility of glycinin in transgenic rice. Biosci Biotechnol Biochem 63(2):314–318. https://doi.org/10.1271/bbb.63.314

Nakamura Y, Sakurai A, Inaba Y, Kimura K, Iwasawa N, Nagamine T (2002) The fine structure of amylopectin in endosperm from Asian cultivated rice can be largely classified into two classes. Starch-Starke 54:117–131. https://doi.org/10.1002/1521-379x(200204)54:3/4<117::Aid-Star117>3.0.Co;2-2

Nakamura Y, Francisco PB Jr, Hosaka Y, Sato A, Sawada T, Kubo A, Fujita N (2005) Essential amino acids of starch synthase IIa differentiate amylopectin structure and starch quality between japonica and indica rice varieties. Plant Mol Biol 58(2):213–227. https://doi.org/10.1007/s11103-005-6507-2

Ndjiondjop MN, Wambugu P, Sangare JR, Dro T, Kpeki B, Gnikoua K (2018) Oryza glaberrima Steud. In: Mondal TK, Henry RJ (eds) The wild Oryza genomes. Springer International Publishing, Cham, pp 105–126. https://doi.org/10.1007/978-3-319-71997-9_9

Nishi A, Nakamura Y, Tanaka N, Satoh H (2001) Biochemical and genetic analysis of the effects of amylose-extender mutation in rice endosperm. Plant Physiol 127(2):459–472. https://doi.org/10.1104/pp.010127

Ohdan T, Francisco PB Jr, Sawada T, Hirose T, Terao T, Satoh H, Nakamura Y (2005) Expression profiling of genes involved in starch synthesis in sink and source organs of rice. J Exp Bot 56(422):3229–3244. https://doi.org/10.1093/jxb/eri292

Peng S, Huang J, Sheehy JE, Laza RC, Visperas RM, Zhong X, Centeno GS, Khush GS, Cassman KG (2004) Rice yields decline with higher night temperature from global warming. Proc Natl Acad Sci U S A 101(27):9971–9975. https://doi.org/10.1073/pnas.0403720101

Purugganan MD, Fuller DQ (2009) The nature of selection during plant domestication. Nature 457(7231):843–848. https://doi.org/10.1038/nature07895

Qiu X, Pang Y, Yuan Z, Xing D, Xu J, Dingkuhn M, Li Z, Ye G (2015) Genome-wide association study of grain appearance and milling quality in a worldwide collection of indica rice germplasm. PLoS One 10(12):e0145577. https://doi.org/10.1371/journal.pone.0145577

Rahman ML, Chu SH, Choi MS, Qiao YL, Jiang W, Piao R, Khanam S, Cho YI, Jeung JU, Jena KK, Koh HJ (2007) Identification of QTLs for some agronomic traits in rice using an introgression line from Oryza minuta. Mol Cells 24(1):16–26

Rakshit S, Rakshit A, Matsumura H, Takahashi Y, Hasegawa Y, Ito A, Ishii T, Miyashita NT, Terauchi R (2007) Large-scale DNA polymorphism study of Oryza sativa and O. rufipogon reveals the origin and divergence of Asian rice. Theor Appl Genet 114(4):731–743. https://doi.org/10.1007/s00122-006-0473-1

Sanchez PL, Wing RA, Brar DS (2013) The wild relative of rice: genomes and genomics. In: Zhang Q, Wing RA (eds) Genetics and genomics of rice. Springer, New York, pp 9–25. https://doi.org/10.1007/978-1-4614-7903-1_2

Sang T, Ge S (2013) Understanding rice domestication and implications for cultivar improvement. Curr Opin Plant Biol 16(2):139–146. https://doi.org/10.1016/j.pbi.2013.03.003

Septiningsih EM, Trijatmiko KR, Moeljopawiro S, McCouch SR (2003) Identification of quantitative trait loci for grain quality in an advanced backcross population derived from the Oryza sativa variety IR64 and the wild relative O. rufipogon. Theor Appl Genet 107(8):1433–1441. https://doi.org/10.1007/s00122-003-1376-z

Shao Y, Jin L, Zhang G, Lu Y, Shen Y, Bao J (2011) Association mapping of grain color, phenolic content, flavonoid content and antioxidant capacity in dehulled rice. Theor Appl Genet 122(5):1005–1016. https://doi.org/10.1007/s00122-010-1505-4

Shen Y, Jin L, Xiao P, Lu Y, Bao JS (2009). Total phenolics, flavonoids, antioxidant capacity in rice grain and their relations to grain color, size and weight. Journal of Cereal Science 49:106–111. https://doi.org/10.1016/j.jcs.2008.07.010

Shomura A, Izawa T, Ebana K, Ebitani T, Kanegae H, Konishi S, Yano M (2008) Deletion in a gene associated with grain size increased yields during rice domestication. Nat Genet 40(8):1023–1028. https://doi.org/10.1038/ng.169

Si L, Chen J, Huang X, Gong H, Luo J, Hou Q, Zhou T, Lu T, Zhu J, Shangguan Y, Chen E, Gong C, Zhao Q, Jing Y, Zhao Y, Li Y, Cui L, Fan D, Lu Y, Weng Q, Wang Y, Zhan Q, Liu K, Wei X, An K, An G, Han B (2016) OsSPL13 controls grain size in cultivated rice. Nat Genet 48(4):447–456. https://doi.org/10.1038/ng.3518

Siebenmorgen TJ, Grigg BC, Lanning SB (2013) Impacts of preharvest factors during kernel development on rice quality and functionality. Annu Rev Food Sci Technol 4(1):101–115. https://doi.org/10.1146/annurev-food-030212-182644

Song XJ, Huang W, Shi M, Zhu MZ, Lin HX (2007) A QTL for rice grain width and weight encodes a previously unknown RING-type E3 ubiquitin ligase. Nat Genet 39(5):623–630. https://doi.org/10.1038/ng2014

Stein JC, Yu Y, Copetti D, Zwickl DJ, Zhang L, Zhang C, Chougule K, Gao D, Iwata A, Goicoechea JL et al (2018) Genomes of 13 domesticated and wild rice relatives highlight genetic conservation, turnover and innovation across the genus Oryza. Nat Genet 50(2):285–296. https://doi.org/10.1038/s41588-018-0040-0

Swamy B, Kaladhar K, Anuradha K, Batchu A, Longvah T, Viraktamath B, Sarla N 2011 Enhancing iron and zinc concentration in rice grains using wild species. In: CCMB, Hyderabad, 23–25 Feb, p 71

Swamy BP, Kaladhar K, Shobha Rani N, Prasad GS, Viraktamath BC, Reddy GA, Sarla N (2012) QTL analysis for grain quality traits in 2 BC2F2 populations derived from crosses between Oryza sativa cv Swarna and 2 accessions of O. nivara. J Hered 103(3):442–452. https://doi.org/10.1093/jhered/esr145

Swamy BP, Kaladhar K, Reddy GA, Viraktamath BC, Sarla N (2014) Mapping and introgression of QTL for yield and related traits in two backcross populations derived from Oryza sativa cv. Swarna and two accessions of O. nivara. J Genet 93(3):643–654

Sweeney M, McCouch S (2007) The complex history of the domestication of rice. Ann Bot 100(5):951–957. https://doi.org/10.1093/aob/mcm128

Sweeney MT, Thomson MJ, Pfeil BE, McCouch S (2006) Caught red-handed: Rc encodes a basic helix-loop-helix protein conditioning red pericarp in rice. Plant Cell 18(2):283–294. https://doi.org/10.1105/tpc.105.038430

Tan YF, Xing YZ, Li JX, Yu SB, Xu CG, Zhang QF (2000) Genetic bases of appearance quality of rice grains in Shanyou 63, an elite rice hybrid. Theor Appl Genet 101:823–829. https://doi.org/10.1007/s001220051549

Tan L, Li X, Liu F, Sun X, Li C, Zhu Z, Fu Y, Cai H, Wang X, Xie D, Sun C (2008) Control of a key transition from prostrate to erect growth in rice domestication. Nat Genet 40(11):1360–1364. https://doi.org/10.1038/ng.197

Tang G, Qin J, Dolnikowski GG, Russell RM, Grusak MA (2009) Golden rice is an effective source of vitamin A. Am J Clin Nutr 89(6):1776–1783. https://doi.org/10.3945/ajcn.2008.27119

Tikapunya T, Fox G, Furtado A, Henry R (2017a) Grain physical characteristic of the Australian wild rices. Plant Genet Resour: Charact Utilization 15(5):409–420. https://doi.org/10.1017/S1479262116000083

Tikapunya T, Zou W, Yu W, Powell PO, Fox GP, Furtado A, Henry RJ, Gilbert RG (2017b) Molecular structures and properties of starches of Australian wild rice. Carbohydr Polym 172:213–222. https://doi.org/10.1016/j.carbpol.2017.05.046

Tikapunya T, Henry RJ, Smyth H (2018) Evaluating the sensory properties of unpolished Australian wild rice. Food Res Int 103:406–414. https://doi.org/10.1016/j.foodres.2017.10.037

Umemoto T, Aoki N (2005) Single-nucleotide polymorphisms in rice starch synthase IIa that alter starch gelatinisation and starch association of the enzyme. Funct Plant Biol 32(9):763–768. https://doi.org/10.1071/fp04214

Umemoto T, Yano M, Satoh H, Shomura A, Nakamura Y (2002) Mapping of a gene responsible for the difference in amylopectin structure between japonica-type and indica-type rice varieties. Theor Appl Genet 104(1):1–8. https://doi.org/10.1007/s001220200000

Umemoto T, Aoki N, Lin HX, Nakamura Y, Inouchi N, Sato Y, Yano M, Hirabayashi H, Maruyama S (2004) Natural variation in rice starch synthase IIa affects enzyme and starch properties. Funct Plant Biol 31(7):671–684. https://doi.org/10.1071/Fp04009

Underwood BA (2000) Overcoming micronutrient defi ciencies in developing countries: is there a role for agriculture? Food Nutr Bull 21(4):356–360

Varavinit S, Shobsngob S, Varanyanond W, Chinachoti P, Naivikul O (2003) Effect of amylose content on gelatinization, retrogradation and pasting properties of flours from different cultivars of Thai rice. Starch-Starke 55(9):410–415. https://doi.org/10.1002/star.200300185

Varshney RK, Singh VK, Kumar A, Powell W, Sorrells ME (2018) Can genomics deliver climate-change ready crops? Curr Opin Plant Biol 45:205–211. https://doi.org/10.1016/j.pbi.2018.03.007

Vaughan DA, Lu BR, Tomooka N (2008a) The evolving story of rice evolution. Plant Sci 174(4):394–408. https://doi.org/10.1016/j.plantsci.2008.01.016

Vaughan DA, Lu BR, Tomooka N (2008b) Was Asian rice (Oryza sativa) domesticated more than once? Rice 1:16–24. https://doi.org/10.1007/s12284-008-9000-0

Wambugu PW, Brozynska M, Furtado A, Waters DL, Henry RJ (2015) Relationships of wild and domesticated rices (Oryza AA genome species) based upon whole chloroplast genome sequences. Sci Rep 5:13957. https://doi.org/10.1038/srep13957

Wambugu P, Ndjiondjop MN, Furtado A, Henry R (2018) Sequencing of bulks of segregants allows dissection of genetic control of amylose content in rice. Plant Biotechnol J 16(1):100–110. https://doi.org/10.1111/pbi.12752

Wan XY, Wan JM, Weng JF, Jiang L, Bi JC, Wang CM, Zhai HQ (2005) Stability of QTLs for rice grain dimension and endosperm chalkiness characteristics across eight environments. Theor Appl Genet 110(7):1334–1346. https://doi.org/10.1007/s00122-005-1976-x

Wang E, Wang J, Zhu X, Hao W, Wang L, Li Q, Zhang L, He W, Lu B, Lin H, Ma H, Zhang G, He Z (2008) Control of rice grain-filling and yield by a gene with a potential signature of domestication. Nat Genet 40(11):1370–1374. https://doi.org/10.1038/ng.220

Wang S, Wu K, Yuan Q, Liu X, Liu Z, Lin X, Zeng R, Zhu H, Dong G, Qian Q, Zhang G, Fu X (2012) Control of grain size, shape and quality by OsSPL16 in rice. Nat Genet 44(8):950–954. https://doi.org/10.1038/ng.2327

Wang M, Yu Y, Haberer G, Marri PR, Fan C, Goicoechea JL, Zuccolo A, Song X, Kudrna D, Ammiraju JS, Cossu RM, Maldonado C, Chen J, Lee S, Sisneros N, de Baynast K, Golser W, Wissotski M, Kim W, Sanchez P, Ndjiondjop MN, Sanni K, Long M, Carney J, Panaud O, Wicker T, Machado CA, Chen M, Mayer KF, Rounsley S, Wing RA (2014) The genome sequence of African rice (Oryza glaberrima) and evidence for independent domestication. Nat Genet 46(9):982–988. https://doi.org/10.1038/ng.3044

Wang S, Li S, Liu Q, Wu K, Zhang J, Wang S, Wang Y, Chen X, Zhang Y, Gao C, Wang F, Huang H, Fu X (2015a) The OsSPL16-GW7 regulatory module determines grain shape and simultaneously improves rice yield and grain quality. Nat Genet 47(8):949–954. https://doi.org/10.1038/ng.3352

Wang K, Wambugu PW, Zhang B, Wu AC, Henry RJ, Gilbert RG (2015b) The biosynthesis, structure and gelatinization properties of starches from wild and cultivated African rice species (Oryza barthii and Oryza glaberrima). Carbohydr Polym 129:92–100. https://doi.org/10.1016/j.carbpol.2015.04.035

Waters DL, Henry RJ, Reinke RF, Fitzgerald MA (2006) Gelatinization temperature of rice explained by polymorphisms in starch synthase. Plant Biotechnol J 4(1):115–122. https://doi.org/10.1111/j.1467-7652.2005.00162.x

Weng J, Gu S, Wan X, Gao H, Guo T, Su N, Lei C, Zhang X, Cheng Z, Guo X, Wang J, Jiang L, Zhai H, Wan J (2008) Isolation and initial characterization of GW5, a major QTL associated with rice grain width and weight. Cell Res 18(12):1199–1209. https://doi.org/10.1038/cr.2008.307

Wu W, Liu X, Wang M, Meyer RS, Luo X, Ndjiondjop MN, Tan L, Zhang J, Wu J, Cai H, Sun C, Wang X, Wing RA, Zhu Z (2017) A single-nucleotide polymorphism causes smaller grain size and loss of seed shattering during African rice domestication. Nat Plants 3(6):17064. https://doi.org/10.1038/nplants.2017.64

Yan CJ, Tian ZX, Fang YW, Yang YC, Li J, Zeng SY, Gu SL, Xu CW, Tang SZ, Gu MH (2011) Genetic analysis of starch paste viscosity parameters in glutinous rice (Oryza sativa L.). Theor Appl Genet 122(1):63. https://doi.org/10.1007/s00122-010-1423-5

Yu YH, Li G, Fan YY, Zhang KQ, Min J, Zhu ZW, Zhuang JY (2009) Genetic relationship between grain yield and the contents of protein and fat in a recombinant inbred population of rice. J Cereal Sci 50(1):121–125. https://doi.org/10.1016/j.jcs.2009.03.008

Yun BW, Kim MG, Handoyo T, Kim KM (2014) Analysis of rice grain quality-associated quantitative trait loci by using genetic mapping. Am J Plant Sci 5(09):1125–1132. https://doi.org/10.4236/ajps.2014.59125

Zhang C, Zhou L, Zhu Z, Lu H, Zhou X, Qian Y, Li Q, Lu Y, Gu M, Liu Q (2016) Characterization of grain quality and starch fine structure of two japonica rice (Oryza Sativa) cultivars with good sensory properties at different temperatures during the filling stage. J Agric Food Chem 64(20):4048–4057. https://doi.org/10.1021/acs.jafc.6b00083

Zhang H, Yu C, Hou D, Liu H, Zhang H, Tao R, Cai H, Gu J, Liu L, Zhang Z, Wang Z, Yang J (2018) Changes in mineral elements and starch quality of grains during the improvement of japonica rice cultivars. J Sci Food Agric 98(1):122–133. https://doi.org/10.1002/jsfa.8446

Zhao X, Daygon VD, McNally KL, Hamilton RS, Xie F, Reinke RF, Fitzgerald MA (2016) Identification of stable QTLs causing chalk in rice grains in nine environments. Theor Appl Genet 129(1):141–153. https://doi.org/10.1007/s00122-015-2616-8

Zhou ZK, Robards K, Helliwell S, Blanchard C (2002) Composition and functional properties of rice. Int J Food Sci Technol 37(8):849–868. https://doi.org/10.1046/j.1365-2621.2002.00625.x

Zhou Y, Zheng HY, Wei GC, Zhou H, Han YN, Bai XF, Xing YZ, Han YP (2016) Nucleotide diversity and molecular evolution of the ALK gene in cultivated rice and its wild relatives. Plant Mol Biol Report 34(5):923–930. https://doi.org/10.1007/s11105-016-0975-1

Part VI
Breeding Strategies

Chapter 14
Conventional Breeding for Rice Grain Quality

Priscila Zaczuk Bassinello, Adriano Pereira de Castro, and Tereza Cristina de Oliveira Borba

14.1 Rice Grain Quality

Cereals comprise a great proportion of crop production and are considered an essential food source for humankind (FAO 2018). Plant breeding, since the 1970s, has been responsible for the development of a successful agriculture and can also be considered as a prosperous research (Khush 2005). Improved varieties/cultivars are the raw material for agricultural development, and such fact can be perceived more clearly, when one considers that half of the world's population rely on rice varieties as a main staple food (Lenaerts et al. 2018).

Along the past years, high-yield varieties have been one of the most important contributions from plant breeding (Wang et al. 2015), and each country, which is a rice producer, displayed different strategies to achieve self-sufficiency (Asante 2017). Nevertheless, since rice is consumed mainly as a whole cooked grain, its quality is increasingly becoming a major goal for breeders (Wang et al. 2015; Asante 2017).

According to Juliano and Villareal (1993), in a survey addressed to 11 countries, grain quality was the second breeding goal in national rice programs. However, breeding for grain quality is not trivial, as there is not a clear definition (Asante 2017) because the perception of quality varies according to each culture or country (Cruz and Khush 2000). Therefore, the international rice market is highly segmented due to the specific preferences from different consumers (Cuevas et al. 2016), and a large range of segments can be identified within a diverse set of countries and regions (Fitzgerald et al. 2008). However, according to Calingacion et al. (2014), the most common combination of quality characteristics, all over the world, for rice is both long and slender grains.

P. Z. Bassinello (✉) · A. P. de Castro · T. C. de Oliveira Borba
Embrapa Arroz e Feijão, Santo Antônio de Goías, Goías, Brazil
e-mail: priscila.bassinello@embrapa.br

A. Costa de Oliveira et al. (eds.), *The Future of Rice Demand: Quality Beyond Productivity*, https://doi.org/10.1007/978-3-030-37510-2_14

Despite its broad range of characteristics, grain quality can be linked to physical appearance, cooking and sensory properties, and nutritional value (Unnevehr et al. 1992; Fitzgerald et al. 2009a, Fitzgerald et al. 2009b). Usually, the physical appearance can be decisive to the definition of market prices, and the cooking and sensory properties can support the establishment of a "premium" variety (Fitzgerald 2010). Some authors also consider that besides technical aspects, economic factors are an important share of quality (Unnevehr et al. 1992; Cuevas et al. 2016). And due to a steep decline of prices, in the 1980s, Unnevehr et al. (1992) emphasized that breeders became progressively interested in grain quality.

There are several classes in which it is possible to group rice varieties; those classes can be based on physical aspects and on starch quality (cooking properties) (Champagne et al. 2010; Anacleto et al. 2015). The physical aspects are very important attributes since they affect directly the decision of a consumer to pay for the product and are one of the first characteristics to be selected in breeding programs (Tomlins et al. 2007; Cuevaset al. 2016).

Rice breeding programs adopt several methodologies for screening rice lines/varieties (Fitzgerald et al. 2009b; Vilaplana et al. 2012), but, usually, the evaluated traits are separately considered (Anacleto et al. 2015). In this scenario, sensory analysis (or evaluation) can be crucial because it can combine different characteristics within a single evaluation, but unfortunately it is not established as a high-throughput exploration (Anacleto et al. 2015). Sensory analysis is mostly employed on advanced breeding lines, when just a few genotypes are available (Anacleto et al. 2015).

Regardless the importance of physical and chemical aspects of grain quality, until recently a limited relevance has been displayed toward micronutrient content of rice (Bhullar and Gruissem 2013). But, considering that rice plays an important role as a staple food resource, improving the nutritional value of this cereal can support initiatives to overcome micronutrient deficiency in developing countries (Welch and Graham 2004).

14.2 Rice Grain Quality Traits and Methods

Increasing crop yields does not necessarily translate into greater food security, as whole grain yield cannot be guaranteed after rice milling. Improvement in the quality of rice processing is essential to ensure food security, mitigating the impact of the large postharvest losses (30%) observed due to rice breakage and increasing the recovery of whole yields while ensuring better financial returns to producers.

Factors related to mechanical stress, extrinsic to rice grain, affect rice breakage more than grain intrinsic strength (Bell et al. 2000). In addition, the impact of poor efficiency of improper postharvest practices also falls on the loss of grain culinary quality at any stage of the production chain (Hodges et al. 2011).

Significant advances in rice postharvest processing technology have been achieved toward reducing the incidence of cracking and plastering and increasing whole grain yield (HRY). Harvesting should be properly programmed to coincide with 22–24% moisture content (MC) and preferably combined to avoid delays and stacking of the harvested rice panicles in the field. After threshing, the harvest must be dried immediately, within 24 h. During drying, the temperature should not exceed 43 °C in a flatbed dryer or any other fixed-bed dryer. Alternatively, for drying in the sun, the grains should have a depth between 2 and 4 cm and should be mixed every 30 min to promote a slow drying up to 14% MC, which ensures the minimum breakage of grains. After drying, the grains should be stored safely to prevent pests, water, and re-humidification.

Environmental conditions of temperature, relative humidity, and grain moisture content influence the formation of cracks in the grain both in the preharvest, postharvest, and storage stages. When the MC of the grain falls below 15% (Otto and David 2004) in the field, cracks can form. In the same panicle, there is variation in MC due to the difference in plant grain maturity, so it is obvious that this variation can also be perceived between different plants cultivated in the same area. The tendency of the grain is to find an equilibrium humidity with the environment, and this moisture gradient induces the grain to internal forces that lead it to expand and, consequently, to fissure. Depending on the grain's resistance to this tensile and compressive stress, fissures may appear during the grain-filling phase (maturation) or desiccation (Kunze and Choudhury 1972). Exposure of grains to extreme drying conditions after harvesting should be avoided. Most often, cracked grains tend to significantly increase the percentage of broken grains in processing (Zhang et al. 2005; Iguaz et al. 2006).

Rice starch represents 90% of the grain content and is characterized by a partially crystalline and a partially amorphous structure of the polymeric glucose chain, consisting of essentially linear amylose and highly branched amylopectin. The amylose and amylopectin branch points form the amorphous layer, while the amylopectin outer chains form the crystalline region. The semicrystalline property of starch dynamically changes depending on the temperature and moisture content of the grain (Slade and Levine 1991, 1995).

In rice drying, the grain temperature increases irregularly from the surface to the inside (Sarker et al. 1996; Yang et al. 2002), forming a moisture gradient that results in the glass transition of starch in different parts of the grain. This usually occurs during desiccation and commercial drying of rice (Perdon et al. 2000).

The moisture exchange of the grain with the environment (Banaszek and Siebenmorgen 1993; Chen et al. 1997) can strongly affect the head rice yield (HRY), since the reversible changes of the resulting vitreous and rubbery states of the starch cause the cracks in the grain (Kunze and Choudhury 1972; Kunze 1979).

Low-cost optimal drying methods for freshly harvested rice grains were developed. The Solar Bubble Dryer (SBD) is a low-cost, portable, and innovative sun-drying technology developed by the International Rice Research Institute, Hohenheim University, and GrainPro. Proper and timely drying of rice grains ensures good quality produce, which can command a competitive market price. The traditional sun-drying method is beset with problems like uneven drying or overheating, which can cause grain breakage during milling, spillage, and contamination (e.g., animals and vehicles). Additionally, with climate change, the weather may become unpredictable, thus, making grain drying more difficult for farmers. With SBD, the drying process is faster, more regulated, and cost-efficient as it does not require fuel or the power grid to operate. SBD can process up to one ton of rice compared to mechanical dryer. Besides, the enclosed feature of the system protects the grains from spillage and foreign contaminants.

Grain chalkiness is considered a physical defect, related to a whitish opaque spot on the grain and is another relevant factor affecting the quality and head rice yield. The susceptibility of rice to chalkiness depends on the genotype, and its incidence is related to the position of the spikelet (Fitzgerald and Resurreccion 2009) and is commonly induced by the high temperature in the grain-filling phase (Zakaria et al. 2002; Tabata et al. 2007). Chalk formation prevents or hinders the flow of assimilates (Fitzgerald and Resurreccion 2009) by aborting starch and protein biosynthesis in grain development (Sreenivasulu et al. 2015), causing the grain to have many unfilled air spaces between starch granules, producing opaque points along the translucent grain (Ashida et al. 2009) which are responsible for increased grain brittleness in processing and consequently greater breakage (Lyman et al. 2013). Furthermore, if chalk is formed in the dorsal and the ventral side of grain, it can affect HRY more than the chalk in the core (Butardo and Sreenivasulu 2019). Chalky grains that resist the milling process without breaking will cause a reduction in the commercial value of rice if present in the package, as well as a decrease in consumer acceptance (Fitzgerald et al. 2009a) due to its poor appearance quality (unevenness) and a change in rice cooking and sensory quality (Singh et al. 2003; Cheng et al. 2005; Chun et al. 2009). The chalky region of the grain requires higher temperature to gelatinize the starch, because it has lower gel consistency and higher hardness than normal rice.

Measurement of grain chalkiness has most recently been executed with individual grain imaging equipment (Cervitec, Image J) or statistical rice analyzer (S21) but still needs improvements in both method standardization and differentiation of the kind of chalk. The conventional method discards the broken grains before analysis, which generates a bias in the result, since only the head rice grains are evaluated for chalk and dimension. This characterization reveals only the percentage of chalky breakage-resistant kernels that will affect the physical (appearance), cooking, and eating quality of commercialized rice, but it has no relation with the chalk that caused grain breakage in the milling process and resulted in loss of HRY, because the chalky kernels susceptible to breakage were eliminated from the evaluation. Therefore, there is no way to measure, in this case, the real contribution of the chalk in reducing the HRY. On the other hand, current technology makes it possible to overcome the traditional visual analysis, which was subjective, tiring, and inconsistent, enabling more accurate identification of export quality rice that meets the requirements of the importing country.

Most of the characterization of the rice grain quality is focused on starch (structure and composition), which in turn affects the functional (technological) properties of rice. Starch is the largest component of milled rice grain (80–90% of grain weight), and in addition to influencing the physical, industrial, cooking, eating, and sensory qualities of rice, it also affects the nutritional quality of the product.

The available routine tests are limited to determining apparent amylose content (AAC), starch gelatinization temperature (GT), and gel consistency (GC). Further evaluations focus on the characterization of starch properties such as viscoelasticity, structure and texture, and aroma of rice, generally employed in laboratories that are more sophisticated. However, these analyses are not routinely applied by breeding programs to differentiate rice quality classes that reflect the evolution of consumer preference and market demand.

It is known that AAC is directly related to the proportion and structure of amylose, while GT and GC are amylopectin properties. Low-GT starch has shorter chain amylopectin branches, whereas intermediate/high GT have more long-chain branches. AAC of milled rice grains can be classified as waxy (0–2%), very low (3–9%), low (10–19%), intermediate (20–25%), and high (>25%) (Kumar and Khush 1986). Milled rice can also be classified as non-glutinous (non-waxy) or glutinous (waxy), depending on the presence or absence of amylose chains, respectively.

The original methods of amylose determination had some limitations and interferences in the results, such as the simultaneous detection of longer amylopectin chains given their affinity for the acetate buffer used for amylose extraction and to iodine at wavelength of 595 nm or even at wavelength 620 nm, which leads to an overestimation of amylose values (Takeda et al. 1987; Butardo et al. 2011; Butardo et al. 2017). Since the 1970s, some improvements in methodology have been proposed to reduce errors and sources of variation (Juliano 1971; Perez and Juliano 1978), such as the different brands and sources of commercial amylose standards. The inclusion in the test of a continuous flow injection automatic sampling equipment was one of the modifications validated by several international laboratories and widely adopted by different rice breeding programs, given the advantage of routine evaluation of hundreds of samples simultaneously by daily or annual analysis. Additional improvements were proposed including the use of calibrated rice standards, the use of longer wavelengths (e.g., 720 nm) to minimize the interference of iodine–amylopectin complex (Fitzgerald et al. 2009b), and the replacement of acetate with ammonium buffer to reduce amylopectin–iodine absorbance (Juliano et al. 2012). There are two main officially approved methods for determining amylose in polished rice: one established by the Association for American Cereal Chemists (AACC) International (AACCI Method 61-03.01) and the other by the International Organization for Standardization (ISO Method 6647-1:2015) (AACC 2009; International Organization for Standardization 2007). Several analytical methods have been proposed to more accurately determine the amylose content of rice grains which do not rely on iodine binding, including the use of differential scanning calorimetry (DSC), near-infrared spectroscopy (NIRS), and debranched starch size exclusion chromatography (SEC).

Besides being the major starch fraction, the amylopectin has also to be considered to explain functional properties of rice, once its proportion and fine structure contribute to the quality of cooked rice. Amylopectin chain ratio (ACR) or molar ratio of Σ degree of polymerization (DP) 6–10/ΣDP 6–24 <0.20 of chain length distribution of debranched amylopectin denotes intermediate/high GT, and ACR of >0.23 is low GT (Nakamura et al. 2002).

The gelatinization temperature (GT) test is important to estimate the time and temperature of cooking of milled rice grains. It indicates the temperature at which the semicrystalline structure of starch begins to melt upon cooking of rice flour in a specific ratio of water. It is classified as low (55–69.5 °C), intermediate (70–74 °C), and high (74.5–80 °C) (Juliano 2007). The two popular methods for GT evaluation are based on grain digestion by alkali and differential scanning calorimetry (DSC). The first is an indirect measure which relies on the tendency of starch granules of milled rice to disintegrate when soaked in dilute alkali such as potassium hydroxide, and it is numerically represented as alkali spreading value (ASV) (Little et al. 1958) or alkali score (AS) (Bhattacharya and Sowbhagya 1972). On the other hand, despite its high cost, the DSC method is more accurate and direct because it measures the first peak of the endotherm in real time as the starch granules gelatinize (Biliaderis et al. 1986; Normand and Marshall 1988), which is associated with variations in amylopectin fine structure (Nakamura et al. 2002; Umemoto et al. 2004). However, for routine analysis, the ASV is the feasible choice to screen thousands of lines in the breeding pipeline, while DSC is usually applied for GT validation on advanced breeding lines before their release as new rice varieties.

The gelatinization temperature of rice can also be estimated by measuring the viscosity of starch pastes by Rapid Visco Analyzer (RVA) which offers speed, convenience, and lower sample size requirement. More recently, the AACC International approved an RVA method (AACCI Method 61-04.01) to precisely estimate the GT of milled rice flour based on a recent collaborative interlaboratory study (Butardo and Sreenivasulu 2019).

Most RVA parameters are significantly correlated with amylose content (Bao et al. 2006a, b). For instance, a low setback value is associated with softness after cooking, and a high viscosity breakdown value is related to good palatability (Champagne et al. 1999; Suwannaporn et al. 2007). Protein content can influence pasting properties to explain the differences in cooked rice texture as well (Champagne et al. 2004).

Cooked rice texture is increasingly recognized as an important characteristic that can influence its acceptability and consumption. Its instrumental measurement can be performed by using Instron food tester, texturometer, and viscoelastograph (Blakeney 1979; Suzuki et al. 1979; Laignelet and Feillet 1979; Juliano et al. 1981). The main concern is how to cook rice grains with different starch properties to the same degree that will make them intercomparable. Mestres et al. (2011) demonstrated that firmness decreased and stickiness increased with an increase in the cooking time. Several authors examined the effect of the water-to-rice ratio on the flavor and texture of cooked rice. The hardness of the cooked rice decreases, and its stickiness and the ratio of stickiness to hardness increase with an increase in the ratio of water to rice (Soo-Jeong 1996). Interestingly, an international cooperative testing to measure cooked rice grain texture revealed that hardness and stickiness values measured by textural profiling instruments are more accurate than sensory panel scores (Juliano et al. 1981).

Considering the highly correlated AAC measure for texture, it is reasonable to recommend the use of AAC estimation as the primary screening tool to classify rice based on texture, with the instrumental methods used only for verification of selected advanced breeding lines (Champagne et al. 2004).

The main rice carbohydrate is not digested at equal rate in the human gastrointestinal tract, and its different digestibility resistance can play an interesting role as health-promoting factor by mitigating the impact of noncommunicable diseases. The estimate of starch digestibility can reflect the glycemic index (GI) of rice. Routinely, screening for low GI in rice breeding programs is not feasible due to the high cost of the clinical assays and low throughputs. In vitro digestion methodologies are simpler, faster, and economically viable compared to the in vivo method (Tamura et al. 2016). In vitro digestibility of starch is determined by the rate of glucose released during enzymatic hydrolysis (Ye et al. 2019).

Cooking process that can alter digestibility should also be investigated. This should also be coupled with industrial processes to modify rice digestibility such as parboiling and processing of high amylose-based products. Embrapa Rice and Beans' research team has conducted a project to investigate the in vitro starch digestibility of rice genotypes developed in Brazil, among them there are the whole pigmented, black and red rice and milled and parboiled rice. Starch digestibility was highly influenced by the bran fraction as well as by its constituents, which reduces enzyme access to the granules. Differences were observed in starch hydrolysis between genotypes due to different amylose levels and processing, as well as interference of non-starch components, such as protein interaction and presence of bran (Reis et al. 2019 – under review).

Rice can also be considered an important protein source in the diet whose digestibility is increased by its combination with protein from legume sources. In addition to its nutritional role, rice protein plays a significant role in the culinary quality of grain given its interaction with starch fractions. In the germplasm of common rice (*Oryza sativa*) as well as wild species such as *Oryza glumaepatula* (Santos et al. 2013), variability in quality (amino acid profile) and protein content in the grain was detected, especially for the rice storage protein. Such information may be exploited by breeding programs to improve the nutritional quality of commercial rice, as it has already been accomplished for mineral content, especially iron and zinc, by different tools such as conventional breeding, agronomic breeding (fertilization and soil fertility), biofortification, and transgenesis (golden rice). These initiatives are of great impact to populations where rice is the mainstay of the diet.

Defining or quantifying rice quality is a challenge, mainly because it depends on consumer preference, market demand, and the type of application or use of the product, and it changes from region to region and between different countries or cultures, besides cooking, eating, and textural quality of rice are complex and multifactorial attributes. The classification of a cultivar as "good quality rice" is completely connected to the target consumer or market. Quality evaluation of cooked rice would be better if performed by a panel of trained tasters; however, this alternative is not convenient for large routines due to a number of factors. Sensory evaluation is a time-consuming, expensive, and generally subjective method.

In addition, there remains a demand for specific quantification methodologies that are simple, accurate, robust, and universally applicable to all rice varieties for each desired quality trait, while being easily adopted and reproducible across all players of the value chain. This is critical to distinguishing the low, medium, and premium rice classes. Ideally, molecular markers associated with grain quality traits should also be developed to facilitate multiple screening using molecular techniques.

At Embrapa Rice and Beans (Araújo Silva 2018), we have developed a free software application in the prediction of rice cooking quality, called QualiArroz ("QualiRice"). From a validated database of different physicochemical quality parameters associated with sensory panel evaluation, it was possible to conceive value rules for the app development. The chosen parameters and the models of prediction allowed the prediction of the most likely classification of the culinary quality, without the need of the sensory test in routine analysis (Araújo Silva 2018).

In 2015, the International Rice Research Institute (IRRI) created the International Network for Quality Rice (INQR), of which Embrapa Rice and Beans from Brazil and other research institutes such as CIAT-FLAR from Colombia were a part, with a panel of experts from rice breeding, grain quality, and industry sectors. The international meetings were fundamental to exchange information on methodologies and parameters of grain quality with other rice quality laboratories around the world. Some strategies and recommendations came up by INQR, including the need of comparison of data from commercial, laboratory, and hand mills, to determine adjustment factors, such as moisture, equipment, and brand variability that interfere with milling results. Besides there is also a demand for a better understanding of the genetic basis of kernel hardness/breaking strength, the development and validation of molecular markers for key quality traits (specially grain breakage), and others (Butardo and Sreenivasulu 2019).

14.3 Breeding for Grain Quality in Brazil

Considering industrial and culinary qualities, the preference of the Brazilian market is for vitreous rice (translucent), without chalkiness and stained grains, with high HRY after industrial processing; besides the grains must be fluffy and non-sticky after cooking.

14.3.1 Milling Quality

At the milling procedure, after grain dehulling and polishing, husk and bran are considered as by-products and milled grains as the main product. The higher the proportion of recovered product the better since it represents the share with the highest commercial value and is defined as *benefit income* (whole grains plus broken grains). HRY is the most important component, as it is more variable and influenced by the environmental conditions of crop management.

These characteristics are greatly influenced by the environment and by the genotype. As grains lose moisture in the maturation process, different genotypes respond differently to their grain susceptibility to breakage because of environmental variations. HRY depends mainly on how dried grains are at harvest and under what conditions the process of moisture loss has taken place. In plant breeding, both traits, total milled rice and HRY, should be considered in relation to the maximum potential obtained by each cultivar.

14.3.2 Product Appearance

The Brazilian consumer is very demanding about the appearance of rice. The preference is for the translucent long and slender grain without opaque areas and free of impurities, stains, or imperfections caused by improper handling, insects, or disease. To be classified as long and slender, rice must have compliance dimensions of length and width according to the Ministry of Agriculture regulations (Normative Instruction N°. 06 of 2009). According to Jennings et al. (1979), grain dimensions (length, width, and thickness) are inherited independently from each other, with little environmental influence, high heritability, and apparently with little or no restriction for recombination with other agronomic characteristics important to the rice crop.

The appearance of rice is also greatly affected by the occurrence of opaque white areas in the grain endosperm (chalkiness, white core, or white belly), caused by loose conditioning of starch and protein chains. The presence and intensity of chalkiness have great genetic effect, with marked difference in its expression between genotypes. However, the environment has a strong influence on the expression of this mandatory trait. The determination of the chalkiness is visually performed using polished grains or with a classifying image equipment.

14.3.3 Cooking Quality

In general, Brazilians prefer rice grains that, after cooking, has non-sticky, fluffy, and soft grains, with good volume expansion (or cooking yield) and short cooking time.

The cooking test is usually adopted by breeding programs to evaluate rice cooking time, grain expansion, and texture (hardness and stickiness), because it simulates the consumer preparation of rice at home. However, it depends on a trained sensory panel and a method standardization to analyze the cooking traits (Martínez and Cuevas 1989), and it can deliver incoherent or poorly comparable or reproducible results given its subjectivity.

The rice cooking test is one of the quality parameters most used by breeding programs and processing industries as a way to evaluate the culinary behavior of the cultivars launched and/or new lines under study. As breeding programs deal with a large number of genotypes (samples) and such analysis requires an expressive amount of grains, time, and training, this methodology becomes impractical and relatively costly, especially when numerous lines need to be evaluated. However, when the number of genotypes is appropriate, and they present a satisfactory amount of grains, rice cooking test is essential.

One of the practical ways of assessing cooking performance of a large number of genotypes is by inferring about culinary characteristics through indirect tests. The main test is the amylose content of the starch fraction, which accounts for about 90% of the polished grain composition. Amylose content and gelatinization temperature are traits transmitted with high heritability; therefore, they are quite easily selected and fixed within breeding programs. However, they require laboratory structure to be determined. For more accurate data, other analysis should be performed, such as gel consistency, viscoamylographic profile, cooking time, and volume expansion.

Grain quality sensory parameters that influence consumer preferences are not fully captured by physical and biochemical methods. Sensory characteristics that include texture and aroma are considered important in certain market segments but are not fully implemented as a trait in rice breeding routine due to the lack of high-capacity or precision technologies (Anacleto et al. 2015).

14.3.4 Nutritional Value

Rice is an essentially energetic food and can also be an important source of B complex proteins, minerals, and vitamins. This cereal provides between 20% and 27% of energy in the form of complex carbohydrates (starch) and between 15% and 20% of proteins daily required by humans and stands out for its easy digestion (Bottini 2008; Naves 2007). Rice contains negligible amounts of fat and is cholesterol-free. Important variation in nutritional values can be identified within cultivated rice; however, at present, breeding programs usually do not focus breeding efforts on these traits.

14.3.5 Rice Improvement for Grain Quality: Embrapa Breeding Program

Grain quality analysis depends on the results of a series of direct or indirect data, which together work as indicators of the quality of a particular rice cultivar or line. The chosen analysis will depend fundamentally on the available infrastructure and the objectives of the genetic improvement program (Vieira and Rabelo 2006).

Frequently, the number and complexity of evaluations advance along the process of breeding; the selection pressure also intensifies generation after generation. Breeding programs typically work with numerous genotypes in early generations, but this number decreases as genotypes advance through the selection steps. Therefore, due to a smaller number of genotypes, it is possible to expand both the complexity and range of different analysis.

Embrapa's rice breeding program, for example, applies the following evaluations for grain quality along the different stages of progeny advancement, development, and selection of elite lines:

The first evaluations for grain quality are performed on $F_{2:3}$ progenies (Progeny Observation Trial). In this generation, samples of 100 g of grains from previously selected progenies (at first for agronomic characteristics of interest) are characterized for milling quality, being determined the benefit income, HRY, grain size (length, width, and length/width ratio), and chalkiness or white belly. These data, used simultaneously, are a primordial tool to discard progenies that do not meet the Brazilian market standards.

In the next generation, $F_{2:4}$ progenies (Progeny Yield Trials), which are evaluated in multilocal trials, are checked for the same characteristics as in the previous generation. However, samples from more than one trial guarantee repetition accuracy and greater reliability in the results, since there is environmental influence on the expression of characteristics such as HRY and chalkiness. This method is applied whenever there are multilocation trials on the breeding program. The amylose content and gelatinization temperature are also evaluated, in order to estimate indirectly the grain hardness and stickiness after cooking.

In the next generation, individual plant selection nursery, there are no evaluations for grain quality due to the small amount of available grain. In $F_{5:6}$ generation, when fixed lines are obtained, the same evaluations described for determination of milling performance are carried out, and only the superior lines remain.

The $F_{5:7}$ generation (Preliminary Trials), consisting of multilocal trials, is evaluated for industrial quality and indirectly for culinary quality. The repetition of these evaluations throughout the development process of new cultivars ensures the follow-up of the different characteristics associated with quality, ensuring precision and reliability in the selection of superior genotypes.

In generation $F_{5:8}$ (Regional Assays), the evaluations for grain quality expand, especially regarding the culinary quality of the lines. In addition to evaluations commonly performed for industrial quality, evaluations of gelatinization temperature, amylose content, cooking test, and determination of the viscoamylographic profile are executed. In the cooking test, the samples are cooked at two times (30 and 60 days after harvest) aiming at the evaluation of softness, stickiness, and detection of the existence and duration of postharvest maturation. The analysis of the viscoamylographic profile provides a *profile* of each genotype and compares it to checks known for their premium culinary quality. The results of all these evaluations, combined to the history of each line, support the selection of the best genotypes.

Finally, in the $F_{5:9}$ and $F_{5:10}$ generations, which make up the Value of Cultivation and Use Trials, the last stage of the program, all evaluations described for the Regional Assays are performed, with the addition of the evaluation of HRY stability.

These trials are conducted in different environments, which are representative of the main producing regions, and aims to evaluate the HRY over five harvest periods, reproducing possible harvest delays that may occur in commercial crops. The five harvest periods are spaced in every 7 days from 25 days after panicle emission. The mean data from each line, from different experiments, enable the elimination of susceptible lines to harvest delays and at the same time the identification of the optimum harvest point (% of relative grain moisture that maximizes HRY) of the selected lines.

14.4 Conclusions

If there is no demand for the newly released rice cultivar, the variety will not succeed in penetrating the target market. So, it is fundamental to start the cultivar development by a market research for preference mapping and clustering of consumer demands for various quality and sensory attributes and not only for agronomic traits.

Breeding strategies for improving HRY include optimization of genetic gains for reduced grain breakage with reduced chalk and optimized postharvest handling and storage. For stakeholder strategies, millers should customize machinery and milling procedures. Parboiling can be considered whenever applicable to enhance HRY in long slender grains, but this strategy is strictly limited to consumer segments that prefer parboiled rice.

There is still need for development of innovative methods to estimate grain quality traits to attend the constant changing and rapidly expanding consumer preference and demand. More than a cultivar with high HRY and good cooking quality, rice should be faced as a food with good sensory, nutritional, and functional quality, meeting the current needs of modern consumers.

References

AACC International (2009) Approved methods of analysis. AACC, St. Paul

Anacleto R, Cuevas RP, Jimenez R, Llorente C, Nissila E, Henry R, Sreenivasulu N (2015) Prospects of breeding high-quality rice using post-genomic tools. Theor Appl Genet 128:1449–1466

Araújo Silva M (2018) Nova estratégia de integração de parâmetros de qualidade de grãos com uso de aplicativo em software livre para estimativa da qualidade culinária de arroz branco. Universidade Federal de Goiás, Goiânia, Brasil, 121p

Asante MD (2017). Breeding rice for improved grain quality. In: Jinquan Li (ed.) Advances in international rice research. IntechOpen. p 69–89. https://doi.org/10.5772/66684

Ashida K, Iida S, Yasui T (2009) Morphological, physical, and chemical properties of grain and flour from chalky rice mutants. Cereal Chem 86:225–231. https://doi.org/10.1094/cchem-86-2-0225

Banaszek MM, Siebenmorgen TJ (1993) Individual rice kernel drying curves. Trans ASAE 36:521–528. https://doi.org/10.13031/2013.28368

Bao JS, Corke H, Sun M (2006a) Nucleotide diversity in starch synthase IIa and validation of single nucleotide polymorphisms in relation to starch gelatinization temperature and other physicochemical properties in rice (Oryza sativa L.). Theor Appl Genet 113:1171–1183

Bao JS, Shen S, Sun M, Corke H (2006b) Analysis of genotypic diversity in the starch physicochemical properties of nonwaxy rice: apparent amylose content, pasting viscosity and gel texture. Starch 58:259–267

Bell M, Bakker R, de Padua D, Rickman J (2000) Rice quality management-principles and some lessons. In: ACIAR proceedings, 2000. ACIAR, p 255–263.

Bhattacharya KR (2009) Physicochemical basis of eating quality of rice. Cereal Foods World 54:18–28. https://doi.org/10.1094/CFW-54-1-0018

Bhattacharya KR, Sowbhagya CM (1972) An improved alkali reaction test for rice quality. J Food Technol 7:323–331

Bhattacharya, (2009) Physicochemical Basis of Eating Quality of Rice. Cereal Foods World 54: 18–28

Bhullar NK, Gruissem W (2013) Nutritional enhancement of rice for human health: the contribution of biotechnology. Biotechnol Adv 31:50–57

Biliaderis CG, Page CM, Maurice TJ, Juliano BO (1986) Thermal characterization of rice starches: a polymeric approach to phase transition of granular starch. J Agric Food Chem 34:6–24

Blakeney A (1979) Instron measurement of cooked rice texture. In: Proceedings of the workshop on chemical aspects of rice grain quality 1978. International Rice Research Institute, Los Banos, p 343–353

Bottini RL (2008) Arroz – História, variedades, receitas. SENAC, São Paulo, Brasil, 390p

Butardo Jr VM, Sreenivasulu N (2019) Improving head rice yield and milling quality: state-of-the-art and future prospects. In: Sreenivasulu N (ed) Rice grain quality: methods and protocols, vol. 1892. Springer Science+Business Media, LLC, part of Springer Nature, Switzerland, p 1–18

Butardo VM Jr, Fitzgerald MA, Bird AR, Gidley MJ, Flanagan BM, Larroque O, Resurreccion AP, Laidlaw HK, Jobling SA, Morell MK, Rahman S (2011) Impact of down-regulation of starch branching enzyme IIb in rice by artificial microRNA and hairpin RNA-mediated RNA silencing. J Exp Bot 62:4927–4941

Butardo VM Jr, Anacleto R, Parween S, Samson I, de Guzman K, Alhambra CM, Misra G, Sreenivasulu N (2017) Systems genetics identifies a novel regulatory domain of amylose synthesis. Plant Physiol 173:887–906

Butardo Jr VM, Sreenivasulu N, Juliano BO (2019) Improving rice grain quality: state-of-the-art and future prospects. In: Sreenivasulu N (ed) Rice grain quality: methods and protocols, vol. 1892. Springer Science+Business Media, LLC, part of Springer Nature 2019, Switzerland, p 19–55

Calingacion M, Laborte A, Nelson A, Resurreccion A, Concepcion JC et al (2014) Diversity of global rice markets and the science required for consumer-targeted rice breeding. PLoS One 9(1):e85106. https://doi.org/10.1371/journal.pone.0085106.

Champagne ET, Bett KL, Vinyard BT, McClung AM, Barton FE, Moldenhauer KA, Linscombe S, McKenzie KS (1999) Correlation between cooked rice texture and rapid Visco Analyser measurements. Cereal Chem 76:764–771

Champagne ET, Bett-Garber KL, McClung AM, Bergman CJ (2004) Sensory characteristics of diverse rice cultivars as influenced by genetic and environmental factors. Cereal Chem 81:237–243

Champagne ET, Bett-Garber KL, Fitzgerald MA, Grimm CC, Lea J, Ohtsubo K, Jongdee S, Xie L, Bassinello P, Resurreccion AP, Ahmad R, Habibi F, Reinke RF (2010) Important sensory properties differentiating premium rice varieties. Rice 3:270–281

Chen H, Siebenmorgen TJ, Marks BP (1997) Relating drying rate constant to head rice yield reduction of long-grain rice. Trans ASAE 40:1133–1139. https://doi.org/10.13031/2013.21331

Cheng FM, Zhong LJ, Wang F, Zhang GP (2005) Differences in cooking and eating properties between chalky and translucent parts in rice grains. Food Chem 90:39–46. https://doi.org/10.1016/j.foodchem.2004.03.018
Christian Mestres, Fabienne Ribeyre, Brigitte Pons, Véronique Fallet, Françoise Matencio, (2011) Sensory texture of cooked rice is rather linked to chemical than to physical characteristics of raw grain. Journal of Cereal Science 53 (1):81–89
Chun A, Song J, Kim K-J, Lee H-J (2009) Quality of head and chalky rice and deterio- ration of eating quality by chalky rice. J Crop Sci Biotechnol 12:239–244. https://doi.org/10.1007/s12892-009-0142-4
Cruz ND, Khush GS (2000) Rice grain quality evaluation procedures. In: Singh RK, Singh US, Kush GS. Aromatic rices. Oxford & IBH Publishing Co. Pvt. Ltd, New Delhi, p 15–28
Cuevas RP, Pede VO, McKinley J, Velarde O, Demont M (2016) Rice grain quality and consumer preferences: a case study of two rural towns in the Philippines. PLoS One 11(3):e0150345. https://doi.org/10.1371/journal.pone.0150345
FAO (2018) World food and agriculture – statistical pocketbook 2018. Rome, p 254. Licence: CC BY-NC-SA 3.0 IGO.
Fitzgerald MA (2010) Rice: characteristics and quality requirements. In: Wrigley CW, Batey IL (eds). Cereal grains. Woodhead Publishing Series in Food Science, Technology and Nutrition, Cambridge, p 212–236. doi: https://doi.org/10.1533/9781845699529.2.212
Fitzgerald MA, Resurreccion AP (2009) Maintaining the yield of edible rice in a warming world. Funct Plant Biol 36:1037–1045. https://doi.org/10.1071/fp09055.
Fitzgerald MA, Sackville Hamilton NR, Calingacion MN, Verhoeven HA, Butardo VM (2008) Is there a second fragrance gene in rice? Plant Biotechnol J 6:416–423
Fitzgerald MA, Bergman CJ, Resurreccion AP, Moller J, Jimenez R, Reinke RF, Martin M, Blanco P, Molina F, Chen M-H, Kuri V, Romero MV, Habibi F, Umemoto T, Jongdee S, Graterol E, Reddy KR, Bassinello PZ, Sivakami R, Rani NS, Das S, Wang Y-J, Indrasari SD, Ramli A, Ahmad R, Dipti SS, Xie L, Lang NT, Singh P, Toro DC, Tavasoli F, Mestres C (2009a) Addressing the dilemmas of measuring amylose in rice. Cereal Chem 86:492–498
Fitzgerald MA, McCouch SR, Hall RD (2009b) Not just a grain of rice: the quest for quality. Trends Plant Sci 14:133–139
Hodges RJ, Buzby JC, Bennett B (2011) Postharvest losses and waste in developed and less developed countries: opportunities to improve resource use. J Agric Sci 149:37–45
Iguaz A, Rodríguez M, Vírseda P (2006) Influence of handling and processing of rough rice on fissures and head rice yields. J Food Eng 77:803–809. https://doi.org/10.1016/j.jfoodeng.2005.08.006
International Organization for Standardization (2007) ISO 6647-2:2007 – rice determination of amylose content – part 2: routine methods. International Organization for Standardization, Geneva
Jennings PR, Coffman WR, Kauffman HE (1979) Rice Improvements. International Rice Research Institute, Los Banos, Philippines, 186p.
Juliano BO (1971) A simplified assay for milled-rice amylose. Cereal Sci Today 16:334–338
Juliano BO (2007) Rice chemistry and quality. Philippine Rice Research Institute, Munoz
Juliano BO, Villareal CP (1993) Grain quality evaluation of world rices. International Rice Research Institute, Los Banos, p 205
Juliano BO, Perez CM, Barber S, Blakeney AB, Iwasaki TA, Shibuya N, Keneaster KK, Chung S, Laignelet B, Launay B, del Mundo AM, Suzuki H, Shiki JI, Tsuji S, Yama JT, Tatsumi K, Webb BD (1981) International cooperative comparison of instrument methods for cooked rice texture. J Texture Stud 12:17–38
Juliano BO, Tuaño APP, Monteroso DN, Aoki N, Mestres C, Duldulao JBA, Bergonio KB (2012) Replacement of acetate with ammonium buffer to determine apparent amylose content of milled rice. Cereal Foods World 57:14–19
Khush GS (2005) What it will take to feed 5.0 billion rice consumers in 2030. Plant Mol Biol 59:1–6

Kumar I, Khush GS (1986) Genetics of amylose content in rice (Oryza sativa L). J Genet 65:1–11. https://doi.org/10.1007/Bf02923530

Kunze O (1979) Fissuring of the rice grain after heated air drying. Trans ASEA 22:1197–1201. https://doi.org/10.13031/2013.35183

Kunze O, Choudhury M (1972) Moisture adsorption related to the tensile strength of rice. Cereal Chem 49:684–696

Laignelet B, Feillet P (1979) Use of the visco- elastograph for measuring the texture of cooked rice. In: Chemical aspects of rice grain quality 1978. International Rice Research Institute, Los Banos, pp 355–361

Lenaerts B, de Mey Y, Demont M (2018) Global impact of accelerated plant breeding: evidence from a meta-analysis on rice breeding. PLoS One 13(6):e0199016. https://doi.org/10.1371/journal.pone.0199016

Little RR, Hilder GB, Dawson EH (1958) Differential effect of dilute alkali on 25 varieties of milled white rice. Cereal Chem 35:111–126

Lyman NB, Jagadish KSV, Nalley LL, Dixon BL, Siebenmorgen T (2013) Neglecting rice milling yield and quality underestimates economic losses from high-temperature stress. PLoS One 8(8):e72157. https://doi.org/10.1371/journal.pone.0072157.

Martinéz C, Cuevas F (1989) Evaluación de la calidad culinaria y molinera del arroz: Guía de estudio para ser usada como complemento de la unidad auditutorial sobre el mismo tema. 3 ed. Cali: CIAT, 73p.

Nakamura Y, Sakurai A, Inaba Y, Kimura K, Iwasawa N, Nagamine T (2002) The fine structure of amylopectin in endosperm from Asian cultivated rice can be largely classified into two classes. Starch 54:117–131

Naves MMV (2007) Características químicas e nutricionais do arroz. Boletim do Centro de Pesquisa e Processamento de Alimentos, Curitiba, 25: 51–61.

Normand FL, Marshall WE (1988) Thermal- properties of whole grain milled rice and milled rice flour studied by differential scanning calorimetry. Cereal Foods World 33:663–663

Otto RK, David LC (2004) Rough-rice drying – moisture adsorption and desorption. In: Champagne ET (ed) Rice: chemistry and technology. Grain science references. American Association of Cereal Chemists, Inc., St. Paul, pp 223–268. https://doi.org/10.1094/1891127349.009

Reis O, Batista CS, Aires NGM, Colussi R, Vanier NL, Bassinello PZ (2019) In vitrostarch digestibility of milled rice with different amylose content. In: XI Congresso Brasileiro de Arroz Irrigado. Balneário Camboriú, Santa Catarina, Brasil.

Perez CM, Juliano BO (1978) Modification of the simplified amylose test for milled rice. Starch 30:424 426

Perdon AA, Siebenmorgen TJ, Mauromoustakos A (2000) Glassy state transition and rice drying: development of a brown rice state diagram. Cereal Chem 77:708–713

Santos K, Silveira RDD, Didonet CGM, Brondani C (2013) Storage protein profile and amino acid content in wild rice Oryza glumaepatula. Pesq Agrop Brasileira 48:66–72

Sarker NN, Kunze OR, Strouboulis T (1996) Transient Moisture Gradients in Rough Rice Mapped with Finite Element Model and Related to Fissures After Heated Air Drying. Trans ASAE 39:625–631

Singh N, Sodhi NS, Kaur M, Saxena, SK (2003) Physico -chemical, Morphological, thermal, cooking and texluralproperties of chalky and translucent rice kernels. J. FoodChem. 82 (3): 433–439.

Slade L, Levine H (1991) A food polymer science approach to structure-property relationships in aqueous food systems: non- equilibrium behavior of carbohydrate-water systems. In: Levine H, Slade L (eds) Water relationships in foods, advances in experimental medicine and biology, vol 302. Springer, New York, pp 29–101. https://doi.org/10.1007/978-1-4899-0664-9_3.

Slade L, Levine H (1995) Glass transitions and water-food structure interactions. Adv Food Nutr Res 38:103–179

Soo-Jeong L (1996) Water addition ratio affected texture properties of cooked rice. Korean J. Food & Nutr 25:810–816

Sreenivasulu N, Butardo VM, Misra G, Cuevas RP, Anacleto R, Kavi Kishor PB (2015) Designing climate-resilient rice with ideal grain quality suited for high-temperature stress. J Exp Bot 66:1737–1748. https://doi.org/10.1093/jxb/eru544

Suwannaporn P, Pitiphunpong S, Champangern S (2007) Classification of rice amylose content by discriminant analysis of physicochemical properties. Starch 59:171–177

Suzuki H, Ikehashi H, Kushibuchi K (1979) Rice grain quality evaluation in Japan. In: Proceedings of the workshop on chemical aspects of rice grain quality. International Rice Research Institute, Los Banos, pp 123–127

Tabata M, Hirabayashi H, Takeuchi Y, Ando I, Iida Y, Ohsawa R (2007) Mapping of quantitative trait loci for the occurrence of white back kernels associated with high temperatures during the ripening period of rice(Oryza sativa L.). Breed Sci 57:47–52. https://doi.org/10.1270/Jsbbs.57.47

Takeda Y, Hizukuri S, Juliano BO (1987) Structures of rice amylopectins with low and high affinities for iodine. Carbohydr Res 168:79–88

Tamura M, Singh J, Kaur L, Ogawa Y (2016) Impact of structural characteristics on starch digestibility of cooked rice. Food Chem 191:91–97

Tomlins K, Manful J, Gayin J, Karlton-Senaye B, Tamakloe I (2007) Study of sensory evaluation, consumer acceptability, affordability and market price of rice. J Sci Food Agric 87:1564–1575. https://doi.org/10.1002/jsfa.2889

Umemoto T, Aoki N, Lin H, Nakamura Y, Inouchi N, Sato Y, Yano M, Hirabayashi H, Maruyama S (2004) Natural variation in rice starch synthase IIa affects enzyme and starch properties. Funct Plant Biol 31:671–684

Unnevehr LJ, Duff B, Juliano BO (1992) Consumer demand for rice grain quality: introduction and major findings. In: Duff UB, Juliano BO (eds) Consumer demand for rice grain quality. International Rice Research Institute, Los Baños, pp 5–20

Vilaplana F, Hasjim J, Gilbert RG (2012) Amylose content in starches: toward optimal definition and validating experimental methods. Carbohydr Polym 88:103–111

Vieira NRA, Rabelo RR (2006) Qualidade tecnológica. In: Santos AB, Stone LF and Vieira RA (ed) A cultura do arroz no Brasil. 2 ed, Embrapa Arroz e Feijão, Goiânia, p. 969-1005.

Wang S, Li S, Liu Q, Wu K, Zhang J, Wang S, Wang Y, Chen X, Zhang Y, Gao C, Wang F, Huang H, Fu X (2015) The OsSPL16-GW7 regulatory module determines grain shape and simultaneously improves rice yield and grain quality. Nat Genet 47:949–954

Welch RM, Graham R (2004) Breeding for micronutrients in staple food crops from a human nutrition perspective. J Exp Bot 55:353–364

Yang W, Jia CC, Siebenmorgen TJ, Howell TA, Cnossen AG (2002) Intra-kernel mois- ture responses of rice to drying and temper- ing treatments by finite element simulation. J Long Form Workform 45(4):1037–1044

Ye X, Zhang Y, Qiu C, Corke H, Sui Z (2019) Extraction and characterization of starch granule-associated proteins from rice that affect in vitro starch digestibility. Food Chem 276:754–760

Zakaria S, Matsuda T, Tajima S, Niita Y (2002) Effect of high temperature at ripening stage on the reserve accumulation in seed in some rice cultivars. Plant Prod Sci 5:160–168

Zhang Q, Yang W, Sun Z (2005) Mechanical properties of sound and fissured rice kernels and their implications for rice breakage. J Food Eng 68:65–72

Chapter 15
Mutation Breeding for Rice Grain Quality: Aspects, Considerations, and Promising Results

Viviane Kopp da Luz, Victoria Freitas de Oliveira, Latóia Eduarda Maltzahn, and Eduardo Venske

15.1 Introduction

Genetic diversity is the raw material for both evolution and plant breeding. It is basically the result of mutations on the DNA of the individuals belonging to any gene pool of any living species under consideration. Mutations occur spontaneously, and although at a very low rate, these natural DNA changes, accumulated along millennia, have been the main source of genetic variability for crop improvement up to the current days. However, the discovery that mutations could be intentionally provoked, and that the rate of generation of novelty at genomes could be enormously increased, has also established man-managed mutagenesis as an important breeding approach (Taheri et al. 2017; Ke et al. 2019). Stadler at the late 1920s (Stadler 1928) has contributed with the first steps to this field which has gradually been adopted and improved all around the world, in a constant rate (Oladosu et al. 2016).

Currently, there is a high pressure on plant breeding for fast and effective results, which bring new reasons to consider artificially inducted mutations as an even more important approach. On the one hand, the ever-increasing human population and the ongoing climate change are enormously modifying the magnitudes of the breeding demands to concerning levels (Jain 2010; Ray et al. 2013; Hickey et al. 2019). On the other hand, several evidences point out to the narrowing of the genetic diversity of several crops, including rice, which are contributing to the plateauing of the genetic improvement of these crops (van de Wouw et al. 2009). Artificially induced mutants are a valuable source of novel genetic variability, being an approach which contributes to fight these adverse scenarios. Another especial aspect of induced mutation is that two simultaneous aims may be pursued: development of novel variability for use in breeding programs and also generation of modified individuals for functional genomic research (Jiang et al. 2012).

V. K. da Luz · V. F. de Oliveira · L. E. Maltzahn · E. Venske (✉)
Plant Breeding and Genomics Center, Universidade Federal de Pelotas, Pelotas, Brazil
e-mail: eduardo.venske@yahoo.com.br

A. Costa de Oliveira et al. (eds.), *The Future of Rice Demand: Quality Beyond Productivity*, https://doi.org/10.1007/978-3-030-37510-2_15

Food has been required by the human population in large quantity, but quality – the scope of this book – has rapidly been increasing in importance, demanding fast breeding progress. Rice grain quality comprises several traits, including different aspects of physical appearance, cooking and sensory properties, aroma, and, more importantly, nutritional value (Fitzgerald et al. 2009; Birla et al. 2017). However, it is important to emphasize that there is variability for rice preferences worldwide (Calingacion et al. 2014). Thus, any given "improvement" could be considered a "worsening" depending on the rice market under consideration. This is why Calingacion et al. (2014) have termed that "rice is by no mean a one size fits all crop."

Even though several "site-directed" DNA modification approaches, such as the recent cutting-edge methods of gene or genome editing (including the CRISPR-Cas9 technique) (Ding et al. 2016), can also be considered mutation approaches – and promising results have been obtained (Sun et al. 2017; Tang et al. 2017; Fiaz et al. 2019) – in this chapter, these methods will not be covered, as another chapter of this book is dedicated to this topic. In fact, just the "classic" mutation breeding for rice grain quality deserves itself a whole book giving the wealth of studies already carried out and the different results obtained. Considering the exposed, the aim of this chapter is to briefly discuss the importance, aspects, considerations, and promising results already obtained through mutation breeding for rice grain quality improvement.

15.2 Mutation Breeding for Rice Grain Quality Improvement: Cultivars Released

Over the last eight decades, mutagenesis has been one of the main wheels of plant breeding, harvesting important results (Ahloowalia et al. 2004). As an especial feature, mutation breeding has been largely supported by the Joint Food and Agriculture Organization of the United Nations/International Atomic Energy Agency (FAO/IAEA), which has supported several projects all around the world. This Joint also makes available, through an online database (https://mvd.iaea.org) (FAO/IAEA 2019), a wealth of information regarding mutant varieties improved for several traits and through the aid of different methods. Searching this database, it is possible to verify that mutation breeding has been responsible already for the development and release of 3.303 cultivars/clones all around the world. These numbers are probably underestimated, as a large unaccounted number of mutant cultivars may have been developed but not registered in different countries. From these improved genotypes, ca. 49.5% are cereals, 21.9% are ornamentals, and 15% are leguminous (Taheri et al. 2017).

The largest number of mutant cultivars released are of rice crop, with a total of 824 improved genotypes (~25% of all mutant cultivars registered so far), demonstrating both the importance of rice itself and of mutant induction for breeding this cereal. Importantly, rice is also a role-model crop in which expressive numbers of mutant cultivars are in fact widely adopted in different countries and regions, giving positive economic and social impact to farmers, industry, and consumers (Ahloowalia et al. 2004). Among the traits already improved in rice through this approach, one can list grain quality (including nutrition traits – the database uses to keep these traits separated), with a total of 270 improved genotypes so far (Fig. 15.1). It represents one-third (~33%) of all mutant rice cultivars registered up to now (note – for all classifications made here, a mutant cultivar may have been simultaneously improved in more than one trait). The first registered mutant cultivar dated by 1970 and the latest by 2017.

As already mentioned, rice quality comprises several traits (Fitzgerald et al. 2009; Birla et al. 2017). The grain quality trait most recurrently improved in rice has been culinary quality, which comprises cooking and eating aspects, with 185 cultivars registered; second, special types of grain and reduced anti-nutrients, with 79 cultivars; third, industrial and marketing quality, with 14 cultivars; andfinally, food processing quality with 6 cultivars and macronutrient content with 5 cultivars. Yet, nine cultivars have been improved for other remaining traits, viz., high amino acids, high protein, long and slender grains with golden color, long fine grains, and nutritional quality – micronutrients, slightly aromatic, and colored grain.

According to the database, from all these mutants, 104 have been developed for direct use as cultivars, so the remaining 166 registered cultivars are in fact the result of an indirect use of a mutant line, such as in crosses with other genotypes.

The largest part of these registered improved mutant genotypes has been obtained through physical mutagens, being 209 cultivars. Through chemical agents, 28 cultivars were developed, and for 33 genotypes, this information has not been made available. From the chemical agents, the most used was nitrous methyl urea, with ten cultivars. Among the physical agents most used, gamma rays occupy the first place, being responsible for the release of 167 mutant rice cultivars with improved grain quality.

Different methods can be used to either induce mutations and also detect them. Next, a brief section is devoted in this chapter to shortly review methods of mutation induction and detection. When appropriate, these methods are discussed in the context of inducing mutations for rice grain quality improvement.

15.3 Mutagenic Agents and Detection Methods

15.3.1 Chemical Mutagenic Agents

The effect chemical mutagenic agents on treated DNAs have is in general smoother comparing to the effect of physical agents (Oldach 2011). It may be considered an important point regarding breeding for quality traits, especially for rice, which is a staple food, as the strategy must comprise improving quality without affecting yield

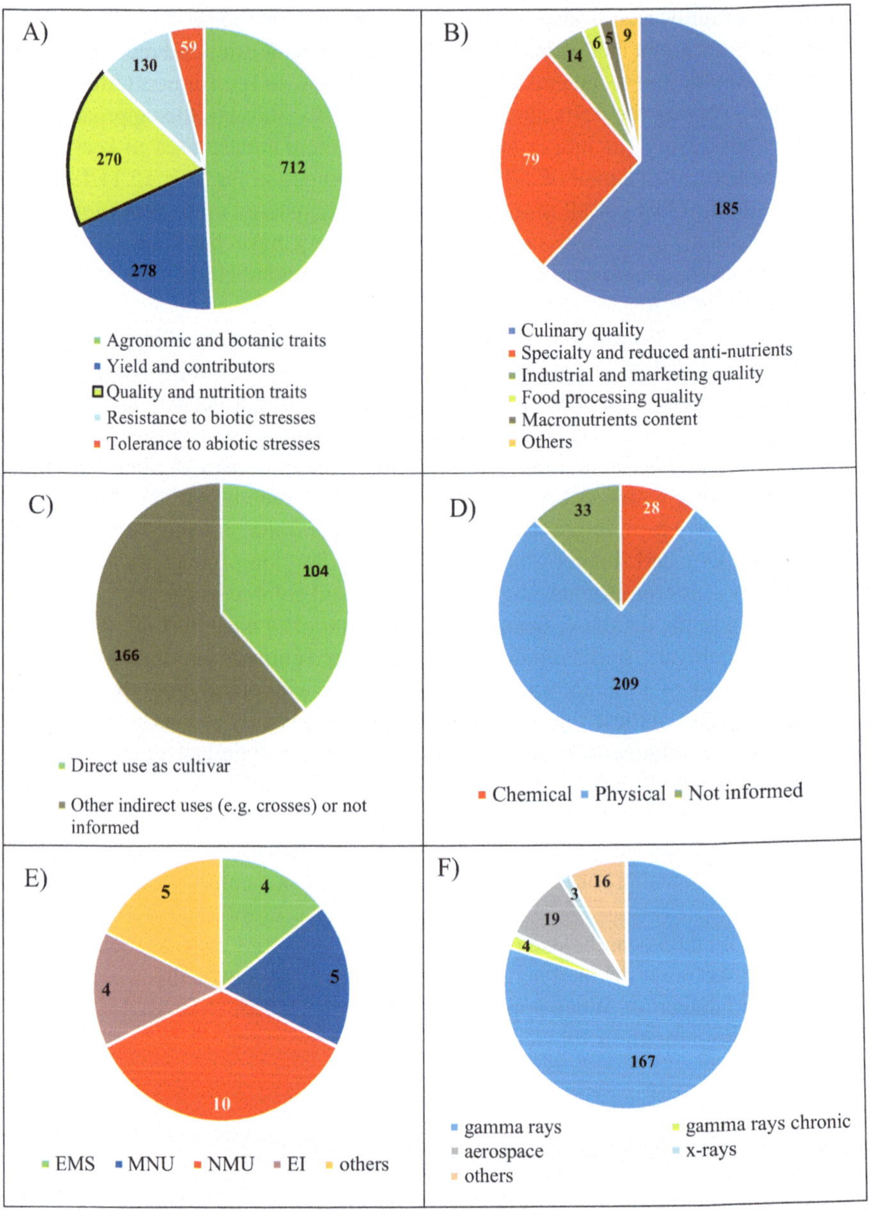

Fig. 15.1 Mutant rice cultivars registered at the Joint FAO/IAEA mutant database (https://mvd.iaea.org). (**a**) Type of improved trait; (**b**) specific grain quality trait; (**c**) mutant use; (**d**) mutagenic agent used; (**e**) chemical mutagenic agent used; and (**f**) physical mutagenic agent used. (Note: For (**a**) and (**b**), a same mutant cultivar may belong to different groups)

potential, or other "field traits," e.g., adaptation, which in the large majority of the cases must come first. Another advantage of the use of chemical mutagenic agents is that there is no need of special facilities, i.e., sophisticated laboratories. Also, the protocols are usually very simple; thus, no special training is required. Summarizing the procedure in very general words, the seeds to be treated are just immersed in a mutagenic solution for a stablished time, and after that they can be sown (Ke et al. 2019). Thus, the process is suitable for rice, a crop which usually receives less financial investment, especially by private companies, comparing to other crops, such as soybean or maize, for example.

There are, however, several factors which affect the results of mutagenesis via chemical agents. They include the concentration of the mutagenic agent in the treatment solution, duration of the exposure, temperature, pre-imbibition of the seeds, pH, the use of catalytic agents (Cu^{+2} and Zn^{+2}), the treated organs (e.g., for rice, usually seeds, but it may also be pollen), the posttreatment handling of the mutated seeds, and even the treated genotype (Ke et al. 2019; Mba et al. 2010; Xin et al. 2008).

Among the different chemical agents (Table 15.1), the ethyl methane sulfonate (EMS) has been considered the most effective and reliable; thus, it has been the most used chemical agent for mutagenesis in plants (Taheri et al. 2017). The treatment with this agent causes nuclear DNA breaks, and during the DNA repair process, mutations are randomly induced (Serrat et al. 2014). In fact, more than 80% of the mutant cultivars registered at the International Atomic Energy Association (IAEA) database were obtained through the use of chemical mutagenesis through alkylating agents. EMS is included in this group, and also 1-methyl-1-nitrosourea (MNU) and 1-ethyl-1-nitrosourea (ENU), being responsible for 64% of these varieties (Oladosu et al. 2016).

The best dose for an effective mutation density shows high variation not only among species, especially due to ploidy levels, but also among genotypes (Parry et al. 2009). Rice is diploid; therefore, there is no difficulty regarding polyploidy buffer effect. A long incubation time with a high EMS concentration provokes a high number of punctual mutations, increasing, however, the lethality (Lee et al. 2017). Thus, previous to the development of a mutant population through the use of any chemical agent, it is important to consider the optimal mutagenesis condition for the agent for the given cultivar, which has to be determined through a pattern able to induce a 50% lethal dose (Talebi et al. 2012; Lee et al. 2017). An advantage of punctual mutations created by chemical mutagens is the potential of generating not only loss of function phenotypes but also gain of function (Oladosu et al. 2016).

15.3.2 Physical Mutagenic Agents

Physical mutagenic agents mainly comprise ionizing radiations (alpha, beta, and gamma rays, fast, and thermic neutrons) and nonionizing radiation (X rays) (Jency et al. 2016; Kayalvizhi et al. 2017). The mutagenic action of the X-rays, discovered by Muller, in 1927, in *Drosophila* and of gamma and X-rays, in 1928, by Stadler, in

Table 15.1 Examples of most commonly used chemical mutagens

Mutagen group	Mutagenic agent
Acridines	Acridine orange
Alkylating agents	Diethyl sulphate (DES)
	Dimethyl sulphate (DMS)
	Ethyl methanesulfonate (EMS)
	Methyl methanesulfonate (MMS)
	N,N-diethylnitrous amide (NDEA)
	N,N-dimethylnitrous amide (NDMA)
	1-Ethyl-1-nitrosourea (ENU)
	1-Ethyl-2-nitro-1-nitrosoguanidine (ENNG)
	1-Methyl-1-nitrosourea (MNU)
	1-Methyl-2-nitro-1-nitrosoguanidine (MNNG)
	Ethyleneimine (EI)
	N-nitrous-N-ethylurea (NEU)
	N-nitrous-N-methylurea (NMU)
Antibiotics	Actinomycin D
	Mitomycin C
	Azaserine
	Streptonigrin
Azide	Sodium azide
Base analogs	Maleic hydrazide
	2-Aminopurine (2AP)
	5-Bromodeoxyuridine
	5-Bromouracil (5-BU)
Hydroxylamine	Hydroxylamine
Nitrous acid	Nitrous acid

Source: Adapted from Oladosu et al. (2016)

barley and maize has opened a new scientific field in genetics (Jency et al. 2016). According to Kovacs and Keresztesa (2002), gamma rays are considered more penetrating in comparison to the other radiations, such as alpha and beta rays. Gamma rays are electromagnetic waves with very short length and are obtained through the disintegration of the radioisotopes ^{60}Co or ^{137}Cs. Radioactive cobalt (^{60}Co) is broadly used, has a high penetration potential, and can be used to irradiate from seeds to whole living plants but also very delicate materials, such as pollen grains. Several mutants have already been developed with the aid of gamma rays (Oladosu et al. 2016). The main advantage of physical in comparison to chemical mutagens is the level of precision and reproducibility, especially the gamma rays (Yamaguchi et al. 2003).

For mutant induction, it is recommended to use two or three radiation doses together with the control (Kodym and Afza 2003). Importantly, there is no linear correlation between radiation dose and mutation quality (proportion of useful mutation) (Acquaah 2006). The neutron-based approaches are known to cause chromosome damages to the treated tissues and have a lower penetration ability. Although their mutagenic effect has been proved to be efficient, particularly for the deletion of large DNA fragments, their use for induced mutagenesis has been limited (Kharkwal 2012). The exposure to ionizing radiation also causes several biological effects on the treated cells, causing mutations due to the break of the chemical links in DNA, deleting nucleotides or replacing them by others (Acquaah 2006). Among the several types of DNA damages, the double-strand breaks (DSBs) are the most severe (Kawamura et al. 2018). The ionizing radiation also provokes the accumulation of reactive oxygen species (ROS) in the irradiated cells. Studies have suggested that after the exposure to a low level of irradiation, the ROS reaction induced has been relatively more significant for mutant induction than the one mediated by direct DNA damages (Kawamura et al. 2018).

The radio-sensibility is influenced by biological factors, such as genetic differences, physiological status, and treated organs, and by environmental modifiers (oxygen, water content, postirradiation store and temperature, as well as the mutagen itself) (Conger et al. 1977; Sarsu et al. 2018). Modifying factors highly affect the mutagenic efficiency and the reproducibility of the results. Oxygen is the main modifying factor, while the water content, temperature, and storage seem to be secondary, interacting with the effect of the oxygen (Revised by Kodym and Afza 2003).

15.3.3 Mutation Detection

Forward and reverse genetics are the two main approaches which connect the genotype to the phenotype. Traditionally, the classical genetic analysis, also called as forward genetics, aims to discover the gene responsible for a known phenotype. In this approach, each phenotype is studied with the aim of determining the DNA sequence responsible for it and also its genetic characteristics. The existence of phenotypic variability is essential for the forward genetics approach, as the comparisons among contrasting phenotypes allows the determination of its genetic basis (Schneider and Leister 2006).

In the reverse genetics, on the other hand, the role of any given gene is identified through the analysis of the phenotypes resulting from the alteration of their sequence (caused by either artificial mutations or natural variations) (Jankowicz-Cieslak and Till 2015). Reverse genetics can employ the targeted mutagenesis to a given locus of interest or, alternatively, generate mutant populations large enough to saturate the genome under study, being the next necessary step the molecular detection and characterization of the induced genomic lesions (Henikoff and Comai 2003).

15.3.3.1 Via Phenotyping

Since the beginning of plant breeding, the identification (phenotyping) and deliberated selection of plants with modifications (i.e., spontaneous or induced mutants) are part of the applied strategies (Sang 2009). The visual/phenotypic assessment is frequently used in selection for yield, plant height, adaptation to soil and climate conditions, earliness, disease resistance, and color changes, among several other traits, including grain quality (Roychowdhury and Tah 2013). When a mutant line seems promising for a pursued trait, the next step is seed multiplication for extensive field assays. In this case, the mother cultivar and other elite genotypes are also included in the assays as controls. The most common methods for these evaluations are the same as for any other type of breeding approach aiming for performance assessment. The aim of these field assays is to evaluate whether the mutant can become a commercial cultivar, improved in comparison to the mother line. Before the release as a commercial type, the promising mutant has to be deeply studied for performance in different traits such as growth habit, yield components, population density, and sown dates, in addition to target trait improved, such as grain quality (Roychowdhury and Tah 2013).

Phenotypic markers are most of the time easy to be used, usually without the requirement for specific tools. They do not require any specific biochemical or molecular techniques; however, the main disadvantages of the visual/phenotypic selection are that they are highly influenced by the developmental stages of the plant and several environmental factors (Eagles et al. 2001). An important aspect to emphasize regarding the assessment of mutant lines for grain quality regards the cost and time required for assessing grain quality traits in comparison to several other morphological traits. Although assessing cooking properties is usually cheap, there is always the need to wait until harvesting to start the assessments, which represents considerable time. Also, assessing the content of trace elements, such as arsenic, requires sophisticated and expensive equipment and reagents. Thus, alternative methods are useful in mutant identification regarding rice grain quality.

15.3.3.2 Molecular Markers

The classical mutation breeding may present some disadvantages, such as the need of a very large mutant population to increase the probability of finding a genotype with an improved trait, the occurrence of chimeras, and the heterozygosity of the mutated loci. Biotechnological approaches can overcome part of these limitations. Several advances in genetics, since the discovery of the DNA and the development of the polymerase chain reaction and genome sequencing, allowed the development in methods to improve the efficiency in mutation detection, leading to the development of the reverse genetics, i.e., from the genotype to the phenotype, in which the effect of the phenotype in the manipulation of any given gene or gene product is assayed (Taheri et al. 2017).

Considering that another chapter of this book is dedicated to the use of molecular markers for grain quality improvement in rice, only a brief discussion on the topic will be presented here, in the context of mutation breeding. DNA-based molecular markers can be used to investigate the mutagenic effects of any method applied. Molecular methods to identify mutations in known target genes may allow for the selection of candidate mutants in the initial stages of a mutation breeding program, which can significantly increase the efficiency in the mutant selection. It is especially important in the context of rice grain quality, whose component traits have to be assayed after harvest (Dreher et al. 2003). Several methods for marker-assisted selection (MAS) have been used to select for favorable alleles (Charcosset and Moreau 2004). Currently, the most common markers are SNPs (*single nucleotide polymorphisms*) and SSRs (simple sequence repeats or microsatellites) (Tadele 2016; Nadeem et al. 2018). Several studies have shown promising results for the use of MAS for grain quality improvement (Yi et al. 2009; Wang et al. 2009; Ni et al. 2011; Lau et al. 2017).

15.3.3.3 Genome Sequencing

Genome sequencing of mutants is a simple way to identify mutations which cause phenotypes of interest. However, due to the numerous polymorphisms which may be generated through mutagenesis not related to the phenotype under investigation which segregates together (in linkage disequilibrium) with the actual causative DNA alteration, in most cases the results obtained itself may represent a weak evidence. Thus, some genetic analysis to purify the chromosomal location are usually needed, such as methods that attempt to map by sequencing through next-generation sequencing (Schneeberger and Weigel 2011; Schneeberger 2014).

SNP discovery is one of the main applications of NGS in the context of mutation breeding. Once SNPs in linkage disequilibrium with the causal variation of a mutant type of interest are discovered, these loci can be used in marker-assisted selection, which is the most practical use of molecular markers (revised by Taheri et al. 2017). NGS platforms have been successfully used to detect mutations (Rigola et al. 2009; Tsai et al. 2011). Among the advantages of NGS, one can list the lower cost, comparing to older sequencing methods, and the rapid results generated, in comparison with other laboratorial approaches (GWAS) (revised by Nadeem et al. 2018).

15.3.3.4 Tilling

Chemical mutagenesis usually generates punctual mutations, which are difficult of being directly detected via PCR; thus, the TILLING (Targeting Induced Local Lesions IN Genomes) method (Mccallum et al. 2000a, b), which has been originally developed for *Arabidopsis*, constitutes an alternative. TILLING is a targeted mutation identifying approach, highly efficient to identify single-nucleotide mutations in a specific region of a gene of interest. In this approach, a pool of DNA from the mutated plants is amplified via PCR using specific primers for the region on interest. The amplified products are denatured and, next, renatured leading to homodimers and heterodimers between the mutated and wild DNA. The aberrant pairing due to the mutations is recognized and cleaved by an endonuclease specific for this purpose (*mismatch-specific celery endonuclease – CELI*) (Oleykowski et al. 1998). The presence of a mutation in the DNA pool is identified during the separation of the products of the digestion with *CELI* in a denaturing gel (Till et al. 2006).

Based on previous studies, TILLING can be applied to any plant species, independently of the genome size and ploidy level, and has proved to be an efficient and robust method for the identification of useful mutation lines in several crops, including rice (Taheri et al. 2017; Tadele 2016). Several TILLING platforms have been already stablished for several crops, with the list of the mutant population and the description of the phenotypes (Kurowska et al. 2011; Chen et al. 2014). Usually 3000–5000 M_2 individuals compose the platform, even though up to 10,000 plants have been already included in platforms. The large majority of the TILLING populations have been developed using chemical mutagenic agents, and among them, the alkylating agent EMS has been the most frequently applied (Kurowska et al. 2011). Using EMS, an elevated mutation rate can be reached, widely and randomly spread along genomes, without excessive DNA damage (Gilchrist and Haughn 2005). Important studies have already been carried out applying TILLING in the aim of improving grain quality in several crops, such as oat, sorghum, wheat, and even rice, decreasing, for example, the phytic acid in the grain (Xin et al. 2008; Chawade et al. 2010; Slade et al. 2012; Kim and Tai 2014).

15.3.3.5 MutMap (Mapping-by-Sequencing)

The traditional genetic mapping approach is usually a complex, time-consuming, and expensive approach, which requires large mapping populations and an abundant number of molecular markers widely distributed on all chromosomes of the studied species. In the same regard, generating mutant population usually also requires time and work. Recently novel approaches for genetic mapping using next-generation sequencing (NGS) have been developed, facilitating the identification of mutations related to traits of interest (Nakata et al. 2018).

A sequencing-based method has been applied for genetic mapping, termed MutMap, which included the use of a backcross population and applying genome sequencing to identify the position of mutations induced by EMS, showing that the mutations were responsible for improvements in important agronomic traits, such as semidwarf height (Abe et al. 2012). This method may also be based on the cross of a mutant for a trait of interest in the M_3–M_5 to the wild type, following the self-pollination of the F_1, allowing for a clear separation of the phenotypic differences already at the second generation. Next, DNA samples composed of 20 individuals (showing a mutant phenotype) are sequenced, aiming to detect the presence of SNPs which differ from the reference genome (Taheri et al. 2017).

Cao et al. (2019), applying both MutMap and RNA-seq to identify mutations in rice, have evidenced the advantage of this procedure by the identification and isolation of a promising mutant showing very low cadmium content at grains. Due to its high applicability and low sequencing cost, MutMap approach can aid the discovery of improvements for grain quality in rice (Abe et al. 2012).

15.4 Improvements in Rice Grain Quality Through Mutagenesis

In the literature, there are significant number of reports on rice grain quality improvements through the use of different mutation breeding methods. In addition to the improved mutation lines obtained, most of these studies also aimed to depict the result of these mutations in biochemical, physiological, technological, industrial terms. (Table 15.2).

Table 15.2 Summary of examples of improvements in rice grain quality obtained through mutation breeding approaches

Trait	Improvement	Method for mutant development and/or assay	References
Gelatinization properties	Varying textures	MutMapPlus	Nakata et al. (2018)
Amylose content	High amylose content	Sodium azide	Suzuki et al. (2008)
	Low amylose content	EMS	Wu et al. (2015)
Starch properties	Resistant starch	n.i.	Yang et al. (2006)
Improved protein or amino acids	Increased glutenin and prolamin	N-methyl-N-nitrosourea	Kumamaru et al. (1988)
	High-lysine	N-methyl-N-nitrosourea	Kumamaru et al. (1997)
Mineral biofortification	Increased content of Fe and Zn	Sodium azide	Jeng et al. (2012)
Reduced content of toxic elements	Low cadmium	Ion beam; EMS + MutMap and RNA-seq	Ishikawa et al. (2012); Cao et al. (2019)
	Low arsenic	Ion beam	Ishikawa et al. (2016)
Reduced content of anti-nutritional compounds	Low phytic acid	γ-Irradiation and NaN_3; sodium azide and methyl nitrosourea and TILLING	Liu et al. (2007), Kim and Tai (2014)
Flour making	Decreased starch damage and fine particles when milled	γ-Irradiation and EMS	Ashida et al. (2009)
Bread making	High-quality bread	γ-Irradiation and EMS	Ashida et al. (2010)
Low protein	Low content of glutelin and a high content of prolamin	Ethyleneimine	Iida et al. (1993)

15.4.1 Starch Properties

By screening a mutant population obtained from the cultivar Nipponbare, two mutants showing moderate changes in starch gelatinization property were selected, and a combination of conventional genetic analyses and the MutMapPlus method revealed that both of these lines harbor novel independent mutant alleles of the starch branching enzyme *IIb* (*BEIIb*) gene (Nakata et al. 2018). Production of a series of mutant lines by combining these alleles in another background established steps for the alteration of the physicochemical properties of starch including apparent amylose content, thermal property, digestibility by alpha-amylase, and branched structures of amylopectin. Sensory evaluation demonstrated that these mutants varied in texture without losing overall palatability.

Resistant starch (RS) in rice after cooking is of high interest for the dietary prevention of diabetes and hyperlipidemia. A mutant showing resistant starch character has been obtained, whose starch has been proved to be slowly digested (Yang et al. 2006).

15.4.2 Amylose Content

There is an expressive variation for preferences regarding content of amylose at rice grains; thus, depending on the country/region, breeding programs could either put effort on increasing amylose or decreasing it. By screening a mutant population obtained through sodium azide treatment on the low-amylose cultivar "Snow Pearl," a mutant line was selected, as presenting enhanced amylose content (6–7% higher than "Snow Pearl") without affecting amylopectin structure (Suzuki et al. 2008). On the other hand, three novel alleles of *FLOURY ENDOSPERM2* (*FLO2*) which confer dull grains with low amylose content in rice have also been obtained recently, just giving an example of a study seeking the opposite: lower amylose in rice (Wu et al. 2015).

15.4.3 Protein Content and Quality

Birla et al. (2017) has been fortuned to state that it is an irony that rice, a staple food for more than half of the human population, is a poor source of nutrition constituents. Protein content and quality are part of this group of nutrition factors that rice scarcely provides. By screening a mutant population of 3000 lines obtained through N-methyl-N-nitrosourea, 17 lines were selected due to their differential protein content (Kumamaru et al. 1988). Determination of extracted protein in the starchy endosperm of the mutants revealed changes in the content of prolamin and glutelin but not globulin. Two out of these 17 lines presented high globulin to a certain level to be considered promising breeding materials to improve rice regarding protein quality. Another three mutants have also shown interesting variation in polypeptide composition.

In terms of protein, quality may matter more than quantity; thus, in order to improve the nutritive value of rice protein, it is necessary to increase the level of lysine. Ten high-lysine mutants were selected from N-methyl-N-nitrosourea-induced mutant lines, using high-performance liquid chromatography for amino acid analysis (Kumamaru et al. 1997). The lysine content of these mutants ranged from 5.10% to 6.38% of total amino acids, while that of the original varieties was only about 4%. The increase in lysine content was detected in the albumin/globulin fraction. All high-lysine mutants also had a greatly improved histidine content.

Although rice is already considered a "starchy" food, decreasing even more its amount of protein, or bioavailable protein, would be of interest for consumers which present kidney disease. Also, protein content and rice quality (cooking and eating) may negatively correlate. A mutant line has been found to have a lower content of glutelin and a higher content of prolamin in its grain protein than other Japanese cultivars assayed (Iida et al. 1993). Although glutelin accounts for ~80% of the total protein found in rice, this line controversially presented a higher content of total protein comparing to the parental cultivar. Successfully, a backcross line (backcrossing to the parental cultivar) was selected showing the same level of total protein that of the parental cultivar but kept the altered proportions of glutelin and prolamin. As an important fraction of the prolamin (i.e., the 13-kDa) is indigestible for humans, this line represents a successful improvement regarding decreasing "effective" amount of grain protein.

15.4.4 *Biofortification*

Biofortification of staple foods such as rice through genetic improvement is the most efficient and also strategic approach to fight hidden hunger. The most targeted elements in this regard are iron (Fe) and zinc (Zn), given their importance to human health and also frequent deficiency among poorest communities. Mutagenesis has shown promising results. From a set of 254 sodium azide-induced mutants originated from IR64, 2 lines have shown increased levels of Fe in polished grains, and 3 had improved Zn content, also in polished rice (Jeng et al. 2012). It has to be emphasized that these mutant lines performed several times better than the parental cultivar, i.e., the lines improved for Fe content presented more than 27 mg kg^{-1} of the element in the grains, while IR64 just 3.9 mg kg^{-1}, which is almost a sevenfold increase. For Zn the gain has been lower, however still expressive, in which the mutant lines showed a Zn concentration above 26 mg kg^{-1}, while IR64 presented 16 mg kg^{-1}.

15.4.5 Decreasing the Content of Toxic Elements

Nutritional improvement of foods include increasing the content of beneficial elements but also decreasing the levels of the toxic ones. Rice is an important source of cadmium (Cd) through diet, which is a highly toxic element. Carbon ion beam irradiation has produced three rice mutants with <0.05 mg Cd kg^{-1} in the grain compared with a mean of 1.73 mg Cd kg^{-1} in the parent, Koshihikari (Ishikawa et al. 2012). In addition to the development of this valuable source of genetic variability, the gene responsible for the trait was identified, and a strategy for marker-assisted selection has been stablished for use in molecular breeding. Later, this mutant line, already called as a cultivar ("Koshihikari Kan No.1"), was assayed for arsenic (As) accumulation at grains, when grown at different soil conditions which are known to affect As bioavailability (Ishikawa et al. 2016). The authors found that the cultivar can simultaneously produce grains with low Cd and As, especially when grown at aerobically soil conditions.

In another study, a mutant generated through EMS treatment displaying an extremely low Cd content at grains (*lcd1*) was selected for gene function study (Cao et al. 2019). Through MutMap, candidate SNP markers associated to the trait were discovered, and through RNA-seq, transcriptomic changes between the mutant line and the wild type brought evidences that a Pro236Leu amino acid substitution in the highly conserved region of *OsNRAMP5* is likely responsible for low Cd accumulation in the *lcd1* mutant. In addition to the elucidation of the molecular basis of the improved phenotype, this study has provided molecular markers for subsequent selection for this allele.

Phytic acid is commonly regarded as an anti-nutritional factor. A total of five independent nonlethal mutants for the gene *lpa* (low phytic acid) have been obtained through both chemical and physical mutagens, and their phytic acid content at brown grains ranged from 34% to 64% of the content found in the parental cultivars (Liu et al. 2007). In another study, four mutants were identified in a population obtained through a chemical mutagen through TILLING method in which the reduction in phytic acid ranged from 20% to 68% (Kim and Tai 2014).

15.4.6 Flour and Bread Making

Most rice produced all around the world is consumed as whole grain; however, flour is also a relatively common product obtained from rice. A set of milky-white chalky mutant lines derived from the cultivar Koshihikari have shown especial properties for flour production, proper for dry milling, such as decreased starch damage and finer particle size (Ashida et al. 2009). Later, these mutant lines were assayed for further baking parameters and presented low pasting temperature, high final viscosity, large setback, and long peak time, thus making this flour suitable for bread making (Ashida et al. 2010).

It is important to emphasize that even though some achievements obtained through mutagenesis may seem redundant when taking into account the phenotypes obtained in different studies, e.g., different studies reaching a similar content of iron in the rice grain, for a breeding point of view, the availability of a variability of sources for any giving trait is of high importance, always taking into account risks of genetic vulnerability when overusing few genes or alleles. It has to be highlighted that similar phenotypes may be obtained through the mutation of different genes, thus, with potential for gene pyramiding strategies.

15.5 Final Considerations and Future Perspectives

There is an increasing demand for high-quality food all around the world. It includes physical aspects, eating and coking quality, aroma, and others but with especial regard to nutritional value, which is also true for rice. In fact, the quest for high-quality rice definitely makes sense, as the cereal is the staple food for more than half of the human population, which includes the poorest fraction. Several breeding methods can and should be applied for this purpose; however, approaches which led to the development of novel variation, such as the classic mutation breeding, discussed in this chapter, must also be applied. The main reason for that is the fact that any breeding method which relies on the already existing genetic variability within the crop gene pool will, by definition, gradually exhaust the sources of genes and alleles for breeding. Mutation breeding creates novelty. Also, how naturally variable can be this starchy cereal for grain quality, especially for mineral and protein content? In order to reach the nutrition quality demanded by rice consumers in several countries, breeders would definitely need to artificially accelerate the mutation rate within rice genomes.

It is undeniable that the very recent methods of gene editing, such as the CRISPR-Cas9 system, represent a powerful tool for precise gene modification. The classic mutagenic methods, however, still present advantages. Mutation induction, especially through chemical agents, is fast and cheaper. Applying physical methods is also cheap; once in most cases, the procedure simply consists in sending the seeds to an institution which has the irradiation equipment, which may ask for a small fee. Also, in most cases there is no knowledge regarding which candidate gene should be modified to get an improvement on a desired phenotype, which is a requirement for gene editing but not necessarily for most classic mutation approaches. The main advantage, however, of the classic mutation breeding is the possibility of generating novel variability for uncountable genes responsible for several other traits and through inducing several different alleles, which can be considered a source of raw material for breeding future crops. It is decisive in the current scenario in which, in addition to quality, the yield potential and adaptation to the ongoing climate change have also to be prioritized. This is why the almost centenary approach of mutation breeding will continue to be applied for the decades to come.

Several achievements have been obtained through mutation breeding regarding improving rice grain quality, for the very distinct examples presented and discussed in this chapter, which does mean that this approach can be even more successfully used for this goal, especially in conjunction with other methods. These achievements, however, are just the first steps of a long way in direction to a high grain quality rice.

References

Abe A, Kosugi S, Yoshida K, Natsume S, Takagi H, Kanzaki H, Matsumura H, Yoshida K, Mitsuoka C, Tamiru M, Innan H, Cano L, Kamoun S, Terauchi R (2012) Genome sequencing reveals agronomically important loci in rice using MutMap. Nat Biotech 30:174–178. https://doi.org/10.1038/nbt.2095

Acquaah G (2006) Principles of plant genetics and breeding. Wiley-Blackwell, Chichester

Ahloowalia B, Maluszynski M, Nichterlein K (2004) Global impact of mutation-derived varieties. Euphytica 135:187–204. https://doi.org/10.1023/B:EUPH.0000014914.85465.4f

Ashida K, Iida S, Yasui T (2009) Morphological, physical, and chemical properties of grain and flour from chalky rice mutants. Cereal Chem 86:225–231. https://doi.org/10.1094/CCHEM-86-2-0225

Ashida K, Araki E, Iida S, Yasui T (2010) Flour properties of milky-white rice mutants in relation to specific loaf volume of rice bread. Food Sci Technol Res 16:305–312. https://doi.org/10.3136/fstr.16.305

Birla DS, Malik K, Sainger M, Chaudhary D, Jaiwal R, Jaiwal PK (2017) Progress and challenges in improving the nutritional quality of rice (Oryza sativa L.). Crit Rev Food Sci Nutr 57(11):2455–2481. https://doi.org/10.1080/10408398.2015.1084992

Calingacion M et al (2014) Diversity of global rice markets and the science required for consumer-targeted rice breeding. PLoS One 9:e85106. https://doi.org/10.1371/journal.pone.0085106

Cao ZZ, Lin XY, Yang YJ, Guan MY, Xu P, Chen MX (2019) Gene identification and transcriptome analysis of low cadmium accumulation rice mutant (lcd1) in response to cadmium stress using MutMap and RNA-seq. BMC Plant Biol 19:250. https://doi.org/10.1186/s12870-019-1867-y

Charcosset A, Moreau L (2004) Use of molecular markers for the development of new cultivars and the evaluation of genetic diversity. Euphytica 137:81–94. https://doi.org/10.1023/B:EUPH.0000040505.65040.75

Chawade A, Sikora P, Brautigam M, Larsson M, Vivekanand V, Nakash MA, Chen TS, Olsson O (2010) Development and characterization of an oat TILLING-population and identification of mutations in lignin and beta-glucan biosynthesis genes. BMC Plant Biol 10:86. https://doi.org/10.1186/1471-2229-10-86

Chen L, Hao LG, Parry MAJ, Phillips AL, Hu YG (2014) Progress in TILLING as a tool for functional genomics and improvement of crops. J Integr Plant Biol 56:425–443. https://doi.org/10.1111/jipb.12192

Conger BV, Konzak CF, Nilan RA (1977) Radiation sensitivity and modifying factors and methods of applying pre- and post-treatments, in manual on mutation breeding. IAEA Techn. Reports Series, No. 119, 2nd ed, pp 40–50

Ding Y, Li H, Chen LL, Xie K (2016) Recent advances in genome editing using CRISPR/Cas9. Front Plant Sci 7:1–12. https://doi.org/10.3389/fpls.2016.00703

Dreher K, Khairallah M, Ribaut J, Morris M (2003) Money matters (I): costs of field and laboratory procedures associated with conventional and marker-assisted maize breeding at CIMMYT. Mol Breed 11:221–234. https://doi.org/10.1023/A:1022820520673

Eagles HA, Bariana HS, Ogbonnaya FC, Rebetzke GJ, Hollamby GJ, Henry RJ, Henschke P, Carter M (2001) Implementation of markers in Australian wheat breeding. Aust J Agric Res 52:1349–1356. https://doi.org/10.1071/AR01067

FAO/IAEA- Food of Agriculture Organization of the United Nations and International Atomic Energy Agency 2019 Mutant variety database. Accessed July. Available from: http://mvd.iaea.org/

Fiaz S, Ahmad S, Noor MA, Wang X, Younas A, Riaz A, Riaz A, Ali F (2019) Applications of the CRISPR/Cas9 system for rice grain quality improvement: perspectives and opportunities. Int J Mol Sci 20:888. https://doi.org/10.3390/ijms20040888

Fitzgerald MA, McCouch SR, Hall RD (2009) Not just a grain of rice: the quest for quality. Trends Plant Sci 14:133–139. https://doi.org/10.1016/j.tplants.2008.12.004

Gilchrist EJ, Haughn GW (2005) TILLING without a plough: a new method with applications for reverse genetics. Curr Opin Plant Biol 8:211–215. https://doi.org/10.1016/j.pbi.2005.01.004

Henikoff S, Comai L (2003) Single-nucleotide mutations for plant functional genomics. Annu Rev Plant Biol 54:375–401. https://doi.org/10.1146/annurev.arplant.54.031902.135009

Hickey LT, Hafeez AN, Robinson H, Jackson SA, Leal-Bertioli SCM, Tester M, Gao C, Godwin ID, Hayes BJ, Wulff BBH (2019) Breeding crops to feed 10 billion. Nat Biotechnol 37:744–754. https://doi.org/10.1038/s41587-019-0152-9

Iida S, Amano E, Nishio T (1993) A rice (Oryza sativa L.) mutant having a low content of glutelin and a high content of prolamine. Theor Appl Genet 87:374–378. https://doi.org/10.1007/BF01184926

Ishikawa S, Ishimaru Y, Igura M, Kuramata M, Abe T, Senoura T, Hase Y, Arao T, Nishizawa NK, Nakanishi H (2012) Ion-beam irradiation, gene identification, and marker-assisted breeding in the development of low-cadmium rice. Proc Natl Acad Sci U S A 109:19166–19171. https://doi.org/10.1073/pnas.1211132109

Ishikawa S, Makino T, Ito M, Harada K, Nakada H, Nishida I, Nishimura M, Tokunaga T, Shirao K, Yoshizawa C, Matsuyama M, Abe T, Arao T (2016) Low-cadmium rice (Oryza sativa L.) cultivar can simultaneously reduce arsenic and cadmium concentrations in rice grains. Soil Sci Plant Nutr 62:327–339. https://doi.org/10.1080/00380768.2016.1144452

Jain SM (2010) Mutagenesis in crop improvement under the climate change. Rom Biotechnol Lett 15:88–106

Jankowicz-Cieslak J, Till BJ (2015) Forward and reverse genetics in crop breeding. In: Al-Khayri JM, Jain SM, Johnson DV(eds) Advances in plant breeding strategies: breeding, biotechnology and molecular tools. Springer International Publishing, Switzerland. p 215–240. https://doi.org/10.1007/978 3 319-22521-0

Jency JP, Ravikesavan R, Sumathi P, Raveendran M (2016) Determination of lethal dose and effect of physical mutagen on germination percentage and seedling parameters in kodomillet variety CO_3. Electron J Plant Breed 7:1122–1126. https://doi.org/10.5958/0975-928X.2016.00155.1

Jeng TL, Lin YW, Wang CS, Sung JM (2012) Comparisons and selection of rice mutants with high iron and zinc contents in their polished grains that were mutated from the indica type cultivar IR64. J Food Compos Anal 28:149–154. https://doi.org/10.1016/j.jfca.2012.08.008

Jiang Y, Cai Z, Xie W, Long T, Yu H, Zhang Q (2012) Rice functional genomics research: progress and implications for crop genetic improvement. Biotechnol Adv 30:1059–1070. https://doi.org/10.1016/j.biotechadv.2011.08.013

Kawamura K, Qi F, Kobayashi J (2018) Potential relationship between the biological effects of low-dose irradiation and mitochondrial ROS production. J Radiat Res 59:ii91–ii97. https://doi.org/10.1093/jrr/rrx091

Kayalvizhi K, Kannan M, Ganga M (2017) Effect of physical and chemical mutagens on morphological characters in M1V2 generation of tuberose Polianthes tuberosa L. Int J Curr Microbiol App Sci 6:2492–2499. https://doi.org/10.20546/ijcmas.2017.604.290

Ke C, Guan W, Bu S, Li X, Deng Y, Wei Z, Wu W, Zheng Y (2019) Determination of absorption dose in chemical mutagenesis in plants. PLoS One 14:e0210596. https://doi.org/10.1371/journal.pone.0210596

Kharkwal MC (2012) A brief history of plant mutagenesis. In: Shu QY, Forster BP, Nakagawa H (eds) Plant mutation breeding and biotechnology. CABI, Wallingford, pp 21–30

Kim SI, Tai TH (2014) Identification of novel rice low phytic acid mutations via TILLING by sequencing. Mol Breed 34:1717–1729. https://doi.org/10.1007/s11032-014-0127-y

Kodym A, Afza R (2003) Physical and chemical mutagenesis. In: Grotewold E (ed) Plant functional genomics. Humana Press, Totowa, pp 189–203

Kovacs E, Keresztesa A (2002) Effect of gamma and UV-B/C radiation on plant cell. Micron 33:199–210. https://doi.org/10.1016/S0968-4328(01)00012-9

Kumamaru T, Sato H, lwata N, Omura T, Ogawa M, Tanaka K (1988) Mutants for rice storage proteins 1 Screening of mutants for rice storage proteins of protein bodies in the starchy endosperm. Theor Appl Genet 76:11–16. https://doi.org/10.1007/BF00288825

Kumamaru T, Sato H, Satoh H (1997) High-lysine mutants of rice, Oryza sativa L. Plant Breed 116:245–249. https://doi.org/10.1111/j.1439-0523.1997.tb00990.x

Kurowska M, Daszkowska-Golec A, Gruszka D, Marzec M, Szurman M, Szarejko I, Maluszynski M (2011) TILLING: a shortcut in functional genomics. J Appl Genet 52:371–390. https://doi.org/10.1007/s13353-011-0061-1

Lau WCP, Rafii MY, Ismail MR, Puteh A, Latif MA, Asfaliza R, Miah G (2017) Development of advanced fragrant rice lines from MR269 × basmati 370 through marker-assisted backcrossing. Euphytica 213:11. https://doi.org/10.1007/s10681-016-1794-z

Lee DK, Kim YS, Kim JK (2017) Determination of the optimal condition for ethylmethane sulfonate-mediated mutagenesis in a Korean commercial rice, Japonica cv. Dongjin Appl Biol Chem 60:241–247. https://doi.org/10.1007/s13765-017-0273-0

Liu Q-L, Xu XH, Ren X-L, Fu HW, Wu D-X, Shu Q-Y (2007) Generation and characterization of low phytic acid germplasm in rice (Oryza sativa L.). Theor Appl Genet 114:803–814. https://doi.org/10.1007/s00122-006-0478-9

Mba C, Afza R, Bado S, Jain SM (2010) Induced mutagenesis in plants using physical and chemical agents. In: Davey MR, Anthony P (eds) Plant cell culture: essential methods. Wiley, Chichester, pp 111–130

Mccallum CM, Comai L, Greene EA, Henikoff S (2000a) Targeted screening for induced mutations. Nat Biotechnol 18:455–457. https://doi.org/10.1038/74542

Mccallum CM, Comai L, Greene EA, Henikoff S (2000b) Targeting induced local lesions in genomes (TILLING) for plant functional genomics. Plant Physiol 123:439–442. https://doi.org/10.1104/pp.123.2.439

Nadeem MA, Nawaz MA, Shahid MQ, Doğan Y, Comertpay G, Yıldız M, Hatipoğlu R, Ahmad F, Alsaleh A, Labhane N, Özkan H, Chung G, Baloch FS (2018) DNA molecular markers in plant breeding: current status and recent advancements in genomic selection and genome editing. Biotechnol Biotechnol Equip 32:261–285. https://doi.org/10.1080/13102818.2017.1400401

Nakata M, Miyashita T, Kimura R, Nakata Y, Takagi H, Kuroda M, Yamaguchi T, Umemoto T, Yamakawa H (2018) MutMapPlus identified novel mutant alleles of a rice starch branching enzyme IIb gene for fine-tuning of cooked rice texture. Plant Biotechnol J 16:111–123. https://doi.org/10.1111/pbi.12753

Ni D, Zhang S, Chen S, Xu Y, Li L, Li H, Wang Z, Cai X, Li Z, Yang J (2011) Improving cooking and eating quality of Xieyou57, an elite indica hybrid rice, by marker-assisted selection of the Wx locus. Euphytica 179:355–362. https://doi.org/10.1007/s10681-011-0377-2

Oladosu Y, Rafii MY, Abdullah N, Hussin G, Ramli A, Rahim HA, Miah G, Usman M (2016) Principle and application of plant mutagenesis in crop improvement: a review. Biotechnol Biotechnol Equip 30:1–16. https://doi.org/10.1080/13102818.2015.1087333

Oldach KH (2011) Mutagenesis. In: Pratap A, Kumar J (eds) Biology and breeding of food legumes. CABI, Wallingford, pp 208–219

Oleykowski CA, Mullins CRB, Godwin AK, Yeung AT (1998) Mutation detection using a novel plant endonuclease. Nucleic Acids Res 26:4597–4602. https://doi.org/10.1093/nar/26.20.4597

Parry MAJ, Madgwick PJ, Bayon C, Tearall K, Hernandez-Lopez A, Baudo M, Rakszegi M, Hamada W, Al-Yassin A, Ouabbou H, Labhilili M, Phillips AL (2009) Mutation discovery for crop improvement. J Exp Bot 60:2817–2825. https://doi.org/10.1093/jxb/erp189

Ray DK, Mueller ND, West PC, Foley JA (2013) Yield trends are insufficient to double global crop production by 2050. PLoS One 8(6):e66428. https://doi.org/10.1371/journal.pone.0066428
Rigola D, Van Oeveren J, Janssen A, Bonne A, Schneiders H, Van Der Poel HJ, Van Orsouw NJ, Hogers RC, de Both MT, Van Eijk MJ (2009) High-throughput detection of induced mutations and natural variation using key point technology. PLoS One 4(3):e4761. https://doi.org/10.1371/journal.pone.0004761
Roychowdhury R, Tah J (2013) Mutagenesis a potential approach for crop improvement. In: Hakeem KR, Ahmad P, Ozturk M (eds) Crop improvement: new approaches and modern techniques. Springer, New York, pp 149–187
Sang T (2009) Genes and mutations underlying domestication transitions in grasses. Plant Physiol 149:63–70. https://doi.org/10.1104/pp.108.128827
Sarsu F, Spencer-Lopes MM, Sangwan R, Penna S (2018) Specific techniques for increasing efficiency of mutation breeding (a) in vitro methods in plant mutation breeding. In: Spencer-Lopes MM, Forster BP, Jankuloski L (eds) FAO/IAEA. Manual on mutation breeding, 3rd edn. Food and Agriculture Organization of the United Nations, Rome, p 301
Schneeberger K (2014) Using next-generation sequencing to isolate mutant genes from forward genetic screens. Nature Rev Genet 15:662–676. https://doi.org/10.1038/nrg3745
Schneeberger K, Weigel D (2011) Fast-forward genetics enabled by new sequencing technologies. Trends Plant Sci 16:282–288. https://doi.org/10.1016/j.tplants.2011.02.006
Schneider A, Leister D (2006) Forward genetic screening of insertional mutants. In: Arabidopsis protocols. Humana Press, Switzerland. 2006. p. 147–161. https://doi.org/10.1385/1-59745-003-0:147
Serrat X, Esteban R, Guibourt N, Moysset L, Nogués S, Lalanne E (2014) EMS mutagenesis in mature seed-derived rice calli asa new method for rapidly obtaining TILLING mutant populations. Plant Methods 10:5. https://doi.org/10.1186/1746-4811-10-5
Slade AJ, McGuire C, Loeffler D, Mullenberg J, Skinner W, Fazio G, Holm A, Brandt KM, Steine MN, Goodstal JF, Knauf VC (2012) Development of high amylose wheat through TILLING. BMC Plant Biol 12:69. https://doi.org/10.1186/1471-2229-12-69
Stadler LJ (1928) Genetic effects of x-rays in maize. Proc Natl Acad Sci U S A 14:69–75. https://doi.org/10.1073/pnas.14.1.69
Sun YW, Jiao GA, Liu ZP, Zhang X, Li JY, Guo XP, Du WM, Du JL, Francis F, Zhao YD, Xia LQ (2017) Generation of high-amylose rice through CRISPR/Cas9-mediated targeted mutagenesis of starch branching enzymes. Front Plant Sci 8:298. https://doi.org/10.3389/fpls.2017.00298
Suzuki Y, Sano Y, Ise K, Matsukura U, Aoki N, Sato H (2008) A rice mutant with enhanced amylose content in endosperm without affecting amylopectin structure. Breed Sci 58:209–215. https://doi.org/10.1270/jsbbs.58.209
Tadele Z (2016) Mutagenesis and TILLING to dissect gene function in plants. Curr Genomics 17:499–508. https://doi.org/10.2174/1389202917666160520104158
Taheri S, Abdullah TL, Jain SM, Sahebi M, Azizi P (2017) TILLING, high-resolution melting (HRM), and next generation sequencing (NGS) techniques in plant mutation breeding. Mol Breed 37:40. https://doi.org/10.1007/s11032-017-0643-7
Talebi AB, Talebi AB, Shahrokhifar B (2012) Ethyl methane sulphonate (EMS) induced mutagenesis in Malaysian rice (cv.MR219) for lethal dose determination. Am J Plant Sci 3:1661–1665. https://doi.org/10.4236/ajps.2012.312202
Tang L, Mao B, Li Y, Lv Q, Zhang L et al (2017) Knockout of OsNramp5 using the CRISPR/Cas9 system produces low cd-accumulating indica rice without compromising yield. Sci Rep 7:14438. https://doi.org/10.1038/s41598-017-14832-9
Till BJ, Zerr T, Comai L, Henikoff S (2006) A protocol for TILLING and Ecotilling in plants and animals. Nat Protoc 1:2465–2477. https://doi.org/10.1038/nprot.2006.329
Tsai H, Howell T, Nitcher R, Missirian V, Watson B, Ngo KJ, Lieberman M, Fass J, Uauy C, Tran RK, Khan AA, Filkov V, Tai TH, Dubcovsky J, Comai L (2011) Discovery of rare mutations in populations: TILLING by sequencing. Plant Physiol 156:1257–1268. https://doi.org/10.1104/pp.110.169748
Van de Wouw M, Kik C, Van Hintum T, Van Treuren R, Visser B (2009) Genetic erosion in crops: concept, research results and challenges. Plant Gen Resour: Charact Utilization 8:1–15. https://doi.org/10.1017/S1479262109990062

Wang C, Zhang Y, Zhu Z, Chen T, Zhao L, Lin J, Zhou L (2009) Development of a new japonica rice variety Nan-jing 46 with good eating quality by marker assisted selection. Mol Plant Breed 7:1070–1076. https://doi.org/10.5376/rgg.2010.01.0003

Wu Y-P, Pu C-H, Lin H-Y, Huang H-Y, Huang Y-C, Hong C-Y, Chang M-C, Lin Y-R (2015) Three novel alleles of FLOURY ENDOSPERM2 (FLO2) confer dull grains with low amylose content in rice. Plant Sci 233:44–52. https://doi.org/10.1016/j.plantsci.2014.12.011

Xin ZG, Wang ML, Barkley NA, Burow G, Franks C, Pederson G, Burke J (2008) Applying genotyping (TILLING) and phenotyping analyses to elucidate gene function in a chemically induced sorghum mutant population. BMC Plant Biol 8:103. https://doi.org/10.1186/1471-2229-8-103

Yamaguchi H, Nagatomi S, Morishita T, Degi K, Tanaka A, Shikazono N, Hase Y (2003) Mutation induced with ion-beam irradiation in rose. Nucl Instr Meth Phys Res 206:561–564. https://doi.org/10.1016/S0168-583X(03)00825-5

Yang CZ, Shu XL, Zhang LL, Wang XY, Zhao HJ, Ma CX, Wu DX (2006) Starch properties of mutant rice high in resistant starch. J Agric Food Chem 54:523–528. https://doi.org/10.1021/jf0524123

Yi M, Nwe KT, Vanavichit A, Chai-arree W, Toojinda T (2009) Marker assisted backcross breeding to improve cooking quality traits in Myanmar rice cultivar Manawthukha. F Crop Res 113:178–186. https://doi.org/10.1016/j.fcr.2009.05.006

Chapter 16
Gene Delivery Systems for Conventional Genetic Engineering and Genome Editing to Improve Grain Quality

Vera Quecini, Vívian Ebeling Viana, Camila Pegoraro, and Antonio Costa de Oliveira

16.1 Introduction

Grain quality is an increasing demand to rice (*Oryza sativa* L.) breeding programs to guarantee the acceptance of novel cultivars by consumers and growers alike. The trait is complex and defined by several features, including grain appearance, milling properties, eating and cooking qualities, and sensory and nutritional composition (reviewed by Mukamuhirwa et al. 2018; Butardo Jr. et al. 2019), which are genetically and environmentally controlled. The combination of high yield and grain quality is difficult to achieve; however, the relationship between productivity and quality can be balanced considering that starch type and contents determine rice quality (Xu et al. 2015).

Starch consists of the polymers amylose and amylopectin and corresponds to 90% of the rice grain. Amylose contents influence cooking, eating, and processing features of the rice and are directly associated with water absorption, volume expansion, fluffiness, and separability of the grains during cooking, which are inversely correlated to cohesiveness, tenderness, and glossiness (reviewed by Asanti 2016). Grain aspect (size and length) and milling quality (brown, milled and head rice recovery) are also important traits in rice quality breeding programs. Thus, the genetic framework of starch and quality traits have been intensively studied and are summarized in Bao (2014) and Asanti (2016).

Nutritional quality is critical to rice breeding programs, since the grain is a staple food for low-income populations in developing countries. The grain is naturally limited in nutrients, and many studies have identified genes and genomic regions

V. Quecini (✉)
Embrapa Uva e Vinho, Bento Gonçalves, RS, Brazil
e-mail: vera.quecini@embrapa.br

V. Ebeling Viana · C. Pegoraro · A. Costa de Oliveira
Universidade Federal de Pelotas, Campus Capão do Leão, Pelotas, RS, Brazil

A. Costa de Oliveira et al. (eds.), *The Future of Rice Demand: Quality Beyond Productivity*, https://doi.org/10.1007/978-3-030-37510-2_16

associated with the synthesis and accumulation of minerals, vitamins, and essential amino acids (Mahender et al. 2016).

The increasing availability of genomic and post-genomic tools has allowed the use of precision breeding technologies, such as genetic engineering and genome editing, to improve the existing and introduce novel traits in plants (Hahne et al. 2019). These methods are particularly effective for complex traits, such as quality of the grain in rice (Birlaa et al. 2017). In the present chapter, we present and discuss the available systems for gene delivery, used in conventional transformation and genome editing, of rice and the regeneration pathways employed in the production of whole plants expressing the novel phenotypes associated to grain quality (Fig. 16.1). We also discuss the interaction between the gene delivery system and its biological target in recovering transformed and/or edited plants.

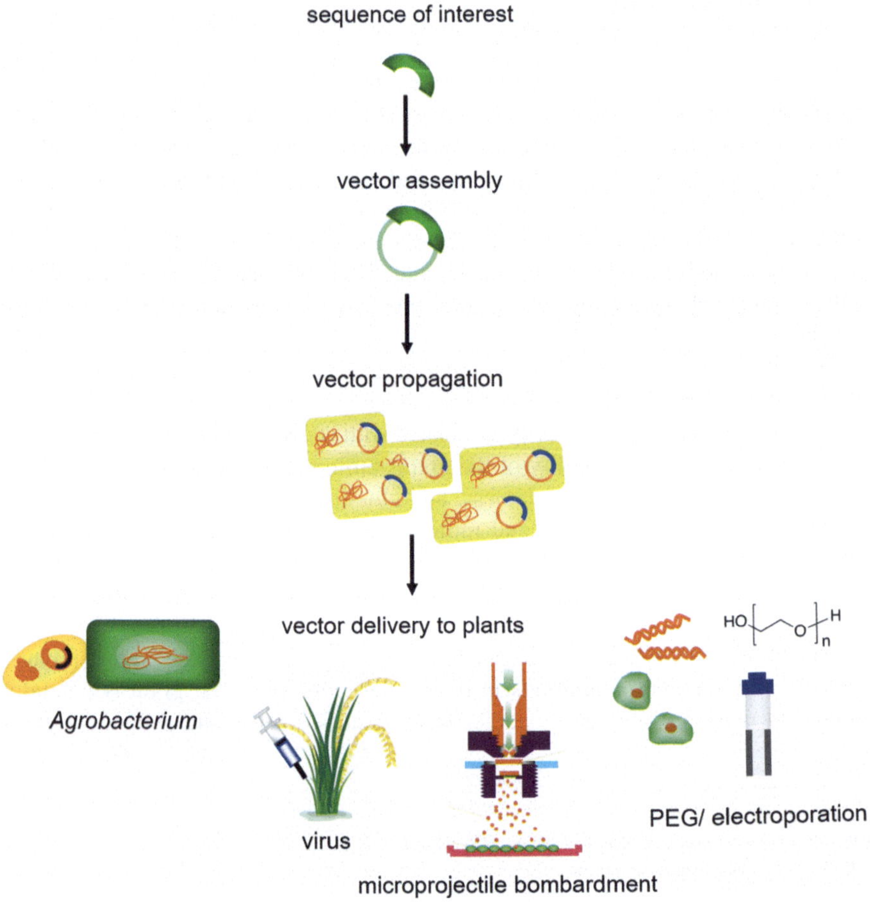

Fig. 16.1 Schematic representation of the principal steps in the most commonly used gene delivery systems to rice

16.2 Vectors for Gene Delivery

The genes of interest in plant genetic engineering and genome editing, along with other sequences of interest, are transferred to the host via delivery vectors based on natural systems, such as plasmids, viruses, and artificial chromosomes. The sequences of interest in the vectors used to deliver genes to plant genomes consist of regulatory regions and other sequences, employed in nucleic acid transfer and insertion processes, or at later stages, in transformant identification (Weising et al. 1988).

Regulatory sequences include natural or artificial promoters, subcellular targeting sequences, and terminator regions and are essential to control temporally and spatially the transcription of the introduced sequences in conventional genetic engineering (Liu et al. 2013; Pandiarajan and Grover 2018). In contrast, in precision breeding technologies based on sequence-specific nucleases (SSN), regulatory elements in the delivery vector drive the expression of the SSN and other sequences required for gene targeting (Huang and Puchta 2019). The introduction of large DNA segments into plant genomes via artificial chromosomes also requires specific structural features of the delivery vector, such as an origin of replication, a centromere, and telomeres (Liu et al. 2013). The mobilization of the vector to the plant host cell(s)/tissue(s) can use biological intermediates, such as *Agrobacterium*, or employ physical or chemical forces to introduce the genetic material directly into the host (Weising et al. 1988). Vectors used in direct gene transfer do not need special sequence features, other than those used in its assembly, whereas *Agrobacterium*-mediated DNA delivery requires several virulence (*vir*) loci in the bacterial tumor-inducing (Ti) plasmid and the transferred DNA (T-DNA) borders (Gelvin 2003). In SSN-based technologies, gene targeting requires positioning of the endonuclease-binding site in the delivery vector to allow excision of the target site in the host genome but concomitantly prevent recognition of the region by DNA repair systems while in the vector and later in the edited genome (Huang and Puchta 2019). The methods of gene delivery to plants require the identification and selection of the cells where the modification is expressed, transiently or stably. Vector-encoded selectable markers and reporter genes allow the identification and isolation of the cells expressing the introduced character and to monitor it in future generations (Miki and McHugh 2004).

The sequences in gene delivery vectors can be physically assembled by restriction/ligation cloning or homology-dependent methods, where adjacent DNA fragments sharing identical sequences are spliced by homologous recombination in vivo or by annealing and extension of the homologous ends in vitro (Liu et al. 2013; Patron 2014; Kim et al. 2015). Traditional subcloning requires the presence of restriction sites flanking the insert and, in the vector, multiple cloning site (MCS) and absence from elsewhere in the sequence of interest and vector (Fig. 16.2a).

Distinct convenient restriction sites on the 3′ and 5′ of a sequence of interest allow directional subcloning and prevents the vector from self-ligating. When suitable restriction sites are naturally not present, they can be created or introduced at

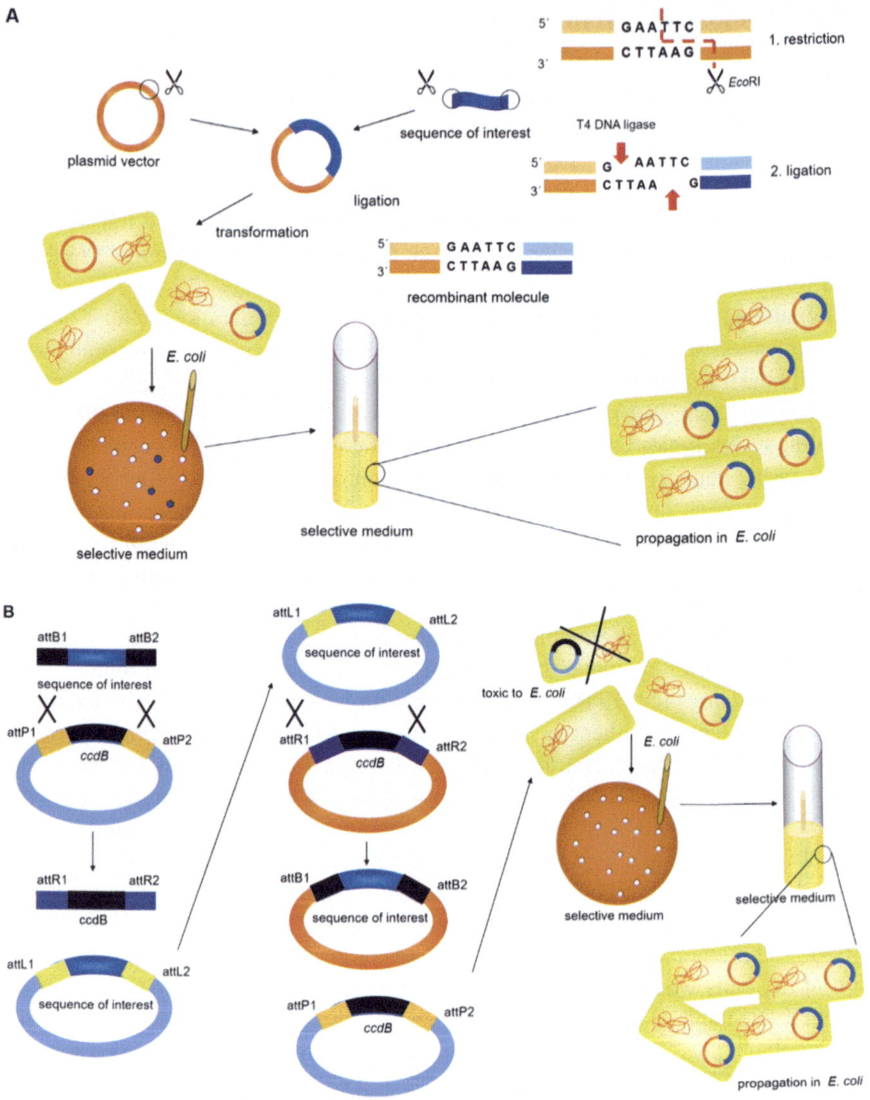

Fig. 16.2 Restriction/ligation (**a**) and recombination-based (**b**) cloning schemes in vector assembly for gene delivery

the extremity of the sequence of interest by polymerase chain reaction (PCR)-based cloning and adding sequences recognized by the desired restriction enzymes to PCR oligonucleotides. The optimal reaction condition of the restriction enzymes can also be considered to allow combined digestion steps. Recombination cloning is based on the reversible integration of λ phage in *E. coli*, where specific recombination sites and enzymatic activities allow the phage DNA to be integrated and removed from the host genome. Recombination cloning requires the sequence of interest to

Table 16.1 Summary of high-throughput cloning strategies used in the construction of gene delivery vectors to plant cells

Cloning system	Principles	Reference
Gateway cloning	Nondisclosed (proprietary) – Thermo Fisher Scientific	Hartley et al. (2000)
Recombineering	*E. coli* strains expressing *RecET* or *Redαβγ*	Ellis et al. (2001)
GoldenGate	Type IIS endonuclease, DNA ligase – New England Biolabs	Engler et al. (2008)
gBlocks Gene Fragments	Nondisclosed (proprietary) – IDT DNA	–
GeneArt Seamless Cloning	Nondisclosed (proprietary) – Thermo Fisher Scientific	–
NEBuilder HiFi DNA assembly	Nondisclosed (proprietary) – New England Biolabs	–
Gibson assembly	T5 DNA exonuclease, Phusion	Gibson et al. (2009)
	DNA polymerase, Taq DNA	
	Ligase	
In-Fusion	Vaccinia virus DNA polymerase – Clontech	–

be flanked by a specific recombination site (*attB*) that interacts with the recombination site (*attP*) in a donor vector, producing a transient co-integrate plasmid, which is subsequently resolved into one molecule containing sequence of interest in the new vector backbone flanked by recombination site *attL* (Fig. 16.2b). The other resultant molecule carries the *ccdB* gene that encodes a toxin responsible for the post-segregational killing (PSK) of the cells. Thus, only the *E. coli* carrying the molecule with the sequence of interest will be viable.

High-throughput cloning by restriction/ligation relies on methods of standardized restriction enzyme assembly that uses type II enzymes, such as *BsaI*, *BsmBI*, and *BbsI*, to produce interchangeable DNA fragments that are ligated simultaneously and directionally by T4 DNA ligase (Liu et al. 2013). In contrast, overlap assembly techniques in vitro use specific recombinases or the method known as "chew back and anneal," where one DNA strand is digested back to produce overhangs at the ends of fragments that are able to anneal without ligation (Liu et al. 2013; Kim et al. 2015). High-throughput seamless cloning methods are summarized in Table 16.1.

The most common features of the vectors used to deliver genes to plant genomes are presented in Table 16.2.

16.2.1 Tumor Inducing (Ti) Plasmid-Based Vectors

A large number of gene delivery vectors was designed based on the tumor-inducing (Ti) plasmid of *Agrobacterium tumefaciens* (Zambryski et al. 1989). Under natural conditions, *Agrobacterium* colonizes the plant and promotes the insertion of a delimited region of the Ti plasmid, the transferred DNA (T-DNA), to the host

Table 16.2 Principal features of gene delivery vectors used in rice conventional genetic engineering and genome editing techniques

Vector type	Origin	Main characteristics	Reference
Cointegrated	*Agrobacterium* Ti plasmid	Linked *vir* genes, T-DNA borders, ColE1, bacteria selectable marker, ColE1 *bom*, ColE1 *mob* (*in trans*)	Hooykaa et al. (1977), Deblaere et al. 1985
Binary	*Agrobacterium* Ti plasmid	Unlinked *vir* genes, T-DNA borders, wide host origin of replication, bacteria selectable marker,	Hoekema et al. (1983)
Viral	Wheat streak mosaic virus (WSMV), barley stripe mosaic virus (BSMV), tobacco rattle virus (TRV), Begomoviruses	Engineered virus, integrated into binary vectors	
Artificial chromosome	Bacteria (BIBAC, TAC)	Transfer of large sequences, requires helper plasmid	He (2016)
Genome editing (SSN)	Based on binary and/or viral vectors	Transient expression and/or stable insertion	Mahas et al. (2019), Wu et al. (2018)

BIBAC binary bacterial artificial chromosome, *ColE1 E. coli* origin of replication, *ColE1bom* ColE1 basis of mobilization, *ColE1mob* ColE1 mobility, *SSN* sequence-specific nucleases, *TAC* transformation-competent binary bacterial artificial chromosome

genome (Zambryski et al. 1989; Gelvin 2003). The delivery and insertion of exogenous sequences via *Agrobacterium* require three genetic bacterial components: (i) the T-DNA, flanked at both ends by imperfect direct 25 bp repeats, called the T-DNA borders; (ii) the virulence (*vir*) region in the Ti plasmid, consisting of seven principal loci *virA*, *virB*, *virC*, *virD*, *virE*, *virG*, and *virH*; and (iii) the chromosome virulence genes *chv* (*chromosomal virulence*) (Zambryski et al. 1989; Gelvin 2003).

Initially, the vectors designed for plant genetic modifications consisted of large Ti-based plasmids carrying the T-DNA borders and the *vir* region in the same molecule (Walkerpeach and Velten 1994). However, operational problems due to the large size of cointegrated vectors, such as recalcitrance of bacterial transformation and difficulties in cloning steps, coupled with the absence of any specific advantage in maintaining the T-DNA and the *vir* genes in a single regulon prompted the development of binary vectors (Hoekema et al. 1983; Walkerpeach and Velten 1994).

The binary system consists of two independent plasmids autonomously replicating in *Agrobacterium*: a small binary vector carrying the sequences of interest flanked by T-DNA borders and a larger Ti plasmid with *vir* genes Hoekema et al. 1983; Walkerpeach and Velten 1994).

16.2.2 Cointegrated Vectors

Cointegrated vectors are no longer in use due to the technical difficulties mentioned previously. Thus, they will be described briefly to provide a historical perspective of the development of the commonly used binary systems.

Large portions of the wild-type Ti plasmid are retained in cointegrated vectors, with exception of the genes responsible for the neoplastic growth in the native *Agrobacterium* pathosystem (Walkerpeach and Velten 1994). Ti plasmids lacking the genes responsible associated to the pathogenic phenotypes of *Agrobacterium* infection are called "disarmed," and by analogy, the strains carrying disarmed Ti plasmids are called "disarmed" agrobacteria (Walkerpeach and Velten 1994).

The removed oncogenes were replaced by a DNA fragment common to several *Escherichia coli* cloning vectors and homologous to T-DNA sequences allowing recombination, mediated by native *Agrobacterium rec* function, to give rise to a hybrid, or cointegrated, Ti plasmid (Walkerpeach and Velten 1994). Standard cointegrate vectors carry (i) a multicloning site for the insertion of the sequence of interest, (ii) antibiotic selectable markers for *E. coli* and *A. tumefaciens* cloning, (iii) a eukaryotic selectable marker gene, and (iv) an *E. coli* origin of replication that is not functional in *Agrobacterium* (Walkerpeach and Velten 1994).

Operationally, the cointegrated vector carrying the sequence of interest is transferred to *Agrobacterium*, and transformed colonies are identified by antibiotic resistance, provided by the selection marker. The origin of replication *ColEI*, used in most cointegrate vectors, does not function in *Agrobacterium*; thus, the maintenance of the linked sequence of interest and antibiotic resistance gene is dependent on the formation of the hybrid (cointegrated) molecule, which, in turn, allows the Ti plasmid origin to replicate both the Ti plasmid and the cointegrate vector plus the linked sequence of interest (Walkerpeach and Velten 1994).

16.2.3 Binary Vectors

The binary system has several advantages over the cointegrated vectors, so it remains the basis for the development of delivery vectors in current genetic transformation and genome editing methods (Zhang et al. 2018). Binary vectors have a functional origin of replication for *Agrobacterium*, so they do not require the recombination event to maintain the vector stable; therefore, transformation is more efficient and faster. As independent replicons, binary vectors are present in high copy number in agrobacteria; their small sizes facilitate in vivo and vitro DNA manipulation and promote higher transformation efficiencies (Walkerpeach and Velten 1994; Lee and Gelvin 2008).

In general, binary vectors carry (i) a multiple cloning site (MCS); (ii) an origin of replication, such as RK2, that is functional in *E. coli* and *A. tumefaciens*; (iii) selectable marker genes for bacteria and plants; (iv) genes required for bacterial conjugation, as *oriV*, *oriT*, and *trfA*; and (v) T-DNA border sequences, although only the right border is absolutely essential (Hoekema et al. 1983; Lee and Gelvin 2008). To increase the fidelity of T-strand synthesis, sequences as overdrive and TSS (T-DNA transfer stimulator sequence) were also incorporated (Walkerpeach and Velten 1994). The *vir* functions, essential to sequence mobilization and delivery, are provided *in trans* by the resident Ti plasmid during the process of plant transformation (Hoekema et al. 1983; Lee and Gelvin 2008). Although binary vectors have features required for *Agrobacterium*-mediated gene transfer, the operational facilities and widespread availability of the system prompt their use in direct gene transfer methods in rice (He et al. 2016; Wu et al. 2009; Lee and Gelvin 2008; Hajdukiewicz et al. 1994).

The binary system has been modified to accomplish specialized functions of genetic modifications in rice, including the identification of transcriptionally active regions in the genome (promoters and enhancers) (Yu et al. 2007; Wu et al. 2015; Bureau et al. 2018), protein tagging and identification of subcellular localization (Ma et al. 2019; He et al. 2016; Gan et al. 2016), large-scale mutant generation (Lo et al. 2016; Wu et al. 2017), and, more recently, genome editing for gene functional characterization (Chiou et al. 2019; Hu et al. 2019; Pérez et al. 2019; Mishra et al. 2018).

16.2.4 Viral Vectors

The intrinsic proximity of plant viruses and their host cells made them attractive targets for gene delivery systems, with special interest in DNA viruses (Weising et al. 1988; Zaid and Mansoor 2017). Viral vectors profit from the autonomous replication, effective genome machinery, and comprehensive structure to deliver nucleic acids to plant host cells (Zaid and Mansoor 2017; Pasin et al. 2019). Moreover, high-throughput sequencing technologies promoted a steady increase in the genomic resources for viral and subviral agents (Pasin et al. 2019).

The most frequently used viral vectors include RNA viruses, such as wheat streak mosaic virus (WSMV), barley stripe mosaic virus (BSMV) and tobacco rattle virus (TRV), and single-stranded DNA viruses, as geminiviruses (family Geminiviridae) and Begomoviruses (genus Begomovirus in the family Geminiviridae) (Zaid and Mansoor 2017). The majority of viral vectors require mechanical inoculation and do not stably integrate in the host genome, making the expression of the introduced trait free of positional effects and not transmissible to the progeny (Zaid and Mansoor 2017).

The mechanical inoculation imposed restrictions to the development of virus-derived vectors, since it limited the possible vectors to those mechanically transmissible and was able to reach a relatively small number of cells at the inoculation site (Peyret and Lomonossoff 2015). Thus, the viral vectors had to allow the introduced sequences to retain their systemic movement ability (Peyret and Lomonossoff 2015). The insertion of viral sequences between the T-DNA borders of binary vectors allowed the viral infection of a mature plant via *Agrobacterium*-mediated gene transfer, a technique later called "agroinfection" and "agroinoculation" (Pasin et al. 2019; Peyret and Lomonossoff 2015). Engineering of viral sequences into binary plasmids expanded the repertoire delivery vectors since it allowed the inclusion of sequences from nonmechanically transmissible viruses and expanded *Agrobacterium* host range to include monocotyledonous species (Pasin et al. 2019; Peyret and Lomonossoff 2015). The use of *Agrobacterium* to initiate the infection by virus-based vectors allowed the removal of unnecessary viral sequences, such as those coding for the viral coat (CP) or movement (MP) proteins and replacing them by sequences of interest. This technical advance has widened the range of viruses used as expression vectors to include tobamoviruses, potexviruses, tobraviruses, geminiviruses, and comoviruses (Peyret and Lomonossoff 2015). Recently, infectious clones and other devices of viral or subviral origins are increasingly used to induce the expression of desirable traits in plants (Pasin et al. 2019).

Viral vectors were reported to induce stable genome alterations in plant cells by the expression of SSN and the use of virus-delivered sequences as template for targeted DNA replacement, indicating their potential in genome editing techniques (Pasin et al. 2019).

16.2.5 Artificial Chromosome Vectors

The stable introduction of large DNA fragments in plant genomes is desired for map-based (positional) gene cloning, since it allows the isolation of linked dominant alleles, and for biotechnological applications that require the simultaneous introduction of several genes, for metabolic engineering of novel pathways, for example (Shibata and Liu 2000). Thus, hybrid vectors combining characteristics of bacterial artificial chromosomes (BACs) and binary vectors for *Agrobacterium*-mediated transformation were developed (Shibata and Liu 2000; He et al. 2016). The resultant vectors have large DNA fragments (≤150 kb) cloned inside the T-DNA borders of binary Ti-based plasmids, allowing them to successfully deliver the exogenous sequences to plant species.

The hybrid chromosome vectors are called binary bacterial artificial chromosomes (BIBACs) or transformation-competent bacterial artificial chromosome (TACs) and have similar components, including (i) positive-selection markers for *E. coli* (*sacB*), (ii) elements for the stable maintenance of foreign DNA in *E. coli* (the *F* factor or *P1* phage replication origin) or *Agrobacterium* (the *Ri* replication origin of *Agrobacterium rhizogenes*), (iii) T-DNA borders, and (iv) a selectable marker active in plants (Shibata and Liu 2000).

Besides the basic shared framework, BIBACs and TACs have distinct features, such as the RK2 oriT sequence and the T7 and SP6 RNA polymerase promoters, the coding sequence of a bifunctional fusion peptide encoding a reporter/selective marker in BIBACs, and P1 lytic replicon to increase DNA yields and the convenient restriction sites for cloning and restriction analyses in TACs (Shibata and Liu 2000; Yu et al. 2016).

16.2.6 Genome Editing Vectors

Plant genome editing has benefited largely from several novel methods based on SSNs, and the ultimate method of choice for a wide range of organisms, including plants, is based on the *Streptococcus pyogenes* CRISPR (clustered regularly interspaced short palindromic repeats)/Cas9 (CRISPR-associated protein 9) system (Puchta 2017; Hsu et al. 2014). The endonuclease Cas9 is driven by a single-guide RNA (sgRNA) to induce a DNA double-strand break (DSB) at a specific genomic site that is complementary to a 20-base sequence in the sgRNA next to a short *pro*tospacer *a*djacent *m*otif (PAM sequence) (Jinek et al. 2012). The failure to repair the induced DSB correctly leads to short insertion/deletions (indels) or base substitutions by nonhomologous end joining (NHEJ). The CRISPR/Cas9 system is also able to promote the deletion of DNA fragments flanked by target sites for two different sgRNAs, leading to the removal of specific sequences in the genome (Puchta 2017; Hsu et al. 2014).

As discussed for other systems, binary vectors based on agrobacteria Ti/Ri plasmids constitute the basis for engineering novel delivery tools used in genome editing in plants. Typically, CRISPR/Cas 9 constructs in plants are transferred as transgenes, via *Agrobacterium*, in binary vectors where the coding sequences of the sgRNA and of the endonuclease Cas9 are driven by plant promoters and cloned between T-DNA borders in binary vectors (Puchta 2017). Alternatively, a modular system, where the sgRNA is delivered by a shuffle-in vector and the Cas9 coding sequence is expressed by a destination binary vector, has been proposed (Wu et al. 2018). The dual module vector system facilitates the generation of transgenic constructs and, later, the identification of transgene-free plants (Wu et al. 2018). The modular system has convenient restriction sites in front of the Cas9 endonuclease open reading frame (ORF) in the binary destination vector to allow the insertion of promoter sequences of interest and the coding sequence of a fluorescent protein driven by a seed coat-specific promoter to allow positive selection of the modified plants in the first generation (T1) and negative in the second (T2) (Wu et al. 2018).

The new CRISPR/Cas9 vectors, as the modular system proposed by Wu et al. (2018), designed for genome editing in plants take advantage of the high-throughput assembly methods summarized in Table 16.1.

16.3 Gene Delivery Targets

The choice of biological targets is crucial to the effectiveness of gene delivery and, later expression of the desired trait, in plant systems. Successful methods should be reproducible, cultivar-/variety-independent, and not be labor-intensive and time-consuming. The compatibility between the biological target, the method of delivery, and the sequence carrying vector are also important.

The delivery of sequences of interest to plants is a cellular event, in general occurring in a single or a limited number of cells, which have to be induced to give rise to complete, viable, modified plants (Motte et al. 2014; Ikeuchi et al. 2016). The regeneration of plants from a single transformed cell is advantageous, since it is more likely to generate uniformly modified plants (Verdeil et al. 2007; Motte et al. 2014).

In vitro tissue culture techniques induce the production of toti- or pluripotent targets for gene delivery methods in plants (Neelakandan and Wang 2012; Motte et al. 2014). However, in vitro manipulations of plant cells and tissues require sterile conditions, are time- and work-consuming, and may cause somatic mutation or somaclonal variation. Alternatively, genes can be delivered to intact plant organs via in planta techniques (Weigel and Glazebrook 2006). In planta transformation does not require in vitro manipulations and is faster than tissue culture-dependent methods, although to obtain heritable mutations, the tissue/organ of choice should be embryogenic or contain germline cells.

In rice, subspecies and genotype are determinant to in vitro performance due to distinct wiring of hormone signaling pathways (Indoliya et al. 2016). These differences have to be taken into account in choosing the biological target to gene delivery. To circumvent tissue culture performance barriers, in planta gene delivery systems can be used (Table 16.3). However, there are other specific requirements for rice target tissues to in planta transformation, which include stronger resistance to pathogens and stress and higher differentiation/regeneration ability in comparison to the in vitro counterparts (Supartana et al. 2005).

16.3.1 In Vitro Targets

The generation of uniformly modified plants requires stable genomic modifications in a single cell, able to generate a complete viable, new plant (Weising et al. 1988). In vitro tissue culture techniques, involving the manipulation of environmental signals and physical stimuli, allow the proliferation and control of the differentiation of plant cells, capable of producing complete, new individuals (Sugimoto et al. 2019; Verdeil et al. 2007). The principal pathways of regeneration consist of a de novo organ regeneration system and of somatic embryogenesis induction (Sugimoto et al. 2019). The first consists of a two-stage process, where an intermediate growing structure (callus) is typically induced by hormonal manipulation, followed by the formation of novel meristems that sprout into shoots (Sugimoto et al. 2019; Verdeil et al. 2007). The plant growth regions (meristems) retain portions of undifferentiated stem cells that can be directly induced into shoots (Sugimoto et al. 2019; Verdeil et al. 2007). Alternatively, new plants can be produced from embryos induced from certain somatic cells, a process termed somatic embryogenesis (Sugimoto et al. 2019; Verdeil et al. 2007). Thus, organogenic meristems, from adult plants or zygotic embryos, and embryogenic calli are adequate biological

Table 16.3 Summary of the biological targets used in gene delivery systems to rice

Cultivar	Biological target	Gene delivery system	Reference
In vitro			
Taipei 309, IR 43	Protoplasts	PEG	Datta et al. (1990), Datta et al. (1992), Goshi-Biswas et al. (1994)
Taipei 309, Basmati-370, IR43	Protoplasts	Electroporation	Toriyama et al. (1988), Zhang et al. (1988), Shimamoto et al. (1989)
	Intact tissue (transient)	Electroporation	Dekeyser et al. (1990)
Asanohikari, Shiokari Akitakomachi, Chiyonishiki, Hinohikari, Asominori, Koshihikari, M202, Hitomebore, Nipponbare, Sasanishiki, Tsukinohikari, Yukihikari, Suweon 258, Nanjing 11, Xingingai 1, IR8, IR72, IR24, IR26, IR36, IR54, IR64, Kasalat, CSR10, PB1, Swarna	SE calli	*Agrobacterium*	Hiei and Komari (2008), Sahoo et al. (2011)
IR24, IR64, IR72, IR57311–95–2-3, Chin Guang	SE calli	Microparticle bombardment	Christou et al. (1991), Li et al. (1993), Sivamani et al. (1996), Zhang et al. (1996), Devi et al. (2012)
Asanohikari, Shiokari Akitakomachi, Chiyonishiki, Hinohikari, Asominori, Koshihikari, M202, Hitomebore, Nipponbare, Sasanishiki, Tsukinohikari, Yukihikari, Suweon 258, Nanjing 11, Xingingai 1, IR8, IR72, IR24, IR26, IR36, IR54, IR64, Kasalat, CSR10, PB1, Swarna	Organogenic calli	*Agrobacterium*	Hiei and Komari (2008), Sahoo et al. (2011)
Chinsurah Boro II, Bengal, cypress, lido, IR72, Koshihikari, Gulfmont	Organogenic calli	Microparticle bombardment	Cooley et al. (1995), Christou (1997), Kohli et al. (1998)

(continued)

Table 16.3 (continued)

Cultivar	Biological target	Gene delivery system	Reference
In planta			
Taipei 309 (japonica)	Inflorescence	Vacuum *Agro*infiltration/ *A. tumefaciens* LBA4404	Dong et al. (2001)
Koshihikari (japonica)	SAM, germinating seed	Piercing/*A. tumefaciens* M-21 and LBA4404	Supartana et al. (2006)
R207 (indica), Teqing (indica)	SAM, germinating seed	Piercing and vacuum *Agro*infiltration/*A. tumefaciens* EHA105	Lin et al. (2009)
Hashemi	SAM, germinating seed	Piercing/*A. tumefaciens* EHA101	Naseri et al. (2012)
Zhong Zuo321 (japonica), Nipponbare (japonica), IR64 (indica)	Leaf (transient)	*Agro*infiltration/strain n.a.	Andrieu et al. (2012)
BRRIdhan 55 (indica)	SAM, germinating seed	Piercing/*A. tumefaciens* LBA4404	Parvin et al. (2015)

SAM shoot apical meristem, *SE* somatic embryogenic, *PEG* polyethylene glycol

targets for gene delivery systems, since new plants are regenerated from a small number of cells. The most important regeneration pathways for rice gene delivery are represented in Fig. 16.3.

Historically, embryogenic and organogenic cultures, as protoplast sources or direct target to microparticle bombardment, constituted the preferred biological explant used to produced genetically altered rice plants (Hiei and Komari 2008; Christou et al. 1991; Datta et al. 1990; Shimamoto et al. 1989). Most recently, immature embryos or calli derived from mature seeds are the favored explants for gene delivery to rice (Hiei and Komari 2008). The efficiency of transformation is higher for immature embryos, especially for indica genotypes, although they are more difficult to obtain and lose viability over time. In contrast, mature seeds are easily obtained and stored. The major constraint of the mature seed method is the efficiency of callus induction, since many indica genotypes and some elite japonica varieties are recalcitrant to the process (Hiei and Komari 2008).

16.3.1.1 Somatic Embryogenic Cultures

The development of bipolar structures from somatic cells via an embryological stage that does not involve the fusion of gametes is called somatic embryogenesis and the structures, somatic embryos (SE) (Sugimoto et al. 2019; Verdeil et al. 2007). In rice, somatic embryogenesis is a feasible regeneration pathway that can be

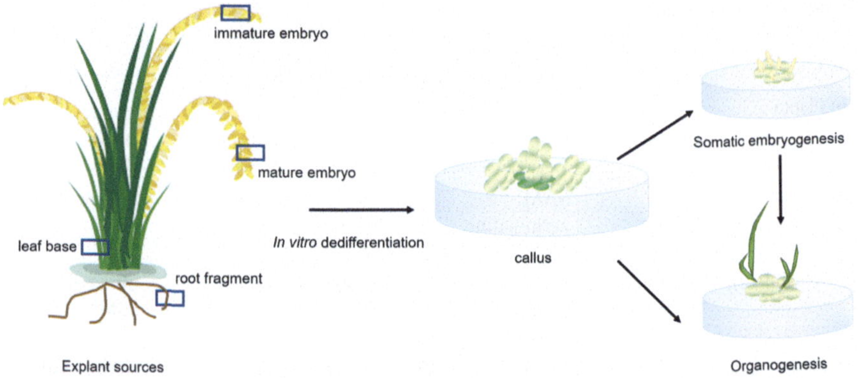

Fig. 16.3 Schematic representation of the most frequently used in vitro regeneration systems to deliver genes to rice

induced in vitro from mature seed explants (Indoliya et al. 2016; Hiei and Komari 2008).

Briefly, mature seeds are in vitro germinated in medium containing a synthetic auxin, usually 2,4-dichlorophenoxyacetic acid (2,4-D), until the isolation of the scutellum, which is, in turn, inoculated to another induction medium until callus formation. Calli are sliced and transferred to proliferation medium until cocultivation with *Agrobacterium* (Hiei and Komari 2008).

16.3.1.2 Organogenic Cultures

Organogenesis consists in the development of adventitious organs or primordia from undifferentiated cell mass in vitro, via dedifferentiation of the original explant, and its subsequent redifferentiation into adventitious organs/primordia (Sugimoto et al. 2019; Verdeil et al. 2007). The mass of dedifferentiated cells is labeled callus and can be induced from in vitro manipulation of the explant hormone, nutrition, osmotic, and environmental growth conditions. Similarly, redifferentiation is also achieved by controlled alterations in the growth environment of the calli (Verdeil et al. 2007). The undifferentiated cells can be induced to take two distinct differentiation pathways, via somatic embryogenesis or organogenesis, although in *Oryza* species, organogenesis is the preferred pathway (Lee and Huang 2013).

Several parts of the rice plant have been used to produce organogenic calli, such as immature or mature embryos, root segments, coleoptiles, and leaf bases (Lee and Huang 2013). Shoot organogenesis from rice calli is dependent on the subspecies, with *indica* genotypes being less amenable to regeneration than the *japonica* ones, although a large difference in the organogenic potential is also observed among the genotypes (Hiei and Komari 2008; Lee and Huang 2013).

In short, organogenic calli can be induced by inducing immature embryos to dedifferentiate in media containing artificial auxin 2,4-D (Hiei and Komari 2008; Lee and Huang 2013). Temperature and osmotic stress treatments, prior to the callus induction, have been reported to increase organogenesis (Hiei and Komari 2008; Lee and Huang 2013).

16.3.2 In Planta Targets

In planta gene delivery methods were developed as an alternative to laborious, expensive, and time-consuming transformation protocols based on tissue culture. An additional advantage of in planta gene delivery is the avoidance of tissue-culture induced epigenetic changes and somaclonal variation. Tissue culture-independent transformation was initially established for model species *Arabidopsis thaliana* (Feldmann and Marks 1987) and, later, expanded to several other species, including rice (Dong et al. 2001; Supartana et al. 2006; Lin et al. 2009; Naseri et al. 2012; Andrieu et al. 2012; Parvin et al. 2015). The main characteristics of the in planta gene delivery systems for *Oryza* are summarized in Table 16.3.

16.4 Gene Delivery Systems

16.4.1 Biological Systems

The most important biological systems used in gene delivery to rice are mediated by *Agrobacterium* and, to a lesser extent, viruses. A survey estimates that more than 80% of the work involving rice genetic modification, published in leading journals, uses *Agrobacterium* (Sah et al. 2014). *Oryza* species, as most monocotyledonous plants, are commonly considered recalcitrant to *Agrobacterium* infection (Gelvin 2003; Altpeter et al. 2016), but several technical advances have elevated the efficiency of gene delivery use to rice (Sah et al. 2014) (Fig. 16.1).

16.4.1.1 *Agrobacterium* spp.

Agrobacterium is the most important method of gene delivery to plants, due to its several advantages, including (i) easiness of the method, (ii) equipment not required to deliver the sequences of interest, (iii) high efficiency in compatible plants, (iv) clean genome insertions, (v) low insertion copy numbers, and (vi) more predictable integration patterns in the host genome (Gelvin 2003; Lee and Gelvin 2008). In nature, *Agrobacterium* displays higher compatibility to dicot plants; thus monocots, such as rice, are considered recalcitrant to infection (Gelvin 2003; Altpeter et al. 2016).

Technological advances, including cocultivation conditions, selection and regeneration methods, and vector development, allowed increased frequencies of T-DNA transfer to monocots, and the method is currently used to in vitro and in planta gene delivery to rice (Sah et al. 2014). A summary of the use of *Agrobacterium* to gene delivery in rice is presented in Table 16.3.

16.4.1.2 Other Bacteria

Besides *Agrobacterium*, other species also display the capacity of integrating DNA segments into plant genomes (Altpeter et al. 2016). Several species of *Rhizobium* have been demonstrated to promote the integration of DNA in plant genomes, including rice (Broothaerts et al. 2005), however at lower efficiency in comparison to *Agrobacterium* (Hooykaas et al. 1977; Broothaerts et al. 2005).

The nitrogen-fixing *Sinorhizobium*, *Ensifer adhaerens*, was also shown to induce transformation events to plants, including two japonica rice varieties, Curinga and Nipponbare, at frequencies similar to those of *Agrobacterium* (Zuniga-Soto et al. 2015). However, the molecular machinery of *Agrobacterium* is still required by the other bacteria to transfer and integrate exogenous DNA into plant genomes (Altpeter et al. 2016). Recently, a novel *Rhizobium* species, encoding its own set of proteins engaged in DNA transfer and integration, has been identified and shown to effectively deliver genes to *Arabidopsis* and tobacco (Lacroix and Citovsky 2016).

16.4.2 Viral Systems

Viruses are interesting systems to shuttle exogenous sequences into plants due to their intracellular nature and propensity to recombine genetic material (Pasin et al. 2019; Catoni et al. 2018). Therefore, several RNA viruses, such as wheat streak mosaic virus (WSMV), barley stripe mosaic virus (BSMV) and tobacco rattle virus (TRV), and single-stranded DNA viruses, as geminiviruses and begomoviruses, are used to deliver genes to plants (Zaid and Mansoor 2017). However, viral-based gene delivery has several restrictions, including (i) the requirement of mechanical inoculation; (ii) the removal of the circulating viruses from the plant body; (iii) the lack of stable integration, in many cases; and (iv) the limited number of cells reached by the viruses at the inoculation site (Pasin et al. 2019; Peyret and Lomonossoff 2015). In contrast, the non-insertional character of many virus-mediated gene delivery systems is promising to the generation of non-transgenic genome-edited plants, where the transient expression of the endonuclease and sgRNA is sufficient to induce the desired mutation (Pasin et al. 2019).

In rice, virus-mediated gene delivery has been used mainly in functional studies (Kant and Dasgupta 2017; Kant et al. 2015; Purkayastha et al. 2010). Recently, self-replicating hybrid DNA molecules between beet curly top Iran virus (BCTIV) and a plant host DNA were demonstrated to become encapsidated in viral particles, and spread systemically throughout infected plants, suggesting a possible novel pathway for virus-mediated horizontal transfer of chromosomal DNA between plant species (Catoni et al. 2018).

16.4.3 Physical Systems

Agrobacterium- and virus-mediated gene delivery to plants is highly dependent on complex biological interactions between the shuttling and the host organisms, which renders the effectiveness of the methods highly specific (Lee and Gelvin 2008; Peyret and Lomonossoff 2015; Altpeter et al. 2016; Pasin et al. 2019). Alternatively, physical forces used to deliver DNA to cells are universal and intend to circumvent compatibility problems (Klein et al. 1992; Frame et al. 1994; Bates et al. 1988). The factors influencing direct systems to deliver nucleic acids to living cells are associated to the damage imposed by physical forces on the biological targets (Southgate et al. 1995, Bates et al. 1988). The main disadvantages of direct DNA delivery methods are associated to the insertion of multiple copies of the sequence of interest, resulting in the generation of genome sites prone to epigenetic modifications and erratic expression of the introduced character (Kohli et al. 1999; Stoger et al. 1998).

16.4.3.1 Microprojectile Bombardment

Nucleic acid-covered microprojectiles can be accelerated to supersonic speed and delivered to a wide range of biological targets (Klein et al. 1992). The main advantages of the method consist in its universal characteristics and in the possibility of delivering several DNA sequences concomitantly (Altpeter et al. 2016; Klein et al. 1992). In rice, an effective protocol using microprojectile bombardment to deliver genes and produce modified plants was developed early on (Christou et al. 1991) and has been widely employed and is still currently used as basis to novel improved methods (Sah et al. 2014). A summary of the use of microparticle bombardment to deliver genes to rice is presented in Table 16.3.

Recently, other methods have been developed to deliver nucleic acid-coated nanoparticles to plants (Mitter et al. 2017; Demirer et al. 2018). In the first study, layered double hydroxide clay nanosheets were loaded with double-stranded RNA to produce the so-called BioClays that were applied to the plant's external surface by spraying tobacco and *Vigna unguiculata* leaves (Mitter et al. 2017). The second work coated carbon nanotube scaffolds with linear and plasmid DNA and infused the complex to arugula and wheat (Demirer et al. 2018). The nanoparticle methods were successful to deliver sequences of interest into non-model plants, in the absence of other unaccounted for exogenous nucleic acid molecules, without requiring tissue culture steps. The potential of these novel technologies remains to be tested in rice.

16.4.3.2 Electroporation

The polar end of membrane phospholipids can be ionized by the passage of a directional electric current (Chen et al. 2006). The polarization of the phospholipids leads to the formation of transient pores in the membrane structure, making it permeable to macromolecules (Chen et al. 2006). In plants, the cell wall functions as a physical barrier preventing the transfer of molecules to the interior of the cells even in the presence of membrane pores. Thus, electroporation is often coupled to the enzymatic removal of the cell wall, given rise to protoplasts, in gene delivery systems to plants. The regeneration of whole, viable plants from electroporated protoplasts is laborious, time-consuming, and highly dependent on the species in vitro performance, making their use more frequent in transient assays (Shen et al. 2014). The production of intact rice plants from protoplasts was considered a major breakthrough to the implementation of modern biotechnological tools to cereals (Marx 1987) and has been continuously used to deliver exogenous nucleic acid molecules in the species (Sah et al. 2014).

Alternatively, electroporation has been used to deliver and transiently express nucleic acids to intact plant tissues, including rice (Dekeyser et al. 1990). The actual mechanisms allowing movement of nucleic acids through the cell wall matrix remain unclear, and the method has restricted use.

16.4.4 Chemical Systems

Protoplasts, isolated plant cells from which the cell wall has been removed, were demonstrated to aggregate with other cells and/or macromolecules in the presence of synthetic polymers (Shen et al. 2014). Polyethylene glycol (PEG), an inert hydrophilic polymer of ethylene oxide, alone or associated with $CaCl_2$, is the most frequently used molecule to induce the aggregation of plant protoplasts and protoplasts/biomolecules (Shen et al. 2014). In comparison to protoplast electroporation, PEG-mediated delivery of nucleic acids is less prejudicial to the cells, although, in general, it is considered to have lower transfection efficiency. An advantage of the method over electroporation is that it does not require special equipment to be used. The constraints involved in the regeneration of complete, viable plants from protoplasts are comparable to those of electroporation.

The polymer has been used with success in rice to produce modified plants expressing the introduced traits (Yang et al. 1988; Peng et al. 1990; Datta et al. 1992).

16.5 Gene Delivery for Quality Improvement

The quality of the grain is complex trait in rice and is associated to starch properties and contents, synthesis, and accumulation of lipids, amino acids and other phytochemicals, seed morphogenesis, and other factors. Table 16.4 presents a summary of quality traits modified in rice by genetic engineering and genome editing.

Table 16.4 Summary of quality traits and delivery systems used to improve grain quality in rice

Cultivar	Delivery system	Trait (gene)	Reference
	Binary vector, *Agrobacterium tumefaciens*	Provitamin A – synthetic genes encoding phytoene synthase, phytoene desaturase, β-carotene ketolase, and β-carotene hydroxylase (*sZmPSY1*, *sPaCrtI*, *sCrBKT*, and *sHpBHY*)	Zhu et al. (2018)
Taipei 309	CRISPR/Cas9, *Agrobacterium tumefaciens*	Oleic acid contents – FATTY ACID DESATURASE 2–1 (FAD 2–1) knock-out	Abe et al. (2018)
ZhongHua 11 (japonica) and HuaGuang1 (Indica)	Binary vector, *Agrobacterium tumefaciens*	Antioxidant compounds – 2 regulatory genes from maize (*ZmLc* (leaf color) and *ZmPl* (purple leaf)), six anthocyanin biosynthesis genes from *Coleus* (*SsCHS* (chalcone synthase), *SsCHI* (chalcone isomerase), *SsF3H* (flavanone 3-hydroxylase), *SsF30H* (flavonoid 30-hydroxylase), *SsDFR* (dihydroflavonol 4-reductase), and *SsANS* (anthocyanidin synthase)	Zhu et al. (2017)
IR64	Binary vector, *Agrobacterium tumefaciens*	Iron contents – rice and soybean ferritin genes	Oliva et al. (2014)
Wuxiangjing 9	Binary vector, *Agrobacterium tumefaciens*	Lysine contents – bacterial lysine feedback-insensitive aspartate kinase (AK) and dihydrodipicolinate synthase (DHPS), RNA interference of rice lysine ketoglutaric acid reductase/saccharopine dehydropine dehydrogenase (LKR/SDH)	Long et al. (2013)
Taipei 309	Binary vector, *Agrobacterium tumefaciens*	Cysteine, glutathione, and methionine contents – *E. coli* serine acetyltransferase (*EcSAT*)	Nguyen et al. (2012)
Nipponbare	Binary vector, *Agrobacterium tumefaciens*	Amylose contents – RNA hairpin for starch branching IIb (SBEIIb)	Butardo et al. (2011)
Nipponbare	Binary vector, *Agrobacterium tumefaciens*	Pro vitamin A – feedback-insensitive alpha subunit of rice anthranilate synthase (*OASA*1D)	Wakasa et al. (2006)
Taipei 309	Binary vector, *Agrobacterium tumefaciens*	Pro vitamin A – maize PHYTOENE SYNTHASE (PSY) and PHYTOENE DESATURASE (CRT I)	Al-Babili et al. (2006)
Taipei 309	Binary vector, microparticle bombardment	Starch quality – *Os*BEIIb (starch branching enzyme)	Tanaka et al. (2004)
Taipei 309	Binary vector, microparticle bombardment	Pro vitamin A – protein contents – *A. thaliana* $tRNA^{lys}$	Wu et al. (2003)
Taipei 309	Binary vector, *Agrobacterium tumefaciens*	Iron contents – *Phaseolus vulgaris* ferritin gene (*PvPFE*)	Lucca et al. (2001)
Taipei 309	Binary vector, *Agrobacterium tumefaciens*	Pro vitamin A – PHYTOENE SYNTHASE (PSY) and PHYTOENE DESATURASE (CRT I)	Ye et al. (2000)
Taipei 309	Binary vector, microparticle bombardment	Pro vitamin A – PHYTOENE SYNTHASE (PSY) from *Narcissus pseudonarcissus*	Bukhardt et al. (1997)

References

Abe K, Araki E, Suzuki Y, Toki S, Saika H (2018) Production of high oleic/low linoleic rice by genome editing. Plant Physiol Biochem 131:58–62. https://doi.org/10.1016/j.plaphy.2018.04.033

Al-Babili S, Hoa TTC, Schaub P (2006) Exploring the potential of the bacterial carotene desaturase *CrtI* to increase the β-carotene content in Golden Rice. J Exp Bot 57:1007–1014. https://doi.org/10.1093/jxb/erj086

Altpeter F, Springer NM, Bartley LE, Blechl AE, Brutnell TP, Citovsky V, Conrad LJ, Gelvin SB, Jackson DP, Kausch AP, Lemaux PG, Medford JI, Orozco-Cárdenas ML, Tricoli DM, Van Eck J, Voytas DF, Walbot V, Wang K, Zhang ZJ, Stewart CN Jr (2016) Advancing crop transformation in the era of genome editing. Plant Cell 28:1510–1520. https://doi.org/10.1105/tpc.16.00196

Andrieu A, Breitler JC, Siré C, Meynard D, Gantet P, Guiderdon E (2012) An *in planta*, *Agrobacterium*-mediated transient gene expression method for inducing gene silencing in rice (*Oryza sativa* L.) leaves. Rice 5:23. https://doi.org/10.1186/1939-8433-5-23

Asanti MD (2016) Breeding rice for improved grain quality. In: Li JQ (ed) Advances in international rice research. InTech. https://doi.org/10.5772/66684

Bao J (2014) Genes and QTLs for rice grain quality improvement, rice – germplasm, genetics and improvement. In: Wengui Yan (ed) InTech. https://doi.org/10.5772/56621

Bates GW, Piastuch W, Riggs CD, Rabussay D (1988) Electroporation for DNA delivery to plant protoplasts. In: Puite KJ, Dons JJM, Huizing HJ, Kool AJ, Koornneef M, Krens FA (eds) Progress in plant protoplast research. Current plant science and biotechnology in agriculture, vol 7. Springer, Dordrecht, pp 305–310

Birlaa DS, Kapil Malika K, Saingera M, Chaudharya D, Jaiwalb R, Jaiwala PK (2017) Progress and challenges in improving the nutritional quality of rice (*Oryza sativa* L.). Crit Rev Food Sci 57(11):2455–2481. https://doi.org/10.1080/10408398.2015.1084992

Broothaerts W, Mitchell HJ, Weir B, Kaines S, Smith LMA, Yang W, Mayer JE, Roa-Rodríguez C, Jefferson RA (2005) Gene transfer to plants by diverse species of bacteria. Nature 433:629–633. https://doi.org/10.1038/nature03309

Bureau C, Lanau N, Ingouff M, Hassan B, Meunier AC, Divol F, Sevilla R, Mieulet D, Dievart A, Périn C (2018) A protocol combining multiphoton microscopy and propidium iodide for deep 3D root meristem imaging in rice: application for the screening and identification of tissue-specific enhancer trap lines. Plant Methods 14:96. https://doi.org/10.1186/s13007-018-0364-x

Burkhardt PK, Beyer P, Wünn J, Klöti A, Armstrong GA, Schledz M, von Lintig J, Potrykus I (1997) Transgenic rice (*Oryza sativa*) endosperm expressing daffodil (*Narcissus pseudonarcissus*) phytoene synthase accumulates phytoene, a key intermediate of provitamin a biosynthesis. Plant J 11(5):1071–1078. https://doi.org/10.1046/j.1365-313X.1997.11051071.x

Butardo VM Jr, Fitzgerald MA, Bird AR, Gidley MJ, Flanagan BM, Larroque O, Resurreccion AP, Laidlaw HKC, Jobling SA, Morel MK, Rahman S (2011) Impact of down-regulation of starch branching enzyme IIb in rice by artificial microRNA-and hairpin RNA-mediated RNA silencing. J Exp Bot 62:4927–4941. https://doi.org/10.1093/jxb/err188

Butardo VM Jr, Sreenivasulu N, Juliano BO (2019) Improving rice grain quality: state-of-the-art and future prospects. In: Sreenivasulu N. (ed) Rice grain quality. Methods in Molecular Biology, vol 1892. Humana Press, New York, NY. https://doi.org/10.1007/978-1-4939-8914-0_2

Catoni M, Noris E, Vaira AM, Jonesman T, Matić S, Soleimani R, Behjatnia SAA, Vinals N, Paszkowski J, Accotto GP (2018) Virus-mediated export of chromosomal DNA in plants. Nat Commun 9:5308. https://doi.org/10.1038/s41467-018-07775

Chen C, Smye S, Robinson M, Evans JA (2006) Membrane electroporation theories: a review. Med Bio Eng Comput 44:5. https://doi.org/10.1007/s11517-005-0020-2

Chiou WY, Kawamoto T, Himi E, Rikiishi, Sugimoto M, Hayashi-Tsugane M, Tsugane K, Maekawa M (2019) *LARGE GRAIN* encodes a putative RNA-binding protein that regulates spikelet hull length in rice. Plant Cell Physiol. https://doi.org/10.1093/pcp/pcz014

Christou P (1997) Rice transformation: bombardment. Plant Mol Biol September 1997 35(1–2):197–203. https://doi.org/10.1023/A:1005791230345

Christou P, Ford TL, Kofron M (1991) Production of transgenic rice (*Oryza sativa* L.) plants from agronomically important *indica* and *japonica* varieties via electric discharge particle acceleration of exogenous DNA into immature zygotic embryos. Bio/Technology 9:957–962. https://doi.org/10.1038/nbt1091-957

Cooley J, Ford T, Christou P (1995) Molecular and genetic characterization of elite transgenic rice plants produced by electric-discharge particle acceleration. Theor Appl Genet 90:97–104. https://doi.org/10.1007/BF00221001

Datta SK, Peterhans A, Datta K, Potrykus I (1990) Genetically engineered fertile *indica*-rice recovered from protoplasts. Bio/Technology 8:736–740. https://doi.org/10.1038/nbt0890-736

Datta SK, Datta K, Soltanifar N, Donn G, Potrykus I (1992) Herbicide-resistant *Indica* rice plants from IRRI breeding line IR72 after PEG-mediated transformation of protoplasts. Plant Mol Biol 20:619–629. https://doi.org/10.1007/BF00046447

Deblaere R, Bytebier B, De Greve H, Deboeck F, Schell J, Van Montagu M, Leemans J (1985) Efficient octopine Ti plasmid-derived vectors for *Agrobacterium*-mediated gene transfer to plants. Nucleic Acids Res 13:4777–4788

Dekeyser RA, Claes B, De Rycke R, Habets ME, Van Montagu MC, Caplan AB (1990) Transient gene expression in intact and organized rice tissues. Plant Cell 2:591–602. https://doi.org/10.1105/tpc.2.7.591

Demirer GS, Zhang H, Matos J, Chang R, Chio L, Staskawicz B, Landry MP (2018) High aspect ratio nanomaterials enable biomolecule delivery and transgene expression or silencing in mature plants. BioRxiv. https://doi.org/10.1101/179549

Devi S, Mishra MK, Elliott M (2012) Regeneration of transgenic rice with bacterial *ipt* gene driven by senescence specific (*SAG12*) promoter by particle bombardment. Trop Life Sci Res 23:39–48

Dong J, Kharb P, Teng W, Hall TC (2001) Characterization of rice transformed via an *Agrobacterium*-mediated inflorescence approach. Mol Breed 7:187–194. https://doi.org/10.1023/A:1011357709073

Ellis HM, Yu D, DiTizio T, Court DL (2001) High efficiency mutagenesis, repair, and engineering of chromosomal DNA using single-stranded oligonucleotides. Proc Natl Acad Sci U S A 98:6742–6746. https://doi.org/10.1073/pnas.121164898

Engler C, Kandzia R, Marillonnet S (2008) A one pot, one step, precision cloning method with high throughput capability. PLoS One 3(11):e3647. https://doi.org/10.1371/journal.pone.0003647

Feldmann KA, Marks MD (1987) *Agrobacterium*-mediated transformation of germinating seeds of *Arabidopsis thaliana*: a non-tissue culture approach. Mol Gen Genet 208:1–9. https://doi.org/10.1007/BF00330414

Frame BR, Drayton PR, Bagnall SV, Lewnau CJ, Bullock WP, Wilson HM, Dunwell JM, Thompson JA, Wang K (1994) Production of fertile transgenic maize plants by silicon carbide whisker-mediated transformation. Plant J 6:941–948. https://doi.org/10.1046/j.1365-313X.1994.6060941.x

Gan L, Wang X, Cheng Z, Liu L, Wang J, Zhang Z, Ren Y, Lei C, Zhao Z, Zhu S, Lin Q, Wu F, Guo X, Wang J, Zhang X, Wan J (2016) *Wax crystal-sparse leaf 3* encoding a β-ketoacyl-CoA reductase is involved in cuticular wax biosynthesis in rice. Plant Cell Rep 35:1687–1698. https://doi.org/10.1007/s00299-016-1983-1

Gelvin SB (2003) *Agrobacterium*-mediated plant transformation: the biology behind the "gene-jockeying" tool. Microbiol Mol Biol Rev 67:16–37. https://doi.org/10.1128/MMBR.67.1.16-37.2003

Ghosh-Biswas GC, Iglesias VA, Datta SK, Potrykus I (1994) Transgenic *Indica* rice (*Oryza sativa* L.) plants obtained by direct gene transfer to protoplasts. J Biotechnol 32:1–10. https://doi.org/10.1016/0168-1656(94)90114-7

Gibson DG, Young L, Chuang RY, Venter JC, Hutchison CA 3rd, Smith HO (2009) Enzymatic assembly of DNA molecules up to several hundred kilobases. Nat Methods 6:343–345. https://doi.org/10.1038/nmeth.1318

Hahne G, Tomlinson L, Nogué F (2019) Precision genetic engineering tools for next-generation plant breeding. Plant Cell Rep 38(4):435–436. https://doi.org/10.1007/s00299-019-02400-6

Hajdukiewicz P, Svab Z, Maliga P (1994) The small, versatile pPZP family of *Agrobacterium* binary vectors for plant transformation. Plant Mol Biol 25:989–994. https://doi.org/10.1007/BF00014672

Hartley JL, Temple GF, Brasch MA (2000) DNA cloning using *in vitro* site-specific recombination. Genome Res 10:1788–1795. https://doi.org/10.1101/gr.143000

He R (2016) Multigene engineering in rice using high-capacity *Agrobacterium tumefaciens* BIBAC vectors. Methods Mol Biol 1385:29–37. https://doi.org/10.1007/978-1-4939-3289-4_2

He F, Chen S, Ning Y, Wang GL (2016) Rice (*Oryza sativa*) protoplast isolation and its application for transient expression analysis. Curr Protoc Plant Biol 1:373–383. https://doi.org/10.1002/cppb.20026

Hiei Y, Komari T (2008) *Agrobacterium*-mediated transformation of rice using immature embryos or calli induced from mature seed. Nat Protoc 3:824. https://doi.org/10.1038/nprot.2008.46

Hoekema A, Hirsch PR, Hooykaas PH, Schilperoort RA (1983) A binary plant vector strategy based on separation of the vir and T-DNA regions of the *Agrobacterium tumefaciens* Ti plasmid. Nature 303:179–180. https://doi.org/10.1038/303179a0

Hooykaas PJJ, Klapwijk PM, Nuti MP, Schilperoort RA, Rorsch A (1977) Transfer of the *Agrobacterium tumefaciens* TI plasmid to avirulent *Agrobacteria* and to *Rhizobium* ex planta. J Gen Microbiol 98:477–484. https://doi.org/10.1099/00221287-98-2-477

Hsu PD, Lander ES, Zhang F (2014) Development and applications of CRISPR-Cas9 for genome engineering. Cell 157:1262–1278. https://doi.org/10.1016/j.cell.2014.05.010

Hu N, Xian Z, Li N, Liu Y, Huang W, Yan F, Su D, Chen J, Li Z (2019) Rapid and user-friendly open-source CRISPR/Cas9 system for single- or multi-site editing of tomato genome. Hortic Res 6:7. https://doi.org/10.1038/s41438-018-0082-6

Huang TK, Puchta H (2019) CRISPR/Cas-mediated gene targeting in plants: finally a turn for the better for homologous recombination. Plant Cell Rep. https://doi.org/10.1007/s00299-019-02379-0

Ikeuchi M, Ogawa Y, Iwase A, Sugimoto K (2016) Plant regeneration: cellular origins and molecular mechanisms. Development 143:1442–1451. https://doi.org/10.1242/dev.134668

Indoliya Y, Tiwari P, Chauhan AS, Goel R, Shri M, Bag SK, Chakrabarty D (2016) Decoding regulatory landscape of somatic embryogenesis reveals differential regulatory networks between *japonica* and *indica* rice subspecies. Sci Rep 6:23050. https://doi.org/10.1038/srep23050

Jinek M, Chylinski K, Fonfara I, Hauer M, Doudna JA, Charpentier E (2012) A programmable dual-RNA-guided DNA endonuclease in adaptive bacterial immunity. Science 337:816–821. https://doi.org/10.1126/science.1225829

Kant R, Dasgupta I (2017) Phenotyping of VIGS-mediated gene silencing in rice using a vector derived from a DNA virus. Plant Cell Rep 36:1159–1170. https://doi.org/10.1007/s00299-017-2156-6

Kant R, Sharma S, Dasgupta I (2015) Virus-induced gene silencing (VIGS) for functional genomics in rice using Rice Tungro bacilliform virus (RTBV) as a vector. Methods Mol Biol 1287:201–217. https://doi.org/10.1007/978-1-4939-2453-0_15

Kim E, Moore BS, Yoon YJ (2015) Reinvigorating natural product combinatorial biosynthesis with synthetic biology. Nat Chem Biol 11:649–659. https://doi.org/10.1038/nchembio.1893

Klein TM, Arentzen R, Lewis PA, Fitzpatrick-McElligott S (1992) Transformation of microbes, plants and animals by particle bombardment. Bio/Technology 10:286–291. https://doi.org/10.1038/nbt0392-286

Kohli A, Leech M, Vain P, Laurie DA, Christou P (1998) Transgene organization in rice engineered through direct DNA transfer supports a two-phase integration mechanism mediated by the establishment of integration hot spots. Proc Natl Acad Sci U S A 95:7203–7208. https://doi.org/10.1073/pnas.95.12.7203

Kohli A, Gahakwa D, Vain P, Laurie DA, Christou P (1999) Transgene expression in rice engineered through particle bombardment: molecular factors controlling stable expression and transgene silencing. Planta 208:88–97. https://doi.org/10.1007/s004250050538

Lacroix B, Citovsky V (2016) A functional bacterium-to-plant DNA transfer machinery of *Rhizobium etli*. PLoS Pathog 12:e1005502. https://doi.org/10.1371/journal.ppat.1005502

Lee L-Y, Gelvin SB (2008) T-DNA binary vectors and systems. Plant Physiol 146:325–332. https://doi.org/10.1104/pp.107.113001

Lee ST, Huang WL (2013) Cytokinin, auxin, and abscisic acid affects sucrose metabolism conduce to de novo shoot organogenesis in rice (*Oryza sativa* L.) callus. Bot Stud 54:5. https://doi.org/10.1186/1999-3110-54-5

Li L, Qu R, Kochko A, Fauquet C, Beachy RN (1993) An improved rice transformation system using the biolistic method. Plant Cell Rep 12:250–255. https://doi.org/10.1007/BF00237129

Lin J, Zhou B, Yang Y, Mei J, Zhao X, Guo X, Huang X, Tang D, Liu X (2009) Piercing and vacuum infiltration of the mature embryo: a simplified method for *Agrobacterium*-mediated transformation of *indica* rice. Plant Cell Rep 28:1065–1074. https://doi.org/10.1007/s00299-009-0706-2

Lo SF, Fan MJ, Hsing YI, Chen LJ, Chen S, Wen IC, Liu YL, Chen KT, Jiang MJ, Lin MK, Rao MY, Yu LC, Ho TH, Yu SM (2016) Genetic resources offer efficient tools for rice functional genomics research. Plant Cell Environ 39:998–1013. https://doi.org/10.1111/pce.12632

Long X, Liu Q, Chan M, Wang Q, Sun SS (2013) Metabolic engineering and profiling of rice with increased lysine. Plant Biotechnol J 11:490–501. https://doi.org/10.1111/pbi.12037

Lucca P, Hurrell R, Potrykus I (2001) Genetic engineering approaches to improve the bioavailability and the level of iron in rice grains. Theor Appl Genet 102:392–397. https://doi.org/10.1007/s001220051659

Liu W, Yuan JS, Stewart CN Jr (2013) Advanced genetic tools for plant biotechnology. Nat Rev Genet 14:781–793. https://doi.org/10.1038/nrg3583

Ma X, Zhang J, Han B, Tang J, Cui D, Han L (2019) *FLA*, which encodes a homolog of UBP, is required for chlorophyll accumulation and development of lemma and palea in rice. Plant Cell Rep 38(3):321–331. https://doi.org/10.1007/s00299-018-2368-4

Mahas A, Ali Z, Tashkandi M, Mahfouz MM (2019) Virus-mediated genome editing in plants using the CRISPR/Cas9 system. Methods Mol Biol 1917:311–326. https://doi.org/10.1007/978-1-4939-8991-1_23

Mahender A, Anandan A, Pradhan SK, Pandit E (2016) Rice grain nutritional traits and their enhancement using relevant genes and QTLs through advanced approaches. Springerplus 5(1):2086. https://doi.org/10.1186/s40064-016-3744-6

Marx JL (1987) Rice plants regenerated from protoplasts: the ability to regenerate rice protoplasts means that for the first time a major cereal will become subject to modern biotechnological methods. Science 235:31–32. https://doi.org/10.1126/science.235.4784.31

Miki B, McHugh S (2004) Selectable marker genes in transgenic plants: applications, alternatives and biosafety. J Biotechnol 107:193–232. https://doi.org/10.1016/j.jbiotec.2003.10.011

Mishra R, Joshi RK, Zhao K (2018) Genome editing in rice: recent advances, challenges, and future implications. Front Plant Sci 9:1361. https://doi.org/10.3389/fpls.2018.01361

Mitter N, Worrall EA, Robinson KE, Li P, Jain RG, Taochy C, Fletcher SJ, Carroll BJ, Lu GQ, Xu ZP (2017) Clay nanosheets for topical delivery of RNAi for sustained protection against plant viruses. Nat Plants 3:16207. https://doi.org/10.1038/nplants.2016.207

Motte H, Vereecke D, Geelen D, Werbrouck S (2014) The molecular path to in vitro shoot regeneration. Biotechnol Adv 32:107–121. https://doi.org/10.1016/j.biotechadv.2013.12.002

Mukamuhirwa A, Hovmalm HP, Ortiz R, Nyamangyoku O, Johansson E (2018) Quality and grain yield attributes of rwandan rice (*Oryza sativa* L.) cultivars grown in a biotron applying two NPK levels. J Food Qual. https://doi.org/10.1155/2018/5134569

Naseri G, Sohani MM, Pourmassalehgou A, Allahi S (2012) In planta transformation of rice (*Oryza sativa*) using thaumatin-like protein gene for enhancing resistance to sheath blight. African J Biotechnol 11:7885–7893. https://doi.org/10.5897/AJB11.3331

Neelakandan AK, Wang K (2012) Recent progress in the understanding of tissue culture-induced genome level changes in plants and potential applications. Plant Cell Rep 31:597–620. https://doi.org/10.1007/s00299-011-1202-z

Nguyen HC, Hoefgen R, Hesse H (2012) Improving the nutritive value of rice seeds: elevation of cysteine and methionine contents in rice plants by ectopic expression of a bacterial serine acetyltransferase. J Exp Bot 63:5991–6001. https://doi.org/10.1093/jxb/ers253

Oliva N, Chadha-Mohanty P, Poletti S, Abrigo E, Atienza G, Torrizo L, Garcia R, Duenas C Jr, Poncio MA, Balindong J, Manzanilla M, Montecillo F, Zaidem M, Barry G, Herve P, Shou H, Slamet- Loedin IH (2014) Large-scale production and evaluation of marker-free indica rice IR64 expressing phytoferritin genes. Mol Breed 33:23–37. https://doi.org/10.1007/s11032-013-9931-z

Pandiarajan R, Grover A (2018) In vivo promoter engineering in plants: are we ready? Plant Sci 277:132–138. https://doi.org/10.1016/j.plantsci.2018.10.011

Parvin S, Biswas S, Razzaque S, Haque T, Elias SM, Tammi RS, Seraj ZI (2015) Salinity and drought tolerance conferred by in planta transformation of SNAC1 transcription factor into a high-yielding rice variety of Bangladesh. Acta Physiol Plant 37:68. https://doi.org/10.1007/s11738-015-1817-8

Pasin F, Menzel W, Daròs JA (2019) Harnessed viruses in the age of metagenomics and synthetic biology: an update on infectious clone assembly and biotechnologies of plant viruses. Plant Biotechnol J. https://doi.org/10.1111/pbi.13084

Patron NJ (2014) DNA assembly for plant biology: techniques and tools. Curr Opin Plant Biol 19:14–19. https://doi.org/10.1016/j.pbi.2014.02.004

Peng J, Lyznik LA, Lee L, Hodges TK (1990) Co-transformation of indica rice protoplasts with *gusA* and *neo* genes. Plant Cell Rep 9:168–172. https://doi.org/10.1007/BF00232097

Pérez L, Soto E, Farré G, Juanos J, Villorbina G, Bassie L, Medina V, Serrato AJ, Sahrawy M, Rojas JA, Romagosa I, Muñoz P, Zhu C, Christou P (2019) CRISPR/Cas9 mutations in the rice *waxy/GBSSI* gene induce allele-specific and zygosity-dependent feedback effects on endosperm starch biosynthesis. Plant Cell Rep 38:417–433. https://doi.org/10.1007/s00299-019-02388-z

Peyret H, Lomonossoff GP (2015) When plant virology met *Agrobacterium*: the rise of the deconstructed clones. Plant Biotechnol J 13:1121–1135. https://doi.org/10.1111/pbi.12412

Puchta H (2017) Applying CRISPR/Cas for genome engineering in plants: the best is yet to come. Curr Opin Plant Biol 36:1–8. https://doi.org/10.1016/j.pbi.2016.11.011

Purkayastha A, Mathur S, Verma V, Sharma S, Dasgupta I (2010) Virus-induced gene silencing in rice using a vector derived from a DNA virus. Planta 232:1531–1540. https://doi.org/10.1007/s00425-010-1273-z

Sah SK, Kaur A, Kaur G, Cheema GS (2014) Genetic transformation of rice: problems, progress and prospects. J Rice Res 3:132. https://doi.org/10.4172/2375-4338.1000132

Sahoo KK, Tripathi AK, Pareek A, Sopory SK, Singla-Pareek SL (2011) An improved protocol for efficient transformation and regeneration of diverse indica rice cultivars. Plant Methods 7(1):49. https://doi.org/10.1186/1746-4811-7-49

Shen J, Fu J, Ma J, Wang X, Gao C, Zhuang C, Wan J, Jiang L (2014) Isolation, culture, and transient transformation of plant protoplasts. Curr Protoc Cell Biol 63:1–17. https://doi.org/10.1002/0471143030.cb0208s63

Shibata D, Liu YG (2000) *Agrobacterium*-mediated plant transformation with large DNA fragments. Trends Plant Sci 5:354–357. https://doi.org/10.1016/S1360-1385(00)01689-7

Shimamoto K, Terada R, Izawa T, Fujimoto H (1989) Fertile transgenic rice plants regenerated from transformed protoplasts. Nature 338:274–276. https://doi.org/10.1038/338274a0

Sivamani E, Shen P, Opalka N, Beachy RN, Fauquet CM (1996) Selection of large quantities of embryogenic calli from *indica* rice seeds for production of fertile transgenic plants using the biolistic method. Plant Cell Rep 15:322–327. https://doi.org/10.1007/BF00232364

Southgate EM, Davey MR, Power JB, Marchant R (1995) Factors affecting the genetic engineering of plants by microprojectile bombardment. Biotechnol Adv 13:631–651. https://doi.org/10.1016/0734-9750(95)02008-X

Stoger E, Williams S, Keen D, Christou P (1998) Molecular characteristics of transgenic wheat and the effect on transgene expression. Transgenic Res 7:463–s471. https://doi.org/10.1023/A:1008833324193

Sugimoto K, Temman H, Kadokura S, Matsunaga S (2019) To regenerate or not to regenerate: factors that drive plant regeneration. Curr Opin Plant Biol 47:138–150. https://doi.org/10.1016/j.pbi.2018.12.002

Supartana P, Shimizu T, Shioiri H, Nogawa M, Nozue M, Kojima M (2005) Development of simple and efficient in planta transformation method for rice (*Oryza sativa* L.) using *Agrobacterium tumefaciens*. J Biosci Bioeng 100:391–397. https://doi.org/10.1263/jbb.100.391

Supartana P, Shimizu T, Nogawa M, Shioiri H, Nakajima T, Haramoto N, Nozue M, Kojima M (2006) Development of simple and efficient in Planta transformation method for wheat (Triticum aestivum L.) using Agrobacterium tumefaciens. J Biosci Bioeng 102(3):162–170. https://doi.org/10.1263/jbb.102.162

Toriyama K, Arimoto Y, Uchimiya H, Hinata K (1988) Transgenic rice plants after direct gene transfer into protoplasts. Bio/Technology 6:1072–1074. https://doi.org/10.1038/nbt0988-1072

Tanaka N, Fujita N, Nishi A, Satoh H, Hosaka Y, Ugaki M, Kawasaki S, Nakamura Y (2004) The structure of starch can be manipulated by changing the expression levels of STARCH BRANCHING ENZYME IIB in rice endosperm. Plant Biotechnol J 2(6):507–516. https://doi.org/10.1111/j.1467-7652.2004.00097.x

Verdeil JL, Alemanno L, Niemenak N, Tranbarger TJ (2007) Pluripotent versus totipotent plant stem cells: dependence versus autonomy? Trends Plant Sci 12:245–252. https://doi.org/10.1016/j.tplants.2007.04.002

Wakasa K, Hasegawa H, Nemoto H, Matsuda F, Miyazawa H, Tozawa Y, Morino K, Komatsu A, Yamada T, Terakawa T, Miyagawa H (2006) High-level tryptophan accumulation in seeds of transgenic rice and its limited effects on agronomic traits and seed metabolite profile. J Exp Bot 57(12):3069–3078. https://doi.org/10.1093/jxb/erl068

Walkerpeach CR, Velten J (1994) *Agrobacterium*-mediated gene transfer to plant cells: cointegrate and binary vector systems. In: Gelvin SB, Schilperoort RA (eds) Plant molecular biology manual. Springer, Dordrecht. https://doi.org/10.1007/978-94-011-0511-8_3

Weigel D, Glazebrook J (2006) In planta transformation of *Arabidopsis*. CSH Protoc (7). pii: pdb.prot4668. https://doi.org/10.1101/pdb.prot4668

Weising K, Schell J, Kahl G (1988) Foreign genes in plants: transfer, structure, expression, and applications. Annu Rev Genet 22:421–477. https://doi.org/10.1146/annurev.ge.22.120188.002225

Wu XR, Chen ZH, Folk WR (2003) Enrichment of cereal protein lysine content by altered tRNAlys coding during protein synthesis. Plant Biotech 1:187–194. https://doi.org/10.1046/j.1467-7652.2003.00017.x

Wu R, Guo WL, Wang XR, Wang XL, Zhuang TT, Clarke JL, Liu B (2009) Unintended consequence of plant transformation: biolistic transformation caused transpositional activation of an endogenous retrotransposon Tos17 in rice ssp. japonica cv. Matsumae. Plant Cell Rep 28:1043–1051. https://doi.org/10.1007/s00299-009-0704-4

Wu J, Zhang Z, Zhang Q, Han X, Gu X, Lu T (2015) The molecular cloning and clarification of a photorespiratory mutant, *oscdm1*, using enhancer trapping. Front Genet 6:226. https://doi.org/10.3389/fgene.2015.00226

Wu HP, Wei FJ, Wu CC, Lo SF, Chen LJ, Fan MJ, Chen S, Wen IC, Yu SM, Ho TD, Lai MH, Hsing YC (2017) Large-scale phenomics analysis of a T-DNA tagged mutant population. Gigascience 6:1–7. https://doi.org/10.1093/gigascience/gix055

Wu R, Lucke M, Jang YT, Zhu W, Symeonidi E, Wang C, Fitz J, Xi W, Schwab R, Weigel D (2018) An efficient CRISPR vector toolbox for engineering large deletions in *Arabidopsis thaliana*. Plant Methods 14:65. https://doi.org/10.1186/s13007-018-0330-7

Xu Q, Chen W, Xu Z (2015) Relationship between grain yield and quality in rice germplasms grown across different growing areas. Breed Sci 65(3):226–232. https://doi.org/10.1270/jsbbs.65.226

Ye X, Al-Babili S, Klöti A, Zhang J, Lucca P, Beyer P, Potrykus I (2000) Engineering the provitamin A (beta-carotene) biosynthetic pathway into (carotenoid-free) rice endosperm. Science 287(5451):303–305. https://doi.org/10.1126/science.287.5451.303

Yang H, Zhang HM, Davey MR, Mulligan BJ, Cocking EC (1988) Production of kanamycin resistant rice tissues following DNA uptake into protoplasts. Plant Cell Rep 7:421–425. https://doi.org/10.1007/BF00269528

Yu SM, Ko SS, Hong CY, Sun HJ, Hsing YI, Tong CG, Ho TH (2007) Global functional analyses of rice promoters by genomics approaches. Plant Mol Biol 65:417–425. https://doi.org/10.1007/s11103-007-9232-1

Yu W, Yau YY, Birchler JA (2016) Plant artificial chromosome technology and its potential application in genetic engineering. Plant Biotechnol J 14:1175–1182. https://doi.org/10.1111/pbi.12466

Zaidi SS, Mansoor S (2017) Viral vectors for plant genome engineering. Front Plant Sci 8:539. https://doi.org/10.3389/fpls.2017.00539

Zambryski P, Tempe J, Schell J (1989) Transfer and function of T-DNA genes from *Agrobacterium* Ti and Ri plasmids in plants. Cell 56:193–201. https://doi.org/10.1016/0092-8674(89)90892-1

Zhang HM, Yang H, Rech E, Golds TJ, Davis AS, Mulligan BJ, Cocking EC, Davey MR (1988) Transgenic rice plants produced by electroporation-mediated plasmid uptake into protoplasts. Plant Cell Rep 7:379–384. https://doi.org/10.1007/BF00269517

Zhang S, Chen L, Qu R, Marmey P, Beachy R, Fauquet C (1996) Regeneration of fertile transgenic *indica* (group 1) rice plants following microprojectile transformation of embryogenic suspension culture cells. Plant Cell Rep 15:465–469. https://doi.org/10.1007/BF00232975

Zhang Y, Massel K, Godwin ID, Gao C (2018) Applications and potential of genome editing in crop improvement. Genome Biol 19:210. https://doi.org/10.1186/s13059-018-1586-y

Zhu Q, Yu S, Zeng D, Liu H, Wang H, Yang Z, Xie X, Shen R, Tan J, Li H, Zhao X, Zhang Q, Chen Y, Guo J, Chen L, Liu Y-G (2017) Development of "purple endosperm Rice" by engineering anthocyanin biosynthesis in the endosperm with a high-efficiency transgene stacking system. Mol Plant 10:918–929. https://doi.org/10.1016/j.molp.2017.05.008

Zhu Q, Zeng D, Yu S, Cui C, Li J, Li H, Chen J, Zhang R, Zhao X, Chen L, Liu Y-G (2018) From Golden Rice to aSTARice: bioengineering astaxanthin biosynthesis in rice endosperm. Mol Plant 11(12):1440–1448. https://doi.org/10.1016/j.molp.2018.09.007

Zuniga-Soto E, Mullins E, Dedicova B (2015) *Ensifer*-mediated transformation: an efficient non-*Agrobacterium* protocol for the genetic modification of rice. Springerplus 4:600. https://doi.org/10.1186/s40064-015-1369-9

Chapter 17
Genome Editing and Rice Grain Quality

Mohamed Abdelrahman and Kaijun Zhao

17.1 Introduction

The world has added approximately one billion people in the span of the last 12 years, according to the most recent estimate by the United Nations (UN 2017). This growth along with other dietary factors is driving up global food demand. As a consequence, global food production should be doubled between 2005 and 2050 to fulfill the needs of the ever- growing population (Tilman et al. 2011). However, global production in the main crops, wheat, rice, soybean, and maize would increase in 2050 by ~67%, ~42%, ~38%, and ~55%, respectively, which is far below what is needed to meet projected demands (Ray et al. 2013). Rice is considered the predominant staple food at most of the developing countries, providing 25% of the caloric needs of the world's population (Kennedy et al. 2003; Kusano et al. 2015). Reduced arable land, water scarcity, and climate change are the main threats facing rice global production especially in developing countries. Hence, besides rice high yielding ability, it is required to develop new varieties with improved traits, such as increased quality, enhanced nutrition, disease resistance, stress tolerance, and reduced resource requirements.

Conventional breeding approaches for crop improvement like hybridization and radiation/chemical mutagenesis breeding have played a vital role in increasing crop productivity. World rice production has witnessed successful significant increase

M. Abdelrahman
National Key Facility for Crop Gene Resources and Genetic Improvement (NFCRI), Institute of Crop Science, Chinese Academy of Agriculture Sciences (CAAS), Beijing, China

Rice Research and Training Center, Field Crops Research Institute, Agricultural Research Center, Kafrelsheikh, Egypt

K. Zhao (✉)
National Key Facility for Crop Gene Resources and Genetic Improvement (NFCRI), Institute of Crop Science, Chinese Academy of Agriculture Sciences (CAAS), Beijing, China
e-mail: Zhaokaijun@caas.cn

A. Costa de Oliveira et al. (eds.), *The Future of Rice Demand: Quality Beyond Productivity*, https://doi.org/10.1007/978-3-030-37510-2_17

during the last half-century mainly due to large-scale adoption of high yielding semidwarf rice varieties (Khush 2005) as well as exploiting heterosis through production of hybrids using the three-line or cytoplasmic male sterile system in 1970s (Yuan 1987). However, the application of chemical pesticides and fertilizers needed for these varieties is costly and causes environmental problems (Abdel-Rahman et al. 2016), in addition to the expensive hybrid seeds. Furthermore, since mid-1980, the rate of increase of rice production has slowed down and currently reached a plateau and no significant increase in rice yield is observed.

In many rice-growing areas, rice breeders' efforts have focused on improving production neglecting grain quality traits mainly due to the difficulty associated with combining high yielding ability with superior grain quality (Peng et al. 2008). However, the adoption of new varieties by farmers is strongly driven by grain quality traits (Wang et al. 2012a; Dalton 2004). Such rice varieties with improved grain quality will be more acceptable to consumers (Fitzgerald et al. 2009; Juliano and Villareal 1993), providing profit to farmers from increased commercial value or higher price of high-quality rice and multiple uses in food processing industries (Hsu et al. 2014). The prime characteristics affecting rice grain quality are cooking and eating parameters, phytochemicals, and micronutrients. Desirable physical properties, cooking, and eating parameters are the most important grain quality characteristics.

In last few decades, powerful genetically modified (GM) crops have held great promise in overcoming the problems of conventional breeding approach and have strengthened the agricultural productivity in many crops and their quality. Golden rice is a well-known example, which is rice with extra three genes allowing rice to produce beta-carotene as a source of vitamin A (Paine et al. 2005) to fight its deficiency especially in the children with malnutrition problems of developing countries. Despite its valuable importance their adoption has been controversial due to social, environmental, and political issues (Lusser et al. 2012) which doesn't accept introducing foreign genes into the food crop.

Identification of molecular markers and their use for direct genotypic identification/selection of traits irrespective of the developmental stage of the plant (marker-assisted selection, MAS) has accelerated the rice breeding (Rao et al. 2014). The work on molecular breeding, i.e., MAS and identification of QTLs for grain quality traits, has been reviewed recently (Brar et al. 2012; Bao 2014; Rao et al. 2014). Further, with the availability of high-quality genomic sequence of rice (Goff et al. 2002; Yu et al. 2002), and recent improvements in rice, genomics research has made significant progress in developing functional genomics resources which have greatly accelerated identification and cloning of novel genes controlling rice yield and grain quality (Duan and Sun 2005; Jiang et al. 2012). Advances in genetic engineering have been dominated by the transfer of one or a few well-characterized desirable genes, affecting mainly the output traits such as herbicide and insect resistance, etc., in very precise and faster way to develop the first-generation genetically modified (GM) rice plants (Bajaj and Mohanty 2005; Kathuria et al. 2007; Dunwell 2014). Current efforts in developing rice with output traits including nutritional enhancement, the second-generation transgenics are on rise and are under advance stage of development. Thus, to overcome the food and nutritional security, there is an urgent need of new high yielding and superior quality rice varieties that are more resilient to stress/climate change and contain higher levels of bioavailable vitamins, essential amino acids, minerals, and phytochemicals that provide nutritional and health benefits (Khush 2005).

Recently, a novel technology with more potential innovation in overcoming the limitations of conventional breeding approaches in introducing targeted mutations in genomes with high efficiency and precision has been developed. Namely, genome editing with engineered nuclease (GEEN) tools that regulate the site of genomic modification by cleavage and DNA rejoining could help to speed up breeding processes by using site-specific endonucleases (SSE) to alter specific genes at a chosen site or by introduction of new traits at a given location. SSE acts to introduce double-strand breaks (DSBs) into the DNA at the target site, such as zinc finger meganucleases, zinc finger nucleases (ZFNs), transcription activator-like effector nucleases (TALENs), and CRISPR (clustered regularly interspaced short palindromic repeats)/Cas9 (CRISPR-associated protein 9). The efficiency has been raised tremendously by introducing each new tool for genome editing, more recently CRISPR/Cpf1 which was recently identified to overcome some limitations in Cas9. GEEN can save years or even decades of time and effort it takes to develop new crop varieties with preferred grain quality characteristics for farmers and consumers. This technology is more precise than GMO processes and keeps getting more reliable without cutting and pasting genes from animals or bacteria into plants. This technology is relatively cost-effective compared to other methods, and more scientists could gain access to it.

In this chapter we will survey research and development that has taken place in genome editing methodologies, reviewing genome editing with more emphasis on grain quality traits in rice. As rice is an economically and socially important crop, increased focus on these studies could result in an increased rate of improved rice discovery. We also speculate the perspectives and future utilization of the genome editing technology in rice grain quality improvement.

17.2 Overview Regarding Genome Editing

Plant biotechnology is now entering a new era where random mutagenesis methods are being dominated by genome editing using engineered nucleases (GEEN) (Belhaj et al. 2015). GEEN is an effective technology that uses artificially engineered sequence-specific nucleases (SSNs), to act as molecular scissors, enabling precise DNA manipulation at specific genomic sequence in a manner to help to speed up breeding processes by introduction of new traits at this site. Several artificially engineered nucleases have emerged in the last decade, including engineered homing endonucleases/meganucleases (MNs; Smith et al. 2006), zinc finger nucleases (ZFNs; Kim et al. 1996), transcription activator-like effector nucleases (TALENs; Boch et al. 2009), CRISPR (clustered regularly interspaced short palindromic

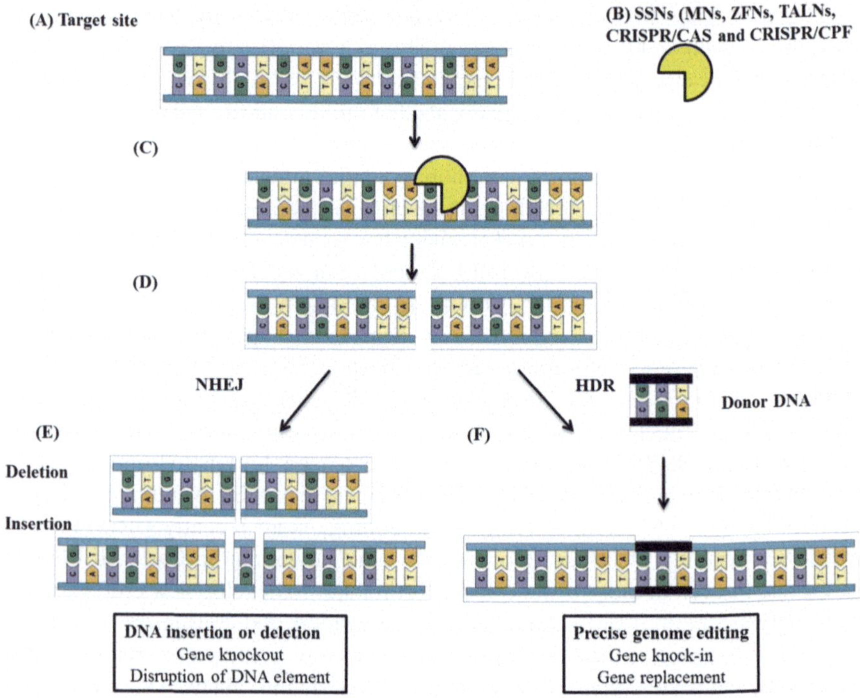

Fig. 17.1 Generating double-strand DNA breaks (DSB) at the target site in the genome using SSNs which is then repaired by either NHEJ or HDR pathways. Targeted genome, (**A**), editing in plants using engineered nucleases (GEEN), (**B**). Engineered nucleases are used to induce a double-strand DNA break (DSB) at a specified locus of the gene of interest (**C** and **D**). DSBs are repaired by either nonhomologous end joining (NHEJ) to introduce deletion or insertion (**E**) or homologous recombination (HR) for gene replacement (**F**)

repeats)/Cas9 (CRISPR-associated protein 9) (Jinek et al. 2012), and CRISPR/Cpf. These techniques are common in generating a double-strand DNA break (DSB) at the target site in the genome which is then repaired by either the error-prone nonhomologous end joining (NHEJ) or the error-free homology-directed repair (HDR) pathways (Fig. 17.1) (Hsu et al. 2013; Miglani 2017; Mishra et al. 2018).

17.2.1 NHEJ and HDR Double-Strand Break Repair

Repair of DSBs created by nucleases is an imperative step in the genome editing process as sequence modifications then occur at the cleaved sites, which can include deletions or insertions that result in gene disruption in the case of NHEJ or integration of exogenous sequences by HDR (Osakabe and Osakabe 2014).

In plants, DSBs are mainly repaired by NHEJ by a variety of enzymes. These enzymes have the ability to directly join the DSB ends without the presence of a homologous repair (Puchta 2004). In this stage error occurs, owing to its error-prone nature, since NHEJ repair often leads to the insertion or deletion (indels) of nucleotides and consequently DNA sequence alterations occurred at the target site (Fig. 17.1E). Deletion or insertion of nucleotide(s) through NHEJ causes a complete loss of gene function as an indels introduced in exons leading to missense or nonsense mutations, this affecting the gene reading frame resulting in gene knock-out (Perkin et al. 2016). NHEJ is the simplest method; all published grain quality gene editing work used the NHEJ pathway for gene loss of function so far. In contrast, HDR pathway requires the existence of a homologue sequence as a template to repair the DSBs for accurate error-free repair (Puchta 2004 and Perkin et al. 2016). Hence, HDR is conducted by giving an exogenously supplied donor DNA that comprises a prefered sequence of concern as a template surrounded by sequences homologous to both sides of the CRISPR/Cas9 DSB point(s), that will allow gene replacement/insertion into the target site (Fig. 17.1F). Unlike NHEJ which occurs throughout the cell cycle in higher eukaryotes and exhibits low fidelity in the repair (Mladenov and Iliakis 2011), HDR occurs mainly during the S/G2 phases of the cell cycle and has a much lower efficiency, making it more challenging in genome editing of plants (Puchta 2004).

17.2.2 Engineered Sequence-Specific Nucleases (SSNs)

The different SSN types MNs, ZFNs, TALENs, and CRISPR are different in the way they target the specific site in order to produce the DSB.

17.2.2.1 Meganucleases (MNs)

MNs are naturally occurring endonucleases from different organisms such as bacteria, fungi, and algae. They are unique enzymes with high activity and long recognition site with variable size differing between 14 and 30 bp resulting in site-specific digestion of target DNA (Epinat et al. 2003; Smith et al. 2006). Naturally occurring meganucleases can be modified by introducing changes to the amino acids found in the recognition site of the protein to adapt for restriction of a specifically chosen sequence. A further study described the synthesis of hybrid enzymes using two meganucleases, I-Cre I and I-Dmo I, that recognize new target sequences (Epinat et al. 2003). Specialized methods of mutagenesis and high-throughput screening have also been used to create novel meganuclease variants that recognize unique sequences and possess improved nuclease activity (Smith et al. 2006; Arnould et al. 2007; Grizot et al. 2009). Initially developed for the animal and human system, it has also been used for genome editing purposes in plants, including *Arabidopsis thaliana* and *Zea mays* (Gao et al. 2010; Antunes et al. 2012).

17.2.2.2 Zinc Finger Nucleases (ZFNs)

ZFNs were considered the first generation of genome editing tools being first exploited to edit plant genomes in 2005 (Lloyd et al. 2005). ZFNs are chimeric proteins composed of a synthetic sequence-specific zinc finger-based DNA-binding domain and a DNA cleavage domain derived from the type II most frequently used restriction endonuclease FokI (Kim et al. 1996). The DNA-binding domain contains several (3–6) linked zinc finger (ZF) motifs, each of which recognizes a 3 bp, allowing binding to a 9–18-bp-long specific DNA sequence (Fig. 17.2). ZFNs can be designed to cleave almost any long stretch of double- strand DNA by modification of the zinc finger DNA-binding domain (Durai et al. 2005; Camenisch et al. 2008). Individual FokI nuclease must dimerize to cleave double-strand DNA. Therefore, two individual ZFNs need to bind to the forward and reverse strands, respectively, and the two target sequences, forward and reverse, must be separated by a 5–7 bp spacer sequence (Bitinaite et al. 1998). The ZFN monomer can cut the target site if the two ZF-binding sites are palindromic. Genome modification using ZFNs has been successful in various organisms including plants, such as corn, tobacco, *Arabidopsis*, and soybean, although not all engineered ZFNs can create DSBs efficiently (Shukla et al. 2009; Townsend et al. 2009; Zhang et al. 2010; Sander et al. 2011).

In 2009, Shukla et al. engineered a zinc finger nuclease (ZFN) to modify *IPK1* gene, which is involved in regulation of biogenesis of phytic acid, aiming to increase kernel quality by declining of phytic acid accumulation in kernels. Phytic acid cannot be easily digested by humans and is considered as an anti-nutritional compound negatively impacting nutrient uptake and it also impacted the environment through its induction in waste stream. With the use of genome editing, significant reduction in phytic acid concentration can be achieved. ZFN has some drawbacks as the design and assembly of ZFNs are technically challenging and outsourcing the construction of ZFN modules to commercial suppliers is expensive (Ramirez et al. 2008). These shortcomings have greatly limited its widespread adoption by the scientific community.

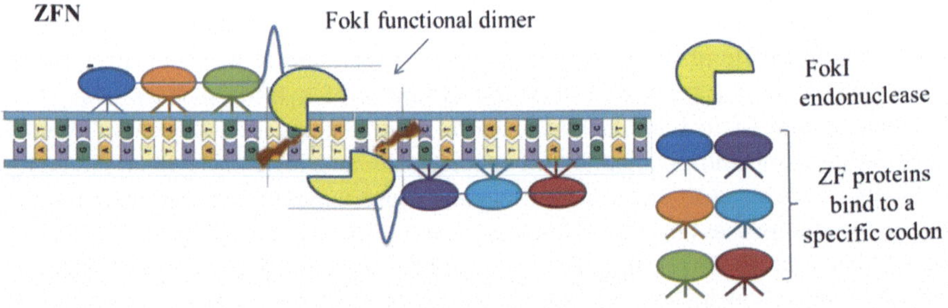

Fig. 17.2 Illustration of a pair of ZFNs bound to DNA. A zinc finger nuclease (ZFN) dimer bound to its target. Each ZFN contains the cleavage domain of FokI linked to an array of three to six zinc fingers (three are shown here) that have been designed to specifically recognize sequences that flank the cleavage site. A small number of bases (typically five or six) separate the ZFN targets

17.2.2.3 Transcription Activator-Like Effector Nucleases (TALENs)

TALEN is a recently developed platform for GEEN. It has rapidly emerged as an alternative to ZFNs for genome editing and introducing targeted DSBs. TALENs are similar to ZFNs and comprise a nonspecific FokI nuclease domain fused to a customizable DNA-binding domain (Joung and Sander 2013; Fig. 17.3). This DNA-binding domain is composed of highly conserved repeats derived from transcription activator-like effectors (TALEs), which are proteins that are secreted by various plant bacterial pathogens of the genus *Xanthomonas*. The *Xanthomonas* pathogens secrete TALEs into the host plant cell during infection. The natural function of TALE is to recognize and bind to a specific DNA sequence in the promoter region of a specific gene of the host genome when infection by the *Xanthomonas* occurred (Boch and Bonas 2010). The DNA binding capability of these proteins was first reported in 2007 (Römer et al. 2007). A typical TALE protein contains an N-terminal translocation signal, a C-terminal acidic transcription activation domain, and a central DNA-binding domain (Bogdanove et al. 2010; Miller et al. 2010; Joung and Sander 2013). The central DNA-binding domain is composed of several tandems of nearly identical 33–35 amino acid repeats which are highly conserved except for two hyper variable amino acid residues at positions 12 and 13, which are called repeat variable di-residues (RVDs) (Bogdanove et al. 2010). Each of the individual repeat units in the DNA- binding domain binds to a single nucleotide, and the RVDs in the repeats show a strong correlation with a specific nucleotide (A, G, C, or T). For example, the RVDs His-Asp (HD), Asn-Ile (NI), Asn-Gly (NG), and Asn-Asn (NN) recognize the nucleotides C, A, T, and G, respectively (Boch et al. 2009; Moscou and Bogdanove 2009; Deng et al. 2012). The co-crystal structures of TALE

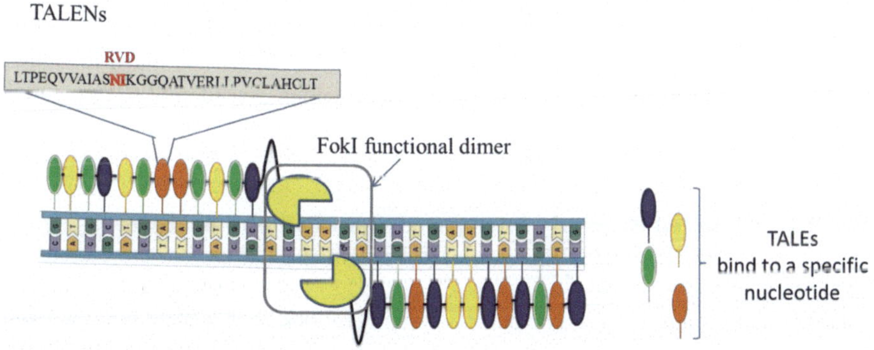

Fig. 17.3 Illustration of a pair of TALENs bound to DNA. Structurally TALE consists of 18 repeats of 33–35 amino acids. The repeat varies at amino acids 12 and 13. These variations are called as RVD (repeat variable di-residue). Different RVDs associate preferentially with different nucleotides, with the four most common RVDs (HD, NG, NI, and NN) accounting for each of the four nucleotides (C, T, A, and G, respectively). By fusing the TALE sequence to a Fok1 nuclease, the synthetic compound is now called a TALEN. However, Fok1 works only in a dimerized form and hence TALENs are always designed as pairs binding opposing strands of the DNA to allow dimerization of FokI in a spacer region that is bridging the two TALE binding sites

DNA-binding domains bound to their recognition sites revealed that individual repeats comprise two-helix v-shaped bundles that stack to form a superhelix around the DNA, and the RVDs are positioned in the DNA major groove. The residues at position eight and position 12 within the same repeat interact with each other, thereby possibly stabilizing the structure of the domain, whereas the residue at position 13 can make base-specific contacts with the DNA (Deng et al. 2012; Mak et al. 2012). This straightforward relationship between protein sequence and DNA recognition allows the engineering of specific DNA-binding proteins by selecting specific combinations of the four prominent repeats containing the appropriate RVDs.

TALENs have generated much interest and excitement over ZFNs because they can be easily and rapidly designed using a simple "protein-DNA code" that relates modular DNA-binding TALE repeat domains to individual bases in a target binding site. This resulted in a wider adoption of the editing method in plants, including *Arabidopsis*, rice, *Brachypodium*, barley, maize, tobacco, soybean, wheat, tomato, potato, and sugarcane (Cermak et al. 2011; Mahfouz et al. 2011; Li et al. 2012a, b; Christian et al. 2013; Shan et al. 2013; Wendt et al. 2013; Zhang et al. 2013; Haun et al. 2014; Liang et al. 2014; Lor et al. 2014; Wang et al. 2014a; Nicolia et al. 2015; Jung and Altpeter 2016).

TALENs have been used for improving quality of soybean oil by improving oleic acid level through slowing down the fatty acid desaturase (FAD2 and FAD3) genes activity (Haun et al. 2014). In their study, they developed mutant genotypes with high amount of healthy oleic acid (60% increase) and decreased amount of poly-saturated fatty acids (4% decrease) compared to the WT. The merits behind editing *FAD2* and *FAD3* genes are to produce plants which are healthier and high-quality oil producing as well as have improved shelf life. In another study, TALENs had been used to improve the characteristic of commercial potato. This was conducted by knocking out vacuolar invertase gene (*VInv*), which encodes a protein that cleaves sucrose to glucose and fructose (Clasen et al. 2015). The knockout lines resulted in lower levels of reducing sugars and lacked foreign DNA. This will allow the long-term storage without the formation of acrylamide which considered as a potential carcinogenic.

17.2.2.4 Clustered Regularly Interspaced Short Palindromic Repeats (CRISPR)/CRISPR-Associated Protein (Cas) 9

CRISPR/Cas9 is a novel genome editing system that has been widely accepted and has become extensively common tool for genome editing, because of its simplicity, efficiency, ease of use, and applicability. CRISPR is adaptive RNA-mediated bacterial and archaeal immune system for sequence-specific detection and silencing of foreign nucleic acids, including viruses and plasmids (Barrangou et al. 2007; Jinek et al. 2012; Cong et al. 2013). It is unique and flexible owing to its dependence on RNA as a guide that targets the nuclease to a desired DNA sequence. In contrast to ZFN and TALEN methods, which use protein-DNA interactions for targeting, RNA-guided nucleases (RGNs) use simple, base- pairing rules between an engineered RNA and the target DNA site. Its discovery was in 1980s when Ishino and coworkers founded a 29-nucleotide repeat sequence in the *E. coli* genome while working on the *iap* gene (Ishino et al. 1987).

The CRISPR/Cas9 system is transcribed from the CRISPR array which encodes the Cas9 protein, CRISPR-derived RNAs (crRNAs), which are complementary to the exogenous DNA, and a trans-acting antisense RNA (tracrRNA). TracrRNA forms an RNA hybrid (crRNA-tracrRNA complex) with complementarity to the crRNAs. Then, crRNA-tracrRNA complex guides the Cas9 nuclease to cleave the DNA and called guide RNA (gRNA) (Datsenko et al. 2012; Jinek et al. 2012; Mishra and Zhao 2018). Moreover, the sequence motif "NGG" downstream of the protospacer PAM (protospacer adjacent motif, a short nucleotide sequence at the 3′ end of the target sequence in the genome; Fig. 17.4) is crucial for binding (Jinek et al. 2012). gRNA-Cas9 complex searches for protospacer sequence by searching the target DNA for the PAM sequence and triggers DSB in the target sequence (Bolotin et al. 2005; Marraffini and Sontheimer 2008; Mojica et al. 2005; Garneau et al. 2010; Cong et al. 2013; Hsu et al. 2014).

Using this system, universal genetic constructs encoding artificial elements of CRISPR/Cas "genome editor" have been produced (Doudna and Charpentier 2014). Moreover, this system has the advantage that the researchers need to make only a single construct for expression or RNA synthesis of gRNA (Jinek et al. 2013, 2014). Therefore, the CRISPR/Cas9 system is able to target and create DSBs in any genomic DNA sequence using a specifically designed gRNA together with a PAM sequence. Therefore, due to its ease, efficacy, and extensive abilities, in a little time

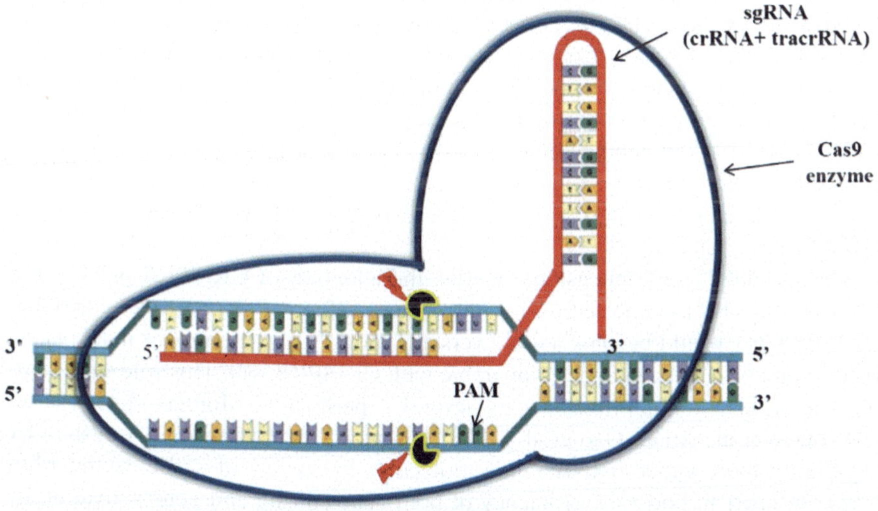

Fig. 17.4 Components of the CRISPR/Cas9 system for directing Cas9 endonuclease to genomic targets. The cleavage site is specified by the protospacer element of the sgRNA (crRNA+ tracrRNA). The gRNA protospacer element recognizes 19 or 20 nt on the opposite strand of the NGG PAM site

CRISPR/Cas9 system has previously found use in numerous fields of fundamental and applied biology, biotechnology, and genetic engineering. Phytic acid was also targeted in maize using CRISPR/Cas9 as Liang et al. (2014) designed two gRNAs targeting the *ZmIPK* (inositol phosphate kinase) gene whose protein catalyzes a key step in PA biosynthetic pathway causing decrease of phytic accumulation in mutant kernels.

New CRISPR system with a single sequence-specific nuclease has recently been identified including CRISPR from *Prevotella* and *Francisella* 1 (Cpf1) effector nucleases (Zetsche et al. 2015; Shmakov et al. 2015; Abudayyeh et al. 2016; Liu et al. 2017). CRISPR/Cpf1 is an RNA-guided, class II CRISPR/Cas system type V endonuclease that is analogous to CRISPR/Cas9. Cpf1 is generally smaller than most of Cas9 orthologs that has a protospacer adjacent motif (PAM) sequence with a high frequency in the genome. In contrast to the G-rich PAM of Cas9, Cpf1 depends on a T-rich PAM. 5′-TTTN-3′ is present in the 5′ end of the target site and requires a single crRNA of ~ 44 nucleotide length for cleavage. In addition, Cpf1 creates a sticky end double-strand break in a staggered manner at the PAM-distal position, which may provide additional advantages for the efficiency of gene insertion particularly for knock-in strategies. Cpf1 provide additional advantages regarding safety issues as its enzymes have been shown to have lower rates of off target activity as compared to Cas9 nucleases (Kleinstiver et al. 2016; Kim et al. 2016, 2017).

Cpf1 system differs in several important ways from the previously described Cas9. In its natural form, the DNA-cutting enzyme Cas9 forms a complex with two small RNAs, both of which are required for the cutting activity. The Cpf1 system is simpler in that it requires only a single RNA. The Cpf1 enzyme is also smaller than the standard SpCas9, making it easier to deliver into cells and tissues. Most significantly, Cpf1 cuts far away from the recognition site (Fonfara et al. 2016), meaning that even if the targeted gene becomes mutated at the cut site, it can likely still be recut, allowing multiple opportunities for correct editing to occur. Cpf1 system was used for multiplexed genome editing in rice by editing four *OsBEL* genes (Wang et al. 2017). More recently, Yang and colleagues have successfully generated stable and heritable mutations in rice by selecting two genome targets in the *OsPDS* and *OsBEL* genes (Xu et al. 2017).

Despite these multidimensional merits, the adoption of CRISPR/Cpf1 has not been as explosive as expected for the past 2 years. Moon et al. (2018) assumed that CRISPR/Cpf1 would become a more versatile tool if its efficiency for indel occurrence could be enhanced to be comparable with CRISPR/Cas9. This sole shortcoming was tackled for improvement by several experiments (Moreno-Mateos et al. 2017; Gao et al. 2018). Gao et al. (2018) optimized the CRISPR/Cpf1 system by expressing more exact crRNA guide molecules by means of self-cleaving ribozymes in order to boost the efficiency of both gene editing and gene activation by CRISPR/Cpf1 systems.

17.3 Genome Editing for Rice Grain Quality Improvement

Rice grain quality is all features and characteristics of rice variety that meets the demand and preferences of the consumer. The market value of rice and the adoption of new rice cultivars by farmers is strongly based on grain yield and grain features (Wang et al. 2012a; Dalton 2004). The concept of grain quality covers consumer preferences for particular physical, cooking and eating, and nutritional characteristics (Fitzgerald et al. 2009; Juliano and Villareal 1993). Figure 17.5 describes the main categories of rice grain quality. Many features of physical properties include uniform grain length, width, weight, color (whiteness and translucence), and chalk content. Desirable cooking and eating characteristics are cooking time (gelatinization temperature and viscosity) (Juliano 2003), the ability of rice to remain soft after cooking (gel consistency) (Philpot et al. 2006), textural properties of the cooked rice (amylose content) (Cagampang et al. 1973; Pandey et al. 2012), and aroma (Fitzgerald et al. 2008). Aroma is the most important quality trait that the consumer focuses on and in turn brings the highest economic gain for the farmer (Fitzgerald et al. 2008; Kovach et al. 2009).

It is a challenge for plant breeders to use conventional breeding in developing high yielding genotypes with premium grain quality traits and has been met with limited success, as it is very time-consuming and sometimes combined with adverse genes due to linkage drag. The more challenge for plant conventional breeders is that varieties must be developed to meet regional quality preferences. The world economic development is changing in many countries and the demand for specific quality traits will be increased. Traditional plant breeding techniques cannot afford this demand. However, our information regarding genetic basis underlying many grain quality traits is increasing (Fitzgerald et al. 2009; Sreenivasulu et al. 2015; Bandyopadhyay et al. 2019). Quantitative trait loci (QTL) controlling different grain qualities have been reported (Sreenivasulu et al. 2015; Peng et al. 2014; Fan et al. 2006; Weng et al. 2008; Kwon et al. 2011; Liu et al. 2011; Isshiki et al. 1998; Wang et al. 2008) and reviewed elsewhere (Fitzgerald et al. 2009; Sreenivasulu et al. 2015; Bradbury et al. 2005). All the available

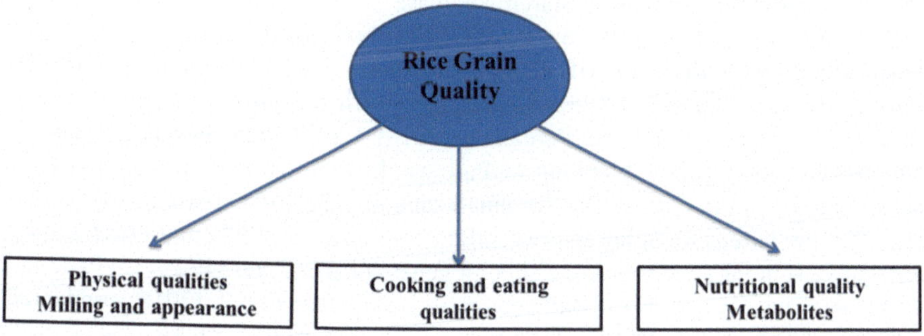

Fig. 17.5 Main features of rice grain quality

knowledge deciphers this challenge for the breeders and further increases the potential to use biotechnology to develop high yielding varieties carrying these traits (Lucca et al. 2001; Bhullar and Gruissem 2013). Moreover, with the current information about rice genome sequence through rice genome annotation project (2005), 3 k rice genome project (2014) and the new cutting-edge bioinformatics tools available make the challenges for molecular breeders then become how best to use this available information. Georges and Ray (2017) describe the differences and limitations underlying the classical plant breeding and genetic modification. At the same time, the major drawbacks with conventional genetic modification technologies are consumer acceptance and the fact that they cannot eliminate the endogenous target gene; thus there is an emerging role for modern tools such as targeted genome editing technologies.

Genome editing tools afford the ability to efficiently and precisely manipulate the genome to induce a desired trait. Back to grain quality and the possibilities to develop high yielding genotypes with desirable grain quality traits will be increased with technology of genome editing.

17.3.1 Eating and Cooking Quality Traits

To date multiple successful genome editing studies have been conducted targeting grain quality traits in rice. Eating and cooking quality traits had been in the focus of the scientists using genome editing techniques; this might be because of its importance as it largely determines consumer appeal and economic value of rice (Hsu et al. 2014). These traits are based on the physicochemical properties of the major component in rice grain endosperm starch. Starch is a mix of two glucans: amylose and amylopectin. Amylose content (AC) refers to the percentage of starch instead of the percentage of total weight in the grain and is arguably the most important quality indicator in rice, especially for cooking and eating quality (Juliano 1998). Based on AC values, rice is commercially classified into five groups: waxy (0–5%), very low (5–12%), low (12–20%), intermediate (20–25%), and high (25–33%) AC (Juliano 1992). Starch metabolism in plants requires the enzymes ADP-glucose pyrophosphorylase (AGPase), starch synthase (SS), starch branching enzyme (SBE), and starch debranching enzyme (SDBE) (Fiaz et al. 2019; Wang et al. 2014b; Dian et al. 2005) with SS and BE playing key roles in starch biosynthesis (Qu et al. 2018). AGPase catalyzes the first step in the starch biosynthesis pathway (Tang et al. 2016; Lee et al. 2016a, b; Pérez et al. 2018). Rather than altering the ratio of amylose and amylopectin, blocking AGPase would therefore prevent starch synthesis all together. In rice, there are two small subunit genes of AGPase (*OsAPS1* and *OsAPS2*, the latter producing the two mRNA variants *OsAPS2a* and *OsAPS2b* by alternative splicing) and four large subunit genes (*OsAPL1*, *OsAPL2*, *OsAPL3*, and *OsAPL4*) (Tuncel et al. 2014; Ballicora et al. 2003). Pérez et al. (2018) conducted genome editing using CRISPR/Cas9 targeting two mutations in *OsAPL2* gene, they observed a significant decline in starch content in the leaves of both mutants, and further increase was noticed in the level of soluble sugars. One of the mutations showed a remarkable decrease in starch content to 2% by weight, which equals around 15% of the normal levels found in wild-type leaves. Pérez and his group, however, conducted their measures on rice leaves, but interestingly their results indicated that starch synthesis in leaves remains dependent on the plastidial AGPase even when cytosolic large and small subunits are expressed. Furthermore, after heading plant starch is transformed to sucrose for translocation to grain, enzymes related to conversion of starch to sucrose are the regulating factors that affect non-structural carbohydrate remobilization in rice (Yang et al. 2001). Later, during grain developing, grain sucrose is converted to starch for grain filling (Yang et al. 2004; Braun et al. 2014). Rapid sucrose to starch conversion enhances also carbohydrate translocation into grains which in turn affect grain physical and chemical properties.

For starch synthesis (SS) controlling enzymes they are diveded into two main classes of enzymes based on their function in building up the starch two main components (amylose and amylopecting). Granule bound synthases (GBSS) are responsible mainly for the synthesis of amylose and contribute to the formation of the extra-long-chain fraction of amylopectin while soluble starch synthases (SSS) are the main enzymes for amylopectin synthesis (Denyer et al. 1999; Lin et al. 2012; Qu et al. 2018). Ma et al. (2015) knocked out *OsWaxy* gene encoding a GBSS enzyme that is responsible for amylose synthesis using CRISPR/Cas9 in Taichung 65 cultivar, causing a decrease in the amylose content from 14.6% in the wild type to 2.6% in the T1 generation mutants, similar to a natural glutinous rice variety. Markedly, Zhang et al. (2018) produced two glutinous and transgene-free lines while maintaining all the desired agronomic traits by introducing a loss-of-function mutation into the *Waxy* gene using CRISPR/Cas9 in two widely cultivated elite japonica varieties, Xiushui134 and Wuyunjing 7. The mutants had a "waxy" appearance, being white and fully opaque in contrast with the typical "non-waxy" translucent appearance of their wild type. Moreover, there was a twofold increase in gel consistency and a marked reduction in gelatinization temperature for CRISPR waxy seeds, compared to their respective wild type. Amylose is an essentially linear glucan that contains very few alpha-1,6-branches and makes up to 20–30% of normal starch, while amylopectin is more highly branched. Branch points are introduced by two classes of starch branching enzymes (SBE I and II), which differ in terms of length of the glucan chains transferred and substrate specificities. SBE plays a major role in determining the fine structure and physical properties of starch. High-amylose rice mutants have been produced by targeting the genes encoding SBEI and SBEIIb using CRISPR/Cas9 (Sun et al. 2017). Biallelic T0 mutants with insertions and deletions at the target sites were generated and the mutations were stably transmitted to progeny. The *OsBEIIb* mutants accumulate higher proportion of amylose and debranched amylopectin in the seeds than normal (Sun et al. 2017).

Interestingly, starch synthesis also involves two types of debranching enzymes, which cleave branch points and are probably involved in tailoring the branched glucans into a form capable of crystallization within the granule. ISA, isoamylase-type debranching enzyme, plays a predominant role in amylopectin synthesis. Chao et al. (2019) generated ISA1-deficient mutants (*isa1*) using the CRISPR/Cas9 genomic editing system. Such mutants display shrunken endosperm, significantly reduced amylose content and increased content of total soluble sugar, altered expression of genes associated to starch synthesis, and grains with reduced weight and altered starch granule morphology.

Furthermore, fragrant rice (*Oryza sativa*) is favored worldwide for the characteristic fragrance of its grains. Its market price particularly that of the Indian Basmati and the Thai Jasmine types is much higher than that of the conventional non-fragrant rice (Bhattacharjee et al. 2002). The presence of a defective *badh2* allele encoding betaine aldehyde dehydrogenase (BADH2) results in the synthesis of 2-acetyl-1-pyrroline (2AP), which is the most abundant and is considered responsible for the fragrance. No apparent 2AP (at least about two orders of magnitude less) is detected in non-fragrant rice (Jezussek et al. 2002; Lorieux et al. 1996). TALENs were engineered to target disruption of the *OsBADH2* gene by Shan et al. (2015). The 2AP content of rice grains of the T1 lines with homozygous mutations increased from 0 to 0.35–0.75 mg/kg, which was similar to the content of the positive control, fragrant rice cultivar Daohuaxiang, harboring the *badh2-E7* gene.

17.3.2 Physical and Appearance Quality Traits

Rice grain morphology, notably grain size or shape (GS), is one of the key determinants of both yield and quality. In general, large grain size is associated with high yield but poor quality, especially the appearance of milled rice, such as chalkiness (Sakamoto and Matsuoka 2008; Harberd 2015). Zhao et al. (2018) identified *GS9* (Grain Shape Gene on Chromosome 9) gene by map-based cloning. Introducing the *gs9* allele into elite rice cultivars significantly improves grain shape and appearance quality. They used CRISPR/Cas9 to produce a lake of function in *GS9* in Nipponbare with a 1 bp insertion and subsequent frameshift mutation in the coding region of *LOC_Os09g27590*. The NPB-cas9-1 mutant grew normally, but presented a slender grain as that of normal *gs9* line with a 4% increase in length, 10% decrease in width, and 23% increase in length-width ratio comparing to their respective wild-type milled grain. In the same study they confirmed the roll of *GS9* and its *gs9* allele slender grain shape effect using the same approach they edited Yandao 8 (Y8), another high-yield cultivar. One homologous null *gs9* mutant, Y8-cas9-1, was identified with a 1 bp insertion, similar with NPB-cas9-1 mutant in Nipponbare background. As expected, the grain shape was significantly slender than that of the non-mutated control (Y8-cas9-WT), with a 5.5% increase in length and 9.1% decrease in width and no changes in grain thickness. More importantly, the milled rice of Y8-cas9-1 was also much slender than that of Y8-cas9-WT and resulted in significant decrease of chalkiness. These findings confirmed the great potential for editing application of null *gs9* allele to improve grain appearance quality in high-yield but poor appearance rice cultivars.

Shen et al. (2018) designed and constructed a CRISPR/Cas9 vector that targets the disruption of both *GS3* (GRAIN SIZE3), the first QTL characterized to regulate grain size (Fan et al. 2006), and *Gn1a* (Grain number 1a Gn1a), the first QTL identified to control the grain number (Ashikari et al. 2005). Five widely cultivated japonica varieties, namely, Nanjing 9108 (N9108), Wuyunjing 27 (W27), Yangjing 4227 (Y4227), Zhejing 22 (Z22), and Zhejing 88 (Z88), were chosen for targeted mutation. They quantified the effects of loss of function of *GS3* and *Gn1a* on grain size and grain number in all of the tested japonica varieties. The grain lengths of the *gs3* and *gs3gn1a* mutants in all of the genetic backgrounds were increased compared to the WT. Similar results were obtained by Li et al. (2016) when they constructed CRISPR/Cas9 with the aim of creating defined deletions in *GS3* region. They found that frameshift mutants of *GS3* in Zhonghua11, *gs3-4* (4-base deletion) and *gs3-5* (5-base deletion), showed larger grain size. The grain length in the mutants was significantly increased to more than 8 mm compared with the wild-type average of 6.4 mm, and consequently the weight of the grain was increased to more than 31 mg on average, in contrast with 25.7 mg in wild type. Similarly, Xu et al. (2016) disrupted the function of three other important genes *GW2*, *GW5*, and *TGW6*, negative regulators of grain weight, through CRISPR/Cas9-mediated multiplex genome editing. The obtained results indicated that the genome editing of these genes significantly increased grain size and thousand grain weight.

17.3.3 Nutritional and Element Composition Quality Traits

Breeding rice cultivars with specific grain element composition is very important to prevent diseases through decreasing some elements as well as to increase nutrient value gained from other beneficial elements. In an attempt to develop new indica rice lines with low Cd accumulation in rice grain, Tang et al. (2017) used CRISPR/Cas9 to knock out the metal transporter gene *OsNramp5*. In this regard they obtained *osnramp5* mutants in the backgrounds of Huazhan (HZ) and Longke 638S (638S), both of which are extensively used indica parental lines in the current two-line hybrid rice breeding. The transgene-free homozygous mutants had extremely low Cd content in grains and did not show stunted growth or reduced yield when grown in highly Cd-contaminated paddy fields. Cd-contaminated paddy field trials demonstrated that Cd concentration in grains (brown rice) was <0.05 mg/kg, compared with an average of 0.33 mg/kg and 2.90 mg/kg in the grains of WT plants. The analysis of other relevant metals in grains showed that mutants accumulate 34–53% less Mn than in WT plants. Unexpectedly, the concentration of Fe was higher in the mutants, but no significant differences in Cu or Zn concentration were observed between mutants and WT plants.

With the aim of increasing β-carotene accumulation in rice endosperm, Yang et al. (2017) used CRISPR/Cas9 to knock out five rice carotenoid catabolic genes (*OsCYP97A4, OsDSM2*, *OsCCD4a*, *OsCCD4b*, and *OsCCD7*). Their results indicated that in the five catabolism genes, mutations failed to boost carotenoid accumulation in rice seeds.

17.4 Conclusion of Current and Future Prospective

The genome editing systems hold great promise in generating crop varieties with improved yield, enhanced grain quality characteristics, and novel agronomic traits which will be beneficial for farmers and consumers. CRISPR/Cas9 system was the most utilized system for enhancing the grain quality traits; TALENs and CRISPR/Cpf1 are also precise approaches to improve traits including enhanced nutritional quality and commercial valuable compounds. Given the high precision nature of genome editing techniques, genes encoding feedback-sensitive enzymes in the amino acid biosynthesis pathways can be edited to become insensitive to feedback inhibition.

Table 17.1 summarizes the progress conducted for rice grain quality improvement by genome editing, which showed that CRISPR/Cas9 dramatically increases the potential to improve grain quality in crops compared to other approaches. It is a powerful tool for introducing heritable, trait-related mutations indistinguishable from natural allelic variants (Schaeffer and Nakata 2015). There are several other identified genes for these traits, and their interactions and functions have been fully characterized, such as *GIF1* (Wang et al. 2008), *qSW5* (Shomura et al. 2008), *GS5* (Li et al. 2011), *qGL3* (Zhang et al. 2012), *GW8* (Wang et al. 2012b), *GS6* (Sun et al. 2013), *GS2* (Hu et al. 2015), *GL7/GW7* (Wang et al. 2015a; Wang et al. 2015b), *OsMAPK6* (Liu et al. 2015), and *GLW7* (Si et al. 2015). Some of them have been fine mapped such as *GW1–1* and *qGRL1.1* (Singh et al. 2012), *GW3* and *GW6* (Guo et al. 2009), *qGL-7* (Bai et al. 2010), and *qGRL7.1* (Singh et al. 2012).

For grain chalkiness, only one gene, *Chalk5*, is cloned (Li et al. 2014). Recently, *qBRR-10* for brown rice recovery (BRR) was narrowed to a 39.5 kb region on chromosome 10 and two candidate genes were determined (Ren et al. 2016). The incorporation of these genes through knock in/out using CRISPR/Cas9, and the assessment of their interactions among each other, can greatly improve our understanding of rice grain appearance and milling quality. Similarly, for eating, cooking, and nutritional grain quality traits, several genes and transcription factors are affecting the efficiency of Wx gene such as a tetratricopeptide domain-containing flo2 protein (Wu et al. 2015) and the transcription factors like *rsr1*, *OsBP-5*, *OsEBP-89*, *OsbZIP58*, and *OsMADS7* (Zhang et al. 2017; Wambugu et al. 2018). On the other hand, Zheng et al. (2010) and Johnson et al. (2011) identified the importance of NAS genes (*OsNAS1*, *OsNAS2*, and *OsNAS3*) in increasing Fe and Zn accumulation in polished rice grain.

CRISPR/Cas9 genome editing system targeting multiple loci provide a straightforward mean of disrupting the functions of different genes of interest simultaneously. Furthermore, crossing or selfing the edited plants can generate transgene-free progeny, which will reduce the potential risks and regulatory requirements that may apply to transgenic crops (Scheben et al. 2017). CRISPR multiplex genome editing system has been well adapted for the rapid generation and pyramiding of beneficial alleles. Pyramiding via multiplex gene editing will strongly facilitate the investigations of complex traits and increase their applications in crop breeding (Xu et al. 2016).

Table 17.1 Summary of genes targeted for genome editing for different grain quality traits

Quality traits	Gene name	Editing tool	Gene function	Resulted phenotype	Reference
Eating and cooking quality	*OsAPL2*	CRISPR/Cas9	AGPase (starch biosynthesis)	Starch depletion in leaves	Pérez et al. (2018)
	OsWaxy	CRISPR/Cas9	GBSS (amylose synthesis)	Decrease in amylose content (glutinous rice production)	Ma et al. (2015)
	OsWaxy	CRISPR/Cas9	GBSS (amylose synthesis)	Decrease in amylose content (glutinous rice production)	Zhang et al. (2018)
	OsBEI and *OsBEIIb*	CRISPR/Cas9	Starch branching enzyme	High amylose content	Sun et al. (2017)
	IAS	CRISPR/Cas9	Starch (isoamylase-type) debranching enzymes	Reduced amylose content and increased total soluble sugar	Chao et al. (2019)
	OsBADH2	TALENs	Betaine aldehyde dehydrogenase (fragrant rice)	Increased 2AP content (fragrance)	Shan et al. (2015)
Physical and appearance quality	*GS9*	CRISPR/Cas9	Grain size-9	Slender grain shape	Zhao et al. (2018)
	GS3/Gn1a	CRISPR/Cas9	Grain size-3/grain number 1a	Increased grain length	Shen et al. (2018)
	GW2/GW5/TGW6	CRISPR/Cas9	Grain weight 2/grain weight 3/thousand-grain weight 6	Increased grain length and width	Xu et al. (2016)
Nutritional and element composition quality	*OsNramp5*	CRISPR/Cas9	Cd accumulation	Low Cd accumulations in grains	Tang et al. (2017)
	OsCYP97A4, *OsDSM2*, *OsCCD4a*, *OsCCD4b* and *OsCCD7*	CRISPR/Cas9	Carotenoid catabolic genes	Mutations failed to boost carotenoid accumulation in rice seeds	Yang et al. (2017)

The CRISPR/Cpf1 system is a recent addition to the toolbox of sequence-specific nucleases. The simplicity and robustness of this system makes it as an attractive genome editing tool for plant biology. It is very promising especially in the studies which aim at gene knock-in. Still it requires some investigations to enhance its mutation efficiency to be comparable to CRISPR/Cas9.

CRISPR/Cas9 will be a great tool for rice biofortification to overcome nutrient starvation especially for poor populations. This could be achieved by knocking out genes causing nutrient deterioration in grains through degradation processes. Also enhancing grain nutrient composition needs more focus in the upcoming years especially as rice is one of the world's most demanding staple food, by developing rice high in nutrient values that will facilitate mitigation of malnutrition in developing countries with an inexpensive and sustainable solution. More efforts are required to work on waxy gene in rice to manipulate AC especially in hybrid rice, by editing waxy gene in either the sterile or restorer lines.

Acknowledgment This research was supported by grants from the China Priority Program-Breeding of Seven Major Crops (2017YFD01100100), the Innovation Program of Chinese Academy of Agricultural Sciences (to K. Zhao), and the Talented Young Scientist Program of China (to M. Abdelrahman).

Conflict of Interest Statement The authors declare that they have no conflict of interest.

Author Contributions The manuscript was written by M. Abdelrahman with the supervision of K. Zhao. The manuscript was critically revized by K. Zhao.

References

3,000 Rice Genomes Project (2014) The 3,000 rice genomes project. GigaScience 3(1). https://doi.org/10.1186/2047-217X-3-7.

Abdel-Rahman MA, El-Denary ME, Ammar M, Abdelkhalik AF, Draz AE, Dora SA (2016) QTL validation for grain yield and nitrogen use efficiency under different nitrogen levels in rice. Egypt J Gen Cytol 44(2). Available at http://journal.esg.net.eg/index.php/EJGC/article/view/191/194.

Abudayyeh OO, Gootenberg JS, Konermann S, Joung J, Slaymaker IM, Cox DB, Shmakov S, Makarova KS, Semenova E, Minakhin L, Severinov K, Regev A, Lander ES, Koonin EV, Zhang F (2016) C2c2 is a single-component programmable RNA-guided RNA- targeting CRISPR effector. Science 353(6299):aaf5573. https://doi.org/10.1126/science.aaf5573

Antunes MS, Smith JJ, Jantz D, Medford JI (2012) Targeted DNA excision in Arabidopsis by a re-engineered homing endonuclease. BMC Biotechnol 12(1):86. https://doi.org/10.1186/1472-6750-12-86

Arnould S, Perez C, Cabaniols JP, Smith J, Gouble A, Grizot S, Epinat JC, Duclert A, Duchateau P, Pâques F (2007) Engineered I-CreI derivatives cleaving sequences from the human XPC gene can induce highly efficient gene correction in mammalian cells. J Mol Biol 371(1):49–65. https://doi.org/10.1016/j.jmb.2007.04.079

Ashikari M, Sakakibara H, Lin S, Yamamoto T, Takashi T, Nishimura A, Angeles E, Qian Q, Kitano H, Matsuoka M (2005) Cytokinin oxidase regulates rice grain production. Science 309(5735):741–745. https://doi.org/10.1126/science.1113373

Bai X, Luo L, Yan W, Kovi MR, Zhan W, Xing Y (2010) Genetic dissection of rice grain shape using a recombinant inbred line population derived from two contrasting parents and fine mapping a pleiotropic quantitative trait locus qGL7. BMC Genet 11:16. https://doi.org/10.1186/1471-2156-11-16

Bajaj S, Mohanty A (2005) Recent advances in rice biotechnologytowards genetically superior transgenic rice. Plant Biotechnol J 3:275–307. https://doi.org/10.1111/j.1467-7652.2005.00130.x.

Ballicora MA, Iglesias AA, Preiss J (2003) ADP-glucose pyrophosphorylase: a regulatory enzyme for plant starch synthesis. Microbiol Mol Biol Rev 67:213–225. https://doi.org/10.1023/B:PRES.0000011916.67519.58

Bandyopadhyay A, Yin X, Biswal A, Coe R, Quick W (2019) CRISPR-Cas9-mediated genome editing of rice towards better grain quality. In: Sreenivasulu N (ed) Rice grain quality. Humana Press, New York, pp 311–336. https://doi.org/10.1007/978-1-4939-8914-0_18

Bao, J. (2014) Genes and QTLs for rice grain quality improvement. In: Yan W (ed) Rice – germplasm, genetics and improvement, ISBN: 978–953–51–1240-2, InTech. Doi: https://doi.org/10.5772/56621. Available at. http://www.intechopen.com/books/rice-germplasm-genetics-andimprovement/genes- and-qtls-for-rice-grain-quality-improvement.

Barrangou R, Fremaux C, Deveau H, Richards M, Boyaval P, Moineau S, Romero DA, Horvath P (2007) CRISPR provides acquired resistance against viruses in prokaryotes. Science 315(5819):1709–1712. https://doi.org/10.1126/science.1138140.

Belhaj K, Chaparro-Garcia A, Kamoun S, Patron NJ, Nekrasov V (2015) Editing plant genomes with CRISPR/Cas9. Curr Opin Biotechnol 32:76–84. https://doi.org/10.1016/j.copbio.2014.11.007

Bhattacharjee P, Singhal RS, Kulkarni PR et al (2002) Basmati rice: a review. Int J Food Sci Technol 37:1–12. https://doi.org/10.1046/j.1365-2621.2002.00541.x

Bhullar NK, Gruissem W (2013) Nutritional enhancement of rice for human health: the contribution of biotechnology. Biotechnol Adv 31:50–57. https://doi.org/10.1016/j.biotechadv.2012.02.001

Bitinaite J, Wah DA, Aggarwal AK, Schildkraut I (1998) FokI dimerization is required for DNA cleavage. Proc Natl Acad Sci U S A 95:10570–10575. https://doi.org/10.1073/pnas.95.18.10570

Boch J, Bonas U (2010) Xanthomonas AvrBs3 family-type III effectors: discovery and function. Annu Rev Phytopathol 48:419–436. https://doi.org/10.1146/annurev-phyto-080508-081936

Boch J, Scholze H, Schornack S, Landgraf A, Hahn S, Kay S, Lahaye T, Nickstadt A, Bonas U (2009) Breaking the code of DNA binding specificity of TAL-type III effectors. Science 326:1509–1512. https://doi.org/10.1126/science.1178811

Bogdanove AJ, Schornack S, Lahaye T (2010) TAL effectors: finding plant genes for disease and defense. Curr Opin Plant Biol 13:394–401

Bolotin A, Quinquis B, Sorokin A, Ehrlich SD (2005) Clustered regularly interspaced short palindrome repeats (CRISPRs) have spacers of extrachromosomal origin. Microbiology 151:2551–2561. https://doi.org/10.1099/mic.0.28048-0

Bradbury LMT, Fitzgerald TL, Henry RJ, Jin Q, Waters DLE (2005) The gene for fragrance in rice. Plant Biotechnol J 3(363–370):30. https://doi.org/10.1111/j.1467-7652.2005.00131.x

Brar DS, Virk PS, Grewal D, Slamet-Loedin I, Fitzgerald M, Khush GS (2012) Breeding rice varieties with improved grain and nutritional quality. Qual Assur Saf Crops Foods 4:137

Braun DM, Wang L, Ruan YL (2014) Understanding and manipulating sucrose phloem loading, unloading, metabolism, and signalling to enhance crop yield and food security. J Exp Bot 65:1713–1735. https://doi.org/10.1093/jxb/ert416

Cagampang GB, Perez CM, Juliano BO (1973) A gel consistency test for eating quality of rice. J Sci Food Agric 24:1589–1594. https://doi.org/10.1002/jsfa.2740241214

Camenisch TD, Brilliant MH, Segal DJ (2008) Critical parameters for genome editing using zinc finger nucleases. Mini Rev Med Chem 8:669–676. https://doi.org/10.2174/138955708784567458

Cermak T, Doyle EL, Christian M, Wang L, Zhang Y, Schmidt C, Baller JA, Somia NV, Bogdanove AJ, Voytas DF (2011) Efficient design and assembly of custom TALEN and other TAL effector based constructs for DNA targeting. Nucleic Acids Res 39:e82. https://doi.org/10.1093/nar/gkr218

Chao S, Cai Y, Feng B, Jiao G, Sheng Z, Luo J, Tang S, Wang J, Wei X, Hu P (2019) Editing of the rice isoamylase gene ISA1 provides insights into its function in starch formation. Rice Sci 26(2):77–87

Christian M, Qi Y, Zhang Y, Voytas DF (2013) Targeted mutagenesis of Arabidopsis thaliana using engineered TAL effector nucleases. G3 3:1697–1705. https://doi.org/10.1534/g3.113.007104

Clasen BM, Stoddard TJ, Luo S, Demorest ZL, Li J, Cedrone F, Tibebu R, Davison S, Ray EE, Daulhac A, Coffman A, Yabandith A, Retterath A, Haun W, Baltes NJ, Mathis L, Voytas DF, Zhang F (2015) Improving cold storage and processing traits in potato through targeted gene knockout. Plant Biotechnol J 14:169–176. https://doi.org/10.1111/pbi.12370

Cong L, Ran FA, Cox D, Lin S, Barretto R, Habib N, Hsu PD, Wu X, Jiang W, Marraffini L, Zhang F (2013) Multiplex genome engineering using CRISPR/Cas systems. Science 3:1231143. https://doi.org/10.1126/science.1231143

Dalton TJ (2004) A household hedonic model of rice traits: economic values from farmers in West Africa. Agric Econ 31:149–159. https://doi.org/10.1111/j.1574-0862.2004.tb00253.x

Datsenko KA, Pougach K, Tikhonov A, Wanner BL, Severinov K, Semenova E (2012) Molecular memory of prior infections activates the CRISPR/Cas adaptive bacterial immunity system. Nat Commun 3:945. https://doi.org/10.1038/ncomms1937

Deng D, Yan C, Pan X, Mahfouz M, Wang J, Zhu J, Shi Y, Yan N (2012) Structural basis for sequence specific recognition of DNA by TAL effectors. Science 335:720–723. https://doi.org/10.1126/science.1215670

Denyer K, Waite D, Motawia S, Moller BL, Smith AM (1999) Granule-bound starch synthase I in isolated starch granules elongates malto-oligosaccharides processively. Biochem J 340:183–191. https://doi.org/10.1042/bj3400183

Dian W, Jiang H, Wu P (2005) Evolution and expression analysis of starch synthase III and IV in rice. Journal of Experimental Botany 6:623–632. https://doi.org/10.1093/jxb/eri065

Doudna JA, Charpentier E (2014) The new frontier of genome engineering with CRISPR-Cas9. Science 346:1258096. https://doi.org/10.1126/science.1258096

Duan M, Sun S (2005) Profiling the expression of genes controlling rice grain quality. Plant Mol Biol 59:165–178. https://doi.org/10.1007/s11103-004-7507-3

Dunwell J (2014) Transgenic cereals: current status and future prospects. J Cereal Sci 59:419–434. https://doi.org/10.1016/j.jcs.2013.08.008

Durai S, Mani M, Kandavelou K, Wu J, Porteus MH, Chandrasegaran S (2005) Zinc finger nucleases: custom-designed molecular scissors for genome engineering of plant and mammalian cells. Nucleic Acids Res 18:5978–5990. https://doi.org/10.1093/nar/gki912

Epinat JC, Arnould S, Chames P, Rochaix P, Desfontaines D, Puzin C, Patin A, Zanghellini A, Pâques F, Lacroix E (2003) A novel engineered meganuclease induces homologous recombination in yeast and mammalian cells. Nucleic Acids Res 31(11):2952–2962. https://doi.org/10.1093/nar/gkg375

Fan C, Xing Y, Mao H, Lu T, Han B, Xu C, Li X, Zhang Q (2006) GS3, a major QTL for grain length and weight and minor QTL for grain width and thickness in rice, encodes a putative transmembrane protein. Theor Appl Genet 112:1164–1171. https://doi.org/10.1007/s00122-006-0218-1

Fiaz S, Ahmad S, Noor MA, Wang X, Younas A, Riaz A, Riaz A, Ali F (2019) Applications of the CRISPR/Cas9 system for rice grain quality improvement: perspectives and opportunities. Int J Mol Sci 20(4):888. https://doi.org/10.3390/ijms20040888

Fitzgerald MA, Sackville Hamilton NR, Calingacion MN, Verhoeven HA, Butardo VM (2008) Is there a second fragrance gene in rice? Plant Biotechnol J 6:416–423. https://doi.org/10.1111/j.1467-7652.2008.00327.x

Fitzgerald MA, McCouch SR, Hall RD (2009) Not just a grain of rice: the quest for quality. Trends Plant Sci 14:133–139. https://doi.org/10.1016/j.tplants.2008.12.004

Fonfara I, Richter H, Bratovič M, Le Rhun A, Charpentier E (2016) The CRISPR- associated DNA-cleaving enzyme Cpf1 also processes precursor CRISPR RNA. Nature 532:517–521. https://doi.org/10.1038/nature17945

Gao H, Smith J, Yang M, Jones S, Djukanovic V, Nicholson MG, West A, Bidney D, Falco SC, Jantz D, Lyznik LA (2010) Heritable targeted mutagenesis in maize using a designed endonuclease. Plant J 61(1):176–187. https://doi.org/10.1111/j.1365-313X.2009.04041.x

Gao Z, Herrera-Carrillo E, Berkhout B (2018) Improvement of the CRISPR-Cpf1 system with ribozyme-processed crRNA. RNA Biol 15(12):1458–1467. https://doi.org/10.1080/15476286.2018.1551703

Garneau JE, Dupuis ME, Villion M, Romero DA, Barrangou R, Boyaval P, Fremaux C, Horvath P, Magadan AH, Moineau S (2010) The CRISPR/Cas bacterial immune system cleaves bacteriophage and plasmid DNA. Nature 468:67–71. https://doi.org/10.1038/nature09523

Georges F, Ray H (2017) Genome editing of crops: a renewed opportunity for food security. GM Crops Food 8(1):1–12. https://doi.org/10.1080/21645698.2016.1270489

Goff SA, Ricke D, Lan TH, Presting G, Wang R, Dunn M, Glazebrook J, Sessions A, Oeller P, Varma H, Hadley D, Hutchison D, Martin C, Katagiri F, Lange BM, Moughamer T, Xia Y, Budworth P, Zhong J, Miguel T, Paszkowski U, Zhang S, Colbert M, Sun WL, Chen L, Cooper B, Park S, Wood TC, Mao L, Quail P, Wing R, Dean R, Yu Y, Zharkikh A, Shen R,

Sahasrabudhe S, Thomas A, Cannings R, Gutin A, Pruss D, Reid J, Tavtigian S, Mitchell J, Eldredge G, Scholl T, Miller RM, Bhatnagar S, Adey N, Rubano T, Tusneem N, Robinson R, Feldhaus J, Macalma T, Oliphant A, Briggs S (2002) A draft sequence of the rice genome (Oryza sativa L. ssp. japonica). Science 296:92–100. https://doi.org/10.1126/science.1068275.

Grizot S, Smith J, Daboussi F, Prieto J, Redondo P, Merino N, Villate M, Thomas S, Lemaire L, Montoya G, Blanco FJ (2009) Efficient targeting of a SCID gene by an engineered single-chain homing endonuclease. Nucleic Acids Res 37(16):5405–5419. https://doi.org/10.1093/nar/gkp548

Guo L, Ma L, Jiang H, Zeng D, Hu J, Wu L, Gao Z, Zhang G, Qian Q (2009) Genetic analysis and fine mapping of two genes for grain shape and weight in rice. J Integr Plant Biol 51:45–51. https://doi.org/10.1111/j.1744-7909.2008.00793.x

Harberd NP (2015) Shaping taste: the molecular discovery of rice genes improving grain size, shape and quality. J Genet Genom 42:597–599. https://doi.org/10.1016/j.jgg.2015.09.008.

Haun W, Coffman A, Clasen BM, Demorest ZL, Lowy A, Ray E, Retterath A, Stoddard T, Juillerat A, Cedrone F, Mathis L, Voytas DF, Zhang F (2014) Improved soybean oil quality by targeted mutagenesis of the fatty acid desaturase 2 gene family. Plant Biotechnol J 12:934–940. https://doi.org/10.1111/pbi.12201

Hsu PD, Lander ES, Zhang F (2014) Development and applications of CRISPR-Cas9 for genome engineering. Cell 157(6):1262–1278. https://doi.org/10.1016/j.cell.2014.05.010

Hu J, Wang Y, Fang Y, Zeng L, Xu J, Yu H, Shi Z, Pan J, Zhang D, Kang S, Zhu L (2015) A rare allele of GS2 enhances grain size and grain yield in rice. Molecular Plantx 8(10):1455–1465. https://doi.org/10.1016/j.molp.2015.07.002

Hsu PD, Scott DA, Weinstein JA, Ran FA, Konermann S, Agarwala V, Li Y, Fine EJ, Wu X, Shalem O, Cradick TJ (2013) DNA targeting specificity of RNA-guided Cas9 nucleases. Nat Biotechnol 31:827–832. https://doi.org/10.1038/nbt.2647

Ishino Y, Shinagawa H, Makino K, Amemura M, Nakata A (1987) Nucleotide sequence of the iap gene, responsible for alkaline phosphatase isozyme conversion in Escherichia coli, and identification of the gene product. J Bacteriol 169(12):5429–5433. https://doi.org/10.1128/jb.169.12.5429-5433.1987

Isshiki M, Morino K, Nakajima M, Okagaki RJ, Wessler SR, Izawa T, Shimamoto K (1998) A naturally occurring functional allele of the rice waxy locus has a GT to TT mutation at the 5′ splice site of the first intron. Plant J 15:133–138. https://doi.org/10.1046/j.1365-313X.1998.00189.x

Jezussek M, Juliano BO, Schieberle P (2002) Comparison of key aroma compounds in cooked brown rice varieties based on aroma extract dilution analyses. J Agric Food Chem 50:1101–1105. https://doi.org/10.1021/jf0108720

Jiang Y, Cai Z, Xie W, Long T, Yu H, Zhang Q (2012) Rice functional genomics research: progress and implications for crop genetic improvement. Biotechnol Adv 30:1059–1070. https://doi.org/10.1016/j.biotechadv.2011.08.013

Jinek M, Chylinski K, Fonfara I, Hauer M, Doudna JA, Charpentier E (2012) A programmable dual-RNA-guided DNA endonuclease in adaptive bacterial immunity. Science 337:816–821. https://doi.org/10.1126/science.1225829

Jinek M, East A, Cheng A, Lin S, Ma E, Doudna J (2013) RNA-programmed genome editing in human cells. elife 2:e00471. https://doi.org/10.7554/eLife.00471.001

Jinek M, Jiang F, Taylor DW, Sternberg SH, Kaya E, Ma E, Anders C, Hauer M, Zhou K, Lin S, et a (2014) Structures of Cas9 Endonucleases Reveal RNA-Mediated Conformational Activation. Science 343:1247997. https://doi.org/10.1126/science.1247997

Johnson A, Kyriacou B, Callahan D, Carruthers L, Stangoulis J, Lombi E, Tester M (2011) Constitutive overexpression of the OsNAS gene family reveals single-gene strategies for effective iron-and zinc-biofortification of rice endosperm. PLoS One 6:e24476. https://doi.org/10.1371/journal.pone.0024476

Joung JK, Sander JD (2013) TALENs: a widely applicable technology for targeted genome editing. Nat Rev Mol Cell Biol 14:49–55. https://doi.org/10.1038/nrm3486

Juliano B (1992) Structure, chemistry, and function of the rice grain and its fractions. Cereal Foods World 37:772–772

Juliano B (1998) Varietal impact on rice quality. Cereal Foods World 43:207–222
Juliano B (2003) Rice chemistry and quality. International Rice Research Institute, Los Baños
Juliano B, Villareal CP (1993) Grain quality evaluation of world rices. International Rice Research Institute, Los Baños
Jung JH, Altpeter F (2016) TALEN mediated targeted mutagenesis of the caffeic acid O methyltransferase in highly polyploid sugarcane improves cell wall composition for production of bioethanol. Plant Mol Biol 92:131–142. https://doi.org/10.1007/s11103-016-0499-y
Kathuria H, Giri J, Tyagi H, Tyagi AK (2007) Advances in transgenic rice biotechnology. Crit Rev Plant Sci 26:65–103. https://doi.org/10.1080/07352680701252809
Kennedy G, Burlingame B, Nguyen VN (2003) Nutritional contribution of rice and impact of biotechnology and biodiversity in rice-consuming countries. In: Proceedings of the 20th session of the international rice commission. Bangkok, Thailand. FAO, Rome, Italy, pp 59–70
Khush GS (2005) What it will take to feed 5.0 billion rice consumers in 2030. Plant Mol Biol 59(1):1–6. https://doi.org/10.1007/s11103-005-2159-5
Kim YG, Cha J, Chandrasegaran S (1996) Hybrid restriction enzymes: zinc finger fusions to Fok I cleavage domain. Proc Natl Acad Sci U S A 93:1156–1160. https://doi.org/10.1073/pnas.93.3.1156
Kim D, Kim J, Hur JK, Been KW, Yoon SH, Kim JS (2016) Genome-wide analysis reveals specificities of Cpf1 endonucleases in human cells. Nat Biotechnol 34(8):863–868. https://doi.org/10.1038/nbt.3609
Kim K, Park SW, Kim JH, Lee SH, Kim D, Koo T, Kim KE, Kim JH, Kim JS (2017) Genome surgery using Cas9 ribonucleoproteins for the treatment of age-related macular degeneration. Genome Res 27:419–426. https://doi.org/10.1101/gr.219089.116
Kleinstiver BP, Tsai SQ, Prew MS, Nguyen NT, Welch MM, Lopez JM, McCaw ZR, Aryee MJ, Joung JK (2016) Genome-wide specificities of CRISPR-Cas Cpf1 nucleases in human cells. Nat Biotechnol 34(8):869–874. https://doi.org/10.1038/nbt.3620
Kovach MJ, Calingacion MN, Fitzgerald MA, McCouch SR (2009) The origin and evolution of fragrance in rice (Oryza sativa L.). Proc Natl Acad Sci 106:14444–14449. https://doi.org/10.1073/pnas.0904077106
Kusano M, Yang Z, Okazaki Y, Nakabayashi R, Fukushima A, Saito K (2015) Using metabolomic approaches to explore chemical diversity in rice. Mol Plant 8(1):58–67. https://doi.org/10.1016/j.molp.2014.11.010
Kwon SW, Cho YC, Lee JH, Suh JP, Kim JJ, Kim MK, Choi IS, Hwang HG, Koh HJ, Kim YG (2011) Identification of quantitative trait loci associated with rice eating quality traits using a population of recombinant inbred lines derived from a cross between two temperate japonica cultivars. Mol Cells 31:437–445. https://doi.org/10.1007/s10059-011-0289-y
Lee J, Chung JH, Kim HM, Kim DW, Kim H (2016a) Designed nucleases for targeted genome editing. Plant Biotechnol J 14:448–462. https://doi.org/10.1111/pbi.12465
Lee SK, Eom JS, Hwang SK, Shin D, An G, Okita TW, Jeon JS (2016b) Plastidic phosphoglucomutase and ADP-glucose pyrophosphorylase mutants impair starch synthesis in rice pollen grains and cause male sterility. J Exp Bot 67:5557–5569. https://doi.org/10.1093/jxb/erw324
Li Y, Fan C, Xing Y, Jiang Y, Luo L, Sun L, Shao D, Xu C, Li X, Xiao J, He Y (2011) Natural variation in GS5 plays an important role in regulating grain size and yield in rice. Nat Genet 43(12):1266. https://doi.org/10.1038/ng.977
Li L, Piatek MJ, Atef A, Piatek A, Wibowo A, Fang X, Sabir JS, Zhu JK, Mahfouz MM (2012a) Rapid and highly efficient construction of TALE-based transcriptional regulators and nucleases for genome modification. Plant Biotechnol J 78:407–416. https://doi.org/10.1007/s11103-012-9875-4
Li T, Liu B, Spalding MH, Weeks DP, Yang B (2012b) High-efficiency TALEN-based gene editing produces disease-resistant rice. Nat Biotechnol 30:390–392. https://doi.org/10.1038/nbt.2199
Li Y, Fan C, Xing Y, Yun P, Luo L, Yan B, Peng B, Xie W, Wang G, Li X, Xiao J (2014) Chalk5 encodes a vacuolar H+-translocating pyrophosphatase influencing grain chalkiness in rice. Nat Genet 46:398–404. https://doi.org/10.1038/ng.2923

Li M, Li X, Zhou Z, Wu P, Fang M, Pan X, Lin Q, Luo W, Wu G, Li H (2016) Reassessment of the four yield-related genes Gn1a, DEP1, GS3, and IPA1 in rice using a CRISPR/Cas9 system. Front Plant Sci 7:377. https://doi.org/10.3389/fpls.2016.00377
Liang Z, Zhang K, Chen K, Gao C (2014) Targeted mutagenesis in Zea mays using TALENs and the CRISPR/Cas system. J Genet Genomics 41(2):63–68. https://doi.org/10.1016/j.jgg.2013.12.001
Lin Q, Huang B, Zhang M, Zhang X, Rivenbark J, Lappe RL, James MG, Myers AM, Hennen-Bierwagen TA (2012) Functional interactions between starch synthase III andisoamylase-type starch-debranching enzyme in maize endosperm. Plant Physiol 158:679–692. https://doi.org/10.1104/pp.111.189704
Liu X, Wan X, Ma X, Wan J (2011) Dissecting the genetic basis for the effect of rice chalkiness, amylose content, protein content, and rapid viscosity analyzer profile characteristics on the eating quality of cooked rice using the chromosome segment substitution line population across eight environments. Genome 54:64–80. https://doi.org/10.1139/G10-070
Liu S, Hua L, Dong S, Chen H, Zhu X, Jiang JE, Zhang F, Li Y, Fang X, Chen F (2015) Os MAPK 6, a mitogen-activated protein kinase, influences rice grain size and biomass production. Plant J 84(4):672–681. https://doi.org/10.1111/tpj.13025
Liu L, Chen P, Wang M, Li X, Wang J, Yin M, Wang Y (2017) C2c1-sgRNA complex structure reveals RNA-guided DNA cleavage mechanism. Mol Cell 65(2):310–322. https://doi.org/10.1016/j.molcel.2016.11.040
Lloyd A, Plaisier CL, Carroll D, Drews GN (2005) Targeted mutagenesis using zinc-finger nucleases in Arabidopsis. Proc Natl Acad Sci 102(6):2232–2237. https://doi.org/10.1073/pnas.0409339102
Lor VS, Starker CG, Voytas DF, Weiss D, Olszewski NE (2014) Targeted mutagenesis of the tomato PROCERA gene using transcription activator-like effector nucleases. Plant Physiol 166:1288–1291. https://doi.org/10.1104/pp.114.247593
Lorieux M, Petrov M, Huang N, Guiderdoni E, Ghesquière A (1996) Aroma in rice: genetic analysis of a quantitative trait. Theor Appl Genet 93:1145–1151. https://doi.org/10.1007/BF00230138
Lucca P, Hurrell R, Potrykus I (2001) Genetic engineering approaches to improve the bioavailability and the level of iron in rice grains. Theor Appl Genet 102:392–397. https://doi.org/10.1007/s001220051659
Lusser M, Parisi C, Plan D, Rodríguez-Cerezo E (2012) Development of new biotechnologies in plant breeding. Nat Biotechnol 30:231–239. https://doi.org/10.1038/nbt.2142
Ma X, Zhang Q, Zhu Q, Liu W, Chen Y, Qiu R, Wang B, Yang Z, Li H, Lin Y (2015) A robust CRISPR/Cas9 system for convenient, high-efficiency multiplex genome editing in monocot and dicot plants. Mol Plant 8:1274–1284. https://doi.org/10.1016/j.molp.2015.04.007
Mahfouz MM, Li L, Shamimuzzaman M, Wibowo A, Fang X, Zhu JK (2011) De novo- engineered transcription activator-like effector TALE hybrid nuclease with novel DNA binding specificity creates double strand breaks. Proc Natl Acad Sci U S A 108:2623–2628. https://doi.org/10.1073/pnas.1019533108
Mak AN, Bradley P, Cernadas RA, Bogdanove AJ, Stoddard BL (2012) The crystal structure of TAL effector PthXo1 bound to its DNA target. Sci Technol Humanit 335(6069):716–719. https://doi.org/10.1126/science.1216211
Marraffini LA, Sontheimer EJ (2008) CRISPR interference limits horizontal gene transfer in staphylococci by targeting DNA. Science 322:1843–1845. https://doi.org/10.1126/science.1165771
Miglani GS (2017) Genome editing in crop improvement: present scenario and future prospects. J Crop Improv 31:453–559. https://doi.org/10.1080/15427528.2017.1333192
Miller JC, Tan S, Qiao G, Barlow KA, Wang J, Xia DF, Meng X, Paschon DE, Leung E, Hinkley SJ, Dulay GP (2010) A TALE nuclease architecture for efficient genome editing. Nat Biotechnol 29(2):143. https://doi.org/10.1038/nbt.1755
Mishra R, Zhao K (2018) Genome editing technologies and their applications in crop improvement. Plant Biotechnol. Rep. 12, 57–68. https://doi.org/10.1007/s11816-018-0472-0
Mishra R, Joshi RK, Zhao K (2018) Genome editing in rice: recent advances, challenges, and future implications. Front Plant Sci 9:1361. https://doi.org/10.3389/fpls.2018.01361

Mladenov E, Iliakis G (2011) Induction and repair of DNA double strand breaks: the increasing spectrum of non-homologous end joining pathways. Mutat Res 711:61–72. https://doi.org/10.1016/j.mrfmmm.2011.02.005

Mojica FJ, Diez-Villasenor C, Garcia-Martinez J, Soria E (2005) Intervening sequences of regularly spaced prokaryotic repeats derive from foreign genetic elements. J Mol Evol 60:174–182. https://doi.org/10.1007/s00239-004-0046-3

Moon SB, Lee JM, Kang JG, Lee NE, Ha DI, Kim SH, Yoo K, Kim D, Ko JH, Kim YS (2018) Highly efficient genome editing by CRISPR-Cpf1 using CRISPR RNA with a uridinylate-rich 3′-overhang. Nat Commun 9(1):3651. https://doi.org/10.1038/s41467-018-06129-w

Moreno-Mateos MA, Fernandez JP, Rouet R, Vejnar CE, Lane MA, Mis E, Khokha MK, Doudna JA, Giraldez AJ (2017) CRISPR-Cpf1 mediates efficient homology-directed repair and temperature-controlled genome editing. Nat Commun 8(1):2024. https://doi.org/10.1038/s41467-017-01836-2

Moscou MJ, Bogdanove AJ (2009) A simple cipher governs DNA recognition by TAL effectors. Science 326:1501. https://doi.org/10.1126/science.1178817

Nicolia A, Proux-Wera E, Ahman I, Onkokesung N, Andersson M, Andreasson E, Zhu LH (2015) Targeted gene mutation in tetraploid potato through transient TALEN expression in protoplasts. J Biotechnol 204:17–24. https://doi.org/10.1016/j.jbiotec.2015.03.021

Osakabe Y, Osakabe K (2014) Genome editing with engineered nucleases in plants. Plant Cell Physiol 56(3):389–400. https://doi.org/10.1093/pcp/pcu170

Paine JA, Shipton CA, Chaggar S, Howells RM, Kennedy MJ, Vernon G, Wright SY, Hinchliffe E, Adams JL, Silverstone AL, Drake R (2005) Improving the nutritional value of Golden Rice through increased pro-vitamin A content. Nat Biotechnol 23(4):482. https://doi.org/10.1038/nbt1082

Pandey MK, Rani NS, Madhav MS, Sundaram RM, Varaprasad GS, Sivaranjani AK, Bohra A, Kumar GR, Kumar A (2012) Different isoforms of starch-synthesizing enzymes controlling amylose and amylopectin content in rice (Oryza sativa L.). Biotechnol Adv 30:1697–1706. https://doi.org/10.1016/j.biotechadv.2012.08.011

Peng S, Khush GS, Virk P, Tang Q, Zou Y (2008) Progress in ideotype breeding to increase rice yield potential. Field Crops Res 108:32–38. https://doi.org/10.1016/j.fcr.2008.04.001

Peng B, Kong H, Li Y, Wang L, Zhong M, Sun L, Gao G, Zhang Q, Luo L, Wang G, Xie W, Chen J, Yao W, Peng Y, Lei L, Lian X, Xiao J, Xu C, Li X, He Y (2014) OsAAP6 functions as an important regulator of grain protein content and nutritional quality in rice. Nat Commun 5:4847. https://doi.org/10.1038/ncomms5847

Pérez L, Soto E, Villorbina G, Bassie L, Medina V, Muñoz P, Capell T, Zhu C, Christou P, Farré G (2018) CRISPR/Cas9-induced monoallelic mutations in the cytosolic AGPase large subunit gene APL2 induce the ectopic expression of APL2 and the corresponding small subunit gene APS2b in rice leaves. Transgenic Res 27(5):423–439. https://doi.org/10.1007/s11248-018-0089-7

Perkin LC, Adrianos SL, Oppert B (2016) Gene disruption technologies have the potential to transform stored product insect pest control. Insects 7(3):46. https://doi.org/10.3390/insects7030046

Philpot K, Martin M, Butardo V, Willoughby D, Fitzgerald M (2006) Environmental factors that affect the ability of amylose to contribute to retrogradation in gels made from rice flour. J Agric Food Chem 54:5182–5190. https://doi.org/10.1021/jf060666y

Puchta H (2004) The repair of double-strand breaks in plants: mechanisms and consequences for genome evolution. J Exp Bot 56:1–14. https://doi.org/10.1093/jxb/eri025

Qu J, Xu S, Zhang Z, Chen G, Zhong Y, Liu L, Zhang R, Xue J, Guo D (2018) Evolutionary, structural and expression analysis of core genes involved in starch synthesis. Sci Rep 8(1):12736. https://doi.org/10.1038/s41598-018-30411-y

Ramirez CL, Foley JE, Wright DA, Müller-Lerch F, Rahman SH, Cornu TI, Winfrey RJ, Sander JD, Fu F, Townsend JA, Cathomen T, Voytas DF, Joung JK (2008) Unexpected failure rates for modular assembly of engineered zinc fingers. Nat Methods 5(5):374. https://doi.org/10.1038/nmeth0508-374

Rao VT, Mohan YC, Bhadru D, Bharathi D, Venkanna V (2014) Genetic variability and association analysis in rice. Intl J Appl Biol Pharma Tech 5:63–65

Ray DK, Mueller ND, West PC, Foley JA (2013) Yield trends are insufficient to double global crop production by 2050. PLoS One 8(6):e66428. https://doi.org/10.1371/journal.pone.0066428

Ren D, Rao Y, Huang L, Leng Y, Hu J, Lu M, Zhang G, Zhu L, Gao Z, Dong G, Guo L (2016) Fine mapping identifies a new QTL for brown rice rate in rice (Oryza Sativa L.). Rice 9:1–10. https://doi.org/10.1186/s12284-016-0076-7

Römer P, Hahn S, Jordan T, Strauß T, Bonas U, Lahaye T (2007) Plant pathogen recognition mediated by promoter activation of the pepper Bs3 resistance gene. Science 318(5850):645–648. https://doi.org/10.1126/science.1144958

Sakamoto T, Matsuoka M (2008) Identifying and exploiting grain yield genes in rice. Curr Opin Plant Biol 11:209–214. https://doi.org/10.1016/j.pbi.2008.01.009

Sander JD, Dahlborg EJ, Goodwin MJ, Cade L, Zhang F, Cifuentes D, Curtin SJ, Blackburn JS, Thibodeau-Beganny S, Qi Y, Pierick CJ (2011) Selection-free zinc- fingernuclease engineering by context-dependent assembly CoDA. Nat Methods 8:67–69. https://doi.org/10.1038/nmeth.1542

Schaeffer SM, Nakata PA (2015) CRISPR/Cas9-mediated genome editing and gene replacement in plants: transitioning from lab to field. Plant Sci 240:130–142. https://doi.org/10.1016/j.plantsci.2015.09.011

Scheben A, Wolter F, Batley J, Puchta H, Edwards D (2017) Towards CRISPR/Cas crops – bringing together genomics and genome editing. New Phytol 216(3):682–698. https://doi.org/10.1111/nph.14702

Shan Q, Wang Y, Chen K, Liang Z, Li J, Zhang Y, Zhang K, Liu J, Voytas DF, Zheng X, Zhang Y, Gao C (2013) Rapid and efficient gene modification in rice and Brachypodium using TALENs. Mol Plant 6:1365–1368. https://doi.org/10.1093/mp/sss162

Shan Q, Zhang Y, Chen K, Zhang K, Gao C (2015) Creation of fragrant rice by targeted knockout of the OsBADH2 gene using TALEN technology. Plant Biotechnol J 13(6):791–800. https://doi.org/10.1111/pbi.12312

Shen L, Wang C, Fu Y, Wang J, Liu Q, Zhang X, Yan C, Qian Q, Wang K (2018) QTL editing confers opposing yield performance in different rice varieties. J Integr Plant Biol 60(2):89–93. https://doi.org/10.1111/jipb.12501

Shmakov S, Abudayyeh OO, Makarova KS, Wolf YI, Gootenberg JS, Semenova E, Minakhin L, Joung J, Konermann S, Severinov K, Zhang F (2015) Discovery and functional characterization of diverse class 2 CRISPR-Cas systems. Mol Cell 60(3):385–397. https://doi.org/10.1016/j.molcel.2015.10.008

Shomura A, Izawa T, Ebana K, Ebitani T, Kanegae H, Konishi S, Yano M (2008) Deletion in a gene associated with grain size increased yields during rice domestication. Nat Genet 40(8):1023. https://doi.org/10.1038/ng.169

Shukla VK, Doyon Y, Miller JC, DeKelver RC, Moehle EA, Worden SE, Mitchell JC, Arnold NL, Gopalan S, Meng X, Choi VM, Rock JM, Wu YY, Katibah GE, Zhifang G, McCaskill D, Simpson MA, Blakeslee B, Greenwalt SA, Butler HJ, Hinkley SJ, Zhang L, Rebar EJ, Gregory PD, Urnov FD (2009) Precise genome modification in the crop species Zea mays using zinc-finger nucleases. Nature 459:437–441. https://doi.org/10.1038/nature07992

Si L, Chen J, Huang X, Gong H, Luo J, Hou Q, Zhou T, Lu T, Zhu J, Shangguan Y, Chen E (2015) OsSPL13 controls grain size in cultivated rice. Nat Genet 47(3):447. https://doi.org/10.1038/ng.3518

Singh R, Singh A, Sharma T, Singh A, Singh N (2012) Fine mapping of grain length QTLs on chromosomes 1 and 7 in Basmati rice (Oryza sativa L.). J. Plant Biochem. Biotechnol 21:157–166. https://doi.org/10.1007/s13562-011-0080-3

Smith J, Grizot S, Arnould S, Duclert A, Epinat J-C, Chames P, Prieto J, Redondo P, Blanco FJ, Bravo J (2006) A combinatorial approach to create artificial homing endonucleases cleaving chosen sequences. Nucleic Acids Res 34:e149. https://doi.org/10.1093/nar/gkl720

Sreenivasulu N, Butardo VM, Misra G, Cuevas RP, Anacleto R, Kavi Kishor PB (2015) Designing climate-resilient rice with ideal grain quality suited for high-temperature stress. J Exp Bot 66:1737–1748. https://doi.org/10.1093/jxb/eru544

Sun L, Li X, Fu Y, Zhu Z, Tan L, Liu F, Sun X, Sun X, Sun C (2013) GS 6, a member of the GRAS gene family, negatively regulates grain size in rice. J Integr Plant Biol 55(10):938–949. https://doi.org/10.1111/jipb.12062

Sun YW, Jiao GA, Liu ZP, Zhang X, Li JY, Guo XP, Du WM, Du JL, Francis F, Zhao YD, Xia LQ (2017) Generation of high-amylose rice through CRISPR/Cas9-mediated targeted mutagenesis of starch branching enzymes. Front Plant Sci 8:298. https://doi.org/10.3389/fpls.2017.00298

Tang XJ, Peng C, Zhang J, Cai Y, You XM, Kong F et al (2016) ADP-glucose pyrophosphorylase large subunit 2 is essential for storage substance accumulation and subunit interactions in rice endosperm. Plant Sci 249:70–83. https://doi.org/10.1016/j.plantsci.2016.05.010

Tang L, Mao B, Li Y, Lv Q, Zhang L, Chen C, He H, Wang W, Zeng X, Shao Y, Pan Y (2017) Knockout of OsNramp5 using the CRISPR/Cas9 system produces low Cd- accumulating indica rice without compromising yield. Scientific reports. 31 7(1):14438. https://doi.org/10.1038/s41598-017-14832-9

Tilman D, Balzer C, Hill J, Befort BL (2011) Global food demand and the sustainable intensification of agriculture. Proc Natl Acad Sci 108(50):20260–20264. https://doi.org/10.1073/pnas.1116437108

Townsend JA, Wright DA, Winfrey RJ, Fu F, Maeder ML, Joung JK, Voytas DF (2009) High-frequency modification of plant genes using engineered zinc-finger nucleases. Nature 459:442–445. https://doi.org/10.1038/nature07845

Tuncel A, Kawaguchi J, Ihara Y, Matsusaka H, Nishi A, Nakamura T et al (2014) The rice endosperm ADP-glucose pyrophosphorylase large subunits essential for optimal catalysis and allosteric regulation of the heterotetrameric enzyme. Plant Cell Physiol 55:1169–1183. https://doi.org/10.1093/pcp/pcu057

United Nations, Department of Economic and Social Affairs, Population Division (2017) World population prospects: the 2017 revision, key findings and advance tables. Working Paper No. ESA/P/WP/248. Available at: https://esa.un.org/unpd/wpp/publications/files/wpp2017_key-findings.pdf.

Wambugu P, Ndjiondjop MN, Furtado A, Henry R (2018) Sequencing of bulks of segregants allows dissection of genetic control of amylose content in rice. Plant Biotechnol J 16:100–110. https://doi.org/10.1111/pbi.12752

Wang E, Wang J, Zhu X, Hao W, Wang L, Li Q, Zhang L, He W, Lu B, Lin H, Ma H, Zhang G, He Z (2008) Control of rice grain-filling and yield by a gene with a potential signature of domestication. Nat Genet 40:1370–1374. https://doi.org/10.1038/ng.220

Wang H, Pandey S, Velarde O (2012a) Synthesis of key results and implications. In: Wang H, Pandey S, Velarde O, Hardy B (eds) Patterns of varietal adoption and economics of rice production in Asia. International Rice Research Institute, Los Baños, pp 1–14

Wang S, Wu K, Yuan Q, Liu X, Liu Z, Lin X, Zeng R, Zhu H, Dong G, Qian Q, Zhang G (2012b) Control of grain size, shape and quality by OsSPL16 in rice. Nat Genet 44(8):950. https://doi.org/10.1038/ng.2327

Wang Y, Cheng X, Shan Q, Zhang Y, Liu J, Gao C, Qiu JL (2014a) Simultaneous editing of three homoeoalleles in hexaploid bread wheat confers heritable resistance to powdery mildew. Nat Biotechnol 32:947–951. https://doi.org/10.1038/nbt.2969

Wang Z, Li W, Qi J, Shi P, Yin Y (2014b) Starch accumulation, activities of key enzyme and gene expression in starch synthesis of wheat endosperm with different starch contents. J Food Sci Technol 51(3):419–429. https://doi.org/10.1007/s13197-011-0520-z

Wang S, Li S, Liu Q, Wu K, Zhang J, Wang S, Wang Y, Chen X, Zhang Y, Gao C, Wang F (2015a) The OsSPL16-GW7 regulatory module determines grain shape and simultaneously improves rice yield and grain quality. Nat Genet 47(8):949. https://doi.org/10.1038/ng.3352

Wang Y, Xiong G, Hu J, Jiang L, Yu H, Xu J, Fang Y, Zeng L, Xu E, Xu J, Ye W (2015b) Copy number variation at the GL7 locus contributes to grain size diversity in rice. Nat Genet 47(8):944. https://doi.org/10.1038/ng.3346

Wang M, Mao Y, Lu Y, Tao X, Zhu JK (2017) Multiplex gene editing in rice using the CRISPR-Cpf1 system. Mol Plant 10(7):1011–1013. https://doi.org/10.1016/j.molp.2017.03.001

Wendt T, Holm PB, Starker CG, Christian M, Voytas DF, Brinch-Pedersen H, Holme IB (2013) TAL effector nucleases induce mutations at a pre-selected location in the genome of primary barley transformants. Plant Mol Biol 83:279–285. https://doi.org/10.1007/s11103-013-0078-4

Weng J, Gu S, Wan X, Gao H, Guo T, Su N, Lei C, Zhang X, Cheng Z, Guo X, Wang J, Jiang L, Zhai H, Wan J (2008) Isolation and initial characterization of GW5, a major QTL associated with rice grain width and weight. Cell Res 18:1199–1209. https://doi.org/10.1038/cr.2008.307

Wu YP, Pu CH, Lin HY, Huang HY, Huang YC, Hong CY, Chang MC, Lin YR (2015) Three novel alleles of FLOURY ENDOSPERM2 (FLO2) confer dull grains with low amylose content in rice. Plant Sci 233:44–52. https://doi.org/10.1016/j.plantsci.2014.12.011

Xu R, Yang Y, Qin R, Li H, Qiu C, Li L, Wei P, Yang J (2016) Rapid improvement of grain weight via highly efficient CRISPR/Cas9-mediated multiplex genome editing in rice. J Genet Genom 43:529–532. https://doi.org/10.1016/j.jgg.2016.07.003

Xu R, Qin R, Li H, Li D, Li L, Wei P, Yang J (2017) Generation of targeted mutant rice using a CRISPR-Cpf1 system. Plant Biotechnol J 15(6):713–717. https://doi.org/10.1111/pbi.12669

Yang J, Zhang J, Wang Z, Zhu Q (2001) Activities of starch hydrolytic enzymes and sucrose-phosphate synthase in the stems of rice subjected to water stress during grain filling. J Exp Bot 52:2169–2179. https://doi.org/10.1093/jexbot/52.364.2169

Yang J, Zhang J, Wang Z, Xu G, Zhu Q (2004) Activities of key enzymes in sucrose-to- starch conversion in wheat grains subjected to water deficit during grain filling. Plant Physiol 135:1621–1629. https://doi.org/10.1104/pp.104.041038

Yang X, Chen L, Yu W (2017) Knocking out of carotenoid catabolic genes in rice fails to boost carotenoid accumulation, but reveals a mutation in strigolactone biosynthesis. Plant Cell Rep 36:1533–1545. https://doi.org/10.1007/s00299-017-2172-6

Yu J, Hu S, Wang J, Wong GK, Li S, Liu B, Deng Y, Dai L, Zhou Y, Zhang X, Cao M, Liu J, Sun J, Tang J, Chen Y, Huang X, Lin W, Ye C, Tong W, Cong L, Geng J, Han Y, Li L, Li W, Hu G, Huang X, Li W, Li J, Liu Z, Li L, Liu J, Qi Q, Liu J, Li L, Li T, Wang X, Lu H, Wu T, Zhu M, Ni P, Han H, Dong W, Ren X, Feng X, Cui P, Li X, Wang H, Xu X, Zhai W, Xu Z, Zhang J, He S, Zhang J, Xu J, Zhang K, Zheng X, Dong J, Zeng W, Tao L, Ye J, Tan J, Ren X, Chen X, He J, Liu D, Tian W, Tian C, Xia H, Bao Q, Li G, Gao H, Cao T, Wang J, Zhao W, Li P, Chen W, Wang X, Zhang Y, Hu J, Wang J, Liu S, Yang J, Zhang G, Xiong Y, Li Z, Mao L, Zhou C, Zhu Z, Chen R, Hao B, Zheng W, Chen S, Guo W, Li G, Liu S, Tao M, Wang J, Zhu L, Yuan L, Yang H (2002) A draft sequence of the rice genome (Oryza sativa L. ssp. indica). Science 296:279–292. https://doi.org/10.1126/science.1068037

Yuan LP (1987) Strategy conception of hybrid rice breeding. Hybrid Rice 1:1–3

Zetsche B, Gootenberg JS, Abudayyeh OO, Slaymaker IM, Makarova KS, Essletzbichler P, Volz SE, Joung J, Van Der Oost J, Regev A, Koonin EV (2015) Cpf1 is a single RNA-guided endonuclease of a class 2 CRISPR-Cas system. Cell 163(3):759–771. https://doi.org/10.1016/j.cell.2015.09.038

Zhang X, Wang J, Huang J, Lan H, Wang C, Yin C, Wu Y, Tang H, Qian Q, Li J, Zhang H (2012) Rare allele of OsPPKL1 associated with grain length causes extra-large grain and a significant yield increase in rice. Proc Natl Acad Sci 109(52):21534–21539. https://doi.org/10.1073/pnas.1219776110

Zhang Y, Zhang F, Li X, Baller JA, Qi Y, Starker CG, Bogdanove AJ, Voytas DF (2013) Transcription activator-like effector nucleases enable efficient plant genome engineering. Plant Physiol 161:20–27. https://doi.org/10.1007/s11103-013-0078-4

Zhang K, Raboanatahiry N, Zhu B, Li M (2017) Progress in genome editing technology and its application in plants. Front Plant Sci 8:177. https://doi.org/10.3389/fpls.2017.00177

Zhang J, Zhang H, Botella JR, Zhu JK (2018) Generation of new glutinous rice by CRISPR/Cas9-targeted mutagenesis of the Waxy gene in elite rice varieties. J Integr Plant Biol 60(5):369–375. https://doi.org/10.1111/jipb.12620

Zhao D, Li Q, Zhang C, Zhang C, Yang Q, Pan L, Ren X, Lu J, Gu M, Liu Q (2018) GS9 acts as a transcriptional activator to regulate rice grain shape and appearance quality. Nat Commun 9(1):1240. https://doi.org/10.1038/s41467-018-03616-y

Zheng L, Cheng Z, Ai C, Jiang X, Bei X, Zheng Y, Glahn R, Welch R, Miller D, Lei X (2010) Nicotianamine, a novel enhancer of rice iron bioavailability to humans. PLoS One 5:e10190. https://doi.org/10.1371/journal.pone.0010190

Part VII
Omics

Chapter 18
Genomics and Molecular Markers for Rice Grain Quality: A Review

Jeanaflor Crystal T. Concepcion, Christopher Proud, and Melissa A. Fitzgerald

18.1 Introduction

Rice is one of the world's most valuable crops, cultivated in more than 100 countries and providing food to a third of the global population (Yu et al. 2011; Calingacion et al. 2014). Rice is mainly produced for direct human consumption, providing 21% of global human energy per capita and 15% of per capita protein (United Nations 2017). Rice belongs to the genus *Oryza* and has two cultivated species *Oryza sativa* and *Oryza glaberrima* and 22 wild species (Kumagai et al. 2018). Possibly the oldest domesticated grain (~10,000 years), *O. sativa*, is grown all over the world (Purugganan 2014). Among the accessions of *O. sativa*, certain types of varieties have been identified as premium or high quality. These rice types include the jasmine, basmati and sadri-style rice that are a premium commodity as well as a staple in many parts of the world. These rice types command a high prize in both domestic and international markets because consumers prefer their attributes (Fitzgerald et al. 2009; Champagne et al. 2010; Calingacion et al. 2014). On international markets, these varieties can attract more than double the price of other types.

The size of the global population is increasing, which might suggest a proportional increase in demand for rice. However, the demography within the population is also changing, with economic growth in many countries. This could mean that the demand is not simply for more rice, but for high-quality rice (Minten et al. 2011;

J. C. T. Concepcion (✉)
School of Agriculture and Food Science, The University of Queensland, Brisbane, QLD, Australia

Department of Crop Sciences, University of Illinois at Urbana-Champaign, Urbana, IL, USA
e-mail: jcconcepcion13@gmail.com

C. Proud · M. A. Fitzgerald
School of Agriculture and Food Science, The University of Queensland, Brisbane, QLD, Australia

A. Costa de Oliveira et al. (eds.), *The Future of Rice Demand: Quality Beyond Productivity*, https://doi.org/10.1007/978-3-030-37510-2_18

Calingacion et al. 2014). The importance of ensuring an abundant supply of high-quality rice that caters to the growing market demand while capturing regional preference of quality requires screening tools that are cheap, easy to use and robust in terms of the trait they target. Therefore, the use of molecular markers stands out as a method of choice for early screening of rice lines, providing confidence in the screening process without the long wait for seed harvest and further phenotyping procedures (Collard et al. 2005). Genotyping technologies and techniques in molecular biology and biotechnology have progressed rapidly in the past few decades, giving rise to a wealth of information on the genome of rice varieties (Li et al. 2014). From the birth of PCR in the mid-1980s (Mullis et al. 1987) to the availability of the rice reference genome (International Rice Genome Sequencing Project 2005) about a decade later, rapidly followed by the release of the sequence of 3000 rice genomes (Li et al. 2014), there is no doubt that the rice science community has taken advantage of the genetic black box of rice. The availability of rice genome sequences has provided great opportunities for rice scientists to further exploit genetic diversity (Li et al. 2014; Kumagai et al. 2018), as evidenced by the interest in genome-wide association study (GWAS) over the conventional biparental population linkage mapping, therefore delivering information on novel genetic understanding of a wide range of economically important traits in rice, including the traits of quality. This surge in technological capacity has benefited the breeding community by allowing verification of known genetic regions or quantitative trait loci (QTLs) and candidate genes across diverse rice collections (Kovach et al. 2009; Huang et al. 2010; Daygon et al. 2017; Misra et al. 2018; Cuevas et al. 2018) and mapping populations (Calingacion et al. 2017; Concepcion et al. 2019). This abundance in genomic data has also provided new insights to maximise the impacts of minor QTLs and genes by estimating the phenotypic outcomes by means of genomic selection.

This chapter reviews the recent progress on molecular marker genotyping and genomics of rice grain quality and discusses further the promise of metabolomics as a phenotyping tool of the future, its advantages, considerations and the anticipated impact on future breeding plans for rice grain quality.

18.2 Molecular Markers for Rice Grain Quality

Breeding for grain quality is the most challenging task for a rice breeder because quality has to be equally considered with yield, tolerance to abiotic stress and resistance to pests. Screening for quality of lines at the later stages of the breeding cycle does not guarantee a desired quality profile, defeating the main goal of the breeding program. Therefore, the use of molecular markers from the early stages of the breeding cycle is the comparatively promising option to ensure that traits of quality- and those of yield-associated traits are captured. Similar to most of the agronomically important traits, grain quality traits are inherited quantitatively and are presumed to be controlled by a large number of genes (polygenes), also known as quantitative trait loci (QTLs) (Tanksley 1993). While many studies have looked into the inheritance of routinely measured traits such as amylose content (Bao et al. 2004; Xu et al. 2015; Chen et al. 2008a), gelatinisation temperature (GT) (Waters et al. 2006; Cuevas et al. 2010), gel consistency (GelCon) (Tran et al. 2011), pasting properties derived from a rapid visco analyser (RVA) (Chen et al. 2008b), cooked rice texture (Meullenet et al. 2000; Champagne et al. 2004) and aroma measured by the concentration of 2AP using gas chromatography-mass spectrometry, the connection between the genetics and the expression of these traits is still poorly understood, either due to unavailability of accurate phenotype tools and/or inadequate markers that are useful for marker-assisted selection (MAS). Therefore, it is essential to have accurate and precise tools to identify the causative gene and protein and to design proper markers that are linked to the traits of interest.

With the decrease in the cost for genotyping either by using the customised single nucleotide polymorphism (SNP) chips or by genotyping by sequencing, the majority of genetic mapping studies use SNPs to genotype their samples (Thomson et al. 2012; Thomson 2014). SNPs are the most abundant source of genetic variation (McCouch et al. 2010; Kim et al. 2016). In rice quality, it is known that SNPs can definitely alter a phenotype, such as gelatinisation temperature (Cuevas et al. 2010).

18.2.1 Physical Traits

The physical attributes of the rice grain are those traits that describe the rice grains after the milling process and are associated with grain length (GL), grain width (GW), length-to-width ratio or the grain shape and chalk. These traits are determinants of quality and head rice yield, and they directly influence acceptance of varieties in the market. Variation in physical quality of the grain is most easily distinguishable, as it is obvious to rice consumers as they browse through markets. Without cooking the rice or tasting the raw kernels, a consumer can assess a good quality rice based on the length, width, and colour of the grains. Figure 18.1 shows

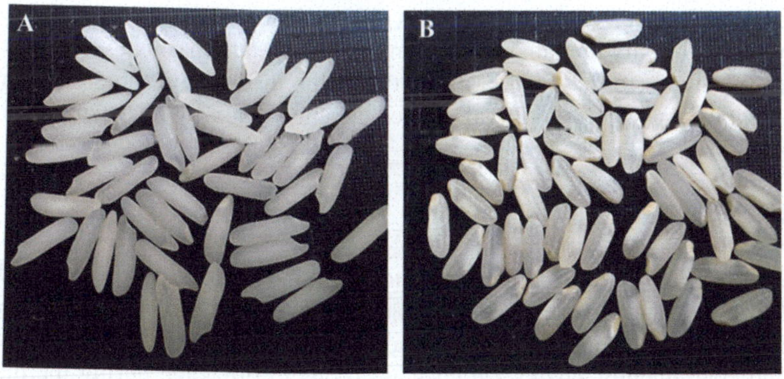

Fig. 18.1 Sample milled rice grains of two traditional Cambodian rice varieties. Phka Rumduol with long, slender and non-chalky grains (**a**); Thmar Krem (TMK) with medium and chalky grains (**b**). These varieties were used as parents for the recombinant inbred lines used for quantitative trait loci mapping for rice physical traits (Concepcion et al. 2019)

the milled rice grains of two *indica* rice varieties that differ in grain dimensions and chalkiness. For brown rice, grain dimensions are also easily assessed. Preferences for grain size and shape vary across different countries and cultures. Therefore, it is essential that breeding programs aim to release varieties of improved agronomic traits to incorporate specific physical grain traits to match consumer preference of a specific rice-consuming area.

18.2.1.1 Grain Dimensions

Grain size is often described by its three-dimensional structures: grain length (GL), grain width (GW) and length-to-width ratio (Calingacion et al. 2014). Several QTLs and functional markers have been reported for GL (Fan et al. 2006; Amarawathi et al. 2008; Wang et al. 2011; Zeng et al. 2016) and GW (Song et al. 2007; Wan et al. 2008; Weng et al. 2008). Using 185 SSR markers and a set of 286 F_8 recombinant inbred lines (RILs) derived from an *indica* × *japonica* cross, Lou et al. (2009) identified 16 QTLs across six rice chromosomes. Of these, a pleiotropic main effect QTL on chromosome 3 revealed the biggest effect on variation in GL (46.0%), whereas four QTLs identified for GW explained only about 10% of the total phenotypic variation in GW.

Of these studies, the genetic basis for GL and GW has been attributed to major genes *GS3* and *GW2*. Using a diverse collection of rice varieties, an association study revealed that a C to A mutation in the second exon of *GS3* (A allele) was associated with enhanced grain length in *Oryza sativa* but was absent from other *Oryza* species (Fan et al. 2009; Takano-Kai et al. 2009). *GW2* encodes a RING-type E3 ubiquitin ligase and associated with grain weight in rice (Song et al. 2007). A 1-bp deletion resulting in a premature stop codon in the *GW2* gene was responsible for increasing seed size (Song et al. 2007). In a more recent QTL mapping study using the 5 K SNP chip, Concepcion et al. (2019) identified a large-effect QTL for GL on rice chromosome 3, which explains 41.72% of the variation in GL, and in addition identified QTL on chromosomes 2, 3, 4, 7 and 9. However, there was no QTL identified on chromosome 2 and instead a large-effect QTL was revealed on chromosome 5, which explains 33.69% in GW variation.

Given the diversity in grain shape, from the long slender basmati varieties to the bold arborio types to the tiny Carnaroli types, the diversity in three-dimensional shape of grains has not been fully described genetically nor by image analysis.

18.2.1.2 Chalk

Chalkiness in rice grains is both an indicator of grain physical quality and head rice yield (Fitzgerald and Resurreccion 2009; Peng et al. 2014; Zhao et al. 2015). Chalky grains are inferior to non-chalky grains and due to the loose packing of starch granules, and consequently, chalky varieties are prone to breakage during the milling process (Cheng et al. 2005). Starch granules in chalky grains are much smaller than those present in translucent parts of the grain (Lisle et al. 2000). Further, in the chalky parts of the grain, the individual granules are not sufficiently built with starch

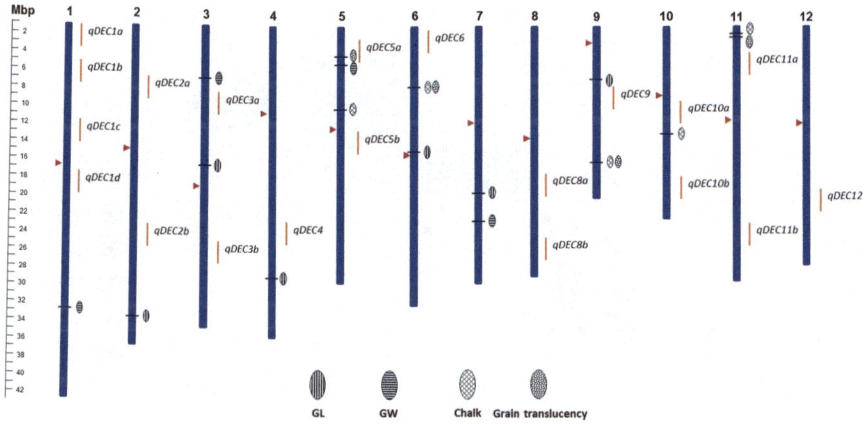

Fig. 18.2 Physical map of QTLs identified for grain length (GL), grain width (GW), chalk and grain translucency mapped on 12 rice chromosomes. These QTLs were identified using GLM and CIM methods (Concepcion et al. 2019). On the left, the scale indicates the mega base pair (Mbp) distance. Arrows (in red) on chromosomal maps represent the position of centromere. Vertical orange lines indicate the QTLs identified for degree of chalk (DEC) by Zhao et al. (2015)

to form the compound granules that are characteristic of rice. Chalk has been shown to diminish the cooking quality of rice as well (Cheng et al. 2005). There are several QTLs identified for chalk (Zhao et al. 2015, Concepcion et al. 2019) (Fig. 18.2), yet further work is still being done to identify the genetic basis of chalk, and perhaps this can be done by exploring alternative methods of deciphering the trait, such as developing mutant lines and/or by fine mapping the QTL regions identified as stable across several environments.

18.2.2 *Textural Traits*

The texture of cooked rice is widely studied; yet it is still not fully characterised because of several reasons related to its multigenic nature. Texture can further be defined by traits that rice consumers easily recognise: softness, stickiness, elasticity of chew and time to break into particles when chewing. Softness on cooking is easily measured by instruments and can be easily measured by a sensory panel. Softness relates to the amount of amylose, possibly also the chain length distribution of amylose and the amount of protein. Stickiness between grains has been shown to be due to small amylopectin molecules and small amylose molecules moving to the surface of the grain during cooking (Li et al. 2016).

Traditionally, rice texture is related to the amount of amylose in the endosperm and is associated with the *granule-bound starch synthase* (*GBSSI*) protein, responsible for amylose synthesis, but it is clear that the size, or chain length distribution of amylose, plays a key role in textural traits. Single nucleotide substitutions at the coding regions of the Waxy gene are associated with variation in the amylose content in rice varieties.

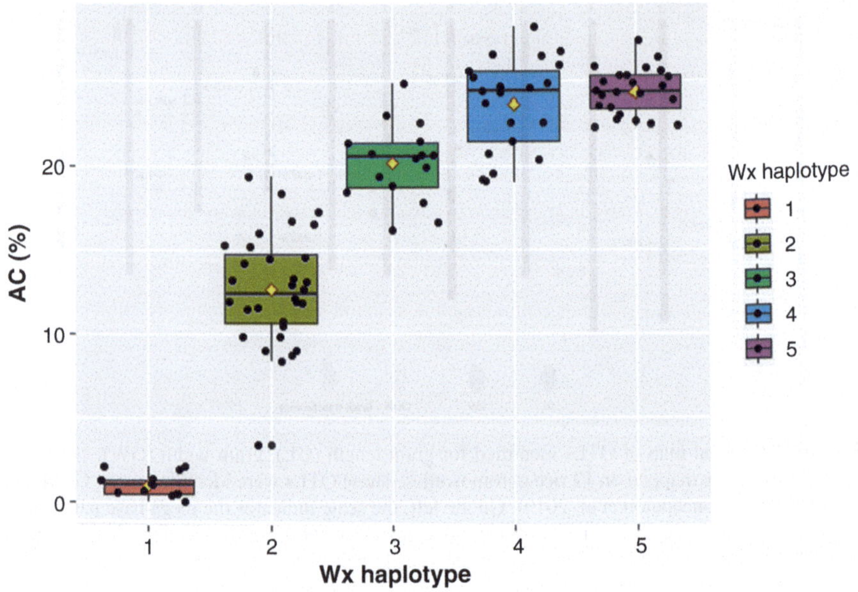

Fig. 18.3 The amylose content (%) across 106 rice varieties classified under five haplotypes of the *Waxy* gene (*Wx*). Haplotypes were grouped based on the mutations in the rice waxy gene: 1 = G/T SNP at intron 1, 2 = 23 duplication at exon 2, 3 = T/C SNP at exon 6, 4 = soft GelCon and T/C SNP at exon 10, hard GelCon and T/C SNP at exon 10. Black horizontal lines and yellow triangles denote the median and the means, respectively

Figure 18.3 shows the AC of milled rice derived from 106 rice varieties classified under five haplotypes of the *Waxy* gene (*Wx*). Haplotypes were grouped based on the mutations *Wx* (Table 18.1). Black horizontal lines and yellow triangles denote the median and the means, respectively. It is worth noting that within each AC class, differentiation between varieties is observed, indicating that other parameters and/or metrics are needed to reveal such variation in the textural phenotype.

To identify further sources of variation within the low AC class, Calingacion et al. (2014) sequenced the *Wx* gene and found no difference in the nucleotide sequences among these low AC rice. This is not surprising considering that more and more genetic mapping studies have been identifying QTLs for AC in other regions of the rice genome, particularly on chromosomes 1 and 4 (Concepcion et al. 2019). In this recent work on a mapping population derived from the cross between two traditional varieties of different AC (PRD and TMK), QTL mapping of a RIL population derived from two varieties of similar GelCon (Concepcion et al. 2019). The eating and cooking quality of PRD is largely explained by the *Waxy* (Yamanaka et al. 2004) and *SSIIa* (Cuevas et al. 2010), genes which regulate the synthesis of amylose and amylopectin. These two genes are known to exhibit mutations that correspond to variation in AC and GT across populations (Tables 18.1 and 18.2). However, the fact that multiple QTLs explain the variation of AC and GT in the population means that the *Wx* and the *SSIIa* markers at chromosome 6 are not sufficient to carry out targeted MAS for these traits.

Table 18.1 Known mutations in the rice *Waxy* gene and the corresponding grain quality phenotype

Wx haplotype	Intron 1	Exon 2	Exon 6	Exon 10	Phenotype
1	T*	+			Waxy
2	T	–			Low AC
3	G	–	C		Intermediate AC
4	G	–	A	C	High AC/soft GelCon
5	G	–	A	T	High AC/hard GelCon

[a]Waxy rice varieties predominantly carry the Wx^b allele; however, there are waxy rice varieties that carry the Wx^a allele. The Wx^a allele was predominant in non-waxy *indica* cultivars, whereas the Wx^b allele was common to the non-waxy *japonica* variety. Waxy rice strains have both Wx^a - and Wx^b -derived alleles, but that the Wx^b-derived allele was predominant, and its distribution was independent of *indica-japonica* differentiation (Yamanaka et al. 2004)

Table 18.2 Known mutations in the rice *Starch Synthase IIa* gene (*SSIIa*) and the corresponding grain quality phenotype

SSIIa haplotype	Sequence	DSC values (°C)[a]	Phenotype
1	G/G/GC	77.5 ± 1.5 ($n = 622$)	High GT
2	A/G/GC	77.3 ± 1.7 ($n = 491$)	High GT
3	A/A/GC	70.6 ± 2.9 ($n = 127$)	Low GT
4	A/G/TT	70.4 ± 2.5 ($n = 303$)	Low GT

[a]Values were derived from the measurements of 1543 traditional rice varieties (Cuevas et al. 2010)

Often considered as alternative method to measure the texture of cooked rice is the rapid visco analysis (RVA) of rice flour. The values for the various points of the RVA curves are peak viscosity (PV), trough viscosity (TV) and final viscosity (FV), and from these inflection points, other traits are derived: breakdown (BD = PV-TV), hot paste consistency (HPC=FV-TV) and setback (SB=FV-PV) (Champagne et al. 1999; Fitzgerald et al. 2003). Peak viscosity, TV and FV are routinely measured to assess starch texture and this is mainly because of their existing significant correlations with amylose content (Bao et al. 2004; Kong et al. 2015).

Using 768 SNPs in a recombinant inbred (RI) *indica-indica* population, several QTLs were identified for RVA-derived traits (Concepcion et al. 2019). Most of the major QTLs in this study were found to co-localise with QTLs for AC and GT. In this study, the most significant QTL for AC, *qAAC6.1*, co-localised with the QTLs for PV and TV and the derived traits of SB, BD and HPC, reaffirming their strong statistical correlations with AC. PV and FV, for instance, are strongly correlated with AC, which shows that pasting properties are impacted by amylose. Amylose leaches from granules during swelling and so restricts granule expansion, and amylose forms a gel on cooling, which shows the different functions of amylose throughout the cooking process, and these are reflected in both PV and FV (Fitzgerald et al. 2003).

The most recent association mapping study done on rice used about 150,000 high-quality SNPs resulting from genotyping data of 700 K high Density Rice Array of the Indica diversity panel of 218 lines (Misra et al. 2018). In this study, 224 significant loci were identified by multilocus GWAS and 97 of these were validated in at least two out of the four multilocus methods. The GWAS results coincide with the very significant negative correlation observed between AC and adhesiveness, which is the degree to which the rice kernels adhere to each other and to the surface of the probe, and the significant correlation between AC and hardness, which is defined as the force required to bite through the sample with the molars (Meullenet et al. 2000).

The novel genetic loci and putative candidate genes influencing textural properties beyond AC will be a useful resource for deployment into the MAS to capture consumer preferences influencing rice texture. Putting it all together, at this stage we understand that rice texture is largely controlled by the major genes *Wx*, *SSIIa* and known minor genes, *SSI* and *SBE* enzymes, and in the future, we expect more functional genes to join the panel that affect texture of cooked rice. The direction of finding new genetic loci that explains the textural variations in cooked rice must be pursued in order to provide more specificity in varietal grading, hence in MAS strategies.

18.2.3 Aroma

Aroma is one of the most sought after quality traits in rice. The most popular high-quality rice varieties such as the jasmine (Daygon et al. 2016; Concepcion et al. 2018) and basmati rice types (Amarawathi et al. 2008; Mahattanatawee and Rouseff, 2014; Hinge et al. 2016) are aromatic. This phenotypic trait is associated with the presence of the volatile organic compound (VOC) 2-acetyl-1-pyrroline (2AP) (Buttery et al. 1983). The earliest genetic mapping on rice aroma was done using restriction fragment length polymorphic (RFLP) markers (Ahn et al. 1992). Currently, the allele-specific marker developed by Bradbury et al. (2005b) for rice fragrance is widely deployed across rice breeding programs that target fragrant rice. The advantage of this marker set is its ability to detect heterozygotes, which therefore offers more precision for marker-assisted selection. However, because this marker is unable to detect the actual differences in the amounts of 2AP among fragrant homozygous lines (Roferos et al. 2008), the bottleneck of further assessment of the fragrance trait is necessary after postharvest processing.

The genetics of 2AP is linked to mutations in the amino aldehyde dehydrogenase 2 gene (*AADH2*) or the *FGR* locus (Chen et al. 2008c; Shi et al. 2008; Kovach et al. 2009). The most widespread of these mutations is the 8-bp deletion and three SNPs at the seventh exon of *AAHD2* (Bradbury et al. 2005a). Varieties carrying *fgr*, the recessive allele, produce a truncated AADH protein which is unable to catalyse the oxidation of 4-aminobutanal to GABA, hence enabling the accumulation of 2AP (Bradbury et al. 2008; Daygon et al. 2017).

Using untargeted metabolomics, the volatile profiles, including 2AP and other nitrogen-containing compounds, were detected in fragrant rice lines and varieties (Daygon et al. 2016; Concepcion et al. 2018). All of these N-containing VOCs are part of the biochemical mechanism of 2AP and provide information on further differentiating the fragrant lines from each other. Because of the apparent benefits of producing high-quality rice, improved techniques in ensuring the expression of the desired traits are necessary. Future work that involves development of markers specific to these other compounds will allow greater precision and accuracy in fragrant line selection and improved classification of rice quality grades (Daygon et al. 2017).

18.3 Metabolism Verifies the Genome

The availability of genomic tools has indeed enhanced the curation of the known and putative genetic loci for rice quality traits (Varshney et al. 2009; Rasheed et al. 2017); however, up to date, not much progress has been made that enables the pool of genetic markers for quality to grow. This is because the major genes for quality are now known and we have reached a limited phenotyping capacity. A wide array of quality traits has emerged to be important for various stakeholders in the rice value chain. However, limited information is known on the genetic and biochemical basis of the variations in the quality of these rice types.

Metabolomics has recently revealed the wealth of biochemical footprints that ultimately differentiate varieties which the current suite of quality evaluation tools is unable to measure. Mumm et al. (2016) screened a diverse set of varieties using four different platforms in metabolomics and identified differences among jasmine and basmati rice types in terms of their metabolic profiles. Calingacion et al. (2015) identified characterising volatile compounds between two non-aromatic *indica* varieties Apo and IR64, which are also of the same amylose class. Daygon et al. (2016) identified characterising volatile compounds for aromatic varieties and related these compounds to the sensory qualities of the varieties. Concepcion et al. (2018) identified significant differences in the amount of unsaturated fatty acids in the milled rice grains of fragrant and non-fragrant varieties. These unsaturated fatty acids are prone to oxidation, a process which produces secondary metabolites such as volatile organic compounds that emit characteristic aromas.

From these sets of independent metabolomics experiments, there are several directions to which further research can be pursued. One of the areas that can be explored is lipidomics, which is a branch of metabolomics that covers that study of the lipidome (Shevchenko and Simons 2010). Lipids are one of the macromolecules that are found in the rice endosperm (Fujino 1978). Lipids constitute less than 1% of the rice endosperm (Resurreccion and Juliano 1975). Often overlooked due to the small quantity relative to starch and proteins, the effect of lipids on rice quality traits has not been well studied. Considering these functional differences and the minute abundance of lipids in the grain, fewer studies have investigated the impact of lipids on rice grain quality.

Rice starches are known to contain amylose-lipid complexes (Godet et al. 1995; Godet et al. 1993). Amylose forms double helices or single helices in the native state (Buléon et al. 1998). Single helices give rise to a central cavity that can be filled with compounds such as iodine, alcohols or fatty acids. Further work is required to identify the lipid species that characterise varieties and, more so, the

A

B

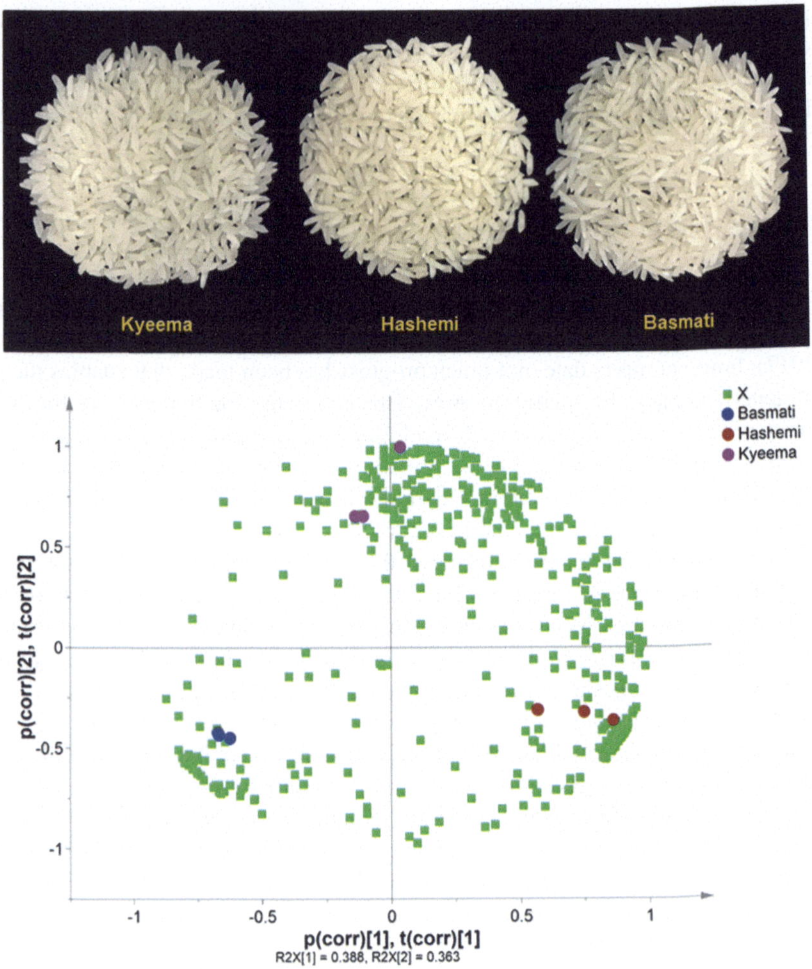

Fig. 18.4 Milled rice grains of three rice varieties Kyeema, Hashemi and basmati (**a**); biplot of the lipidomics features detected in the lipid extracts of these three varieties using the ultra-performance liquid chromatography tandem mass spectrometry. Scores are coloured according to variety and loadings that represent the lipidomics features are shown as green boxes (**b**)

environmental impact on these compounds in order to take advantage of this phenotyping technique. Figure 18.4 shows the milled rice grains of three popular rice varieties basmati from India, Hashemi from Iran and Kyeema from Australia (Fig. 18.4a) and a principal component analysis biplot showing the lipid features in green boxes gathered from lipid extracts of these varieties (shown in coloured circles) (Fig. 18.4b) analysed using an ultra-performance liquid chromatography tandem mass spectrometry (UPLC-MS/MS) (Waters Corporation, USA). Based on this untargeted profiling of lipids in rice, lipidomics can be used as an additional phenotyping platform, and with progress in the identification of lipid species using curated databases (Fahy et al. 2005, 2009; Sud et al. 2007) and synthetic standards, more lipid species can be accurately identified, hence properly used for future rice metabolite QTL mapping studies.

18.4 Genomic Selection for Grain Quality

Quality evaluation is a significant bottleneck within a breeding program in terms of the time required to measure samples but also the quantities of paddy required for these evaluations not being available until later generations, thus increasing the time of a breeding cycle and limiting the rates of genetic gains within a breeding program. The use of the large-effect QTLs/genes as described above for marker-assisted selection is an effective strategy in culling unfavourable genotypes without the need for phenotypic evaluation, allowing for greater investment in germplasm of greater relevance. However, there is still the bottleneck regarding the processing time and paddy requirements as mentioned above. On top of these known limitations in phenotyping requirements, MAS has reached a plateau as this process has mainly targeted large-effect QTLs and known genes (Jannink et al. 2010). Therefore, a new strategy such as the use of genomic selection (GS) for rice quality traits is viable option forward in breeding programs.

The advantage of genomic selection (GS) is that it uses individuals with known phenotypic and genotypic information, or a training population, to predict the genetic estimated breeding values (GEBVs), or with some model genetic values (GV), of individuals with unknown phenotype (Meuwissen et al. 2001). In contrast to marker-assisted breeding approaches which fit one marker/QTL, genomic selection uses the training set to develop a prediction equation which estimates all marker effects of a given trait simultaneously (Meuwissen et al. 2001). There have been numerous papers on training population development, genotyping requirements, statistical models and machine learning methods, and breeding strategies for its implementation (Spindel et al. 2015; Sverrisdóttir et al. 2018; Edwards et al. 2019).

More research is needed to show the validity of GS for quality parameters in breeding, but k-fold cross-validation approaches within diverse or breeding panels have demonstrated promise for the likes of AC (Zhang et al. 2014; Suela et al. 2019) and chalkiness (Morais et al. 2017; Morais Júnior et al. 2018). Furthermore, cross-validation has highlighted the importance of selecting the appropriate method, whether a GS approach or MAS, to reflect the genetic architecture of a trait. For example, a random forest method ($r = 0.76$) led to 12% increase in the selection accuracies for chalkiness compared to genomic best linear unbiased prediction (BLUP), which assumes marker effects follow a normal distribution (Morais Júnior et al. 2018). Zhang et al. (2014) noted similar results with a weighted relationship matrix for amylose content, and Spindel et al. (2015) demonstrated a multiple linear regression containing up to the 100 most significant SNPs from a genome-wide association was more informative than a genomic best linear unbiased prediction (GBLUP) model for flowering time. GBLUP is a method that utilises genomic relationships by way of nucleotide information to estimate the genetic contribution of an individual in a population.

Genomic selection is well-suited to a classical recurrent selection program, given their small effective population and high relatedness. We created a GS scheme for rice that removes yield trials, and substantial quality evaluation, before a subsequent breeding cycle (Fig. 18.5). Genetic gains would be achieved by the reduction in the time for the breeding cycle, on the assumption that selection accuracies are not reduced by the same proportion as per the breeder's equation (Cobb et al. 2019). The scheme fits an existing recurrent selection program with approximately 2 years per cycle. Progeny is advanced to the F_4 through rapid generation advancement (Collard et al. 2017), where plants are genotyped for MAS and GS of agronomic and quality traits (i.e. amylose, RVA derived traits, aroma) to provide (a) genomic breeding values (GEBV) to identify parents for subsequent crossing cycles and (b) GV for potential varietal release. Selected individuals are advanced for crossing and phenotypic evaluation, which is also used to update GS models of subsequent breeding cycles. Gezan et al. (2017) demonstrated that advance selection trials could be training populations of subsequent breeding cycles in octoploid strawberries, but the approach needs validation in rice.

In breeding programs, F_4 phenotyping for AC was incorporated to adopt a multivariate GS approach to improve the predictive ability of texture-related traits (i.e. RVA derived traits) as non-destructive sampling becomes more and more feasible. Multivariate GS with AC is hypothesised to improve the predictive accuracies of texture-related traits given their known relationships (Allahgholipour et al. 2006; Blazek and Copeland 2008; Varavinit et al. 2003; Bao et al. 2004; Concepcion et al. 2019). Multivariate GS has been demonstrated to improve selection accuracies for end-use quality traits compared to univariate models in wheat (Michel et al. 2018; Lado et al. 2018). The extent of gains achieved by the inclusion of multi-trait data in GS models relies on genetic correlations between targeted traits and secondary traits, but also the heritability of a given trait.

There are numerous approaches for the deployment of GS and its implementation comes down to the genetic gains per unit cost and time (i.e. it may be more cost-effective to do field evaluations and MAS before genome-wide genotyping for GS). There are numerous tools available to simulate estimated genetic gains and efficiencies of a proposed breeding strategy that can be utilised to provide evidence of their effectiveness (Wang et al. 2007; Denis and Bouvet 2013; Gouy et al. 2013). Genomic selection is indeed one of many tools at a program disposal to improve rates of genetic gains and it is in the best interest of breeding programs to leverage appropriate tools in order to optimise genetic gains with the resources available. This section on GS and the new GS scheme that we present in this current chapter (Fig. 18.5) can be used to further improve the breeding strategy for rice quality improvement.

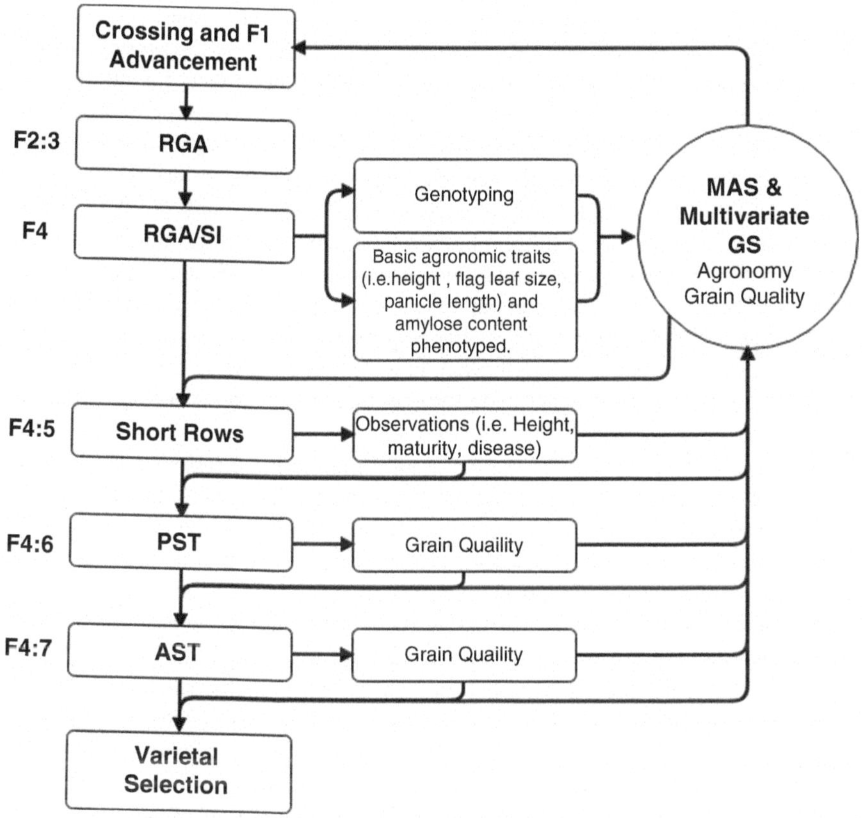

Fig. 18.5 A flow diagram of a potential breeding scheme for an existing recurrent selection program that implements rapid generation advancement (RGA) and genomic selection to reduce the length of a breeding cycle. Arrows indicate the flow of breeding material/information. Abbreviations: SI seed increase, PST preliminary selection trials, AST advanced selection trials

18.5 Post-Genomics Era: Are We There Yet?

Since sequencing technologies have become affordable, they have grown in popularity for genotyping in rice. The main limitations of rice research now are caused by phenotyping tools and the slow gains in those (Concepcion et al. 2015). As much as the idea of a post-genomics era for rice can mean that we have reached the point of fully revealing the genetics of rice, it is still important to note that the area of rice genomics and the current suite of molecular markers for grain quality still require further improvements. In one direction, new traits are being revealed using metabolomics as an additional phenotyping platform, therefore improving the current phenotyping tools for rice grain quality, and in parallel, genomic selection instead of the usual MAS will be incorporated in breeding programs.

The most recent genomic development is the clustered regularly interspaced short palindromic repeats/CRISPR-associated protein (CRISPR/Cas) system, which is a most controversial, but very promising tool that can be used for future trait-targeted crop improvement programs. The tool is an RNA-guided genome editing system (Mali et al. 2013) that has recently emerged as an alternative nuclease-based method for efficient and versatile genome engineering (Mao et al. 2013). In this system, only the 20-nt targeting sequence within the single-guide RNA (sgRNA) needs to be changed to target different genes. The simplicity of the cloning strategy and the few limitations on potential target sites make the CRISPR/Cas system very appealing. There are a small number of reports which have used CRISPR/Cas9 gene-editing technology for grain quality improvement in rice (Zhang et al. 2018; Sun et al. 2017; Shufen et al. 2019; Shan et al. 2014; Yang et al. 2016; Tang et al. 2017). Zhang et al. (2018) generated glutinous rice by CRISPR/Cas9 targeting the first exon of the *Waxy* gene in two elite non-waxy or non-glutinous *japonica* varieties and with the expectation to produce a null mutation that introduces a loss-of-function mutation into the *Waxy* gene. In this study, the induced mutations reduced AC and produced glutinous rice that showed typical physical traits of waxy rice – white, fully opaque – without affecting other desirable agronomic traits. With the goal of developing rice with lower glycaemic index (GI) and therefore improved nutritional properties, Sun et al. (2017) used CRISPR/Cas9 technology to generate targeted mutagenesis in the *STARCH BRANCHING ENZYMES I* and *IIb* (*SBEI* and *SBEIIb*) in rice. This study successfully produced *sbeII* mutants with altered amylopectin structure and with significantly increased AC and resistant starch content. In another study, the *ISOAMYLASE 1* (*ISA1*) gene in rice, which plays a predominant role in amylopectin synthesis, was edited using CRISPR/Cas9 technology via *Agrobacterium*-mediated transformation (Shufen et al. 2019). The mutation events lowered the transcript levels of most starch synthesis-related genes. The total starch, amylose and amylopectin in the endosperm of mutants were significantly reduced, whereas sugar content and starch gel consistency were observably increased compared to the wild-type and the gelatinisation temperature and starch chain length distributions of the mutants were also altered. Shan et al. (2013) demonstrated that CRISPR/Cas9 demonstrated produced higher mutation frequencies on the major locus for rice aroma, *OsBAHD2* among other rice genes, compared to that induced by transcription activator-like effector nucleases (TALENs). These studies have targeted well-known genes for rice texture and aroma but can be later used for altering novel genetic regions for novel traits of quality.

Since the main markers for rice grain quality have been popularised and are currently used across several biparental and diverse populations, and increasingly for developing gene-edited crops using the CRISPR/Cas system, perhaps it is about time to utilise a more dynamic population set such as multi-parent mapping populations and specifically the multi-parent advanced generation intercross (MAGIC) population (Cavanagh et al. 2008). These types of mapping populations present unparalleled genetic and phenotypic diversity that are suitable for high-resolution mapping of complex traits (Huang et al. 2015; Meng et al. 2016; Ponce et al. 2018). Compared to the currently popular GWAS, which also offers high-resolution mapping by exploiting historical recombination events across a natural diverse set of germplasm, MAGIC populations are less prone to extensive population structure and ambiguous kinship that increases the likelihood of discovering false positive marker-trait associations (Huang et al. 2015). Recently, a MAGIC population consisting of 508 F_6 MAGIC lines developed from eight founder *indica* rice lines (Meng et al. 2016) was used for GWAS for rice textural traits (Ponce et al. 2018). In this study, the MAGIC population showed greater phenotypic variation in almost all traits than the eight founder lines and suggested the possibility of selecting specific recombinants based on variations in AC and GelCon.

18.6 Conclusions

Better strategies are needed to reach a perfect variety of rice. With genome sequencing and genetic markers becoming cheaper, the future of rice genomics rests upon its utility to combine the emerging omics technology such as metabolomics and then by using genomic selection to predict traits of interest. This means that greater emphasis must be made on early selection during the breeding cycle instead of waiting until more plots and plants are available. This saves resources and will be more beneficial for rice farmers in the long run.

References

Ahn SN, Bollich CN, Tanksley SD (1992) RFLP tagging of a gene for aroma in rice. Theor Appl Genet 84:825–828

Allahgholipour M, Ali AJ, Alinia F, Nagamine T, Kojima Y (2006) Relationship between rice grain amylose and pasting properties for breeding better quality rice varieties. Plant Breed 125:357–362

Amarawathi Y, Singh R, Singh AK, Singh VP, Mohapatra T, Sharma TR, Singh NK (2008) Mapping of quantitative trait loci for basmati quality traits in rice (Oryza sativa L.). Mol Breed 21:49–65

Bao J, Kong X, Xie J, Xu L (2004) Analysis of genotypic and environmental effects on rice starch: Apparent amylose content, pasting viscosity, and gel texture. J Agric Food Chem 52:6010–6016

Blazek J, Copeland L (2008) Pasting and swelling properties of wheat flour and starch in relation to amylose content. Carbohydr Polym 71:380–387

Bradbury LM, Gillies SA, Brushett DJ, Waters DL, Henry RJ (2008) Inactivation of an aminoaldehyde dehydrogenase is responsible for fragrance in rice. Plant Mol Biol 68:439–449

Bradbury LMT, Fitzgerald TL, Henry RJ, Jin Q, Waters DLE (2005a) The gene for fragrance in rice. Plant Biotechnol J 3:363–370

Bradbury LMT, Henry RJ, Jin Q, Reinke RF, Waters DLE (2005b) A perfect marker for fragrance genotyping in rice. Mol Breed 16:279–283

Buléon A, Colonna P, Planchot V, Ball S (1998) Starch granules: structure and biosynthesis. Int J Biol Macromol 23:85–112

Buttery RG, Ling LC, Juliano BO, Turnbaugh JG (1983) Cooked rice aroma and 2-acetyl-1-pyrroline. J Agric Food Chem 31:823–826

Calingacion M, Mumm R, Tan K, Quiatchon-Baeza L, Concepcion JCT, Hageman JA, Prakash S, Fitzgerald M, Hall RD (2017) A multidisciplinary phenotyping and genotyping analysis of a mapping population enables quality to be combined with yield in rice. Front Mol Biosci 4:1–16

Calingacion MN, Fang L, Quiatchon-Baeza L, Mumm R, Riedel A, Hall RD, Fitzgerald MA (2015) Delving deeper into technological innovations to understand differences in rice quality. Rice 8:6

Calingacion MN, Laborte A, Nelson A, Resurreccion A, Concepcion JCT, Daygon VD, Mumm R, Reinke R, Dipti S, Bassinello PZ (2014) Diversity of global rice markets and the science required for consumer-targeted rice breeding. PLoS One 9:e85106

Cavanagh C, Morell M, Mackay I, Powell W (2008) From mutations to MAGIC: resources for gene discovery, validation and delivery in crop plants. Curr Opin Plant Biol 11:215–221

Champagne E, Bett-Garber K, McClung A, Bergman C (2004) Sensory characteristics of diverse rice cultivars as influenced by genetic and environmental factors. Cereal Chem J 81:237–243

Champagne ET, Bett-Garber KL, Fitzgerald MA, Grimm CC, Lea J, Ohtsubo KI, Jongdee S, Xie L, Bassinello PZ, Resurreccion A, Ahmad R, Habibi F, Reinke R (2010) Important sensory properties differentiating premium rice varieties. Rice 3:270–281

Champagne ET, Bett KL, Vinyard BT, McClung AM, Barton FE II, Moldenhauer K, Linscombe S, McKenzie K (1999) Correlation between cooked rice texture and rapid visco analyser measurements. Cereal Chem 76:764–771

Chen M-H, Bergman C, Pinson S, Fjellstrom R (2008a) Waxy gene haplotypes: associations with apparent amylose content and the effect by the environment in an international rice germplasm collection. J Cereal Sci 47:536–545

Chen M-H, Bergman CJ, Pinson SRM, Fjellstrom RG (2008b) Waxy gene haplotypes: associations with pasting properties in an international rice germplasm collection. J Cereal Sci 48:781–788

Chen S, Yang Y, Shi W, Ji Q, He F, Zhang Z, Cheng Z, Liu X, Xu M (2008c) Badh2, encoding betaine aldehyde dehydrogenase, inhibits the biosynthesis of 2-acetyl-1-pyrroline, a major component in rice fragrance. Plant Cell 20:1850–1861

Cheng FM, Zhong LJ, Wang F, Zhang GP (2005) Differences in cooking and eating properties between chalky and translucent parts in rice grains. Food Chem 90:39–46

Cobb JN, Juma RU, Biswas PS, Arbelaez JD, Rutkoski J, Atlin G, Hagen T, Quinn M, Ng EH (2019) Enhancing the rate of genetic gain in public-sector plant breeding programs: lessons from the breeder's equation. Theor Appl Genet 132:627–645

Collard BCY, Beredo JC, Lenaerts B, Mendoza R, Santelices R, Lopena V, Verdeprado H, Raghavan C, Gregorio GB, Vial L, Demont M, Biswas PS, Iftekharuddaula KM, Rahman MA, Cobb JN, Islam MR (2017) Revisiting rice breeding methods – evaluating the use of rapid generation advance (RGA) for routine rice breeding. Plant Prod Sci 20:337–352

Collard BCY, Jahufer MZZ, Brouwer JB, Pang ECK (2005) An introduction to markers, quantitative trait loci (QTL) mapping and marker-assisted selection for crop improvement: The basic concepts. Euphytica 142:169–196

Concepcion JCT, Ouk M, Zhao D, Fitzgerald MA (2015) The need for new tools and investment to improve the accuracy of selecting for grain quality in rice. Field Crop Res 182:60–67

Concepcion JCT, Ouk S, Ouk M, Fitzgerald MA (2019) Mapping quantitative traits for grain physical and textural quality in Cambodian Jasmine rice PRD. Euphytica 215:135

Concepcion JCT, Ouk S, Riedel A, Calingacion M, Zhao D, Ouk M, Garson MJ, Fitzgerald MA (2018) Quality evaluation, fatty acid analysis and untargeted profiling of volatiles in Cambodian rice. Food Chem 240:1014–1021

Cuevas RP, Daygon VD, Corpuz HM, Nora L, Reinke RF, Waters DLE, Fitzgerald MA (2010) Melting the secrets of gelatinisation temperature in rice. Funct Plant Biol 37:439–447

Cuevas RPO, Domingo CJ, Sreenivasulu N (2018) Multivariate-based classification of predicting cooking quality ideotypes in rice (Oryza sativa L.) indica germplasm. Rice 11:56

Daygon VD, Calingacion M, Forster L, Voss J, Schwartz B, Ovenden B, Alonso D, McCouch S, Garson MJ, Fitzgerald MA (2017) Metabolomics and genomics combine to unravel the pathway for the presence of fragrance in rice. Sci Rep 7:8767–8767

Daygon VD, Prakash S, Calingacion M, Riedel A, Ovenden B, Snell P, Mitchell J, Fitzgerald MA (2016) Understanding the Jasmine phenotype of rice through metabolite profiling and sensory evaluation. Metabolomics 12:1–15

Denis M, Bouvet J-M (2013) Efficiency of genomic selection with models including dominance effect in the context of Eucalyptus breeding. Tree Genet Genomes 9:37–51

Edwards S, Buntjer J, Jackson R, Bentley A, Lage J, Byrne E, Burt C, Jack P, Berry S, Flatman E, Poupard B, Smith S, Hayes C, Gaynor R, Gorjanc G, Howell P, Ober E, Mackay I (2019) The effects of training population design on genomic prediction accuracy in wheat. Theor Appl Genet:1–10

Fahy E, Subramaniam S, Brown HA, Glass CK, Merrill AH, Murphy RC, Raetz CRH, Russell DW, Seyama Y, Shaw W, Shimizu T, Spener F, Van Meer G, Vannieuwenhze MS, White SH, Witztum JL, Dennis EA (2005) A comprehensive classification system for lipids. J Lipid Res 46:839–862

Fahy E, Subramaniam S, Murphy RC, Nishijima M, Raetz CRH, Shimizu T, Spener F, Van Meer G, Wakelam MJO, Dennis EA (2009) Update of the LIPID MAPS comprehensive classification system for lipids. J Lipid Res 50:S9–S14

Fan C, Xing Y, Mao H, Lu T, Han B, Xu C, Li X, Zhang Q (2006) GS3, a major QTL for grain length and weight and minor QTL for grain width and thickness in rice, encodes a putative transmembrane protein. Theor Appl Genet 112:1164–1171

Fan C, Yu S, Wang C, Xing Y (2009) A causal C–A mutation in the second exon of GS3 highly associated with rice grain length and validated as a functional marker. Theor Appl Genet 118:465–472

Fitzgerald MA, Martin M, Ward RM, Park WD, Shead HJ (2003) Viscosity of rice four: a rheological and biological study. J Agric Food Chem 51:2295–2299

Fitzgerald MA, McCouch SR, Hall RD (2009) Not just a grain of rice: the quest for quality. Trends Plant Sci 14:133–139

Fitzgerald MA, Resurreccion AP (2009) Maintaining the yield of edible rice in a warming world. Funct Plant Biol 36:1037–1045

Fujino Y (1978) Rice lipids. Cereal Chem J 55:559–571

Gezan SA, Osorio LF, Verma S, Whitaker VM (2017) An experimental validation of genomic selection in octoploid strawberry. Hortic Res 4:16070

Godet MC, Buléon A, Tran V, Colonna P (1993) Structural features of fatty acid-amylose complexes. Carbohydr Polym 21:91–95

Godet MC, Tran V, Colonna P, Buleon A, Pezolet M (1995) Inclusion/exclusion of fatty acids in amylose complexes as a function of the fatty acid chain length. Int J Biol Macromol 17:405–408

Gouy M, Rousselle Y, Bastianelli D, Lecomte P, Bonnal L, Roques D, Efile J-C, Rocher S, Daugrois J, Toubi L, Nabeneza S, Hervouet C, Telismart H, Denis M, Thong-Chane A, Glaszmann JC, Hoarau J-Y, Nibouche S, Costet L (2013) Experimental assessment of the accuracy of genomic selection in sugarcane. Theor Appl Genet 126:2575–2586

Hinge VR, Patil HB, Nadaf AB (2016) Aroma volatile analyses and 2AP characterization at various developmental stages in basmati and non-basmati scented rice (Oryza sativa L.) cultivars. Rice 9

Huang BE, Verbyla KL, Verbyla AP, Raghavan C, Singh VK, Gaur P, Leung H, Varshney RK, Cavanagh CR (2015) MAGIC populations in crops: current status and future prospects. Theor Appl Genet 128:999–1017

Huang X, Wei X, Sang T, Zhao Q, Feng Q, Zhao Y, Li C, Zhu C, Lu T, Zhang Z, Li M, Fan D, Guo Y, Wang A, Wang L, Deng L, Li W, Lu Y, Weng Q, Liu K, Huang T, Zhou T, Jing Y, Li W, Lin Z, Buckler ES, Qian Q, Zhang Q-F, Li J, Han B (2010) Genome-wide association studies of 14 agronomic traits in rice landraces. Nat Genet 42:961–967

International Rice Genome Sequencing Project (2005) The map-based sequence of the rice genome. Nature 436:793–800

Jannink J-L, Lorenz AJ, Iwata H (2010) Genomic selection in plant breeding: from theory to practice. Brief Funct Genomics 9:166–177

Kim C, Guo H, Kong W, Chandnani R, Shuang L-S, Paterson AH (2016) Application of genotyping by sequencing technology to a variety of crop breeding programs. Plant Sci 242:14–22

Kong X, Zhu P, Sui Z, Bao J (2015) Physicochemical properties of starches from diverse rice cultivars varying in apparent amylose content and gelatinisation temperature combinations. Food Chem 172:433–440

Kovach MJ, Calingacion MN, Fitzgerald MA, McCouch SR (2009) The origin and evolution of fragrance in rice (Oryza sativa L.). Proc Natl Acad Sci U S A 106:14444–14449

Kumagai M, Tanaka T, Ohyanagi H, Hsing Y-IC, Itoh T (2018) Genome Sequences of Oryza Species. In: Sasaki T, Ashikari M (eds) Rice genomics, genetics and breeding. Springer Singapore, Singapore

Lado B, Vázquez D, Quincke M, Silva P, Aguilar I, Gutiérrez L (2018) Resource allocation optimization with multi-trait genomic prediction for bread wheat (Triticum aestivum L.) baking quality. Theor Appl Genet 131:2719–2731

Li H, Prakash S, Nicholson TM, Fitzgerald MA, Gilbert RG (2016) The importance of amylose and amylopectin fine structure for textural properties of cooked rice grains. Food Chem 196:702–711

Li J-Y, Wang J, Zeigler RS (2014) The 3,000 rice genomes project: new opportunities and challenges for future rice research. GigaScience 3:1–3

Lisle AJ, Martin M, Fitzgerald MA (2000) Chalky and translucent rice grains differ in starch composition and structure and cooking properties. Cereal Chem 77:627–632

Lou J, Chen L, Yue G, Lou Q, Mei H, Xiong L, Luo L (2009) QTL mapping of grain quality traits in rice. J Cereal Sci 50:145–151

Mahattanatawee K, Rouseff RL (2014) Comparison of aroma active and sulfur volatiles in three fragrant rice cultivars using GC–Olfactometry and GC–PFPD. Food Chem 154:1–6

Mali P, Yang L, Esvelt KM, Aach J, Guell M, Dicarlo JE, Norville JE, Church GM (2013) RNA-guided human genome engineering via Cas9. Science 339:823–826

Mao Y, Zhang H, Xu N, Zhang B, Gou F, Zhu J-K (2013) Application of the CRISPR-Cas system for efficient genome engineering in plants. Mol Plant 6:2008–2011

McCouch SR, Zhao K, Wright M, Tung C-W, Ebana K, Thomson M, Reynolds A, Wang D, Declerck G, Ali ML, McClung A, Eizenga G, Bustamante C (2010) Development of genome-wide SNP assays for rice. Breed Sci 60:524–535

Meng L, Guo L, Ponce K, Zhao X, Ye G (2016) Characterization of three Indica Rice multiparent advanced generation intercross (MAGIC) populations for quantitative trait loci identification. Plant Genome 9

Meullenet J-F, Champagne ET, Bett KL, McClung AM, Kauffmann D (2000) Instrumental assessment of cooked rice texture characteristics: a method for breeders. Cereal Chem J 77:512–517

Meuwissen THE, Hayes BJ, Goddard ME (2001) Prediction of total genetic value using genome-wide dense marker maps. Genetics 157:1819–1829

Michel S, Kummer C, Gallee M, Hellinger J, Ametz C, Akgöl B, Epure D, Löschenberger F, Buerstmayr H (2018) Improving the baking quality of bread wheat by genomic selection in early generations. Theor Appl Genet 131:477–493

Minten, B., Murshid, K. A. S. & Reardon, T (2011) The quiet revolution in agrifood value chains in Asia: the case of increasing quality in rice markets in Bangladesh. IFPRI - Discussion Papers, vi + 25 pp.

Misra G, Badoni S, Domingo CJ, Cuevas RPO, Llorente C, Mbanjo EGN, Sreenivasulu N (2018) Deciphering the genetic architecture of cooked rice texture. Front Plant Sci 9:1405–1405

Morais Júnior OP, Batista Duarte J, Breseghello F, Coelho ASG, Borba TCO, Aguiar J, Neves PCF, Morais OP (2017) Relevance of additive and nonadditive genetic relatedness for genomic prediction in rice population under recurrent selection breeding. Genet Mol Res 16:1–16

Morais Júnior OP, Breseghello F, Duarte JB, Coelho ASG, Borba TCO, Aguiar JT, Neves PCF, Morais OP (2018) Assessing prediction models for different traits in a rice population derived from a recurrent selection program. Crop Sci 58:2347–2359

Mullis, K. B., Erlich, H. A., Arnheim, N., Horn, G. T., Saiki, R. K. & Scharf, S. J. 1987. One of the first polymerase chain reaction (PCR) patents. Google Patents.

Mumm R, Hageman JA, Calingacion MN, De Vos RCH, Jonker HH, Erban A, Kopka J, Hansen TH, Laursen KH, Schjoerring JK, Ward JL, Beale MH, Jongee S, Rauf A, Habibi F, Indrasari SD, Sakhan S, Ramli A, Romero M, Reinke RF, Ohtsubo K, Boualaphanh C, Fitzgerald MA, Hall RD (2016) Multi-platform metabolomics analyses of a broad collection of fragrant and non-fragrant rice varieties reveals the high complexity of grain quality characteristics. Metabolomics 12:38

Peng B, Wang LQ, Fan CC, Jiang GH, Luo LJ, Li YB, He YQ (2014) Comparative mapping of chalkiness components in rice using five populations across two environments. BMC Genet 15:14

Ponce KS, Ye G, Zhao X (2018) QTL identification for cooking and eating quality in indica rice using multi-parent advanced generation intercross (MAGIC) population. Front Plant Sci 9:1–9

Purugganan M (2014) An evolutionary genomic tale of two rice species. Nat Genet 46:931–932

Rasheed A, Hao Y, Xia X, Khan A, Xu Y, Varshney RK, He Z (2017) Crop breeding chips and genotyping platforms: progress, challenges, and perspectives. Mol Plant 10:1047–1064

Resurreccion AP, Juliano BO (1975) Fatty acid composition of rice oils. J Sci Food Agric 26:437–439

Roferos, L. T., Butardo JR, V. M., Fitzgerald, M. A. & Juliano, B. O. 2008. Association between alleles of the Waxy gene and traits of grain quality in Philippine seed board rice varieties. Philipp Agric Sci, 91, 335.

Shan Q, Wang Y, Li J, Gao C (2014) Genome editing in rice and wheat using the CRISPR/Cas system. Nat Protoc 9:2395

Shan Q, Wang Y, Li J, Zhang Y, Chen K, Liang Z, Zhang K, Liu J, Xi JJ, Qiu J-L, Gao C (2013) Targeted genome modification of crop plants using a CRISPR-Cas system. Nat Biotechnol 31:686

Shevchenko A, Simons K (2010) Lipidomics: coming to grips with lipid diversity. Nat Rev Mol Cell Biol 11:593–598

Shi W, Yang Y, Chen S, Xu M (2008) Discovery of a new fragrance allele and the development of functional markers for the breeding of fragrant rice varieties. Mol Breed 22:185–192

Shufen C, Yicong C, Baobing F, Guiai J, Zhonghua S, Ju L, Shaoqing T, Jianlong W, Peisong H, Xiangjin W (2019) Editing of rice isoamylase gene ISA1 provides insights into its function in starch formation. Rice Sci 26:77–87

Song X-J, Huang W, Shi M, Zhu M-Z, Lin H-X (2007) A QTL for rice grain width and weight encodes a previously unknown RING-type E3 ubiquitin ligase. Nat Genet 39:623–630

Spindel J, Begum H, Akdemir D, Virk P, Collard B, Redoña E, Atlin G, Jannink J-L, McCouch SR (2015) genomic selection and association mapping in rice (Oryza sativa): effect of trait genetic architecture, training population composition, marker number and statistical model on accuracy of rice genomic selection in elite, tropical rice breeding lines. PLoS Genet 11:e1004982

Sud M, Fahy E, Cotter D, Brown A, Dennis EA, Glass CK, Merrill AH Jr, Murphy RC, Raetz CR, Russell DW, Subramaniam S (2007) LMSD: LIPID MAPS structure database. Nucleic Acids Res 35:D527–D532

Suela MM, Lima LP, Azevedo CF, Resende MDVD, Nascimento M, SILVA FFE (2019) Combined index of genomic prediction methods applied to productivity. Ciência Rural 49

Sun Y, Jiao G, Liu Z, Zhang X, Li J, Guo X, Du W, Du J, Francis F, Zhao Y, Xia L (2017) Generation of high-amylose rice through CRISPR/Cas9-mediated targeted mutagenesis of starch branching enzymes. Front Plant Sci 8:1–15

Sverrisdóttir E, Sundmark EHR, Johnsen HØ, Kirk HG, Asp T, Janss L, Bryan G, Nielsen KL (2018) The value of expanding the training population to improve genomic selection models in tetraploid potato. Front Plant Sci 9:1–4

Takano-Kai N, Jiang H, Kubo T, Sweeney M, Matsumoto T, Kanamori H, Padhukasahasram B, Bustamante C, Yoshimura A, Doi K, McCouch S (2009) Evolutionary history of GS3, a gene conferring grain length in rice. Genetics 182:1323–1334

Tang L, Mao B, Li Y, Lv Q, Zhang L, Chen C, He H, Wang W, Zeng X, Shao Y, Pan Y, Hu Y, Peng Y, Fu X, Li H, Xia S, Zhao B (2017) Knockout of OsNramp5 using the CRISPR/Cas9 system produces low Cdaccumulating indica rice without compromising yield. Sci Rep 7:14438

Tanksley SD (1993) Mapping polygenes. Annu Rev Genet 27:205–233
Thomson MJ (2014) High-throughput SNP genotyping to accelerate crop improvement. Plant Breed Biotechnol 2:195–212
Thomson MJ, Zhao K, Wright M, McNally K, Rey J, Tung C-W, Reynolds A, Scheffler B, Eizenga G, McClung A, Kim H, Ismail A, De Ocampo M, Mojica C, Reveche MY, Dilla-Ermita C, Mauleon R, Leung H, Bustamante C, McCouch S (2012) High-throughput single nucleotide polymorphism genotyping for breeding applications in rice using the BeadXpress platform. Mol Breed 29:875–886
Tran NA, Daygon VD, Resurreccion AP, Cuevas RP, Corpuz HM, Fitzgerald MA (2011) A single nucleotide polymorphism in the Waxy gene explains a significant component of gel consistency. Theor Appl Genet 123:519–525
United Nations, D. O. E. A. S. A., Population Division 2017. World population prospects: the 2017 revision, key findings and advance tables. Working paper No. ESA/P/WP/248.
Varavinit S, Shobsngob S, Varanyanond W, Chinachoti P, Naivikul O (2003) Effect of amylose content on gelatinization, retrogradation and pasting properties of flours from different cultivars of thai rice. Starch - Stärke 55:410–415
Varshney RK, Nayak SN, May GD, Jackson SA (2009) Next-generation sequencing technologies and their implications for crop genetics and breeding. Trends Biotechnol 27:522–530
Wan X, Weng J, Zhai H, Wang J, Lei C, Liu X, Guo T, Jiang L, Su N, Wan J (2008) Quantitative trait loci (QTL) analysis for rice grain width and fine mapping of an identified QTL allele gw-5 in a recombination hotspot region on chromosome 5. Genetics 179:2239–2252
Wang C, Chen S, Yu S (2011) Functional markers developed from multiple loci in GS3 for fine marker-assisted selection of grain length in rice. Theor Appl Genet 122:905–913
Wang J, Chapman SC, Bonnett DG, Rebetzke GJ, Crouch J (2007) Application of population genetic theory and simulation models to efficiently pyramid multiple genes via marker-assisted selection. Crop Sci 47:582–588
Waters DLE, Henry RJ, Reinke RF, Fitzgerald MA (2006) Gelatinization temperature of rice explained by polymorphisms in starch synthase. Plant Biotechnol J 4:115–122
Weng J, Gu S, Wan X, Gao H, Guo T, Su N, Lei C, Zhang X, Cheng Z, Guo X, Wang J, Jiang L, Zhai H, Wan J (2008) Isolation and initial characterization of GW5, a major QTL associated with rice grain width and weight. Cell Res 18:1199–1209
Xu F, Sun C, Huang Y, Chen Y, Tong C, Bao J (2015) QTL mapping for rice grain quality: a strategy to detect more QTLs within sub-populations. Mol Breed 35:105
Yamanaka S, Nakamura I, Watanabe KN, Sato Y-I (2004) Identification of SNPs in the waxy gene among glutinous rice cultivars and their evolutionary significance during the domestication process of rice. Theor Appl Genet 108:1200–1204
Yang Q-Q, Zhang C-Q, Chan M-L, Zhao D-S, Chen J-Z, Wang Q, Li Q-F, Yu H-X, Gu M-H, Sun SS-M, Liu Q-Q (2016) Biofortification of rice with the essential amino acid lysine: molecular characterization, nutritional evaluation, and field performance. J Exp Bot 67:4285–4296
Yu G, Olsen KM, Schaal BA (2011) Molecular evolution of the endosperm starch synthesis pathway genes in rice (Oryza sativa L.) and its wild ancestor, O. rufipogon L. Mol Biol Evol 28:659–671
Zeng Y, Ji Z, Wen Z, Liang Y, Yang C (2016) Combination of eight alleles at four quantitative trait loci determines grain length in rice. PLoS One 11:e0150832
Zhang J, Zhang H, Botella JR, Zhu J-K (2018) Generation of new glutinous rice by CRISPR/Cas9-targeted mutagenesis of the Waxy gene in elite rice varieties. J Integr Plant Biol 60:369–375
Zhang Z, Ober U, Erbe M, Zhang H, Gao N, He J, Li J, Simianer H (2014) Improving the accuracy of whole genome prediction for complex traits using the results of genome wide association studies. PLoS One 9:e93017
Zhao X, Daygon VD, McNally KL, Hamilton RS, Xie F, Reinke RF, Fitzgerald MA (2015) Identification of stable QTLs causing chalk in rice grains in nine environments. Theor Appl Genet:1–13

Chapter 19
Transcriptomics Applied to Rice Grain Quality

Carlos Busanello, Railson Schreinert dos Santos, and Vívian Ebeling Viana

19.1 Introduction

Demand for quality in agricultural production has been growing steadily over time, and for rice (*Oryza sativa* L.), which is a staple food for a large part of the world's population, it is no different. Quality is defined by the International Organization for Standardization (ISO 8402: 1986) as the totality of features and characteristics of a product or service that determine its ability to satisfy explicit or implicit needs, depending heavily on the type of use that will be given to the product and the consumer market.

When defining the quality characteristics that the consumers take into consideration, we are trying to reach some economic advantages through genetic breeding to serve certain markets, since quality indicators change according to the country and many of them establish their own protocols for certain aspects of quality. In addition, there are still differences between the various standardization organizations such as the International Organization for Standardization (ISO), the Association of Analytical Communities International (AOAC), or the American Association of Cereal Chemists International (AACCI) (Juliano 1993; Bao 2014).

It has been some time since it was understood that the greater the supply of a product on the market, the higher the requirement for higher standards of superior quality, and it is not different for rice. The increased production of this cereal leads to an increased effort to improve their quality traits since the consumers are willing to pay for a product that meets their expectations (Ferreira et al. 2005). It is also important to note that the economic context of consumers is related to quality preferences (reviewed in Cuevas et al. 2016).

C. Busanello (✉) · V. Ebeling Viana
Universidade Federal de Pelotas, Campus Capão do Leão, Pelotas, RS, Brazil
e-mail: carlosbuzza@gmail.com

R. S. dos Santos
Instituto Federal Farroupilha, Alegrete, RS, Brazil

A. Costa de Oliveira et al. (eds.), *The Future of Rice Demand: Quality Beyond Productivity*, https://doi.org/10.1007/978-3-030-37510-2_19

At the point we are dealing specifically with in this chapter, cultivar seems to be one of the main factors that contribute to differences in grain quality, although several factors unrelated to genetics may also influence (Juliano 1993). However, quality is determined by several factors, including the selection of the genotype (cultivar), cultivation conditions, as well as harvesting and postharvest procedures (Juliano 1990; Kibanda and Luzi-Kihupi 2007; Chen et al. 2012). Developing genotypes able to provide a high-quality product depends on the breeding research, which is highly dependent on different areas of knowledge (Moose and Mumm 2008; Perez-de-Castro et al. 2012).

All this together can affect aspects such as milling, appearance, cooking behavior, as well as sensorial and nutritional characteristics. Several quantitative trait loci (QTLs) and genes related to grain quality have been identified (Bao 2014). In addition, it has been emphasized that further studies of these regions toward understanding the transcriptional regulation of target genes may lead to a better understanding of how grain quality is defined in genetic terms, a knowledge that will be an important auxiliary tool in rice breeding.

19.2 Biotechnology in Rice Quality

It can be said that, in general, the two main contributions of biotechnology to plant breeding are the use of molecular markers for the detection of genotypic variation associated with phenotype as well as the use of genetic engineering in its various forms (Braun and Virchow 2001).

It is easy to perceive that molecular biology techniques help to reduce difficulties and problems that hinder plant breeding, especially when studying some agronomic characters of quantitative inheritance. In rice breeding, plant size, yield components, and grain quality comprehend the target polygenic traits (reviewed by Fukuoka et al. 2010; Kamara et al. 2017). Nowadays, difficulties of phenotypic selection due to strong environmental effects are present, which harm both the choice of parents that will be used in controlled crosses and the selection of plants in segregating populations (Brammer 2000).

For the correct understanding of plant physiology and for efficient design of processes that make it possible to obtain improved plants, it is necessary to characterize different genes to understand both their function in the metabolism and the control of their expression. It is interesting to note that there is still little material available in the literature that provides information on gene expression at both the transcriptional and translational levels, as more information on this point would be needed. Many studies have arisen in order to identify and understand the genes involved, metabolic pathways, and processes that comprehend the entire grain quality information set in rice (Liu et al. 2010; Wang et al. 2010; Sato et al. 2011; Sharma et al. 2012; Gao et al. 2013; Sun et al. 2015; Biselli et al. 2015; Ishimaru et al. 2019; Bello et al. 2019).

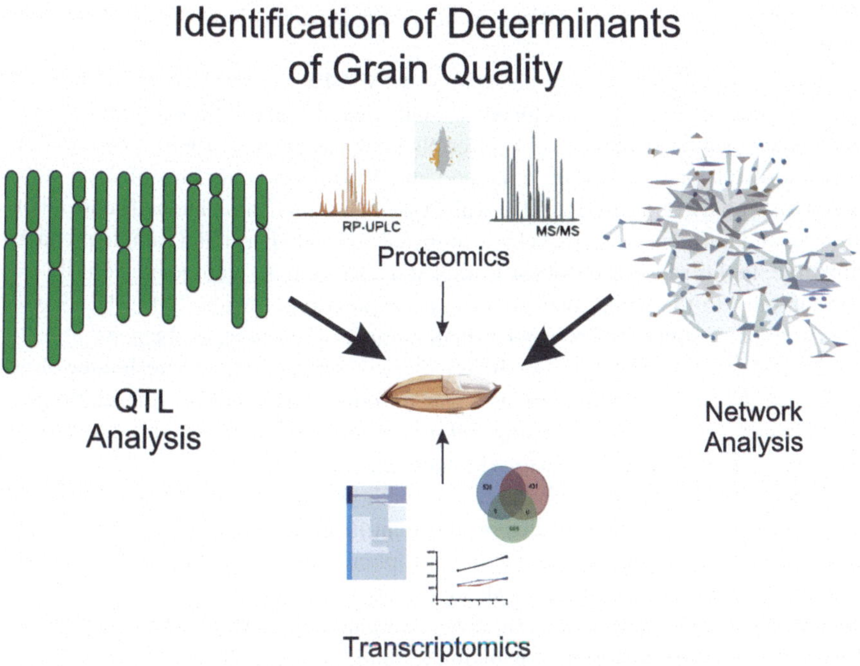

Fig. 19.1 The studies about rice grain quality are becoming increasingly complex, bringing together a large amount of data obtained by different research groups on different varieties. Currently the ability to gather data from different fronts to obtain information is of great importance for the adequate understanding of how each of the characteristics that influence quality is controlled physiologically

Thus, new studies have tried to unite information of different types to facilitate the identification of genes probably involved with different aspects of quality (Fig. 19.1). This includes searching for different QTLs related to each trait, which are generally found through genotyping and phenotyping, linking this knowledge to the transcriptional expression of genes within these QTLs under different conditions. After this, the in silico analysis of promoters of these genes and identification of mutants for them can also be performed.

19.3 Transcriptomics and the Quality of Rice Grains

Transcriptomics refers to the study of the complete set of transcripts of a given cell, tissue, or organism at any given time. The profile of transcripts may vary according to physiological state, physical, chemical, biological, or disease stimuli. Understanding it means understanding the first phases of the interaction between genotype and environment, which may be of great value in devising useful strategies for modifying these plants (Wang et al. 2009; Monteverde et al. 2019; Kadam et al. 2019).

Initially, serial analysis of gene expression (SAGE), a method for transcript analysis by sequencing, was developed to analyzing the absolute and simultaneous abundance of a large number of transcripts (Velculescu et al. 1995). Later, SAGE was improved for LongSAGE, a method able to support grater tags of 21pb compared to 14pb used in SAGE (Saha et al. 2002). However, these methods depend on cloning and purification steps, which contribute to higher cost and sample preparation time. In this sense, the Robust-LongSAGE (RL-SAGE) method has contributed substantially to the serial analysis of gene expression (Gowda 2004). RL-SAGE uses a small amount of mRNA and a small number of reactions to obtain a complete library and also a partial digestion before cloning, which makes the technique more robust. In the same year, SuperSAGE was introduced as a tool for transcriptomics of pathogen–host interaction using 26bp tags allowing simultaneous analysis of host and pathogen gene expression (Matsumura et al. 2004).

Almost simultaneously with the SAGE methods, microarray arises, a technique that came to dominate transcription profile identification. In microarrays, the nucleic acids of the samples are identified with fluorophores and then hybridized in a matrix with probes for specific sequences already known, something that will be discussed later in this chapter (Southern 2001; McGettigan 2013). More recently, after the emergence of new-generation sequencing technologies or high-throughput sequencing (NGS/HTS), the RNA-sequencing-based transcriptomics has been gaining ground, with microarray studies being used less frequently. The use of NGS allows to analyze several molecules and may even discover new forms of RNA transcribed in a given sample (Wang et al. 2009). Although most sequencing methods allow only the reading of relatively short fragments or regions at the edges of certain fragments, several processes can be used to select specific RNA sets. One of the first reports applying sequence by synthesis of RNA used the 454/Roche technology, although with the use of long (110bp) reads, the low number of reads resulted in modest processing results (Bainbridge et al. 2006). The advent of technology came later with the Illumina/Solexa having shorter reads but with many replicates which generates many data in each run. The preparation of the most accessible library also improved the Illumina technology acceptance (reviewed by McGettigan 2013).

Several new publications on the possibilities of the use of RNA-seq have been made available, giving an overview about its possibility of use as well as providing protocols for simplifying the analysis of the data obtained by this type of technique (Djebali et al. 2017).

Currently, transcriptomics has addressed not only messenger RNAs (mRNAs) but also noncoding RNAs (ncRNAs), including small RNAs such as microRNAs and even larger ones, such as long noncoding RNAs (lnRNAs). Also, splicing profile and posttranscriptional changes are being analyzed. These have gained space with the new sequencing technologies and while RNA-seq is expanding over microarray (Wang et al. 2015; He et al. 2015; Singh et al. 2017; Shin et al. 2018) (Table 19.1).

Table 19.1 Comparison of global detection: mRNA methods, microarray, and RNA-seq

Technology	Microarray	RNA-seq
Principle	Hybridization	High-throughput sequencing
Resolution	From several to 100pb	Single base
Throughput	High	High
Background noise	High	Low
Dynamic range to quantify gene expression level	Up to a few hundred times	>8000 times
Ability to distinguish different isoforms	Limited	Yes
Ability to distinguish allelic expression	Limited	Yes
Required amount of RNA	High	Low
Cost for mapping transcripts from large genomes	High	Relatively low

Adapted from Wang et al. (2009)

The general transcript profile using microarray or RNA-seq is generally, since it is relatively accessible and very informative, the first approach applied and generates information about gene expression and also about transcribed isoforms (alternative splicing) (McGettigan 2013).

19.4 Main Steps in Each Analysis

As discussed, different types of experiments can be performed for transcriptomic studies, with microarray and RNA-seq being the two most important ones. Even within these experiments, there are many variations, with several protocols both for wet lab steps of the experiment and for bioinformatic analyses. Here we present these steps in a simplified way both for microarray and for RNA-seq analysis, with several references that will help in performing those experiments.

RNA Isolation

Transcriptomic methods require that RNA be previously isolated from the experimental organism. Methods for RNA isolation involve cell disruption and precipitation of RNA with chemical solvents. The mRNA is an unstable molecule and prone to degradation if exposed to RNases; therefore, for the maintenance of RNA integrity, immediate freezing is recommended. Also, the use of low temperatures is recommended during and after sample processing.

Many different methods for total RNA extraction have been applied in rice grain quality studies (Table 19.2), among them, widely used methods for total RNA extraction from general purpose such as vegetative tissues. However, a method for isolation of total RNA from endosperm derived from seeds containing high starch contents has been reported (Li and Trick 2005). Li and Trick (2005) reported that samples containing high starch levels may affect total RNA extraction when phenol–

Table 19.2 RNA isolation methods used for transcript global analyses of tissues related with rice grain quality

Extraction methods	Sample	Reference
TRIzol® (Invitrogen)	Endosperm	Liu et al. (2010), and Gao et al. (2013)
	Spikelet, endosperm, and panicle	Wang et al. (2010)
	Panicle and seeds	Sharma et al. (2012)
	Seeds	Sun et al. (2015), and Bello et al. (2019)
RNeasy® Plant Mini Kit (QIAGEN)	Palea, lemma, ovary, embryo, and endosperm	Sato et al. (2011)
Chang et al. (1993)	Endosperm	Ishimaru et al. (2019)
Li and Trick (2005)	Seeds	Biselli et al. (2015)

chloroform buffer, TRIzol® Reagent (Invitrogen, Carlsbad, CA, USA), and the RNeasy® Plant Mini Kit (QIAGEN, Valencia, CA, USA) are used. The method of Li and Trick (2005) is able to extract high-quality RNA from 50 to 100 mg of starchy seed samples. Isolated RNA must be DNase digested to remove possible contaminating genomic DNA. After isolation, it is necessary to verify the integrity and quality of RNA in agarose gel electrophoresis and spectrophotometry.

cDNA synthesis
Following DNase treatment, mRNA is converted to cDNA (complementary DNA) by the reverse transcriptase enzyme. The method for the first strand of cDNA synthesis needs to be chosen according to the objectives of the research (Green and Sambrook 2019). Oligo(dt) method binds to the endogenous poly(A) tails of eukaryotic mRNAs and can be used as universal primer to bind generally mRNA. Also, with the same purpose to bind the mRNAs in general, the random hexanucleotides can be used acting as oligonucleotides that bind sequences in different exons. This method is recommended for RNAs extremely long or with secondary structures which commit the oligo(dt) method. On the other hand, it is also possible to obtain a cDNA from a specific target RNA by using a synthetic antisense oligonucleotide for a chosen region in a particular or for a family of mRNAs.

19.4.1 Microarray

Microarray, also known as DNA chip, biochip, or gene array, consists in a collection of microscopic spots of DNA probes for specific target molecules in a solid surface; these probes are usually for known genes and those that are mapped on a slide (Southern 2001; Bumgarner 2013). Each of these spots will hybridize to molecules obtained from the biological samples being tested; these should be labeled with proper fluorophores before hybridization. A strong attachment between the probe and the molecule will result in an increase in fluorescence that can be measured by

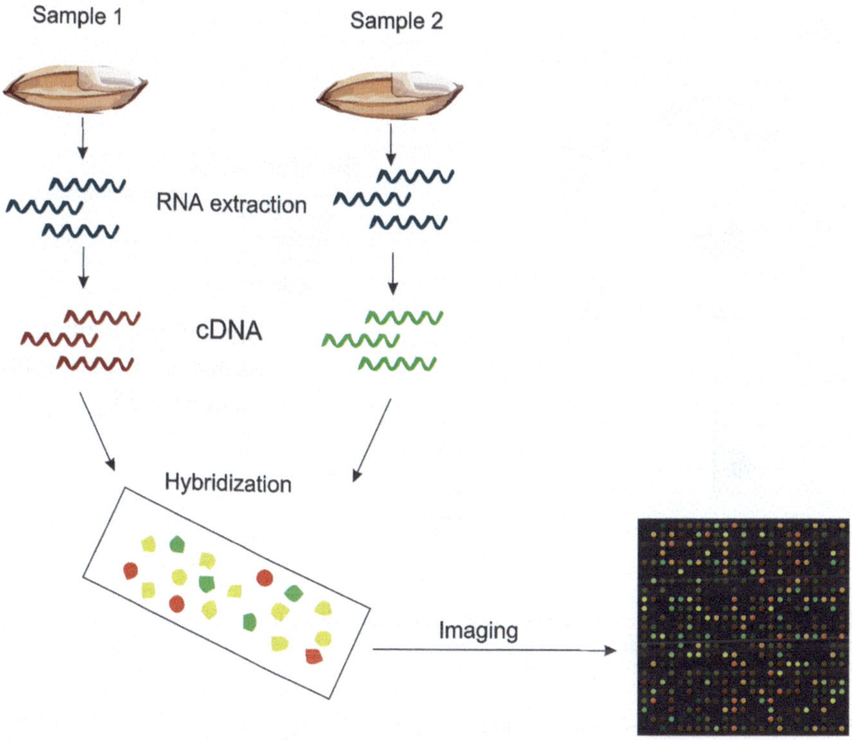

Fig. 19.2 A synthesis of a microarray analysis

a scanner and analyzed by different bioinformatics methods (Fig. 19.2). Generally, the technique is used for analysis of gene expression, our focus in this chapter, and of point mutations.

The use of microarray allowed a fast and relatively accessible way to observe the expression of numerous transcripts being the accepted standard method. However, this approach has some limitation of detection (Duggan et al. 1999) and also limited to genome coverage due to a lack of accurate genetic annotation (Kikuchi et al. 2007).

Wet Lab (Sample Collection, Nucleic Acid Preparation, Hybridization, and Data Collection)

A microarray experiment, just as for most experiments in the area of molecular biology, begins in what we call "wet lab," dealing directly with biological material. This first part goes from preparing the experiment itself and collecting samples to obtaining the slide image by the scanner, generating the data that will be used later in the "dry lab" (Fig. 19.3).

Before we even start the study, we need, in the case of the microarray, to already have in mind the information we want to collect, which genes we want to measure the transcription, etc. Some commercial arrays for rice are already available such as Affymetrix Gene Chip rice® which contains probe sets representing more than

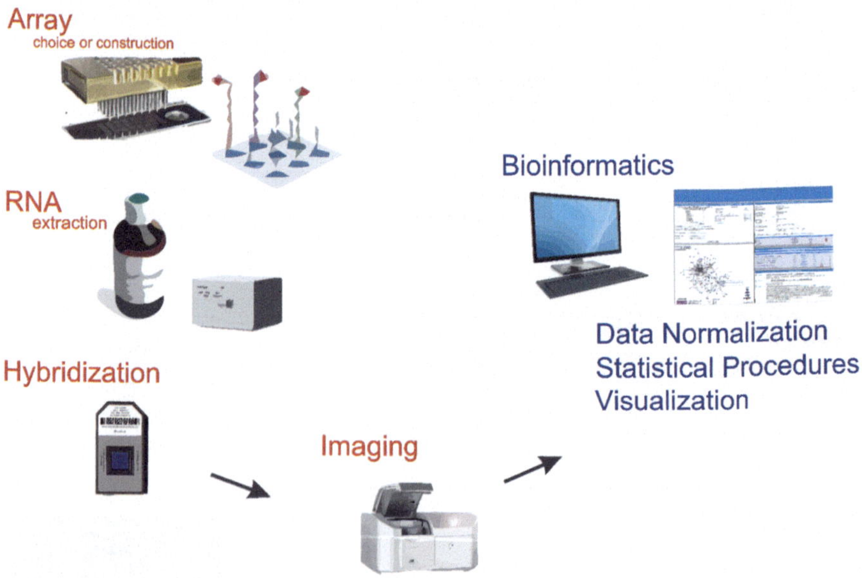

Fig. 19.3 Main steps in a typical microarray experiment. Steps in red refer to the wet lab procedures, while the steps in and blue to dry lab procedures

50,000 transcripts (Bruce et al. 2009). Agilent also offers the Rice Gene Expression Microarray® chip. Some companies offer, in addition to the standard clip for analysis, a customization, depending on the need for some special type, with any particular information, so you can build your own chip (Dalton 2000).

To obtain good RNA samples for the study of rice quality, it is important to have well-designed protocols for extracting nucleic acids mainly from flowers and rice grains, as well as from their different tissues (Zeng et al. 2017; Wang et al. 2018b, 2019b). This RNA is then used to obtain nucleic acids (i.e., complementary DNA or cDNA) labeled with different fluorophores, such as Cyanine-3 (Cy3) and Cyanine-5 (Cy5) (for charge-coupled device (CCD) scanners) onto the microarray slides.

There are two different types of scanners, the CCD and the laser scanning. The first reads the excited slides with white light all the way up, getting photographs from the fluorescence of Cy3 and Cy5, while the second one scans the slide with a laser (Vukojević et al. 2011).

After this procedure, the probe can hybridize with the array, and this matrix is then washed and scanned twice for specific wavelengths of the fluorescent markers used, one for each sample (Püffeld et al. 2019). The labeled cDNA will emit a fluorescence at the site where it hybridized, which is proportional to the amount of cDNA for that specific target in the sample. Such fluorescence will be captured by the device, generating a self-reading file that will be used in computational analysis, the dry lab. Details and specifics of the entire process can be found at Püffeld et al. (2019).

Dry Lab (Bioinformatics)
After obtaining the files with the reads, generated by the sequencer, it is time to start the statistical analyses. The generated images are stored in the device and must undergo a process of segmentation of the signal/spot, identifying what is foreground and what is background fluorescence, something important for subsequent calculations, that measures the brightness intensity. Different forms of data normalization within and between arrays can be applied, and then several statistical procedures will detect differences in gene expression between samples.

Different methods can be used to carry them out, many of them quite simple, and are already available in a very didactic way in the Internet. The Bioconductor is a tool to analyze high-throughput genomic data. This tool uses an R statistical programming language and is an open source and development (Frazee et al. 2015). Another way to analyze the microarray data is following the steps described by Babu (2004) and Püffeld et al. (2019)

1. Image processing and analysis: processes that convert microarray digitalized images into quantifying values. This file contains the primary data collection
2. Quality control: used to perform diagnostic charts in order to identify problematic samples
3. Normalization: intended to preserve biological variations and control technical variations of the assay
4. Differential expression analysis: aims to identify genes that have different expression in relation to study conditions (i.e., control vs. stress)

19.4.2 RNA-seq

In RNA-seq protocols, different ways of obtaining data are available. In this type of analysis, different from what occurs in the microarray, there is no need to have prior knowledge of the genes that will have their expression measured, although there are some advantages we can take from some previous knowledge. Through RNA-seq, it is also possible to detect alternative forms of splicing and even new genes, which is why this type of approach has been gaining ground on microarray technology (Wang et al. 2009; Ozsolak and Milos 2011; He et al. 2015).

Wet Lab (Sample Collection, Nucleic Acid Preparation)
As for the microarray, the experiment begins with the biological part itself, requiring a good experimental design, a well-cared for experiment, and adequate collection of samples, which must be placed at extremely low temperatures (<−80 °C) soon after the parts (leaves, flowers, etc.) are excised.

The RNA can then be extracted from these samples, which are usually macerated at very low temperatures. Different reagents and/or commercial kits may be used, depending on the purpose. It is important to remember that these extracts should be evaluated for integrity in appropriate equipment (second step of Fig. 19.4).

After obtaining high-quality RNAs, we can proceed to cDNA synthesis and sequencing in following the instructions of the available apparatus. It is important to notice that the type of sequencing impacts how we treat the data later, and the vast majority of the protocols available today are focused on Illumina, NextSeq (2x150pb), HiSeq 4000 (2x150pb), and NovasSeq 6000 (2v250pb – specification based on dual-flow cell run of SP flow cells).

Dry Lab (Bioinformatics)

Once the sequences are obtained, we can start the "dry lab" analyses, which include the different bioinformatics steps shown in Fig. 19.4. Different programs, from public to paid, are available for reading, editing, and analyzing sequences. Software such as the CLC Genomics Workbench 6.5 (CLC bio-QIAGEN, Aarhus, Denmark) can be a great help for a simple, quick, and very informative analysis. Yet different free-to-use computer programs are available and can provide a tremendous amount of information if you apply a little effort to your proper management.

Basically, the assembly steps of these transcripts go through an evaluation of the quality of readings with removal of the ones that have below our threshold and of sequences that should not be used, like adapters. Subsequently the genes can be compiled de novo, possibly providing a range of information of great value, as alternative forms of splicing. It can also be directly mapped in a reference genome or transcriptome. Genes are identified by comparison with the genome itself or with similar sequences of other genomes; then these transcripts are quantified by evaluating the number of reads that cover them; this coupled with varied statistical methods is able to provide a wide range of information.

Among the most used programs are the R DESeq in Bioconductor (Anders and Huber 2010), to assemble and understand these transcripts, and MapMan (see Sect. 3.5.1) (Usadel et al. 2009) for annotation. Also, DEXSeq-Count tool can be applied to count the number of reads/fragments per exon of each RNA-seq sample

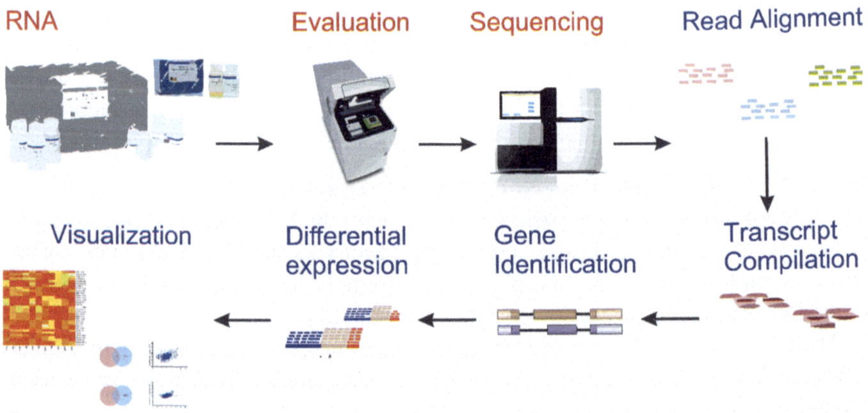

Fig. 19.4 Main steps for a RNA-seq experient. Steps in red refer to the wet lab procedures, while the steps in and blue to dry lab procedures

(Anders et al. 2012). To assemble transcript sequences from Illumina RNA-seq data, the Trinity tool can be used (Grabherr et al. 2011). RNA STAR and TopHat2 are tools to align sequences (Kim et al. 2013; Dobin et al. 2013), and for transcript abundance quantification obtained from RNA-seq data, the Cufflinks, Kallisto quant, Sailfish, and Salmon tools can be used (Trapnell et al. 2010; Patro et al. 2014; Patro et al. 2017; Bray et al. 2016).

19.5 Applications of Global mRNA Analysis Methods to Rice Grain Quality Traits

Although many rice studies focus on global transcript identification, these studies are related to gene activation in response to environmental stimulus, such as biotic and abiotic stresses. There are still few studies related to global identification of transcripts related to grain quality (Table 19.3).

Table 19.3 Studies developed for global analysis of transcripts related to grain quality in rice

Technology	Target trait and tissue	Reference
Microarray	Seed endosperm15 days after fertilization	Liu et al. (2010)
Microarray	39 Tissues collected throughout the life cycle of the rice plant including spikelet, endosperm, and five stages of panicle development	Wang et al. (2010)
Microarray	48 Tissues were used throughout the rice life cycle, including palea and lemma, ovary, embryo, and endosperm	Sato et al. (2011)
Microarray	Tissues of five stages of panicle development, from early to late stage and seed development, from fertilization to maturation, were analyzed	Sharma et al. (2012)
Microarray	Endosperm-specific tissues, central starch endosperm (milky-white-type chalk grain), lateral starch endosperm, dorsal aleurone layer, and dorsal starch endosperm (white-black-type chalk grain) obtained by laser microdissection plants under heat stress and under control conditions	Ishimaru et al. (2019)
RNA-seq	Rice endosperm development at 3, 6, and 10 days after fertilization	Gao et al. (2013)
RNA-seq	Developing seeds with 14 days after fertilization were used, an intermediate time for the accumulation of nutritional compounds in the grains	Biselli et al. (2015)
RNA-seq	Upper and lower grains at 10, 15, 21, and 27 days after pollination were analyzed	Sun et al. (2015)
RNA-seq	Developing seeds with 7 days after fertilization from NF-Ys transcription factors (involved in grain quality)	Bello et al. (2019)

19.6 Identification of Target Gene Expression: A Focus on Quantitative PCR

After global analysis of transcripts, gene expression is usually validated by differentiated PCR techniques. Polymerase chain reaction (PCR), described by Kary Mullis in 1984, is a sensitive and specific method for detecting nucleic acids and quantifying the amount of a specific nucleic acid in a sample. Using the PCR and the RNA to cDNA conversion by the reverse transcriptase enzyme, the RNA can also be amplified. Researchers have adapted the technique to precisely quantify the amount of DNA or RNA in a sample with specificity for a single target gene.

However, the biggest challenge for a target RNA quantification is the PCR reaction efficiency, since after many PCR cycles, cDNA is amplified millions of times which can provide a source of error if amplification efficiency is not equal in all samples (VanGuilder et al. 2008). These high-precision RNA quantification approaches used to confirm transcriptome results, whether obtained by microarray or RNA-seq as well as the constrains, will be discussed hereafter. In this chapter we will focus on the quantitative PCR technique.

19.6.1 Quantitative PCR: An Introduction

Quantitative PCR (qPCR) was originally developed for DNA quantification, for example, to identify viral load in a sample (Syvänen et al. 1988; Abbott et al. 1988), and was then adapted for mRNA quantification (Becker-André and Hahlbrock 1989; Wang et al. 1989). PCR associated with reverse transcriptase (RT)-PCR allowed the analysis of minimal amounts of nucleic acids from which different qPCR and RT-PCR methods were developed.

In general, the method is based on quantifying the PCR product in the log phase, before the plateau, and requires that each sample presents an equal amount of nucleic acid and that all sample under analysis amplify with equal efficiency until the quantitative analysis point A reference gene is used as a normalizer for amplification efficiency (reviewed in Heid et al. 1996). The quantitative competitive (QC)-PCR method was developed and depended on an internal control concurrent with known quantity in each reaction (Gilliland et al. 1990). For relative quantitation, the unknown target PCR product is compared to the known concurrent PCR product. However, the success depends on the development of an internal control that amplifies with the same efficiency as the target (Heid et al. 1996).

The qPCR reached an advanced level of development with the introduction of qPCR methods in real time (also called here qPCR) for both DNA and RNA detection (Higuchi et al. 1993; Chiang et al. 1996; Gibson et al. 1996; Heid et al. 1996). Real-time detection methods use fluorophores as a detection system and do not require post-reaction PCR product handling as previously described methods. In these methods, it is possible to identify the PCR fluorescence progressively in real time during the exponential phase of the reaction and thus to quantify the nucleic acid in the sample.

The first real-time qPCR method used the ethidium bromide (DNA intercalant) during the PCR process and a modified thermocycler to irradiate the samples with ultraviolet light for subsequent fluorescent signal detection with a charged-coupled device (CCD) camera (Higuchi et al. 1993). The fluorescence signal was plotted against the PCR cycle number indicating the amount of PCR product generated during each PCR cycle. However, the limitation, besides the use of carcinogenic ethidium bromide, was that the nonspecific PCR products were equally detected and included in the quantification (Ginzinger 2002).

The identification of the 5′-3′ exonuclease activity of *Thermus aquaticus* (*Taq*) DNA polymerase thermostable enzyme and demonstrating that cleavage of a target probe during PCR by *Taq* polymerase 5′ nuclease activity could be used to detect amplification of the specific target product (Holland et al. 1991) revolutionized the qPCR detection system. Holland and collaborators (1991) used a ^{32}P-labeled probe in the 5′ end region and nonextendable in the 3′ end region to ensure that the probe did not act as a primer. The probe annealing to the PCR product during amplification forms a substrate for the *Taq* 5′-3′ exonuclease activity. During amplification, as the enzyme extends from the primer to the probe target region, *Taq* degrades the probe into small fragments; the dependence of polymerization for probe degrading ensures the breakage occurs only if the target sequence is amplified. Following PCR, probe cleavage was measured using chromatography to separate the degraded from intact probes. Later, in order to eliminate the post-PCR step, a dual-labeled oligonucleotide was introduced to PCR (Lee et al. 1993). The probe had a fluorescent dye reporter in the 5′ end and a quencher dye in the 3′ end. When the probe is intact, the proximity of the quencher significantly decreases the fluorescence emitted by the reporter dye. The fluorescence signal is only emitted after probe degradation. Other studies have demonstrated the efficiency of the double-labeled probe in PCR product detection, the most appropriate reporter and quencher dyes, and the acceptable distance between them for a suitable 5′-3′ nuclease activity performance (Livak et al. 1995a, b; Bassler et al. 1995).

Nowadays, two general chemistries are available as qPCR detector, the double-stranded (ds) DNA-intercalating agents (DNA-binding dyes) and fluorescent probes. As a dsDNA intercalating, the SYBR Green I is the most used, and their fluorescence is higher when intercalated into dsDNA (Arya et al. 2005). Dispite being the most cost-effective method, SYBR Green I is not specific since binding to all dsDNA formed during the PCR (primer dimers and nonspecific products). On the other hand, the dual-labeled TaqMan® fluorogenic nonextendable probe, which is based on *Taq* 5′-3’exonuclease activity, is highly precisioned since it is specific for a target sequence and only generates fluorescence when degraded by the *Taq* nuclease activity (Heid et al. 1996). TaqMan probe presents a reporter (5′ end) and a quencher (3′ end) which are closer. The quencher absorbs the fluorescence emitted by the reporter until broken by the 5′-3′ nuclease activity of the DNA polymerase during amplification.

To apply fluorescence as a detection system, a modified thermocycler able to detect the progress of the amplification in real time was need. The first machine was commercialized by Applied Biosystems (ABI) in 1996, after other companies added new instruments to the market (Tajadini et al. 2014).

19.6.2 Fluorescence Detection and Basics Concepts of Real-Time PCR

Using any of the developed chemicals, the increase in fluorescence emission during PCR can be detected in real time by a modified thermocycler. Computer software develops amplification graphs using fluorescence emission data collected during PCR amplification (Arya et al. 2005). Amplification graphs use specific terms.

Baseline Baseline is defined as the PCR cycle in which a reporter fluorescent signal is accumulating but is below the detection limit of the instrument, generally, between the cycles 3 and 15.

ΔRn The computer software calculates ΔRn using the equation Rn = Rnf-Rnb, where Rnf is the fluorescence emission of the product at each time point and Rnb is the baseline fluorescence emission. The ΔRn values are plotted versus the cycle number. During the first PCR amplification cycles, the ΔRn values do not exceed the baseline.

Threshold An arbitrary threshold is chosen by the computer software based on baseline variability. The threshold is calculated ten times the standard deviation of the mean fluorescence signal emitted by the baseline between cycles 3 and 15. The fluorescent signal detected above the threshold is considered the real signal and is used to define the threshold cycle (Ct) of a sample.

Ct It is defined as the PCR cycle number at which the reporter's fluorescence is greater than the minimum detection level. The Ct value will always be obtained at the exponential phase of amplification, which occurs during the initial PCR cycles. As the reaction components become limited, the amplification rate decreases until the PCR reaction no longer generates exponential rate (plateau phase) presenting a small or no increase in the PCR product. This is the main reason why Ct is a more reliable to measure initial copy number than a final measure of the amount of accumulated PCR product.

19.6.3 Basic Principles of qPCR

All qPCR technologies are based on fluorescence measurement during PCR. The amount of fluorescence emitted is proportional to the amount of PCR product which helps monitor the PCR reaction. The resulting PCR curve is used to define the exponential phase of the reaction and is used for the accurate calculation of the initial nucleic acid copy number at the beginning of the reaction (Klein 2002).

qPCR is guided by basic principles (reviewed by Klein 2002; Gachon et al. 2004):

I. qPCR is used to accurately quantify different amounts of the product of a target gene present in independent samples. In the case of qRT-PCR, samples comprise cDNA reverse transcribed from RNA.
II. Each sample is subjected to PCR reaction together with a fluorophore present in the reaction, which emits fluorescence only when the specific product is recognized or synthesized. The threshold is determined on the linear part of the curve, corresponding to the phase with the greater amplification efficiency. Ct is defined as the number of cycles required to reach the fluorescence threshold.
III. A calibration curve is designed from parallel runs using an initial amount of the specific target. This assists in the calculation of the amount of target product in a sample of interest.

19.6.4 Fundamental Aspects to Perform qPCR Experiments

The appropriate performance and interpretation of qPCR assays are critical to ensure the reliability and reproducibility of the generated data. Therefore, Bustin et al. (2009) developed a guide, MIQE guidelines, containing 85 parameters to ensure the quality of qPCR results. Later a research was reported based on MIQE guidelines demonstrating the application in establishing a solid experiment (Taylor et al. 2010).

Here, we review some important aspects that must be taken into account following Bustin et al. (2009) and Taylor et al. (2010).

The *experimental design* is crucial for any gene expression experiment and must be carefully established to ensure reproducibility. Considering that the transcription of the mRNA is sensitive to external stimuli that are not related to the studied processes, it is important to stablish a rigid work control. Time is needed to define experimental conditions, control groups, type and number of replicates, and the use of well-established sample manipulation methods (Taylor et al. 2010).

Validation of the qPCR assay assists in identifying the ideal range of primer annealing temperature and specificity and reaction efficiency. Validation ensures that reaction conditions, buffers, and primers have been optimized and that cDNA samples are not contaminated with *Taq* polymerase inhibitors. Taylor et al. (2010) reviews the validation conditions, determination of temperature, and primer specificity and amplification efficiency. To identify the optimal annealing temperature of the primer, it is ideal to test temperatures using a thermocycler able to promote temperature gradient. Also, at this stage it is important to verify the specificity of the primer by analyzing the PCR product. The melting curve that is generated at the end of the PCR cycles confirms the specificity of primer annealing. The melting curve should show only one peak.

To determine reaction efficiency, a standard curve is used. Reaction efficiency is assessed by the rate that DNA polymerase converts the reagents (dNTPs, oligonucleotides, and template) to amplicon. A 100% efficient reaction has a maximum amplicon increase of twofold per reaction cycle. Inefficient reactions (<90%) may be caused by contamination with DNA polymerase inhibitors, nonoptimal annealing temperature, old or inactive DNA polymerase, inefficient primers, or by amplicons with secondary structure. On the other hand, reactions with very high efficiency (>110%) are usually due to the occurrence of primers dimers or unspecific amplicons.

To obtain the standard curve, reactions with serial cDNA dilutions are performed. Initial tenfold dilutions at more than eight points are recommended, starting from the most concentrated cDNA sample ensuring that the standard curve will cover the full concentration potential of the template that can be obtained. For validation by the standard curve, each dilution should be performed in triplicate for all primer pairs used in the experiment. The standard curve is constructed by plotting the log of the initial template amount against the Ct value obtained. A linear regression by Pearson's correlation coefficient (r) or determination coefficient (r^2) can be used to assess whether qPCR is optimized. Serial dilutions should produce amplification curves with grouped technical replicates that are spaced between dilutions. The r or r^2 values of the standard curve represent how well the experimental data fit the regression line demonstrating how linear the data are. Linearity demonstrates the variability between replicates and whether the amplification efficiency is the same for the different initial template concentrations. How much more different the Ct values between replicates smaller will be the r and r^2 values. Values of $r > 0.990$ or $r^2 > 0.980$ are desirable for qRT-PCR experiments.

Control reactions should be added to qPCR experiments, as described by Bustin et al. (2009). Controls should be added during the treatment of the RNA sample with DNase. Following digestion with the specific DNase, a reaction on the thermocycler should be performed containing digested samples not yet converted into cDNA and a positive control, a genomic DNA. Negative control reactions (NTCs) must be included on each plate or batch of samples. NTCs detect contamination when probes are used and can distinguish undesirable amplification products, such as primer dimers in reactions with SYBR Green. Positive controls containing nucleic acids are useful for monitoring assay variation over time and are essential when the run does not generate a calibration curve.

Normalization is essential for a reliable qPCR assay as it controls variations in nucleic acid extraction yield, reverse transcription yield, and amplification efficiency, allowing comparison of mRNA concentrations in different samples (Bustin et al. 2009). Using reference genes as internal control is the most common method to normalizing mRNA data; however, the reference gene should be experimentally validated for specific tissues, cells, and conditions. Normalization demonstrates the mRNA concentration of a target gene relative to the mRNA concentration of reference gene. Therefore, reference genes should be expressed in stable manner. Normalization with only one reference gene is not acceptable unless the author demonstrates clear evidence of invariable expression under the experimental condition used.

Proper *statistical design* is critical to the reliability of a qPCR assay. Rieu and Powers (2009) review the key factors that should be considered in a qPCR experiment. An experiment should consist of at least three independent biological triplicates of each treatment. For each biological replicate, at least two technical replicates are required. Each sample has the material for the target and reference gene reactions; ideally both genes should be tested on the same plate (typically 96 wells). Experiments composed of many treatments and/or target and references genes require the use of multiple plaques. In this sense, the ideal is to use the "treatment maximization" strategy, where the plate constitutes a statistical block and the biological replicates are completely separated (i.e., a biological replicate from each treatment) in each plate, constituting a randomized block experiment.

19.6.5 Quantification Methods

Quantitation of mRNA by qPCR can be performed in two ways, absolute quantification and relative quantification. Absolute quantification of the amount of a target in an unknown sample evaluates Ct using a standard curve to determine the initial copy quantity; in this method no comparisons or references are required (Pfaffl 2001). This method is applied to circumstances where absolute quantification is critical, i.e., by quantifying viral load to correlate with disease progression (reviewed by Arya et al. (2005). Plasmids are one of the most widely used known sample (standard) sources; in addition, synthetic single-stranded oligonucleotides for the complete amplicon and cell line with known copy number or expression level of the gene of interest may also be used (Arya et al. 2005).

Relative quantitation determines expression compared to a reference sample; the target gene expression is compared to expression of a reference gene (Pfaffl 2001). This method eliminates the need for standard curves, and mathematical equations are used to calculate the relative expression of the target relative to a control or calibrator sample and normalized to a reference gene (Arya et al. 2005). The most common method for relative quantification is $2^{-\Delta\Delta Ct}$ which was tested by Livak and Schmittgen (2001). The amount of the target gene in a sample normalized to a reference gene and relative to the calibrator is given by $2^{-\Delta\Delta Ct}$ where $\Delta\Delta Ct = \Delta Ct$ (sample) – ΔCt (calibrator); ΔCt is the Ct of the target gene subtracted from the Ct of the reference gene. To apply this method, the amplification efficiencies of the target and the reference genes must be close to and/or above 90%. The amplification efficiency (E) can be calculated after the qPCR; a linear regression is performed with the Ct mean from replicates with the log of the cDNA serial dilutions to obtain a slope. The slope is applied to the equation proposed by Pfaffl (2001) to obtain the amplification efficiency, $E = 10^{-1/\text{slope}}$. A slope of −3.32 indicates a PCR reaction with 100% efficiency (Applied Biosystems 2004).

19.6.6 qPCR Applications in Rice Grain Quality

The introduction of qPCR revolutionized molecular biology studies helping the detection of a target nucleic acid in a rapid, specific, and sensitive manner. Nevertheless, many studies have emerged regarding gene expression related to rice development and grain quality.

Reference Genes

The reserve accumulation in rice endosperm has a direct effect on grain yield and quality and is related to seed development. However, this is a complex process and is controlled by a dynamic of signaling pathways, from genetic programming to hormonal and metabolic signals (Li et al. 2010). In this sense, for qPCR studies of genes related to seed development processes require the identification of appropriate reference genes to ensure the accuracy of quantification of target gene expression. However, since the beginning of the use of qPCR, only few reports arisen focus on identifying suitable reference genes to be used in analysis of rice grain quality aspects (Table 19.4).

Based on the abovementioned records, it can be observed that although qPCR is over 10 years old, few studies have focused on identifying reference genes for tissue studies and conditions related to seed development. In addition, the lack of information regarding the increased accuracy of qPCR assays focused on rice grain quality aspects is also evident. Many studies have been performed to identify the expression of target genes involved in starch biosynthesis; however, the identification of more specific reference genes may help in understanding this highly complex process.

Target Genes

As previouly stated, qPCR has been applied to validate the global data obtained from microarray and RNA-seq assay, as can be observed in Table 19.3. However, despite the few reports regarding the identification of reference genes to apply for rice grain quality traits, there are many records related to expression of targeted genes. Taking into account the high quantity of reports, we summarize here the published works from the past 3 years (Table 19.5). It is important to observe that the reports in general perform qPCR applying SYBR Green as a detection system and that the data is normalized against one reference gene.

19.7 Conclusion and Future Prospects

The union of different methods of analysis in the understanding of the interaction "genetic vs. environment" is extremely important in the identification of factors that affect the different components of rice grain quality. Some of them have already been studied at some considerable degree; others have a certain delay in the available information; however, the lack of transcriptomic studies for a more in-depth

Table 19.4 Reference genes identified for qPCR-aiming studies related to rice grain quality aspects

Gene name	Accession no.	Gene description	Primer sequence (5′-3′) F/R	Tissue/condition	Detection method	Reference
eEF-1a	AK061464	Eukaryotic elongation factor 1-alpha	TTTCACTCTTGGTGTGAAGCAGAT GACTTCCTTCACGATTTCATCGTAA	Post-fertilized flower/ mature seed	qPCR/geNorm	Jain et al. (2006)
UBQ5	AK061988	Ubiquitin 5	ACCACTTCGACCGCCACTACT	Post-fertilized flower/ mature seed	qPCR/geNorm	Jain et al. (2006)
			ACGCCTAAGCCTGCTGGTT	Endosperm/ endosperm under heat	qPCR/ geNorm-NormFinder	Xu et al. (2015)
ACT1	AK100267	Actin 1	CTTCATAGGAATGGAAGCTGCGGGTA CGACCACCTTGATCTTCATGCTGCTA	Panicle (6 stages)/ seed (5 stages)	qPCR/geNorm	Li et al. (2010)
eIF-4a	AK073620	Eukaryotic initiation factor 4a	TTGTGCTGGATGAAGCTGATG	Panicle (6 stages)/ seed (5 stages)	qPCR/ geNorm	Li et al. (2010)
			GGAAGGAGCTGGAAGATATCATAGA	Endosperm/ endosperm under heat	qPCR/ geNorm-NormFinder	Xu et al. (2015)
Fb15	LOC_ Os02g07910	Fiber protein Fb15	GAGCGGCAACACTAAGGATG AATCTCCTTGACGGCACAGA	Endosperm/ endosperm under heat	qPCR/ geNorm-NormFinder	Xu et al. (2015)

Table 19.5 qPCR assays performed in target genes related to rice grain quality traits from the past 3 years

Gene	Trait	Reference gene	Detection system	Reference
PFP, OsAGPS1, OsAGPS2a, OsAGPS2b, sAGPL1, OsAGPL2, OsAGPL3, OsAGPL4, OsSSI, OsSSIIa, OsSSIIb, OsSSIIc, OsSSIIIb, OsSSIVa, OsSSIVb, OsGBSSI, OsGBSSII, OsBEI, OsBEIIa, OsBEIIb, OsISA1, OsISA2, OsISA3, OsPUL, OsPHOL, OsPHOH, OsDPE2, Flo2, Flo4, SS1, SS2, SS3, α-amylase 3C, α-amylase 3E, PGIa, and *PGIb*	Expression in *fructose-6-phosphate 1-phosphotransferase (PFP)* mutant, which regulates carbon metabolism during grain filling	*Ubiquitin*	SYBR Green	Duan et al. (2016)
LOC_Os09g126620, LOC_Os09g126630, LOC_Os09g126640, LOC_Os09g126650, and *LOC_Os09g12660*	Expression of candidate genes in qACE9 QTL involved in chalkiness trait	*Actin*	SYBR Green	Gao et al. (2016)
HMA4	Expression in shoots and roots *OsHMA4* prevents cooper accumulation in grains	*HistoneH3*	SYBR Green	Huang et al. (2016)
MADS1, MADS14, MADS15, MADS16, DL, SNB, MFS1, ASP1, and *G1*	Expression in young panicles from 05 to 5 cm in *mads34* mutant MADS34 determine glumes fate, grain yield, and quality	–	SYBR Green	Ren et al. (2016b)
LOC_Os10g32124, LOC_Os10g32140, LOC_Os10g321250, LOC_Os10g32160, LOC_Os10g32170, and *LOC_Os10g32190*	Expression of genes in a new QTL qBRR-10 controlling brown rice rate	–	SYBR Green	Ren et al. (2016a)
SSIIIa, SSI, and *GBSSI*	Expression in grains upon ozone treatment	*Acin 1*	SYBR Green	Sawada et al. (2016)
GL3.1, srs3, GW2, GS5, gw8, GW5, LP, Ghd7, sh4, qSH1, TGW6, and *Gn1a*	Expression of genes related to panicle traits in lines Overexpressing *OsGRF4* and in *osgrf4* mutant OsGRF4 controls grain shape, panicle length, and seed shattering	*Ubiquitin 5*	SYBR Green	Sun et al. (2016)

(continued)

Table 19.5 (continued)

Gene	Trait	Reference gene	Detection system	Reference
OsSuSy2, *OsAGPS2b*, *OsAGPL2*, *OsGBSSI*, *OsBEIIb*, and *OsAmy3E*	Expression of the starch synthesis and degradation related genes in grains submitted to heat stress	*Actin*	SYBR Green	Tanamachi et al. (2016)
Os01g08350, *Os01g58010*, *Os12g14930*, *Os12g34018*, *Os12g34108*, *Os01g41720*, *Os11g47520*, *Os02g48790*, *Os06g40704*, and *Os03g08460*	Expression of genes differentially expressed obtained from RNA-seq in aleurone layer of NF-YB1-RNAi lines. NF-YB1 regulates grain filling and endosperm development	*Actin*	SYBR Green	Xu et al. (2016)
HMA3	Expression in roots. HMA3 is involved with cadmium accumulation in shoots and grains	*Histone H3*	SYBR Green	Yan et al. (2016)
SBEI, *SBEIIa*, and *SBEIIb*	Expression in seeds of wild type and *sbe3-rs* mutant	*Actin1*	SYBR Green	Yan et al. (2016)
SSIIIa	Expression in spikelets 15 Days after fertilization in starch-resistant lines	*Actin*	EvaGreen	Zhou et al. (2016)
NRAMP5	Expression of NRAMP5, responsible for manganese and cadmium uptake, in flag leaves at the grain-filling stage	*Actin 1*	EvaGreen	Liu et al. (2017)
OsLG3	Expression in young panicle from 3 to 22 cm in length. OsLG3 is a positive regulator of grain length	*Acin 1*	SYBR Green	Yu et al. (2017)
MADS55, *MADS34*, *PIN1a*, *ARF14*, *KANADI2*, and *KANADI4*	Expression in lines combining the *OsMADS*^{lgye} allele, *dep1-1*, and *gs3* which improve productivity and quality	*Acin 3*	SYBR Green	Liu et al. (2018)
WB1, *APS1*, *APL1*, *PPDKB*, *SSIIIa*, *SSG4*, *FLO7*, *FLO6*, and *bZIP68*	Expression of genes associated with starch synthesis during endosperm development in wild type and *wb1* mutant *WB1* is regulator of endosperm development	–	SYBR Green	Wang et al. (2018a)

(continued)

Table 19.5 (continued)

Gene	Trait	Reference gene	Detection system	Reference
NF-YB1, *NF-YC12*, and *bHLH144*	Panicle and developing seeds from 0 to 25 days after fertilization. The complex of these three genes activates Wx regulating grain quality	*Ubiquitin*	SYBR Green	Bello et al. (2019)
Os11g0605500, *Os05g0132100*, *Os08g0508800*, *Os03g0738600*, *Os08g0509100*, *Os06g0604400*, *Os02g0676000*, *Os03g0826600*, and *Os09g0543100*	Expression of genes involved in flavor metabolic pathway in seeds treated with different storage temperatures and humidity	*Ubiquitin 5*	SYBR Green	Biao et al. (2019)
SSI, *SSIII-2*, *GBSS1*, *AGPS2*, and *AGLP1*	Expression of the starch synthase genes in endosperm of seeds from OsSUS3-transgenic plants	*Polyubiquitin*	SYBR Green	Fan et al. (2019)
IPA1, *MOC1*, *TAD1*	Expression of genes involved in tillering in young panicle at differentiation stage V in the wild type and *gs9-1* mutant	*Actin*	SYBR Green	Jiang et al. (2019)
qPE9-1	Expression in lines overexpressing different truncated qPE9-1 proteins qPE9-1 affects grain length and weight	*Actin*	SYBR Green	Li et al. (2019)
GS5, *TGW6*, *GW5*, *GS2*, *GL3.1*, *GL7*, *SPL13*, and *WAK11*	Expression of genes controlling grain size and weight in wild type and *Ossnb* mutant	*Actin*	SYBR Green	Ma et al. (2019)
NF-YC12	Expression in endosperm and aleurone layer NF-YC12 is a regulator of accumulation of seed storage substances	*Ubiquitin*	–	Xiong et al. (2019)
GBP	Expression in roots, stems, leaves, panicles, and developing grains at 9, 12, and 15 days after fertilization GBP effects grain quality and yield	*Ubiquitin*	–	Wang et al. (2019b)
GL6	Expression in vegetative and developing panicles. GL6 affects grain length and number	*Ubiquitin 5*	SYBR Green	Wang et al. (2019a)

understanding of these processes is notorious. In particular, when observing rice grain quality traits, it is possible to note that the transcript analyses are still scarce.

Researchers have been mindful of the understanding of the molecular mechanisms underlying grain quality traits in rice. The knowledge related to transcriptional activation, also known as gene expression, has gained special attention. We look forward to more applications of global transcript analysis tools so that they can assist in identifying individual target genes that can be applied for improving grain quality in rice.

References

Abbott MA, Poiesz BJ, Byrne BC et al (1988) Enzymatic gene amplification: qualitative and quantitative methods for detecting proviral DNA amplified in vitro. J Infect Dis 158:1158–1169. https://doi.org/10.1093/infdis/158.6.1158

Anders S, Huber W (2010) Differential expression analysis for sequence count data. Genome Biol 11:R106. https://doi.org/10.1186/gb-2010-11-10-r106

Anders S, Reyes A, Huber W (2012) Detecting differential usage of exons from RNA-seq data. Genome Res 22:2008–2017. https://doi.org/10.1101/gr.133744.111

Applied Biosystems (2004) Guide to performing relative quantitation of gene expression using real-time quantitative PCR. Springer Berlin Heidelberg, Berlin, Heidelberg

Arya M, Shergill IS, Williamson M et al (2005) Basic principles of real-time quantitative PCR. Expert Rev Mol Diagn 5:209–219. https://doi.org/10.1586/14737159.5.2.209

Babu MM (2004) Introduction to microarray data analysis. Comput Genomics Theory Appl 225:249

Bainbridge MN, Warren RL, Hirst M et al (2006) Analysis of the prostate cancer cell line LNCaP transcriptome using a sequencing-by-synthesis approach. BMC Genomics 7:246. https://doi.org/10.1186/1471-2164-7-246

Bao J (2014) Genes and QTLs for rice grain quality improvement. In: Rice – germplasm, genetics and improvement. InTech, Vienna, p 64

Bassler HA, Flood SJA, Livak KJ et al (1995) Use of a fluorogenic probe in a PCR-based assay for the detection of Listeria monocytogenes. Appl Environ Microbiol 61:3724–3728

Becker-André M, Hahlbrock K (1989) Absolute mRNA quantification using the polymerase chain reaction (PCR). A novel approach by a P CR aided t ranscipt t itration assay (PATTY). Nucleic Acids Res 17:9437–9446. https://doi.org/10.1093/nar/17.22.9437

Bello BK, Hou Y, Zhao J et al (2019) NF-YB1-YC12-bHLH144 complex directly activates Wx to regulate grain quality in rice (Oryza sativa L.). Plant Biotechnol J 17:1222–1235. https://doi.org/10.1111/pbi.13048

Biao Y, Chanjuan Z, Ming Y et al (2019) Influence of gene regulation on rice quality: impact of storage temperature and humidity on flavor profile. Food Chem 283:141–147. https://doi.org/10.1016/j.foodchem.2019.01.042

Biselli C, Bagnaresi P, Cavalluzzo D et al (2015) Deep sequencing transcriptional fingerprinting of rice kernels for dissecting grain quality traits. BMC Genomics 16:1091. https://doi.org/10.1186/s12864-015-2321-7

Brammer SP (2000) Marcadores moleculares: princípios básicos e uso em programas de melhoramento genético vegetal

Bray NL, Pimentel H, Melsted P, Pachter L (2016) Near-optimal probabilistic RNA-seq quantification. Nat Biotechnol 34:525–527. https://doi.org/10.1038/nbt.3519

Bruce M, Hess A, Bai J et al (2009) Detection of genomic deletions in rice using oligonucleotide microarrays. BMC Genomics 10:129. https://doi.org/10.1186/1471-2164-10-129

Bumgarner R (2013) Overview of DNA microarrays: types, applications, and their future. Curr Protoc Mol Biol Chapter 22:Unit-22.1. https://doi.org/10.1002/0471142727.mb2201s101

Bustin SA, Benes V, Garson JA et al (2009) The MIQE guidelines: minimum information for publication of quantitative real-time PCR experiments. Clin Chem 55:611–622. https://doi.org/10.1373/clinchem.2008.112797

Chen Y, Wang M, Ouwerkerk PBF (2012) Molecular and environmental factors determining grain quality in rice. Food Energy Secur 1:111–132. https://doi.org/10.1002/fes3.11

Chiang PW, Song WJ, Wu KY et al (1996) Use of a fluorescent-PCR reaction to detect genomic sequence copy number and transcriptional abundance. Genome Res 6:1013–1026. https://doi.org/10.1101/gr.6.10.1013

Cuevas RP, Pede VO, McKinley J et al (2016) Rice grain quality and consumer preferences: a case study of two rural towns in the Philippines. PLoS One 11:e0150345. https://doi.org/10.1371/journal.pone.0150345

Dalton R (2000) DIY microarrayers promise DNA chips with everything. Nature 403:234–234. https://doi.org/10.1038/35002159

Djebali S, Wucher V, Foissac S, et al (2017) Bioinformatics Pipeline for Transcriptome Sequencing Analysis. Methods in molecular biology, 1468:201–19. https://doi.org/10.1007/978-1-4939-4035-6_14

Dobin A, Davis CA, Schlesinger F et al (2013) STAR: ultrafast universal RNA-seq aligner. Bioinformatics 29:15–21. https://doi.org/10.1093/bioinformatics/bts635

Duan E, Wang Y, Liu L et al (2016) Pyrophosphate: fructose-6-phosphate 1-phosphotransferase (PFP) regulates carbon metabolism during grain filling in rice. Plant Cell Rep 35:1321–1331. https://doi.org/10.1007/s00299-016-1964-4

Duggan DJ, Bittner M, Chen Y et al (1999) Expression profiling using cDNA microarrays. Nat Genet 21:10–14. https://doi.org/10.1038/4434

Fan C, Wang G, Wang Y et al (2019) Sucrose synthase enhances hull size and grain weight by regulating cell division and starch accumulation in transgenic rice. Int J Mol Sci 20:4971. https://doi.org/10.3390/ijms20204971

Ferreira CM, Pinheiro B da S, Sousa ISF de, de Morais OP (2005) Evolução e Padronização Embrapa Arroz e Feijão. Embrapa Arroz e Feijão, Santo Antônio de Goiás

Frazee AC, Pertea G, Jaffe AE et al (2015) Ballgown bridges the gap between transcriptome assembly and expression analysis. Nat Biotechnol 33:243–246. https://doi.org/10.1038/nbt.3172

Fukuoka S, Ebana K, Yamamoto T, Yano M (2010) Integration of genomics into rice breeding. Rice 3:131–137. https://doi.org/10.1007/s12284-010-9044-9

Gachon C, Mingam A, Charrier B (2004) Real-time PCR: what relevance to plant studies? J Exp Bot 55:1445–1454. https://doi.org/10.1093/jxb/erh181

Gao Y, Xu H, Shen Y, Wang J (2013) Transcriptomic analysis of rice (Oryza sativa) endosperm using the RNA-Seq technique. Plant Mol Biol 81:363–378. https://doi.org/10.1007/s11103-013-0009-4

Gao Y, Liu C, Li Y et al (2016) QTL analysis for chalkiness of rice and fine mapping of a candidate gene for qACE9. Rice 9:41. https://doi.org/10.1186/s12284-016-0114-5

Gibson UEM, Heid CA, Williams PM (1996) A novel method for real time quantitative RT-PCR. Genome Res 6:995–1001. https://doi.org/10.1101/gr.6.10.995

Gilliland G, Perrin S, Blanchard K, Bunn HF (1990) Analysis of cytokine mRNA and DNA: detection and quantitation by competitive polymerase chain reaction. Proc Natl Acad Sci 87:2725–2729. https://doi.org/10.1073/pnas.87.7.2725

Ginzinger DG (2002) Gene quantification using real-time quantitative PCR. Exp Hematol 30:503–512. https://doi.org/10.1016/S0301-472X(02)00806-8

Gowda M (2004) Robust-LongSAGE (RL-SAGE): a substantially improved LongSAGE method for gene discovery and transcriptome analysis. Plant Physiol 134:890–897. https://doi.org/10.1104/pp.103.034496

Grabherr MG, Haas BJ, Yassour M et al (2011) Full-length transcriptome assembly from RNA-Seq data without a reference genome. Nat Biotechnol 29:644–652. https://doi.org/10.1038/nbt.1883

Green MR, Sambrook J (2019) Amplification of cDNA generated by reverse transcription of mRNA: two-step reverse transcription-polymerase chain reaction (RT-PCR). Cold Spring Harb Protoc 2019:pdb.prot095190. https://doi.org/10.1101/pdb.prot095190

He F, Liu Q, Zheng L et al (2015) RNA-Seq analysis of rice roots reveals the involvement of post-transcriptional regulation in response to cadmium stress. Front Plant Sci 6:1–16. https://doi.org/10.3389/fpls.2015.01136

Heid CA, Stevens J, Livak KJ, Williams PM (1996) Real time quantitative PCR. Genome Res 6:986–994. https://doi.org/10.1101/gr.6.10.986

Higuchi R, Fockler C, Dollinger G, Watson R (1993) Kinetic PCR analysis: real-time monitoring of DNA amplification reactions. Nat Biotechnol 11:1026–1030. https://doi.org/10.1038/nbt0993-1026

Holland PM, Abramson RD, Watson R, Gelfand DH (1991) Detection of specific polymerase chain reaction product by utilizing the 5′----3′ exonuclease activity of Thermus aquaticus DNA polymerase. Proc Natl Acad Sci 88:7276–7280. https://doi.org/10.1073/pnas.88.16.7276

Huang X-Y, Deng F, Yamaji N et al (2016) A heavy metal P-type ATPase OsHMA4 prevents copper accumulation in rice grain. Nat Commun 7:12138. https://doi.org/10.1038/ncomms12138

Ishimaru T, Parween S, Saito Y et al (2019) Laser microdissection-based tissue-specific transcriptome analysis reveals a novel regulatory network of genes involved in heat-induced grain chalk in rice endosperm. Plant Cell Physiol 60:626–642. https://doi.org/10.1093/pcp/pcy233

Jain M, Nijhawan A, Tyagi AK, Khurana JP (2006) Validation of housekeeping genes as internal control for studying gene expression in rice by quantitative real-time PCR. Biochem Biophys Res Commun 345:646–651. https://doi.org/10.1016/j.bbrc.2006.04.140

Jiang L, Li G, Chern M et al (2019) Whole-genome sequencing identifies a rice grain shape mutant, gs9–1. Rice 12:52. https://doi.org/10.1186/s12284-019-0308-8

Juliano BO (1990) Rice grain quality: problems and challenges. Cereal Foods World 35:245–253

Juliano BO (1993) Rice in human nutrition, 1° edn. FAO, Rome

Kadam NN, Krishna Jagadish SV, Struik PC et al (2019) Incorporating genome-wide association into eco-physiological simulation to identify markers for improving rice yields. J Exp Bot. https://doi.org/10.1093/jxb/erz120

Kamara N, Asante MD, Akromah R (2017) Inheritance studies of aroma and yield attributes in rice (Oryza sativa L.) genotypes. 12:3461–3469. https://doi.org/10.5897/AJAR2017.12583

Kibanda JM, Luzi-Kihupi A (2007) Influence of genetic and genotype x environment interaction on quality of rice grain. African Crop Sci J 15:173–182

Kikuchi S, Wang G-L, Li L (2007) Genome-wide RNA expression profiling in rice. In: Upadhyaya NM (org) rice functional genomics, 1° edn. Springer New York, p 500

Kim D, Pertea G, Trapnell C et al (2013) TopHat2: accurate alignment of transcriptomes in the presence of insertions, deletions and gene fusions. Genome Biol 14:R36. https://doi.org/10.1186/gb-2013-14-4-r36

Klein D (2002) Quantification using real-time PCR technology: applications and limitations. Trends Mol Med 8:257–260. https://doi.org/10.1016/S1471-4914(02)02355-9

Lee LG, Connell CR, Bloch W (1993) Allelic discrimination by nick-translation PCR with fluorgenic probes. Nucleic Acids Res 21:3761–3766. https://doi.org/10.1093/nar/21.16.3761

Li Z, Trick HN (2005) Rapid method for high-quality RNA isolation from seed endosperm containing high levels of starch. BioTechniques 38:872–876. https://doi.org/10.2144/05386BM05

Li Q-F, Sun SSM, Yuan D-Y et al (2010) Validation of candidate reference genes for the accurate normalization of real-time quantitative RT-PCR data in rice during seed development. Plant Mol Biol Report 28:49–57. https://doi.org/10.1007/s11105-009-0124-1

Li X, Tao Q, Miao J et al (2019) Evaluation of differential qPE9-1/DEP1 protein domains in rice grain length and weight variation. Rice 12:5. https://doi.org/10.1186/s12284-019-0263-4

Liu X, Guo T, Wan X et al (2010) Transcriptome analysis of grain-filling caryopses reveals involvement of multiple regulatory pathways in chalky grain formation in rice. BMC Genomics 11:730. https://doi.org/10.1186/1471-2164-11-730

Liu C, Chen G, Li Y et al (2017) Characterization of a major QTL for manganese accumulation in rice grain. Sci Rep 7:17704. https://doi.org/10.1038/s41598-017-18090-7

Liu Q, Han R, Wu K et al (2018) G-protein βγ subunits determine grain size through interaction with MADS-domain transcription factors in rice. Nat Commun 9:852. https://doi.org/10.1038/s41467-018-03047-9

Livak KJ, Schmittgen TD (2001) Analysis of relative gene expression data using real-time quantitative PCR and the 2−ΔΔCT method. Methods 25:402–408. https://doi.org/10.1006/meth.2001.1262

Livak KJ, Flood SJA, Marmaro J et al (1995a) Oligonucleotides with fluorescent dyes at opposite ends provide a quenched probe system useful for detecting PCR product and nucleic acid hybridization. Genome Res 4:357–362. https://doi.org/10.1101/gr.4.6.357

Livak KJ, Marmaro J, Todd JA (1995b) Towards fully automated genome–wide polymorphism screening. Nat Genet 9:341–342. https://doi.org/10.1038/ng0495-341

Ma X, Feng F, Zhang Y et al (2019) A novel rice grain size gene OsSNB was identified by genome-wide association study in natural population. PLoS Genet 15:e1008191. https://doi.org/10.1371/journal.pgen.1008191

Matsumura H, Ito A, Saitoh H et al (2004) SuperSAGE. Cell Microbiol 7:11–18. https://doi.org/10.1111/j.1462-5822.2004.00478.x

McGettigan PA (2013) Transcriptomics in the RNA-seq era. Curr Opin Chem Biol 17:4–11. https://doi.org/10.1016/j.cbpa.2012.12.008

Monteverde E, Gutierrez L, Blanco P et al (2019) Integrating molecular markers and environmental covariates to interpret genotype by environment interaction in rice (Oryza sativa L.) grown in subtropical areas. G3 (Bethesda) g3.400064.2019. https://doi.org/10.1534/g3.119.400064

Moose SP, Mumm RH (2008) Molecular plant breeding as the foundation for 21st century crop improvement. Plant Physiol 147:969–977. https://doi.org/10.1104/pp.108.118232

Ozsolak F, Milos PM (2011) RNA sequencing: advances, challenges and opportunities. Nat Rev Genet 12:87–98. https://doi.org/10.1038/nrg2934

Patro R, Mount SM, Kingsford C (2014) Sailfish enables alignment-free isoform quantification from RNA-seq reads using lightweight algorithms. Nat Biotechnol 32:462–464. https://doi.org/10.1038/nbt.2862

Patro R, Duggal G, Love MI et al (2017) Salmon provides fast and bias-aware quantification of transcript expression. Nat Methods 14:417–419. https://doi.org/10.1038/nmeth.4197

Perez-de-Castro AM, Vilanova S, Canizares J et al (2012) Application of genomic tools in plant breeding. Curr Genomics 13:179–195. https://doi.org/10.2174/138920212800543084

Pfaffl MW (2001) A new mathematical model for relative quantification in real-time RT-PCR. Nucleic Acids Res 29:45e–45. https://doi.org/10.1093/nar/29.9.e45

Püffeld M, Seiler C, Kuhlmann M et al (2019) Analysis of developing rice grain transcriptome using the agilent microarray platform. In: Sreenivasulu N (org) rice grain quality. Springer New York, New York, pp 277–300

Ren D, Rao Y, Huang L et al (2016a) Fine mapping identifies a new QTL for Brown rice rate in rice (Oryza Sativa L.). Rice 9:4. https://doi.org/10.1186/s12284-016-0076-7

Ren D, Rao Y, Leng Y et al (2016b) Regulatory role of OsMADS34 in the determination of glumes fate, grain yield, and quality in rice. Front Plant Sci 7:1–15. https://doi.org/10.3389/fpls.2016.01853

Rieu I, Powers SJ (2009) Real-time quantitative RT-PCR: design, calculations, and statistics. Plant Cell 21:1031–1033. https://doi.org/10.1105/tpc.109.066001

Saha S, Sparks AB, Rago C et al (2002) Using the transcriptome to annotate the genome. Nat Biotechnol 20:508–512. https://doi.org/10.1038/nbt0502-508

Sato Y, Antonio B, Namiki N et al (2011) Field transcriptome revealed critical developmental and physiological transitions involved in the expression of growth potential in japonica rice. BMC Plant Biol 11:10. https://doi.org/10.1186/1471-2229-11-10

Sawada H, Tsukahara K, Kohno Y et al (2016) Elevated ozone deteriorates grain quality of Japonica Rice cv. Koshihikari, even if it does not cause yield reduction. Rice 9:7. https://doi.org/10.1186/s12284-016-0079-4

Sharma R, Agarwal P, Ray S et al (2012) Expression dynamics of metabolic and regulatory components across stages of panicle and seed development in indica rice. Funct Integr Genomics 12:229–248. https://doi.org/10.1007/s10142-012-0274-3

Shin S-Y, Jeong JS, Lim JY et al (2018) Transcriptomic analyses of rice (Oryza sativa) genes and non-coding RNAs under nitrogen starvation using multiple omics technologies. BMC Genomics 19:532. https://doi.org/10.1186/s12864-018-4897-1

Singh U, Khemka N, Rajkumar MS et al (2017) PLncPRO for prediction of long non-coding RNAs (lncRNAs) in plants and its application for discovery of abiotic stress-responsive lncRNAs in rice and chickpea. Nucleic Acids Res 45:e183–e183. https://doi.org/10.1093/nar/gkx866

Southern EM (2001) DNA microarrays: history and overview. In: DNA Arrays. Humana Press, New Jersey, pp 1–15

Sun H, Peng T, Zhao Y et al (2015) Dynamic analysis of gene expression in rice superior and inferior grains by RNA-Seq. PLoS One 10:e0137168. https://doi.org/10.1371/journal.pone.0137168

Sun P, Zhang W, Wang Y et al (2016) OsGRF4 controls grain shape, panicle length and seed shattering in rice. J Integr Plant Biol 58:836–847. https://doi.org/10.1111/jipb.12473

Syvänen A-C, Bengtström M, Jukka T, Söderlund H (1988) Quantification of polymerase chain reaction products by affinity-based hybrid collection. Nucleic Acids Res 16:11327–11338. https://doi.org/10.1093/nar/16.23.11327

Tajadini M, Panjehpour M, Javanmard S (2014) Comparison of SYBR Green and TaqMan methods in quantitative real-time polymerase chain reaction analysis of four adenosine receptor subtypes. Adv Biomed Res 3:85. https://doi.org/10.4103/2277-9175.127998

Tanamachi K, Miyazaki M, Matsuo K et al (2016) Differential responses to high temperature during maturation in heat-stress-tolerant cultivars of Japonica rice. Plant Prod Sci 19:300–308. https://doi.org/10.1080/1343943X.2016.1140007

Taylor S, Wakem M, Dijkman G et al (2010) A practical approach to RT-qPCR—Publishing data that conform to the MIQE guidelines. Methods 50:S1–S5. https://doi.org/10.1016/j.ymeth.2010.01.005

Trapnell C, Williams BA, Pertea G et al (2010) Transcript assembly and quantification by RNA-Seq reveals unannotated transcripts and isoform switching during cell differentiation. Nat Biotechnol 28:511–515. https://doi.org/10.1038/nbt.1621

Usadel B, Poree F, Nagel A et al (2009) A guide to using MapMan to visualize and compare Omics data in plants: a case study in the crop species, Maize. Plant Cell Environ 32:1211–1229. https://doi.org/10.1111/j.1365-3040.2009.01978.x

VanGuilder HD, Vrana KE, Freeman WM (2008) Twenty-five years of quantitative PCR for gene expression analysis. BioTechniques 44:619–626. https://doi.org/10.2144/000112776

Velculescu VE, Zhang L, Vogelstein B, Kinzler KW (1995) Serial analysis of gene expression. Science (80-) 270:484–487. https://doi.org/10.1126/science.270.5235.484

von Braun J, Virchow D (2001) Villages in the future. Springer Berlin Heidelberg, Berlin, Heidelberg

Vukojević V, Gräslund A, Bakalkin G (2011) Fluorescence Imaging with Single-Molecule Sensitivity and Fluorescence Correlation Spectroscopy of Cell-Penetrating Neuropeptides. Methods in Molecular Biology, 789:147-70. https://doi.org/10.1007/978-1-61779-310-3_9

Wang AM, Doyle MV, Mark DF (1989) Quantitation of mRNA by the polymerase chain reaction. Proc Natl Acad Sci 86:9717–9721. https://doi.org/10.1073/pnas.86.24.9717

Wang Z, Gerstein M, Snyder M (2009) RNA-Seq: a revolutionary tool for transcriptomics. Nat Rev Genet 10:57–63. https://doi.org/10.1038/nrg2484

Wang L, Xie W, Chen Y et al (2010) A dynamic gene expression atlas covering the entire life cycle of rice. Plant J 61:752–766. https://doi.org/10.1111/j.1365-313X.2009.04100.x

Wang H, Niu Q-W, Wu H-W et al (2015) Analysis of non-coding transcriptome in rice and maize uncovers roles of conserved lncRNAs associated with agriculture traits. Plant J 84:404–416. https://doi.org/10.1111/tpj.13018

Wang H, Zhang Y, Sun L et al (2018a) WB1, a regulator of endosperm development in rice, is identified by a modified MutMap method. Int J Mol Sci 19:2159. https://doi.org/10.3390/ijms19082159

Wang J, Long Y, Zhang J et al (2018b) Combined analysis and miRNA expression profiles of the flowering related genes in common wild rice (oryza rufipogon Griff.). Genes Genomics 40:835–845. https://doi.org/10.1007/s13258-018-0688-y

Wang A, Hou Q, Si L, et al (2019a) The PLATZ Transcription Factor GL6 Affects Grain Length and Number in Rice. Plant Physiol 180:2077–2090. https://doi.org/10.1104/pp.18.01574

Wang W, Wei X, Jiao G, et al (2019b) GBSS-BINDING PROTEIN , encoding a CBM48 domain-containing protein, affects rice quality and yield. J Integr Plant Biol jipb.12866. https://doi.org/10.1111/jipb.12866

Xiong Y, Ren Y, Li W et al (2019) NF-YC12 is a key multi-functional regulator of accumulation of seed storage substances in rice. J Exp Bot 70:3765–3780. https://doi.org/10.1093/jxb/erz168

Xu H, Bao J-D, Dai J-S et al (2015) Genome-wide identification of new reference genes for qRT-PCR normalization under high temperature stress in rice endosperm. PLoS One 10:e0142015. https://doi.org/10.1371/journal.pone.0142015

Xu J-J, Zhang X-F, Xue H-W (2016) Rice aleurone layer specific OsNF-YB1 regulates grain filling and endosperm development by interacting with an ERF transcription factor. J Exp Bot 67:6399–6411. https://doi.org/10.1093/jxb/erw409

Yan J, Wang P, Wang P et al (2016) A loss-of-function allele of OsHMA3 associated with high cadmium accumulation in shoots and grain of Japonica rice cultivars. Plant Cell Environ 39:1941–1954. https://doi.org/10.1111/pce.12747

Yu J, Xiong H, Zhu X et al (2017) OsLG3 contributing to rice grain length and yield was mined by Ho-LAMap. BMC Biol 15:28. https://doi.org/10.1186/s12915-017-0365-7

Zeng F, Biligetu B, Coulman B et al (2017) RNA-Seq analysis of gene expression for floral development in crested wheatgrass (Agropyron cristatum L.). PLoS One 12:e0177417. https://doi.org/10.1371/journal.pone.0177417

Zhou H, Wang L, Liu G et al (2016) Critical roles of soluble starch synthase SSIIIa and granule-bound starch synthase Waxy in synthesizing resistant starch in rice. Proc Natl Acad Sci 113:12844–12849. https://doi.org/10.1073/pnas.1615104113

Chapter 20
Proteomic for Quality: Mining the Proteome as a Strategy to Elucidate the Protein Complex Applied for Quality Improvement

Marouane Baslam and Toshiaki Mitsui

20.1 State of the Art

Cereals are the most important crops in the world, with total global annual grain production exceeding 2700 million tons (mt), compared with less than 250 mt for legume seeds (including pulses, soybean, and groundnut) (FAO 2019). The big three cereal species (rice, maize, and wheat) account for over 70% of the total production, being rice (*Oryza sativa* L.) alone the staple crop for more than half of the world's population. Other cereals, which include barley, sorghum, millets, oats, and rye, are also quite significant. Plants are the major source of dietary proteins for humans and animals since these proteins – building blocks are amino acids – are one of the primary components of the living cells. Humans can synthesize half of the 20 common amino acids needed for life; the rest considered as essential amino acids must be obtained via diet (Le et al. 2016). Cereal grains have low protein (10–12% dry wt) content compared to legume seeds. Thus, ranges of 5.8–7.7% of protein on a dry weight basis have been quoted for rice (Champagne et al. 2004), 8–15% for barley (Shewry and Halford 2002), and 9–11% for maize (Darrah et al. 2019). These ranges almost certainly reflect the impacts of the environment as well as genotype. Nevertheless, they provide over 200 mt of protein for the nutrition of humans and livestock, which is about three times the amount derived from the more protein-rich (20–40%) legume seeds (Shewry and Halford 2002). In this sense, one of the

M. Baslam (✉)
Department of Applied Biological Chemistry, Faculty of Agriculture, Niigata University, Niigata, Japan
e-mail: mbaslam@agr.niigata-u.ac.jp

T. Mitsui
Department of Applied Biological Chemistry, Faculty of Agriculture, Niigata University, Niigata, Japan

Graduate School of Science and Technology, Niigata University, Niigata, Japan
e-mail: t.mitsui@agr.niigata-u.ac.jp

A. Costa de Oliveira et al. (eds.), *The Future of Rice Demand: Quality Beyond Productivity*, https://doi.org/10.1007/978-3-030-37510-2_20

challenges in identifying grains proteins is their relatively high content (~90–95%) of carbohydrate (starch) and low protein and lipid (~1%) fractions.

Yields are the most noticeable characteristic for farmers while the crop is in the ground; however, the quality becomes a crucial determinant when the product of crop reaches the market. The traits of cereal grain quality are common to every market, while the metric of the traits is not. Thus, quality analysis should have the capacity to measure the important traits. In cereals, the main quality traits are milling properties, appearance, nutritional value, and cooking quality. All these qualities are many associated with the storage starch and protein in the endosperm. The inner endosperm is assigned as starchy endosperm and, as the name indicates, contains the totality of energy-yielding starch of the grain, which is used by the embryo as the major source of nutrition for development. In contrast to its high concentration of carbohydrates, this tissue is poor in proteins, fats, dietary fibers, and minerals as compared to the embryo and aleurone. The distribution of proteins and weight is not equal among different grain tissues. In starchy endosperm, three types of cell are reported differing according to their size, shape, and sites of occurrence. The storage and/or metabolic protein synthesis in seeds is timing, tissue specificity, and nutrient availability regulated manner (Shewry 2004). Indeed, the synthesis and accumulation of proteins takes place during the subsequent expansion of the cells of the storage tissue(s) and ends when these cells mature and desiccate (Bewley and Nonogaki 2017). The onset and duration of the phases are largely affected by the environmental scenario. More stressful conditions experienced in the field (drought, salt stress, high light, etc.) directly affect seed maturation and compromise their vigorousness or ability to maintain viability during storage (Finch-Savage and Bassel 2016). For instance, synthesis may be initiated earlier in plants grown in high temperature, dry climates (Shewry 2004). While the grain qualitative protein composition depends on the genotype, its quantitative composition is largely determined by environmental factors.

The climate condition has a great impact on seed growth, including protein synthesis, with higher temperatures that generally increase the rate of development. The storage compound (proteins, lipids, and/or carbohydrates) accumulation and water intake by cotyledon or endosperm cells influence the growth and expansion of cell size and hence the seed size and weight (Dante et al. 2014). The proteins are to a great extent responsible for the nutritional quality and processing properties of the grain. Environmentally induced changes in grain composition are associated with the altered expression of genes and proteins in response to signals. Under stressful conditions, metabolic adjustment is achieved through the accumulation of osmolytes such as sugars and amino acids in plant organs and tissue. The oxidative stress caused might result in enhanced oxidative and glycoxidative modification of seed proteins affecting negatively the quality of seeds (Frolov et al. 2018). To achieve superior quality, highly nutritious, and safe grain, the omics-based research including genomics, transcriptomics, epigenetics, proteomics, peptidomics, and/or metabolomics has emerged recently. These technologies aim to understand genome structure and the overall expression of genes, proteins, peptides, and metabolites in a functionally relevant context and provide insights into the molecular basis of various fundamental processes involved in grain formation and their interaction with the environment. Proteins play the functional units of a cell. All cellular states in grain development, whether normal or altered (chalky), depend ultimately on protein expression and regulation. Knowledge of protein expression and its dynamic alterations are therefore paramount to understanding the biology of grain quality. The proteome of a cell consists of the proteins set expressed in the cell, while the examination of the protein-protein interactions on a genome-wide scale in an attempt to dissect functional pathways is designed as "functional proteome." The large-scale analysis of proteins unravels and identifies the full complement of proteins expressed in a cell, tissue or organism, as well as their changes under a given set of conditions to understand cellular mechanisms underlying biological processes. The proteomics provides valuable information on changes in protein structure occurring at determined amino acid residues during processing events and helps to predict the quality of grains. It includes not only the identification and quantification of proteins but also the determination of their localization, modifications, interactions, activities, and, ultimately, their function. The proteomic approaches can be grouped into six categories: expression proteomics, protein-protein interactions, functional proteomics, structural proteomics, proteome mining, and posttranslational modifications (PTMs) (Carbonaro 2004). The functions of a protein depend on its nucleotide sequence and also on PTMs and its interaction with other proteins.

20.2 Proteomics for Quality

Rice proteomics had an earlier beginning applying one-dimensional gel electrophoresis (1-DGE) and two-DGE (2-DGE) coupled with electroblotting and N-terminal amino acid sequencing. The availability of the rice genomes along with various genetic resources of rice has accelerated the rice proteomics. Recently, a mass spectrometry (MS)-based quantitative proteomics featuring stable isotope labeling is becoming indispensable for gaining insights into the biological systems at the molecular level. MS measures mass-to-charge ratio (m/z), yielding the molecular weight and the fragmentation pattern of peptides derived from proteins. MS represents a general method for all modifications that change the molecular weight (Roepstorff 1997). Accordingly, MS has long been used to map the complete primary structure of individual proteins (Mann and Jensen 2003). The shotgun proteomics is MS-based proteomics for identifying proteins in a single spot or gel band slice from a simple experiment and used to identify as many proteins in a single experiment as possible. The new advances in mass spectrometry (MS) permitted the development of methods which are more sensitive and are not hindered by modifications or by a blocked N-terminus. This technique cleaved the proteins enzymatically using proteases and the resulting peptides are analyzed by MS. The trypsin is commonly used to digest the proteins and it cleaves proteins at the C-terminal side of arginine and lysine residues. Other used proteases cleave at the C-terminus such as Lys-C with lysine and Glu-C with glutamate residues and Asp-N which cleaves at the N-terminus of aspartate residues. Digesting proteins to produce peptides overcomes problems that protein analysis presents such as solubility and instability under the conditions required for MS. The MS/MS is now the most commonly used technology. Isobaric labeling techniques such as tandem mass tags (TMT) (Thompson et al. 2003) and isobaric tags for relative and absolute quantification (iTRAQ) (Ross et al. 2004) achieve multiplex quantitation without increasing spectral complexity. As a result, isobaric labeling approaches offer greater multiplexing capacity up to eight samples in a single run with high coverage and analytical throughput compared to mass difference labeling. The iTRAQ method screens for global proteomic changes for identifying differentially regulated proteins and the activated transduction pathways. The iTRAQ approach has several advantages compared to other quantitative proteomic techniques such as cleavable isotope-coded affinity tags (cICAT) or differential in-gel electrophoresis (DIGE) in regard, on the one hand, to the detection limit (Wu et al. 2006) having a good dynamic range detecting up to 24-fold changes and, on the other hand, to the great potential to identify and quantify low-abundant (Li et al. 2017) proteins which make this technique a good promise having a great potential in discovery studies. iTRAQ method is used, among all, to: (1) quantify the phosphorylation status of plant cells (Nühse et al. 2007; Inomata et al. 2018; Zhao et al. 2018) which is based on the retention and preservation in the labeled peptides of other important structural features such as posttranslational modifications, (2) determine the reversible reduction-oxidation (redox) thiol modifications of cysteines (such as disulfide bond formation, S-glutathionylation, S-nitrosylation, and S-sulfenylation), which provide fundamental posttranslational "switches" in the regulatory mechanisms of metabolism, growth, and development to adjust to continuously changing environmental constraints (Hu et al. 2015; Keech et al. 2017; Ameztoy et al. 2019), and (3) the subcellular localization of proteins by isotope tagging (LOPIT) to increase our understanding of the functions of different organelles (Dunkley et al. 2006). Recently, a new multiplex isobaric tag, the 12-plex DiLeu reagent set, represents a powerful tool for large-scale quantitative experiments and the highest multiplexing capacity currently available in an isobaric labeling experiment, enabling the triplicate analysis of four samples in a single experiment without increasing mass spectral complexity (Frost et al. 2015). The potential candidate proteins extracted from these approaches may then be utilized in elucidating the molecular mechanism explaining the response of grains to a particular environmental condition. This field of research aims to identify molecular features that can be developed as biomarkers for crop improvement and provides genetic resources underlying rice grain chalkiness, one of the principal targets for the improvement of rice characteristics.

As mention above, the proteomics analysis is leaping ahead directly in the field of rice grain quality evaluation in order to improve the genetic engineering and cross-breeding to realize desired phenotypes in response to certain environments. To circumvent quantitative and qualitative losses of crop production by forthcoming global warming, studies concerning physiological effects of environmental changes such as elevated temperature on grain filling are indispensable. In this context, knowledge of the proteome profile of important crop plants, as well as its dynamics in response to environmental and biological stressors, will pave the way to impact the improvement of their quality and nutritional properties. Furthermore, MS-based proteomics exclusively provides tools to investigate large-scale variations at the level of PTMs whose role is crucial to study the dynamics of modifications to improve the quality or understand the signaling under environmental stresses. Investigations at the level of PTMs hold great promise to provide new understandings on protein populations. Phosphoproteomics, for instance, contributed to examining the key proteins such as kinases, transcription factors, and ubiquitin ligase function (Hamel et al. 2012), the downstream signaling networks that regulate the (sub)cellular activities (van Wijk et al. 2014; Inomata et al. 2018) and the identification and functional analysis of novel phosphorylation sites (Lasalde et al. 2014).

Sample preparation for proteomic analysis remains one of the most critical steps for plant cells owing to the rigidity of plant cell walls and to the presence of large amounts of secondary compounds such as polyphenols and polysaccharides, accumulated in the central vacuole, which can lead to protein precipitation when tissues are disrupted. In cereal grains, one of the challenges in identifying the proteins is their relatively high content (~90–95%) of carbohydrate (starch) and low protein (~4–10%) and lipid (~1%) fractions. In order to overcome the problem of starchy endosperm, studies have successfully optimized the conditions for rice grain proteome analysis by iTRAQ LC-MS/MS (Baslam et al., “unpublished data”).

20.3 Impact of Rice Proteins in Its Quality

20.3.1 Rice Grain Protein Quality

Rice quality parameters are wide depending on its uses. The different kernel parts have several uses; for instance, bran (and extracts) is used for feed or recently as superfood owing to its richness of bioactive components, germ as a functional food supplement with broad health benefits, and the endosperm for cooking. Quality criteria in general terms refer to the characteristics of the endosperm. Grain quality is not just dependent on rice species and varieties, but quality also depends on the crop production environment, harvesting, processing, and milling systems. The physical quality (i.e., weight per unit volume, kernel weight, size, shape, hardness, and grain condition (chalkiness and whiteness), color, milling degree, head rice) and chemical characteristics (amylose content, gelatinization temperature, protein content and quality, alpha-amylase activity, gel consistency) are for quality determination. The protein content of rice ranges from 4% to 10% (6.8 g per 100 g) depending upon variety and environmental conditions during growth. The low quality of proteins in cereals, especially in rice, is owing to the limitations in the amounts of essential amino acids, notably lysine. Proteins are useful markers of quality (Payne 1987).

The rice grain proteins are classified into metabolic protein, water-soluble albumin, and storage proteins. While the storage proteins in the protein bodies (PBs) (prolamin in PB-I, globulin that accumulates in PB-II, and glutelin) act as a storage reserve for N and C, the metabolic protein, albumin, plays a key role in growth and structure of the seed and the biochemical pathways of storage product biosynthesis. Several studies indicate that protein composition may explain some differences in rice grain quality and hence eating quality (Martin and Fitzgerald 2002; Baxter et al. 2004; Baxter et al. 2010; Baxter et al. 2014). The low prolamin contents were related to softer rice grains and could promote more digestible rice grains (Baxter et al. 2004). The globulin content of grains was associated with head rice yield (Baxter et al. 2014). Further, other seminal works have reported that the total protein content is related to the texture of cooked rice (Derycke et al. 2005; Xie et al. 2008), with high protein content related to harder cooked rice, which in turn relates to perceived sensory texture characteristics (Lyon et al. 1999). Furukawa et al. (2006) showed that prolamin enrichment increased the hardness of cooked rice and glutelin enrichment degraded the appearance of cooked rice.

The nutritional value of rice could thus be raised by deregulating its protein content. Studies focused to obtain rice lines with varying levels of proteins have been conducted and mutants with high or low levels of seed protein have been selected (Kumamaru et al. 1988; Kawakatsu et al. 2010a, b; Ren et al. 2014). High-protein rice is likely to increase human nutrition in poor families, especially where rice is a staple food. Cooked rice prolamin is remarkably resistant to digestion and the protein inclusion bodies pass through the human gut essentially intact (Takaiwa et al. 1999). The easily digestible rice glutelins are important protein sources for human consumption. Thus, obtaining high-glutelin rice, owing to its greater digestive capacity and lysine content, would be of great interest especially in Asia. Previous studies have identified rice 57H mutants that accumulate relatively high levels of 57 kDa pro-glutelin and have floury/opaque endosperm phenotype (Wang et al. 2010). Schaeffer and Sharpe (1987) altered the lysine pathways leading to increasing the lysine levels and protein synthesis in rice plants. Other strategy was applied to increase essential amino acid and hence the rice protein by modifying biosynthetic and catabolic fluxes by overexpressing aspartate kinase and dihydrodipicolinate synthase and silencing lysine ketoglutaric acid reductase/saccharopine dehydropine dehydrogenase (LKR/SDH) by RNAi (Long et al. 2013).

In the same line, increasing the content of free tryptophan, one of the essential amino acids for forming proteins, in rice seeds would be likely to increase their acceptability for consumption by humans and farm animals. Thus, expression in rice seeds of a mutated feedback-insensitive form of the endogenous biosynthetic enzyme anthranilate synthase resulted in increases in free tryptophan of between 55- and 431-fold (Wakasa et al. 2006). In addition, genetic manipulation of seed storage proteins has also been employed to improve protein quality and quantity in rice by overexpressing the endogenous rice lysine-rich histone proteins, RLRH1 and RLRH2 (Wong et al. 2015). On the contrary, other studies have been carried out to identify low protein lines. Low-protein rice is required for the diet of patients with kidney disease. Also, rice with low glutelin is preferred by the sake industry

(Takaiwa et al. 1999). Altogether, proteomics should be on the urgent need for further efforts to identify high/low protein lines for their nutritional value.

20.3.2 *High-Throughput Proteomics as a Tool for Understanding and Improving Quality*

Proteomics plays a central role in systems biology as it complements transcriptome and metabolome data. Several works on grain have been performed to describe proteome profiling during development in cereal crops such as rice (Thelen 2009; Zhang et al. 2012; Yang et al. 2013), wheat (Skylas et al. 2005; Dong et al. 2012; Ma et al. 2014; Zhang et al. 2015), maize (Méchin et al. 2007), and barley (Finnie et al. 2002). The use and integration of proteomics as a selection tool in rice engineering and molecular breeding programs will provide guidance in selecting desired target grain qualities. This will help the rice industry to breed with greater accuracy and produce rice with higher grain quality and better nutritional value. The analysis of proteins is the most direct approach to define the gene function. The proteomic characterization could provide valuable information for dissection of the molecular and the genetic basis of yield-related traits and insights in the biology of middle and late grain development in order to improve the quality.

Proteome analysis also could determine the proteins for subcellular compartment localization and identify novel components involved in the components' biogenesis. Proteomics of rice embryo and endosperm (Deng et al. 2013; Galland et al. 2017), root (Koller et al. 2002; Komatsu and Konishi 2005; Paul et al. 2015), green and etiolated shoot (Komatsu et al. 1999), cultured suspension cells (Komatsu et al. 1998), anther (Sarhadi et al. 2012; Kim et al. 2015; Mu et al. 2017), leaf sheath (Shen et al. 2002; Ali and Komatsu 2006), and cellular organelles (Heazlewood et al. 2003; Huang et al. 2009, b; Huang et al. 2013; Oikawa et al. 2018) has opened an avenue to critically understand the functions of rice proteins and improve its quality. Yang et al. (2013) identified 302 proteins in the proteome of rice seeds representing 252 gene products and classified into 12 functional categories. Jia et al. (2013) apply the proteomic analysis to better understand which proteins are altered in mutants with enhanced nutritional value. These analyses showed that the accumulation of some lysine-rich proteins (i.e., sorbitol dehydrogenase and glyceraldehyde 3-phosphate dehydrogenase) was upregulated in mature kernels contributing probably to the lysine content of endosperm. Further, deregulation of some defense (i.e., beta-glucosidase) and several starch biosynthesis proteins contributed to the generation of soft and highly crystalline starchy endosperm. Further, the proteomic analysis of mature seeds helps to identify specific proteins that contribute disproportionately to the increased amino acid (i.e., lysine and tryptophan) content in order to relate these more abundant gene products to gene expression in maturing endosperm. Proteomic analysis of protein accumulation during grain development is generally consistent with the pattern of gene expression observed by microarray analysis (Damerval and Guilloux 1998). Accordingly, bypassing feedback inhibition by tryptophan biosynthesis pathway was demonstrated to be feasible to improve dietary tryptophan and protein quality in the seed of rice (Zhang et al. 2014). As well, several anthranilate synthase alpha subunit (*OASA1D*)-overexpressed rice lines were significantly correlated with overaccumulation of tryptophan in rice (Dubouzet et al. 2007). Other studies focused on increasing the lysine – first limiting essential amino acid in rice – content in grains. Indeed, overaccumulation of lysine-rich binding protein (BiP) in the rice endosperm resulted in a relative increase in nutritionally balanced non-seed storage proteins and offered a unique strategy to improve the lysine content of cereal grains (Kawakatsu et al. 2010a, b). Other transgenic Lys-rich rice lines have been developed by transforming Lys-rich protein gene from the seeds of *Psophocarpus* (Wang et al. 2011).

Moreover, the proteomic approach helps to check the fine composition of embryo and endosperm and better understand the compartmentalization of nutritionally relevant pathways between the endosperm and embryo (nutritional quality). These pathways could be later on improved through metabolic engineering or molecular breeding systems. Galland et al. (2017) using LC-MS/MS technique showed that the embryo proteome is much more diversified and equipped with proteins involved in the translation initiation machinery than the endosperm proteome. Further, determining the protein profile in seeds during storage would be of great interest owing to the involvement of different proteins in catalyzing the complex processes during seed aging, in which a series of physiological and biochemical reactions occur, including signal transduction, reactivation of metabolism, and redox homeostasis regulation (Parkhey et al. 2012). Using iTRAQ method, Xu et al. (2015) provided comprehensive information at the protein level and improve the understanding of the mechanism of rice seed aging and the roles played by relevant enzymes. These authors showed that the suppression of OsLOX3 expression in the rice endosperm improved resistance to seed aging and hence preserves rice grain quality during storage with no impact on grain yield (Xu et al. 2015). Additionally, Zhang et al. (2016) have proposed different proteins to be involved in during rice seed aging and in this way provide new knowledge to improve and maintain seed quality. Indeed, using the proteomic approach, the endosperm and embryo were differentially affected by seed aging in particular regarding glycolytic enzymes that decreased in abundance in the endosperm while increasing in the embryo. It is worthy to note that previous proteome studies identified a specific subset of LEA proteins, including the dehydrin family, detected exclusively in the embryo that were associated with the seed longevity and dry storage survival (Chatelain et al. 2012; Galland et al. 2017).

Dynamic proteomic analysis during eight sequential rice grain filling stages revealed a switch between central carbon metabolism and alcoholic fermentation in a coordinated manner consistent with the transition from cell growth and differentiation. These proteins were involved in different cellular and metabolic processes with a prominently functional skew toward metabolism and protein synthesis/destination (Xu et al. 2008). These studies provided several insights into the mechanism of reserve accumulation in grains. Given the fact that the endosperm is the prominent part of cereal seeds and its functional skew to starch accumulation, the proteomic analysis showed that the starch synthesis-related proteins were upregulated to the maximal level after the middle phase of grain filling. Starch in rice endosperm is mainly synthesized in the amyloplast by a series of enzymes. Identification of these proteins would shed light into mechanisms of starch synthesis, structure of starch granules, and regulation of grain quality. In light of the proteomics features conditioning regulatory approaches of grain quality, the ability to manipulate storage starch structures depends on understanding the biosynthetic pathway and, in particular, how the many components of the pathway are coordinated and regulated. Recent advances point to the implication of the PTM lysine acetylation in the starch metabolism through the regulation of protein interactions, the catalytic activity, and protein localization (Choudhary et al. 2014; Friso and van Wijk 2015). For instance, FLOURY ENDOSPERM2 (FLO2) overexpressing rice lines produce larger and heavier grains coincident with starch accumulation compared with WT plants, providing evidence that FLO2 overexpression is a good strategy for increasing starch content and yield in rice (She et al. 2010). FLO2 was found co-regulating positively the expression of genes coding for many starch synthesis enzymes, and in maize the floury mutants (fl1 and fl2) result in soft and starchy endosperm with a reduced amount of prolamin proteins (Holding et al. 2007).

Other study reported that natural variation at the DENSE AND ERECT PANICLE1 (DEP1) locus, including a gain-of-function mutation causing truncation of the encoded phosphatidylethanolamine binding protein-like domain protein, increases grain yield and quality in rice (Huang et al. 2009a, b). The plastidial starch phosphorylase (Pho) accounts for over 95% of the total phosphorylase activity in developing rice seeds and it is involved in starch biosynthesis in storage tissues. Rice generated mutants of the Pho1 altered the endosperm starch resulting in smaller starch granules and in modified amylopectin structure (Satoh et al. 2008). Other studies influence the strength of sink productivity in rice grains by deregulating a calmodulin-like domain protein kinase (SPK). As a result, the SPK phosphorylates sucrose synthase to enhance the degrading activity of sucrose in endosperm to fuel the biosynthetic pathway of storage starch in developing rice seed (Asano et al. 2002). Other strategies targeting protein manipulation in starch metabolism focused on enhancing AGP activity. The ectopic expression of allosterically insensitive *E. coli* AGP in rice resulted in enhanced starch synthesis and, in turn, higher seed weights (Smidansky et al. 2003; Nagai et al. 2009). Bahaji et al. (2014) review the rational design of genetic engineering traits and potential biotechnological strategies aimed at increasing starch in crop species. Future increases in starch yield and properties to grain quality will be driven by the isolation of natural or insertion mutants, breeding, or through the genetic engineering approaches. Underpinning these techniques requires identification and characterization of known or novel protein complexes and associations (and genes) directly or indirectly associated with starch metabolism in crops. Future works are required to elucidate starch metabolic proteins and signaling cues involved in metabolic regulation of starch-storing crops, in conjunction with protein complex formation in plastid.

It is worthy to note that microbial volatile-mediated overaccumulation of starch via promotion of posttranslational reductive activation of starch biosynthetic enzymes is a recently discovered mechanism for the induction of plant carbohydrate metabolism by microbes and used as tools to complement molecular breeding and biotechnology programs aimed at improving grain yield and quality (Sánchez-López et al. 2016a; Sánchez-lópez et al. 2016b; García-Gómez et al. 2019; Ameztoy et al. 2019).

Storage proteins in cereal seeds are delivered by the secretory pathway and deposited in protein bodies. Still, the molecular machinery of the protein bodies determining the pathway of storage protein trafficking and deposition remain to be clarified. Recently, a proteomic study employing combined LC MS/MS technology (LTQ Orbitrap XL and Q Exactive mass spectrometry) of rice Golgi membrane has shed light on the dynamic function of the Golgi complex (Oikawa et al. 2018). Previous studies have shown that α-amylase and NPPs traffic between endoplasmic reticulum-Golgi and the plastidial compartment (Nanjo et al. 2006; Kitajima et al. 2009). Thus, the use of a model glycoprotein enables us to visualize and characterize the vacuolar compartments in the endosperm and to determine the fate of the protein vacuole in the developmental changes in protein trafficking route and distribution (Arcalis et al. 2010).

Advances in mass spectrometry-based proteomics have allowed the ability to measure the ubiquitous changes that occur when autophagy (self-eating) is initiated and progresses in the cell in order to get insight into the functions of various proteins linked to the autophagic machinery, the conserved cellular process by which cells recycle their contents either to maintain basal homeostasis or in response to external stimuli. Autophagy impacts seed production due to its important role during senescence, nutrient remobilization, and by functioning during seed reproductive development. Autophagy plays also a role in the transport of seed storage proteins. Previous studies showed that prolamins were transported from the ER to protein storage vacuoles through an atypical autophagic process, suggesting that this may be a common mechanism for the vacuolar delivery of prolamins in cereals (Reyes et al. 2011) and non-cereal (Di Berardino et al. 2018) seeds. Several labeling and label-free proteomics approaches were applied to monitor and explore autophagy, such as stable isotope labeling by amino acids in cell culture (SILAC), bioorthogonal noncanonical amino acid tagging (BONCAT), and isobaric tagging for relative and absolute quantification (iTRAQ) (for more details see Cudjoe et al. 2017). Emerging proteome data established the role of autophagy in clearing mitochondria, peroxisomes, Golgi bodies, ER, chloroplasts, proteasome, ribosomes, and aberrant protein aggregates and found the possible role of autophagy in regulating fatty acid catabolism (McLoughlin et al. 2018). In the absence of autophagosome trafficking, proteases and their substrates could not meet and cause the overaccumulation of non-degraded proteins resulting in turn to protease activity stimulation.

Applying global and activity-based proteomic analyses, Havé et al. (2018) identified the nature of the overabundant proteases – cysteine proteases – in autophagy mutants. They showed that proteasome activity was increased in the autophagy mutants, playing a backup role in nutrient recycling and remobilization or act to promote cell death. Combined techniques of proteomic and lipidomic analyses revealed significant accumulations of phospholipids and ceramides and changes in glycosyl inositol phosphoryl ceramides in the mutants indicating large modifications in endomembrane-lipid and especially plasma membrane-lipid composition (Havé et al. 2019). Such disorders indicate the role of autophagy in the control of plant lipid metabolism and catabolism, influencing both lipid homeostasis and endomembrane composition, and nutrient allocation to grains. In different study, overexpressing *OsATG8a* significantly increased the level of autophagy, the number of effective tillers in the transgenic rice, the nitrogen uptake efficiency, nitrogen use efficiency, and the N in grains (Yu et al. 2019). Recent research showed that *OsATG8b*-mediated autophagy is involved in N recycling to grains and contributes to grain quality (Fan et al. 2019). Emerging data such as the involvement of autophagy in reproductive development are increasing our understanding of autophagy, but much work focusing on the proteome profile to identify new regulatory mechanisms is a critical field for future studies.

Proteomic analyses-based characterization of candidate proteins involved in different biological processes controlling the metabolism functions will be crucially important to rationally design molecular strategies aimed at enhancing the grain quality of main crops. The efforts were made to deregulate the protein and starch contents in rice grain, but significant and stable improvement could not be achieved due to the involvement of many small-effect genes/quantitative trait loci substantially affected by the environment (Chattopadhyay et al. 2019). Seed storage protein is quantitatively inherited and is affected by the growing environment (Shewry 2007).

20.3.3 Proteomics to Decipher the Environmental Impacts on Crops Quality

Proteins and peptides are significant macromolecules carrying out fundamental functions and are susceptible to various spontaneous covalent modifications, especially under stress and aging conditions. The proteome analysis is a tool that can be used both to visualize and compare complex mixtures of proteins of complex cellular process of crops under different environmental scenarios and to gain a large amount of information about the individual proteins involved in specific biological responses. Proteomics is becoming an increasingly powerful tool for the investigation since it provides an essential link between the transcriptome and metabolome and complementing genomics research. Also, the use of the techniques of proteome analysis can address a specific biological question involved in stress/defense such as the role of jasmonate in the rice self-defense signaling (Rakwal and Komatsu 2000), to identify plant-specific proteins phosphorylated in response to bacterial and fungal elicitors (Peck et al. 2001) or under climate change scenario (Kaneko et al. 2016). Among abiotic stresses, high temperature, particularly during grain filling, is the major factor responsible for causing deleterious effects (chalky grains) on the yield and quality of grain products and hence grain end-use value. Grain chalkiness is a complex trait adversely affecting appearance and milling quality and has been one of the principal goals of crop, especially rice, improvement. All scientific efforts have been in vain to eliminate chalkiness from rice owing to the complex interaction genotype-environment as well as the lack of overview of the molecular mechanisms governing the grain chalkiness formation. Although several works on plants proteomic have been conducted, there are relatively scarce studies specifically addressing the grain proteome. Few studies employing proteomics approaches have focused on grain development to high temperature during grain filling stages (Koller et al. 2002; Skylas et al. 2002; Lin et al. 2005; Balmer et al. 2006; Xu et al. 2008; Hurkman et al. 2009; Laino et al. 2010; Lee and Koh 2011; Majoul-Haddad et al. 2013; Z. Lin et al. 2014; Branlard et al. 2015). The abnormally elevated temperature during endosperm development and grain filling periods modify the chemical ingredients of rice caryopses (i.e., starch, storage proteins, fatty acid content), causing a decrease in grain yield and quality. The expression analyses of functionally categorized protein groups revealed changes in regulatory metabolism during grain development. Indeed, proteins involved in glycolysis, TCA cycle, lipid metabolism, and proteolysis pathways were upregulated at higher levels in mature grain compared to developing stages (Lee and Koh 2011).

High temperature during the early stage of caryopsis development downregulates the expression of starch granule-bound starch synthase (Wx) and prolamin, but upregulates dnaK-type hsp70 and glutelin expressions (Lin et al. 2005). In the same line, several starch synthesis-related genes, such as GBSS, BEI, and BEIIb, were found to be downregulated in rice grains developed under high-temperature stress resulting in chalky grain texture, milky appearance, and reduction of grain weight (Yamakawa et al. 2007). Thus, comprehensive poteome exploration could shed light on tissue- and time- specifics and subcellular compartement of the presence of distinct regulatory mechanisms involved in the biosynthesis and breakdown of separate starch pools and are tissue, subcellular compartment, and time specifics. The environmental temperature at the grain filling stage has been reported to influence the starch-associated protein composition in rice grains and hence their quality (Mitsui et al. 2016). Indeed, follow-up study employing a comparative proteomic analysis by iTRAQ suggested that unusual starch degradation instead of starch synthesis is involved in the occurrence of chalky grains of rice under the high-temperature stress during grain filling period (Kaneko et al. 2016). As an illustration, RNAi-mediated suppression of the starch-hydrolyzing enzyme α-amylase (EC 3.2.1.1) genes mitigated chalky grain production (Yamakawa and Hakata 2010). Suppressing the expression of genes for a single α-amylase was sufficient to reduce chalky grains under high temperature exposure (Hakata et al. 2012). Yamakawa et al. (2017) reported that increasing the activity of α-amylase isoform Amy3E failed to produce chalky grains, very likely owing to its low specific activity with intact starch, corroborating that the heat-induced expression of α-amylases in developing endosperm triggers grain chalkiness. These studies dealing with the deregulation of starch proteins propose a new strategy to develop high temperature-tolerant rice plants. Comparison study of the grain filling rate between high temperature-sensitive and high temperature-tolerant cultivars suggested that the occurrence of milky white kernels caused by the stress is the result of limited assimilate supply during the active grain filling period and not a reduction of sink capacity (Kobata et al. 2011).

More successful attempts of proteomics research have been emerged to substantially improve the grain quality against a background of changing global climate and shifting arable land. The protein analysis showed that rice aldehyde dehydrogenase7 (*OsALDH7*) – responsible for the detoxification of the lipid peroxidation by-product malondialdehyde as a consequence of lipid degradation during storage – is needed for seed maturation and viability under stress condition and is essential for desiccation tolerance (Shin et al. 2009). During seed maturation, orthodox seeds acquire desiccation tolerance and longevity as defined by the ability to withstand extreme water loss and to survive in the dry state. Further, the proteomic studies revealed one of the important mechanisms through which orthodox seeds acquire stress tolerance is the accumulation of specific protective proteins, such as late embryogenesis abundant proteins (LEA), heat shock proteins (HSPs), etc. In general, these protective proteins are linked to stress adaptation and longevity and are accumulated in the maturation phase during seed development. In fact, it has been demonstrated the involvement of HSP family to stabilize rice soluble proteins from heat denaturation and would primarily serve to protect the embryo against desiccation injuries during late maturation program. (Kaur et al. 2015). HSPs are known to function as (i) stabilizing neosynthesized proteins to facilitate correct folding and prevent irreversible protein aggregation (ii) protecting proteins against oxidative damage and (iii) helping to refold proteins that have been altered during seed storage.

Additionally, the protein repair systems were found to reconstitute damages of antioxidative enzymes and protective proteins involved in the restriction of ROS accumulation, lipid peroxidation, etc., in seed, particularly during stress condition and aging, and thus contribute to seed vigor and longevity. Proteomic analysis study carried out by Shiraya et al. (2015) revealed the involvement of a novel Golgi/plastid-type manganese superoxide dismutase (MnSOD) protein in the maintenance of normal grain filling and production of perfect grains of rice under heat stress. In fact, the ectopic expression of MSD1 ($MSD1^{OE}$) increased the expression of ROS scavengers, chaperonins, calreticulin, proteasome, molecular chaperones, and the quality control system in rice developing seeds under elevated temperature. In contrast, glutelin, prolamin, and allergen family proteins were strongly downregulated (Shiraya et al. 2015). Studies using the protein-repairing enzyme mutants, PROTEIN l-ISOASPARTYL O-METHYLTRANSFERASE (PIMT) deficient and PIMT over-expressing, revealed the role of PIMT in maintaining the longevity and survivability of seeds, particularly under various stresses, including salt, dehydration, paraquat, and cold (Kindrachuk et al. 2003; Petla et al. 2016). Moreover, the proteomic analysis also revealed several other categories of chaperone proteins (i.e., annexins, lipocalins, and caseinolytic protease) involved in lipid protection, which is critical for stress adaptation (Charron et al. 2005; Galland et al. 2017). Several proteins engaged in tocopherol biosynthesis pathway were specifically found in the embryo such as the 4-hydroxyphenylpyruvate dioxygenase (HPPD) and OsVTE proteins, which are involved in the production of plastoquinone and tocopherol essential for plant survival.

Proteome analyses on dry seeds at two physiological states revealed that seed storage proteins can also buffer the oxidative stress caused by seed aging and indicated the role of antioxidant systems, notably vitamin E, and indicated that protection and maintenance of the translation machinery and energy pathways are essential for seed longevity (Nguyen et al. 2015). Under oxidative stress, Arc et al. (2011) showed that PTMs of proteins in dry seeds play a central role in dormancy release, metabolism resumption, and aging processes. The accumulation of oxidized proteins in the dry seed is associated with aging and might induce loss of function of proteins and enzymes. Proteome diversity and plasticity resulting from PTMs give to seeds an important adaptive capacity to their environment (Arc et al. 2011).

In plants, autophagy has been shown to function under several environmental stresses such as nutrient or oxygen starvation, drought, salt, and heat. These functions in stress responses and seed production suggest a potential for autophagy as a target for manipulation that may lead to agronomic benefits owing to the homeostatic role and maintaining levels of proteins and energy in the cell conducted by autophagy. The proteomics analysis into autophagy allowed the global identification and characterization of range changes to the cellular proteome and secretome. As a result, autophagy-defective mutants are more sensitive to stress, indicating that autophagy contributes to plant survival under stress conditions. Rice atg7 and atg9 mutants showed a reduction in the vegetative growth and early senescence and they are unable to produce seeds due to male sterility (Kurusu et al. 2014; Wada et al. 2015). In addition, *atg* mutants produce fewer seeds compared to WT plants, while ATG-overexpressing plants have enhanced autophagy, suppressed aging, stimulated stress resistance, and increased seed production and oil accumulation (Li et al. 2015; Minina et al. 2018; Tang and Bassham 2018). Therefore, any environmental condition that could amplify ER stress is believed to amplify metabolic disorders if autophagy machinery is malfunctioning, leading to plant susceptibility and jeopardizing the plant and grain qualities.

20.4 Perspectives: Proteomics Approach to Nutritional Improvement

The studies discussed above showed a potential to use proteomics technique to improve grain quality in crops. Collectively, emphasis in rice breeding should not only be on the concentration but also on the quality of rice protein owing that some proteins possess some allergenicity potential. Exploring the role and regulation of proteins in grains under stress condition will not only elucidate the molecular aspects of grain quality but can pave the way to the possibility of protein-mediated improvement in seed adaptation, vigor, and longevity during stressful conditions and storability of this global food crop. Thus, it will have a huge impact on the agricultural economy.

Further, the emerging peptidome, phosphoproteomics, and redox proteomics approaches could provide interesting insights into the compositional differences among cultivars and treatments and can give further pieces of evidence for models of molecular interactions and for the prediction of protein function.

Acknowledgments This work was supported by KAKENHI Grants-in-Aid for Scientific Research (A) (15H02486) from Japan Society for the Promotion of Sciences, Strategic International Collaborative Research Program by the Japan Science and Technology Agency (JPMJSC16C5), and Grant for Promotion of KAAB Projects (Niigata University) from the Ministry of Education, Culture, Sports, Science, and Technology, Japan.

References

Ali GM, Komatsu S (2006) Proteomic analysis of rice leaf sheath during drought stress. J Proteome Res 5:396–403. https://doi.org/10.1021/pr050291g

Ameztoy K, Baslam M, Sánchez-López ÁM et al (2019) Plant responses to fungal volatiles involve global posttranslational thiol redox proteome changes that affect photosynthesis. Plant Cell Environ 42:pce.13601. https://doi.org/10.1111/pce.13601

Arc E, Galland M, Cueff G et al (2011) Reboot the system thanks to protein post-translational modifications and proteome diversity: how quiescent seeds restart their metabolism to prepare seedling establishment. Proteomics 11:1606–1618. https://doi.org/10.1002/pmic.201000641

Arcalis E, Stadlmann J, Marcel S et al (2010) The changing fate of a secretory glycoprotein in developing maize endosperm. Plant Physiol 153:693–702. https://doi.org/10.1104/pp.109.152363

Asano T, Kunieda N, Omura Y et al (2002) Rice SPK, a calmodulin-like domain protein kinase, is required for storage product accumulation during seed development: phosphorylation of sucrose synthase is a possible factor. Plant Cell 14:619–628. https://doi.org/10.1105/tpc.010454

Bahaji A, Li J, Sánchez-López ÁM et al (2014) Starch biosynthesis, its regulation and biotechnological approaches to improve crop yields. Biotechnol Adv 32:87–106. https://doi.org/10.1016/J.BIOTECHADV.2013.06.006

Balmer Y, Vensel WH, Cai N et al (2006) A complete ferredoxin/thioredoxin system regulates fundamental processes in amyloplasts. Proc Natl Acad Sci U S A 103:2988–2993. https://doi.org/10.1073/pnas.0511040103

Baxter G, Blanchard C, Zhao J (2004) Effects of prolamin on the textural and pasting properties of rice flour and starch. J Cereal Sci 40:205–211. https://doi.org/10.1016/J.JCS.2004.07.004

Baxter G, Zhao J, Blanchard C (2010) Albumin significantly affects pasting and textural characteristics of rice flour. Cereal Chem J 87:250–255. https://doi.org/10.1094/CCHEM-87-3-0250

Baxter G, Blanchard C, Zhao J (2014) Effects of glutelin and globulin on the physicochemical properties of rice starch and flour. J Cereal Sci 60:414–420. https://doi.org/10.1016/J.JCS.2014.05.002

Bewley JD, Nonogaki H (2017) Seed maturation and germination. Ref Modul Life Sci. https://doi.org/10.1016/B978-0-12-809633-8.05092-5

Branlard G, Lesage VS, Bancel E et al (2015) Coping with wheat quality in a changing environment: proteomics evidence for stress caused by environmental changes. In: Advances in Wheat Genetics. From genome to field. Springer, Tokyo, pp 255–264

Carbonaro M (2004) Proteomics: present and future in food quality evaluation. Trends Food Sci Technol 15:209–216. https://doi.org/10.1016/J.TIFS.2003.09.020

Champagne ET, Wood DF, Juliano BO, Bechtel DB (2004) The rice grain and its gross composition. In: Champagne ET (ed) Rice: chemistry and technology, 3rd edn. AACC, St. Paul, pp 77–107. https://doi.org/10.1094/grain, https://doi.org/10.1094/1891127349

Charron J-BF, Ouellet F, Pelletier M et al (2005) Identification, expression, and evolutionary analyses of plant lipocalins. Plant Physiol 139:2017–2028. https://doi.org/10.1104/pp.105.070466

Chatelain E, Hundertmark M, Leprince O et al (2012) Temporal profiling of the heat-stable proteome during late maturation of Medicago truncatula seeds identifies a restricted subset of late embryogenesis abundant proteins associated with longevity. Plant Cell Environ 35:1440–1455. https://doi.org/10.1111/j.1365-3040.2012.02501.x

Chattopadhyay K, Behera L, Bagchi TB et al (2019) Detection of stable QTLs for grain protein content in rice (Oryza sativa L.) employing high throughput phenotyping and genotyping platforms. Sci Rep 9:3196. https://doi.org/10.1038/s41598-019-39863-2

Choudhary C, Weinert BT, Nishida Y et al (2014) The growing landscape of lysine acetylation links metabolism and cell signalling. Nat Rev Mol Cell Biol 15:536–550. https://doi.org/10.1038/nrm3841

Cudjoe EK, Saleh T, Hawkridge AM, Gewirtz DA (2017) Proteomics insights into autophagy. Proteomics 17:1700022. https://doi.org/10.1002/pmic.201700022

Damerval C, Le Guilloux M (1998) Characterization of novel proteins affected by the o2 mutation and expressed during maize endosperm development. Mol Gen Genet MGG 257:354–361. https://doi.org/10.1007/s004380050657

Dante RA, Larkins BA, Sabelli PA (2014) Cell cycle control and seed development. Front Plant Sci 5:493. https://doi.org/10.3389/fpls.2014.00493

Darrah LL, McMullen MD, Zuber MS (2019) Breeding, genetics and seed corn production. Cornea:19–41. https://doi.org/10.1016/B978-0-12-811971-6.00002-4

Deng ZY, Gong CY, Wang T (2013) Use of proteomics to understand seed development in rice. Proteomics 13:1784–1800. https://doi.org/10.1002/pmic.201200389

Derycke V, Veraverbeke WS, Vandeputte GE et al (2005) Impact of proteins on pasting and cooking properties of nonparboiled and parboiled rice. Cereal Chem J 82:468–474. https://doi.org/10.1094/CC-82-0468

Di Berardino J, Marmagne A, Berger A et al (2018) Autophagy controls resource allocation and protein storage accumulation in Arabidopsis seeds. J Exp Bot 69:1403–1414. https://doi.org/10.1093/jxb/ery012

Dong K, Ge P, Ma C et al (2012) Albumin and globulin dynamics during grain development of elite Chinese wheat cultivar Xiaoyan 6. J Cereal Sci 56:615–622. https://doi.org/10.1016/J.JCS.2012.08.016

Dubouzet JG, Ishihara A, Matsuda F et al (2007) Integrated metabolomic and transcriptomic analyses of high-tryptophan rice expressing a mutant anthranilate synthase alpha subunit. J Exp Bot 58:3309–3321. https://doi.org/10.1093/jxb/erm179

Dunkley TPJ, Hester S, Shadforth IP et al (2006) Mapping the Arabidopsis organelle proteome. Proc Natl Acad Sci U S A 103:6518–6523. https://doi.org/10.1073/pnas.0506958103

Fan T, Yang W, Zeng X et al (2019) A rice autophagy gene OsATG8b is involved in nitrogen remobilization and grain quality. BMC Plant Biol. https://doi.org/10.21203/RS.2.12993/V1

FAO (2019). World food situation. http://www.fao.org/worldfoodsituation/csdb/en/

Finch-Savage WE, Bassel GW (2016) Seed vigour and crop establishment: extending performance beyond adaptation. J Exp Bot 67:567–591. https://doi.org/10.1093/jxb/erv490

Finnie C, Melchior S, Roepstorff P, Svensson B (2002) Proteome analysis of grain filling and seed maturation in barley. Plant Physiol 129:1308–1319. https://doi.org/10.1104/pp.003681

Friso G, van Wijk KJ (2015) Posttranslational protein modifications in plant metabolism. Plant Physiol 169:1469–1487. https://doi.org/10.1104/pp.15.01378

Frolov A, Mamontova T, Ihling C et al (2018) Mining seed proteome: from protein dynamics to modification profiles. Biol Commun 63:43–58. https://doi.org/10.21638/spbu03.2018.106

Frost DC, Greer T, Li L (2015) High-resolution enabled 12-plex DiLeu isobaric tags for quantitative proteomics. Anal Chem 87:1646–1654. https://doi.org/10.1021/ac503276z

Furukawa S, Tanaka K, Masumura T et al (2006) Influence of rice proteins on eating quality of cooked rice and on aroma and flavor of sake. Cereal Chem J 83:439–446. https://doi.org/10.1094/CC-83-0439

Galland M, He D, Lounifi I et al (2017) An integrated "Multi-Omics" comparison of embryo and endosperm tissue-specific features and their impact on rice seed quality. Front Plant Sci 8:1984. https://doi.org/10.3389/fpls.2017.01984

García-Gómez P, Almagro G, Sánchez-López ÁM et al (2019) Volatile compounds other than CO2 emitted by different microorganisms promote distinct posttranscriptionally regulated responses in plants. Plant Cell Environ. https://doi.org/10.1111/pce.13490

Hakata M, Kuroda M, Miyashita T et al (2012) Suppression of α-amylase genes improves quality of rice grain ripened under high temperature. Plant Biotechnol J 10:1110–1117. https://doi.org/10.1111/j.1467-7652.2012.00741.x

Hamel L-P, Nicole M-C, Duplessis S, Ellis BE (2012) Mitogen-activated protein kinase signaling in plant-interacting fungi: distinct messages from conserved messengers. Plant Cell 24:1327–1351. https://doi.org/10.1105/tpc.112.096156

Havé M, Balliau T, Cottyn-Boitte B et al (2018) Increases in activity of proteasome and papain-like cysteine protease in Arabidopsis autophagy mutants: back-up compensatory effect or cell-death promoting effect? J Exp Bot 69:1369–1385. https://doi.org/10.1093/jxb/erx482

Havé M, Luo J, Tellier F et al (2019) Proteomic and lipidomic analyses of the Arabidopsis atg5 autophagy mutant reveal major changes in endoplasmic reticulum and peroxisome metabolisms and in lipid composition. New Phytol 223:1461–1477. https://doi.org/10.1111/nph.15913

Heazlewood JL, Howell KA, Whelan J, Millar AH (2003) Towards an analysis of the rice mitochondrial proteome. Plant Physiol 132:230–242. https://doi.org/10.1104/pp.102.018986

Holding DR, Otegui MS, Li B et al (2007) The maize floury1 gene encodes a novel endoplasmic reticulum protein involved in zein protein body formation. Plant Cell 19:2569–2582. https://doi.org/10.1105/tpc.107.053538

Hu J, Huang X, Chen L et al (2015) Site-specific nitrosoproteomic identification of endogenously S-nitrosylated proteins in Arabidopsis. Plant Physiol 167:1731–1746. https://doi.org/10.1104/pp.15.00026

Huang X, Qian Q, Liu Z et al (2009a) Natural variation at the DEP1 locus enhances grain yield in rice. Nat Genet 41:494–497. https://doi.org/10.1038/ng.352

Huang S, Taylor NL, Narsai R et al (2009b) Experimental analysis of the rice mitochondrial proteome, its biogenesis, and heterogeneity. Plant Physiol 149:719. https://doi.org/10.1104/PP.108.131300

Huang S, Shingaki-Wells RN, Taylor NL, Millar AH (2013) The rice mitochondria proteome and its response during development and to the environment. Front Plant Sci 4:16. https://doi.org/10.3389/fpls.2013.00016

Hurkman WJ, Vensel WH, Tanaka CK et al (2009) Effect of high temperature on albumin and globulin accumulation in the endosperm proteome of the developing wheat grain. J Cereal Sci 49:12–23. https://doi.org/10.1016/J.JCS.2008.06.014

Inomata T, Baslam M, Masui T et al (2018) Proteomics analysis reveals non-controlled activation of photosynthesis and protein synthesis in a rice npp1 mutant under high temperature and elevated CO2 conditions. Int J Mol Sci. https://doi.org/10.3390/ijms19092655

Jia M, Wu H, Clay KL et al (2013) Identification and characterization of lysine-rich proteins and starch biosynthesis genes in the opaque2 mutant by transcriptional and proteomic analysis. BMC Plant Biol 13:60. https://doi.org/10.1186/1471-2229-13-60

Kaneko K, Sasaki M, Kuribayashi N et al (2016) Proteomic and glycomic characterization of rice chalky grains produced under moderate and high-temperature conditions in field system. Rice. https://doi.org/10.1186/s12284-016-0100-y

Kaur H, Petla BP, Kamble NU et al (2015) Differentially expressed seed aging responsive heat shock protein OsHSP18.2 implicates in seed vigor, longevity and improves germination and seedling establishment under abiotic stress. Front Plant Sci 6:713. https://doi.org/10.3389/fpls.2015.00713

Kawakatsu T, Hirose S, Yasuda H, Takaiwa F (2010a) Reducing rice seed storage protein accumulation leads to changes in nutrient quality and storage organelle formation. Plant Physiol 154:1842–1854. https://doi.org/10.1104/pp.110.164343

Kawakatsu T, Wang S, Wakasa Y, Takaiwa F (2010b) Increased lysine content in rice grains by over-accumulation of BiP in the endosperm. Biosci Biotechnol Biochem 74:2529–2531. https://doi.org/10.1271/bbb.100619

Keech O, Gardeström P, Kleczkowski LA, Rouhier N (2017) The redox control of photorespiration: from biochemical and physiological aspects to biotechnological considerations. Plant Cell Environ 40:553–569. https://doi.org/10.1111/pce.12713

Kim M, Kim H, Lee W et al (2015) Quantitative shotgun proteomics analysis of rice anther proteins after exposure to high temperature. Int J Genomics 2015:1–9. https://doi.org/10.1155/2015/238704

Kindrachuk J, Parent J, Davies GF et al (2003) Overexpression of L-isoaspartate O-methyltransferase in Escherichia coli increases heat shock survival by a mechanism independent of methyltransferase activity. J Biol Chem 278:50880–50886. https://doi.org/10.1074/jbc.M308423200

Kitajima A, Asatsuma S, Okada H et al (2009) The rice alpha-amylase glycoprotein is targeted from the Golgi apparatus through the secretory pathway to the plastids. Plant Cell 21:2844–2858. https://doi.org/10.1105/tpc.109.068288

Kobata T, Miya N, Anh NT (2011) High risk of the formation of milky white rice kernels in cultivars with higher potential grain growth rate under elevated temperatures. Plant Prod Sci 14:359–364. https://doi.org/10.1626/pps.14.359

Koller A, Washburn MP, Lange BM et al (2002) Proteomic survey of metabolic pathways in rice. Proc Natl Acad Sci U S A 99:11969–11974. https://doi.org/10.1073/pnas.172183199

Komatsu S, Konishi H (2005) Proteome analysis of rice root proteins regulated by gibberellin. Genomics Proteomics Bioinformatics 3:132–142. https://doi.org/10.1016/S1672-0229(05)03019-6

Komatsu S, Rakwal R, Li Z (1998) Separation and characterization of proteins in rice (Oryza sativa) suspension cultured cells. Plant Cell Tissue Organ Cult 55:183–192. https://doi.org/10.1023/A:1006145716701

Komatsu S, Muhammad A, Rakwal R (1999) Separation and characterization of proteins from green and etiolated shoots of rice (Oryza sativa L.): towards a rice proteome. Electrophoresis 20:630–636. https://doi.org/10.1002/(SICI)1522-2683(19990301)20:3<630::AID-ELPS630>3.0.CO;2-Z

Kumamaru T, Satoh H, Iwata N et al (1988) Mutants for rice storage proteins. Theor Appl Genet 76:11–16. https://doi.org/10.1007/BF00288825

Kurusu T, Koyano T, Hanamata S et al (2014) OsATG7 is required for autophagy-dependent lipid metabolism in rice postmeiotic anther development. Autophagy 10:878–888. https://doi.org/10.4161/auto.28279

Laino P, Shelton D, Finnie C et al (2010) Comparative proteome analysis of metabolic proteins from seeds of durum wheat (cv. Svevo) subjected to heat stress. Proteomics 10:2359–2368. https://doi.org/10.1002/pmic.200900803

Lasalde C, Rivera AV, León AJ et al (2014) Identification and functional analysis of novel phosphorylation sites in the RNA surveillance protein Upf1. Nucleic Acids Res 42:1916–1929. https://doi.org/10.1093/nar/gkt1049

Le DT, Chu HD, Le NQ (2016) Improving nutritional quality of plant proteins through genetic engineering. Curr Genom 17:220–229

Lee J, Koh H-J (2011) A label-free quantitative shotgun proteomics analysis of rice grain development. Proteome Sci 9:61. https://doi.org/10.1186/1477-5956-9-61

Li F, Chung T, Pennington JG et al (2015) Autophagic recycling plays a central role in maize nitrogen remobilization. Plant Cell 27:1389–1408. https://doi.org/10.1105/tpc.15.00158

Li H, Han J, Pan J et al (2017) Current trends in quantitative proteomics – an update. J Mass Spectrom 52:319–341. https://doi.org/10.1002/jms.3932

Lin S-K, Chang M-C, Tsai Y-G, Lur H-S (2005) Proteomic analysis of the expression of proteins related to rice quality during caryopsis development and the effect of high temperature on expression. Proteomics 5:2140–2156. https://doi.org/10.1002/pmic.200401105

Lin Z, Zhang X, Yang X et al (2014) Proteomic analysis of proteins related to rice grain chalkiness using iTRAQ and a novel comparison system based on a notched-belly mutant with white-belly. BMC Plant Biol 14:163. https://doi.org/10.1186/1471-2229-14-163

Long X, Liu Q, Chan M et al (2013) Metabolic engineering and profiling of rice with increased lysine. Plant Biotechnol J 11:490–501. https://doi.org/10.1111/pbi.12037

Lyon BG, Champagne ET, Vinyard BT et al (1999) Effects of degree of milling, drying condition, and final moisture content on sensory texture of cooked rice. Cereal Chem J 76:56–62. https://doi.org/10.1094/CCHEM.1999.76.1.56

Ma C, Zhou J, Chen G et al (2014) iTRAQ-based quantitative proteome and phosphoprotein characterization reveals the central metabolism changes involved in wheat grain development. BMC Genomics 15:1029. https://doi.org/10.1186/1471-2164-15-1029

Majoul-Haddad T, Bancel E, Martre P et al (2013) Effect of short heat shocks applied during grain development on wheat (Triticum aestivum L.) grain proteome. J Cereal Sci 57:486–495. https://doi.org/10.1016/J.JCS.2013.02.003

Mann M, Jensen ON (2003) Proteomic analysis of post-translational modifications. Nat Biotechnol 21:255–261. https://doi.org/10.1038/nbt0303-255

Martin M, Fitzgerald MA (2002) Proteins in rice grains influence cooking properties. J Cereal Sci 36:285–294. https://doi.org/10.1006/JCRS.2001.0465

McLoughlin F, Augustine RC, Marshall RS et al (2018) Maize multi-omics reveal roles for autophagic recycling in proteome remodelling and lipid turnover. Nat Plants 4:1056–1070. https://doi.org/10.1038/s41477-018-0299-2

Méchin V, Thévenot C, Le Guilloux M et al (2007) Developmental analysis of maize endosperm proteome suggests a pivotal role for pyruvate orthophosphate dikinase. Plant Physiol 143:1203–1219. https://doi.org/10.1104/pp.106.092148

Minina EA, Moschou PN, Vetukuri RR et al (2018) Transcriptional stimulation of rate-limiting components of the autophagic pathway improves plant fitness. J Exp Bot 69:1415–1432. https://doi.org/10.1093/jxb/ery010

Mitsui T, Yamakawa H, Kobata T (2016) Molecular physiological aspects of chalking mechanism in rice grains under high-temperature stress. Plant Prod Sci 19:22–29. https://doi.org/10.1080/1343943X.2015.1128112

Mu Q, Zhang W, Zhang Y et al (2017) iTRAQ-based quantitative proteomics analysis on rice anther responding to high temperature. Int J Mol Sci. https://doi.org/10.3390/ijms18091811

Nagai YS, Sakulsingharoj C, Edwards GE et al (2009) Control of starch synthesis in cereals: metabolite analysis of transgenic rice expressing an up-regulated cytoplasmic ADP-Glucose Pyrophosphorylase in developing seeds. Plant Cell Physiol 50:635–643. https://doi.org/10.1093/pcp/pcp021

Nanjo Y, Oka H, Ikarashi N et al (2006) Rice plastidial N-glycosylated nucleotide pyrophosphatase/phosphodiesterase is transported from the ER-golgi to the chloroplast through the secretory pathway. Plant Cell 18:2582–2592. https://doi.org/10.1105/tpc.105.039891

Nguyen T-P, Cueff G, Hegedus DD et al (2015) A role for seed storage proteins in Arabidopsis seed longevity. J Exp Bot 66:6399–6413. https://doi.org/10.1093/jxb/erv348

Nühse TS, Bottrill AR, Jones AME, Peck SC (2007) Quantitative phosphoproteomic analysis of plasma membrane proteins reveals regulatory mechanisms of plant innate immune responses. Plant J 51:931–940. https://doi.org/10.1111/j.1365-313X.2007.03192.x

Oikawa K, Inomata T, Hirao Y et al (2018) Proteomic analysis of rice golgi membranes isolated by floating through discontinuous sucrose density gradient. Methods Mol Biol. https://doi.org/10.1007/978-1-4939-7411-5_6

Parkhey S, Naithani SC, Keshavkant S (2012) ROS production and lipid catabolism in desiccating Shorea robusta seeds during aging. Plant Physiol Biochem 57:261–267. https://doi.org/10.1016/J.PLAPHY.2012.06.008

Paul S, Gayen D, Datta SK, Datta K (2015) Dissecting root proteome of transgenic rice cultivars unravels metabolic alterations and accumulation of novel stress responsive proteins under drought stress. Plant Sci 234:133–143. https://doi.org/10.1016/j.plantsci.2015.02.006

Payne PI. Nightingale MA. Krattiger AF. and Holt LM (1987) The relationship between HMW glutenin subunit composition and the bread-making quality of British-grown wheat varieties. J. Sci. Agric. 40:51

Peck SC, Nühse TS, Hess D et al (2001) Directed proteomics identifies a plant-specific protein rapidly phosphorylated in response to bacterial and fungal elicitors. Plant Cell 13:1467–1475. https://doi.org/10.1105/tpc.13.6.1467

Petla BP, Kamble NU, Kumar M et al (2016) Rice PROTEIN l -ISOASPARTYL METHYLTRANSFERASE isoforms differentially accumulate during seed maturation to restrict deleterious isoAsp and reactive oxygen species accumulation and are implicated in seed vigor and longevity. New Phytol 211:627–645. https://doi.org/10.1111/nph.13923

Rakwal R, Komatsu S (2000) Role of jasmonate in the rice (Oryza sativa L.) self-defense mechanism using proteome analysis. Electrophoresis 21:2492–2500. https://doi.org/10.1002/1522-2683(20000701)21:12<2492::AID-ELPS2492>3.0.CO;2-2

Ren Y, Wang Y, Liu F et al (2014) GLUTELIN PRECURSOR ACCUMULATION3 encodes a regulator of post-Golgi vesicular traffic essential for vacuolar protein sorting in rice endosperm. Plant Cell 26:410–425. https://doi.org/10.1105/tpc.113.121376

Reyes FC, Chung T, Holding D et al (2011) Delivery of prolamins to the protein storage vacuole in maize aleurone cells. Plant Cell 23:769–784. https://doi.org/10.1105/tpc.110.082156

Roepstorff P (1997) Mass spectrometry in protein studies from genome to function. Curr Opin Biotechnol 8:6–13. https://doi.org/10.1016/S0958-1669(97)80151-6

Ross PL, Huang YN, Marchese JN et al (2004) Multiplexed protein quantitation in Saccharomyces cerevisiae using amine-reactive isobaric tagging reagents. Mol Cell Proteomics 3:1154–1169. https://doi.org/10.1074/mcp.M400129-MCP200

Sánchez-López ÁM, Baslam M, de Diego N et al (2016a) Volatile compounds emitted by diverse phytopathogenic microorganisms promote plant growth and flowering through cytokinin action. Plant Cell Environ 39(12):2592–2608. https://doi.org/10.1111/pce.12759

Sánchez-López ÁM, Bahaji A, De Diego N et al (2016b) Arabidopsis responds to Alternaria alternata volatiles by triggering plastid phosphoglucose isomerase-independent mechanisms. Plant Physiol. https://doi.org/10.1104/pp.16.00945

Sarhadi E, Bazargani MM, Sajise AG et al (2012) Proteomic analysis of rice anthers under salt stress. Plant Physiol Biochem 58:280–287. https://doi.org/10.1016/J.PLAPHY.2012.07.013

Satoh H, Shibahara K, Tokunaga T et al (2008) Mutation of the plastidial alpha-glucan phosphorylase gene in rice affects the synthesis and structure of starch in the endosperm. Plant Cell 20:1833–1849. https://doi.org/10.1105/tpc.107.054007

Schaeffer GW, Sharpe FT (1987) Increased lysine and seed storage protein in rice plants recovered from calli selected with inhibitory levels of lysine plus threonine and S-(2-Aminoethyl)cysteine. Plant Physiol 84:509–515. https://doi.org/10.1104/pp.84.2.509

She K-C, Kusano H, Koizumi K et al (2010) A novel factor FLOURY ENDOSPERM2 is involved in regulation of rice grain size and starch quality. Plant Cell 22:3280–3294. https://doi.org/10.1105/tpc.109.070821

Shen S, Matsubae M, Takao T et al (2002) A proteomic analysis of leaf sheaths from rice. J Biochem 132:613–620. https://doi.org/10.1093/oxfordjournals.jbchem.a003264

Shewry PR (2004) Protein synthesis and deposition. Encycl Grain Sci:472–480. https://doi.org/10.1016/B0-12-765490-9/00131-2

Shewry PR, Halford NG (2002) Cereal seed storage proteins: structures, properties and role in grain utilization. J Exp Bot 53:947–958. https://doi.org/10.1093/jexbot/53.370.947

Shewry PR (2007) Improving the protein content and composition of cereal grain. 46 239–250. https://doi.org/10.1016/j.jcs.2007.06.006

Shin J-H, Kim S-R, An G (2009) Rice aldehyde dehydrogenase7 is needed for seed maturation and viability. Plant Physiol 149:905–915. https://doi.org/10.1104/pp.108.130716

Shiraya T, Mori T, Mruyama T et al (2015) Golgi/plastid-type manganese superoxide dismutase involved in heat-stress tolerance during grain filling of rice. Plant Biotechnol J 13:1251–1263. https://doi.org/10.1111/pbi.12314

Skylas DJ, Cordwell SJ, Hains PG et al (2002) Heat Shock of wheat during grain filling: proteins associated with heat-tolerance. J Cereal Sci 35:175–188. https://doi.org/10.1006/JCRS.2001.0410

Skylas DJ, Van Dyk D, Wrigley CW (2005) Proteomics of wheat grain. J Cereal Sci 41:165–179. https://doi.org/10.1016/J.JCS.2004.08.010

Smidansky ED, Martin JM, Hannah LC et al (2003) Seed yield and plant biomass increases in rice are conferred by deregulation of endosperm ADP-glucose pyrophosphorylase. Planta 216:656–664. https://doi.org/10.2307/23387686

Takaiwa F, Ogawa M, Okita TW (1999) Rice glutelins. In: Shewry P, Casey R (eds) Seed proteins. Springer, Netherlands, pp 401–425

Tang J, Bassham DC (2018) Autophagy in crop plants: what's new beyond Arabidopsis ? Open Biol 8:180162. https://doi.org/10.1098/rsob.180162

Thelen JJ (2009) Proteomics tools and resources for investigating protein allergens in oilseeds. Regul Toxicol Pharmacol 54:S41–S45. https://doi.org/10.1016/J.YRTPH.2009.01.005

Thompson A, Schäfer J, Kuhn K et al (2003) Tandem mass tags: a novel quantification strategy for comparative analysis of complex protein mixtures by MS/MS. https://doi.org/10.1021/AC0262560

van Wijk KJ, Friso G, Walther D, Schulze WX (2014) Meta-analysis of Arabidopsis thaliana phospho-proteomics data reveals compartmentalization of phosphorylation motifs. Plant Cell 26:2367–2389. https://doi.org/10.1105/tpc.114.125815

Wada S, Hayashida Y, Izumi M et al (2015) Autophagy supports biomass production and nitrogen use efficiency at the vegetative stage in rice. Plant Physiol 168:60–73. https://doi.org/10.1104/pp.15.00242

Wakasa K, Hasegawa H, Nemoto H et al (2006) High-level tryptophan accumulation in seeds of transgenic rice and its limited effects on agronomic traits and seed metabolite profile. J Exp Bot 57:3069–3078. https://doi.org/10.1093/jxb/erl068

Wang Y, Ren Y, Liu X et al (2010) OsRab5a regulates endomembrane organization and storage protein trafficking in rice endosperm cells. Plant J 64:812–824. https://doi.org/10.1111/j.1365-313X.2010.04370.x

Wang J, Chen L, Liu QQ et al (2011) Transformation of LRP gene into Brassica napus mediated by Agrobacterium tumefaciens to enhance lysine content in seeds. Genetika 47:1616–1621

Wong HW, Liu Q, Sun SSM (2015) Biofortification of rice with lysine using endogenous histones. Plant Mol Biol 87:235–248. https://doi.org/10.1007/s11103-014-0272-z

Wu L, Fan J, Belasco JG (2006) From the Cover: microRNAs direct rapid deadenylation of mRNA. Proc. Natl Acad. Sci. USA 103, 4034–4039 https://doi.org/10.1073/pnas.0510928103

Xie L, Chen N, Duan B et al (2008) Impact of proteins on pasting and cooking properties of waxy and non-waxy rice. J Cereal Sci 47:372–379. https://doi.org/10.1016/J.JCS.2007.05.018

Xu SB, Li T, Deng ZY et al (2008) Dynamic proteomic analysis reveals a switch between central carbon metabolism and alcoholic fermentation in rice filling grains. Plant Physiol 148:908–925. https://doi.org/10.1104/pp.108.125633

Xu H, Wei Y, Zhu Y et al (2015) Antisense suppression of LOX3 gene expression in rice endosperm enhances seed longevity. Plant Biotechnol J 13:526–539. https://doi.org/10.1111/pbi.12277

Yamakawa H, Hakata M (2010) Atlas of rice grain filling-related metabolism under high temperature: joint analysis of metabolome and transcriptome demonstrated inhibition of starch accumulation and induction of amino acid accumulation. Plant Cell Physiol 51:795–809. https://doi.org/10.1093/pcp/pcq034

Yamakawa H, Hirose T, Kuroda M, Yamaguchi T (2007) Comprehensive expressionp of rice grain filling-related genes under high temperature using DNA microarray. Plant Physiol 144:258–277. https://doi.org/10.1104/pp.107.098665

Yamakawa H, Hirai-Kimura R, Nakata Y et al (2017) An activity-staining method on filtration paper enables high-throughput screening of temperature-sensitive and inactive mutations of rice α-Amylase for improvement of rice grain quality. Plant Cell Physiol 58:658–667. https://doi.org/10.1093/pcp/pcx030

Yang Y, Dai L, Xia H et al (2013) Protein profile of rice (Oryza sativa) seeds. Genet Mol Biol 36:87–92. https://doi.org/10.1590/S1415-47572013000100012

Yu J, Zhen X, Li X et al (2019) Increased autophagy of rice can increase yield and Nitrogen Use Efficiency (NUE). Front Plant Sci 10:584. https://doi.org/10.3389/fpls.2019.00584

Zhang C, Yin Y, Zhang A et al (2012) Comparative proteomic study reveals dynamic proteome changes between superhybrid rice LYP9 and its parents at different developmental stages. J Plant Physiol 169:387–398. https://doi.org/10.1016/j.jplph.2011.11.016

Zhang Y, Schernthaner J, Labbé N et al (2014) Improved protein quality in transgenic soybean expressing a de novo synthetic protein, MB-16. Transgenic Res 23:455–467. https://doi.org/10.1007/s11248-013-9777-5

Zhang N, Chen F, Huo W, Cui D (2015) Proteomic analysis of middle and late stages of bread wheat (Triticum aestivum L.) grain development. Front Plant Sci 6:735. https://doi.org/10.3389/fpls.2015.00735

Zhang Y-X, Xu H-H, Liu S-J et al (2016) Proteomic analysis reveals different involvement of embryo and endosperm proteins during aging of Yliangyou 2 hybrid rice seeds. Front Plant Sci 7:1394. https://doi.org/10.3389/fpls.2016.01394

Zhao Q, Chen W, Bian J et al (2018) Proteomics and phosphoproteomics of heat stress-responsive mechanisms in Spinach. Front Plant Sci 9:800. https://doi.org/10.3389/fpls.2018.00800

Chapter 21
Metabolomics for Rice Grain Quality

Fabio Clasen Chaves and Corey D. Broeckling

21.1 Introduction

Rice and rice-derived products make up a significant amount of the daily diet throughout the world. Rice is considered a staple food in several countries, and rice-derived products such as sake, shochu, and makgeoli have a social component and are highly appreciated in certain communities. In addition, products such as brown rice, germinated brown rice, rice bran, rice bran oils, and pigmented rice are consumed for their health beneficial contents (Fitzgerald et al. 2009).

Rice grain processing usually follows dehulling, which turns rice into brown rice. Brown rice is then milled and polished, removing the bran (pericarp, tegument, and embryo), and turning it into white rice, which contains the aleurone and endosperm layers only. Red and black rice variants contain pigments in the pericarp and aleurone layers, and therefore, depending on the degree on milling, more or less of the pigments will be removed (Huang and Lai 2016). During the dehulling, milling and polishing of rice, 20% is separated as husk (hull), 10% as bran, and the remaining 70% constitutes the milled or polished white rice, approximately 95% of which is starch (Fitzgerald et al. 2009). For the purpose of this chapter, only small organic molecules (<1200 Da), whether volatile, nonvolatile, polar, or apolar, with known involvement in rice grain quality are discussed. We refer to the full collection of small organic molecules found in rice grain as rice metabolome. "The term 'rice grain quality' is inclusive of parameters such as grain appearance (color, shape, chalkiness), aroma, texture, nutritional and functional properties, and cooking time". However, the physiochemical properties of starch, a large carbohydrate polymer, largely define qualities such as shape, chalkiness, texture, and cooking time. Given that starch falls outside our definition of a small molecule, we do not discuss

F. C. Chaves (✉) · C. D. Broeckling
Proteomics and Metabolomics Facility, Colorado State University, Fort Collins, CO, USA
e-mail: fabio.clasen_chaves@colostate.edu

A. Costa de Oliveira et al. (eds.), *The Future of Rice Demand: Quality Beyond Productivity*, https://doi.org/10.1007/978-3-030-37510-2_21

these components of grain quality extensively in this chapter. Conversely, color, aroma, nutritional and functional properties are directly associated with small organic compounds. Rice grain quality is dependent on its chemical components, which vary according to the variety; growing conditions, including environmental factors; management practices; and post-harvest storage and processing steps and conditions (Yanjie et al. 2018).

Phytochemicals associated with quality attributes of rice grain include volatiles responsible for aroma and flavor; amino acids; vitamins such as tocopherols, which contribute to nutritional quality; and specialized metabolites such as phenolic compounds abundant in red and black rice, which contribute to the color and functional or health attributes, all of which can be determined by metabolomics analyses (Fitzgerald et al. 2009). Among the functional and health-promoting aspects of compounds present in rice products are anti-cancer properties, anti-atherogenic properties, and properties controlling blood cholesterol and sugar (Mohd Esa and Ling 2016). Two comprehensive lists of compounds identified in rice can be found at (1) the Food Database project, supported by the Canadian Institutes of Health Research, Canada Foundation for Innovation, and by The Metabolomics Innovation Centre (http://foodb.ca/), and at (2) the Plant Metabolic Network project, funded by the National Science Foundation, the Department of Energy and executed at the Carnegie Institution for Science, Department of Plant Biology (https://pmn.plantcyc.org/organism-summary?object=ORYZA). We provide a comprehensive review of the literature on metabolomic-quality relationships in section three of this chapter.

Metabolomics is an analytical strategy used to study small molecule composition in biological materials. Metabolomic approaches can be broadly divided into targeted and untargeted, which differ in the data acquisition approach and breadth of metabolite coverage (Ortmayr et al. 2016). Untargeted metabolomics offers broad metabolite coverage, and the resulting data can be used either for hypothesis testing or generation. No prior knowledge of treatment responsive compounds is necessary, and the results can be used as a screen to guide more targeted interrogation of compounds and/or pathways of interest (Lee et al. 2017a). In contrast, targeted metabolomics is more purely hypothesis driven, as the analytical assay is defined before data acquisition begins to measure relatively few metabolites in a highly targeted manner (Mahdavi et al. 2015).

Mass spectrometry (MS) and nuclear magnetic resonance (NMR) are the primary analytical techniques that have been used to support metabolomics studies. MS provides evidence for compound identification based on the mass-to-charge ratio (*m/z*) of formed ions, and the quantification is based on ion signal intensity, which can be converted to absolute concentration through comparison to appropriate authentic and internal standards. In NMR, compound identification is based on chemical shift, and quantification is based on comparison of integrals (signal intensity) with appropriate standards (Wishart 2008).

MS-based metabolomics usually, but not always, employs a coupled separation step which facilitates compound identification and quantitation. Capillary electrophoresis (CE) (Sato et al. 2004), gas-chromatography (GC), high-performance liquid chromatography (HPLC), ultra-high-pressure liquid chromatography (UHPLC), and their use in tandem or 2-D (Daygon et al. 2016; Navarro-Reig et al. 2017a) are among the different separation techniques that can be hyphenated to mass spectrometers. We further describe metabolomics instrumentation and technologies in section two of this chapter.

This chapter will collectively introduce metabolomics approaches as applied to the analysis of rice grain, with particular emphasis on quality-related metabolites.

21.2 Metabolomics Analysis

21.2.1 Instrumentation

Mass spectrometers are basically constituted of an ionization source, a mass analyzer, and a detector. Compounds are ionized (charged) in the ionization source. Following ionization, the gas-phase ions are accelerated by electromagnetic fields and filtered or sorted by the mass-to-charge (m/z) ratio in the mass analyzer. Following separation by m/z, the detector registers the signal in a manner that reflects the mass analyzer separation. The different mass analyzers available vary in resolving power and range from low resolution instruments such as single or triple quadrupoles (Q, QQQ) to high-resolution time-of-flight (TOF) to ultrahigh-resolution Fourier-transform ion cyclotron resonance (FT-ICR) and Orbitrap (Akram et al. 2017). While low-resolution instruments are used in targeted metabolomics approaches because of their greater selectivity and sensitivity (Lim et al. 2017a), high-resolution instruments are used in untargeted metabolomics because they provide evidence for compound annotation (exact mass, isotope profile, and fragmentation).

Nuclear magnetic resonance (NMR) instruments are primarily composed of magnet, probe, shim coils, signal transmitters, and amplifiers. NMR spectroscopy uses radiofrequency wave pulses to obtain compound structural and quantitative information based on chemical shifts, coupling constants, and signal intensity of magnetic nuclei (^{1}H, ^{13}C, ^{19}F, or ^{31}P) (Barding et al. 2013). Most NMR spectroscopy is performed on samples in solution resuspended in deuterated solvents whose signal do not interfere with the signals of the nuclei of interest and also serve to calibrate their frequencies. The sample placed inside the probe under a strong electromagnetic field receives energy pulses at radio frequency range applied by the surrounding shim coil. The magnetic nuclei absorb this energy going into a higher energy spin state. When pulse is stopped and nuclei relax, the released energy is detected as free induction decay in the time domain, which is converted by Fourier Transformation to the frequency domain (Bothwell and Griffin 2011).

21.2.2 Sample Preparation

Quenching of metabolic activities is performed by flash freezing collected samples in liquid nitrogen. Water is then removed, preferably by freeze-drying. Samples are stored at low temperature (−80 °C) prior to sample extraction to assure limited changes to compound profile and abundance until the sample extraction for analysis.

Compound coverage is determined by the extraction solvent hydrophilicity or hydrophobicity. A solvent mixture, such as chloroform, methanol, and water (1:3:1), can be utilized to widen the metabolite extraction range, including hydrophilic and hydrophobic compounds (Akram et al. 2017). By altering their proportions, it becomes a two-phase extraction system in which compounds with different polarities will be partitioned into their respective phases. For example, different proportions of a mixture of methanol, methyl tert-butyl ether (MTBE), and water are optimized for separate analyses of rice metabolomics and lipidomic by LC-Q-TOF-MS (Chang et al. 2014). Phase-separated compounds can, however, be analyzed within the same run by utilizing a stacked injection method (Broeckling and Prenni 2018).

In order to improve compound extraction efficiency, different techniques may be utilized, including tissue grinding to reduce particle size and increase surface area, microwave, sonication, pressurized solvent extraction, and heating. A study investigated the use of different extraction conditions for metabolomic analysis by GC-MS and LC-MS for determining metabolite markers that could differentiate white rice from Korea and China (Lim et al. 2018a).

Interfering compounds co-extracted from the sample matrix with the analytes of interest can cause signal suppression during analysis (matrix effect) and may be selectively removed by precipitation, centrifugation, filtration, solvent extraction, partition, or the use of solid-phase extraction. In a GC-MS metabolomics analysis of pesticide residues in brown rice and other crops, different extraction conditions using solid-phase extraction were tested and matrix effects were investigated (Sugitate et al. 2012). Similarly, solid-phase extraction, followed by a direct infusion-tandem mass spectrometry (DI-MS/MS)-based metabolomics approach, was used to differentiate rice products of different origin (Lim et al. 2017b). Interfering compounds are even more problematic when the analytes are in comparatively lower concentration, such as the case of plant hormones. LC-MS analysis of brassinosteroid in rice samples followed selective removal of matrix interferents by solid-phase extraction and purification steps (Ding et al. 2013).

Liquid–liquid extraction methods are more time consuming and tend to be a greater source of error than solid-phase extraction, which is more prone to automation. Automated liquid handlers and robotized handling of solid-phase extraction steps are advances which expedite sample preparation and limit variation. A high-throughput extraction of plant hormones from rice samples used microscale ball-milling for sample grinding and an automatic liquid handling system for solid-phase extraction using 96 well plates (Kojima et al. 2009).

The GC-MS platform is utilized for the analysis of naturally volatile compounds and polar compounds from general metabolism, including amino acids, organic acids, and sugars, following sample derivatization steps. Derivatization steps such as methylation, oxymation, and trimethylsilylation are the most usual reactions which optimize compound analysis by making compounds more volatile and less labile, limit the formation of isomers, and improve peak shape and intensity (Frenzel et al. 2002; Long et al. 2013). Inter-laboratory reproducibility of GC-TOF-MS methods was assessed for the analysis of rice grain and other matrices (Allwood et al. 2009).

21.2.3 Data Processing

Mass Spectrometry

In order to ensure repeatability, samples in a batch should be run in randomized order with quality control (QC) samples, consisting of pooled aliquots of equal amounts of all different samples, interspersed in the beginning, middle, and the end of the batch. QC runs can be utilized to measure variation and serve as a parameter to discard an outlier run or to monitor and identify instrumental failure (Dunn et al. 2011).

Raw mass spectrometry chromatograms consist of 3-dimensional data (histograms) with the following basic features: mass-to-charge ratio (*m/z*), intensities, and retention time. Depending on the analytical platform, ample files need to be converted to the appropriate extension before they can be processed. Automated data processing is possible utilizing vendor proprietary or open source software such as XCMS, MetAlign, or MZmine (Smith et al. 2006; Lommen 2009; Pluskal et al. 2010). Outlier detection serves to eliminate peaks or runs that present large deviation when compared across a set of replicate runs. Noise filtering and baseline correction are followed by the peak picking or feature detection step. Features are ions of a specific *m/z* detected from the continuous raw signal which are converted into centroid data represented by peaks eluting at a specified retention time. Although peak area can be utilized for compound quantitative analysis, it is dependent on the compound's ionization capacity under specified analytical parameters, so peak area must be calibrated using authentic standards to obtain absolute concentration. GC-MS data deconvolution for separating co-eluting analytes is usually performed using an external software, such as the freely available Automated Mass Spectral Deconvolution and Identification System (AMDIS) (National Institute of Standards and Technology, NIST). The algorithm detects mass differences in a spectra of neighboring data points separating co-eluting peaks. The alignment step promotes a retention time correction due to run-to-run variation by grouping peaks representing the same feature across different runs. In LC-MS analyses, a compound can be ionized and detected as different species, including its protonated or de-protonated form, by the formation of adducts with salts, mobile phase modifiers, solvent ions, fragment ions, dimers, and trimers. Therefore, the number of detected features is usually greater the actual number of compounds present in the sample. In order to associate a precursor ion with its fragment ions or adducts, different algorithms cluster these ions together based on their elution time and consistent proportional abundance relative to each other across different runs (Broeckling et al. 2014). A normalization step then removes unwanted systematic variation between ion intensities across runs. The output of the processing steps is a two-dimensional matrix consisting of the samples and the compounds representing their retention time and mass-to-charge ratio (or mass spectrum) and intensity. Several peak detection filtering and processing steps were utilized and described in detail in the untargeted metabolite profiling study of natural rice variants (Akram et al. 2017).

NMR

NMR spectral analysis is based on calibration against an internal standard, usually 4,4-dimethyl-4-silapentane-1-sulfonic acid (DSS) added at a specified concentration, which serves to adjust the chemical shift and for quantification purposes by comparing DSS integrals (signal area) to sample compound integrals. Deuterated water or deuterium oxide (D_2O) is utilized as a field lock to ensure that resonance frequencies do not drift while data is being collected, and often times samples are spiked with a buffer solution for maintaining the pH stable (Sotelo and Slupsky 2013). Less polar deuterated solvents such as chloroform ($CDCl_3$) and dimethyl sulfoxide (DMSO) might be utilized depending on the sample. Other relevant parameters to be considered prior to data acquisition include temperature setting, probe tuning and matching, magnet shimming, pulse calibration, and receiver gain adjustment (Nagana Gowda and Raftery 2019).

NMR data processing can either be performed by untargeted or targeted approaches. The untargeted profiling approach uses spectral binning, in which the spectrum is integrated between 0.0 and 10.0 ppm and divided into smaller regions of specific widths (bins). Typically, untargeted NMR data is imported and converted into the appropriate format, aligned, binned, and a multivariate analysis follows. The area under a spectral bin is used for the multivariate statistical analysis, performed using packages running on R software, for example, and for compound identification by comparing to spectral databases (Mochida et al. 2009). In the targeted profiling approach, spectral deconvolution and peak picking are followed by statistical analysis comparing the concentrations of the targeted compounds. Targeted compound identity is confirmed by comparison to a reference library. Many vendor proprietary and open-source software are available for processing and analyzing NMR metabolomics data (Lewis et al. 2009). In addition, several NMR spectral databases containing chemical shift and coupling constants are available, such as the Human Metabolome Database (www.hmdb.ca), the Madison Metabolomic Consortium Database (mmcd.nmrfam.wisc.edu), the Biological Magnetic Resonance Data Bank (www.bmrb.wisc.edu), the Magnetic Resonance Metabolomics Database (www.liu.se/hu/mdl/main), PubChem (http://pubchem.ncbi.nlm.nih.gov), NMRShiftDB (nmrshiftdb.ice.mpg.de), Spectral Database for Organic Compounds (riodb01.ibase.aist.go.jp/sdbs/cgi-bin/cre_index.cgi), and SpecInfo (cds.dl.ac.uk/cds/datasets/spec/specinfo/specinfo.html). Identification of novel metabolites usually requires two-dimensional NMR experiments, including total correlation spectroscopy (TOCSY) and $^{1}H–^{13}C$ heteronuclear single quantum correlation (HSQC) among others (Song et al. 2016). 2D NMR experiments can also be utilized when signal overlap in 1D NMR limits the quantitation of less abundant compounds (Chae and Kim 2016).

21.2.4 Data Analysis

GC-MS metabolite annotation is based on comparison of retention index and fragmentation profiles to those available in compound libraries (NIST) or databases (Golm) (Kopka et al. 2005; Kind et al. 2009). High-resolution mass spectrometry provides accurate mass measurements and isotope profiles, from which a molecular formula can be calculated. Heuristic rules can also be applied to narrow down potential molecular formula (Kind and Fiehn 2007). Tandem mass spectrometry provides fragmentation patterns which add another level of confidence to compound identification. Fragmentation is dependent on the applied collision energy, and publicly available databases provide experimental data with compound fragmentation as well as predicted fragmentation profiles according to specific applied collisional energy (Koellensperger et al. 2018).

The identification of isomers with the same elemental composition but different structures depends on the resolving power of chiral chromatography separation, tandem mass spectrometry, or the application of ion-mobility mass-spectrometry. Yet, unequivocal confirmation of compound identity depends on comparison to reference material or pure standards. When standards are not available, compound purification followed by structure elucidation by NMR is the alternative (Yang et al. 2014). Compound purification can be obtained by fraction collection of the LC or GC eluate. Several runs might be necessary to obtain enough pure compounds, and while the LC solvent requires concentration by evaporation, the GC eluate requires a cold trap for the collection of the separated compounds.

The Metabolomics Standards Initiative recommends that compound identification be reported as compound annotation or putative identification when the only few parameters for identification are available, such as comparison to database information. Matching authentic standard data to experimental data run under the same analytical condition is necessary for unequivocal compound identification (Sumner et al. 2007; Salek et al. 2013).

Relative quantification based on normalizing metabolite signal intensity to that of an internal standard (IS) is typically used in untargeted metabolomics. The IS serves to account for losses due to variation originating during sample preparation or analysis, and the peak area of the analyte is normalized by the peak area of the IS. In targeted metabolomics, absolute quantification uses external standards or internal isotopically labeled standards.

Data normalization by Pareto scaling or other transformation is usually necessary due to the wide dynamic range of the multitude of compounds present in the matrix. Multivariate unsupervised principal component analysis (PCA) is the predominant statistical analysis applied to metabolomics data. It allows for visualizing trends of clustering or grouping of individuals based on the contribution of the variance observed for all measured variables (scores plot). It also allows for the identification of which variables most contributed to each grouping (loadings plot). Partial least squares-discriminant analysis (PLS-DA) is a supervised multivariate approach that allows to achieve maximum group separation by rotating PCA components. Alternative supervised learning methods include k-nearest neighbors, support vector machine, random forest, and neural network (Lim et al. 2017a). From the PLS-DA discriminatory analysis, variable importance in projection (VIP) scores are assigned to the compounds (markers) most contributing to group discrimination. PCA and PLS-DA are the most commonly employed chemometric for untargeted data analysis (Wishart 2008). A partial least squares discriminant analysis (PLS-DA) model exhibited a good classification of white rice between Korea and China (Lim et al. 2018b). Analysis of variance (ANOVA) and means comparison by T test, however, are the definitive parameters to assign compounds differentially accumulated in response to treatment effects. A false discovery rate (FDR) filter is also applied to assure that a peak is only set as significantly influenced by treatment when meeting a minimal threshold (Benajmini and Hochberg 1995).

The functional analysis of metabolomics data helps with the visualization and interpretation of a complex data set. Available software such as Metaboanalyst maps annotated compounds within their respective biological pathways by matching to databases such as KEGG (Kyoto Encyclopedia of Genes and Genomes) and the Human Metabolome (Chagoyen and Pazos 2013; Misra and Mohapatra 2019; Xia and Wishart 2011). Enrichment analysis in addition to mapping the compounds to a pathway considers their abundance (concentration) and allows statistical comparisons (Chagoyen and Pazos 2013). In a study on rice plants applied with an insecticide, a functional analysis of the GC-MS data was applied to assist with the visualization of its impact on different metabolic pathways (Mahdavi et al. 2016). The effect of iron deficiency in rice plants metabolism was reported using a functional analysis of LC-MS metabolomics data (Selby-Pham et al. 2017).

21.3 Applications of Different Metabolomics Platforms to Assess Rice Grain Quality

Genotype

Sensory, nutritional, and functional aspects of rice grain quality are influenced by several quantitative trait loci (QTL) and by biotic and abiotic environmental factors. The genetic analysis of the rice metabolome can associate QTLs and the abundance of compounds to establish how metabolic expression is regulated. With that information, identified markers can be utilized in breeding programs in a marker-assisted selection strategy focused on quality traits.

Metabolite quantitative trait analysis was used to associate genetic loci with metabolite abundance. Accessions from the world rice collection were analyzed by ^{1}H-NMR in order to try establish correlations between metabolic and genetic diversity (Mochida et al. 2009).

GC-TOF-MS metabolomics analysis of rice accessions from a Chinese germplasm collection revealed nutritionally relevant compounds (amino acids, sugars, organic acids, fatty acids, sterols, and polyols) linked to molecular markers analyzed by association maps (Lou et al. 2011). A combined one- and two-dimensional GC (GC × GC)-TOF-MS approach was used to determine the metabolic phenotyping of natural rice variants from the world rice core collection (Kusano et al. 2007).

High yielding rice individuals producing volatile organic compounds (VOC) associated with high grain quality were selected from a crossing population between two commercial varieties (Apo and IR64). QTLs identified by a genomics approach associated with both yield and positive VOC profile can be utilized in breeding strategies (Calingacion et al. 2017).

The two major rice subspecies *japonica* and *indica* have distinct differences in genome sequences as well as quality traits. Untargeted UHPLC-MS/MS and GC-MS metabolomic analyses of grains of 100 *japonica* and *indica* cultivars revealed species-specific diversified metabolomes (Hu et al. 2014).

A genome-wide association study with hundreds of rice genotypes, including accessions from both *indica* and *oryza* subspecies, was matched to a widely targeted HPLC-MS/MS metabolomics. Candidate genes responsible for metabolic traits were identified (Chen et al. 2014). Similar association studies have been utilized with targeted and untargeted HPLC-MS approaches (Gong et al. 2013; Matsuda et al. 2015). Given the metabolite coverage limitation possibly achieved by the use of single platforms, combining different platforms allows for complementarity and increased coverage. A study linked genome-wide genotyping with complementary metabolomic analysis using ^{1}H-NMR for polar compounds, GC-TOF-MS for VOC, and inductively coupled plasma (ICP)-MS for the analysis of minerals and compared commercially available rice varieties in the Lao People's Democratic Republic (Calingacion et al. 2012).

The use of complementary metabolomics platforms GC-TOF-MS, HPLC-Q-TOF-MS, CE-TOF-MS, and HPLC-ion trap (IT)-TOF-MS in the analysis of brown rice grains allowed for establishing correlations between quality traits and metabolites. Integrated data sets yielded better overall prediction in comparison to single platforms. Multivariate regression analysis showed that several quality traits could be predicted from the metabolic composition, and the correlation network of the trait-associated metabolites showed correlations between fatty acids and amylose ratio and between tryptophan and ear emergence day (Redestig et al. 2011).

In order to determine the effect of QTLs on metabolic traits, grains from backcrossed inbred lines of selected cultivars with superior flavor and high yield were analyzed by different platforms, such as HPLC-Q-TOF-MS, CE-TOF-MS, GC-TOF-MS, and HPLC-IT-TOF-MS, in order to obtain a wide range of metabolite coverage. Predominantly, compound abundance was strongly influenced by environmental factors and weakly associated with mQTLs (Matsuda et al. 2012).

Metabolic diversity studies comparing rice varieties by GC-MS metabolomics found sugar, sugar alcohol, organic acid, amino acid, fatty acid, phenol, and sterol contents, all of which impact rice flavor and nutritional quality, influencing genotypes discrimination (De and Nag 2014; Ranjitha et al. 2019).

Pigmented and nonpigmented rice possess a different profile of compounds responsible for their color. Multivariate data analysis of targeted anthocyanin HPLC-MS analysis and untargeted GC-MS analysis of white, red, and black *indica* cultivars separated cultivars with different color background based on their chemical profile (Frank et al. 2012). Untargeted GC-MS analysis identified organic acids, sugars, sugar alcohols, and unsaturated and saturated fatty acids that contributed to differentiate among traditional Thai rice varieties, including white, red, and black genotypes (Wongsa et al. 2018). GC-TOF-MS-based metabolic profiling of black and white rice cultivars revealed correlations between the abundances of general and specialized metabolites (Kim et al. 2013b). ^{1}H NMR-based metabolomics using decision tree chemometric data analysis was able to discriminate among one white and three pigmented (a red, a purple, and a black) rice cultivars (Du et al. 2019).

Ultimately, rice quality is expressed in the ready-to-eat product, such as cooked rice. An untargeted UPLC-Q-TOF-MS metabolomics compared open-boiled cooked rice of different varieties. Multivariate analysis of metabolite profiles lead to the clustering of samples according to their subspecies classifications (Heuberger et al. 2010).

A prominent quality attribute of cooked rice is aroma, which makes fragrant rice varieties highly valued. Aromatic cultivars have been bred for that characteristic, and yet environmental factors also play a role in the rice aroma profile. Solvent-extracted VOC responsible for cooked rice aroma from scented and unscented cultivars produced in different locations were analyzed by GC-MS and GC-olfactometry (GC-O). While several odor-active compounds were identified among samples, oct-1-en-3-one and 2-acetyl-1-pyrroline (2AP) were perceived in all samples. Among the compounds responsible for differentiating cultivars and growing regions were lipid degradation and cinnamic acid–derived compounds (Maraval et al. 2008).

2AP, the main compound responsible for the aroma of fragrant rice varieties, was thought to be formed during processing, but it is actually pre-formed in the grains of selected lines. In order to gain further understanding of its biosynthetic pathway, an untargeted metabolomics approach used a static headspace collection of volatiles, followed by 2D-GC-TOF-MS analysis in combination with a genome-wide association study (Daygon et al. 2017). Pyrrole, 2-acetylpyrrole 1-pyrroline, and 6-methyl-5-oxo-2,3,4,5-tetrahydropyridine were found associated with the 2AP pathway and, along with 2AP, contribute to the floral, popcorn, and toasted aromas of rice (Daygon et al. 2017).

Aroma volatiles of fragrant rice types, Basmati, Jasmati, and Jasmine, after cooking were collected by solid-phase micro extraction (SPME) and analyzed by a combination of techniques including GC–O and GC–MS. Aroma active compounds were identified, including sulfur-containing compounds active only in Jasmine rice. A principal component analysis of the sulfur-volatile data was able to group separately fragrant rice by type (Mahattanatawee and Rouseff 2014).

VOC analysis may follow solvent extraction methods or trapping of volatilized compounds using static or dynamic collection approaches. Cooked black and white rice VOC were trapped in a dynamic headspace system using Tenax adsorbing resins and analyzed by GC-MS and GC-O. 2AP, guaiacol, indole, and p-xylene, more abundant in cooked black rice, were the compounds that most contributed to distinguish between the two rice varieties (Dong et al. 2008).

Basmati varieties require a long storage period for the development of its typical flavor, some of which derive from lipid breakdown. However, inappropriate storage may lead to the formation of off-flavors with negative impact on consumer acceptance (Fitzgerald et al. 2009). Flavor is influenced not only by volatile but also nonvolatile compounds. Untargeted LC-Orbitrap-MS metabolomic analysis of nonvolatile compounds was used to differentiate samples of aromatic rice varieties from the Basmati and non-Basmati groups (Akram et al. 2017).

Untargeted metabolite profiling of volatile compounds collected by static headspace extraction and analyzed by 2D-GC-TOF-MS revealed hexanal and 2AP as the most abundant compounds in all samples and a multivariate analysis of the data separated fragrant *indica* varieties from japonica varieties (Daygon et al. 2016).

NMR-based metabolomics was used to evaluate the influence of growing season on grain quality of cultivars with different maturation cycle lengths. The differentially accumulated metabolite profile revealed greater sucrose, amino acid, and free fatty acid contents in grains of early maturing cultivars (Song et al. 2018).

Waxy or sticky rice contains very little amylose content, and its starch fraction is made up almost exclusively of amylopectin, which is responsible for the glutinous characteristic upon cooked. NMR-based metabolomics of sticky and nonsticky cultivars revealed that their differences extend to fatty acids, phospholipids, glycerophosphocholine, glutamate, aspartate, asparagine, and sucrose contents, which were more abundant in grains of waxy rice cultivars in comparison to nonwaxy ones (Song et al. 2016).

Genetic Modification

Genetic engineering is a tool that has been successfully employed to rice breeding with positive impacts on rice quality attributes and holds the potential of delivering nutritionally and functionally rich rice grains (Fitzgerald et al. 2009). However, its wide application and the launch of commercial cultivars depends on the assurance of the absence of undesirable and unexpected effects derived from the genetic modification. Regulatory agencies assess the safety of a genetically modified (GM) food by comparing its chemical composition to its untransformed background. Untargeted metabolomics is the ideal approach to evaluate unintended, unexpected, and potentially negative changes in the metabolome given its unbiased and encompassing scope (Frenzel et al. 2002).

Although grain quality is the focus, it is important to realize that the rice seed, which will become the processed grain, is a sink tissue whose composition is dependent on the precursors shuttled from source tissues such as mature leaves and roots. Rice has been genetically modified to accumulate more lysine, an essential amino acid of limited natural abundance in rice. Untargeted GC-MS and targeted

LC-FT-MS analysis were used to monitor specific and broader metabolic changes associated with the transformation. Levels of asparagine and glutamine, which are the major amino acids transported from rice leaves into the seeds, remained unaltered in transgenic plants. Although lysine accumulation was observed in the leaves of transgenic lines, its accumulation in the seeds was not observed (Long et al. 2013).

GM rice seedlings overexpressing the essential amino acid tryptophan were compared to the untransformed lines by untargeted LC-MS metabolomics. Transformed seedlings had increased accumulation of tryptophan and other minor indole metabolites. Meanwhile, the abundance of other metabolites remained unaltered (Matsuda et al. 2010).

Flavonoids are known health-promoting compounds abundantly present in many foods but of limited occurrence in unpigmented rice grains, with the exception of tricin, a flavonoid known to accumulate on rice bran with colon cancer cell growth inhibitory potential. GM rice lines accumulating the flavonoids naringenin, kaempferol, genistein, and apigenin in seeds have been developed and their contents were monitored by targeted HPLC-MS analysis (Ogo et al. 2013).

Resveratrol is a polyphenol (stilbene) present in grapes and wine as well as other food products which has been associated with many health benefits, such as the reduction of oxidative stress and inflammation and prevention of cardiovascular disease. Targeted GC-MS metabolite profiling of several polar and apolar compounds did not indicate any metabolic changes promoted by the genetic modifications that lead to a resveratrol-enriched rice with tolerance to the herbicide glufosinate (Kim et al. 2017b).

Rice lines overexpressing *Arabidopsis thaliana* folate biosynthetic genes achieved 100-fold increase in folate content found in milled rice when analyzed by targeted HPLC-QQQ-MS in comparison to the untransformed counterpart. One hundred gram of the folate-rich rice contained four times the daily folate requirement of an adult person (Storozhenko et al. 2007).

Consumption of carotenoid-rich food contributes to reducing the incidence of several chronic-degenerative diseases, including cancers and cardiovascular diseases; in addition, many carotenoids are precursors of vitamin A, which is essential for human health. A genetically modified rice which accumulates beta-carotene and nontransgenic white, red, and black rice cultivars were analyzed using a targeted GC-TOF-MS to measure lipophilic compounds, including tocopherols, tocotrienols, sterols, and policosanols. Transformation did not lead to any undesirable or unintended effect regarding changes in the targeted metabolites. While PCA analysis did not separate GM rice from the non-GM rice, pigmented rice did separate from nonpigmented ones (Kim et al. 2012b). Untargeted GC-TOF-MS similarly did not identify major differences between the beta-carotene-fortified transgenic line and its precursor line but was able to differentiate between pigmented and white rice (Kim et al. 2013a).

An *Arabidopsis* NAD kinase involved in promoting carbon and nitrogen assimilation was introduced to rice and led to increased NADP(H) content. A comprehensive targeted CE-MS metabolomic analysis of the general metabolites present in rice leaves revealed elevated levels of amino acids and sugar phosphates. The comparative differential metabolic changes observed in the transgenic lines were accompanied by enhanced tolerance to oxidative stress and greater electron transport and CO_2 assimilation rates suggesting a role for NADP in those metabolisms (Takahara et al. 2010). Several other studies employing capillary and microchip electrophoresis coupled to MS analysis of transformants have been compiled in a review (Vega and Marina 2014).

Flooded rice grown in paddy fields uses primarily ammonium as nitrogen source and converts it to glutamine. A rice plant lacking a glutamine synthetase gene analyzed by GC-TOF-MS metabolomics had overaccumulation of free ammonium in leaves and roots, limited shoot growth, and imbalanced sugar, amino acid, organic acid, and specialized metabolites contents confirming the role of that gene in nitrogen fixation and its impact on plant metabolism (Kusano et al. 2011). Another approach to assess gene function is by overexpressing genes of interest. Rice plants overexpressing *Os-LBD37/ASL39* evaluated by GC-TOF-MS metabolomics and transcriptomics had significant changes in the contents of metabolites and mRNA of genes associated with nitrogen metabolism (Albinsky et al. 2010).

Rice phosphoenolpyruvate carboxylase is an abundant enzyme encoded by *Osppc4*, highly expressed in leaf tissues. Knockdown of *Osppc4* expression resulted in stunting when rice plants were grown with ammonium as the nitrogen source. Leaf metabolome analysis by GC-TOF-MS revealed a reduction in organic acid content, which indicated that the knockdown suppressed ammonium assimilation, amino acid synthesis, and growth by reducing the carbon source for the biosynthetic pathway (Masumoto et al. 2010).

GM rice containing *cry1Ac*, a *Bacillus thuringiensis* gene (protects mainly against lepidopteran pests), and sck, a cowpea trypsin inhibitor gene (protects mainly against coleopteran pests), analyzed by GC-MS metabolomics showed a different metabolite profile than the non-GM counterpart (Zhou et al. 2009). Follow-up studies examining the unintended effects of the referred modifications by UPLC-Q-TOF-MS metabolomics revealed that phytosphingosine, palmitic acid, 5-hydroxy-2-octadenoic acid, and three other unidentified metabolites contents were altered by the transformation (Chang et al. 2012). The same group investigated time course changes induced upon insecticide treatment in the same GM rice using LC–MS metabolomics and observed that while ferulic acid and sinapic acid were downregulated by insecticide treatment, salicylic acid and nicotinamide were upregulated (Chang et al. 2014). In another study, a pseudo-targeted GC-MS metabolomics analysis of a backcross breeding of the same insect-resistant GM rice revealed changes in the contents of sugars and amino acids of rice seeds (Zhao et al. 2015). The pseudo-targeted approach consisted in transforming untargeted data into targeted data by using retention time locking (to avoid drift) and selected ion monitoring (SIM). Typically in a targeted GC-MS analysis, data is acquired in SIM mode to achieve greater sensitivity by reducing background noise, in addition to avoiding the interference of co-eluting compounds (Li et al. 2012). In the pseudo-targeted approach, data is acquired in the scan mode, deconvoluted, then an algorithm is used to quantify all detected compounds by SIM (Li et al. 2012).

An UHPLC-QQQ-MS pseudo-targeted metabolomics approach was applied for the investigation of metabolic variation between seeds of insect resistant GM lines and nontransformed lines grown in two locations (Zhang et al. 2016). The pseudo-targeted method consisted of analyzing a pooled rice sample by untargeted UHPLC-Q-TOF-MS to establish a list of candidate ion pairs to be used in the pseudo-targeted method. The pooled sample was then analyzed under the same UHPLC conditions in a QQQ-MS in dynamic MRM mode. Cultivar growing site and genetic transformation all influenced differential metabolite accumulation (Zhang et al. 2016). In a similar analytical strategy, a widely targeted HPLC-QQQ-MS metabolomics approach was used to quantify 277 metabolites in the leaves of contrasting rice genotypes for drought tolerance submitted water restriction treatments (Chen et al. 2013). Abscisic acid polyamines and C-glycosylated flavones were the compounds found contributing to the differentiation between contrasting species that might be associated with stress response mechanism.

Environment
Environmental conditions can affect not only rice yield but quality aspects. Different growing regions with different edaphoclimatic conditions can potentially lead to rice grains with different metabolic contents. However, the geographical influence cannot be isolated because it results from a combination of several environmental factors, including biotic and abiotic stressors.

Phospholipids are the main components of rice grain lipids and have been shown to impact quality attributes. These compounds seem to be strongly influenced by environmental conditions, which are distinct for each rice-growing country. Therefore, high-throughput lipidomic methods by direct infusion-tandem mass spectrometry (MS/MS) have been utilized for target analysis of phospholipids in white rice of different origins (Lim et al. 2017a, c; Long et al. 2017). In addition to phospholipids, sugars, sugar alcohols, and fatty acids were identified as markers of the geographical origins of rice when using both GC-MS and LC-MS-untargeted metabolomics approaches (Lim et al. 2018c).

One- and two dimensional NMR metabolomics were able to differentiate among cultivars and geographical origins. In one case, rice from the region known to produce the most flavorful rice contained the greatest sugar and amino acid contents (Chae and Kim 2016; Huo et al. 2017).

Targeted HPLC-Q-TOF-MS analysis of phenolic compounds in two black rice genotypes grown in six different locations revealed that both genotype and geographical origin influenced the phenolic composition of black rice (Dittgen et al. 2019).

VOC biomarkers for origin discrimination between Korea and China were identified using SPME/GC-MS, among them, hexanal and 1-hexanol known to be influenced by environmental factors (Lim et al. 2018b).

Abiotic Stressors
^{1}H NMR and GC-TOF-MS metabolomics were utilized to understand the role of a specific gene (*SUB1A*) associated with response to submergence stress. Cultivars containing or not the gene of interest had comparatively different accumulation of

compounds associated with carbohydrate, organic acid, and amino acid metabolisms upon submergence stress conditions (Barding et al. 2012, 2013).

Rice plants exposed to different irrigation regimes and harvested at different time points were analyzed by untargeted metabolomics using comprehensive two-dimensional [hydrophilic interaction liquid chromatography (HILIC) and reverse-phase (RP)] HPLC-Q-TOF-MS (Navarro-Reig et al. 2017a). The analytical method consisted of a preliminary compression step along the mass spectrometry spectral dimension based on the selection of the regions of interest, followed by a further data compression along the chromatographic dimension by wavelet transforms. In a secondary step, a multivariate curve resolution alternating least squares method was applied to the compressed data sets obtained in the simultaneous analysis of multiple 2D-HPLC-Q-TOF-MS runs from multiple samples. Irrigation treatment affected the concentrations of flavonoids and hormones, while harvest time influenced sugar content (Navarro-Reig et al. 2017a).

Heat and drought stress conditions can result in reduced rice grain yield and quality. Rice cultivars with different tolerance backgrounds submitted to combined drought and heat stress were analyzed by GC-MS metabolomics. While some metabolites were differentially accumulated in response to stress regardless of cultivar tolerance, other responses were cultivar specific (Lawas et al. 2019).

Indica and japonica rice cultivars with different tolerance to drought stress submitted to control and stress conditions were analyzed for gene expression and by GC-MS metabolomics (Degenkolbe et al. 2013). Major differences were observed due to stress rather than cultivar. While glutamine and glutamic acid contents increased upon stress, sugar phosphates contents decreased under stress conditions (Degenkolbe et al. 2013).

NMR and GC-MS metabolomics of well-watered and drought-stressed transgenic rice plants (expressing an Arabidopsis P450) compared to the wild type revealed sugars and amino acids to be differentially accumulated in the grains of drought-tolerant plants under stress and potentially involved in drought-tolerance mechanism (Nam et al. 2016).

While Apo is a drought-tolerant variety, IR64 is susceptible. Both varieties experienced yield reductions when submitted to drought conditions. GC-MS metabolomics and sensory analysis revealed different volatile and aroma profiles for the two varieties, meanwhile the stress only affected the volatile and aroma profile of the drought-susceptible IR64. Sweet aromatic aroma was perceived from irrigated IR64, while and metallic, astringent, sour silage, sewer animal and hay-like/musty aromas were perceived from irrigated Apo. Aroma of drought-stressed IR64 was metallic, astringent, and sour/silage. Sensory perception was matched by measured VOC profile (Calingacion et al. 2015).

NMR metabolomics revealed the accumulation of amino acids and sugars in shoots and roots of rice cultivars under drought and salt stress conditions (Fumagalli et al. 2009). Untargeted GC-MS metabolomics of contrasting rice lines for tolerance to salt stress revealed compounds in roots and leaves associated with salt stress tolerance. While sugars and amino acids increased in leaves and roots upon stress conditions, organic acids increased in roots and decreased in leaves (Zhao et al. 2014).

GC-TOF-MS and CE-MS metabolomics and gene expression analyses indicated that several genes encoding enzymes involved in starch degradation, sucrose metabolism, and the glyoxylate cycle were upregulated in rice plants exposed to cold or dehydration stress conditions, and these changes were correlated with the accumulation of glucose, fructose, and sucrose. Phytohormone accumulation (targeted HPLC-MS) and gene transcript accumulation analysis of rice under cold and dehydration stress conditions revealed an inverse relationship between abscisic acid and cytokinin. While abscisic acid was upregulated, cytokinin was downregulated (Maruyama et al. 2014).

Non-GM and GM rice expressing an enzyme that promotes cytokinin biosynthesis, under the control of a stress-inducible promoter, grown under well-watered and water-stress conditions were analyzed by HPLC-MS metabolomics. While stress conditions impaired photosynthetic capacity and nitrogen assimilation and reduced the contents of proteins and free amino acids in wild-type plants, stress-induced cytokinin accumulation in the GM line promoted tolerance to water stress (Reguera et al. 2013).

Strigolactones are plant hormones that regulate plant growth in response to environmental cues. *Arabidopsis* strigolactone biosynthetic genes overexpressed in rice lead to the accumulation of this hormone measured by targeted LC-MS (Cardoso et al. 2014). In general, plant response to stress is associated with hormonal signaling. Brassinosteriod is another group of plant hormone with various effects on plant growth, and development has been analyzed by targeted LC-MS in rice (Ding et al. 2013). A CE-TOF-MS method has been developed for the determination of plant hormones in rice (Chen et al. 2011). Gibberellin (GA), auxin, and abscisic acid (ABA) detection was improved by derivatization with bromocholine by adding a quaternary ammonium functional group prior to UPLC-QQQ-MS analysis. This method has been utilized for the analysis of GA rice mutants (Kojima et al. 2009).

An untargeted metabolomics approach by HPLC-Q-TOF-MS was utilized to differentiate between organic and conventionally grown rice samples (Xiao et al. 2018). The two crop management systems differ in the source of input applied. While organic crop production relies exclusively on the use of fertilizers and pesticides of natural origin, in conventional crop production, the use of synthetic products is common practice. Even though the study suggested the identification of markers to determine whether a sample had originated from an organically or conventionally grown field, their use may be limited since many uncontrolled variables associated with a chosen management practice might influence the results.

Crop yields are limited by phosphorous availability and plants utilize adaptive mechanisms to cope with this nutrient limitation. LC-IT-TOF-MS lipidomics revealed that rice lipid metabolism was modulated under phosphorous-depleted conditions and involved the accumulation of glucuronosyldiacylglycerol, a membrane lipid associated with the sulfoquinovosyldiacylglycerol biosynthetic pathway (Okazaki et al. 2013).

Soil incorporated with titanium oxide nanoparticles led to improved phosphorous uptake and increased plant growth, and GC-MS-based metabolomics also revealed increased levels of amino acids, palmitic acid, and glycerol contents in grains of treated plants in comparison to the respective controls (Zahra et al. 2017).

Nitrogen deficiency is considered an environmental stress that leads to reduced yields and changes in rice grain quality. A rice hybrid was analyzed for responses to nitrogen supply treatments by UHPLC-TOF-TOF-TOF. Amino sugar and nucleotide sugar contents increased during the period of nitrogen deficiency, while after nitrogen was compensated, lipid biosynthesis was upregulated (Shen et al. 2019b).

GC-MS metabolomics revealed that when grown under low nitrogen conditions, a low nitrogen-tolerant rice genotype accumulated more pyroglutamate, glutamate, 2-oxoglutarate, sorbose, glycerate-2-phosphate, and phosphoenolpyruvic acid than a less-tolerant line (Zhao et al. 2018).

Cadmium and copper are toxic metals readily absorbed by roots that translocate and can potentially accumulate in rice grains. Rice metabolome changes induced by Cd and Cu exposure were evaluated by untargeted LC-TOF-MS (Navarro-Reig et al. 2015). Although responses to Cd seemed more intense, both metals impacted specialized metabolism and amino acid, purine, carbon, and glycerolipid metabolisms in addition to limiting plant growth and development (Navarro-Reig et al. 2017b). HPLC-orbitrap-MS analysis of rice exposed to cadmium revealed accumulation of phytochelatins (cysteine-rich oligopeptides) involved in tolerance and detoxification (Mou et al. 2016).

Ozone has been shown to significantly reduce rice yields. CE-MS-based metabolomic profiling within an integrated omics approach revealed differential accumulation of amino acids, gamma-aminobutyric acid, and glutathione in ozone-treated seedlings in comparison to controls (Cho et al. 2008).

Biotic Stressors

Brown planthopper (BPH; *Nilaparvata lugens* Stal; Hemiptera: Delphacidae) is an aphid pest of rice that has caused significant crop losses. NMR-based metabolomics and gene expression analyses of BPH-susceptible and resistant rice infested with BPH indicated that GABA and shikimate pathways are potentially associated with insect resistance (Liu et al. 2010). Meanwhile, GC-MS metabolomics revealed that BPH feeding led to the differential accumulation of compounds involved in fatty acid oxidation, the glyoxylate cycle, gluconeogenesis, and the GABA shunt in susceptible plants, while insect resistant plants exhibited a differential accumulation of compounds associated with glycolysis and the shikimate pathway (Peng et al. 2016). Another study using untargeted GC-MS and LC-MS metabolomics analysis revealed that cyanoamino acids, lipids, thiamine, taurine, hypotaurine, quercetin, and spermidine were differentially accumulated between varieties and upon BPH infestation (Kang et al. 2019).

Rice sheath blight caused by *Rhizoctonia solani* is a widespread disease that can lead to 50% yield loss. GM rice carry a PR gene and wild-type rice lines infected and noninfected with *R. solani* were evaluated by GC-MS metabolomics.

Compounds differentially accumulated between the lines upon pathogen infection are potentially involved in the resistance mechanism (Karmakar et al. 2019).

Blast caused by the pathogen *Magnaporthe grisea* is also a widespread and economically relevant disease in rice. NMR, GC-MS, and LC-Q-TOF-MS metabolomics were used to identify metabolic changes derived from the plant-pathogen interactions. Diseased plants infected with a pathogenic strain had greater alanine, malate, glutamine, proline, cinnamate contents than plants inoculated with a non-pathogenic strain (Jones et al. 2011).

Mycotoxins and Pesticides

Mycotoxins are heat stable, potentially carcinogenic, mutagenic, hepatotoxic, nephrotoxic, and immunosuppressive compounds produced by filamentous fungi, common contaminants of grain cereals such as rice. In order to protect consumers, regulatory agencies have set limits for their presence in foods. In particular, the European Union has the strictest limits and imposes constraints on imports from producing countries. Pesticides have a wide range of active ingredients with several targets, including more commonly herbicides, insecticides, and fungicides with various levels of toxicity but also with strictly regulated maximum residue limits.

Several analytical methods have been developed, optimized, and validated for the simultaneous or independent measurement of pesticides and mycotoxins in different types of rice samples either by GC-MS or LC-MS (Mondal et al. 2017). Given the complex grain matrix, studies have employed various sample extraction and cleanup methods to remove interferents (Sugitate et al. 2012). Pesticide and mycotoxin extraction methods vary from the so-called dilute and shoot (da Silva et al. 2019) to modified QuEChERS (Quick, Easy, Cheap, Effective, Rugged and Safe) and predominantly utilize the more sensitive targeted LC-QQQ platform (Viera et al. 2017).

Different extraction solvents and types of cleanup, including solid-phase extraction, dispersive solid-phase, QuEChERS, and immunoaffinity columns, have been used for the analysis of mycotoxins and pesticides in rice (Cabrera et al. 2016; Kim et al. 2017a; Solfrizzo et al. 2018). Cleanup steps after extraction, although improve compound detection may lead to lengthy analytical protocols. A simple ultrasound-assisted extraction of several mycotoxins from rice sample without sample cleanup used data analysis algorithms to avoid problems such as base line drift and co-eluting interferents (Liu et al. 2017b).

In multiresidue analysis by low-resolution QQQ-MS, analytical parameters need to be optimized for each analyte, and it is not the ideal choice for screening unknown suspected contamination. High-resolution MS such as TOF, Q-TOF, orbital ion trap, or ion trap or quadrupole orbital ion trap (Q-orbital trap) MS can be used for screening multiple contaminants in full-scan mode with high mass accuracy. It therefore permits the combination of target analysis with screening of nontarget compounds. Q-orbitrap MS has been used for the identification of mycotoxins in finished rice grain products (Liao et al. 2015) and of fungicides in rice bran protein powder (Wu et al. 2018).

Compound quantification usually follows the isotope-dilution method adding known amounts of the isotopically labelled compound as internal standard (Li et al. 2017a). Quantification of neonicotinoid insecticide in rice samples (Nakamura et al. 2019) and of mycotoxins in infant rice cereals have used this method (Al-Taher et al. 2017).

Another aspect of method development widely explored is the choice of stationary phase for optimum compound separation. While reverse phase (C18) stationary phases are the most frequently used, hydrophilic interaction liquid chromatography column (HILIC) for pesticides (Li et al. 2017a), silica hydride–based columns for mycotoxin (Pesek et al. 2017), cation-exchange column for herbicides (Nardin et al. 2017), and ion pack column for glyphosate residue in rice (Santilio et al. 2019) are just a few alternatives in the literature.

Several LC-MS methods for simultaneous analysis of multiple mycotoxins in rice matrices have been developed and validated (Koesukwiwat et al. 2014; Lim et al. 2015; Marley et al. 2015; Jettanajit and Nhujak 2016; Miró-Abella et al. 2017; Zhao et al. 2017; Sinphithakkul et al. 2019).

The LC-MS/MS QTRAP method was applied for the simultaneous measurement of mycotoxins in domestic rice in Iran. All samples were contaminated with at least one mycotoxin. The most prevalent were brevianamide F (81.5%), emodin (46.1%), and tryptophol (43.1%). Aflatoxin B1 (AFB1), ochratoxin A (OTA), zearalenone (ZEN), and fumonisin B1 (FB1) were 21.5%, 4.6%, 29.2%, and 9.2%, respectively. The regulated mycotoxin content was lower than the maximum limit (Nazari et al. 2014).

Red koji rice (angkak) is a product resulting from *Monascus purpureus* fermentation, which promotes rice pigmentation (Wild et al. 2003). However, along with the pigments, *M. purpureus* produces the mycotoxin citrinin. An LC-MS/MS method has been developed for the analysis of citrinin in red koji rice products (Ji et al. 2015).

Rice false smut is a destructive fungal disease in rice which leads to the accumulation of ustiloxins cyclopeptide mycotoxins. These compounds have been isolated and analyzed by 1D and 2D nuclear magnetic resonance (NMR) and high-resolution electrospray ionization-mass spectrometry (Wang et al. 2017).

Phenamacril, a broad-spectrum fungicide commonly used in rice, was analyzed by modified QuEChERS UHPLC-MS/MS (Sun et al. 2019).

While the number of simultaneously monitored mycotoxins in rice has not reached 20, the number of monitored pesticides has passed 300, in particular due to development of novel active ingredients to avoid weed, pathogen, pest resistance. Methods for the simultaneous analysis of hundreds of pesticide multiresidues in rice have been developed (Lee et al. 2018a).

Insecticides are commonly applied to rice, and their extraction from different rice matrices has been optimized by the use of dynamic microwave coupled with matrix solid-phase dispersion (Zhang et al. 2017a), accelerated solvent extraction (Teló et al. 2017), following HPLC-QQQ-MS analysis (Li et al. 2017b).

Insecticide chlorpyrifos and carbosulfan residues applied at a pre-harvest interval (PHI) of 28 days were reduced by 75% upon hulling and milling.

Household-simulated processing by washing twice and high-pressure cooking reduced the amount of insecticide on polished rice to below maximum residue levels, lowering the risk of dietary exposure (Ma et al. 2019)

Clothianidin insecticide residue at PHI 28 days was not detected in brown rice at the tested field rates when measured by QUECHERS-HPLC-MS (Zhang et al. 2017b).

Rice hull is utilized as support in the manufacture of Chinese liquor. In order to survey the presence of pesticides in the hull and the beverage, a method for the simultaneous extraction, cleanup, and analysis of dozens of pesticides have been developed (Han et al. 2018).

Several HPLC-QQQ-MS methods developed for herbicide analysis in different rice matrices used modified QuEChERS extraction (Lee et al. 2016; Rebelo et al. 2016; Hu et al. 2017; Ni et al. 2018; Guo et al. 2019).

Several methods either using GC-MS/MS or LC-MS/MS for the targeted simultaneous analysis of large numbers of pesticides in different rice matrices have been developed (Grande-Martínez et al. 2015a, b; Han et al. 2017; Lee et al. 2017b; Amirahmadi et al. 2018; Grande-Martínez et al. 2015a).

Not only the actual pesticide but its metabolized products need to be monitored. The dissipation rates of the herbicide mesotrione and two metabolites were monitored in rice tissues by HPLC-QQQ-MS. At harvest, no residues were detected (Du et al. 2017). Fifty pesticides and eight related metabolites were measured using a cleanup adsorbent, which reduced sample preparation time by 30% increasing throughput (Liu et al. 2017a). Simultaneous determination of three pesticides (two insecticides and one fungicide) and their metabolites in unprocessed foods was achieved using LC-MS/MS (Rong et al. 2018).

Free fatty acid content serves as a quality indicator for milled rice since it becomes oxidized and leads to rancid flavor upon prolonged storage (shelf life up to 6 months). A single rice grain ambient electrospray ionization method was developed by setting a small chamber in line with an ESI-linear trap quadrupole (LTQ)-MS that allowed for measuring free fatty acids and pesticides without any sample preparation steps (Shen et al. 2019a).

Chiral pesticides account for 30% of commercial pesticides. A chiral LC-HRMS–method was developed for the study of the degradation and residue of the enantiomers of the insecticide indoxacarb in rice plants, rice hulls, and brown rice (Shi et al. 2018).

Postharvest Processing

Depending on the moment of harvest, the location of the drying silo, the amount of rice, and due to unexpected machinery down time, the period between harvest and drying might vary. The moisture content of the grain at harvest and the time period from harvest until the beginning of drying can influence the quality aspects of the grain due to endogenous and microbial-induced metabolic changes. Rice harvested with different moisture content and either dried immediately or after 48 h were analyzed for volatile compound production by SPME/GC-MS. Volatile production in rice harvested with moisture ranging from 17% to 21% did not change over the

48 h until drying. However, rice harvested with 24% or greater moisture content had a significant increase in microbial activity related VOC if 48 h passed until drying. Sensory analysis of the same samples indicated they had off-flavor notes of sour/silage and alfalfa/grassy/green bean (Champagne et al. 2004).

Changes in rice seed composition during a 2-year storage period was investigated by untargeted GC-MS metabolomics for hybrid rice cultivars with contrasting storability capacities to try to understand the metabolic basis for the differential reduction in germination rate and viability. During the 24-month storage period, the contents of free amino acids and soluble sugars were differentially altered in the two cultivars (Yan et al. 2018).

Not only seeds loose quality when stored over time, rice grains destined for consumption also lose quality over storage depending on the storage temperature and whether the grains have been milled. SPME/GC-MS metabolomics was utilized to evaluate changes in the volatile aroma profile of milled and unmilled aromatic black rice stored at different temperatures (Choi et al. 2019). The greater the storage temperature, the greater the increase in volatile compound production associated with lipid oxidation. Milled rice had better stability than unmilled rice during a 3-month storage, but after 6 months, unmilled rice had better lipid stability (Choi et al. 2019).

SPME/GC-MS analysis of rice volatiles assessed the effects of drying conditions and storage time on the aroma of the fragrant Jasmine rice. Drying methods employing the lowest temperatures yielded rice with higher amounts of the positive key aroma and lower amounts of the off-flavor compounds regardless of the storage time (Wongpornchai et al. 2004).

In order to maintain and prolong quality of a specialty product, such as organically produced aromatic red rice, grains are immediately dried, dehulled, and packaged preferably under low oxygen conditions. Pigmented rice is usually not milled in order to keep the bioactive compounds responsible for its color. SPME/GC-MS analysis was used to monitor changes in volatile compounds from organic red fragrant rice vacuum packed, stored at different temperatures for 12 months. Independent of packaging material and storage temperature, lipid oxidation–associated VOC production increased over storage time. Although positive aroma compounds decreased over the storage period, lower storage temperature (15 °C) and gas and water impermeable packaging limited its reduction. In addition, rice packaged in a material with no oxygen or water permeability had less volatiles from lipid oxidation (Tananuwong and Lertsiri 2010).

In a descriptive qualitative SPME/GC-MS analysis of VOC from aromatic and nonaromatic cultivars stored over 3 months, storage time did not influence the profile of volatiles; meanwhile, qualitative differences were seen between aromatic and nonaromatic cultivars (Bryant and McClung 2011).

Excess bran present on milled rice contributes to off-flavors due to fatty acid oxidation. Free unsaturated fatty acids present on the bran layer tend to be more readily oxidized into hydroperoxides, which break down into volatile compounds. Storage temperature and relative humidity (RH) influence this fatty acid degradation. Partially milled rice with 50% bran removed and fully milled (polished) rice with more than 95% of the bran removed stored over 50 days (37 °C and 70% RH) were analyzed by SPME/GC-MS (Lam and Proctor 2003). The qualitative volatile profile did not change over the storage period. Octanal (fatty) and 2-nonenal (rancid) were the compounds that most contributed to milled rice aroma. Hexanal and 2-nonenal, secondary oxidation products of linoleic acid, increased over the storage period. Partially milled rice produced fatty acid–derived volatiles in greater amounts than fully milled rice (Lam and Proctor 2003).

Cooking quality and phenolic composition of black and red rice grains stored during 6 months at different temperatures (16, 24, 32, and 40 °C) were evaluated by HPLC-Q-TOF-MS. Cooking time and phenolic content changed over time depending on the storage temperature (Ziegler et al. 2018).

Parboiled rice has improved nutritional properties due to compound migration from the outer bran layer into the inner layer, which is not removed during milling. However, parboiled rice is usually darker in color due to Maillard reaction occurring during the pressure steaming treatment of the parboiling process. A by-product of this reaction is 5-hydroxymethyl-2-furaldehyde (HMF). A modified QuEChERS combined with dispersive liquid–liquid microextraction was developed for the HPLC-QQQ-MS determination of HMF in cooked japonica rice (Feng et al. 2019). Parboiled rice submitted to different treatments to avoid color development was also analyzed for free lysine content by LC-Q-TOF-MS to monitor proteolytic activity and understand the role of this amino acid in sugar–amino acid interactions (Villanova et al. 2017).

Brown rice has greater nutritional value than white rice but it does not cook as easily. Germinated brown rice, however, is more easily cooked and yields a softer texture. Germinated brown rice has been shown to possess antihyperlipidemic and antihypertensive potential and the capability of reducing the risk of some chronic diseases, such as diabetes, cancer, cardiovascular diseases, and Alzheimer's disease (Patil and Khan 2011). A time course GC-MS metabolomic analysis of germinated rice grains helped determine the ideal moment when maximum accumulation of compounds associated with nutritional quality occurred (Shu et al. 2008).

NMR metabolomics was able to differentiate between germinated brown rice of black, red, and white cultivars, in particular, due to the compound abundance in the black cultivar (Pramai et al. 2018).

Rice intake has been associated with health improvement. In an in vivo study, mice who consumed brown rice had lower total-cholesterol levels and increased levels of plasma growth hormone IGF-1 than those who consumed white rice. In addition, an untargeted GC-TOF-MS metabolomics analysis revealed that livers of mice fed with brown rice had greater L-valine, L-glutamine, and D-mannose contents compared to those who ate white rice (Yang et al. 2016). These differentially accumulated compounds are possibly associated with IGF-1 synthesis involved in promoting cellular growth in mice.

Untargeted NMR metabolomics of fermented brown rice and rice bran was able to differentiate good quality products from defective and unfermented samples. A set of 25 fermented brown rice and rice bran when analyzed by NMR did not separate in the multivariate analysis, indicating their chemical profile is consistently similar, reflecting their standardization and quality (Horie et al. 2019).

Commercial rice bran is a mixture of rice bran (pericarp and tegument) and germ (embryo). Tocopherols and tocotrienols that make up vitamin E and ferulic acid esters of sterols and triterpenoids that make up gamma-oryzanol present in the germ and bran components of commercial bran were analyzed by liquid chromatography/mass spectrometry/mass spectrometry using both positive- and negative-ionization modes. While germ vitamin E content was greater than bran, gamma-oryzanol content in bran was greater than in rice germ (Yu et al. 2007).

Flavonoids present in rice bran extracts of seven black Thai rice varieties have been identified and quantified by HPLC-MS (Sriseadka et al. 2012).

Rice bran is a by-product of rice milling from which rice bran oils can be extracted and utilized in different products in the food industry. A targeted LC-Q-ion trap-MS metabolomics analysis was used to discriminate between cold-pressed rice bran oils produced from two different cultivars of *Oryza sativa* L. ssp. *indica* in Thailand. Fatty acids and gamma-oryzanol contents allowed to differentiate the oils from the different cultivars (Charoonratana et al. 2015).

Untargeted rice bran metabolomics by UPLC-MS/MS was used to compare cultivars from seven countries. Among the several compounds that contributed to the chemical diversity among cultivars were amino acids, sugars, vitamins, lipids, nucleotides, and specialized metabolites (Zarei et al. 2018).

Rice bran volatile analysis by GC-MS followed different extraction/collection methods, SPME, static headspace extraction, accelerated solvent extraction, and simultaneous distillation extraction. GC-MS data analysis was performed by heuristic evolving latent projections to resolve mass spectra with the aid of AMDIS software and temperature-programmed retention indices, while quantitative analysis was conducted using the overall volume integration technique (Zeng et al. 2012).

Due to the high lipid content and the abundance of active lipases, rice bran needs to be stabilized to limit lipid hydrolysis, which facilitates the development of oxidative rancidity. The metabolome of heat-stabilized rice bran from three rice cultivars was characterized by untargeted metabolomic analysis by UPLC-MS/MS and GC–MS (Zarei et al. 2017).

Metabolomics approaches have also been utilized to monitor the influence of a heat-stabilized rice bran diet on human metabolism. The stool microbiome and metabolome (analyzed by GC-MS) of patients who were on a heat-stabilized rice bran diet had greater contents of branched chain fatty acids, secondary bile acids, and 11 other putative microbial metabolites when compared to control group (Sheflin et al. 2015).

Altered blood cholesterol content may serve as an early marker for the prediction of cardiovascular disease development (CVD). In a clinical study, a combined untargeted GC-MS and UPLC-MS/MS metabolomics analysis was used to compare the plasma metabolome of children with abnormal cholesterol levels and of a control group on 4 weeks of daily intake of navy bean, rice bran, or a combination of navy bean + rice bran. Among the differently accumulated metabolites were amino acids, phytochemicals, and plasma lipids, with implications for reducing CVD risk in children (Li et al. 2018).

Saccharomyces cerevisiae var. *boulardii* is a probiotic microorganism that can reduce the antinutritional components and promote the accumulation of health-beneficial components of foods. A combined targeted and untargeted GC-MS analysis of unfermented and *S. cerevisiae* var. *boulardii* fermented rice bran from different rice varieties in combination with human B lymphocyte cell assays lead to the identification of several functional components of this health-beneficial rice product (Ryan et al. 2011).

Untargeted metabolomics by GC–TOF–MS, UHPLC–LTQ–IT–MS/MS and UPLC–Q–TOF–MS revealed how the degree of milling affected the composition of grains used in koji fermentation (Lee et al. 2018b). Greater amylase, β-glucosidase, and protease activities were observed in grains with 50% and 30% of the bran layer removed when compared to unmilled rice. The amount of the bran layer left in rice after different degrees of milling influenced the enzyme activities during rice koji making because it affected the contents of sugar alcohols, organic acids, phenolic acids, fatty acids, and lysophospholipids.

Koji rice derives from the inoculation of polished steamed grains with *Aspergillus* spp., which secretes enzymes to break down starch into fermentable sugars. Koji rice is then fermented by yeast to produce beverages such as sake (Japanese rice wine), shochu (Japanese liquor), and makgeoli (Korean rice wine). Untargeted metabolomics by GC-MS was used to identify the ideal *Aspergillus* spp. strain and saccharification conditions by measuring changes in compounds such as sugars, amino acids, and organic acids (Kim et al. 2012a).

Untargeted metabolite profiling by GC-TOF-MS and UHPLC-LTQ-IT-MS/MS for koji-fermentative bioprocess using soybean, wheat, and rice and *Aspergillus oryzae* and *Bacillus amyloliquefaciens* unraveled a differential compound accumulation based on substrate and inocula (Seo et al. 2018).

While yellow koji mold (*Aspergillus oryzae*) has been used for the production of different fermented foods, white koji mold (*Aspergillus kawachii*) and black koji mold (*Aspergillus awamori*) have been used exclusively in the manufacture of shochu, in particular, due to their capacity of producing acids that can lower the pH of the mash. However, not much is known about their contribution to aroma components. A large volume static headspace collection of koji volatiles was followed by their concentration and GC-MS analysis. This collection method has the advantage of picking up trace compounds. Principal component analysis of VOC data from koji produced from all three mold types allowed for the separation between yellow mold koji from black and white mold koji. However, the profile from the white and black mold koji was very similar, not allowing for their separate grouping (Yoshizaki et al. 2010).

Saccharomyces cerevisiae fermentation of polished and steamed rice saccharified by *Aspergillus oryzae* leads to sake. Untargeted GC-MS metabolomics of derivatized nonvolatile and nonderivatized volatiles compounds compared 40 different sake samples produced under different manufacturing conditions and from different origins. Sensory analysis data was correlated with the chemical analysis and allowed for identifying compounds contributing for each flavor attribute of sake (Mimura et al. 2014). A different analytical approach by 2D-GC-TOF-MS was also utilized to find correlations between sake compounds and its flavor characteristics assessed by a sensory panel (Takahashi et al. 2016).

The greater the amount of bran removed during polishing, the greater tends to be the quality of the product due to the removal of lysophosphatidylcholines (LPC), which negatively impact sake flavor. LPC species were visualized via targeted analysis by matrix-assisted laser desorption/ionization mass spectrometry imaging (MALDI-MSI). MALDI-MSI can determine not only compound identity but its location. Thin slices of frozen or carboxymethyl embedded rice tissues were attached to adhesive films and sprayed with 2,5-dihydroxybenzoic acid as the matrix prior to analysis (Zaima et al. 2010). While LPC 16:0 was present throughout the entire rice endosperm, LPC 18:0 was localized in the core of the endosperm. Meanwhile, LPC (18:2) and LPC (18:1) were present in the outer region of the endosperm (Zaima et al. 2014). Therefore, polishing removes unsaturated fatty acid containing LPC, which have been shown to inhibit the alcohol acetyltransferase enzyme that produced isoamyl acetate, a compound with banana-like flavor important for sake flavor. In addition, MALDI-MSI localized phosphatidylcholine (PC), gamma-oryzanol, and phytic acid in the bran layer, while alpha-tocopherol was distributed in the germ (Zaima et al. 2010). Anthocyanins and their localization in black rice pericarp have also been revealed by MALDI-MSI (Yoshimura et al. 2012).

Fermentation of brown, white, and giant embryo-rice koji was monitored by GC-TOF-MS and UHPLC-LTQ-IT-MS/MS metabolomics to determine the effect of the starting material on the quality of the end product. Giant embryo rice had greater sugars and flavonoid aglycons contents than the other koji. Over the time course of fermentation, phenolic acids, fatty acids, and vitamins contents increased and were greater in brown rice koji while white rice koji had the greatest lysophospholipids and amino acids contents (Lee et al. 2017a).

21.4 Concluding Remarks and Prospects

Here we presented and discussed several analytical and processing tools and approaches used for the assessment of the rice grain metabolome in response to several factors which ultimately reflect grain quality.

All comprehensive metabolomics approaches discussed thus far provide a snapshot of all rice small molecule concentration at a given moment. Even if complicated and highly controlled experimental designs are devised, it is not possible to know definitively whether compound abundant at a specific moment is a result of increased synthesis or lack of downstream turnover. In order to be able to confidently interpret the dynamic changes of metabolite concentration within metabolic pathways, a fluxomics approach is needed. Metabolic flux analysis based on isotope tracer experiments is used for measuring the changing metabolome. Pathway precursor compounds isotopically labeled are fed to plants, and the distribution of those labelled atoms in the metabolome is then measured after a certain time by either MS or 1 or 2-D NMR. When using ^{13}C-labelled precursors, the analysis of the general metabolism is favored given that carbon flow through that pathway is greater when compared to man-specialized metabolic pathways (Islam et al. 2018). ^{15}N-labelled precursors have also been successfully utilized (Millard et al. 2017).

In spite of the sensitive and informative detection methods available, compound identification without having commonly expensive authentic standards is still tentative. Another level of evidence to support compound annotation comes from the currently evolving instruments possessing ion mobility mass spectrometry (IMS) in tandem with high-resolution MS. IMS separates ions by drift time based on their mass, shape, and size as well as their collisions with a buffer gas under an electric field. A collision cross-section (CCS) value is determined from the drift time. CCS values are smaller for tri-dimensionally smaller molecules than for extended ones. CCS is another descriptor that can be used to support compound identity (Bijlsma et al. 2017). Different IMS separation methods available that allow to obtain CCS information, which include drift-time (DTIMS), traveling-wave IMS (TWIMS), and trapped IMS (TIMS) (Righetti et al. 2018).

Metabolomic technologies continue to develop, and each new development will enable increased resolution of the complex chemistry rice, enabling improved understanding of the relationship between small molecule composition and rice grain quality.

References

Akram MI, Vincent IM, Siddiqui AJ, Musharraf SG (2017) Polymeric hydrophilic interaction liquid chromatography coupled with Orbitrap mass spectrometry and chemometric analysis for untargeted metabolite profiling of natural rice variants. J Cereal Sci 73:165–173. https://doi.org/10.1016/j.jcs.2017.01.001

Albinsky D, Kusano M, Higuchi M et al (2010) Metabolomic screening applied to rice FOX arabidopsis lines leads to the identification of a gene-changing nitrogen metabolism. Mol Plant 3:125–142. https://doi.org/10.1093/mp/ssp069

Allwood JW, Erban A, de Koning S et al (2009) Inter-laboratory reproducibility of fast gas chromatography-electron impact-time of flight mass spectrometry (GC-EI-TOF/MS) based plant metabolomics. Metabolomics 5:479–496. https://doi.org/10.1007/s11306-009-0169-z

Al-Taher F, Cappozzo J, Zweigenbaum J et al (2017) Detection and quantitation of mycotoxins in infant cereals in the U.S. market by LC-MS/MS using a stable isotope dilution assay. Food Control 72:27–35. https://doi.org/10.1016/j.foodcont.2016.07.027

Amirahmadi M, Yazdanpanah H, Kobarfard F et al (2018) Exposure assessment for some pesticides through rice consumption in Iran using a multiresidue analysis by GC-MS. Iran J Pharm Res 17:124–139

Barding GA, Fukao T, Béni S et al (2012) Differential metabolic regulation governed by the rice SUB1A gene during submergence stress and identification of alanylglycine by 1H NMR spectroscopy. J Proteome Res 11:320–330. https://doi.org/10.1021/pr200919b

Barding GA, Béni S, Fukao T et al (2013) Comparison of GC-MS and NMR for metabolite profiling of rice subjected to submergence stress. J Proteome Res 12:898–909. https://doi.org/10.1021/pr300953k

Benajmini Y, Hochberg Y (1995) Controlling the false discovery rate: a practical and powerful approach to multiple testing. J R Stat Soc B 57:289–300. https://doi.org/10.2307/2346101

Bijlsma L, Bade R, Celma A et al (2017) Prediction of collision cross-section values for small molecules: application to pesticide residue analysis. Anal Chem 89:6583–6589. https://doi.org/10.1021/acs.analchem.7b00741

Bothwell JHF, Griffin JL (2011) An introduction to biological nuclear magnetic resonance spectroscopy. Biol Rev 86:493–510. https://doi.org/10.1111/j.1469-185X.2010.00157.x

Broeckling CD, Prenni JE (2018) Stacked injections of biphasic extractions for improved metabolomic coverage and sample throughput. Anal Chem 90:1147–1153. https://doi.org/10.1021/acs.analchem.7b03654

Broeckling CD, Afsar FA, Neumann S et al (2014) RAMClust: a novel feature clustering method enables spectral-matching-based annotation for metabolomics data. Anal Chem 86:6812–6817. https://doi.org/10.1021/ac501530d

Bryant RJ, McClung AM (2011) Volatile profiles of aromatic and non-aromatic rice cultivars using SPME/GC-MS. Food Chem 124:501–513. https://doi.org/10.1016/j.foodchem.2010.06.061

Cabrera LDC, Caldas SS, Prestes OD et al (2016) Evaluation of alternative sorbents for dispersive solid-phase extraction clean-up in the QuEChERS method for the determination of pesticide residues in rice by liquid chromatography with tandem mass spectrometry. J Sep Sci 39:1945–1954. https://doi.org/10.1002/jssc.201501204

Calingacion MN, Boualaphanh C, Daygon VD et al (2012) A genomics and multi-platform metabolomics approach to identify new traits of rice quality in traditional and improved varieties. Metabolomics 8:771–783. https://doi.org/10.1007/s11306-011-0374-4

Calingacion M, Fang L, Quiatchon-Baeza L et al (2015) Delving deeper into technological innovations to understand differences in rice quality. Rice 8. https://doi.org/10.1186/s12284-015-0043-8

Calingacion M, Mumm R, Tan K et al (2017) A multidisciplinary phenotyping and genotyping analysis of a mapping population enables quality to be combined with yield in rice. Front Mol Biosci 4. https://doi.org/10.3389/fmolb.2017.00032

Cardoso C, Zhang Y, Jamil M et al (2014) Natural variation of rice strigolactone biosynthesis is associated with the deletion of two MAX1 orthologs. Proc Natl Acad Sci U S A 111:2379–2384. https://doi.org/10.1073/pnas.1405833111

Chae YK, Kim SH (2016) Discrimination of rice products by geographical origins and cultivars by two-dimensional NMR spectroscopy. Bull Kor Chem Soc 37:1612–1617. https://doi.org/10.1002/bkcs.10914

Chagoyen M, Pazos F (2013) Tools for the functional interpretation ofmetabolomic experiments. Brief Bioinform 14:737–744. https://doi.org/10.1093/bib/bbs055

Champagne ET, Thompson JF, Bett-Garber KL et al (2004) Impact of storage of freshly harvested paddy rice on milled white rice flavor. Cereal Chem 81:444–449. https://doi.org/10.1094/CCHEM.2004.81.4.444

Chang Y, Zhao C, Zhu Z et al (2012) Metabolic profiling based on LC/MS to evaluate unintended effects of transgenic rice with cry1Ac and sck genes. Plant Mol Biol 78:477–487. https://doi.org/10.1007/s11103-012-9876-3

Chang Y, Zhang L, Lu X et al (2014) A simultaneous extraction method for metabolome and lipidome and its application in cry1Ac and sck-transgenic rice leaf treated with insecticide based on LC MS analysis. Metabolomics 10:1197–1209. https://doi.org/10.1007/s11306-014-0658-6

Charoonratana T, Songsak T, Sakunpak A, Pathompak P (2015) Using liquid chromatography-mass spectrometry based metabolomics to discriminate between cold pressed rice bran oils produced from two different cultivars of *Oryza sativa* L. ssp. indica in Thailand. Chin J Chromatogr 33:966–973

Chen ML, Huang YQ, Liu JQ et al (2011) Highly sensitive profiling assay of acidic plant hormones using a novel mass probe by capillary electrophoresis-time of flight-mass spectrometry. J Chromatogr B Anal Technol Biomed Life Sci 879:938–944. https://doi.org/10.1016/j.jchromb.2011.03.003

Chen W, Gong L, Guo Z et al (2013) A novel integrated method for large-scale detection, identification, and quantification of widely targeted metabolites: application in the study of rice metabolomics. Mol Plant 6:1769–1780. https://doi.org/10.1093/mp/sst080

Chen W, Gao Y, Xie W et al (2014) Genome-wide association analyses provide genetic and biochemical insights into natural variation in rice metabolism. Nat Genet 46:714–721. https://doi.org/10.1038/ng.3007

Cho K, Shibato J, Agrawal GK et al (2008) Integrated transcriptomics, proteomics, and metabolomics analyses to survey ozone responses in the leaves of rice seedling. J Proteome Res 7:2980–2998. https://doi.org/10.1021/pr800128q

Choi S, Seo HS, Lee KR et al (2019) Effect of milling and long-term storage on volatiles of black rice (*Oryza sativa* L.) determined by headspace solid-phase microextraction with gas chromatography–mass spectrometry. Food Chem 276:572–582. https://doi.org/10.1016/j.foodchem.2018.10.052

da Silva LP, Madureira F, de Azevedo VE et al (2019) Development and validation of a multi-analyte method for quantification of mycotoxins and pesticides in rice using a simple dilute and shoot procedure and UHPLC-MS/MS. Food Chem 270:420–427. https://doi.org/10.1016/j.foodchem.2018.07.126

Daygon VD, Prakash S, Calingacion M et al (2016) Understanding the Jasmine phenotype of rice through metabolite profiling and sensory evaluation. Metabolomics 12:1–15. https://doi.org/10.1007/s11306-016-0989-6

Daygon VD, Calingacion M, Forster LC et al (2017) Metabolomics and genomics combine to unravel the pathway for the presence of fragrance in rice. Sci Rep 7:1–12. https://doi.org/10.1038/s41598-017-07693-9

De B, Nag G (2014) Metabolic diversity in the grains of Indian varieties of rice. Comptes Rendus – Biol 337:283–293. https://doi.org/10.1016/j.crvi.2014.01.006

Degenkolbe T, Do PT, Kopka J et al (2013) Identification of drought tolerance markers in a diverse population of rice cultivars by expression and metabolite profiling. PLoS One 8. https://doi.org/10.1371/journal.pone.0063637

Ding J, Mao LJ, Yuan BF, Feng YQ (2013) A selective pretreatment method for determination of endogenous active brassinosteroids in plant tissues: double layered solid phase extraction combined with boronate affinity polymer monolith microextraction. Plant Methods 9:1–9. https://doi.org/10.1186/1746-4811-9-13

Dittgen CL, Hoffmann JF, Chaves FC et al (2019) Discrimination of genotype and geographical origin of black rice grown in Brazil by LC-MS analysis of phenolics. Food Chem 288:297–305. https://doi.org/10.1016/j.foodchem.2019.03.006

Dong SY, Lee KS, Jeong OY et al (2008) Characterization of volatile aroma compounds in cooked black rice. J Agric Food Chem 56:235–240. https://doi.org/10.1021/jf072360c

Du P, Wu X, Xu J et al (2017) Determination and dissipation of mesotrione and its metabolites in rice using UPLC and triple-quadrupole tandem mass spectrometry. Food Chem 229:260–267. https://doi.org/10.1016/j.foodchem.2017.02.079

Du H, Huo Y, Liu H et al (2019) Fast nutritional characterization of different pigmented rice grains using a combination of NMR and decision tree analysis. CyTA – J Food 17:128–136. https://doi.org/10.1080/19476337.2018.1545800

Dunn WB, Broadhurst D, Begley P et al (2011) Procedures for large-scale metabolic profiling of serum and plasma using gas chromatography and liquid chromatography coupled to mass spectrometry. Nat Protoc 6:1060–1083. https://doi.org/10.1038/nprot.2011.335

Feng T, Zhang Q, Wang X et al (2019) Determination of 5-hydroxymethyl-2-furaldehyde in Cooked japonica rice using a modified QuEChERS method combined with dispersive liquid-liquid microextraction followed by UPLC-ESI-MS/MS. Food Anal Methods 12:1838–1848. https://doi.org/10.1007/s12161-019-01533-4

Fitzgerald MA, McCouch SR, Hall RD (2009) Not just a grain of rice: the quest for quality. Trends Plant Sci 14:133–139. https://doi.org/10.1016/j.tplants.2008.12.004

Frank T, Reichardt B, Shu Q, Engel KH (2012) Metabolite profiling of colored rice (*Oryza sativa* L.) grains. J Cereal Sci 55:112–119. https://doi.org/10.1016/j.jcs.2011.09.009

Frenzel T, Miller A, Engel KH (2002) Metabolite profiling – a fractionation method for analysis of major and minor compounds in rice grains. Cereal Chem 79:215–221. https://doi.org/10.1094/CCHEM.2002.79.2.215

Fumagalli E, Baldoni E, Abbruscato P et al (2009) NMR techniques coupled with multivariate statistical analysis: tools to analyse *Oryza sativa* metabolic content under stress conditions. J Agron Crop Sci 195:77–88. https://doi.org/10.1111/j.1439-037X.2008.00344.x

Gong L, Chen W, Gao Y et al (2013) Genetic analysis of the metabolome exemplified using a rice population. Proc Natl Acad Sci U S A 110:20320–20325. https://doi.org/10.1073/pnas.1319681110

Grande-Martínez Á, Arrebola-Liébanas FJ, Martínez-Vidal JL et al (2015a) Optimization and validation of a multiresidue pesticide method in rice and wheat flour by modified QuECHERS and GS-MS/MS. Food Anal Methods 9:548–563. https://doi.org/10.1007/s12161-015-0214-7

Grande-Martínez Á, Arrebola FJ, Moreno LD et al (2015b) Development and validation of a multiresidue method for the determination of pesticides in dry samples (rice and wheat flour) using liquid chromatography/triple quadrupole tandem mass spectrometry. J AOAC Int 98:1186–1198. https://doi.org/10.5740/jaoacint.SGE4-Grande-Martinez

Guo T, Wang X, Wang H et al (2019) Determination of phenoxy acid herbicides in cereals using high-performance liquid chromatography-tandem mass spectrometry. J Food Prot 82:1160–1165. https://doi.org/10.4315/0362-028X.JFP-18-558

Han Y, Song L, Zou N et al (2017) Rapid multiplug filtration cleanup method for the determination of 124 pesticide residues in rice, wheat, and corn. J Sep Sci 40:878–884. https://doi.org/10.1002/jssc.201600978

Han Y, Song L, Liu S et al (2018) Simultaneous determination of 124 pesticide residues in Chinese liquor and liquor-making raw materials (sorghum and rice hull) by rapid multi-plug filtration cleanup and gas chromatography–tandem mass spectrometry. Food Chem 241:258–267. https://doi.org/10.1016/j.foodchem.2017.08.103

Heuberger AL, Lewis MR, Chen MH et al (2010) Metabolomic and functional genomic analyses reveal varietal differences in bioactive compounds of cooked rice. PLoS One 5:1–10. https://doi.org/10.1371/journal.pone.0012915

Horie Y, Nemoto H, Fujita J et al (2019) Studies on the quality control of fermented brown rice and rice bran with *Aspergillus oryzae* by metabolome analysis using ^{1}H-NMR. Nippon Shokuhin Kagaku Kogaku Kaishi 66:139–146. https://doi.org/10.3136/nskkk.66.139

Hu C, Shi J, Quan S et al (2014) Metabolic variation between japonica and indica rice cultivars as revealed by non-targeted metabolomics. Sci Rep 4. https://doi.org/10.1038/srep05067

Hu M, Chen W, Liu Y et al (2017) Determination of herbicide pyraclonil residue in rice, soil and water using high-performance liquid chromatography/tandem mass spectrometry. Anal Methods 9:4790–4796. https://doi.org/10.1039/c7ay01483j

Huang YP, Lai HM (2016) Bioactive compounds and antioxidative activity of colored rice bran. J Food Drug Anal 24:564–574. https://doi.org/10.1016/j.jfda.2016.01.004

Huo Y, Kamal GM, Wang J et al (2017) 1H NMR-based metabolomics for discrimination of rice from different geographical origins of China. J Cereal Sci 76:243–252. https://doi.org/10.1016/j.jcs.2017.07.002

Islam MM, Al-Siyabi A, Saha R, Obata T (2018) Dissecting metabolic flux in C4 plants: experimental and theoretical approaches. Phytochem Rev 17:1253–1274. https://doi.org/10.1007/s11101-018-9579-8

Jettanajit A, Nhujak T (2016) Determination of mycotoxins in brown rice using QuEChERS sample preparation and UHPLC-MS-MS. J Chromatogr Sci 54:720–729. https://doi.org/10.1093/chromsci/bmv244

Ji X, Xu J, Wang X et al (2015) Citrinin determination in red fermented rice products by optimized extraction method coupled to liquid chromatography tandem mass spectrometry (LC-MS/MS). J Food Sci 80:T1438–T1444. https://doi.org/10.1111/1750-3841.12900

Jones OAH, Maguire ML, Griffin JL et al (2011) Using metabolic profiling to assess plant-pathogen interactions: an example using rice (*Oryza sativa*) and the blast pathogen Magnaporthe grisea. Eur J Plant Pathol 129:539–554. https://doi.org/10.1007/s10658-010-9718-6

Kang K, Yue L, Xia X et al (2019) Comparative metabolomics analysis of different resistant rice varieties in response to the brown planthopper *Nilaparvata lugens* Hemiptera: Delphacidae. Metabolomics 15:1–13. https://doi.org/10.1007/s11306-019-1523-4

Karmakar S, Datta K, Molla KA et al (2019) Proteo-metabolomic investigation of transgenic rice unravels metabolic alterations and accumulation of novel proteins potentially involved in defence against *Rhizoctonia solani*. Sci Rep 9:1–16. https://doi.org/10.1038/s41598-019-46885-3

Kim AJ, Choi JN, Kim J et al (2012a) Metabolite profiling and bioactivity of rice koji fermented by *Aspergillus* strains. J Microbiol Biotechnol 22:100–106. https://doi.org/10.4014/jmb.1106.06033

Kim JK, Ha SH, Park SY et al (2012b) Determination of lipophilic compounds in genetically modified rice using gas chromatography-time-of-flight mass spectrometry. J Food Compos Anal 25:31–38. https://doi.org/10.1016/j.jfca.2011.06.002

Kim JK, Park SY, Lee SM et al (2013a) Unintended polar metabolite profiling of carotenoid-biofortified transgenic rice reveals substantial equivalence to its non-transgenic counterpart. Plant Biotechnol Rep 7:121–128. https://doi.org/10.1007/s11816-012-0231-6

Kim JK, Park SY, Lim SH et al (2013b) Comparative metabolic profiling of pigmented rice (Oryza sativa L.) cultivars reveals primary metabolites are correlated with secondary metabolites. J Cereal Sci 57:14–20. https://doi.org/10.1016/j.jcs.2012.09.012

Kim DH, Hong SY, Kang JW et al (2017a) Simultaneous determination of multi-mycotoxins in cereal grains collected from South Korea by LC/MS/MS. Toxins (Basel) 9:1–13. https://doi.org/10.3390/toxins9030106

Kim MS, Baek SH, Park SU et al (2017b) Targeted metabolite profiling to evaluate unintended metabolic changes of genetic modification in resveratrol-enriched rice (Oryza sativa L.). Appl Biol Chem 60:205–214. https://doi.org/10.1007/s13765-017-0265-0

Kind T, Fiehn O (2007) Seven golden rules for heuristic filtering of molecular formulas obtained by accurate mass spectrometry. BMC Bioinformatics 8:1–20. https://doi.org/10.1186/1471-2105-8-105

Kind T, Wohlgemuth G, Lee DY et al (2009) FiehnLib: mass spectral and retention index libraries for metabolomics based on quadrupole and time-of-flight gas chromatography/mass spectrometry. Anal Chem 81:10038–10048. https://doi.org/10.1021/ac9019522

Koellensperger G, Guijas C, Benton HP et al (2018) METLIN: a technology platform for identifying knowns and unknowns. Anal Chem 90:3156–3164. https://doi.org/10.1021/acs.analchem.7b04424

Koesukwiwat U, Sanguankaew K, Leepipatpiboon N (2014) Evaluation of a modified QuEChERS method for analysis of mycotoxins in rice. Food Chem 153:44–51. https://doi.org/10.1016/j.foodchem.2013.12.029

Kojima M, Kamada-Nobusada T, Komatsu H et al (2009) Highly sensitive and high-throughput analysis of plant hormones using ms-probe modification and liquid chromatography tandem mass spectrometry: an application for hormone profiling in oryza sativa. Plant Cell Physiol 50:1201–1214. https://doi.org/10.1093/pcp/pcp057

Kopka J, Schauer N, Krueger S et al (2005) GMD@CSB.DB: the Golm metabolome database. Bioinformatics 21:1635–1638. https://doi.org/10.1093/bioinformatics/bti236

Kusano M, Fukushima A, Kobayashi M et al (2007) Application of a metabolomic method combining one-dimensional and two-dimensional gas chromatography-time-of-flight/mass spectrometry to metabolic phenotyping of natural variants in rice. J Chromatogr B Anal Technol Biomed Life Sci 855:71–79. https://doi.org/10.1016/j.jchromb.2007.05.002

Kusano M, Tabuchi M, Fukushima A et al (2011) Metabolomics data reveal a crucial role of cytosolic glutamine synthetase 1;1 in coordinating metabolic balance in rice. Plant J 66:456–466. https://doi.org/10.1111/j.1365-313X.2011.04506.x

Lam HS, Proctor A (2003) Milled rice oxidation volatiles and odor development. J Food Sci 68:2676–2681. https://doi.org/10.1111/j.1365-2621.2003.tb05788.x

Lawas LMF, Li X, Erban A et al (2019) Metabolic responses of rice cultivars with different tolerance to combined drought and heat stress under field conditions. Gigascience 8:1–21. https://doi.org/10.1093/gigascience/giz050

Lee YJ, Rahman MM, Abd El-Aty AM et al (2016) Detection of three herbicide, and one metabolite, residues in brown rice and rice straw using various versions of the QuEChERS method and liquid chromatography-tandem mass spectrometry. Food Chem 210:442–450. https://doi.org/10.1016/j.foodchem.2016.05.005

Lee DE, Lee S, Singh D et al (2017a) Time-resolved comparative metabolomes for Koji fermentation with brown-, white-, and giant embryo-rice. Food Chem 231:258–266. https://doi.org/10.1016/j.foodchem.2017.03.119

Lee J, Kim L, Shin Y et al (2017b) Rapid and simultaneous analysis of 360 pesticides in brown rice, spinach, orange, and potato using microbore GC-MS/MS. J Agric Food Chem 65:3387–3395. https://doi.org/10.1021/acs.jafc.7b00576

Lee J, Shin Y, Lee J et al (2018a) Simultaneous analysis of 310 pesticide multiresidues using UHPLC-MS/MS in brown rice, orange, and spinach. Chemosphere 207:519–526. https://doi.org/10.1016/j.chemosphere.2018.05.116

Lee S, Lee DE, Singh D, Lee CH (2018b) Metabolomics reveal optimal grain preprocessing (milling) toward rice koji fermentation. J Agric Food Chem 66:2694–2703. https://doi.org/10.1021/acs.jafc.7b05131

Lewis IA, Schommer SC, Markley JL (2009) rNMR: open source software for identifying and quantifying metabolites in NMR spectra. Magn Reson Chem 47. https://doi.org/10.1002/mrc.2526

Li Y, Ruan Q, Li Y et al (2012) A novel approach to transforming a non-targeted metabolic profiling method to a pseudo-targeted method using the retention time locking gas chromatography/mass spectrometry-selected ions monitoring. J Chromatogr A 1255:228–236. https://doi.org/10.1016/j.chroma.2012.01.076

Li H, Jiang Z, Cao X et al (2017a) Simultaneous determination of three pesticide adjuvant residues in plant-derived agro-products using liquid chromatography-tandem mass spectrometry. J Chromatogr A 1528:53–60. https://doi.org/10.1016/j.chroma.2017.10.075

Li R, Liu T, Cui S et al (2017b) Residue behaviors and dietary risk assessment of dinotefuran and its metabolites in *Oryza sativa* by a new HPLC–MS/MS method. Food Chem 235:188–193. https://doi.org/10.1016/j.foodchem.2017.04.181

Li KJ, Borresen EC, Jenkins-Puccetti N et al (2018) Navy bean and rice bran intake alters the plasma metabolome of children at risk for cardiovascular disease. Front Nutr 4. https://doi.org/10.3389/fnut.2017.00071

Liao CD, Wong JW, Zhang K et al (2015) Multi-mycotoxin analysis of finished grain and nut products using ultrahigh-performance liquid chromatography and positive electrospray ionization-quadrupole orbital ion trap high-resolution mass spectrometry. J Agric Food Chem 63:8314–8332. https://doi.org/10.1021/jf505049a

Lim CW, Yoshinari T, Layne J, Chan SH (2015) Multi-mycotoxin screening reveals separate occurrence of aflatoxins and ochratoxin A in Asian rice. J Agric Food Chem 63:3104–3113. https://doi.org/10.1021/acs.jafc.5b00471

Lim DK, Long NP, Mo C et al (2017a) Combination of mass spectrometry-based targeted lipidomics and supervised machine learning algorithms in detecting adulterated admixtures of white rice. Food Res Int 100:814–821. https://doi.org/10.1016/j.foodres.2017.08.006

Lim DK, Mo C, Long NP et al (2017b) A rapid and reliable method for discriminating rice products from different regions using MCX-based solid-phase extraction and DI-MS/MS-based metabolomics approach. J Chromatogr B Anal Technol Biomed Life Sci 1061–1062:185–192. https://doi.org/10.1016/j.jchromb.2017.07.020

Lim DK, Mo C, Long NP et al (2017c) Simultaneous profiling of lysoglycerophospholipids in rice (*Oryza sativa* L) using direct infusion-tandem mass spectrometry with multiple reaction monitoring. J Agric Food Chem 65:2628–2634. https://doi.org/10.1021/acs.jafc.7b00148

Lim DK, Long NP, Mo C et al (2018a) Optimized mass spectrometry-based metabolite extraction and analysis for the geographical discrimination of white rice (*Oryza sativa* L.): a method comparison study. J AOAC Int 101:498–506. https://doi.org/10.5740/jaoacint.17-0158

Lim DK, Mo C, Lee DK et al (2018b) Non-destructive profiling of volatile organic compounds using HS-SPME/GC–MS and its application for the geographical discrimination of white rice. J Food Drug Anal 26:260–267. https://doi.org/10.1016/j.jfda.2017.04.005

Lim DK, Mo C, Lee JH et al (2018c) The integration of multi-platform MS-based metabolomics and multivariate analysis for the geographical origin discrimination of *Oryza sativa* L. J Food Drug Anal 26:769–777. https://doi.org/10.1016/j.jfda.2017.09.004

Liu C, Hao F, Hu J et al (2010) Revealing different systems responses to brown planthopper infestation for pest susceptible and resistant rice plants with the combined metabonomic and gene-expression analysis. J Proteome Res 9:6774–6785. https://doi.org/10.1021/pr100970q

Liu Z, Qi P, Wang X et al (2017a) Multi-pesticides residue analysis of grains using modified magnetic nanoparticle adsorbent for facile and efficient cleanup. Food Chem 230:423–431. https://doi.org/10.1016/j.foodchem.2017.03.082

Liu Z, Wu HL, Xie LX et al (2017b) Chemometrics-enhanced liquid chromatography-full scan-mass spectrometry for interference-free analysis of multi-class mycotoxins in complex cereal samples. Chemom Intell Lab Syst 160:125–138. https://doi.org/10.1016/j.chemolab.2016.12.003

Lommen A (2009) Metalign: interface-driven, versatile metabolomics tool for hyphenated full-scan mass spectrometry data preprocessing. Anal Chem 81:3079–3086. https://doi.org/10.1021/ac900036d

Long X, Liu Q, Chan M et al (2013) Metabolic engineering and profiling of rice with increased lysine. Plant Biotechnol J 11:490–501. https://doi.org/10.1111/pbi.12037

Long NP, Lim DK, Mo C et al (2017) Development and assessment of a lysophospholipid-based deep learning model to discriminate geographical origins of white rice. Sci Rep 7:1–10. https://doi.org/10.1038/s41598-017-08892-0

Lou Q, Ma C, Wen W et al (2011) Profiling and association mapping of grain metabolites in a subset of the core collection of Chinese rice Germplasm (*Oryza sativa* L.). J Agric Food Chem 59:9257–9264. https://doi.org/10.1021/jf201602g

Ma Y, Zhan L, Yang H et al (2019) Dissipation of two field-incurred pesticides and three degradation products in rice (*Oryza sativa* L.) from harvest to dining table. J Sci Food Agric 99:4602–4608. https://doi.org/10.1002/jsfa.9699

Mahattanatawee K, Rouseff RL (2014) Comparison of aroma active and sulfur volatiles in three fragrant rice cultivars using GC-Olfactometry and GC-PFPD. Food Chem 154:1–6. https://doi.org/10.1016/j.foodchem.2013.12.105

Mahdavi V, Farimani MM, Fathi F, Ghassempour A (2015) A targeted metabolomics approach toward understanding metabolic variations in rice under pesticide stress. Anal Biochem 478:65–72. https://doi.org/10.1016/j.ab.2015.02.021

Mahdavi V, Ghanati F, Ghassempour A (2016) Integrated pathway-based and network-based analysis of GC-MS rice metabolomics data under diazinon stress to infer affected biological pathways. Anal Biochem 494:31–36. https://doi.org/10.1016/j.ab.2015.10.017

Maraval I, Mestres C, Pernin K et al (2008) Odor-active compounds in cooked rice cultivars from Camargue (France) analyzed by GC-O and GC-MS. J Agric Food Chem 56:5291–5298. https://doi.org/10.1021/jf7037373

Marley E, Brown P, Mackie J et al (2015) Analysis of sterigmatocystin in cereals, animal feed, seeds, beer and cheese by immunoaffinity column clean-up and HPLC and LC-MS/MS quantification. Food Addit Contam – Part A Chem Anal Control Expo Risk Assess 32:2131–2137. https://doi.org/10.1080/19440049.2015.1100331

Maruyama K, Urano K, Yoshiwara K et al (2014) Integrated analysis of the effects of cold and dehydration on rice metabolites, phytohormones, and gene transcripts. Plant Physiol 164:1759–1771. https://doi.org/10.1104/pp.113.231720

Masumoto C, Miyazawa SI, Ohkawa H et al (2010) Phosphoenolpyruvate carboxylase intrinsically located in the chloroplast of rice plays a crucial role in ammonium assimilation. Proc Natl Acad Sci U S A 107:5226–5231. https://doi.org/10.1073/pnas.0913127107

Matsuda F, Ishihara A, Takanashi K et al (2010) Metabolic profiling analysis of genetically modified rice seedlings that overproduce tryptophan reveals the occurrence of its inter-tissue translocation. Plant Biotechnology 27:17–27

Matsuda F, Okazaki Y, Oikawa A et al (2012) Dissection of genotype-phenotype associations in rice grains using metabolome quantitative trait loci analysis. Plant J 70:624–636. https://doi.org/10.1111/j.1365-313X.2012.04903.x

Matsuda F, Nakabayashi R, Yang Z et al (2015) Metabolome-genome-wide association study dissects genetic architecture for generating natural variation in rice secondary metabolism. Plant J 81:13–23. https://doi.org/10.1111/tpj.12681

Millard P, Cahoreau E, Heuillet M et al (2017) ^{15}N-NMR-based approach for amino acids-based ^{13}C-metabolic flux analysis of metabolism. Anal Chem 89:2101–2106. https://doi.org/10.1021/acs.analchem.6b04767

Mimura N, Isogai A, Iwashita K et al (2014) Gas chromatography/mass spectrometry based component profiling and quality prediction for Japanese sake. J Biosci Bioeng 118:406–414. https://doi.org/10.1016/j.jbiosc.2014.04.006

Miró-Abella E, Herrero P, Canela N et al (2017) Determination of mycotoxins in plant-based beverages using QuEChERS and liquid chromatography–tandem mass spectrometry. Food Chem 229:366–372. https://doi.org/10.1016/j.foodchem.2017.02.078

Misra BB, Mohapatra S (2019) Tools and resources for metabolomics research community: a 2017–2018 update. Electrophoresis 40:227–246. https://doi.org/10.1002/elps.201800428

Mochida K, Furuta T, Ebana K et al (2009) Correlation exploration of metabolic and genomic diversity in rice. BMC Genomics 10:1–10. https://doi.org/10.1186/1471-2164-10-568

Mohd Esa N, Ling TB (2016) By-products of rice processing: an overview of health benefits and applications. Rice Res Open Access 4:1–11. https://doi.org/10.4172/jrr.1000107

Mondal R, Kole RK, Bhattacharyya A (2017) Validation of multiresidue method for analysis of 31 pesticides in rice using gas chromatography-tandem mass spectrometry. J AOAC Int 100:1094–1101. https://doi.org/10.5740/jaoacint.16-0377

Mou RX, Cao ZY, Lin XY et al (2016) Characterization of the phytochelatins and their derivatives in rice exposed to cadmium based on high-performance liquid chromatography coupled with data-dependent hybrid linear ion trap orbitrap mass spectrometry. Rapid Commun Mass Spectrom 30:1891–1900. https://doi.org/10.1002/rcm.7669

Nagana Gowda GA, Raftery D (2019) NMR-based metabolomics, methods and protocols. Springer, New York

Nakamura K, Otake T, Hanari N, Takatsu A (2019) Evaluation of the impact of matrix effects in LC/MS measurement on the accurate quantification of neonicotinoid pesticides in food by isotope-dilution mass spectrometry. J Environ Sci Heal – Part B Pestic Food Contam Agric Wastes 54:467–474. https://doi.org/10.1080/03601234.2019.1607134

Nam KH, Shin HJ, Pack IS et al (2016) Metabolomic changes in grains of well-watered and drought-stressed transgenic rice. J Sci Food Agric 96:807–814. https://doi.org/10.1002/jsfa.7152

Nardin T, Barnaba C, Abballe F et al (2017) Fast analysis of quaternary ammonium pesticides in food and beverages using cation-exchange chromatography coupled with isotope-dilution high-resolution mass spectrometry. J Sep Sci 40:3928–3937. https://doi.org/10.1002/jssc.201700579

Navarro-Reig M, Jaumot J, García-Reiriz A, Tauler R (2015) Evaluation of changes induced in rice metabolome by Cd and Cu exposure using LC-MS with XCMS and MCR-ALS data analysis strategies. Anal Bioanal Chem 407:8835–8847. https://doi.org/10.1007/s00216-015-9042-2

Navarro-Reig M, Jaumot J, Baglai A et al (2017a) Untargeted comprehensive two-dimensional liquid chromatography coupled with high-resolution mass spectrometry analysis of rice metabolome using multivariate curve resolution. Anal Chem 89:7675–7683. https://doi.org/10.1021/acs.analchem.7b01648

Navarro-Reig M, Jaumot J, Piña B et al (2017b) Metabolomic analysis of the effects of cadmium and copper treatment in: *Oryza sativa* L. using untargeted liquid chromatography coupled to high resolution mass spectrometry and all-ion fragmentation. Metallomics 9:660–675. https://doi.org/10.1039/c6mt00279j

Nazari F, Sulyok M, Yazdanpanah H et al (2014) A survey of mycotoxins in domestic rice in Iran by liquid chromatography tandem mass spectrometry. Toxicol Mech Methods 24:37–41. https://doi.org/10.3109/15376516.2013.844752

Ni Y, Yang H, Zhang H et al (2018) Analysis of four sulfonylurea herbicides in cereals using modified quick, easy, cheap, effective, rugged, and safe sample preparation method coupled with liquid chromatography–tandem mass spectrometry. J Chromatogr A 1537:27–34. https://doi.org/10.1016/j.chroma.2018.01.017

Ogo Y, Ozawa K, Ishimaru T et al (2013) Transgenic rice seed synthesizing diverse flavonoids at high levels: a new platform for flavonoid production with associated health benefits. Plant Biotechnol J 11:734–746. https://doi.org/10.1111/pbi.12064

Okazaki Y, Otsuki H, Narisawa T et al (2013) A new class of plant lipid is essential for protection against phosphorus depletion. Nat Commun 4. https://doi.org/10.1038/ncomms2512

Ortmayr K, Causon TJ, Hann S, Koellensperger G (2016) Increasing selectivity and coverage in LC-MS based metabolome analysis. TrAC - Trends Anal Chem 82:358–366. https://doi.org/10.1016/j.trac.2016.06.011

Patil SB, Khan MK (2011) Germinated brown rice as a value added rice product: a review. J Food Sci Technol 48:661–667. https://doi.org/10.1007/s13197-011-0232-4

Peng L, Zhao Y, Wang H et al (2016) Comparative metabolomics of the interaction between rice and the brown planthopper. Metabolomics 12:1–15. https://doi.org/10.1007/s11306-016-1077-7

Pesek JJ, Matyska MT, Hoffmann JF et al (2017) Liquid chromatography with mass spectrometry analysis of mycotoxins in food samples using silica hydride based stationary phases. J Sep Sci 40:1953–1959. https://doi.org/10.1002/jssc.201601267

Pluskal T, Castillo S, Villar-Briones A, Orešič M (2010) MZmine 2: modular framework for processing, visualizing, and analyzing mass spectrometry-based molecular profile data. BMC Bioinformatics 11. https://doi.org/10.1186/1471-2105-11-395

Pramai P, Abdul Hamid NA, Mediani A et al (2018) Metabolite profiling, antioxidant, and α-glucosidase inhibitory activities of germinated rice: nuclear-magnetic-resonance-based metabolomics study. J Food Drug Anal 26:47–57. https://doi.org/10.1016/j.jfda.2016.11.023

Ranjitha HP, Gowda R, Nethra N et al (2019) Biochemical and metabolomics on rice cultivars. Rice Sci 26:189–194. https://doi.org/10.1016/j.rsci.2018.08.007

Rebelo AM, Dolzan MD, Heller M et al (2016) Simultaneous determination of herbicides in rice by QuEChERS and LC-MS/MS using matrix-matched calibration. J Braz Chem Soc 27:186–193. https://doi.org/10.5935/0103-5053.20150269

Redestig H, Kusano M, Ebana K et al (2011) Exploring molecular backgrounds of quality traits in rice by predictive models based on high-coverage metabolomics. BMC Syst Biol 5. https://doi.org/10.1186/1752-0509-5-176

Reguera M, Peleg Z, Abdel-Tawab YM et al (2013) Stress-induced cytokinin synthesis increases drought tolerance through the coordinated regulation of carbon and nitrogen assimilation in rice. Plant Physiol 163:1609–1622. https://doi.org/10.1104/pp.113.227702

Righetti L, Bergmann A, Galaverna G et al (2018) Ion mobility-derived collision cross section database: application to mycotoxin analysis. Anal Chim Acta 1014:50–57. https://doi.org/10.1016/j.aca.2018.01.047

Rong L, Wu X, Xu J et al (2018) Simultaneous determination of three pesticides and their metabolites in unprocessed foods using ultraperformance liquid chromatography-tandem mass spectrometry. Food Addit Contam – Part A Chem Anal Control Expo Risk Assess 35:273–281. https://doi.org/10.1080/19440049.2017.1398419

Ryan EP, Heuberger AL, Weir TL et al (2011) Rice bran fermented with *Saccharomyces boulardii* generates novel metabolite profiles with bioactivity. J Agric Food Chem 59:1862–1870. https://doi.org/10.1021/jf1038103

Salek RM, Steinbeck C, Viant MR et al (2013) The role of reporting standards for metabolite annotation and identification in metabolomic studies. Gigascience 2:13. https://doi.org/10.1186/2047-217X-2-13

Santilio A, Pompili C, Giambenedetti A (2019) Determination of glyphosate residue in maize and rice using a fast and easy method involving liquid chromatography–mass spectrometry (LC/MS/MS). J Environ Sci Heal – Part B Pestic Food Contam Agric Wastes 54:205–210. https://doi.org/10.1080/03601234.2018.1550306

Sato S, Soga T, Nishioka T, Tomita M (2004) Simultaneous determination of the main metabolites in rice leaves using capillary electrophoresis mass spectrometry and capillary electrophoresis diode array detection. Plant J 40:151–163. https://doi.org/10.1111/j.1365-313X.2004.02187.x

Selby-Pham J, Lutz A, Moreno-Moyano LT et al (2017) Diurnal changes in transcript and metabolite levels during the iron deficiency response of rice. Rice 10. https://doi.org/10.1186/s12284-017-0152-7

Seo HS, Lee S, Singh D et al (2018) Untargeted metabolite profiling for koji-fermentative bioprocess unravels the effects of varying substrate types and microbial inocula. Food Chem 266:161–169. https://doi.org/10.1016/j.foodchem.2018.05.048

Sheflin AM, Borresen EC, Wdowik MJ et al (2015) Pilot dietary intervention with heat-stabilized rice bran modulates stool microbiota and metabolites in healthy adults. Nutrients 7:1282–1300. https://doi.org/10.3390/nu7021282

Shen S, Zhang H, Huang K et al (2019a) Differentiation of cultivation areas and crop years of milled rice using single grain mass spectrometry. New J Chem 43:2118–2125. https://doi.org/10.1039/c8nj02740d

Shen T, Xiong Q, Zhong L et al (2019b) Analysis of main metabolisms during nitrogen deficiency and compensation in rice. Acta Physiol Plant 41. https://doi.org/10.1007/s11738-019-2860-7

Shi L, Gui T, Zhao S et al (2018) Degradation and residues of indoxacarb enantiomers in rice plants, rice hulls and brown rice using enriched S-indoxacarb formulation and enantiopure formulation. Biomed Chromatogr 32:1–9. https://doi.org/10.1002/bmc.4301

Shu XL, Frank T, Shu QY, Engel KH (2008) Metabolite profiling of germinating rice seeds. J Agric Food Chem 56:11612–11620. https://doi.org/10.1021/jf802671p

Sinphithakkul P, Poapolathep A, Klangkaew N et al (2019) Occurrence of multiple mycotoxins in various types of rice and barley samples in Thailand. J Food Prot 82:1007–1015. https://doi.org/10.4315/0362-028X.JFP-18-456

Smith CA, Want EJ, O'Maille G et al (2006) XCMS: processing mass spectrometry data for metabolite profiling using nonlinear peak alignment, matching, and identification. Anal Chem 78:779–787. https://doi.org/10.1021/ac051437y

Solfrizzo M, Gambacorta L, Bibi R et al (2018) Multimycotoxin analysis by LC-MS/MS in cereal food and feed: comparison of different approaches for extraction, purification, and calibration. J AOAC Int 101:647–657. https://doi.org/10.5740/jaoacint.17-0339

Song EH, Kim HJ, Jeong J et al (2016) A ^{1}H HR-MAS NMR-Based metabolomic study for metabolic characterization of rice grain from various *Oryza sativa* L. cultivars. J Agric Food Chem 64:3009–3016. https://doi.org/10.1021/acs.jafc.5b05667

Song EH, Jeong J, Park CY et al (2018) Metabotyping of rice (*Oryza sativa* L.) for understanding its intrinsic physiology and potential eating quality. Food Res Int 111:20–30. https://doi.org/10.1016/j.foodres.2018.05.003

Sotelo J, Slupsky CM (2013) Metabolomics using nuclear magnetic resonance (NMR). In: Metabolomics in food and nutrition. Woodhead Publishing, Cambridge, England, pp 29–43

Sriseadka T, Wongpornchai S, Rayanakorn M (2012) Quantification of flavonoids in black rice by liquid chromatography- negative electrospray ionization tandem mass spectrometry. J Agric Food Chem 60:11723–11732. https://doi.org/10.1021/jf303204s

Storozhenko S, De Brouwer V, Volckaert M et al (2007) Folate fortification of rice by metabolic engineering. Nat Biotechnol 25:1277–1279. https://doi.org/10.1038/nbt1351

Sugitate K, Saka M, Serino T et al (2012) Matrix behavior during sample preparation using metabolomics analysis approach for pesticide residue analysis by GC-MS in agricultural products. J Agric Food Chem 60:10226–10234. https://doi.org/10.1021/jf3029557

Sumner LW, Amberg A, Barrett D et al (2007) Proposed minimum reporting standards for chemical analysis. Metabolomics 3:211–221. https://doi.org/10.1007/s11306-007-0082-2

Sun M, Yu L, Tong Z et al (2019) Determination of phenamacril residues in flour and rice based on Z-Sep+ using ultra-high-performance liquid chromatography–tandem mass spectrometry. Biomed Chromatogr. https://doi.org/10.1002/bmc.4688

Takahara K, Kasajima I, Takahashi H et al (2010) Metabolome and photochemical analysis of rice plants overexpressing *Arabidopsis* NAD kinase gene. Plant Physiol 152:1863–1873. https://doi.org/10.1104/pp.110.153098

Takahashi K, Kabashima F, Tsuchiya F (2016) Comprehensive two-dimensional gas chromatography coupled with time-of-flight mass spectrometry reveals the correlation between chemical compounds in Japanese sake and its organoleptic properties. J Biosci Bioeng 121:274–280. https://doi.org/10.1016/j.jbiosc.2015.06.016

Tananuwong K, Lertsiri S (2010) Changes in volatile aroma compounds of organic fragrant rice during storage under different conditions. J Sci Food Agric 90:1590–1596. https://doi.org/10.1002/jsfa.3976

Teló GM, Senseman SA, Marchesan E et al (2017) Accelerated solvent extraction of insecticides from rice hulls, rice bran, and polished rice grains. J AOAC Int 100:566–572. https://doi.org/10.5740/jaoacint.16-0119

Vega ED, Marina ML (2014) Characterization and study of transgenic cultivars by capillary and microchip electrophoresis. Int J Mol Sci 15:23851–23877. https://doi.org/10.3390/ijms151223851

Viera MS, Rizzetti TM, de Souza MP et al (2017) Multiresidue determination of pesticides in crop plants by the quick, easy, cheap, effective, rugged, and safe method and ultra-high-performance liquid chromatography tandem mass spectrometry using a calibration based on a single level standard addition in the sample. J Chromatogr A 1526:119–127. https://doi.org/10.1016/j.chroma.2017.10.048

Villanova FA, Vanier NL, de Avila MN et al (2017) Improvement of the quality of parboiled rice by using anti-browning agents during parboiling process. Food Chem 235:51–57. https://doi.org/10.1016/j.foodchem.2017.05.053

Wang X, Wang J, Lai D et al (2017) Ustiloxin G, a new cyclopeptide mycotoxin from rice false smut balls. Toxins (Basel) 9:1–9. https://doi.org/10.3390/toxins9020054

Wild D, Tóth G, Humpf HU (2003) New *Monascus* metabolites with a pyridine structure in red fermented rice. J Agric Food Chem 51:5493–5496. https://doi.org/10.1021/jf030213i

Wishart DS (2008) Metabolomics: applications to food science and nutrition research. Trends Food Sci Technol 19:482–493. https://doi.org/10.1016/j.tifs.2008.03.003

Wongpornchai S, Dumri K, Jongkaewwattana S, Siri B (2004) Effects of drying methods and storage time on the aroma and milling quality of rice (*Oryza sativa* L.) cv. Khao Dawk Mali 105. Food Chem 87:407–414. https://doi.org/10.1016/j.foodchem.2003.12.014

Wongsa P, Landberg R, Rattanapanone N (2018) Chemical compositions and metabolite profiling of rice varieties from Chiang Rai Province, Thailand. Chiang Mai J Sci 45:2703–2714

Wu M, Liu M, Deng F et al (2018) Simple solvent extraction coupled with liquid chromatography-high-resolution mass spectrometry for the analysis of pesticide residues in rice bran protein powder. Food Anal Methods 11:2368–2375. https://doi.org/10.1007/s12161-018-1221-2

Xia J, Wishart DS (2011) Web-based inference of biological patterns, functions and pathways from metabolomic data using MetaboAnalyst. Nat Protoc 6:743–760. https://doi.org/10.1038/nprot.2011.319

Xiao R, Ma Y, Zhang D, Qian L (2018) Discrimination of conventional and organic rice using untargeted LC-MS-based metabolomics. J Cereal Sci 82:73–81. https://doi.org/10.1016/j.jcs.2018.05.012

Yan S, Huang W, Gao J et al (2018) Comparative metabolomic analysis of seed metabolites associated with seed storability in rice (*Oryza sativa* L.) during natural aging. Plant Physiol Biochem 127:590–598. https://doi.org/10.1016/j.plaphy.2018.04.020

Yang Z, Nakabayashi R, Okazaki Y et al (2014) Toward better annotation in plant metabolomics: isolation and structure elucidation of 36 specialized metabolites from Oryza sativa (rice) by using MS/MS and NMR analyses. Metabolomics 10:543–555. https://doi.org/10.1007/s11306-013-0619-5

Yang SO, Wu C, So MY et al (2016) Effects of brown rice on cellular growth and metabolic changes in mice. Food Res Int 84:33–40. https://doi.org/10.1016/j.foodres.2016.03.009

Yanjie X, Yining Y, Shuhong O et al (2018) Factors affecting sensory quality of cooked japonica rice. Rice Sci 25:330–339. https://doi.org/10.1016/j.rsci.2018.10.003

Yoshimura Y, Zaima N, Moriyama T, Kawamura Y (2012) Different localization patterns of anthocyanin species in the pericarp of black rice revealed by imaging mass spectrometry. PLoS One 7. https://doi.org/10.1371/journal.pone.0031285

Yoshizaki Y, Yamato H, Takamine K et al (2010) Analysis of volatile compounds in shochu koji, sake koji, and steamed rice by gas chromatography-mass spectrometry. J Inst Brew 116:49–55. https://doi.org/10.1002/j.2050-0416.2010.tb00397.x

Yu S, Nehus ZT, Badger TM, Fang N (2007) Quantification of vitamin E and γ-oryzanol components in rice germ and bran. J Agric Food Chem 55:7308–7313. https://doi.org/10.1021/jf071957p

Zahra Z, Waseem N, Zahra R et al (2017) Growth and metabolic responses of rice (*Oryza sativa* L.) cultivated in phosphorus-deficient soil amended with TiO_2 nanoparticles. J Agric Food Chem 65:5598–5606. https://doi.org/10.1021/acs.jafc.7b01843

Zaima N, Goto-Inoue N, Hayasaka T, Setou M (2010) Application of imaging mass spectrometry for the analysis of *Oryza sativa* rice. Rapid Commun Mass Spectrom 24:2723–2729. https://doi.org/10.1002/rcm.4693

Zaima N, Yoshimura Y, Kawamura Y, Moriyama T (2014) Distribution of lysophosphatidylcholine in the endosperm of *Oryza sativa* rice. Rapid Commun Mass Spectrom 28:1515–1520. https://doi.org/10.1002/rcm.6927

Zarei I, Brown DG, Nealon NJ, Ryan EP (2017) Rice bran metabolome contains amino acids, vitamins & cofactors, and phytochemicals with medicinal and nutritional properties. Rice 10. https://doi.org/10.1186/s12284-017-0157-2

Zarei I, Luna E, Leach JE et al (2018) Comparative rice bran metabolomics across diverse cultivars and functional rice gene–Bran metabolite relationships. Metabolites 8. https://doi.org/10.3390/metabo8040063

Zeng M, Zhang L, He Z et al (2012) Determination of flavor components of rice bran by GC-MS and chemometrics. Anal Methods 4:539–545. https://doi.org/10.1039/c2ay05671b

Zhang J, Zhao C, Zeng Z et al (2016) Sample-directed pseudotargeted method for the metabolic profiling analysis of rice seeds based on liquid chromatography with mass spectrometry. J Sep Sci 39:247–255. https://doi.org/10.1002/jssc.201500858

Zhang P, Ding J, Hou J et al (2017a) Dynamic microwave assisted extraction coupled with matrix solid phase dispersion for the determination of chlorfenapyr and abamectin in rice by LC-MS/MS. Microchem J 133:404–411. https://doi.org/10.1016/j.microc.2017.04.006

Zhang ZY, Zheng ZT, Zhu GY et al (2017b) Validation of analytical method and evaluation of clothianidin residues in rice in a typical Chinese field ecosystem. J Agric Sci 155:1371–1380. https://doi.org/10.1017/S0021859617000491

Zhao X, Wang W, Zhang F et al (2014) Comparative metabolite profiling of two rice genotypes with contrasting salt stress tolerance at the seedling stage. PLoS One 9:1–7. https://doi.org/10.1371/journal.pone.0108020

Zhao Y, Zhang L, Zhao C et al (2015) Metabolic responses of rice leaves and seeds under transgenic backcross breeding and pesticide stress by pseudotargeted metabolomics. Metabolomics 11:1802–1814. https://doi.org/10.1007/s11306-015-0834-3

Zhao Y, Huang J, Ma L, Wang F (2017) Development and validation of a simple and fast method for simultaneous determination of aflatoxin B1 and sterigmatocystin in grains. Food Chem 221:11–17. https://doi.org/10.1016/j.foodchem.2016.10.036

Zhao X, Wang W, Xie Z et al (2018) Comparative analysis of metabolite changes in two contrasting rice genotypes in response to low-nitrogen stress. Crop J 6:464–474. https://doi.org/10.1016/j.cj.2018.05.006

Zhou J, Ma C, Xu H et al (2009) Metabolic profiling of transgenic rice with cryIAc and sck genes: an evaluation of unintended effects at metabolic level by using GC-FID and GC-MS. J Chromatogr B Anal Technol Biomed Life Sci 877:725–732. https://doi.org/10.1016/j.jchromb.2009.01.040

Ziegler V, Ferreira CD, Hoffmann JF et al (2018) Cooking quality properties and free and bound phenolics content of brown, black, and red rice grains stored at different temperatures for six months. Food Chem 242:427–434. https://doi.org/10.1016/j.foodchem.2017.09.077

Index

A. Costa de Oliveira et al. (eds.), *The Future of Rice Demand: Quality Beyond Productivity*, https://doi.org/10.1007/978-3-030-37510-2

S

The manufacturer's authorised representative in the EU is Springer Nature Customer Service Centre GmbH, Europaplatz 3, 69115 Heidelberg, Germany. If you have any concerns regarding our products, please contact ProductSafety@springernature.com

Printed and bound by CPI Group (UK) Ltd, Croydon, CR0 4YY

30/09/2026

02225647-0015